FOR STUDENTS

- Career Opportunities
- Career Fitness Program
- Becoming an Electronics Technician
- Free On-Line Study Guides (companion web sites)

FOR FACULTY*

- Supplements
- On-Line Product Catalog
- Electronics Technology Journal
- Prentice Hall Book Advisor

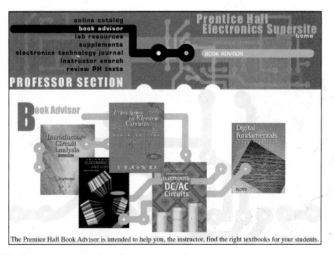

The Prentice Hall Book Advisor is intended to help you, the instructor, find the right textbooks for your students.

* Please contact your Prentice Hall representative for passcode

FUNDAMENTALS
OF ANALOG CIRCUITS

Second Edition

Thomas L. Floyd

David Buchla
Yuba College

Upper Saddle River, New Jersey   Columbus, Ohio

Library of Congress Cataloging-in-Publication Data

Floyd, Thomas L.
    Fundamentals of analog circuits / Thomas L. Floyd, David Buchla.—2nd ed.
      p. cm.
    ISBN 0–13–060619–7 (alk. paper)
    1. Linear integrated circuits. 2. Electronic circuits.   I. Buchla, David.   II. Title.

TK7874 .F5899   2002
621.3815—dc21   00-067601

Vice President and Editor in Chief: Stephen Helba
Product Manager: Scott J. Sambucci
Associate Editor: Kate Linsner
Production Editor: Rex Davidson
Design Coordinator: Robin G. Chukes
Cover Designer: Jason Moore
Cover Photo: David Buchla
Troubleshooting Photo: Copyright 2000, Tektronix, Inc. Reprinted with permission. All rights reserved.
System Application Photo: Reproduced with permission of the John Fluke Mfg. Co., Inc.
Production Manager: Pat Tonneman

This book was set in Times Roman and Optima by The Clarinda Company and was printed and bound by
R. R. Donnelley & Sons Company. The cover was printed by Phoenix Color Corp.

Prentice-Hall International (UK) Limited, *London*
Prentice-Hall of Australia Pty. Limited, *Sydney*
Prentice-Hall Canada, Inc., *Toronto*
Prentice-Hall Hispanoamericana, S. A., *Mexico*
Prentice-Hall of India Private Limited, *New Delhi*
Prentice-Hall of Japan, Inc., *Tokyo*
Prentice-Hall Singapore Pte. Ltd.
Editora Prentice-Hall do Brasil, Ltda., *Rio de Janeiro*

10 9 8 7 6 5 4 3 2 1
ISBN: 0-13-060619-7

Thanks to our wives for their affectionate support. Like fine red wine, they get better with age.

Preface

Fundamentals of Analog Circuits, Second Edition, presents an introduction to discrete linear devices and circuits followed by a thorough coverage of operational amplifiers and other linear integrated circuits. Also, this textbook provides extensive troubleshooting and applications coverage. Applications are shown with a realistic printed circuit board format in the last section of each chapter. They include a Troubleshooter's Bench exercise that presents a troubleshooting problem with the system. In addition to the Troubleshooter's Bench, troubleshooting sections are found in many parts of the text.

This second edition updates and improves coverage of the various operational amplifiers and other analog circuits introduced in the first edition. Each device was reviewed; older devices were replaced with newer ones, and a reference to the manufacturer's Internet site has been added to expedite finding additional information. In some cases, the explanation was streamlined or improved. For example, a brief discussion of triggering SCRs and triacs by microcontrollers was added in Chapter 15.

Two new features of this text include identifying key terms and adding a Troubleshooter's Quiz. Key terms are presented in the chapter opener and highlighted in color in the text with a margin icon. The Troubleshooter's Quiz reinforces critical thinking and troubleshooting skills for circuits introduced in the chapter. The Troubleshooter's Quiz consists of 8 to 12 multiple-choice questions that require students to consider how a given fault will affect voltage, current, gain, and so forth (increase, decrease, no change). Answers to the Troubleshooter's Quiz are found at the end of each chapter.

In addition, circuits have been prepared for many of the examples using Electronics Workbench™/Multisim to enable changes or troubles to be investigated. Electronics Workbench/Multisim is a computer-simulation program that is useful for testing circuits and observing the effect of parameter changes or troubles with the circuit. It uses a graphical interface to place components on a "workbench" and simulated instruments to view the results. These circuits are available on CD-ROM (ISBN: 0-13-060944-7).

Current in *Fundamentals of Analog Circuits, Second Edition,* is indicated by a meter notation rather than by directional arrows. This unique approach accomplishes two things. First, it eliminates the need to distinguish between conventional flow and electron flow because it indicates current direction by polarity signs, just as an actual ammeter does. Users can interpret current direction based on the meter polarity in accordance with their particular preference. Second, in addition to current direction, the meter notation provides relative magnitudes of the currents in a given circuit by observing the number of bars.

Overview

The first five chapters provide a fundamental coverage of basic concepts, diodes, transistors, and amplifiers. The last ten chapters focus on integrated circuit op-amps, active filters, oscillators, power supplies, special amplifiers, communications circuits, data conversion circuits, and measurement and control circuits.

Discrete Devices and Circuits   The first part of the text consists of five chapters as follows: Chapter 1 presents an introduction to analog electronics, analog signals, amplifiers, and troubleshooting. Chapter 2 covers diodes, rectifiers, and regulators. Chapter 3 introduces bipolar junction transistors and BJT amplifiers. Chapter 4 gives a basic treatment of field-effect transistors and FET amplifiers. Chapter 5 deals with multistage amplifiers, radio-frequency (RF) amplifiers, and power amplifiers.

Analog Integrated Circuits   The second part of the text consists of ten chapters that cover analog integrated circuits as follows: Chapter 6 provides an introduction to operational amplifiers. Op-amp frequency response is covered in Chapter 7, and basic op-amp circuits (comparators, summing amplifiers, integrators, and differentiators) is the topic of Chapter 8. Active op-amp filters are covered in Chapter 9, and oscillators and timers are introduced in Chapter 10. Power supplies are covered in Chapter 11. Special amplifiers (instrumentation amplifiers, isolation amplifiers, operational transconductance amplifiers (OTAs), and log/antilog amplifiers) are introduced in Chapter 12. Communication circuits (AM and FM receivers, linear multipliers, mixers, and phase-locked loops) are studied in Chapter 13. Data conversion circuits such as analog switches, sample-and-hold circuits, digital-to-analog and analog-to-digital converters, and voltage-to-frequency and frequency-to-voltage converters are among the topics in Chapter 14. Finally, Chapter 15 covers various types of transducers and associated measurement circuits.

Features

Fundamentals of Analog Circuits, Second Edition, is innovative in four areas:

❏ Current in a circuit is indicated by a polarized meter symbol that allows the user to apply the direction of preference. Also, current meters show relative current magnitude in a circuit. See Figure P–1.

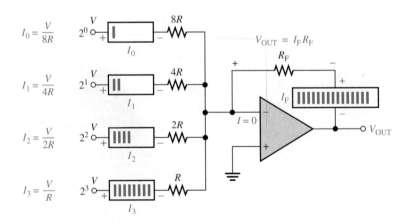

FIGURE P–1
Example of art showing meter symbols.

❑ Emphasis is on analog integrated circuits, but also there is a good coverage of discrete circuits.

❑ Topics that are generally not found in competing textbooks, such as RF amplifiers and transducers, are covered.

❑ System applications with Troubleshooter's Bench exercises incorporate realistic printed circuit boards, and a related full-color insert section is included.

Other features are as follows:

❑ Extensive troubleshooting material, including new Troubleshooter's Quiz

❑ Extensive use of examples

❑ Practice exercise in each numbered example

❑ Standard component values

❑ Two-page chapter openers with introduction, section list, objectives, key terms, and system application preview

❑ Section openers with overview and objectives

❑ End-of-section review questions

❑ Glossary terms boldfaced in text

❑ Answers to practice exercises for examples, section review questions, self-test, and Troubleshooter's Quiz at end of chapter

❑ Minimal mathematics, with important equations numbered

❑ A summary, key formula list, glossary, multiple-choice self-test, Troubleshooter's Quiz, and section problems for each chapter

❑ Key terms in color and with a margin icon in each chapter

❑ End-of-book derivations, manufacturers' specifications sheets, and answers to odd-numbered problems

❑ References to manufacturers' home pages for integrated circuits

❑ Comprehensive end-of-book glossary that includes all the terms from the end-of-chapter glossaries

❑ Lab Exercises manual written by David Buchla

❑ Instructor's Resource Manual that includes transparency masters, System Application worksheets, and test item file

❑ Visit the companion website to this text at www.prenhall.com/floyd.

Chapter Pedagogy

Chapter Opener Each chapter begins with a two-page spread, as indicated in Figure P–2.

Section Opener and Section Review Questions As illustrated in Figure P–3, each section within a chapter begins with an opening introduction and list of section objectives. Each section ends with a set of review questions that focus on key concepts. Answers to review questions are given at the end of the chapter.

Key Terms Certain terms are in color and are identified by a margin icon. These key terms, as well other bold terms, are defined in the end-of-chapter glossary and in the end-of-book glossary.

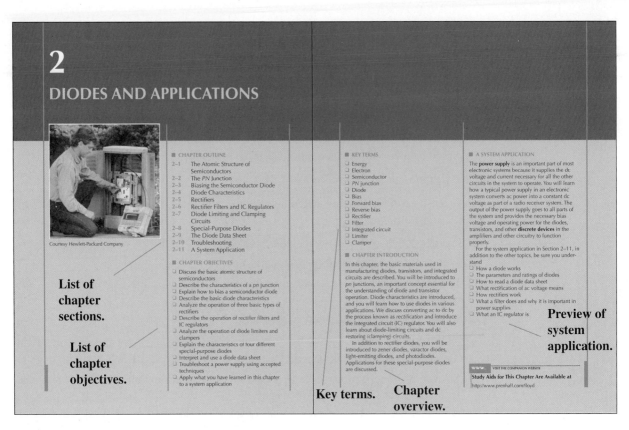

2

DIODES AND APPLICATIONS

Courtesy Hewlett-Packard Company

■ CHAPTER OUTLINE

2–1   The Atomic Structure of Semiconductors
2–2   The *PN* Junction
2–3   Biasing the Semiconductor Diode
2–4   Diode Characteristics
2–5   Rectifiers
2–6   Rectifier Filters and IC Regulators
2–7   Diode Limiting and Clamping Circuits
2–8   Special-Purpose Diodes
2–9   The Diode Data Sheet
2–10   Troubleshooting
2–11   A System Application

List of chapter sections.

■ CHAPTER OBJECTIVES

❏ Discuss the basic atomic structure of semiconductors
❏ Describe the characteristics of a pn junction
❏ Explain how to bias a semiconductor diode
❏ Describe the basic diode characteristics
❏ Analyze the operation of three basic types of rectifiers
❏ Describe the operation of rectifier filters and IC regulators
❏ Analyze the operation of diode limiters and clampers
❏ Explain the characteristics of four different special-purpose diodes
❏ Interpret and use a diode data sheet
❏ Troubleshoot a power supply using accepted techniques
❏ Apply what you have learned in this chapter to a system application

List of chapter objectives.

■ KEY TERMS

❏ Energy
❏ Electron
❏ Semiconductor
❏ *PN* junction
❏ Diode
❏ Bias
❏ Forward bias
❏ Reverse bias
❏ Rectifier
❏ Filter
❏ Integrated circuit
❏ Limiter
❏ Clamper

Key terms.

■ CHAPTER INTRODUCTION

In this chapter, the basic materials used in manufacturing diodes, transistors, and integrated circuits are described. You will be introduced to pn junctions, an important concept essential for the understanding of diode and transistor operation. Diode characteristics are introduced, and you will learn how to use diodes in various applications. We discuss converting ac to dc by the process known as *rectification* and introduce the integrated circuit (IC) regulator. You will also learn about diode-limiting circuits and dc restoring (clamping) circuits.

In addition to rectifier diodes, you will be introduced to zener diodes, varactor diodes, light-emitting diodes, and photodiodes. Applications for these special-purpose diodes are discussed.

Chapter overview.

■ A SYSTEM APPLICATION

The **power supply** is an important part of most electronic systems because it supplies the dc voltage and current necessary for all the other circuits in the system to operate. You will learn how a typical power supply in an electronic system converts ac power into a constant dc voltage as part of a radio receiver system. The output of the power supply goes to all parts of the system and provides the necessary bias voltage and operating power for the diodes, transistors, and other **discrete devices** in the amplifiers and other circuitry to function properly.

For the system application in Section 2–11, in addition to the other topics, be sure you understand

❏ How a diode works
❏ The parameters and ratings of diodes
❏ How to read a diode data sheet
❏ What rectification of ac voltage means
❏ How rectifiers work
❏ What a filter does and why it is important in power supplies
❏ What an IC regulator is

Preview of system application.

www.   VISIT THE COMPANION WEBSITE
Study Aids for This Chapter Are Available at
http://www.prenhall.com/floyd

FIGURE P–2
Chapter opener.

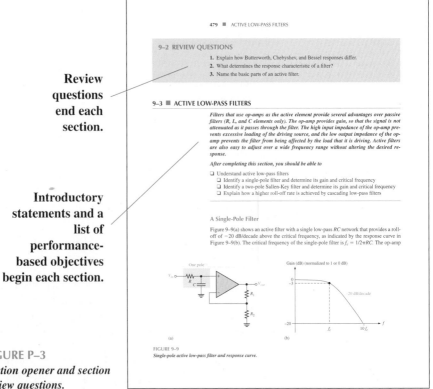

9–2 REVIEW QUESTIONS

1. Explain how Butterworth, Chebyshev, and Bessel responses differ.
2. What determines the response characteristic of a filter?
3. Name the basic parts of an active filter.

Review questions end each section.

9–3 ■ ACTIVE LOW-PASS FILTERS

Filters that use op-amps as the active element provide several advantages over passive filters (R, L, and C elements only). The op-amp provides gain, so that the signal is not attenuated as it passes through the filter. The high input impedance of the op-amp prevents excessive loading of the driving source, and the low output impedance of the op-amp prevents the filter from being affected by the load that it is driving. Active filters are also easy to adjust over a wide frequency range without altering the desired response.

After completing this section, you should be able to

❏ Understand active low-pass filters
   ❏ Identify a single-pole filter and determine its gain and critical frequency
   ❏ Identify a two-pole Sallen-Key filter and determine its gain and critical frequency
   ❏ Explain how a higher roll-off rate is achieved by cascading low-pass filters

Introductory statements and a list of performance-based objectives begin each section.

A Single-Pole Filter

Figure 9–9(a) shows an active filter with a single low-pass *RC* network that provides a roll-off of −20 dB/decade above the critical frequency, as indicated by the response curve in Figure 9–9(b). The critical frequency of the single-pole filter is $f_c = 1/2\pi RC$. The op-amp

FIGURE 9–9
Single-pole active low-pass filter and response curve.

FIGURE P–3
Section opener and section review questions.

Examples and Practice Exercises   Worked-out examples are used to illustrate and clarify topics covered in the text. At the end of every example and within the example box is a practice exercise that either reinforces the example or focuses on a related topic. Answers to the practice exercises are given at the end of the chapter. This feature is illustrated in Figure P–4.

System Application   As illustrated in Figure P–5, the last section of each chapter (except Chapter 1) is a system application of devices and circuits related to the chapter coverage. The Troubleshooter's Bench sections provide a series of activities with a practical slant to simulate "on-the-job" situations. These activities include relating a schematic to a realistic printed circuit board, making measurements, troubleshooting, and writing reports. Three selected system applications are related to the full-color insert as indicated by a special logo. Results are provided in the Instructor's Resource Manual.

    The system application is an optional feature which if omitted will not affect the coverage of any other topics. The variety of "systems" is intended to give students an appreciation for the wide range of applications for electronic devices and to provide motivation to learn the basic concepts of each chapter. The system application sections can be used as:

❑ A part of each chapter for the purpose of relating devices to a realistic application and for establishing a useful purpose for devices covered. All or selected activities can be assigned and discussed in class or turned in for a grade.

**Examples
are contained
within a
ruled box.**

**Each example
contains an
exercise related
to the example.**

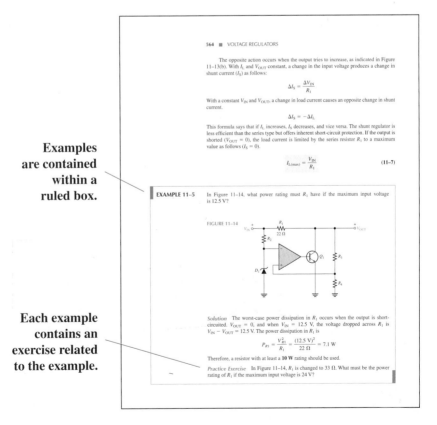

FIGURE P–4
Typical example and practice exercise.

Realistic PC board provides visual information related to the assignment.

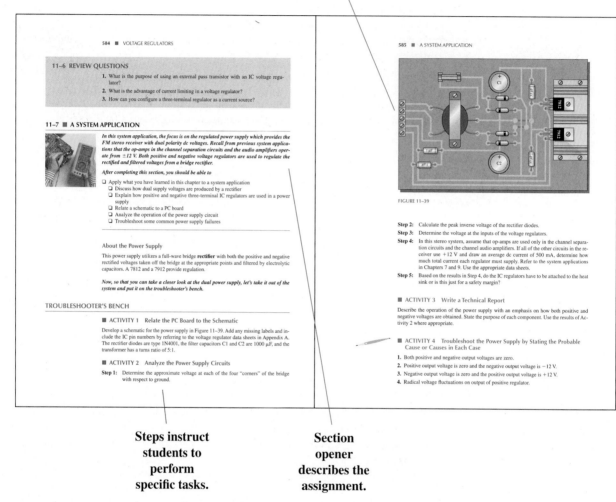

FIGURE P–5
A typical system application.

Steps instruct students to perform specific tasks.

Section opener describes the assignment.

❑ A separate out-of-class assignment to be turned in for extra credit.

❑ An in-class activity to promote and stimulate discussion and interaction among students and between students and instructor.

❑ An illustration to help answer the question that many students have: "Why do I need to know this?"

Chapter End Matter A summary, key formula list, glossary, self-test, Troubleshooter's Quiz, and sectionalized problem sets are found at the end of each chapter. The answers to practice exercises for examples, section review questions, self-test, and Troubleshooter's Quiz are also provided.

To the Student

Any career training requires hard work, and electronics is no exception. The best way to learn new material is by reading, thinking, and doing. This text is designed to help you along the way by providing an overview and objectives for each section, numerous worked-out examples, practice exercises, and review questions with answers.

Don't expect every concept to be crystal clear after a single reading. Read each section of the text carefully and think about what you have read. Work through the example problems step-by-step before trying the practice exercise that goes with the example. Sometimes more than one reading of a section will be necessary. After each section, check your understanding by answering the section review questions.

Review the chapter summary, glossary, and formula list. Take the multiple-choice self-test. Finally, work the problems at the end of the chapter. Check your answers to the self-test and the odd-numbered problems against those provided. Working problems is the most important way to check your comprehension and solidify concepts.

One of the best ways to reinforce text material is through the actual construction of circuits in the laboratory. You will become a better troubleshooter as well if you "learn by doing." Circuit construction reinforces troubleshooting skills because you will find that many times a simple wiring error or other fault is accidentally introduced in your experiment. Making a circuit work correctly involves analysis of the circuit as well as logical thinking. The sort of thinking that goes into lab work is also simulated on Electronics Workbench/Multisim. Another way to develop skill in troubleshooting is to take the Troubleshooter's Quiz, located at the back of each chapter; answers are provided to check your understanding.

Milestones in Electronics

Before you begin your study of analog circuits, let's briefly look at some of the important developments that led to electronics technology as we have today. The names of many of the early pioneers in electricity and electromagnetics still live on in terms of familiar units and quantities. Names such as Ohm, Ampere, Volta, Farad, Henry, Coulomb, Oersted, and Hertz are some of the better known examples. More widely known names such as Franklin and Edison are also significant in the history of electricity and electronics because of their tremendous contributions.

The Beginning of Electronics   Early experiments with electronics involved electric currents in vacuum tubes. Heinrich Geissler (1814–1879) removed most of the air from a glass tube and found that the tube glowed when there was current through it. Later, Sir William Crookes (1832–1919) found the current in vacuum tubes seemed to consist of particles. Thomas Edison (1847–1931) experimented with carbon filament bulbs with plates and discovered that there was a current from the hot filament to a positively charged plate. He patented the idea but never used it.

Other early experimenters measured the properties of the particles that flowed in vacuum tubes. Sir Joseph Thompson (1856–1940) measured properties of these particles, later called *electrons.*

Although wireless telegraphic communication dates back to 1844, electronics is basically a 20th century concept that began with the invention of the vacuum tube amplifier. An early vacuum tube that allowed current in only one direction was constructed by John A. Fleming in 1904. Called the Fleming valve, it was the forerunner of vacuum tube diodes. In 1907, Lee deForest added a grid to the vacuum tube. The new device, called the

FIGURE P–6
The invention of the bipolar junction transistor. Photo copyright by Bell Laboratories. All rights reserved. Used with permission.

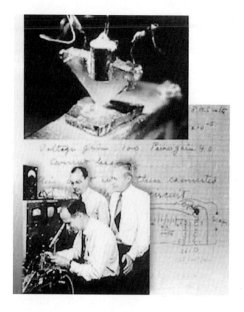

audiotron, could amplify a weak signal. By adding the control element, deForest ushered in the electronics revolution. It was with an improved version of his device that made transcontinental telephone service and radios possible. In 1912, a radio amateur in San Jose, California, was regularly broadcasting music!

In 1921, the secretary of commerce, Herbert Hoover, issued the first license to a broadcast radio station; within two years over 600 licenses were issued. By the end of the 1920s radios were in many homes. A new type of radio, the superheterodyne radio, invented by Edwin Armstrong, solved problems with high-frequency communication. In 1923, Vladimir Zworykin, an American researcher, invented the first television picture tube, and in 1927 Philo T. Farnsworth applied for a patent for a complete television system.

The 1930s saw many developments in radio, including metal tubes, automatic gain control, "midgit sets," directional antennas, and more. Also started in this decade was the development of the first electronic computers. Modern computers trace their origins to the work of John Atanasoff at Iowa State University. Beginning in 1937, he envisioned a binary machine that could do complex mathematical work. By 1939, he and graduate student Clifford Berry had constructed a binary machine called ABC, (for Atanasoff-Berry Computer) that used vacuum tubes for logic and condensers (capacitors) for memory. In 1939, the magnetron, a microwave oscillator, was invented in Britain by Henry Boot and John Randall. In the same year, the klystron microwave tube was invented in America by Russell and Sigurd Varian.

During World War II, electronics developed rapidly. Radar and very high-frequency communication were made possible by the magnetron and klystron. Cathode ray tubes were improved for use in radar. Computer work continued during the war. By 1946, John von Neumann had developed the first stored program computer, the Eniac, at the University of Pennsylvania. The decade ended with one of the most important inventions ever, the transistor.

Solid-State Electronics   The crystal detectors used in early radios were the forerunners of modern solid-state devices. However, the era of solid-state electronics began with the invention of the transistor in 1947 at Bell Labs. The inventors were Walter Brattain, John Bardeen, and William Shockley, shown in Figure P–6. PC (printed circuit) boards were introduced in 1947, the year the transistor was invented. Commercial manufacturing of transistors began in Allentown, Pennsylvania, in 1951.

The most important invention of the 1950s was the integrated circuit. On September 12, 1958, Jack Kilby, at Texas Instruments, made the first integrated circuit, for which he was awarded a Nobel prize in the fall of 2000. This invention literally created the modern computer age and brought about sweeping changes in medicine, communication, manufacturing, and the entertainment industry. Many billions of "chips"—as integrated circuits came to be called—have since been manufactured.

The 1960s saw the space race begin and spurred work on miniaturization and computers. The space race was the driving force behind the rapid changes in electronics that followed. The first successful "op-amp" was designed by Bob Widlar at Fairchild Semiconductor in 1965. Called the μA709, it was very successful but suffered from "latch-up" and other problems. Later, the most popular op-amp ever, the 741, was taking shape at Fairchild. This op-amp became the industry standard and influenced design of op-amps for years to come. Precursors to the Internet began in the 1960s with remote networked computers. Systems were in place within Lawrence Livermore National Laboratory that connected over 100 terminals to a computer system (colorfully called the "Octopus system" and used by one of this text's authors). In an experiment in 1969 with very remote computers, an exchange took place between researchers at UCLA and Stanford. The UCLA group hoped to connect to a Stanford computer and began by typing the word "login" on its terminal. A separate telephone connection was set up and the following conversation occurred.

The UCLA group asked over the phone, "Do you see the letter L?"

"Yes, we see the L."

The UCLA group typed an O. "Do you see the letter O?"

"Yes, we see the O."

The UCLA group typed a G. At this point the system crashed. Such was technology, but a revolution was in the making.

By 1971, a new company that had been formed by a group from Fairchild introduced the first microprocessor. The company was Intel and the product was the 4004 chip, which had the same processing power as the Eniac computer. Later in that same year, Intel announced the first 8-bit processor, the 8008. In 1975, the first personal computer was introduced by Altair, and *Popular Science* magazine featured it on the cover of the January, 1975, issue. The 1970s also saw the introduction of the pocket calculator and new developments in optical integrated circuits.

By the 1980s, half of all U.S. homes were using cable hookups instead of television antennas. The reliability, speed, and miniaturization of electronics continued throughout the 1980s, including automated testing and calibrating of PC boards. The computer became a part of instrumentation and the virtual instrument was created. Computers became a standard tool on the workbench.

The 1990s saw a widespread application of the Internet. In 1993, there were only 130 websites; by the start of the new century (in 2001) there were over 24 million. In the 1990s, companies scrambled to establish a home page and many of the early developments of radio broadcasting had parallels with the Internet. (The bean counters still want to know how it's going to make money!) The exchange of information and e-commerce fueled the tremendous economic growth of the 1990s. The Internet became especially important to scientists and engineers, becoming one of the most important scientific communication tools ever.

In 1995, the FCC allocated spectrum space for a new service called Digital Audio Radio Service. Digital television standards were adopted in 1996 by the FCC for the nation's next generation of broadcast television. As the 20th century drew toward a close, historians could only breathe a sigh of relief. As one wag put it, "I'm all for new technologies, but I wish they'd let the old ones wear out first."

The 21st century dawned in January 2001 (although most people celebrated the new century the previous year, known as "Y2K"). The major technology story was the continued explosive growth of the Internet. Traffic on the Internet doubles every 100 days with no end in sight. The future of technology looks brighter than ever.

Acknowledgments

This textbook is the result of not only the authors' collaboration, but also the skills and efforts of all those at Prentice Hall who were involved in this project. We would particularly like to express our appreciation to Rex Davidson, Scott Sambucci, Kate Linsner, and Steve Helba. Lois Porter did another outstanding job of manuscript editing from beginning to end, and Jane Lopez and Steve Botts deserve our admiration for their work on the art. Thanks also to Gary Snyder and Chuck Garbinski for their superb work in checking accuracy. We also appreciate feedback from various users; we have made several improvements as a result of these suggestions.

Tom Floyd

Dave Buchla

Contents

COLOR INSERT: CIRCUIT BOARDS AND INSTRUMENTATION FOR SELECTED SYSTEM APPLICATIONS

This special 8-page full-color insert, which follows Appendix B, page 892, provides realistic printed circuit boards and test instruments for special Troubleshooter's Bench assignments related to certain System Applications.

1

BASIC CONCEPTS OF ANALOG CIRCUITS AND SIGNALS

■ **CHAPTER OUTLINE**

■ **CHAPTER OBJECTIVES**

❏ Discuss the basic characteristics of analog electronics
❏ Describe analog signals
❏ Analyze signal sources
❏ Explain the characteristics of an amplifier
❏ Describe the process for troubleshooting an analog circuit

■ **KEY TERMS**

❏ Characteristic curve
❏ Analog signal
❏ Digital signal
❏ Period (T)
❏ Cycle
❏ Phase angle
❏ Frequency
❏ Thevenin's theorem
❏ Load line
❏ Transducer
❏ Amplifier
❏ Gain
❏ Decibel (dB)
❏ Attenuation

■ CHAPTER INTRODUCTION

With the influence of computers and other digital devices, it's easy to overlook the fact that virtually all natural phenomena that we measure (for example, pressure, flow rate, and temperature) originate as analog signals. In electronics, transducers are used to convert these analog quantities into voltage or current. Usually amplification or other processing is required for these signals. Depending on the application, either digital or analog techniques may be more efficient for processing. Analog circuits are found in nearly all power supplies, in many "real-time" applications (such as motor-speed controls), and in high-frequency communication systems. Digital processing is more effective when mathematical operations must be performed and has major advantages in reducing the noise inherent in processing analog signals. In short, the two sides of electronics (analog and digital) complement each other, and the competent technician needs to be knowledgeable of both.

1–1 ■ ANALOG ELECTRONICS

The field of electronics can be subdivided into various categories for study. The most basic division is to categorize signals between those that are represented by binary numbers (digital) and those that are represented by continuously variable quantities (analog). Digital electronics includes all arithmetic and logic operations such as performed in computers and calculators. Analog electronics includes virtually all other (nondigital) signals. Analog electronics includes signal-processing functions such as amplification, differentiation, and integration.

After completing this section, you should be able to

❑ Discuss the basic characteristics of analog electronics
  ❑ Contrast the characteristic curve for a linear component with that of a nonlinear component
  ❑ Explain what is meant by a characteristic curve
  ❑ Compare dc and ac resistance and explain how they differ
  ❑ Explain the difference between conventional current and electron flow

Modern electronics had its beginnings in 1907 when Lee deForest first inserted a metallic grid in a vacuum tube and was able to control the current in a circuit. Today, electronic systems still control voltages and currents but use solid-state devices. Basic electronic components, such as resistors or diodes, can be represented with graphs that show their characteristics in a more intuitive manner than mathematical equations. In this section, you will examine graphs representing resistors and diodes. In Chapter 3, you will see how the addition of a control element (like deForest's grid) can also be illustrated with graphs to provide a graphical picture of circuit operation.

Linear Equations

In basic algebra, a linear equation is one that plots a straight line between the variables and is usually written in the following form:

$$y = mx + b$$

where y = the dependent variable
  x = the independent variable
  m = the slope
  b = the y-axis intercept

If the plot of the equation goes through the origin, then the y-axis intercept is zero, and the equation reduces to

$$y = mx$$

which has the same form as Ohm's law.

$$I = \frac{V}{R} \tag{1–1}$$

As written here, the dependent variable in Ohm's law is current (I), the independent variable is voltage (V), and the slope is the reciprocal of resistance ($1/R$). Recall from your dc/ac course that this is simply the conductance, (G). By substitution, the linear form of Ohm's law is more obvious; that is,

$$I = GV$$

A **linear component** is one in which an increase in current is proportional to the applied voltage as given by Ohm's law. In general, a plot that shows the relationship between two variable properties of a device defines a **characteristic curve**. For most electronic devices, a characteristic curve refers to a plot of the current, I, plotted as a function of voltage, V. For example, resistors have an IV characteristic described by the straight lines given in Figure 1–1. Notice that current is plotted on the y-axis because it is the dependent variable.

FIGURE 1–1

IV characteristic curve for two resistors.

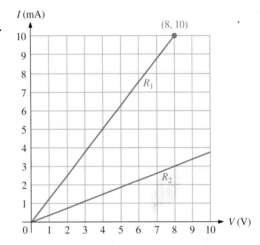

EXAMPLE 1–1

Figure 1–1 shows the IV characteristic curve for two resistors. What are the conductance and resistance of R_1?

Solution Find the conductance, G_1, by measuring the slope of the IV characteristic curve for R_1. The slope is the change in the y variable (written Δy) divided by the corresponding change in the x variable (written Δx).

$$\text{slope} = \frac{\Delta y}{\Delta x}$$

Choosing the point ($x = 8$ V, $y = 10$ mA) from Figure 1–1 and the origin, ($x = 0$ V, $y = 0$ mA), you can find the slope and therefore the conductance as

$$G_1 = \frac{10 \text{ mA} - 0 \text{ mA}}{8.0 \text{ V} - 0 \text{ V}} = \textbf{1.25 mS}$$

For a straight line, the slope is constant so you can use any two points to determine the conductance. The resistance is the reciprocal of the conductance.

$$R_1 = \frac{1}{G_1} = \frac{1}{1.25 \text{ mS}} = \textbf{0.8 k}\boldsymbol{\Omega}$$

*Practice Exercise** Find the conductance and resistance of R_2.

---

* Answers are at the end of the chapter.

AC Resistance

As you have seen, the graph of the characteristic curve for a resistor is a straight line that passes through the origin. The slope of the line is constant and represents the conductance of the resistor; the reciprocal of the slope represents resistance. The ratio of voltage at some point to the corresponding current at that point is referred to as *dc resistance.* DC resistance is defined by Ohm's law, $R = V/I$.

Many devices studied in analog electronics have a characteristic curve for which the current is not proportional to the voltage. These devices are nonlinear by nature but are included in the study of analog electronics because they take on a continuous range of input signals.

Figure 1–2 shows an *IV* characteristic curve for a diode, a nonlinear analog device. (Diodes are discussed in Chapter 2.) Generally, it is more useful to define the resistance of a nonlinear analog device as a small *change* in voltage divided by the corresponding *change* in current, that is, $\Delta V/\Delta I$. The ratio of a small change in voltage divided by the corresponding small change in current is defined as the **ac resistance** of an analog device.

$$r_{ac} = \frac{\Delta V}{\Delta I}$$

This internal resistance (indicated with a lowercase italic *r*) is also called the *dynamic, small signal,* or *bulk resistance* of the device. The ac resistance depends on the particular point on the *IV* characteristic curve where the measurement is made.

FIGURE 1–2

An IV characteristic curve for a diode.

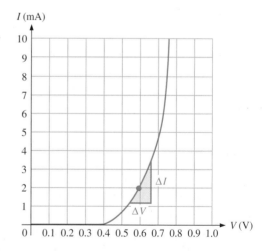

For the diode in Figure 1–2, the slope varies dramatically; the point where the ac resistance is measured needs to be specified with any measurement. For example, the slope at the point $x = 0.6$ V, $y = 2$ mA is found by computing the ratio of the change in current to the change in voltage as defined by the small triangle shown in the figure. The change in current, ΔI, is 3.4 mA − 1.2 mA = 2.2 mA and the change in voltage, ΔV, is 0.66 V − 0.54 V = 0.12 V. The ratio of $\Delta I/\Delta V$ is 2.2 mA/0.12 V = 18.3 mS. This represents the conductance, g, at the specified point. The internal ac resistance is the reciprocal of this value:

$$r = \frac{1}{g} = \frac{1}{18.3 \text{ mS}} = 54.5 \ \Omega$$

Conventional Current Versus Electron Flow

From your dc/ac circuits course, you know that current is the rate of flow of charge. The original definition of current was based on Benjamin Franklin's belief that electricity was an unseen substance that moved from positive to negative. *Conventional current* assumes for analysis purposes that current is out of the positive terminal of a voltage source, through the circuit, and into the negative terminal of the source. Engineers use this definition and many textbooks show current with arrows drawn with this viewpoint.

Today, it is known that in solid metallic conductors, the moving charge is actually negatively charged electrons. Electrons move from the negative to the positive point, opposite to the defined direction of conventional current. The movement of electrons in a conductor is called *electron flow*. Many schools and textbooks show electron flow with current arrows drawn out of the negative terminal of a voltage source.

Unfortunately, the controversy between whether it is better to show conventional current or electron flow in representing circuit behavior has continued for many years and does not appear to be subsiding. It is not important which direction you use to form a mental picture of current. In practice, there is only one correct direction to connect a dc ammeter to make current measurements. Throughout this text, the proper polarity for dc meters is shown when appropriate. Current paths are indicated with special bar meter symbols. In a given circuit, larger or smaller currents are indicated by the relative number of bars shown on a bar graph meter.

1–1 REVIEW QUESTIONS*

1. What is a characteristic curve for a component?
2. How does the characteristic curve for a large resistor compare to the curve for a smaller resistor?
3. What is the difference between dc resistance and ac resistance?

* Answers are at the end of the chapter.

1–2 ■ ANALOG SIGNALS

A signal is any physical quantity that carries information. It can be an audible, visual, or other indication of information. In electronics, the term signal refers to the information that is carried by electrical waves, either in a conductor or as an electromagnetic field.

After completing this section, you should be able to

❏ Describe analog signals
  ❏ Compare an analog signal with a digital signal
  ❏ Define *sampling* and *quantizing*
  ❏ Apply the equation for a sinusoidal wave to find the instantaneous value of a voltage or current
  ❏ Find the peak, rms, or average value, given the equation for a sinusoidal wave
  ❏ Explain the difference between the time-domain signal and the frequency-domain signal

Analog and Digital Signals

Signals can be classified as either continuous or discrete. A continuous signal changes smoothly, without interruption. A discrete signal can have only certain values. The terms *continuous* and *discrete* can be applied either to the amplitude or to the time characteristic of a signal.

In nature, most signals take on a continuous range of values within limits; such signals are referred to as **analog signals**. For example, consider a potentiometer that is used as a shaft encoder as shown in Figure 1–3(a). The output voltage can be continuously varied within the limit of the supply voltage, resulting in an analog signal that is related to the angular position of the shaft.

On the other hand, another type of encoder has a certain number of steps that can be selected as shown in Figure 1–3(b). When numbers are assigned to these steps, the result is called a **digital signal**.

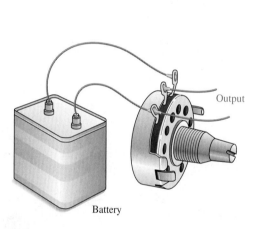

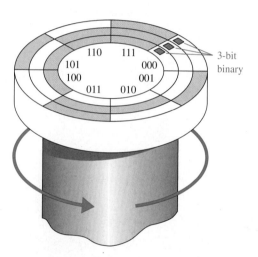

(a) Analog shaft encoder　　　　　(b) Digital shaft encoder

FIGURE 1–3
Analog and digital shaft encoders.

Analog circuits are generally simple, have high speed and low cost, and can readily simulate natural phenomena. They are often used for operations such as performing linearizing functions, waveshaping, transforming voltage to current or current to voltage, multiplying, and mixing. By contrast, digital circuits have high noise immunity, no drift, and the ability to process data rapidly and to perform various calculations. In many electronic systems, a mix of analog and digital signals are required to optimize the overall system's performance or cost.

Many signals have their origin in a natural phenomenon such as a measurement of pressure or temperature. Transducer outputs are typically analog in nature; a microphone, for example, provides an analog signal to an amplifier. Frequently, the analog signal is converted to digital form for storing, processing, or transmitting.

Conversion from analog to digital form is accomplished by a two-step process: sampling and quantizing. **Sampling** is the process of breaking the analog waveform into time "slices" that approximate the original wave. This process always loses some information; however, the advantages of digital systems (noise reduction, digital storage, and processing) outweigh the disadvantages. After sampling, the time slices are assigned a numeric value. This process, called **quantizing,** produces numbers that can be processed by digital computers or other digital circuits. Figure 1–4 illustrates the sampling and quantizing process.

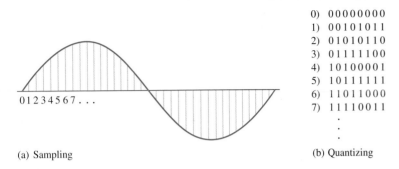

0)	0 0 0 0 0 0 0 0
1)	0 0 1 0 1 0 1 1
2)	0 1 0 1 0 1 1 0
3)	0 1 1 1 1 1 0 0
4)	1 0 1 0 0 0 0 1
5)	1 0 1 1 1 1 1 1
6)	1 1 0 1 1 0 0 0
7)	1 1 1 1 0 0 1 1

0 1 2 3 4 5 6 7 . . .

(a) Sampling   (b) Quantizing

FIGURE 1–4
Digitizing an analog waveform.

Frequently, digital signals need to be converted back to their original analog form to be useful in their final application. For instance, the digitized sound on a CD must be converted to an analog signal and eventually back to sound by a loudspeaker.

Periodic Signals

To carry information, some property such as voltage or frequency of an electrical wave needs to vary. Frequently, an electrical signal repeats at a regular interval of time. Repeating waveforms are said to be **periodic.** The **period (T)** represents the time for a periodic wave to complete one cycle. A **cycle** is the complete sequence of values that a waveform exhibits before another identical pattern occurs. The period can be measured between any two corresponding points on successive cycles.

Periodic waveshapes are used extensively in electronics. Many practical electronic circuits such as oscillators generate periodic waves. Most oscillators are designed to produce a particular shaped waveform—either a sinusoidal wave or nonsinusoidal waves such as the square, rectangular, triangle, and sawtooth waves.

The most basic and important periodic waveform is the sinusoidal wave. Both the trigonometric sine and cosine functions have the shape of a sinusoidal wave. The term *sine wave* usually implies the trigonometric function, whereas the term *sinusoidal wave* means a waveform with the shape of a sine wave. A sinusoidal waveform is generated as the natural waveform from many ac generators and in radio waves. Sinusoidal waves are also present in physical phenomena from the generation of laser light, the vibration of a tuning fork, or the motion of ocean waves.

A **vector** is any quantity that has both magnitude and direction. A sinusoidal curve can be generated by plotting the projection of the end point of a rotating vector that is turning with uniform circular motion, as illustrated in Figure 1–5. Successive revolutions of the point generate a periodic curve which can be expressed mathematically as

$$y(t) = A \sin(\omega t \pm \phi) \tag{1–2}$$

where $y(t)$ = vertical displacement of a point on the curve from the horizontal axis. The bracketed quantity (t) is an optional indicator, called *functional notation,*

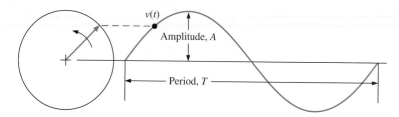

FIGURE 1–5
Generation of a sinusoidal waveform from the projection of a rotating vector.

to emphasize that the signals vary with time. Functional notation is frequently omitted when it isn't important to emphasize the time relationship but is introduced to familiarize you with the concept when it is shown.

A = amplitude. This is the maximum displacement from the horizontal axis.

ω = angular frequency of the rotating vector in radians per second.

t = time in seconds to a point on the curve.

ϕ = phase angle in radians. The **phase angle** is simply a fraction of a cycle that a waveform is shifted from a reference waveform of the same frequency. It is positive if the curve begins before $t = 0$ and is negative if the curve starts after $t = 0$.

Equation (1–2) illustrates that the sinusoidal wave can be defined in terms of three basic parameters. These are the frequency, amplitude, and phase angle.

Frequency and Period When the rotating vector has made one complete cycle, it has rotated through 2π radians. The number of complete cycles generated per second is called the **frequency**. Dividing the angular frequency (ω, in rad/s) of the rotating vector by the number of radians in one cycle (2π rad/cycle) gives the frequency in hertz.[1]

$$f \text{ (Hz)} = \frac{\omega \text{ (rad/s)}}{2\pi \text{ (rad/cycle)}} \qquad (1\text{–}3)$$

One cycle per second is equal to 1 Hz. The frequency (f) of a periodic wave is the number of cycles in one second and the period (T) is the time for one cycle, so it is logical that the reciprocal of the frequency is the period and the reciprocal of the period is the frequency.

$$T = \frac{1}{f} \qquad (1\text{–}4)$$

and

$$f = \frac{1}{T} \qquad (1\text{–}5)$$

---

[1] The unit of frequency was cycles per second (cps) prior to 1960 but was renamed the hertz (abbreviated Hz) in honor of Heinrich Hertz, a German physicist who demonstrated radio waves. The old unit designation was more descriptive of the definition of frequency.

For example, if a signal repeats every 10 ms, then its period is 10 ms and its frequency is

$$f = \frac{1}{T} = \frac{1}{10 \text{ ms}} = 0.1 \text{ kHz}$$

Instantaneous Value of a Sinusoidal Wave   If the sinusoidal waveform shown in Figure 1–5 represents a voltage, Equation (1–2) is written

$$v(t) = V_p \sin(\omega t \pm \phi)$$

In this equation, $v(t)$ is a variable that represents the voltage. Since it changes as a function of time, it is often referred to as the *instantaneous voltage.*

Peak Value of a Sinusoidal Wave   The amplitude of a sinusoidal wave is the maximum displacement from the horizontal axis as shown in Figure 1–5. For a voltage waveform, the amplitude is called the peak voltage, V_p. When making voltage measurements with an oscilloscope, it is often easier to measure the peak-to-peak voltage, V_{pp}. The peak-to-peak voltage is twice the peak value.

Average Value of a Sinusoidal Wave   During one cycle, a sinusoidal waveform has equal positive and negative excursions. Therefore, the mathematical definition of the average value of a sinusoidal waveform must be zero. However, the term *average value* is generally used to mean the average over a cycle without regard to the sign. That is, the average is usually computed by converting all negative values to positive values, then averaging. The average voltage is defined in terms of the peak voltage by the following equation:

$$V_{avg} = \frac{2V_p}{\pi}$$

Simplifying,

$$V_{avg} = 0.637V_p \tag{1–6}$$

The average value is useful in certain practical problems. For example, if a rectified sinusoidal waveform is used to deposit material in an electroplating operation, the quantity of material deposited is related to the average current:

$$I_{avg} = 0.637I_p$$

Effective Value (rms Value) of a Sinusoidal Wave   If you apply a dc voltage to a resistor, a steady amount of power is dissipated in the resistor and can be calculated using the following power law:

$$P = IV \tag{1–7}$$

where V = dc voltage across the resistor (volts)
$\qquad I$ = dc current in the resistor (amperes)
$\qquad P$ = power dissipated (watts)

A sinusoidal waveform transfers maximum power at the peak excursions of the curve and no power at all at the instant the voltage crosses zero. In order to compare ac and dc voltages and currents, ac voltages and currents are defined in terms of the equivalent heating value of dc. This equivalent heating value is computed with calculus, and the result is called the rms (for *root-mean-square*) voltage or current. The rms voltage is related to the peak voltage by the following equation:

$$V_{rms} = 0.707V_p \qquad \qquad \textbf{(1–8)}$$

Likewise, the effective or rms current is

$$I_{rms} = 0.707I_p$$

EXAMPLE 1–2

A certain voltage waveform is described by the following equation:

$$v(t) = 15 \text{ V } \sin(600t)$$

(a) From this equation, determine the peak voltage and the average voltage. Give the angular frequency in rad/s.

(b) Find the instantaneous voltage at a time of 10 ms.

Solution

(a) The form of the equation is

$$y(t) = A \sin(\omega t)$$

The peak voltage is the same as the amplitude (A).

$$V_p = \textbf{15 V}$$

The average voltage is related to the peak voltage.

$$V_{avg} = 0.637V_p = 0.637(15 \text{ V}) = \textbf{9.56 V}$$

The angular frequency, ω, is **600 rad/s.**

(b) The instantaneous voltage at a time of 10 ms is

$$v(t) = 15 \text{ V } \sin(600t) = 15 \text{ V } \sin(600)(10 \text{ ms}) = \textbf{−4.19 V}$$

Note the negative value indicates that the waveform is below the axis at this point.

Practice Exercise   Find the rms voltage, the frequency in hertz, and the period of the waveform described in the example.

Time-Domain Signals

Thus far, the signals you have looked at vary with time, and it is natural to associate time as the independent variable. Some instruments, such as the oscilloscope, are designed to record signals as a function of time. Time is therefore the independent variable. The values assigned to the independent variable are called the **domain.** Signals that have voltage, current, resistance, or other quantity vary as a function of time are called *time-domain* signals.

Frequency-Domain Signals

Sometimes it is useful to view a signal where frequency is represented on the horizontal axis and the signal amplitude (usually in logarithmic form) is plotted along the vertical axis. Since frequency is the independent variable, the instrument works in the *frequency domain,* and the plot of amplitude versus frequency is called a **spectrum.** The spectrum analyzer is an instrument used to view the spectrum of a signal. These instruments are extremely useful in radio frequency (RF) measurements for analyzing the frequency response of a circuit, testing for harmonic distortion, checking the percent modulation from transmitters, and many other applications.

You have seen how the sinusoidal wave can be defined in terms of three basic parameters. These are the amplitude, frequency, and phase angle. A continuous sinusoidal wave can be shown as a time-varying signal defined by these three parameters. The same sinusoidal wave can also be shown as a single line on a frequency spectrum. The frequency-domain representation gives information about the amplitude and frequency, but it does not show the phase angle. These two representations of a sinusoidal wave are compared in Figure 1–6. The height of the line on the spectrum is the amplitude of the sinusoidal wave.

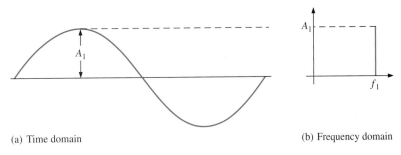

(a) Time domain   (b) Frequency domain

FIGURE 1–6
Time-domain and frequency-domain representations of a sinusoidal wave.

Harmonics   A nonsinusoidal repetitive waveform is composed of a fundamental frequency and harmonic frequencies. The fundamental frequency is the basic repetition rate of the waveform, and the **harmonics** are higher-frequency sinusoidal waves that are multiples of the fundamental. Interestingly, these multiples are all related to the fundamental by integers (whole numbers).

Odd harmonics are frequencies that are odd multiples of the fundamental frequency of a waveform. For example, a 1 kHz square wave consists of a fundamental of 1 kHz and odd harmonics of 3 kHz, 5 kHz, 7 kHz, and so on. The 3 kHz frequency in this case is called the third harmonic, the 5 kHz frequency is called the fifth harmonic, and so on.

Even harmonics are frequencies that are even multiples of the fundamental frequency. For example, if a certain wave has a fundamental of 200 Hz, the second harmonic is 400 Hz, the fourth harmonic is 800 Hz, and the sixth harmonic is 1200 Hz.

Any variation from a pure sinusoidal wave produces harmonics. A nonsinusoidal wave is a composite of the fundamental and certain harmonics. Some types of waveforms have only odd harmonics, some have only even harmonics, and some contain both. The shape of the wave is determined by its harmonic content. Generally, only the fundamental and the first few harmonics are important in determining the waveshape. For example, a square wave is formed from the fundamental and odd harmonics, as illustrated in Figure 1–7.

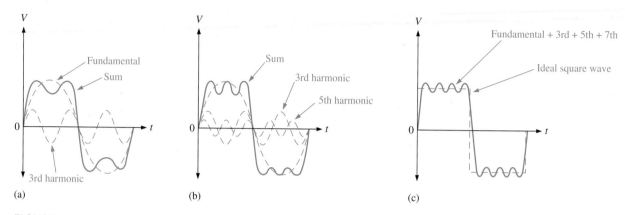

FIGURE 1–7

Odd harmonics combine to produce a square wave.

Fourier Series   All periodic waves except the sinusoidal wave itself are complex wave-forms composed of a series of sinusoidal waves. Jean Fourier, a French mathematician interested in problems of heat conduction, formed a mathematical series of trigonometry terms to describe periodic waves. This series is appropriately called the Fourier series.[2] With the Fourier series, one can mathematically determine the amplitude of each of the sinusoidal waves that compose a complex waveform.

The frequency spectrum developed by Fourier is often shown as an amplitude spectrum with units of voltage or power on the *y*-axis plotted against Hz on the *x*-axis. Figure 1–8(a) illustrates the amplitude spectrum for several different periodic waveforms. Notice that all spectrums for periodic waves are depicted as lines located at harmonics of the fundamental frequency. These individual frequencies can be measured with a spectrum analyzer.

Nonperiodic signals such as speech, or other transient waveforms, can also be represented by a spectrum; however, the spectrum is no longer a series of lines as in the case of repetitive waves. Transient waveforms are computed by another method called the *Fourier transform*. The spectrum of a transient waveform contains a continuum of frequencies rather than just harmonically related components. A representative Fourier pair of signals for a nonrepetitive pulse are shown in Figure 1–8(b).

1–2 REVIEW QUESTIONS

1. What is the difference between an analog signal and a digital signal?
2. Describe the spectrum for a square wave.
3. How does the spectrum for a repetitive waveform differ from that of a nonrepetitive waveform?

---

[2] Although Fourier's work was significant and he was awarded a prize, his colleagues were uneasy about it. The famous mathematician, Legrange, argued in the French Academy of Science that Fourier's claim was impossible. For further information, see Scientific American, June 1989, p. 86.

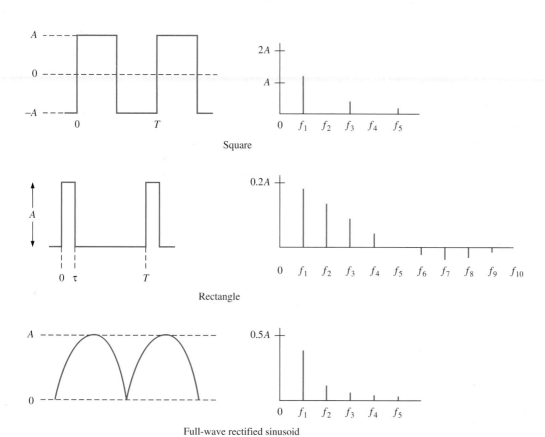

Square

Rectangle

Full-wave rectified sinusoid

(a) Examples of time-domain and frequency-domain representations of repetitive waves

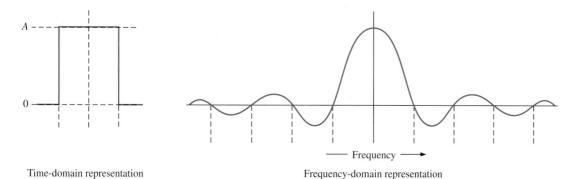

Time-domain representation   Frequency-domain representation

(b) Examples of the frequency spectrum of a nonrepetitive pulse waveform

FIGURE 1–8

Comparison of the frequency spectrum of repetitive and nonrepetitive waves.

1–3 ■ SIGNAL SOURCES

You may recall from basic electronics that Thevenin's theorem allows you to replace a complicated linear circuit with a single voltage source and a series resistance. The circuit is viewed from the standpoint of two output terminals. Likewise, Norton's theorem allows you to replace a complicated two-terminal, linear circuit with a single current source and a parallel resistance. These important theorems are useful for simplifying the analysis of a wide variety of circuits and should be thoroughly understood.

After completing this section, you should be able to

❑ Analyze signal sources
    ❑ Define two types of independent sources
    ❑ Draw a Thevenin or Norton equivalent circuit for a dc resistive circuit
    ❑ Show how to draw a load line for a Thevenin circuit
    ❑ Explain the meaning of Q-point
    ❑ Explain how a passive transducer can be modeled with a Thevenin equivalent circuit

Independent Sources

Signal sources can be defined in terms of either voltage or current and may be defined for either dc or ac. An ideal independent voltage source generates a voltage which does not depend on the load current. An ideal independent current source produces a current in the load which does not depend on the voltage across the load.

The value of an ideal independent source can be specified without regard to any other circuit parameter. Although a truly ideal source cannot be realized, in some cases, (such as a regulated power supply), it can be closely approximated. Actual sources can be modeled as consisting of an ideal source and a resistor (or other passive component for ac sources).

Thevenin's Theorem

Thevenin's theorem allows you to replace a complicated, two-terminal linear circuit with an ideal independent voltage source and a series resistance as illustrated in Figure 1–9. The source can be either a dc or ac source (a dc source is shown). **Thevenin's theorem** provides an equivalent circuit from the standpoint of the two output terminals. That is, the original circuit and the Thevenin circuit will produce exactly the same voltage and current in any load. Thevenin's theorem is particularly useful for analysis of linear circuits such as amplifiers, a topic that will be covered in Section 1–4.

FIGURE 1–9
Thevenin's equivalent for a dc circuit.

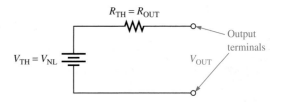

Only two quantities are needed to determine the equivalent Thevenin circuit—the Thevenin voltage and the Thevenin resistance. The Thevenin voltage, V_{TH}, is the open circuit (no load, NL) voltage from the original circuit. The Thevenin resistance, R_{TH}, is the resistance from the point of view of the output terminals with all voltage or current sources replaced by their internal resistance.

EXAMPLE 1–3   Find the equivalent Thevenin circuit for the dc circuit shown in Figure 1–10(a). The output terminals are represented by the open circles.

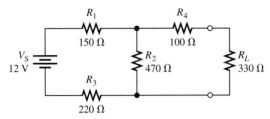

(a) Original circuit with load resistor, R_L

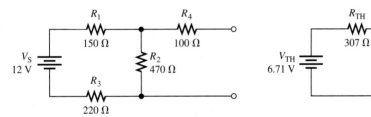

(b) Original circuit with load resistor, R_L, removed   (c) Thevenin equivalent of original circuit

FIGURE 1–10
Simplifying a circuit with Thevenin's theorem.

Solution   Find the Thevenin voltage by computing the voltage on the output terminals *as if the load were removed* as shown in Figure 1–10(b). With no load, there is no path for current through R_4. Therefore, there is no current and no voltage drop will appear across it. The output (Thevenin) voltage must be the same as the drop across R_2. Applying the voltage-divider rule for the equivalent series combination of R_1, R_2, and R_3, the voltage across R_2 is

$$V_{TH} = V_2 = V_S\left(\frac{R_2}{R_1 + R_2 + R_3}\right)$$

$$= 12 \text{ V}\left(\frac{470 \text{ }\Omega}{150 \text{ }\Omega + 470 \text{ }\Omega + 220 \text{ }\Omega}\right) = \textbf{6.71 V}$$

The Thevenin resistance is the resistance from the perspective of the output terminals with sources replaced with their internal resistance. The internal resistance of a voltage source is zero (ideally). Replacing the source with zero resistance places R_1 and R_3 in series and the combination in parallel with R_2. The equivalent resistance of these three resistors is in series with R_4. Thus, the Thevenin resistance for this circuit is

$$R_{TH} = [(R_1 + R_3) \parallel R_2] + R_4$$
$$= [(150 \text{ }\Omega + 220 \text{ }\Omega) \parallel 470 \text{ }\Omega] + 100 \text{ }\Omega = \textbf{307 }\Omega$$

The Thevenin circuit is shown in Figure 1–10(c).

Practice Exercise   Use the Thevenin circuit to find the voltage across the 330 Ω load resistor.

Thevenin's theorem is a useful way of combining linear circuit elements to form an equivalent circuit that can be used to answer questions with respect to various loads. The requirement that the Thevenin circuit elements are linear places some restrictions on the use of Thevenin's theorem. In spite of this, if the circuit to be replaced is approximately linear, Thevenin's theorem may produce useful results. This is the case for many amplifier circuits that we will investigate later.

Norton's Theorem

Norton's theorem provides another equivalent circuit similar to the Thevenin equivalent circuit. Norton's equivalent circuit can also replace any two-terminal linear circuit with a reduced equivalent. Instead of a voltage source, the Norton equivalent circuit uses a current source in parallel with a resistance, as illustrated in Figure 1–11.

FIGURE 1–11
Norton circuit. The arrow in the current source symbol always points to the positive side of the source.

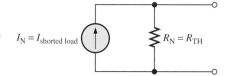

The magnitude of the Norton current source is found by replacing the load with a short and determining the current in the load. The Norton resistance is the same as the Thevenin resistance.

EXAMPLE 1–4   Find the equivalent Norton circuit for the dc circuit shown in Figure 1–12(a). The output terminals are represented by the open circles.

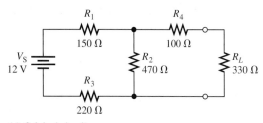

(a) Original circuit

(b) R_L replaced with a short

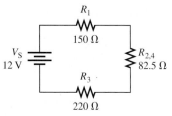

(c) R_2 and R_4 form an equivalent parallel resistor.

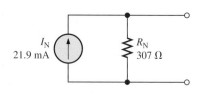

(d) The current in the short is equal to the Norton current.

FIGURE 1–12
Simplifying a circuit with Norton's theorem.

Solution   Find the Norton current by computing the current in the load *as if it were replaced by a short* as shown in Figure 1–12(b). The shorted load causes R_4 to appear in parallel with R_2 as shown in Figure 1–12(b). The total current in the equivalent circuit of Figure 1–12(c) can be found by applying Ohm's law to the total resistance.

$$I = \frac{V_S}{R_T} = \frac{12\ \text{V}}{R_1 + R_{2,4} + R_3} = \frac{12.0\ \text{V}}{452.5\ \Omega} = 26.5\ \text{mA}$$

The current (I_{SL}) in the shorted load is found by applying the current-divider rule to the R_2 and R_4 junction in the circuit of Figure 1–12(b).

$$I_{SL} = I_T \left(\frac{R_2}{R_2 + R_4} \right) = 26.5\ \text{mA} \left(\frac{470\ \Omega}{470\ \Omega + 100\ \Omega} \right) = \textbf{21.9 mA}$$

The current in the shorted load is the Norton current. The Norton resistance is equal to the Thevenin resistance, as found in Example 1–3. Notice that the Norton resistance is in parallel with the Norton current source. The equivalent circuit is shown in Figure 1–12(d).

Practice Exercise   Use Norton's theorem to find the voltage across the 330 Ω load resistor. Show that Norton's theorem gives the same result as Thevenin's theorem for this circuit (see Practice Exercise in Example 1–3).

Load Lines

An interesting way to obtain a "conceptual picture" of circuit operation is through the use of a load line for the circuit. Load lines are introduced here and will be explored further in Chapter 3.

Imagine a linear circuit that has an equivalent Thevenin circuit as shown in Figure 1–13. Let's see what happens if various loads are placed across the output terminals. First, assume there is a short (zero resistance). In this case, the voltage across the load is zero and the current is given by Ohm's law.

$$I_L = \frac{V_{TH}}{R_{TH}} = \frac{10\ \text{V}}{1.0\ \text{k}\Omega} = 10\ \text{mA}$$

Now assume the load is an open (infinite resistance). In this case, the load current is zero, and the voltage across the load is equal to the Thevenin voltage.

FIGURE 1–13

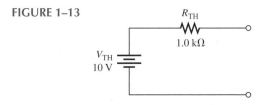

The two tested conditions represent the maximum and minimum current in the load. Table 1–1 shows the results of trying some more points to see what happens with different loads. Plotting the data as shown in Figure 1–14 establishes an *IV* curve for the Thevenin circuit. Because the circuit is a linear circuit, *any load that is placed across the output*

TABLE 1–1

Various load conditions for the circuit in Figure 1–13.

R_L	V_L	I_L
0 Ω	0.0 V	10.0 mA
250 Ω	2.00 V	8.00 mA
500 Ω	3.33 V	6.67 mA
750 Ω	4.29 V	5.72 mA
1.0 kΩ	5.00 V	5.00 mA
2.0 kΩ	6.67 V	3.33 mA
4.0 kΩ	8.00 V	2.00 mA
open	10.0 V	0.00 mA

FIGURE 1–14

Load line for the circuit in Figure 1–13.

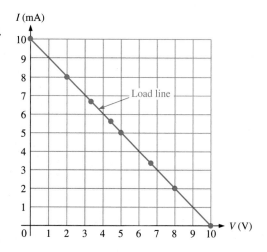

terminals falls onto the same straight line. This line is called the **load line** for the circuit and describes the driving circuit (in this case, the Thevenin circuit), not the load itself. Since the load line is a straight line, the first two calculated conditions (a short and an open load) are all that are needed to establish it.

Before we leave the topic of load lines, consider one more idea. Recall that a resistor (or any other device) has its own characteristic that can be described by its *IV* curve. The characteristic curve for a resistor represents all of the possible operating points for the device, whereas the load line represents all of the possible operating points for the circuit. Combining these ideas, you can superimpose the *IV* curve for a resistor on the plot of the load line for the Thevenin circuit. The intersection of these two lines gives the operating point for the combination.

Figure 1–15(a) shows an 800 Ω load resistor added to the Thevenin circuit from Figure 1–13. The load line for the Thevenin circuit and the characteristic curve for resistor R_1 from Figure 1–1 are shown in Figure 1–15(b). R_1 now serves as a load resistor, R_L. The intersection of the two lines represents the operating point, or **quiescent point,** commonly referred to as the Q-point. Note that the load voltage (4.4 V) and load current (5.6 mA) can be read directly from the graph. In Chapter 3, you will see that this idea can be extended to transistors and other devices to give a graphical tool for understanding circuit operation.

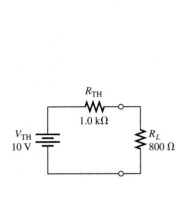

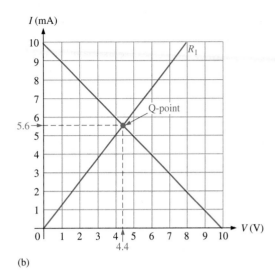

(a)

(b)

FIGURE 1–15
Load line and a resistor IV curve showing the Q-point.

Transducers

Analog circuits are frequently used in conjunction with a measurement that needs to be made. A **transducer** is a device that converts a physical quantity (such as position, pressure, or temperature) from one form to another; for electronic systems, input transducers convert a physical quantity to be measured into an electrical quantity (voltage, current, resistance). Transducers will be covered further in Chapter 15.

The signal from transducers is frequently very small, requiring amplification before being suitable for further processing. Passive transducers, such as strain gages, require a separate source of electrical power (called *excitation*) to perform their job. Others, such as thermocouples, are active transducers; they are self-generating devices that convert a small portion of the quantity to be measured into an electrical signal. Both passive and active transducers are often simplified to a Thevenin or Norton equivalent circuit for analysis.

In order to choose an appropriate amplifier, it is necessary to consider both the size of the source voltage and the size of the equivalent Thevenin or Norton resistance. When the equivalent resistance is very small, Thevenin's equivalent circuit is generally more useful because the circuit approximates an ideal voltage source. When the equivalent resistance is large, Norton's theorem is generally more useful because the circuit approximates an ideal current source. When the source resistance is very high, such as the case with a pH meter, a very high input impedance amplifier must be used. Other considerations, such as the frequency response of the system or the presence of noise, affect the choice of amplifier.

EXAMPLE 1–5

A piezoelectric crystal is used in a vibration monitor. Assume the output of the transducer should be a 60 mV rms sine wave with no load. When a technician connects an oscilloscope with a 10 MΩ input impedance across the output, the voltage is observed to be only 40 mV rms. Based on these observations, draw the Thevenin equivalent circuit for this transducer.

Solution   The open circuit ac voltage is the Thevenin voltage; thus, V_{th} = **60 mV.** The Thevenin resistance can be found indirectly using the voltage-divider rule. The oscilloscope input impedance is considered the load resistance, R_L, in this case. The voltage across the load is

$$V_{R_L} = V_{th}\left(\frac{R_L}{R_L + R_{th}}\right)$$

Rearranging terms,

$$\frac{R_L + R_{th}}{R_L} = \frac{V_{th}}{V_{R_L}}$$

Now solving for R_{th} and substituting the given values,

$$R_{th} = R_L\left(\frac{V_{th}}{V_{R_L}} - 1\right) = 10\ \text{M}\Omega\left(\frac{60\ \text{mV}}{40\ \text{mV}} - 1\right) = \textbf{5.0 M}\Omega$$

The equivalent transducer circuit is shown in Figure 1–16.

FIGURE 1–16

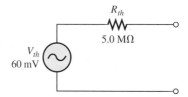

Practice Exercise   Draw the Norton's equivalent circuit for the same transducer.

1–3 REVIEW QUESTIONS

1. What is an independent source?
2. What is the difference between a Thevenin and a Norton circuit?
3. What is the difference between a passive and an active transducer?

1–4 ■ AMPLIFIERS

Before processing, most signals require amplification. Amplification is simply increasing the magnitude of a signal (either voltage, current, or both) and is one of the most important operations in electronics. Other operations in the field of linear electronics include signal generation (oscillators), waveshaping, frequency conversion, modulation, and many other processes. In addition to strictly linear or strictly digital circuits, many electronic circuits involve a combination of digital and linear electronics. These include an important class of interfacing circuits that convert analog-to-digital and digital-to-analog. These circuits will be considered in Chapter 14.

After completing this section, you should be able to

❑ Explain the characteristics of an amplifier
  ❑ Write the equations for voltage gain and power gain
  ❑ Draw the transfer curve for an amplifier
  ❑ Show how an amplifier can be modeled as Thevenin or Norton equivalent circuits to represent the input circuit and the output circuit
  ❑ Describe how an amplifier can be formed by cascading stages
  ❑ Determine the loading effect of one amplifier stage on another
  ❑ Use a calculator to find the logarithm or antilog of a given number
  ❑ Compute decibel voltage and power gain for an amplifier or circuit

Linear Amplifiers

The previous discussion on linear circuits can be extended to **amplifiers**. Linear amplifiers produce a magnified replica **(amplification)** of the input signal in order to produce a useful outcome (such as driving a loudspeaker). The concept of an *ideal amplifier* means that the amplifier introduces no noise or distortion to the signal; the output varies in time and replicates the input exactly.

Amplifiers are designed primarily to amplify either voltage or power. For a voltage amplifier, the output signal, $V_{out}(t)$, is proportional to the input signal, $V_{in}(t)$, and the ratio of output voltage to input voltage is voltage **gain**. To simplify the gain equation, you can omit the functional notation, (t), and simply show the ratio of the output signal voltage to the input signal voltage as

$$A_v = \frac{V_{out}}{V_{in}} \qquad (1\text{-}9)$$

where A_v = voltage gain
  V_{out} = output signal voltage
  V_{in} = input signal voltage

A useful way of looking at any circuit is to show the output for a given input. This plot, called a **transfer curve,** shows the response of the circuit. An ideal amplifier is characterized by a straight line that goes to infinity. For an actual linear amplifier, the transfer curve is a straight line until saturation is reached as shown in Figure 1–17. From this plot, the output voltage can be read for a given input voltage.

All amplifiers have certain limits, beyond which they no longer act as ideal. The output of the amplifier illustrated in Figure 1–17 eventually cannot follow the input; at this point the amplifier is no longer linear. Additionally, all amplifiers must operate from a source of energy, usually in the form of a dc power supply. Essentially, amplifiers convert some of this dc energy from the power supply into signal power. Thus, the output signal has larger power than the input signal. Frequently, block diagrams and other circuit representations omit the power supply, but it is understood to be present.

FIGURE 1–17

Transfer curve for a linear amplifier.

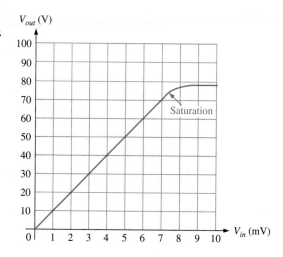

The Nonlinear Amplifier

Amplifiers are frequently used in situations where the output is not intended to be a replica of the input. These amplifiers form an important part of the field of analog electronics. They include two main categories: waveshaping and switching. A *waveshaping amplifier* is used to change the shape of a waveform. A *switching amplifier* produces a rectangular output from some other waveform. The input can be any waveform, for example, sinusoidal, triangle, or sawtooth. The rectangular output wave is often used as a control signal for some digital applications.

EXAMPLE 1–6

The input and output signals for a linear amplifier are shown in Figure 1–18 and represent an oscilloscope display. What is the voltage gain of the amplifier?

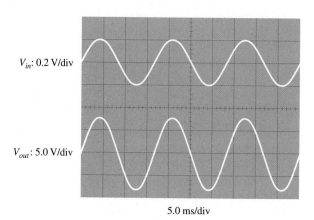

V_{in}: 0.2 V/div

V_{out}: 5.0 V/div

5.0 ms/div

FIGURE 1–18

Oscilloscope display.

Solution The input signal is 2.0 divisions from peak-to peak.

$$V_{in} = 2.0 \text{ div} \times 0.2 \text{ V/div} = 0.4 \text{ V}$$

The output signal is 3.2 divisions from peak-to peak.

$$V_{out} = 3.2 \text{ div} \times 5.0 \text{ V/div} = 16 \text{ V}$$

$$A_v = \frac{V_{out}}{V_{in}} = \frac{16 \text{ V}}{0.4 \text{ V}} = \mathbf{40}$$

Note that voltage gain is a ratio of voltages and therefore has no units. The answer is the same if rms or peak values had been used for both the input and output voltages.

Practice Exercise   The input to an amplifier is 20 mV. If the voltage gain is 300, what is the output signal?

Another gain parameter is power gain, A_p, defined as the ratio of the signal power out to the signal power in. Power is generally calculated using rms values of voltage or current; however, power gain is a ratio so you can use any consistent units. Power gain, shown as a function of time, is given by the following equation:

$$A_p = \frac{P_{out}}{P_{in}} \qquad\qquad \textbf{(1–10)}$$

where A_p = power gain
P_{out} = power out
P_{in} = power in

Power can be expressed by any of the standard power relationships studied in basic electronics. For instance, given the voltage and current of the input and output signals, the power gain can be written

$$A_p = \frac{I_{out}V_{out}}{I_{in}V_{in}}$$

where I_{out} = output signal current to the load
I_{in} = input signal current

Power gain can also be expressed by substituting $P = V^2/R$ for the input and output power.

$$A_p = \left(\frac{V_{out}^2/R_L}{V_{in}^2/R_{in}}\right)$$

where R_L = load resistor
R_{in} = input resistance of the amplifier

The particular equation you choose depends on what information is given.

Amplifier Model

An amplifier is a device that increases the magnitude of a signal for use by a load. Although amplifiers are complicated arrangements of transistors, resistors, and other components, a simplified description is all that is necessary when the requirement is to analyze the source and load behavior. The amplifier can be thought of as the interface between the source and load, as shown in Figure 1–19(a) and 1–19(b). You can apply the concept of equivalent circuits, learned in basic electronics courses, to the more complicated case of an amplifier. By drawing an amplifier as an equivalent circuit, you can simplify equations related to its performance.

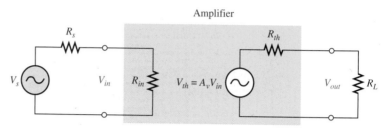

(a) Thevenin output circuit

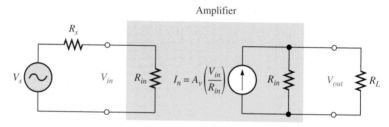

(b) Norton output circuit

FIGURE 1–19
Basic amplifier models showing the equivalent input resistance and dependent output circuits.

The input signal from a source is applied to the input terminals of the amplifier, and the output is taken from a second set of terminals. (Terminals are represented by open circles on a schematic.) The amplifier's input terminals present an input resistance, R_{in}, to the source. This input resistance affects the input voltage to the amplifier because it forms a voltage divider with the source resistance.

The output of the amplifier can be drawn as either a Thevenin or Norton source, as shown in Figure 1–19. The magnitude of this source is dependent on the unloaded gain (A_v) and the input voltage; thus, the amplifier's output circuit (drawn as a Thevenin or Norton equivalent) is said to contain a *dependent* source. The value of a dependent source always depends on voltage or current elsewhere in the circuit.[3] The voltage or current values for the Thevenin and Norton cases are shown in Figure 1–19.

Cascaded Stages

The Thevenin and Norton models reduce an amplifier to its "bare-bones" for analysis purposes. In addition to considering the simplified model for source and load effects, the simplified model is also useful to analyze the internal loading when two or more stages are cascaded to form a single amplifier. Consider two stages cascaded as shown in Figure 1–20. The overall gain is affected by loading effects from each of the three loops. The loops are simple series circuits, so voltages can easily be calculated with the voltage-divider rule.

---

[3] The relationship between the dependent source and its reference cannot be broken. The superposition theorem, which allows sources to be treated separately, does not apply to dependent sources.

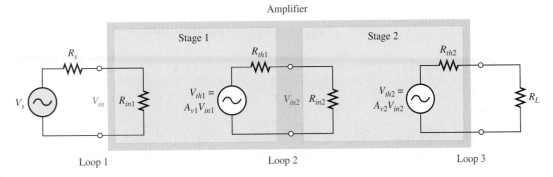

FIGURE 1–20
Cascaded stages in an amplifier.

EXAMPLE 1–7   Assume a transducer with a Thevenin (unloaded) source, V_s, of 10 mV and a Thevenin source resistance, R_s, of 50 kΩ is connected to a two-stage cascaded amplifier, as shown in Figure 1–21. Compute the voltage across a 1.0 kΩ load.

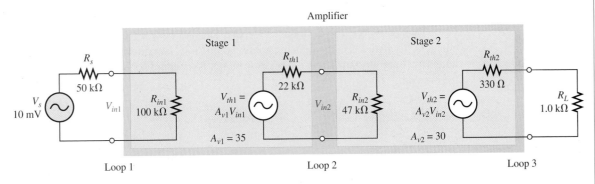

FIGURE 1–21
Two-stage cascaded amplifier.

Solution   Compute the input voltage to stage 1 from the voltage-divider rule applied to loop 1.

$$V_{in1} = V_s\left(\frac{R_{in1}}{R_{in1} + R_s}\right) = 10 \text{ mV}\left(\frac{100 \text{ k}\Omega}{100 \text{ k}\Omega + 50 \text{ k}\Omega}\right) = 6.67 \text{ mV}$$

The Thevenin voltage for stage 1 is

$$V_{th1} = A_{v1}V_{in1} = (35)(6.67 \text{ mV}) = 233 \text{ mV}$$

Compute the input voltage to stage 2 again from the voltage-divider rule, this time applied to loop 2.

$$V_{in2} = V_{th1}\left(\frac{R_{in2}}{R_{in2} + R_{th1}}\right) = 233 \text{ mV}\left(\frac{47 \text{ k}\Omega}{47 \text{ k}\Omega + 22 \text{ k}\Omega}\right) = 159 \text{ mV}$$

The Thevenin voltage for stage 2 is

$$V_{th2} = A_{v2}V_{in2} = (30)(159 \text{ mV}) = 4.77 \text{ V}$$

Apply the voltage-divider rule one more time to loop 3. The voltage across the 1.0 kΩ load is

$$V_{R_L} = V_{th2}\left(\frac{R_L}{R_L + R_{th2}}\right) = 4.77 \text{ V}\left(\frac{1.0 \text{ k}\Omega}{1.0 \text{ k}\Omega + 330 \text{ }\Omega}\right) = \textbf{3.59 V}$$

Practice Exercise Assume a transducer with a Thevenin source voltage of 5.0 mV and a source resistance of 100 kΩ is connected to the same amplifier. Compute the voltage across the 1.0 kΩ load.

Logarithms

A widely used unit in electronics is the *decibel,* which is based on logarithms. Before defining the decibel, let's quickly review logarithms (sometimes called *logs*). A **logarithm** is simply an exponent. Consider the equation

$$y = b^x$$

The value of y is determined by the exponent of the base (b). The exponent, x, is said to be the logarithm of the number represented by the letter y.

Two bases are in common use—base ten and base e (discussed in mathematics courses). To distinguish the two, the abbreviation "log" is written to mean base ten, and the letters "ln" are written to mean base e. Base ten is standard for work with decibels. Thus, for base ten,

$$y = 10^x$$

Solving for x,

$$x = \log_{10}y$$

The subscript 10 can be omitted because it is implied by the abbreviation "log."

Logarithms are useful when you multiply or divide very large or small numbers. When two numbers written with exponents are multiplied, the exponents are simply added. That is,

$$10^x \times 10^y = 10^{x+y}$$

This is equivalent to writing

$$\log xy = \log x + \log y$$

This concept will be applied to problems involving multiple stages of amplification or attenuation.

EXAMPLE 1–8

(a) Determine the logarithm (base ten) for the numbers 2, 20, 200, and 2000.
(b) Find the numbers whose logarithms are 0.5, 1.5, and 2.5.

Solution

(a) Determine the logarithms by entering each number in a calculator and pressing the $\boxed{\text{log}}$ key. The results are

$$\log 2 = \textbf{0.30103} \qquad \log 20 = \textbf{1.30103}$$
$$\log 200 = \textbf{2.30103} \qquad \log 2000 = \textbf{3.30103}$$

Notice that each factor-of-ten increase in y is an increase of 1.0 in the log.

(b) Find the number whose logarithm is a given value by entering the given value in a calculator and pressing the $\boxed{10^x}$ function (or $\boxed{\text{INV}}$ $\boxed{\text{log}}$). The results are

$$10^{0.5} = \textbf{3.16228} \qquad 10^{1.5} = \textbf{31.6228} \qquad 10^{2.5} = \textbf{316.228}$$

Notice that each increase of 1 in x (the logarithm) is a factor-of-10 increase in the number.

Practice Exercise

(a) Find the logarithms for the numbers 0.04, 0.4, 4, and 40.
(b) What number has a logarithm of 4.8?

Decibel Power Ratios

Power ratios are often very large numbers. Early in the development of telephone communication systems, engineers devised the decibel as a means of describing large ratios of gain or attenuation (a signal reduction). The **decibel (dB)** is defined as 10 multiplied by the logarithmic ratio of the power gain.

$$dB = 10 \log\left(\frac{P_2}{P_1}\right) \tag{1–11}$$

where P_1 and P_2 are the two power levels being compared.

Previously, power gain was introduced and defined as the ratio of power delivered from an amplifier to the power supplied to the amplifier. To show power gain, A_p, as a decibel ratio, we use a prime in the abbreviation.

$$A_p' = 10 \log\left(\frac{P_{out}}{P_{in}}\right) \tag{1–12}$$

where A_p' = power gain expressed as a decibel ratio
P_{out} = power delivered to a load
P_{in} = power delivered to the amplifier

The decibel (dB) is a dimensionless quantity because it is a ratio. Any two power measurements with the same ratio are the same number of decibels. For example, the power ratio between 500 W and 1 W is 500:1, and the number of decibels this ratio represents is 27 dB. There is exactly the same number of decibels between 100 mW and 0.2 mW (500:1) or 27 dB. When the power ratio is less than 1, there is a power loss or **attenuation**. The decibel ratio is *positive* for power gain and *negative* for power loss.

One important power ratio is 2:1. This ratio is the defining power ratio for specifying the cutoff frequency of instruments, amplifiers, filters, and the like. By substituting into Equation (1–11), the dB equivalent of a 2:1 power ratio is

$$dB = 10 \log\left(\frac{P_2}{P_1}\right) = 10 \log\left(\frac{2}{1}\right) = 3.01 \text{ dB}$$

This result is usually rounded to 3 dB.

Since 3 dB represents a doubling of power, 6 dB represents another doubling of the original power (a power ratio of 4:1). Nine decibels represents an 8:1 ratio of power and so forth. If the ratio is the same, but P_2 is smaller than P_1, the decibel result remains the same except for the sign.

$$dB = 10 \log\left(\frac{P_2}{P_1}\right) = 10 \log\left(\frac{1}{2}\right) = -3.01 \text{ dB}$$

The negative result indicates that P_2 is less than P_1.

Another useful ratio is 10:1. Since the log of 10 is 1, 10 dB equals a power ratio of 10:1. With this in mind, you can quickly estimate the overall gain (or attenuation) in certain situations. For example, if a signal is attenuated by 23 dB, it can be represented by two 10 dB attenuators and a 3 dB attenuator. Two 10 dB attenuators are a factor of 100 and another 3 dB represents another factor of 2 for an overall attenuation ratio of 1:200.

EXAMPLE 1–9

Compute the overall power gain of the amplifier in Example 1–7. Express the answer as both power gain and decibel power gain.

Solution   The power delivered to the amplifier is

$$P_{in1} = \frac{V_{in1}^2}{R_{in1}} = \frac{(6.67 \text{ mV})^2}{100 \text{ k}\Omega} = 445 \text{ pW}$$

The power delivered to the load is

$$P_{out} = \frac{V_{R_L}^2}{R_L} = \frac{(3.59 \text{ V})^2}{1.0 \text{ k}\Omega} = 12.9 \text{ mW}$$

The power gain, A_p, is the ratio of P_{out}/P_{in1}.

$$A_p = \frac{P_{out}}{P_{in1}} = \frac{12.9 \text{ mW}}{445 \text{ pW}} = 29.0 \times 10^6$$

Expressed in dB,

$$A_p' = 10 \log 29.0 \times 10^6 = \textbf{74.6 dB}$$

Practice Exercise   Compute the power gain (in dB) for an amplifier with an input power of 50 μW and a power delivered to the load of 4 W.

It is common in certain applications of electronics (microwave transmitters, for example) to combine several stages of gain or attenuation. When working with several stages of gain or attenuation, the total voltage gain is the product of the gains in absolute form.

$$A_{v(tot)} = A_{v1} \times A_{v2} \times \cdots \times A_{vn}$$

Decibel units are useful when combining these gains or losses because they involve just addition or subtraction. The algebraic addition of decibel quantities is equivalent to multiplication of the gains in absolute form.

$$A'_{v(tot)} = A'_{v1} = A'_{v2} + \cdots + A'_{vn}$$

EXAMPLE 1–10

Assume the transmitted power from a radar is 10 kW. A directional coupler (a device that samples the transmitted signal) has an output that represents -40 dB of attenuation. Two 3 dB attenuators are connected in series to this output, and the attenuated signal is terminated with a 50 Ω terminator (load resistor). What is the power dissipated in the terminator?

Solution

$$dB = 10 \log\left(\frac{P_2}{P_1}\right)$$

The transmitted power is attenuated by 46 dB (sum of the attenuators). Substituting,

$$-46 \text{ dB} = 10 \log\left(\frac{P_2}{10 \text{ kW}}\right)$$

Divide both sides by 10 and remove the log function.

$$10^{-4.6} = \frac{P_2}{10 \text{ kW}}$$

Therefore,

$$P_2 = \textbf{251 mW}$$

Practice Exercise Assume one of the 3 dB attenuators is removed.
(a) What is the total attenuation?
(b) What is the new power dissipated in the terminator?

Although decibel power ratios are generally used to compare two power levels, they are occasionally used for absolute measurements when the reference power level is understood. Although different standard references are used depending on the application, the most common absolute measurement is the dBm. A **dBm** is the power level when the reference is understood to be 1 mW developed in some assumed load impedance. For radio frequency systems, this is commonly 50 Ω; for audio systems, it is generally 600 Ω. The dBm is defined as

$$dBm = 10 \log\left(\frac{P_2}{1 \text{ mW}}\right)$$

The dBm is commonly used to specify the output level of signal generators and is used in telecommunications to simplify the computation of power levels.

Decibel Voltage Ratios

Since power is given by the ratio of V^2/R, the decibel power ratio can be written as

$$dB = 10 \log\left(\frac{V_2^2/R_2}{V_1^2/R_1}\right)$$

where R_1, R_2 = resistances in which P_1 and P_2 are developed
V_1, V_2 = voltages across the resistances R_1 and R_2

If the resistances are equal, they cancel.

$$dB = 10 \log\left(\frac{V_2^2}{V_1^2}\right)$$

A property of logarithms is

$$\log x^2 = 2 \log x$$

Thus, the decimal voltage ratio is

$$dB = 20 \log\left(\frac{V_2}{V_1}\right)$$

When V_2 is the output voltage (V_{out}) and V_1 is the input voltage (V_{in}) for an amplifier, the equation defines the decibel voltage gain. By substitution,

$$A_v' = 20 \log\left(\frac{V_{out}}{V_{in}}\right) \tag{1–13}$$

where A_v' = voltage gain expressed as a decibel ratio
V_{out} = voltage delivered to a load
V_{in} = voltage delivered to the amplifier

Equation (1–13) gives the decibel voltage gain, a logarithmic ratio of amplitudes. The equation was originally derived from the decibel power equation when both the input and load resistances are the same (as in telephone systems).

Both the decibel voltage gain equation and decibel power gain equation give the same ratio if the input and load resistances are the same. However, it has become common practice to apply the decibel voltage equation to cases where the resistances are *not* the same. When the resistances are not equal, the two equations do not give the same result.[4]

In the case of decibel voltage gain, note that if the amplitudes have a ratio of 2:1, the decibel voltage ratio is very close to 6 dB (since 20 log 2 = 6). If the signal is attenuated by a factor of 2 (ratio = 1:2), the decibel voltage ratio is −6 (since 20 log 1/2 = −6). Another useful ratio is when the amplitudes have a 10:1 ratio; in this case, the decibel voltage ratio is 20 dB (since 20 log 10 = 20).

---

[4] The *IEEE Standard Dictionary of Electrical and Electronic Terms* recommends that a specific statement accompany this application of decibels to avoid confusion.

EXAMPLE 1–11   An amplifier with an input resistance of 200 kΩ drives a load resistance of 16 Ω. If the input voltage is 100 μV and the output voltage is 18 V, calculate the decibel power gain and the decibel voltage gain.

Solution   The power delivered to the amplifier is

$$P_{in} = \frac{V_{in}^2}{R_{in}} = \frac{(100 \ \mu V)^2}{200 \ k\Omega} = 5 \times 10^{-14} \ W$$

The output power (delivered to the load) is

$$P_{out} = \frac{V_{out}^2}{R_L} = \frac{(18 \ V)^2}{16 \ \Omega} = 20.25 \ W$$

The decibel power gain is

$$A'_p = 10 \ \log\left(\frac{P_{out}}{P_{in}}\right) = 10 \ \log\left(\frac{20.25 \ W}{5 \times 10^{-14} \ W}\right) = \textbf{146 dB}$$

The decibel voltage gain is

$$A'_v = 20 \ \log\left(\frac{V_{out}}{V_{in}}\right) = 20 \ \log\left(\frac{18 \ V}{100 \ \mu V}\right) = \textbf{105 dB}$$

Practice Exercise   A video amplifier with an input resistance of 75 Ω drives a load of 75 Ω.
(a) How do the power gain and voltage gains compare?
(b) If the input voltage is 20 mV and the output voltage is 1.0 V, what is the decibel voltage gain?

1–4 REVIEW QUESTIONS

1. What is an ideal amplifier?
2. What is a dependent source?
3. What is a decibel?

1–5 ■ TROUBLESHOOTING ANALOG CIRCUITS

Technicians must diagnose and repair malfunctioning circuits or systems. Troubleshooting is the application of logical thinking to correct the malfunctioning circuit or system. Troubleshooting skills will be emphasized throughout the text.

After completing this section, you should be able to

❑ Describe the process for troubleshooting an analog circuit
  ❑ Explain what is meant by *half-splitting*
  ❑ Cite basic rules for replacing a part in a printed circuit (PC) board
  ❑ Describe basic bench test equipment for troubleshooting

Analysis, Planning, and Measuring

When troubleshooting any circuit, the first step is to analyze the clues (symptoms) of a failure. The analysis can begin by determining the answer to several questions: Has the circuit ever worked? If so, under what conditions did it fail? What are the symptoms of a failure? What are possible causes of this failure? The process of asking these questions is part of the analysis of a problem.

After analyzing the clues, the second step in the troubleshooting process is forming a logical plan for troubleshooting. A lot of time can be saved by planning the process. As part of this plan, you must have a working understanding of the circuit you are troubleshooting. Take the time to review schematics, operating instructions, or other pertinent information if you are not certain how the circuit should operate. It may turn out that the failure was that of the operator, not the circuit! A schematic with proper voltages or waveforms marked at various test points is particularly useful for troubleshooting.

Logical thinking is the most important tool of troubleshooting but rarely can solve the problem by itself. The third step is to narrow the possible failures by making carefully thought-out measurements. These measurements usually confirm the direction you are taking in solving the problem or point to a new direction. Occasionally, you may find a totally unexpected result!

The thinking process that is part of analysis, planning, and measuring is best illustrated with an example. Suppose you have a string of 16 decorative lamps connected in series to a 120 V source as shown in Figure 1–22. Assume that this circuit worked at one time and stopped after moving it to a new location. When plugged in, the lamps fail to turn on. How would you go about finding the trouble?

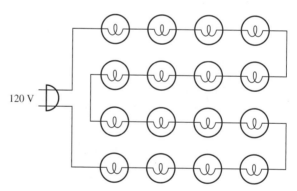

FIGURE 1–22
A series of lights. Is one of them open?

You might think like this: Since the circuit worked before it was moved, the problem could be that there is no voltage at this location. Or perhaps the wiring was loose and pulled apart when moved. It's possible a bulb burned out or became loose. This reasoning has considered possible causes and failures that could have occurred. The fact that the circuit was once working eliminates the possibility that the original circuit may have been incorrectly wired. In a series circuit, the possibility of two open paths occurring together is unlikely. You have analyzed the problem and now you are ready to plan the troubleshooting approach.

The first part of your plan is to measure (or test) for voltage at the new location. If voltage is present, then the problem is in the light string. If voltage is not present, check the circuit breakers at the input panel to the house. Before resetting breakers, you should think about why a breaker may be tripped.

The second part of your plan assumes voltage is present and the string is bad. You can disconnect power from the string and make resistance checks to begin isolating the problem. Alternatively, you could apply power to the string and measure voltage at various points. The decision whether to measure resistance or voltage is a toss-up and can be made based on the ease of making the test. Seldom is a troubleshooting plan developed so completely that all possible contingencies are included. The troubleshooter will frequently need to modify the plan as tests are made. You are ready to make measurements.

Suppose you have a digital multimeter (DMM) handy. You check the voltage at the source and find 120 V present. Now you have eliminated one possibility (no voltage). You know the problem is in the string, so you proceed with the second part of your plan. You might think: Since I have voltage across the entire string, and apparently no current in the circuit (since no bulb is on), there is almost certainly an open in the path—either a bulb or a connection. To eliminate testing each bulb, you decide to break the circuit in the middle and to check the *resistance* of each half of the circuit.

Now you are using logical thinking to reduce the effort needed. The technique you are using is a common troubleshooting procedure called *half-splitting*. By measuring the resistance of half the bulbs at once, you can reduce the effort required to find the open. Continuing along these lines, by half-splitting again, will lead to the solution in a few tests.

Unfortunately, most troubleshooting is more difficult than this example. However, analysis and planning are important for effective troubleshooting. As measurements are made, the plan is modified; the experienced troubleshooter narrows the search by fitting the symptoms and measurements into a possible cause.

Soldering

When repairing circuit boards, sooner or later the technician will need to replace a soldered part. When you replace any part, it is important to be able to remove the old part without damaging the board by excessive force or heat. Transfer of heat for removal of a part is facilitated with a chisel tip (as opposed to a conical tip) on the soldering iron.

Before installing a new part, the area must be clean. Old solder should be completely removed without exposing adjacent devices to excess heat. A degreasing cleaner or alcohol is suggested for cleaning (remember—solder won't stick to a dirty board!). Solder must be a resin core type (acid solder is never used in electronic circuits and shouldn't even be on your workbench!). Solder is applied to the joint (not to the iron). As the solder cools, it must be kept still. A good solder connection is a smooth, shiny one and the solder *flows* into the printed circuit trace. A poor solder connection looks dull. During repair, it is possible for excessive solder to short together two parts or two pins on an integrated circuit (this rarely happens when boards are machine soldered). This is called a solder bridge, and the technician must be alert for this type of error when repairing boards. After the repair is completed, any flux must be removed from the board with alcohol or other cleaner.

Basic Test Equipment

The ability to troubleshoot effectively requires the technician to have a set of test equipment available and to be familiar with the operation of the instruments. An oscilloscope, DMM, and power supply are basic instruments for troubleshooting. These instruments are shown in Figure 1–23. No one instrument is best for all situations, so it is important to understand the limitations of the test equipment at hand. All electronic measuring instruments become part of the circuit they are measuring and thus affect the measurement itself (an effect called *instrument loading*). In addition, instruments are specified for a range of frequencies and must be properly calibrated if readings are to be trusted. An expert troubleshooter must consider these effects when making electronic measurements.

For general-purpose troubleshooting of analog circuits, all technicians need access to an oscilloscope and a DMM. The oscilloscope needs to be a good two-channel scope, fast enough to spot noise or ringing when it occurs. A set of switchable probes, with the ability to switch between $\times 1$ and $\times 10$ is useful for looking at large or small signals. (Note that in the $\times 1$ position, the scope loses bandwidth.)

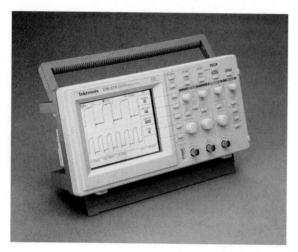

(a) Oscilloscope

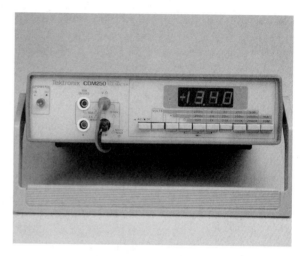

(b) Digital multimeter

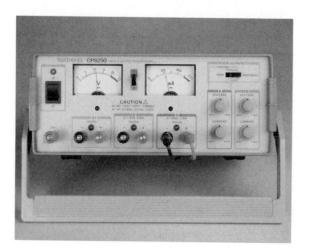

(c) Triple output power supply

FIGURE 1–23

Test instruments. (Copyright 2000, Tekronix, Inc. Reprinted with permission. All rights reserved.)

The DMM is a general-purpose meter that has the advantage of very high input impedance but may have error if used in circuits with frequencies above a few kilohertz. Many new DMMs offer special features, such as continuity testing and diode checking, and may include capacitance and frequency measurements. While DMMs are excellent test instruments, the VOM (volt-ohm-milliammeter) has some advantages (for example, spotting trends faster than a digital meter). Although generally not as accurate as a DMM, a VOM has very small capacitance to ground, and it is isolated from the line voltage. Also, because a VOM is a passive device, it will not tend to inject noise into a circuit under test.

Many times the circuit under test needs to have a test signal injected to simulate operation in a system. The circuit's response is then observed with a scope or other instrument. This type of testing is called *stimulus-response testing* and is commonly used when a portion of a complete system is tested. For general-purpose troubleshooting, the function generator is used as the stimulus instrument. All function generators have a sine wave, square wave, and triangle wave output; the frequency range varies widely, from a low frequency of 1 μHz to a high of 50 MHz (or more) depending on the generator. Higher-quality function generators offer the user a choice of other waveforms (pulses and ramps, for example) and may have triggered or gated outputs as well as other features.

The basic function generator waveforms (sine, square, and triangle) are used in many tests of electronic circuits and equipment. A common application of a function generator is to inject a sine wave into a circuit to check the circuit's response. The signal is capacitively coupled to the circuit to avoid upsetting the bias network; the response is observed on an oscilloscope. With a sine wave, it is easy to ascertain if the circuit is operating properly by checking the amplitude and shape of the sine wave at various points or to look for possible troubles such as high-frequency oscillation.

A common test for wide-band amplifiers is to inject a square wave into a circuit to test the frequency response. Recall that a square wave consists of the fundamental frequency and an infinite number of odd harmonics (as discussed in Section 1–2). The square wave is applied to the input of the test circuit and the output is monitored. The shape of the output square wave indicates if specific frequencies are selectively attenuated.

Figure 1–24 illustrates square wave distortions due to selective attenuation of low or high frequencies. A good amplifier should show a high-quality replica of the input. If the square wave sags, as in Figure 1–24(b), low frequencies are not being passed properly by the circuit. The rising edge contains mostly higher-frequency harmonics. If the square wave rolls over before reaching the peak, as in Figure 1–24(c), high frequencies are being attenuated. The rise time of the square wave is an indirect measurement of the bandwidth of the circuit.

For testing dc voltages or providing power to a circuit under test, a multiple output power supply, with both positive and negative outputs, is necessary. The outputs should be variable from 0 to 15 V. A separate low voltage supply is also handy for powering logic circuits or as a dc source for analog circuits.

For certain situations and applications, there are specialized measuring instruments designed for the application. Some of this specialized equipment is designed for a specific frequency range or for a specific application, so they won't be discussed here. If available, a digital storage scope can be of help. It has some particular advantages for troubleshooting because it can be used to store and compare waveforms from a known good unit or to capture a failure that occurs intermittently. It also has the ability to display events that occur before and after the trigger event, a feature that is invaluable with intermittent problems.

FIGURE 1–24
Square-wave response of wide-band amplifiers.

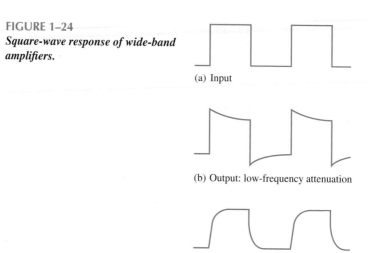

(a) Input

(b) Output: low-frequency attenuation

(c) Output: high-frequency attenuation

A complete list of "nice to have" accessories could be quite long indeed, but another handy set of instruments is a pulser and pulse tracer. These tools are useful for tracing a short such as one from the power supply to ground. The pulser stimulates the circuit with a series of very short pulses. The current tracer can follow the path of the current and lead right to the short. These tools are useful for both digital and analog circuits.

Other Troubleshooting Materials

In general, some materials that are useful for general-purpose troubleshooting that fall under the "must have" category include the following:

❑ A basic set of hand tools for electronics, including long-nose pliers, diagonal wire cutters, wire strippers, screwdrivers (especially jeweler's screwdrivers), and a small flashlight.

❑ Soldering and desoldering tools, including solder wick and a magnifying glass for inspecting work or looking for hairline cracks, solder splashes, or other problems.

❑ A collection of spare parts (resistors, capacitors, transistors, diodes, switches, ICs). In this category, you will also need extra clip leads, cables with various connectors, banana to alligator converters, heat shrink, and the like.

❑ A capacitor and a resistor substitution box. This is a useful tool for various tests such as changing the time constant in a circuit under test.

❑ A hair dryer and freeze spray for testing thermal effects of a circuit.

❑ A static safe wrist strap (and static-free work station, if possible) to prevent damaging static-sensitive circuits.

1–5 REVIEW QUESTIONS

1. What is the first step in troubleshooting a circuit?
2. What is meant by half-splitting?
3. What is meant by instrument loading?

■ SUMMARY

- A linear component is one in which an increase in current is proportional to the applied voltage.
- An analog signal takes on a continuous range of values within limits. A digital signal is a discrete signal that can have only certain values. Many circuits use a combination of analog and digital circuits.
- Waveforms that repeat in a certain interval of time are said to be periodic. A cycle is the complete sequence of values that a waveform exhibits before an identical pattern occurs. The period is the time interval for one cycle.
- Signals that have voltage, current, resistance, or other quantity vary as a function of time are called time-domain signals. When the frequency is made the independent variable, the result is a frequency-domain signal. Any signal can be observed in either the time domain or the frequency domain.
- Thevenin's theorem replaces a complicated, two-terminal, linear circuit with an ideal independent voltage source and a series resistance. The Thevenin circuit is equivalent to the original circuit for any load that is connected to the output terminals.
- Norton's theorem replaces a complicated, two-terminal, linear circuit with an ideal independent current source and a parallel resistance. The Norton circuit is equivalent to the original circuit for any load that is connected to the output terminals.
- A transducer is a device that converts a physical quantity from one form to another; for electronic systems, input transducers convert a physical quantity to an electrical quantity (voltage, current, resistance).
- An ideal amplifier increases the magnitude of an input signal in order to produce a useful outcome. For a voltage amplifier, the output signal, $v_{out}(t)$, is proportional to the input signal, $v_{in}(t)$. The ratio of the output voltage to the input voltage is called the voltage gain, A_v.
- The decibel is a dimensionless number that is ten times the logarithmic ratio of two powers. Decibel gains and losses are combined by algebraic addition.
- Troubleshooting begins with analyzing the symptoms of a failure; then forming a logical plan. Carefully thought-out measurements are made to narrow the search for the cause of the failure. These measurements may modify or change the plan.
- For general-purpose troubleshooting, a reasonable fast, two-channel oscilloscope and a DMM are the principal measuring instruments. The most common stimulus instruments are a function generator and a regulated power supply.

■ GLOSSARY

Key terms are in color. All terms are included in the end-of-book glossary.

ac resistance   The ratio of a small change in voltage divided by a corresponding change in current for a given device; also called *dynamic, small-signal,* or *bulk* resistance.

Amplification   The process of producing a larger voltage, current, or power using a smaller input signal as a "pattern."

Amplifier   An electronic circuit having the capability of amplification and designed specifically for that purpose.

Analog signal   A signal that can take on a continuous range of values within certain limits.

Attenuation   The reduction in the level of power, current, or voltage.

Characteristic curve   A plot which shows the relationship between two variable properties of a device. For most electronic devices, a characteristic curve refers to a plot of the current, I, plotted as a function of voltage, V.

Cycle   The complete sequence of values that a waveform exhibits before another identical pattern occurs.

dBm   Decibel power level when the reference is understood to be 1 mW (see *Decibel*).

Decibel   A dimensionless quantity that is 10 times the logarithm of a power ratio or 20 times the logarithm of a voltage ratio.

Digital signal   A noncontinuous signal that has discrete numerical values assigned to specific steps.

Domain   The values assigned to the independent variable. For example, frequency or time are typically used as the independent variable for plotting signals.

Frequency   The number of repetitions per unit of time for a periodic waveform.

Gain   The amount of amplification. Gain is a ratio of an output quantity to an input quantity (e.g., voltage gain is the ratio of the output voltage to the input voltage).

Harmonics   Higher-frequency sinusoidal waves that are integer multiples of a fundamental frequency.

Linear component   A component in which an increase in current is proportional to the applied voltage.

Load line   A straight line plotted on a current versus voltage plot that represents all possible operating points for an external circuit.

Logarithm   In mathematics, the logarithm of a number is the power to which a base must be raised to give that number.

Norton's theorem   An equivalent circuit that replaces a complicated two-terminal linear circuit with a single current source and a parallel resistance.

Period (T)   The time for one cycle of a repeating wave.

Periodic   A waveform that repeats at regular intervals.

Phase angle (in radians)   The fraction of a cycle that a waveform is shifted from a reference waveform of the same frequency.

Quantizing   The process of assigning numbers to sampled data.

Quiescent point   The point on a load line that represents the current and voltage conditions for a circuit with no signal (also called *operating* or *Q-point*). It is the intersection of a device characteristic curve with a load line.

Sampling   The process of breaking the analog waveform into time "slices" that approximate the original wave.

Spectrum   A plot of amplitude versus frequency for a signal.

Thevenin's theorem   An equivalent circuit that replaces a complicated two-terminal linear circuit with a single voltage source and a series resistance.

Transducer   A device that converts a physical quantity from one form to another; for example, a microphone converts sound into voltage.

Transfer curve   A plot of the output of a circuit or system for a given input.

Vector   Any quantity that has both magnitude and direction.

■ KEY FORMULAS

(1–1)	$I = \dfrac{V}{R}$	Ohm's law
(1–2)	$y(t) = A \sin(\omega t \pm \phi)$	Instantaneous value of a sinusoidal wave
(1–3)	$f\,(\text{Hz}) = \dfrac{\omega\ (\text{rad/s})}{2\pi\ (\text{rad/cycle})}$	Conversion from radian frequency (rad/s) to hertz (Hz)
(1–4)	$T = \dfrac{1}{f}$	Conversion from frequency to period
(1–5)	$f = \dfrac{1}{T}$	Conversion from period to frequency
(1–6)	$V_{avg} = 0.637 V_p$	Conversion from peak voltage to average voltage for a sinusoidal wave
(1–7)	$P = IV$	Power law
(1–8)	$V_{rms} = 0.707 V_p$	Conversion from peak voltage to rms voltage for a sinusoidal wave
(1–9)	$A_v = \dfrac{V_{out}}{V_{in}}$	Voltage gain
(1–10)	$A_p = \dfrac{P_{out}}{P_{in}}$	Power gain
(1–11)	$dB = 10 \log\!\left(\dfrac{P_2}{P_1}\right)$	Definition of the decibel
(1–12)	$A'_p = 10 \log\!\left(\dfrac{P_{out}}{P_{in}}\right)$	Decibel power gain
(1–13)	$A'_v = 20 \log\!\left(\dfrac{V_{out}}{V_{in}}\right)$	Decibel voltage gain

■ SELF-TEST

Answers are at the end of the chapter.

1. The graph of a linear equation
   - (a) always has a constant slope
   - (b) always goes through the origin
   - (c) must have a positive slope
   - (d) answers (a), (b), and (c)
   - (e) none of these answers

2. AC resistance is defined as
   - (a) voltage divided by current
   - (b) a change in voltage divided by a corresponding change in current
   - (c) current divided by voltage
   - (d) a change in current divided by a corresponding change in voltage

3. A discrete signal
   - (a) changes smoothly
   - (b) can take on any value
   - (c) is the same thing as an analog signal
   - (d) answers (a), (b), and (c)
   - (e) none of these answers

4. The process of assigning numeric values to a signal is called
   - (a) sampling
   - (b) multiplexing
   - (c) quantizing
   - (d) digitizing

5. The reciprocal of the repetition time of a periodic signal is the
   (a) frequency (b) angular frequency (c) period (d) amplitude

6. If a sinusoidal wave has a peak amplitude of 10 V, the rms voltage is
   (a) 0.707 V (b) 6.37 V (c) 7.07 V (d) 20 V

7. If a sinusoidal wave has a peak-to-peak amplitude of 325 V, the rms voltage is
   (a) 103 V (b) 115 V (c) 162.5 V (d) 460 V

8. Assume the equation for a sinusoidal wave is $v(t) = 200 \sin(500t)$. The peak voltage is
   (a) 100 V (b) 200 V (c) 400 V (d) 500 V

9. A harmonic is
   (a) an integer multiple of a fundamental frequency
   (b) an unwanted signal that adds noise to a system
   (c) a transient signal
   (d) a pulse

10. A Thevenin circuit consists of a
    (a) current source in parallel with a resistor   (b) current source in series with a resistor
    (c) voltage source in parallel with a resistor   (d) voltage source in series with a resistor

11. A Norton circuit consists of a
    (a) current source in parallel with a resistor   (b) current source in series with a resistor
    (c) voltage source in parallel with a resistor   (d) voltage source in series with a resistor

12. A load line is a plot that describes
    (a) the IV characteristic curve for a load resistor   (b) a driving circuit
    (c) both (a) and (b)   (d) neither (a) nor (b)

13. The intersection of an IV curve with the load line is called the
    (a) transfer curve (b) transition point (c) load point (d) Q-point

14. Assume a certain amount of power is attenuated by 20 dB. This is a factor of
    (a) 10 (b) 20 (c) 100 (d) 200

15. Assume an amplifier has a decibel voltage gain of 100 dB. The output will be larger than the input by a factor of
    (a) 100 (b) 1000 (c) 10,000 (d) 100,000

16. An important rule for soldering is
    (a) always use a good acid-based solder
    (b) always apply solder directly to the iron, never to the parts being soldered
    (c) wiggle the solder joint as it cools to strengthen it
    (d) answers (a), (b), and (c)
    (e) none of these answers

TROUBLESHOOTER'S QUIZ *Answers are at the end of the chapter.*

Refer to Figure 1–22.

❑ Assume the circuit is plugged in and operating normally. If a bulb is then removed,

1. The voltage across the socket of the removed bulb will
   (a) increase (b) decrease (c) not change

2. The voltage across all other bulbs will
   (a) increase (b) decrease (c) not change

3. The voltage to the circuit will
   (a) increase (b) decrease (c) not change

❏ Assume one of the sockets in the circuit is shorted with the bulb out. but the other bulbs are on.

4. As a result of the short, the voltage across each of the other bulbs will

   (a) increase   **(b)** decrease   **(c)** not change

5. The total voltage applied to the circuit will

   (a) increase   **(b)** decrease   **(c)** not change

6. The light output from the other bulbs will

   (a) increase   **(b)** decrease   **(c)** not change

❏ Assume the circuit is disconnected from the source and the resistance is measured between the prongs.

7. If one of the sockets is shorted, the total resistance will

   (a) increase   **(b)** decrease   **(c)** not change

8. If one of the bulbs is open, the total resistance will

   (a) increase   **(b)** decrease   **(c)** not change

■ PROBLEMS

Answers to odd-numbered problems are at the end of the book.

SECTION 1–1 Analog Electronics

1. What is the conductance of a 22 kΩ resistor?

2. How does the ac resistance of a diode change as the voltage increases?

3. Compute the ac resistance of the diode in Figure 1–2 at the point $V = 0.7$ V, $I = 5.0$ mA.

4. Sketch the shape of an IV curve for a device that has a decreasing ac resistance as voltage increases.

SECTION 1–2 Analog Signals

5. Assume a sinusoidal wave is described by the equation $v(t) = 100$ V $\sin(200t + 0.52)$.

   (a) From this expression, determine the peak voltage, the average voltage, and the angular frequency in rad/s.

   (b) Find the instantaneous voltage at a time of 2.0 ms. (Reminder: the angles are in radians in this equation).

6. Determine the frequency (in Hz) and the period (in s) for the sinusoidal wave described in Problem 5.

7. An oscilloscope shows a wave repeating every 27 μs. What is the frequency of the wave?

8. A DMM indicates the rms value of a sinusoidal wave. If a DMM indicates a sinusoidal wave is 3.5 V, what peak-to-peak voltage would you expect to observe on an oscilloscope?

9. The ratio of the rms voltage to the average voltage for any wave is called the *form factor* (used occasionally to convert meter readings). What is the form factor for a sinusoidal wave?

10. What is the fifth harmonic of a 500 Hz triangular wave?

11. What is the only type of harmonics found in a square wave?

SECTION 1–3 Signal Sources

12. Draw the Thevenin equivalent circuit for the circuit shown in Figure 1–25. Show values on your drawing.

13. Assume a 1.0 kΩ, 2.7 kΩ and 3.6 kΩ load resistor are connected, one at a time, across the output terminals of the circuit in Figure 1–25. Determine the voltage across each load.

14. Draw the Norton equivalent circuit for the circuit shown in Figure 1–25. Show values on your drawing.

15. Draw a graph showing the load line for the Thevenin circuit shown in Figure 1–26. On the same graph, show the IV curve for a 150 kΩ resistor. Show the Q-point on your plot.

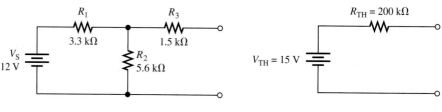

FIGURE 1–25   FIGURE 1–26

16. Assume the output of a transducer is a 10 mV ac signal with no load and drops to 5 mV when a 100 kΩ load is connected to the outputs. Based on these observations, draw the Thevenin equivalent circuit for this transducer.

17. Draw the Norton equivalent circuit for the transducer circuit described in Problem 16.

SECTION 1–4 Amplifiers

18. For the amplifier described by the transfer curve in Figure 1–17, what is the voltage gain in the linear region? What is the largest output voltage before saturation?

19. The input to an amplifier is 80 μV. If the voltage gain of the amplifier is 50,000, what is the output signal?

20. Assume a transducer with a Thevenin (unloaded) voltage of 5.0 mV and a Thevenin resistance of 20 kΩ is connected to a two-stage cascaded amplifier with the following specifications:

$R_{in1} = 50\ \text{k}\Omega$
$A_{v1}\ (\text{unloaded}) = 50$
$R_{th1} = 5\ \text{k}\Omega$
$R_{in2} = 10\ \text{k}\Omega$
$A_{v2}\ (\text{unloaded}) = 40$
$R_{th2} = 1.0\ \text{k}\Omega$

Draw the amplifier model and compute the voltage across a 2.0 kΩ load.

21. Compute the decibel voltage gain for the amplifier in Problem 20.

22. Compute the decibel power gain for the amplifier in Problem 20.

23. Assume you want to attenuate the voltage from a signal generator by a factor of 1000. What is the decibel attenuation required?

24. (a) What is the power dissipated in a 50 Ω load resistor when 20 V is across the load?
    (b) Express your answer in dBm.

25. A certain instrument is limited to 2 watts of input power dissipated in its internal 50 Ω input resistor.
    (a) How much attenuation (in dB) is required in order to connect a 20 W source to the instrument?
    (b) What is the maximum allowable voltage at the input?

SECTION 1–5   Troubleshooting Analog Circuits

26. Figure 1–27 shows a small system consisting of four microphones connected to a two-channel amplifier through a selector switch (SW1). Either the *A* set or the *B* set of microphones is selected and amplified. The output of the amplifier is connected to two speakers. Power to the amplifier is supplied by a single power supply that furnishes dc voltages to the amplifier and two batteries that provide power to two each of the microphones as shown.

    Assume no sound is heard when the system is plugged in and turned on. Outline a basic troubleshooting plan by indicating the tests you would make to isolate the trouble to either the power supply, amplifier, a microphone, microphone battery, switch, speaker, or other fault.

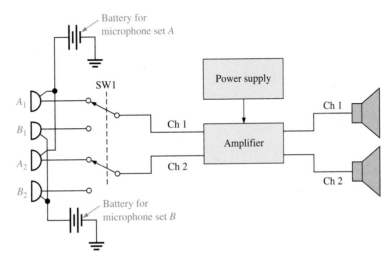

FIGURE 1–27
A small system consisting of a two-channel amplifier and four microphones.

27. For the system described in Problem 26, outline a basic troubleshooting plan for the case where Channel 1 operates normally but no sound is heard from Channel 2. Indicate the tests you would make to isolate the trouble. (Can you think of a method to do half-splitting?)

28. What information is obtained when a square wave calibration signal is used as the input to an oscilloscope?

29. How can you protect a static-sensitive circuit from damage when you are working on it?

30. Cite two important advantages of a digital storage oscilloscope over an analog oscilloscope.

■ ANSWERS TO REVIEW QUESTIONS

Section 1–1

1. A characteristic curve is a graph of the relationship between current and voltage for a component.

2. The slope of curve is lower for larger resistors.

3. DC resistance is the voltage divided by the current. AC resistance is the *change* in voltage divided by the *change* in current.

Section 1–2

1. An analog signal takes on a continuous range of values; a digital signal represents information that has a discrete number of codes.

2. The spectrum is a line spectrum with a single line at the fundamental and other lines at the odd harmonic frequencies. See Figure 1–8(a).

3. The spectrum for the repetitive waveform is a line spectrum; the spectrum for the nonrepetitive waveform is a continuous spectrum.

Section 1–3

1. An independent source is a voltage or current source that can be specified without regard to any other circuit parameter.

2. A Thevenin circuit consists of a series voltage source and resistor that duplicates the performance of a more complicated circuit for any given load. A Norton circuit consists of a parallel current source and resistor that duplicates the performance of a more complicated circuit for any given load.

3. Passive transducers require a separate source of electrical power; active transducers are self-generating devices.

Section 1–4

1. An ideal amplifier is one that introduces no noise or distortion to the signal; the output varies in time and replicates the input exactly.

2. A dependent source is one whose value depends on voltage or current elsewhere in the circuit.

3. A decibel is a dimensionless number that is 10 multiplied by the logarithmic ratio of two powers.

Section 1–5

1. Analyzing the symptoms of a failure by asking questions: Has the circuit ever worked? If so, under what conditions did it fail? What are the symptoms of a failure? What are possible causes of this failure?

2. Half-splitting divides a troubleshooting problem into halves and determines which half of the circuit is likely to have the problem.

3. Instrument loading is the effect of changing circuit voltages due to the process of connecting an instrument.

■ **ANSWERS TO PRACTICE EXERCISES FOR EXAMPLES**

1–1 $G_2 = 375$ mS, $R_2 = 2.67$ kΩ

1–2 $V_{rms} = 10.6$ V, $f = 95$ Hz, $T = 10.5$ ms

1–3 $V_{R_L} = 3.48$ V

1–4 $V_{R_L} = 3.48$ V

1–5   See Figure 1–28.

FIGURE 1–28

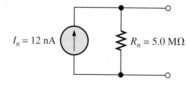

$I_n = 12$ nA $R_n = 5.0$ MΩ

1–6   6 V

1–7   1.34 V

1–8   (a) $\log 0.04 = -1.398$ (b) $10^{4.8} = 63{,}096$
$\log 0.4 = -0.398$
$\log 4.0 = 0.602$
$\log 40 = 1.602$

1–9   49 dB

1–10   (a) -43 dB   (b) 503 mW

1–11   (a) The decibel power gain is one-half the decibel voltage gain.   (b) 34 dB

■ **ANSWERS TO SELF-TEST**

1. (a)	**2.** (b)	**3.** (e)	**4.** (c)	**5.** (a)
6. (c)	**7.** (b)	**8.** (b)	**9.** (a)	**10.** (d)
11. (a)	**12.** (b)	**13.** (d)	**14.** (c)	**15.** (d)
16. (e)				

■ **ANSWERS TO TROUBLE-SHOOTER'S QUIZ**

1. increase	**2.** decrease	**3.** not change	**4.** increase
5. not change	**6.** increase	**7.** decrease	**8.** increase

2

DIODES AND APPLICATIONS

Courtesy Hewlett-Packard Company

■ CHAPTER OBJECTIVES

❑ Discuss the basic atomic structure of
   semiconductors
❑ Describe the characteristics of a *pn* junction
❑ Explain how to bias a semiconductor diode
❑ Describe the basic diode characteristics
❑ Analyze the operation of three basic types of
   rectifiers
❑ Describe the operation of rectifier filters and
   IC regulators
❑ Analyze the operation of diode limiters and
   clampers
❑ Explain the characteristics of four different
   special-purpose diodes
❑ Interpret and use a diode data sheet
❑ Troubleshoot a power supply using accepted
   techniques
❑ Apply what you have learned in this chapter
   to a system application

■ CHAPTER INTRODUCTION

In this chapter, the basic materials used in manufacturing diodes, transistors, and integrated circuits are described. You will be introduced to *pn* junctions, an important concept essential for the understanding of diode and transistor operation. Diode characteristics are introduced, and you will learn how to use diodes in various applications. We discuss converting ac to dc by the process known as *rectification* and introduce the integrated circuit (IC) regulator. You will also learn about diode-limiting circuits and dc restoring (clamping) circuits.

In addition to rectifier diodes, you will be introduced to zener diodes, varactor diodes, light-emitting diodes, and photodiodes. Applications for these special-purpose diodes are discussed.

■ A SYSTEM APPLICATION

The **power supply** is an important part of most electronic systems because it supplies the dc voltage and current necessary for all the other circuits in the system to operate. You will learn how a typical power supply in an electronic system converts ac power into a constant dc voltage as part of a radio receiver system. The output of the power supply goes to all parts of the system and provides the necessary bias voltage and operating power for the diodes, transistors, and other **discrete devices** in the amplifiers and other circuitry to function properly.

For the system application in Section 2–11, in addition to the other topics, be sure you understand
❏ How a diode works
❏ The parameters and ratings of diodes
❏ How to read a diode data sheet
❏ What rectification of ac voltage means
❏ How rectifiers work
❏ What a filter does and why it is important in power supplies
❏ What an IC regulator is

www.   VISIT THE COMPANION WEBSITE

Study Aids for This Chapter Are Available at

http://www.prenhall.com/floyd

2–1 ■ THE ATOMIC STRUCTURE OF SEMICONDUCTORS

Electronic devices such as diodes and transistors are constructed from special materials called semiconductors. This section lays the foundation for understanding how semiconductor devices function.

After completing this section, you should be able to

❑ Discuss the basic atomic structure of semiconductors
    ❑ Describe the planetary model of the atom
    ❑ Discuss how silicon and germanium atoms bond together to form crystals
    ❑ Compare electron energy levels in a conductor, insulator, and a semiconductor

Electron Shells and Orbits

The electrical properties of materials are explained by their atomic structure. In the classic Bohr model of the atom, electrons orbit the nucleus only in certain discrete (separate and distinct) distances. The nucleus contains positively charged protons and uncharged neutrons. The orbiting electrons are negatively charged. Modern quantum mechanical models of the atom retain much of the ideas of the original Bohr model but have replaced the concept of electron "particles" with mathematical "matter waves"; however, the Bohr model provides a useful mental picture of the structure of an atom.

    The distance from the nucleus determines the electron's **energy**. Electrons near the nucleus have less energy than those in more distant orbits. These discrete orbits mean that only certain energy levels are permitted within the atom. These energy levels are known as **shells**. Each shell has a certain maximum permissible number of electrons. The differences in energy levels within a shell are much smaller than the difference in energy between shells. The shells are designated 1, 2, 3, 4, and so on, with 1 being closest to the nucleus. This concept is illustrated in Figure 2–1.

FIGURE 2–1

Energy levels increase as the distance from the nucleus of the atom increases. The ratio of the radii of electron orbits is proportional to the square of the shell number. A neutral silicon atom (with 14 electrons and 14 protons in the nucleus) is shown.

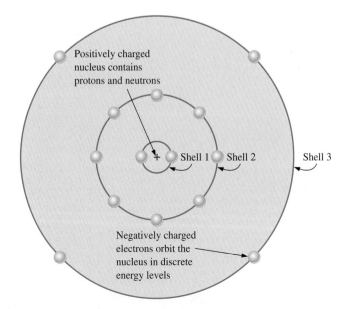

Positively charged nucleus contains protons and neutrons

Shell 1 Shell 2 Shell 3

Negatively charged electrons orbit the nucleus in discrete energy levels

Valence Electrons, Conduction Electrons, and Ions

Electrons in orbits farther from the nucleus are less tightly bound to the atom than those closer to the nucleus. This is because the force of attraction between the positively charged nucleus and the negatively charged electron decreases with increasing distance in accordance with Coulomb's law. Outer-shell electrons are also shielded from the nuclear charge by the inner-shell electrons.

Electrons in the outermost shell, called **valence electrons**, have the highest energy and are relatively loosely bound to their parent atom. For the silicon atom in Figure 2–1, the third shell electrons are the valence electrons. Sometimes, a valence electron can acquire enough energy to break free of its parent atom. This free electron is called a **conduction electron** because it is not bound to any certain atom. When a negatively charged electron is freed from an atom, the rest of the atom is positively charged and is said to be a positive **ion**. In some chemical reactions, the freed electron attaches itself to a neutral atom (or group of atoms), forming a negative ion.

Metallic Bonds

Metals tend to be solids at room temperature. The nucleus and inner-shell electrons of metals occupy fixed lattice positions. The outer valence electrons are held loosely by all of the atoms of the crystal and are free to move about. This "sea" of negatively charged electrons holds the positive ions of the metal together, forming metallic bonding.

With the large number of atoms in the metallic crystal, the discrete energy level for the valence electrons is blurred into a band called the *valence band.* These valence electrons are mobile and account for the thermal and electrical conductivity of metals. In addition to the valence energy band, the next (normally occupied) level from the nucleus in the atom is also blurred into a band of energies called the *conduction band.*

Figure 2–2 compares the energy-level diagram for three types of solids. Notice that for conductors, shown in Figure 2–2(c), the bands are overlapping. Electrons can easily move between the valence and conduction bands by absorbing light. The movement, back and forth, of electrons between the valence band and conduction band accounts for the luster of metals.

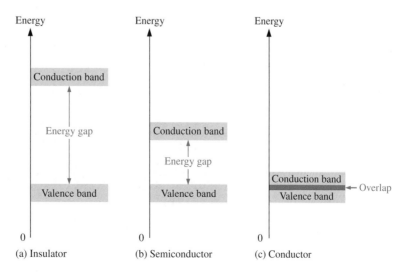

FIGURE 2–2
Energy diagrams for three categories of materials. The upper band is the conduction band; the lower band is the valence band.

Covalent Bonds

Atoms of some solid materials form **crystals**, which are three-dimensional structures held together by strong bonds between the atoms. In diamond, for example, four bonds are formed by the sharing of four valence electrons in each carbon atom with adjacent atoms. This effectively creates eight valence electrons for each atom and produces a state of chemical stability. This sharing of valence electrons produces strong **covalent bonds** that hold the atoms together.

The shared electrons are not mobile; each electron is associated by a covalent bond between the atoms of the crystal. Therefore, there is a large energy gap between the valence band and the conduction band. As a consequence, crystalline materials such as diamond are insulators, or nonconductors, of electricity. Figure 2–2(a) shows the energy bands for a solid insulator.

Electronic devices are constructed from materials called **semiconductors**. The most common semiconductive material is **silicon**; however, **germanium** is sometimes used. At room temperature, silicon forms a covalent crystal. The actual atomic structure is similar to diamond but the covalent bonds in silicon are not as strong as those in diamond. In silicon, each atom shares a valence electron with each of its four neighbors. As in the case of other crystalline materials, the discrete levels are blurred into a valence band and a conduction band, as shown in Figure 2–2(b).

The important difference between a conductor and a semiconductor is the gap that separates the bands. With semiconductors, the gap is narrow; electrons can easily be promoted to the conduction band with the addition of thermal energy. At absolute zero, the electrons in a silicon crystal are all in the valence band, but at room temperature many electrons have sufficient energy to move to the conduction band. The conduction-band electrons are no longer bound to a parent atom within the crystal.

Electrons and Hole Current

When an electron jumps to the conduction band, a vacancy is left in the valence band. This vacancy is called a **hole**. For every electron raised to the conduction band by thermal or light energy, there is one hole left in the valence band, creating what is called an electron-hole pair. **Recombination** occurs when a conduction-band electron loses energy and falls back into a hole in the valence band.

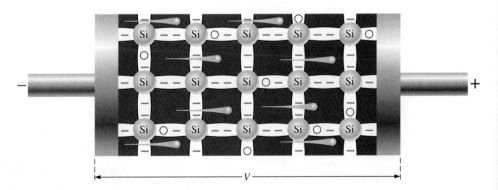

FIGURE 2–3
Electron current in intrinsic silicon is produced by thermally generated electrons.

A piece of **intrinsic** (pure) silicon at room temperature has, at any instant, a number of conduction-band (free) electrons that are unattached to any atom and are essentially drifting randomly throughout the material. Also, an equal number of holes are created in the valence band when these electrons jump into the conduction band.

When a voltage is applied across a piece of intrinsic silicon, as shown in Figure 2–3, the thermally generated free electrons in the conduction band are easily attracted toward the positive end. This movement of free electrons is one type of current in a semiconductor and is called *electron current.*

Another type of current occurs at the valence level, where the holes created by the free electrons exist. Electrons remaining in the valence band are still attached to their atoms and are not free to move randomly in the crystal structure. However, a valence electron can move into a nearby hole, with little change in its energy level, thus leaving another hole where it came from. Effectively, the hole has moved from one place to another in the crystal structure, as illustrated in Figure 2–4. This current is called *hole current.*

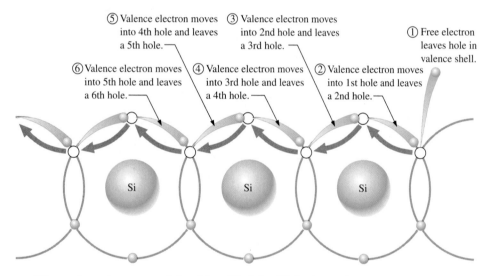

When a valence electron moves left to right to fill a hole while leaving another hole behind, a hole has effectively moved from right to left. Gray arrows indicate effective movement of a hole.

FIGURE 2–4
Hole current in intrinsic silicon.

2–1 REVIEW QUESTIONS*

1. In an intrinsic semiconductor, in which energy band do free electrons exist? In which band do holes exist?

2. How are holes created in an intrinsic semiconductor?

3. Why is current established more easily in a semiconductor than in an insulator?

* Answers are at the end of the chapter.

2–2 ■ THE *PN* JUNCTION

Intrinsic silicon (or germanium) is not a good conductor. It must be modified by increasing either the free electrons or the holes to increase its conductivity. If a pentavalent impurity is added to pure silicon, an n-material is formed; if a trivalent impurity is added, a p-material is formed. During manufacture, these materials can be joined and form a boundary called the pn junction. Amazingly, it is the characteristics of the pn junction that allow diodes and transistors to work.

After completing this section, you should be able to

❏ Describe the characteristics of a *pn* junction
    ❏ Compare *p*-type and *n*-type semiconductive materials
    ❏ Give examples of donor and acceptor materials
    ❏ Describe the formation of a *pn* junction

Doping

The conductivity of silicon (or germanium) can be drastically increased by the controlled addition of impurities to the pure (intrinsic) semiconductive material. This process, called **doping**, increases the number of current carriers (electrons or holes), thus increasing the conductivity and decreasing the resistivity. The two categories of impurities are *n*-type and *p*-type.

To increase the number of conduction-band electrons in pure silicon, a controlled number of pentavalent impurity atoms called *donors* are added to the silicon crystal. These are atoms with five valence electrons, such as arsenic, phosphorus, and antimony. Each pentavalent atom forms covalent bonds with four adjacent silicon atoms, leaving one extra electron. This extra electron becomes a conduction (free) electron because it is not bonded to any atom in the crystal. The electrons in these *n* materials are called the *majority carriers;* the holes are called *minority carriers.*

To increase the number of holes in pure silicon, trivalent impurity atoms called *acceptors* are added. These are atoms with only three valence electrons, such as aluminum, boron, and gallium. Each trivalent atom forms covalent bonds with four adjacent silicon atoms. All three of the impurity atom's valence electrons are used in the covalent bonds. However, since four electrons are required in the crystal structure, a hole is formed with each trivalent atom added. With *p* materials, the acceptor causes extra holes in the valence band; the majority carrier in *p* materials is holes, the minority carrier is electrons.

It is important to note that the process of creating *n*-type or *p*-type materials retains the overall electrical neutrality. With *n*-type materials, the extra electron in the crystal is balanced by the additional positive charge of the donor's nucleus.

The *PN* Junction

When a piece of intrinsic silicon is doped so that half is *n* type and the other half is *p* type, a *pn* **junction** is formed between the two regions. The *n* region has many free electrons (majority carriers) and only a few thermally generated holes (minority carriers). The *p* region has many holes (majority carriers) and only a few thermally generated free electrons (minority carriers). The *pn* junction forms a basic diode and is fundamental to the operation of all solid-state devices. *A diode is a device that allows current in only one direction.*

The Depletion Region When the *pn* junction is formed, some of the conduction electrons near the junction drift across into the *p* region and recombine with holes near the junction as shown in Figure 2–5(a). For each electron that crosses the junction and recombines with a hole, a pentavalent atom is left with a net positive charge in the *n* region near the junction. Also, when the electron recombines with a hole in the *p* region, a trivalent atom acquires a net negative charge. As a result, positive ions are found on the *n* side of the junction and negative ions are found on the *p* side of the junction. The existence of the positive and negative ions on opposite sides of the junction creates a **barrier potential** (V_B) across the depletion region. The barrier potential depends on temperature, but it is approximately 0.7 V for silicon and 0.3 V for germanium at room temperature. Since germanium diodes are rarely used, 0.7 V is normally found in practice and will be assumed in this text.

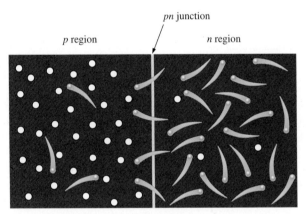

(a) At the instant of junction formation, free electrons in the *n* region near the *pn* junction begin to diffuse across the junction and fall into holes near the junction in the *p* region.

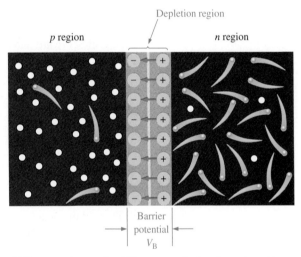

(b) For every electron that diffuses across the junction and combines with a hole, a positive charge is left in the *n* region and a negative charge is created in the *p* region, forming a barrier potential, V_B. This action continues until the voltage of the barrier repels further diffusion.

FIGURE 2–5
Formation of the pn junction.

Conduction electrons in the *n* region must overcome both the attraction of the positive ions and the repulsion of the negative ions in order to migrate into the *p* region. After the ion layers build up, the area on both sides of the junction becomes essentially depleted of any conduction electrons or holes and is known as the **depletion region**. This condition is illustrated in Figure 2–5(b). Any further movement of charge across the boundary requires that the barrier potential be overcome.

2–2 REVIEW QUESTIONS

1. How is an *n*-type semiconductor formed?
2. How is a *p*-type semiconductor formed?
3. What is a *pn* junction?
4. What is the value of the barrier potential for silicon?

2–3 ■ BIASING THE SEMICONDUCTOR DIODE

A single pn junction forms a semiconductor diode. There is no current across a pn junction at equilibrium. The primary usefulness of the semiconductor diode is its ability to allow current in only one direction as determined by the bias. There are two bias conditions for a pn junction—forward and reverse. Either of these conditions is created by connecting an external dc voltage in the proper direction across the pn junction.

After completing this section, you should be able to

❏ Explain how to bias a semiconductor diode
   ❏ Describe forward and reverse bias of a diode
   ❏ Describe avalanche breakdown

Forward Bias

The term **bias** in electronics refers to a fixed dc voltage that sets the operating conditions for a semiconductor device. **Forward bias** is the condition that permits current across a *pn* junction.

Figure 2–6 shows the polarity required from a dc source to forward-bias the semiconductor diode. The negative side of a source is connected to the *n* region (at the cathode **terminal**), and the positive side of a source is connected to the *p* region (at the anode terminal). When the semiconductor diode is forward-biased, the **anode** is the more positive terminal and the **cathode** is the more negative terminal.[1]

This is how forward bias works: When a dc source is connected to forward-bias the diode, the negative side of the source pushes the conduction electrons in the *n* region toward the junction because of electrostatic repulsion. The positive side pushes the holes in the *p* region also toward the junction. When the external bias voltage is sufficient to overcome the barrier potential, electrons have enough energy to penetrate the depletion region and cross the junction, where they combine with the *p* region holes. As electrons leave the *n* region, more electrons flow in from the negative side of the source. Thus, current through the *n* region is formed by the movement of conduction electrons (majority carriers) toward the junction. When the conduction electrons enter the *p* region and combine with holes, they become valence electrons. Then they move as valence electrons from hole to hole toward the positive anode connection. The movement of these valence electrons essentially creates a movement of holes in the opposite direction. Thus, current in the *p* region is formed by the movement of holes (majority carriers) toward the junction.

---

[1] Chemists define anode and cathode in terms of the type of chemical reaction that occurs in electrochemical cells. For electrochemistry, the anode is the terminal that acts as an electron donor; the cathode is the terminal that acts as an electron acceptor.

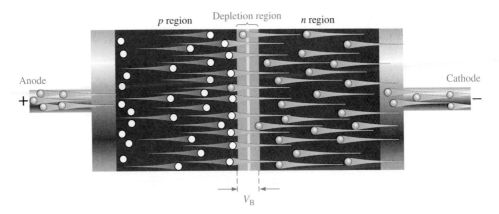

FIGURE 2–6
Electron flow in a forward-biased semiconductor diode.

Reverse Bias

Reverse bias is the bias condition that prevents current across the *pn* junction. Figure 2–7(a) shows the polarity required from a dc source to reverse-bias the semiconductor diode. Notice that the negative side of the source is connected to the *p* region, and the positive side to the *n* region. When the semiconductor diode is reverse-biased, the anode is the more negative terminal and the cathode is the more positive terminal.

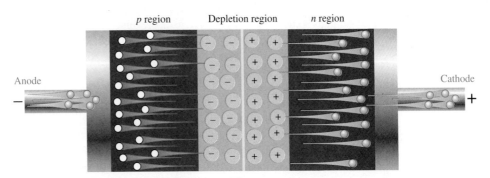

(a) Transient current at initial application of reverse bias

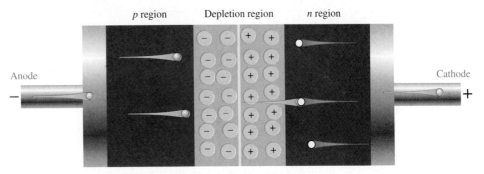

(b) Current ceases when the barrier potential equals the bias voltage.

FIGURE 2–7
Reverse bias.

This is how reverse bias works: The negative side of the source attracts holes in the *p* region away from the *pn* junction, while the positive side of the source attracts electrons away from the junction due to the attraction of opposite charges. As electrons and holes move away from the junction, the depletion region begins to widen; more positive ions are created in the *n* region, and more negative ions are created in the *p* region. The depletion region widens until the potential difference across it is equal to the external bias voltage, as shown in Figure 2–7(b). The depletion region effectively acts as an insulator between the layers of oppositely charged ions when the diode is reverse-biased.

Peak Inverse Voltage (PIV)   When a diode is reverse-biased, it must be able to withstand the maximum value of reverse voltage that is applied or it will break down. The maximum rated voltage for a diode is designated as *peak inverse voltage (PIV)*. The required PIV depends on the application; for most cases with ordinary diodes, the PIV rating should be higher than the reverse voltage.

Reverse Breakdown   If the external reverse-bias voltage is increased to a large enough value, *avalanche breakdown* occurs. Here is what happens: Assume that one minority conduction-band electron acquires enough energy from the external source to accelerate it toward the positive end of the diode. During its travel, it collides with an atom and imparts enough energy to knock a valence electron into the conduction band. There are now two conduction-band electrons. Each will collide with an atom, knocking two more valence electrons into the conduction band. There are now four conduction-band electrons which, in turn, knock four more into the conduction band. This rapid multiplication of conduction-band electrons, known as an *avalanche effect,* results in a rapid buildup of reverse current.

Most diode circuits are not designed to operate in reverse breakdown, and the diode may be destroyed if it is. By itself, reverse breakdown will not harm a diode, but current limiting must be present to prevent excessive heating. One type of diode, the zener diode, is specially designed for reverse-breakdown operation if sufficient current limiting is provided. (Zeners are discussed in Section 2–8.)

2–3 REVIEW QUESTIONS

1. What are the two bias conditions?
2. Which bias condition produces majority carrier current?
3. Which bias condition produces a widening of the depletion region?
4. What is avalanche breakdown?

2–4 ■ DIODE CHARACTERISTICS

In this section, you will learn that the characteristic curve graphically shows the current-voltage relationship for a diode. Three diode models are discussed. Each model represents the diode at a different level of accuracy so that you can use the one most appropriate for a given situation. In some cases, the lowest level of accuracy is all that is needed and additional details only complicate the situation. In other cases, you need the highest level of accuracy so that all factors can be taken into account.

After completing this section, you should be able to

❑ Describe the basic diode characteristics
  ❑ Describe the diode characteristic *IV* curve
  ❑ Explain how to plot the diode characteristic *IV* curve on an oscilloscope
  ❑ Describe three models that are used to simplify diode circuits

Diode Symbol

Figure 2–8(a) shows the standard schematic symbol for a general-purpose diode. The two terminals of the diode are the anode and cathode, labeled A and K on the figure. The arrow always points toward the cathode.

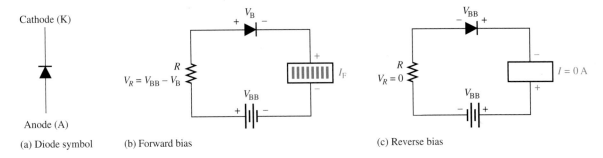

(a) Diode symbol (b) Forward bias (c) Reverse bias

FIGURE 2–8
Diode schematic symbol and bias circuits. V_{BB} is the bias battery voltage, and V_B is the barrier potential. The resistor limits the forward current to a safe value.

Figure 2–8(b) shows a forward-biased diode connected to a source through a current-limiting resistor. The anode is positive with respect to the cathode, causing the diode to conduct as indicated by the ammeter symbol. Remember that when the diode is forward-biased, the barrier potential, V_B, always appears between the anode and cathode, as indicated. The voltage across the resistor, V_R, is V_{BB} less the barrier potential, V_B.

Figure 2–8(c) shows the diode with reverse bias. The anode is negative with respect to the cathode, and the diode does not conduct as indicated by the ammeter symbol. The entire bias voltage, V_{BB}, appears across the diode. There is no voltage across the resistor because there is no current in the circuit. Notice that the bias voltage, V_{BB}, is not the same as the barrier potential, V_B.

Some typical diodes are shown in Figure 2–9 to illustrate common packaging. The letter A is used to identify the anode; K is used to identify the cathode.

Diode Characteristic Curve

Figure 2–10 is a graph of diode current versus voltage. The upper right quadrant of the graph represents the forward-biased condition. As you can see, there is essentially no forward current (I_F) for forward voltages (V_F) below the barrier potential. As the forward voltage approaches the value of the barrier potential (typically 0.7 V for silicon and 0.3 V for germanium), the current begins to increase. Once the forward voltage reaches the barrier

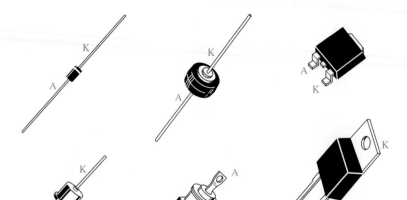

FIGURE 2–9
Typical diode packages and terminal identification.

FIGURE 2–10
Diode characteristic curve.

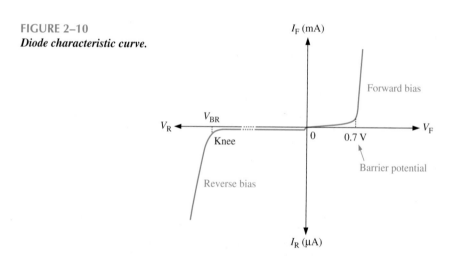

potential, the current increases drastically and must be limited by a series resistor. The voltage across the forward-biased diode remains approximately equal to the barrier potential, but increases slightly with forward current. For a forward-biased diode, this barrier voltage is often referred to as a *diode drop.*

The lower left quadrant of the graph represents the reverse-biased condition. As the reverse voltage increases to the left, the current remains near zero until the breakdown voltage is reached. When breakdown occurs, there is a large reverse current which, if not limited, can destroy the diode.[2] Typically, the breakdown voltage is greater than 50 V for most rectifier diodes. Most applications for ordinary diodes do not include operation in the reverse-breakdown region.

---

[2] With proper current limiting, operation in the reverse-breakdown region does not harm the diode.

Plotting the Characteristic Curve on an Oscilloscope   You can plot the diode's forward characteristic on your oscilloscope by connecting the circuit shown in Figure 2–11. The signal is a 5 V peak-to-peak triangle that is centered about zero volts. This causes the diode to be alternately forward-biased and then reverse-biased. Channel 1 senses the voltage drop across the diode; channel 2 shows a signal that is proportional to the current. The scope is placed in the X-Y mode. The common lead on the signal generator *must not* be the same as the scope ground. Channel 2 must be inverted to display the signal in the proper orientation.

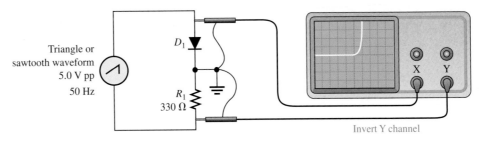

FIGURE 2–11
Plotting the IV curve for a diode on an oscilloscope. The oscilloscope is placed in the X-Y mode and the Y channel is inverted.

Testing Diodes with an Ohmmeter or a Multimeter

The internal battery in most analog ohmmeters can forward-bias or reverse-bias a diode, permitting a quick relative check of the diode. To check the diode with an analog ohmmeter, select the $R \times 100$ range (to limit current through the diode), connect the meter leads to the diode, then reverse the leads. The meter's internal voltage source will tend to forward-bias the diode in one direction and reverse-bias it in the other. As a result, the resistance will read a lower value in one direction than the other. Look for a high ratio between the forward and reverse readings (typically 1000 or more). The actual reading depends on the internal voltage of the meter, the range selected, and the type of diode, so this is only a relative test.

Many digital multimeters have a diode test position that will indicate the forward diode voltage when a good diode is placed across the test leads. The meter will show an overload when the leads are reversed.

Diode Models

The Ideal Model   The simplest way to visualize diode operation is to think of it as a switch. When forward-biased, the diode ideally acts as a closed (on) switch; and when reverse-biased, it acts as an open (off) switch, as shown in Figure 2–12. The characteristic curve for this model is also shown. Note that the forward voltage and the reverse current are always zero in the ideal case. This ideal model, of course, neglects the effect of the barrier potential, the internal resistance, and other effects. However, in many cases, it is accurate enough, particularly when the bias voltage is at least ten times greater than the barrier potential.

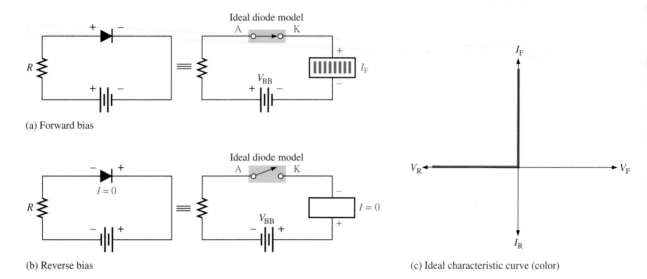

(a) Forward bias

(b) Reverse bias

(c) Ideal characteristic curve (color)

FIGURE 2–12
Ideal model of a diode as a switch.

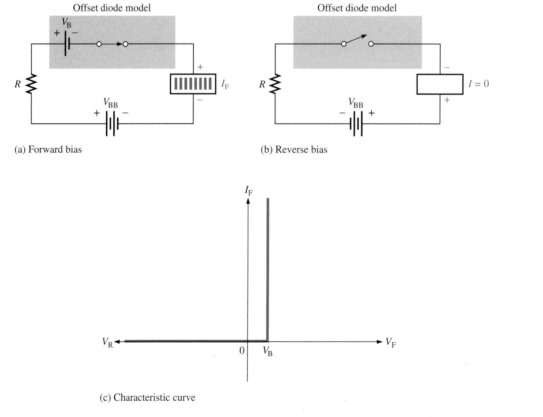

(a) Forward bias

(b) Reverse bias

(c) Characteristic curve

FIGURE 2–13
The offset model for a diode. The barrier potential is included in this model.

The Offset Model   The next higher level of accuracy is the offset model. It includes the barrier potential of the diode. In this model, the forward-biased diode is represented as a closed switch in series with a small "battery" equal to the barrier potential V_B (0.7 V for Si), as shown in Figure 2–13(a). The positive end of the equivalent battery is toward the anode. Keep in mind that the barrier potential is *not* a voltage source and cannot be measured with a voltmeter; rather it only has the effect of an offsetting battery when forward bias is applied because the forward-bias voltage, V_{BB}, must overcome this barrier potential before the diode begins to conduct. The reverse-biased diode is represented by an open switch, as in the ideal case, because the barrier potential does not affect reverse bias, as shown in Figure 2–13(b). The characteristic curve for the offset model is shown in Figure 2–13(c). In this textbook, this model is used for analysis unless otherwise stated.

The Offset-Resistance Model   Figure 2–14(a) shows the forward-biased diode model with both the barrier potential and the low forward (bulk) resistance. The forward resistance is actually an ac resistance (see Section 1–1). The forward resistance varies (depending on where it is measured) but is shown here with a straight-line approximation.

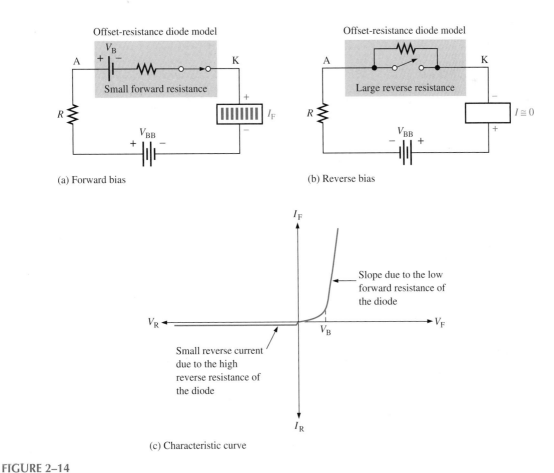

(a) Forward bias

(b) Reverse bias

(c) Characteristic curve

FIGURE 2–14

The offset-resistance model for a diode. The barrier potential and forward ac resistance is included in this model.

The reversed-biased condition is represented in the offset-resistance model with a very high parallel resistance. This results in an extremely small reverse current. Figure 2–14(b) shows how the high reverse resistance affects the reverse-biased model. The characteristic curve is shown in Figure 2–14(c). There are other small-scale effects (such as junction capacitance) that are not included in this model. For these cases, computer modeling is normally done.

2–4 REVIEW QUESTIONS

1. What are the two conditions under which the diode is operated?
2. What region of the diode characteristic curve is not part of normal diode operation?
3. What is the simplest way to visualize a diode?
4. What two approximations are included in the offset-resistance model of a diode?

2–5 ■ RECTIFIERS

Because of their ability to conduct current in one direction and block current in the other direction, diodes are used in circuits called rectifiers that convert ac voltage into dc voltage. Rectifiers are found in all dc power supplies that operate from an ac voltage source. Power supplies are an essential part of all electronic systems from the simplest to the most complex. In this section, you will study three basic types of rectifiers—the half-wave, center-tapped full wave, and full-wave bridge rectifiers.

After completing this section, you should be able to

❏ Analyze the operation of three basic types of rectifiers
  ❏ Recognize a half-wave rectifier and explain how it works
  ❏ Recognize a center-tapped full-wave rectifier and explain how it works
  ❏ Recognize a full-wave bridge rectifier and explain how it works

Half-Wave Rectifiers

A **rectifier** is an electronic circuit that converts ac into pulsating dc. Figure 2–15 illustrates the process called *half-wave rectification*. In a **half-wave rectifier**, shown in part (a), an ac source is connected in series with a diode and the load resistor. When the sinusoidal input voltage goes positive, the diode is forward-biased and conducts current to the load resistor, as shown in part (b). The output voltage is equal to the peak voltage less one diode drop.

$$V_{p(out)} = V_{p(in)} - 0.7 \text{ V} \qquad \text{(2–1)}$$

The current produces a voltage across the load, which has the same shape as the positive half-cycle of the input voltage. When the input voltage goes negative during the sec-

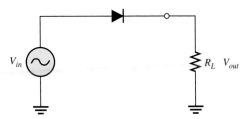

(a) Half-wave rectifier

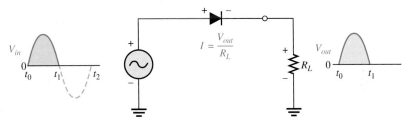

(b) Operation during positive alternation of the input voltage; diode conducts.

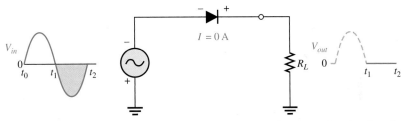

(c) Operation during negative alternation of the input voltage. Diode does not conduct; therefore, the output voltage is zero.

(d) Half-wave output voltage for three input cycles

FIGURE 2–15
Operation of half-wave rectifier. The diode is considered ideal.

ond half of its cycle, the diode is reverse-biased. There is no current, so the voltage across the load resistor is zero, as shown in part (c). The net result is that only the positive half-cycles of the ac input voltage, less one diode drop, appear across the load, making the output a pulsating dc voltage, as shown in part (d). Notice that during the negative cycle, the diode must withstand the negative peak voltage from the source without breaking down.

In working with diode circuits, it is sometimes practical to neglect the diode drop when the peak value of the applied voltage is much greater than the barrier potential. This is equivalent to using the ideal model.

EXAMPLE 2–1 Determine the peak output voltage and the peak inverse voltage (PIV) of the rectifier in Figure 2–16 for the indicated input voltage. Sketch the waveforms you should observe across the diode and the load resistor.

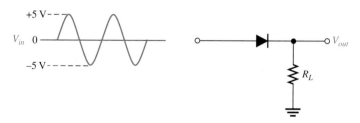

FIGURE 2–16

Solution The peak half-wave output voltage is

$$V_p = 5 \text{ V} - 0.7 \text{ V} = \textbf{4.3 V}$$

The PIV is the maximum voltage across the diode when it is reverse-biased. The PIV is the maximum voltage during the negative half cycle.

$$\text{PIV} = V_p = \textbf{5 V}$$

Waveforms are shown in Figure 2–17. Notice that if you add the load resistor voltage to the diode voltage, you will obtain the input voltage.

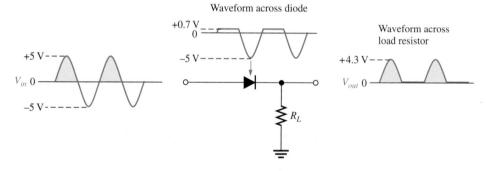

FIGURE 2–17

*Practice Exercise** Determine the peak output voltage and the PIV for the rectifier in Figure 2–16 if the peak input is 3 V.

———————————

* Answers are at the end of the chapter.

Full-Wave Rectifiers

The difference between full-wave and half-wave rectification is that a **full-wave rectifier** allows unidirectional current to the load during the entire input cycle, and the half-wave rectifier allows current only during one-half of the cycle. The result of full-wave rectification is a dc output voltage that pulsates every half-cycle of the input, as shown in Figure 2–18.

FIGURE 2–18
Full-wave rectification.

The Center-tapped Full-wave Rectifier The **center-tapped** (CT) full-wave rectifier uses two diodes connected to the secondary of a center-tapped transformer, as shown in
Figure 2–19. The input signal is coupled through the transformer to the secondary. Half of
the total secondary voltage appears between the center tap and each end of the secondary
winding as shown.

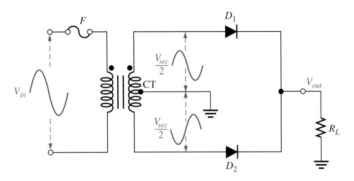

FIGURE 2–19
A center-tapped (CT) full-wave rectifier.

For a positive half-cycle of the input voltage, the polarities of the secondary voltages
are as shown in Figure 2–20(a). This condition forward-biases the upper diode D_1 and
reverse-biases the lower diode D_2. The current path is through D_1 and the load resistor, as
indicated in color.

For a negative half-cycle of the input voltage, the voltage polarities on the secondary
are as shown in Figure 2–20(b). This condition reverse-biases D_1 and forward-biases D_2.
The current path is through D_2 and the load resistor, as indicated in color.

Because the current during both the positive and the negative portions of the input
cycle is in the same direction through the load, the output voltage developed across the
load is a full-wave rectified dc voltage.

Effect of the Turns Ratio on the Full-Wave Output Voltage If the turns ratio of
a transformer is 1, the peak value of the rectified output voltage equals half the peak value
of the primary input voltage less one diode drop. This value occurs because half of the input voltage appears across each half of the secondary winding.

In order to obtain a peak output voltage equal to the peak input voltage (less the barrier potential), you must use a step-up transformer with a turns ratio of 2 (1:2). In this case,
the total secondary voltage is twice the primary voltage, so the voltage across each half of
the secondary is equal to the input.

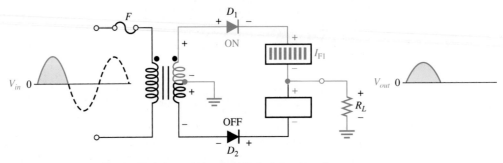

(a) During positive half-cycles, D_1 is forward-biased and D_2 is reverse-biased.

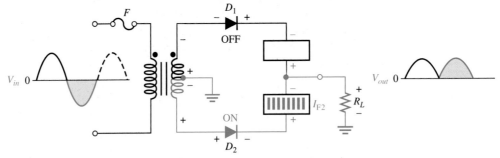

(b) During negative half-cycles, D_2 is forward-biased and D_1 is reverse-biased.

FIGURE 2–20

Conducting paths in the secondary are shown in color.

Peak Inverse Voltage (PIV)   Each diode in the full-wave rectifier is alternately forward-biased and then reverse-biased. The maximum reverse voltage that each diode must withstand is the peak value of the total secondary voltage (V_{sec}). The peak inverse voltage across either diode in the center-tapped full-wave rectifier is

$$\text{PIV} = V_{p(out)}$$

EXAMPLE 2–2

(a) Show the voltage waveforms across the secondary winding and across R_L when a 25 V peak sine wave is applied to the primary winding in Figure 2–21.

(b) What minimum PIV rating must the diodes have?

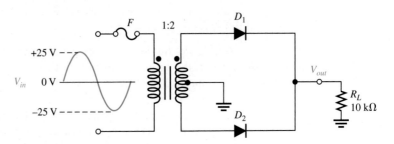

FIGURE 2–21

Solution

(a) The waveforms are shown in Figure 2–22.

(b) The total peak secondary voltage is

$$V_{p(sec)} = \left(\frac{N_{sec}}{N_{pri}}\right)V_{p(in)} = 2(25) \text{ V} = 50 \text{ V}$$

There is a 25 V peak across each half of the secondary. Using the ideal model, one diode is a short while the other diode has the full secondary voltage across it. Each diode must have a minimum PIV rating of **50 V.**

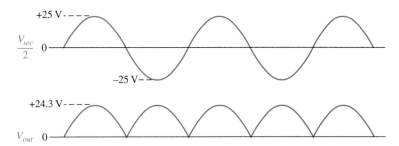

FIGURE 2–22

Practice Exercise What diode PIV rating is required to handle a peak input of 160 V in Figure 2–21?

Bridge Rectifiers

The bridge rectifier uses four diodes, as shown in Figure 2–23 and is the most popular arrangement for power supplies because it does not require a center-tapped transformer. The four diodes are available in a single package, already wired in a bridge configuration. The bridge rectifier is a type of full-wave rectifier because each half of the sine wave contributes to the output.

This is how the bridge rectifier works: When the input cycle is positive as in Figure 2–23(a), diodes D_1 and D_2 are forward-biased and conduct current as shown by the colored path. A voltage is developed across R_L which looks like the positive half of the input cycle. During this time, diodes D_3 and D_4 are reverse-biased. When the input cycle is negative, as in Figure 2–23(b), diodes D_3 and D_4 are forward-biased and conduct as shown by the colored path. A voltage is again developed across R_L in the same direction as during the positive half-cycle. During the negative half-cycle, D_1 and D_2 are reverse-biased. A full-wave rectified output voltage appears across R_L as a result of this action.

Bridge Output Voltage Neglecting the diode drops, the total secondary voltage, V_{sec}, appears across the load resistor. Thus,

$$V_{out} = V_{sec}$$

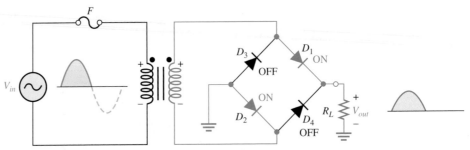

(a) During positive half-cycle of the input, D_1 and D_2 are forward-biased and conduct current. D_3 and D_4 are reverse-biased.

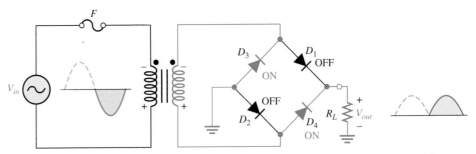

(b) During negative half-cycle of the input, D_3 and D_4 are forward-biased and conduct current. D_1 and D_2 are reverse-biased.

FIGURE 2–23

Operation of the full-wave rectifier. Conducting paths in the secondary are shown in color.

As you can see in Figure 2–23, two diodes are always in series with the load resistor during both the positive and the negative half-cycles. If these diode drops are taken into account, the output voltage (with silicon diodes) is

$$V_{out} = V_{sec} - 1.4 \text{ V} \qquad (2\text{–}2)$$

Peak Inverse Voltage When D_1 and D_2 are forward-biased, the reverse voltage is across D_3 and D_4. Visualizing D_1 and D_2 as shorts (ideally), the peak inverse voltage is equal to the peak secondary voltage.

$$\text{PIV} = V_{p(out)}$$

2–5 REVIEW QUESTIONS

1. Which type of rectifier (half-wave, full-wave, or bridge) has the greatest output voltage for the same input voltage and transformer turns ratio?
2. For a given output voltage, is the PIV for bridge rectifier diodes less than, the same as, or greater than the PIV for center-tapped rectifier diodes?
3. At what point on the input cycle does the PIV occur in a half-wave rectifier that has a positive output?
4. For a half-wave rectifier, there is current through the load for approximately what percentage of the input cycle?

2–6 ■ RECTIFIER FILTERS AND IC REGULATORS

A power supply filter greatly reduces the fluctuations in the output voltage of a rectifier and produces a nearly constant-level dc voltage. The reason for filtering is that electronic circuits require a constant source of dc voltage and current to provide power and biasing for proper operation. Filtering is generally done using large capacitors. To improve the filtering action even more, the capacitor-input filter is followed by a regulator. Today, inexpensive but effective regulators are available as integrated circuits (ICs). IC regulators are introduced here and will be covered in more detail in Chapter 11.

After completing this section, you should be able to

❏ Describe the operation of rectifier filters and IC regulators
  ❏ Give examples of IC regulators and show how they are connected to the output of a rectifier
  ❏ Compute the ripple from an IC regulator given the ripple rejection and the input ripple
  ❏ Compute the load regulation given the loaded and unloaded output voltage
  ❏ Compute the line regulation given the change in output voltage for a given change in the input voltage

In most power supply applications, the standard 60 Hz ac power line voltage must be converted to a nearly constant dc voltage. The 60 Hz pulsating dc output of a half-wave rectifier or the 120 Hz pulsating output of a full-wave or bridge rectifier must be filtered to reduce the large voltage variations. Filtering can be accomplished by a capacitor, an inductor, or a combination of these. The capacitor-input **filter** is the least expensive and most widely used type, by far.

Capacitor-Input Filter

A half-wave rectifier with a capacitor-input filter is shown in Figure 2–24. We will use the half-wave rectifier to illustrate the filtering principle and then expand the concept to the full-wave rectifier.

During the positive first quarter-cycle of the input, the diode is forward-biased, allowing the capacitor to charge to within a diode drop of the input peak, as illustrated in Figure 2–24(a). When the input begins to decrease below its peak, as shown in Figure 2–24(b), the capacitor retains its charge and the diode becomes reverse-biased. During the remaining part of the cycle and the beginning of the next cycle, the capacitor can discharge only through the load resistance at a rate determined by the *RC* time constant. The larger the time constant, the less the capacitor will discharge.

During the peak of the next cycle, as illustrated in Figure 2–24(c), the diode again will become forward-biased when the input voltage exceeds the capacitor voltage by approximately a diode drop.

Ripple Voltage   As you have seen, the capacitor quickly charges at the beginning of a cycle and slowly discharges through R_L after the positive peak (when the diode is reverse-biased). The variation in the capacitor voltage due to the charging and discharging is called the **ripple voltage**. The smaller the ripple voltage, the better the filtering action.

For a given input frequency, the output frequency of a full-wave rectifier is twice that of a half-wave rectifier. As a result, a full-wave rectifier is easier to filter because of the shorter time between peaks. When filtered, the full-wave rectified voltage has less ripple voltage than does a half-wave voltage for the same load resistance and filter capacitor val-

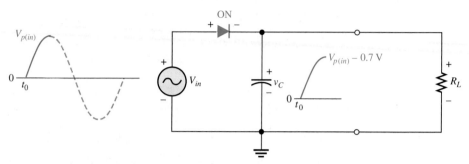

(a) Initial charging of capacitor (diode is forward-biased) happens only once when power is turned on.

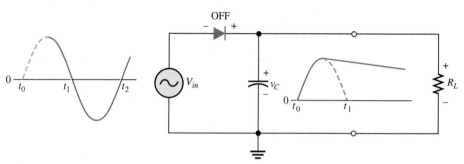

(b) The capacitor discharges through R_L after peak of positive alternation when the diode is reverse-biased. This discharging occurs during the portion of the input voltage indicated by the solid colored curve.

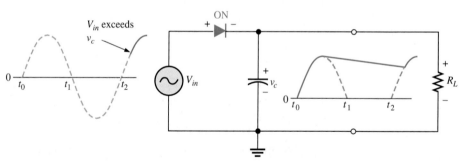

(c) The capacitor charges back to peak of input when the diode becomes forward-biased. This charging occurs during the portion of the input voltage indicated by the solid colored curve. Notice that the diode is not forward-biased on the second cycle until the capacitor voltage is overcome.

FIGURE 2–24

Operation of a half-wave rectifier with a capacitor-input filter.

ues. Less ripple voltage occurs because the capacitor discharges less during the shorter interval between full-wave pulses, as shown in Figure 2–25.

Surge Current in the Capacitor-Input Filter   When the power is first applied to a power supply, the filter capacitor is uncharged. At the instant the switch is closed, voltage is connected to the rectifier and the uncharged capacitor appears as a short. This case is illustrated for a bridge circuit in Figure 2–26(a). An initial surge of current is produced through the forward-biased diodes. The worst-case situation occurs when the switch is closed at a peak of the secondary voltage and a maximum surge current is produced.

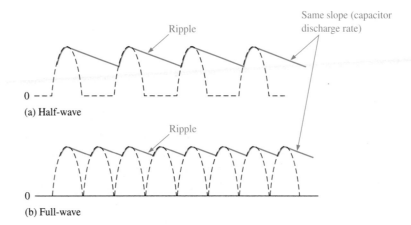

FIGURE 2–25
Comparison of ripple voltages for half-wave and full-wave rectifier outputs with the same filter capacitor and derived from the same sinusoidal input.

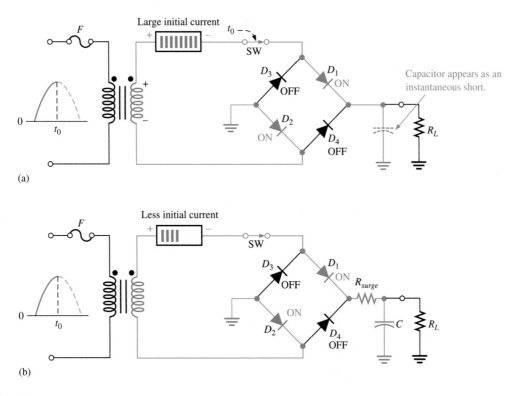

FIGURE 2–26
Surge current in a capacitor-input filter follows the path drawn in color.

It is possible that the surge current could destroy the diodes, and for this reason a surge-limiting resistor, R_{surge}, is sometimes connected, as shown in Figure 2–26(b). The value of this resistor must be small to avoid a significant voltage drop across it. Also, the diodes must have a forward current rating such that they can withstand the momentary surge of current.

IC Regulators

While filters can reduce the ripple from power supplies to a low value, the most effective filter is a combination of a capacitor-input filter used with an IC regulator. In general, an IC (**integrated circuit**) is a complete functional circuit constructed on a single, tiny chip of silicon. An integrated circuit **regulator** is an IC that is connected to the output of a rectifier and maintains a constant output voltage (or current) despite changes in the input, the load current, or the temperature. The capacitor filter reduces the input ripple to the regulator to an acceptable level. The combination of a large capacitor and an IC regulator is inexpensive and helps produce an excellent small power supply.

The most popular IC regulators have three terminals—an input terminal, an output terminal, and a reference (or adjust) terminal. The input to the regulator is first filtered with a capacitor to reduce the ripple to <10%. The regulator reduces the ripple to a negligible amount. In addition, most regulators have an internal voltage reference, short-circuit protection, and thermal shutdown circuitry. They are available in a variety of voltages, including positive and negative outputs, and can be designed for variable outputs with a minimum of external components. Typically, IC regulators can furnish a constant output of one or more amps of current with high ripple rejection. IC regulators are available that can supply load currents of over 5 A.

Three-terminal regulators designed for a fixed output voltage require only external capacitors to complete the regulation portion of the power supply, as shown in Figure 2–27. Filtering is accomplished by a large-value capacitor between the input voltage and ground. Sometimes a second smaller-value input capacitor is connected in parallel, especially if the filter capacitor is not close to the IC regulator, to prevent oscillation. This capacitor needs to be located close to the IC regulator. Finally, an output capacitor (typically 0.1 μF to 1.0 μF) is placed in parallel with the output to improve the transient response.

Examples of fixed three-terminal regulators are the 78XX and 79XX series[3] of regulators that are available with various output voltages and can supply up to 1 A of load current (with adequate heat sinking). Appendix A has the manufacturer's specification sheets for these series of regulators. The last two digits of the number stand for the output voltage; thus, the 7812 has a +12 V output. The negative output versions of the same regulator are numbered as the 79XX series; the 7912 has a −12 V output. The output voltage from these regulators is within 3% of the nominal value but will hold a nearly constant output despite changes in the input voltage or output load. A basic fixed +5 V power supply with a 7805 regulator is shown in Figure 2–28.

As an example of the reduction in ripple that can be obtained from a 7812 regulator, note the typical ripple rejection specification, *RR,* in the data sheet. For the 7812, the typical ripple rejection is 72 dB (refer to Section 1–4 for a review of decibels). This means that the output ripple is 72 dB less than the input ripple, a very significant reduction as illustrated in the next example.

---

[3] Data sheets for MC7800 and MC7900 available at http://www.onsemi.com.

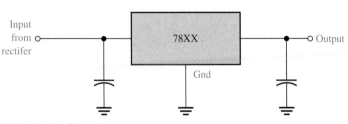

(a) Standard configuration

Type number	Output voltage
7805	+5.0 V
7806	+6.0 V
7808	+8.0 V
7809	+9.0 V
7812	+12.0 V
7815	+15.0 V
7818	+18.0 V
7824	+24.0 V

(b) The 7800 series

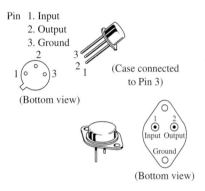

Pin 1. Input
     2. Output
     3. Ground

(Case connected to Pin 3)

(Bottom view)

(Bottom view)

Pins 1 and 2 are electrically isolated from case.
Case is third electrical connection.

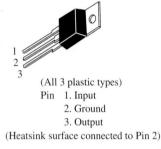

(All 3 plastic types)
Pin 1. Input
     2. Ground
     3. Output
(Heatsink surface connected to Pin 2)

Pin 1. Output
     2. Ground
     3. Input

Pin	1. V_{OUT}	5. NC
	2. Gnd	6. Gnd
	3. Gnd	7. Gnd
	4. NC	8. V_{IN}

(c) Typical metal and plastic packages

FIGURE 2–27
The 7800 series three-terminal fixed positive voltage regulators.

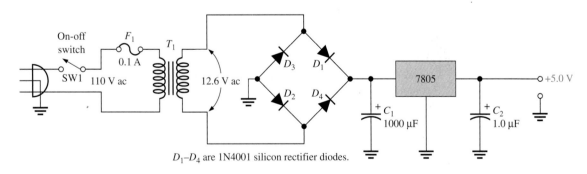

D_1–D_4 are 1N4001 silicon rectifier diodes.

FIGURE 2–28
A basic +5.0 V power supply.

EXAMPLE 2–3

Assume the input ripple to a MC7812B regulator is 100 mV. What is the typical output ripple? From the data sheet, the typical ripple rejection is 60 dB.

Solution   The decibel voltage ratio is

$$dB = 20 \log\left(\frac{V_{out}}{V_{in}}\right)$$

Since 60 dB is an attenuation, it is shown with a negative sign.

$$-60 \text{ dB} = 20 \log\left(\frac{V_{out}}{V_{in}}\right)$$

Dividing by 20,

$$-3.0 = \log\left(\frac{V_{out}}{100 \text{ mV}}\right)$$

and eliminating the log results in

$$10^{-3.0} = \frac{V_{out}}{100 \text{ mV}}$$

$$V_{out} = (100 \text{ mV})1.0 \times 10^{-3} = \mathbf{100 \ \mu V}$$

Alternatively, this result can also be obtained by computing the factor that the output ripple has been reduced: 60 dB is a factor of 1000. The calculator sequence to find this is

The calculator will show 1000. This means the output ripple has been reduced by a factor of 1000 from the input ripple. Thus, the output ripple is

$$\text{ripple} = \frac{100 \text{ mV}}{1000} = \mathbf{100 \ \mu V}$$

Practice Exercise   Find the output ripple for an MC7805B, using the typical value specified on the data sheet in Appendix A.

Another type of three-terminal regulator is an adjustable regulator. Figure 2–29 shows a power supply circuit with an adjustable output, controlled by the variable resistor, R_2. Note that R_2 is adjustable from zero to 1.0 kΩ. The LM317 regulator keeps a constant 1.25 V between the output and adjust terminals. This produces a constant current in R_1 of 1.25 V/240 Ω = 52 mA. Neglecting the very small current through the adjust terminal, the current in R_2 is the same as the current in R_1. The output is taken across both R_1 and R_2 and is found from the equation,

$$V_{out} = 1.25 \text{ V}\left(\frac{R_1 + R_2}{R_1}\right)$$

Notice that the output voltage from the power supply is the regulator's 1.25 V multiplied by a ratio of resistances. For the case shown in Figure 2–29, when R_2 is set to the minimum (zero) resistance, the output is 1.25 V. When R_2 is set to the maximum, the output is nearly 6.5 V.

FIGURE 2–29

A basic power supply with a variable output voltage (from 1.25 V to 6.5 V).

Percent Regulation

The regulation expressed as a percentage is a figure of merit used to specify the performance of a voltage regulator. It can be in terms of input (line) regulation or load regulation. **Line regulation** specifies how much change occurs in the output voltage for a given change in the input voltage. It is typically defined as a ratio of a change in output voltage for a corresponding change in the input voltage expressed as a percentage.

$$\text{Line regulation} = \left(\frac{\Delta V_{\text{OUT}}}{\Delta V_{\text{IN}}}\right)100\% \qquad \textbf{(2–3)}$$

Load regulation specifies how much change occurs in the output voltage over a certain range of load current values, usually from minimum current (no load, NL) to maximum current (full load, FL). It is normally expressed as a percentage and can be calculated with the following formula:

$$\text{Load regulation} = \left(\frac{V_{\text{NL}} - V_{\text{FL}}}{V_{\text{FL}}}\right)100\% \qquad \textbf{(2–4)}$$

where V_{NL} is the output voltage with no load and V_{FL} is the output voltage with full (maximum) load. Line and load regulation are discussed further in Section 11–1.

EXAMPLE 2–4

Assume a certain MC7805B regulator has a measured no-load output voltage of 5.185 V and a full-load output of 5.152 V. What is the load regulation expressed as a percentage? Is this within the manufacturer's specification?

Solution

$$\text{Load regulation} = \left(\frac{V_{\text{NL}} - V_{\text{FL}}}{V_{\text{FL}}}\right)100\% = \left(\frac{5.185 \text{ V} - 5.152 \text{ V}}{5.152 \text{ V}}\right)100\% = \textbf{0.64\%}$$

The data sheet (see the MC7805B in Appendix A) indicates a maximum variation of the output voltage (with a load current from 5 mA to 1.0 A) of 100 mV. This represents a maximum load regulation of 2% (typical is 0.4%). The measured percent regulation is within specifications.

Practice Exercise   If the no-load output voltage of a regulator is 24.8 V and the full-load output is 23.9 V, what is the load regulation expressed as a percentage?

The preceding discussion concentrated on the popular three-terminal regulators. Three-terminal regulators can be adapted to a number of specialized applications or requirements such as current sources or automatic shutdown, current limiting, and the like. For certain other applications (high current, high efficiency, high voltage), more complicated regulators are constructed from integrated circuits and discrete transistors. Chapter 11 discusses some of these applications in more detail.

2–6 REVIEW QUESTIONS

1. What causes the ripple voltage on the output of a capacitor-input filter?

2. The load resistance of a capacitor-filtered full-wave rectifier is reduced. What effect does this reduction have on the ripple voltage?

3. What advantages are offered by a three-terminal regulator?

4. What is the difference between input (line) regulation and load regulation?

2–7 ■ DIODE LIMITING AND CLAMPING CIRCUITS

Diode circuits, called limiters or clippers, are sometimes used to clip off portions of signal voltages above or below certain levels. Another type of diode circuit, called a clamper, is used to restore a dc level to an electrical signal.

After completing this section, you should be able to

❏ Analyze the operation of diode limiters and clampers
  ❏ Explain how a diode limiter works and determine the clipping level for a given circuit
  ❏ Explain how a diode clamping circuit works
  ❏ Cite applications for diode limiting and clamping circuits

Diode Limiters

Figure 2–30(a) shows a diode circuit called a **limiter** that limits or clips off the positive part of the input signal. As the input signal goes positive, the diode becomes forward-biased. Because the cathode is at ground potential (0 V), the anode cannot exceed 0.7 V (assuming a silicon diode). Thus, point *A* is clipped at +0.7 V when the input exceeds this value.

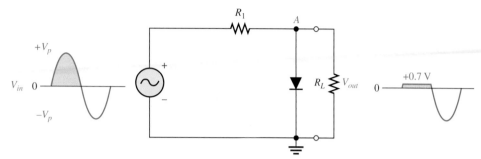

(a) Limiting of the positive alternation; diode conducts on positive alternation.

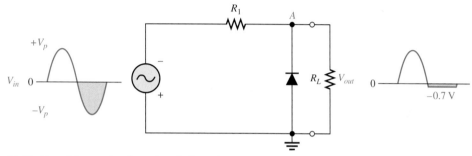

(b) Limiting of the negative alternation; diode conducts on negative alternation.

FIGURE 2–30
Diode limiting (clipping circuits).

Whenever the input is below 0.7 V, the diode is reverse-biased and appears as an open. The output voltage looks like the negative part of the input, but with a magnitude determined by the voltage divider formed by R_1 and R_L, as follows:

$$V_{out} = \left(\frac{R_L}{R_1 + R_L}\right)V_{in}$$

If R_1 is small compared to R_L, then $V_{out} \cong V_{in}$.

Turn the diode around, as in Figure 2–30(b), and the negative part of the input is clipped off. When the diode is forward-biased during the negative part of the input, point *A* is held at −0.7 V by the diode drop. When the input goes above −0.7 V, the diode is no longer forward-biased; and a voltage appears across R_L proportional to the input.

A Limiter Application Figure 2–31 shows an application of a limiter. Suppose you wanted to use the power line to synchronize a computer operation to the ac line. In the case shown, a half-wave rectifier is connected to a 6.3 V output from a transformer. The peak signal from the rectifier is approximately 9 V, too large for a computer input. Computers and other logic circuits are designed for a specific voltage maximum (typically +5.0 V) that cannot be exceeded without risking serious damage to the computer. The limiter shown in the figure prevents the signal to the computer from exceeding 4.7 V.

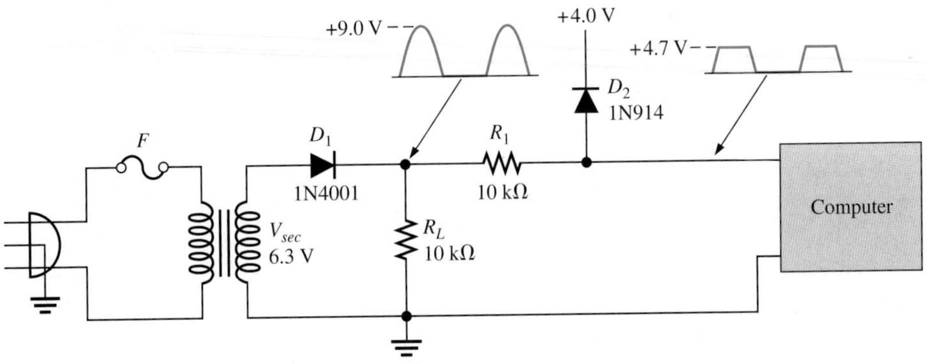

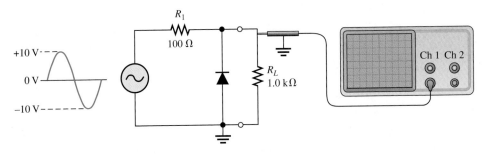

FIGURE 2–31
Limiting the signal into a computer.

EXAMPLE 2–5

What would you expect to see displayed on an oscilloscope connected as shown in Figure 2–32? The time base on the scope is set to show one and one-half cycles.

FIGURE 2–32

Solution   The diode is forward-biased and conducts when the input voltage goes below −0.7 V. Thus, a negative limiter with a peak output voltage across R_L can be determined by the following equation:

$$V_{p(out)} = \left(\frac{R_L}{R_1 + R_L} \right) V_{p(in)} = \left(\frac{1.0 \text{ k}\Omega}{1.1 \text{ k}\Omega} \right) 10 \text{ V} = \textbf{9.1 V}$$

The scope will display an output waveform as shown in Figure 2–33.

FIGURE 2–33
Waveforms for Figure 2–32.

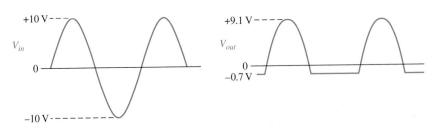

Practice Exercise   Describe the output waveform for Figure 2–32 if R_L is changed to 680 Ω.

Adjustment of the Limiting Level   To adjust the level at which a signal voltage is limited, a bias voltage can be added in series with the diode, as shown in Figure 2–34. The voltage at point *A* must equal V_{BB} + 0.7 V before the diode will become forward-biased and conduct. Once the diode begins to conduct, the voltage at point *A* is limited to V_{BB} + 0.7 V so that all input voltage above this level is clipped off, as shown in the figure.

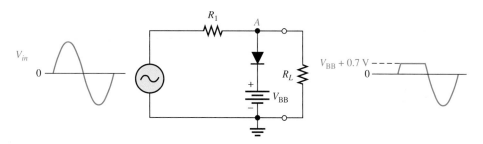

FIGURE 2–34
A positive limiter with positive bias.

If the bias voltage is varied up or down, the clipping level changes correspondingly, as shown in Figure 2–35. If the polarity of the bias voltage is reversed, as in Figure 2–36, voltages above $-V_{BB}$ + 0.7 V are clipped, resulting in an output waveform as shown. The diode is reverse-biased only when the voltage at point *A* goes below $-V_{BB}$ + 0.7 V.

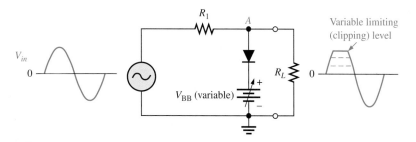

FIGURE 2–35
A positive limiter with variable positive bias.

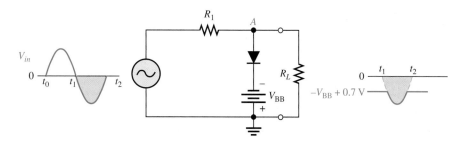

FIGURE 2–36
A positive limiter with negative bias. Notice that the positive side of the waveform is limited above $-V_{BB}$ + 0.7 V.

If it is necessary to clip the voltage below a specified negative level, then the diode and bias voltage must be connected as in Figure 2–37. In this case, the voltage at point A must go below $-V_{BB} - 0.7$ V to forward-bias the diode and initiate clipping action, as shown.

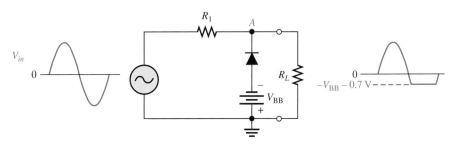

FIGURE 2–37
A negative limiter with negative bias.

EXAMPLE 2–6

Figure 2–38 shows a circuit combining a positive-biased limiter with a negative-biased limiter. Determine the output waveform.

FIGURE 2–38

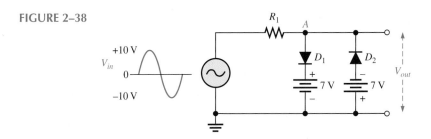

Solution   When the voltage at point A reaches $+7.7$ V, diode D_1 conducts and limits the waveform at $+7.7$ V. Diode D_2 does not conduct until the voltage reaches -7.7 V. Therefore, positive voltages above $+7.7$ V and negative voltages below -7.7 V are clipped. The resulting output waveform is shown in Figure 2–39.

FIGURE 2–39
Output waveform for Figure 2–38.

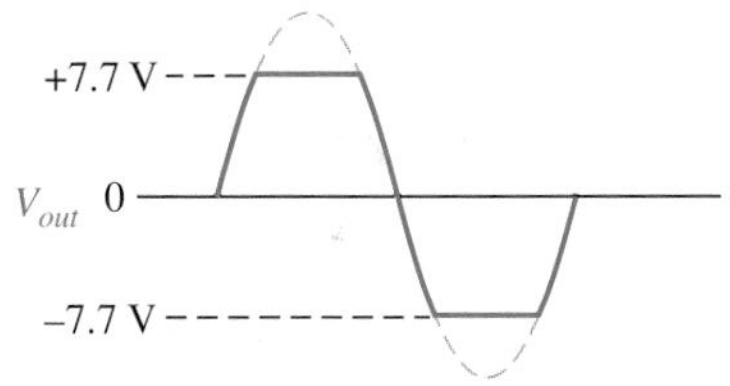

$+7.7$ V ----

V_{out} 0

-7.7 V ---------

Practice Exercise   Determine the output waveform in Figure 2–38 if both dc sources are 10 V and the input has a peak value of 20 V.

Diode Clampers

A diode **clamper** adds a dc level to an ac signal. Clampers are sometimes known as *dc re-storers.* Figure 2–40 shows a diode clamper that inserts a positive dc level in the output waveform. To understand the operation of this circuit, consider the first negative half-cycle of the input voltage. When the input initially goes negative, the diode is forward-biased, allowing the capacitor to charge to near the peak of the input ($V_{p(in)} - 0.7$ V), as shown in Figure 2–40(a). Just past the negative peak, the diode becomes reverse-biased. This is because the cathode is held near $V_{p(in)}$ by the charge on the capacitor.

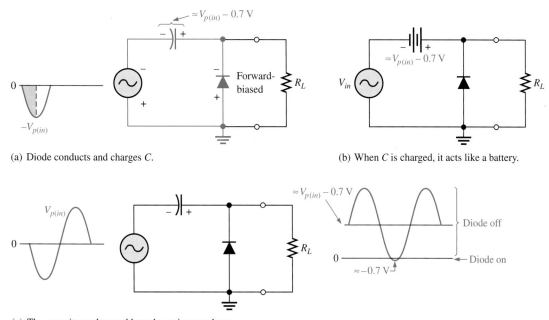

(a) Diode conducts and charges C.

(b) When C is charged, it acts like a battery.

(c) The capacitor voltage adds to the ac input voltage.

FIGURE 2–40

Positive clamping. The diode allows the capacitor to charge rapidly. The capacitor can discharge only through R_L.

The capacitor can discharge only through the high resistance of R_L. Thus, from the peak of one negative half-cycle to the next, the capacitor discharges very little. The amount that is discharged, of course, depends on the value of R_L. For good clamping action, the *RC* time constant should be at least ten times the period of the input signal.

The net effect of the clamping action is that the capacitor retains a charge approximately equal to the peak value of the input less the diode drop. The capacitor voltage acts essentially as a battery in series with the input signal, as shown in Figure 2–40(b). The dc voltage of the capacitor adds to the ac input voltage by superposition, as shown in Figure 2–40(c).

If the diode is turned around, a negative dc voltage is added to the input signal to produce the output signal, as shown in Figure 2–41. If necessary, the diode can be biased to adjust the clamping level.

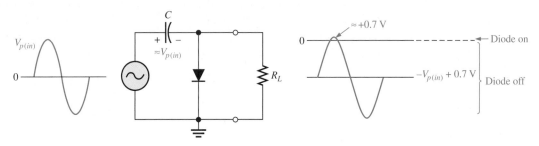

FIGURE 2–41
Negative clamping.

A Clamper Application   A clamping circuit is often used in television receivers as a dc restorer. The incoming composite video signal is normally processed through capacitively coupled amplifiers that eliminate the dc component, thus losing the black and white reference levels and the blanking level. Before being applied to the picture tube, these reference levels must be restored. Figure 2–42 illustrates this process in a general way.

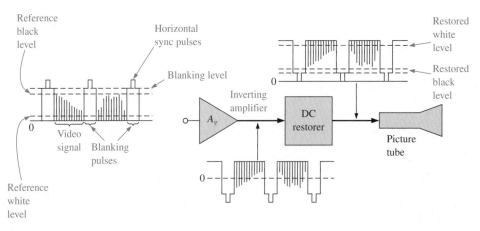

FIGURE 2–42
Clamping circuit (dc restorer) in a TV receiver.

EXAMPLE 2–7   What is the output voltage that you would expect to observe across R_L in the clamping circuit of Figure 2–43? Assume that RC is large enough to prevent significant capacitor discharge.

FIGURE 2–43

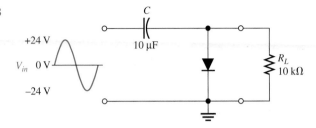

Solution   Ideally, a negative dc value equal to the input peak less the diode drop is inserted by the clamping circuit.

$$V_{DC} \cong -(V_{p(in)} - 0.7\ V) = -(24\ V - 0.7\ V) = \mathbf{-23.3\ V}$$

Actually, the capacitor will discharge slightly between peaks, and, as a result, the output voltage will have an average value of slightly less than that calculated previously. The output waveform goes to approximately 0.7 V above ground, as shown in Figure 2–44.

FIGURE 2–44
Output waveform for Figure 2–43.

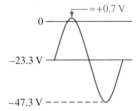

Practice Exercise   What output voltage would you observe across R_L in Figure 2–43 if the polarity of the diode and the polarity of the capacitor were reversed?

2–7 REVIEW QUESTIONS

1. Discuss how diode limiters and diode clampers differ in terms of their function.
2. What happens if the diode is reversed in a limiter?
3. To limit the output of a positive limiter to +5 V when a +10 V peak input is applied, what value must the bias voltage be?
4. What component in a clamper circuit effectively acts as a battery?

2–8 ■ SPECIAL-PURPOSE DIODES

The preceding discussion of diodes has focused on applications that exploit the fact that a diode is a one-way conductor. A number of diodes are designed for other applications. In this section, several special-purpose diodes, namely the zener diode, the varactor diode, the photodiode, and the light-emitting diode will be considered.

After completing this section, you should be able to

❑ Explain the characteristics of four different special-purpose diodes
   ❑ Describe the characteristic curve for a zener diode
   ❑ Show how a zener diode can be used as a basic regulator
   ❑ Explain how a varactor diode is used as a variable capacitor
   ❑ Discuss the basic principles of light-emitting diodes (LEDs) and photodiodes

Zener Diodes

Figure 2–45 shows the schematic symbol for a **zener diode.** The zener diode is a silicon *pn* junction device that differs from the rectifier diode in that it is designed for operation in the reverse-breakdown region. Zeners with breakdown voltages of 1.8 V to 200 V are commercially available. The breakdown voltage is set by carefully controlling the doping level during manufacture. From the discussion of the diode characteristic curve in Section 2–4, recall that when a diode reaches reverse breakdown, its voltage remains almost constant even though the current may change drastically. This volt-ampere characteristic is shown again in Figure 2–46.

FIGURE 2–45
Zener symbol.

Cathode (K)

Anode (A)

FIGURE 2–46
Diode IV characteristic curve.

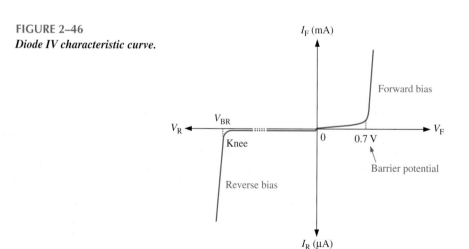

The principal applications of zener diodes are as a voltage reference and for voltage regulators in low-current applications. As a regulator, zeners have limitations: they do not have the high ripple rejection of integrated circuit regulators (discussed in Section 2–6) and they can't handle large load current changes. By combining a zener diode with a transistor or op-amp, better regulators can be constructed.

Figure 2–47 shows the reverse portion of the characteristic curve of a zener diode. Notice that as the reverse voltage (V_R) is increased, the reverse current (I_R) remains extremely small up to the knee of the curve. At this point, the breakdown effect begins; the internal zener ac resistance begins to decrease as the reverse current increases rapidly. This resistance is generally shown on a data sheet as impedance, Z_Z. From the bottom of the knee, the zener breakdown voltage (V_Z) remains essentially constant although it increases slightly as I_Z increases. This constant voltage region of the characteristic curve accounts for the zener's ability to regulate.

FIGURE 2–47

Reverse characteristic of a zener diode. V_Z is usually specified at the test current, I_{ZT}, and is designated V_{ZT}.

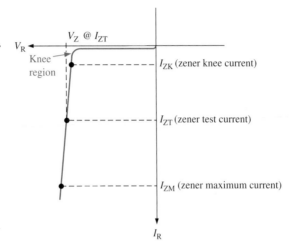

A minimum value of reverse current, I_{ZK}, must be maintained in order to keep the diode in regulation. You can see on the curve in Figure 2–47 that when the reverse current is reduced below the knee of the curve, the voltage decreases drastically and regulation is lost. Also, there is a maximum current, I_{ZM}, above which the diode may be damaged. Thus, basically, the zener diode maintains a nearly constant voltage across its terminals for values of reverse current ranging from I_{ZK} to I_{ZM}. A nominal zener voltage, V_{ZT}, is usually specified on a data sheet at a value of reverse current called the *zener test current, I_{ZT}.*

Zener Equivalent Circuit Figure 2–48(a) shows the ideal approximation of a zener diode in reverse breakdown. It acts simply as a battery having a value equal to the nominal zener voltage. Figure 2–48(b) represents the practical equivalent of a zener, where the zener impedance (Z_Z) is included. Since the actual voltage curve is not ideally vertical, a change in zener current (ΔI_Z) produces a small change in zener voltage (ΔV_Z), as illustrated in Figure 2–48(c).

By Ohm's law, the ratio of ΔV_Z to ΔI_Z is the zener impedance, as expressed in the following equation:

$$Z_Z = \frac{\Delta V_Z}{\Delta I_Z}$$

(2–5)

Normally, Z_Z is specified at I_{ZT}, the zener test current. In most cases, you can assume that Z_Z is constant over the full linear range of zener current values.

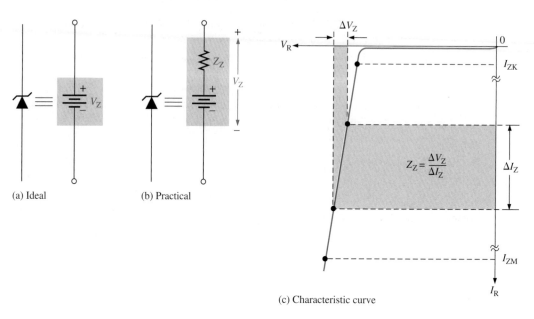

(a) Ideal (b) Practical (c) Characteristic curve

FIGURE 2–48

Zener diode equivalent circuits and the characteristic curve illustrating Z_Z.

EXAMPLE 2–8

A certain zener diode exhibits a 50 mV change in V_Z for a 2 mA change in I_Z on the linear portion of the characteristic curve between I_{ZK} and I_{ZM}. What is the zener impedance?

Solution

$$Z_Z = \frac{\Delta V_Z}{\Delta I_Z} = \frac{50 \text{ mV}}{2 \text{ mA}} = 25 \text{ } \Omega$$

Practice Exercise   Calculate the zener impedance if the zener voltage changes 120 mV for a 15 mA change in zener current.

Zener Voltage Regulation   As mentioned, zener diodes can be used for **voltage regulation** in noncritical applications. Figure 2–49 illustrates how a zener diode can be used to regulate a varying dc input voltage to keep it at a constant level. As you learned earlier, this process is called line regulation. (See Section 2–6.)

As the input voltage varies (within limits), the zener diode maintains a nearly constant voltage across the output terminals. However, as V_{IN} changes, I_Z will change proportionally, and therefore the limitations on the input variation are set by the minimum and maximum current values (I_{ZK} and I_{ZM}) with which the zener can operate and on the condition that $V_{IN} > V_Z$. R is the series current-limiting resistor. The bar graph on the DMM symbols indicates the relative values and trends. Many DMMs provide analog bar graph displays in addition to the digital readout.

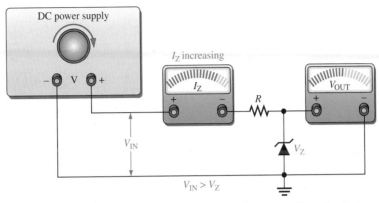

(a) As the input voltage increases, the output voltage remains constant ($I_{ZK} < I_Z < I_{ZM}$).

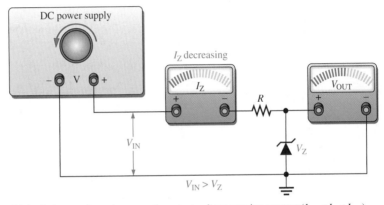

(b) As the input voltage decreases, the output voltage remains constant ($I_{ZK} < I_Z < I_{ZM}$).

FIGURE 2–49
Zener regulation of a varying input voltage.

EXAMPLE 2–9

Figure 2–50 shows a zener diode regulator designed to hold 10 V at the output. Assume the zener impedance is zero and the zener current ranges from 4 mA minimum (I_{ZK}) to 40 mA maximum (I_{ZM}). What are the minimum and maximum input voltages for these currents?

FIGURE 2–50

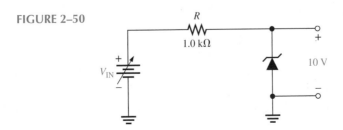

Solution   For the minimum current, the voltage across the 1.0 kΩ resistor is

$$V_R = I_{ZK}R = (4 \text{ mA})(1.0 \text{ k}\Omega) = 4 \text{ V}$$

Since $V_R = V_{IN} - V_Z$,

$$V_{IN} = V_R + V_Z = 4 \text{ V} + 10 \text{ V} = \mathbf{14 \text{ V}}$$

For the maximum zener current, the voltage across the 1.0 kΩ resistor is

$$V_R = (40 \text{ mA})(1.0 \text{ k}\Omega) = 40 \text{ V}$$

Therefore,

$$V_{IN} = 40 \text{ V} + 10 \text{ V} = \mathbf{50 \text{ V}}$$

As you can see, this zener diode can provide line regulation for an input voltage that varies from 14 V to 50 V and maintain approximately a 10 V output. The output will vary slightly from this value because of the zener's impedance.

Practice Exercise   Determine the minimum and maximum input voltages that can be regulated by the zener in Figure 2–51 if the minimum current (I_{ZK}) is 2.5 mA and the maximum (I_{ZM}) is 35 mA.

FIGURE 2–51

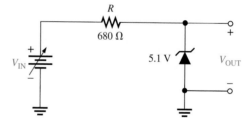

Varactor Diodes

Varactor diodes are also known as variable-capacitance diodes because the junction capacitance varies with the amount of reverse-bias voltage. Varactors are specifically designed to take advantage of this variable-capacitance characteristic. The capacitance can be changed by changing the reverse voltage. These devices are commonly used in electronic tuning circuits used in communications systems.

A **varactor** is basically a reverse-biased *pn* junction that utilizes the inherent capacitance of the depletion region. The depletion region, created by the reverse bias, acts as a capacitor dielectric because of its nonconductive characteristic. The p and n regions are conductive and act as the capacitor plates, as illustrated in Figure 2–52.

Recall that capacitance is determined by the plate area (*A*), dielectric constant (ε), and dielectric thickness (*d*), as expressed in the following formula:

$$C = \frac{A\epsilon}{d}$$

As the reverse-bias voltage increases, the depletion region widens, effectively increasing the dielectric thickness and thus decreasing the capacitance. When the reverse-bias voltage decreases, the depletion region narrows, thus increasing the capacitance. This action is shown in Figure 2–53(a) and (b). A general curve of capacitance versus voltage is shown in Figure 2–53(c).

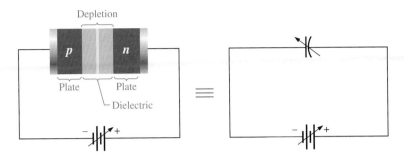

FIGURE 2–52
The reverse-biased varactor diode acts as a variable capacitor.

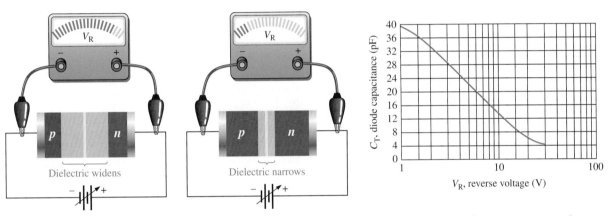

(a) Greater reverse bias, less capacitance   (b) Less reverse bias, greater capacitance   (c) Graph of diode capacitance versus reverse voltage

FIGURE 2–53
Varactor diode capacitance varies with reverse voltage.

In a varactor diode, the capacitance parameters are controlled by the method of doping in the depletion region and the size and geometry of the diode's construction. Varactor capacitances typically range from a few picofarads to a few hundred picofarads.

Figure 2–54(a) shows a common symbol for a varactor, and Figure 2–54(b) shows a simplified equivalent circuit. The internal reverse series resistance is labeled r_s, and the variable capacitance is labeled C_V.

FIGURE 2–54
Varactor diode.

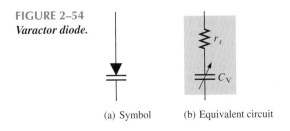

(a) Symbol   (b) Equivalent circuit

Applications   A major application of varactors is in tuning circuits. For example, electronic tuners in TV and other commercial receivers utilize varactors as one of their elements. When used in a resonant circuit, the varactor acts as a variable capacitor, thus allowing the resonant frequency to be adjusted by a variable voltage level, as illustrated in Figure 2–55 where two varactor diodes provide the total variable capacitance in a parallel resonant (tank) circuit. V_C is a variable dc voltage that controls the reverse bias and therefore the capacitance of the diodes.

Recall that the resonant frequency of the tank circuit is

$$f_r \cong \frac{1}{2\pi\sqrt{LC}}$$

This approximation is valid for $Q > 10$.

FIGURE 2–55
Varactors in a resonant circuit.

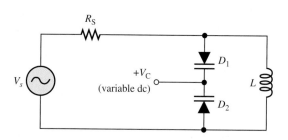

EXAMPLE 2–10

The capacitance of a certain varactor can be varied from 5 pF to 50 pF. The diode is used in a tuned circuit similar to that shown in Figure 2–55. Determine the tuning range for the circuit if $L = 10$ mH.

Solution   The equivalent circuit is shown in Figure 2–56. Notice that the varactor capacitances are in series; the total *minimum* capacitance is the product-over-sum of the individual capacitor's minimum value.

$$C_{T(min)} = \frac{C_{1(min)}C_{2(min)}}{C_{1(min)} + C_{2(min)}} = \frac{(5 \text{ pF})(5 \text{ pF})}{5 \text{ pF} + 5 \text{ pF}} = 2.5 \text{ pF}$$

FIGURE 2–56

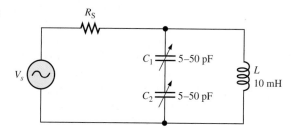

The maximum resonant frequency, therefore, is

$$f_{r(max)} = \frac{1}{2\pi\sqrt{LC_{T(min)}}} = \frac{1}{2\pi\sqrt{(10 \text{ mH})(2.5 \text{ pF})}} \cong \textbf{1 MHz}$$

The maximum total capacitance is

$$C_{T(max)} = \frac{C_{1(max)}C_{2(max)}}{C_{1(max)} + C_{2(max)}} = \frac{(50 \text{ pF})(50 \text{ pF})}{50 \text{ pF} + 50 \text{ pF}} = 25 \text{ pF}$$

The minimum resonant frequency, therefore, is

$$f_{r(min)} = \frac{1}{2\pi\sqrt{LC_{T(max)}}} = \frac{1}{2\pi\sqrt{(10 \text{ mH})(25 \text{ pF})}} \cong \textbf{318 kHz}$$

Practice Exercise   Determine the tuning range for Figure 2–56 if $L = 100$ mH.

Light-Emitting Diodes (LEDs)

As the name implies, the **light-emitting diode** is a light emitter; LEDs are used as indicators (such as in logic probes), in displays such as the familiar seven-segment displays used in many digital clocks, and as sources for optical fiber communication systems. IR-emitting diodes, related to LEDs, are used in optical coupling applications (such as isolating electrocardiogram sensors on a patient from the measuring instrument) and in remote control applications.

The basic operation of the light-emitting diode (LED) is as follows: When the device is forward-biased, electrons cross the *pn* junction from the *n* region and recombine with holes in the *p* region. These free electrons are in the conduction band and are at a higher energy level than the holes in the valence band. When recombination takes place, the recombining electrons release energy in the form of heat and light. A large exposed surface area on one layer of the semiconductive material permits the photons to be emitted as visible light. Figure 2–57 illustrates this process which is called **electroluminescence.**

FIGURE 2–57
Electroluminescence in an LED.

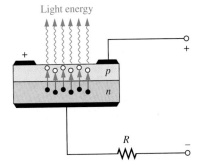

The semiconductive materials used in LEDs are gallium arsenide (GaAs), gallium arsenide phosphide (GaAsP), and gallium phosphide (GaP). Silicon and germanium are not used because they are essentially heat-producing materials and are very poor at producing light. GaAs LEDs emit infrared (IR) radiation, which is invisible. GaAsP produces either red or yellow visible light, and GaP emits red or green visible light. LEDs that emit blue light are also available. Two commonly used LED symbols are shown in Figure 2–58.

FIGURE 2–58
LED symbols.

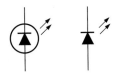

The LED emits light in response to a sufficient forward current (I_F), as shown in Figure 2–59(a). The amount of power output translated into light is directly proportional to the forward current, as indicated in Figure 2–59(b). Typical LEDs are shown in Figure 2–59(c).

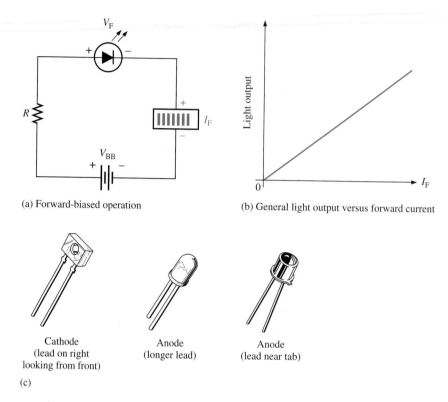

(a) Forward-biased operation

(b) General light output versus forward current

Cathode
(lead on right
looking from front)

Anode
(longer lead)

Anode
(lead near tab)

(c)

FIGURE 2–59
Operation of an LED.

The Photodiode

The **photodiode** is a *pn* junction device that operates in reverse bias, as shown in Figure 2–60(a), where I_λ is the reverse current. Note the schematic symbol for the photodiode. The photodiode has a small transparent window that allows light to strike the *pn* junction. Alternate photodiode symbols are shown in Figure 2–60(b).

FIGURE 2–60
Photodiode.

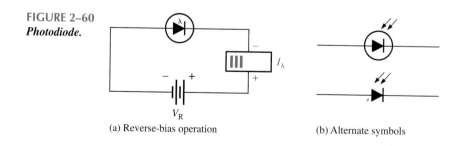

(a) Reverse-bias operation

(b) Alternate symbols

Recall that when reverse-biased, a rectifier diode has a very small reverse leakage current. The same is true for the photodiode. The reverse-biased current is produced by thermally generated electron-hole pairs in the depletion region, which are swept across the junction by the electric field created by the reverse voltage. In a rectifier diode, the reverse leakage current increases with temperature due to an increase in the number of electron-hole pairs.

In a photodiode, the reverse current increases with the light intensity at the *pn* junction. When there is no incident light, the reverse current (I_λ) is almost negligible and is called the *dark current*. An increase in the amount of light energy (measured in lumens per square meter, lm/m^2) produces an increase in the reverse current, as shown by the graph in Figure 2–61(a). For a given value of reverse-bias voltage, Figure 2–61(b) shows a set of characteristic curves for a typical photodiode.

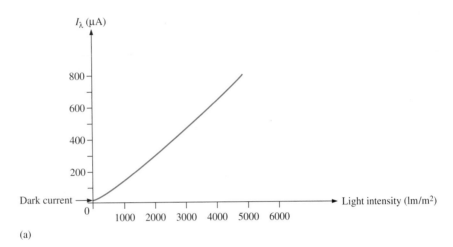

(a)

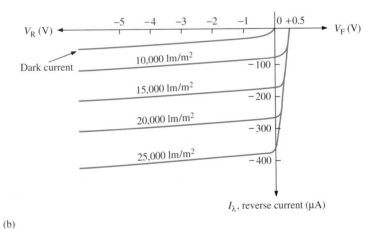

(b)

FIGURE 2–61
Typical photodiode characteristics.

From the characteristic curve in Figure 2–61(b), the dark current for this particular device is approximately 35 μA at a reverse-bias voltage of −3 V. Therefore, the reverse resistance of the device with no incident light is

$$R_R = \frac{V_R}{I_\lambda} = \frac{3\ \text{V}}{35\ \mu\text{A}} = 86\ \text{k}\Omega$$

At 25,000 lm/m^2, the current is approximately 400 μA at −3 V. The resistance under this condition is

$$R_R = \frac{V_R}{I_\lambda} = \frac{3\ \text{V}}{400\ \mu\text{A}} = 7.5\ \text{k}\Omega$$

These calculations show that the photodiode can be used as a variable-resistance device controlled by light intensity.

Figure 2–62 illustrates that the photodiode allows essentially no reverse current (except for a very small dark current) when there is no incident light. When a light beam strikes the photodiode, it conducts an amount of reverse current that is proportional to the light intensity.

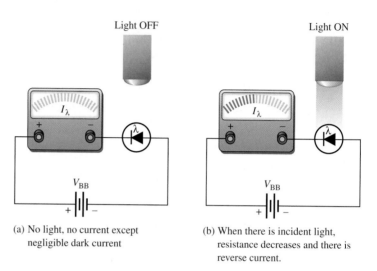

(a) No light, no current except negligible dark current

(b) When there is incident light, resistance decreases and there is reverse current.

FIGURE 2–62
Operation of a photodiode.

2–8 REVIEW QUESTIONS

1. How are zener diodes normally operated?
2. What does the parameter I_{ZM} refer to?
3. What is the purpose of a varactor diode?
4. Based on the general curve in Figure 2–53(c), what happens to the diode capacitance when the reverse voltage is increased?
5. List some semiconductive materials used for LEDs.
6. There is a small current in a photodiode under no-light conditions. What is this current called?

2–9 ■ THE DIODE DATA SHEET

A manufacturer's data sheet gives detailed information on a device so that it can be used properly in a given application. A typical data sheet provides maximum ratings, electrical characteristics, mechanical data, and graphs of various parameters.

After completing this section, you should be able to

❏ Interpret and use a diode data sheet
    ❏ Identify maximum voltage and current ratings
    ❏ Determine the electrical characteristics of a diode

❏ Analyze graphical data
❏ Select an appropriate diode for a given set of specifications

Table 2–1 shows the maximum ratings for a certain series of rectifier diodes (1N4001 through 1N4007[4]). These are the absolute maximum values under which the diode can be operated without damage to the device. For greatest reliability and longer life, the diode should always be operated well under these maximums. Generally, the maximum ratings are specified at 25°C and must be adjusted downward for greater temperatures. (Appendix A has the data sheets for the 1N4001–1N4007 series.)

TABLE 2–1

Rating	Symbol	1N4001	1N4002	1N4003	1N4004	1N4005	1N4006	1N4007	Unit
Peak repetitive reverse voltage Working peak reverse voltage DC blocking voltage	V_{RRM} V_{RWM} V_R	50	100	200	400	600	800	1000	V
Nonrepetitive peak reverse voltage	V_{RSM}	60	120	240	480	720	1000	1200	V
rms reverse voltage	$V_{R(rms)}$	35	70	140	280	420	560	700	V
Average rectified forward current (single-phase, resistive load, 60 Hz, $T_A = 75°C$)	I_O				1.0				A
Nonrepetitive peak surge current (surge applied at rated load conditions)	I_{FSM}				30 (for 1 cycle)				A
Operating and storage junction temperature range	T_J, T_{stg}				−65 to +175				°C

An explanation of some of the parameters from Table 2–1 follows.

V_{RRM} The maximum reverse peak voltage that can be applied repetitively across the diode. Notice that in this case, it is 50 V for the 1N4001 and 1 kV for the 1N4007. This is the same as PIV rating.

V_R The maximum reverse dc voltage that can be applied across the diode.

V_{RSM} The maximum reverse peak value of nonrepetitive (one-cycle) voltage that can be applied across the diode.

I_O The maximum average value of a 60 Hz rectified forward current.

I_{FSM} The maximum peak value of nonrepetitive (one-cycle) forward current. The graph in Figure 2–63 expands on this parameter to show values for more than one cycle at temperatures of 25°C and 175°C. The dashed lines represent values where typical failures occur. Notice what happens on the lower solid line when ten cycles of I_{FSM} are applied. The limit is 15 A rather than the one-cycle value of 30 A.

---

[4] Data sheet for 1N4001–1N4007 available at http://www.onsemi.com

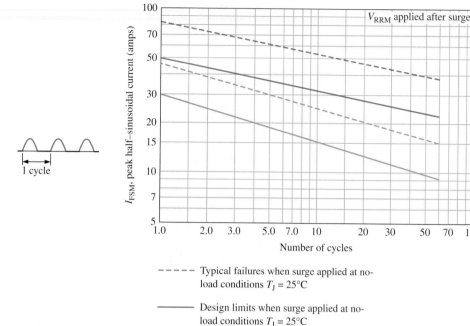

- - - - - Typical failures when surge applied at no-
load conditions $T_J = 25°C$

——— Design limits when surge applied at no-
load conditions $T_J = 25°C$

- - - - - Typical failures when surge applied at rated-
load conditions $T_J = 175°C$

——— Design limits when surge applied at rated-
load conditions $T_J = 175°C$

FIGURE 2–63
Nonrepetitive forward surge current capability.

Table 2–2 lists typical and maximum values of certain electrical characteristics for the 1N4001–1N4007 series. These items differ from the maximum ratings in that they are not selected by design but are the result of operating the diode under specified conditions. A brief explanation of these parameters follows.

TABLE 2–2
Electrical characteristics.

Characteristics and Conditions	Symbol	Typical	Maximum	Unit
Maximum instantaneous forward voltage drop ($I_F = 1$ A, $T_J = 25°C$)	v_F	0.93	1.1	V
Maximum full-cycle average forward voltage drop ($I_O = 1$ A, $T_L = 75°C$, 1 in. leads)	$V_{F(avg)}$	—	0.8	V
Maximum reverse current (rated dc voltage) $T_J = 25°C$ $T_J = 100°C$	I_R	0.05 1.0	10.0 50.0	μA
Maximum full-cycle average reverse current ($I_O = 1$ A, $T_L = 75°C$, 1 in. leads)	$I_{R(avg)}$	—	30.0	μA

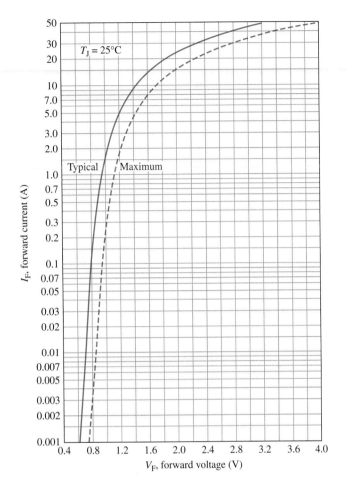

FIGURE 2–64
Forward voltage (V_F) versus forward current (I_F).

v_F The instantaneous voltage across the forward-biased diode when the forward current is 1 A at 25°C. Figure 2–64 shows how the forward voltages vary with forward current.

$V_{F(avg)}$ The maximum forward voltage drop averaged over a full cycle (also shown as V_F on some data sheets).

I_R The maximum current when the diode is reverse-biased with a dc voltage.

$I_{R(avg)}$ The maximum reverse current averaged over one cycle (when reverse-biased with an ac voltage).

T_L The lead temperature.

Figure 2–65 shows a selection of rectifier diodes arranged in order of increasing I_O, I_{FSM}, and V_{RRM} ratings.

	I_O, Average Rectified Forward Current (Amperes)					
	1.0	**1.5**	**3.0**			**6.0**
	59-03 (DO-41) Plastic	59-04 Plastic	60-01 Metal	267-03 Plastic	267-02 Plastic	194-04 Plastic
V_{RRM} (Volts)						
50	1N4001	1N5391	1N4719	MR500	1N5400	MR750
100	1N4002	1N5392	1N4720	MR501	1N5401	MR751
200	1N4003	1N5393 MR5059	1N4721	MR502	1N5402	MR752
400	1N4004	1N5395 MR5060	1N4722	MR504	1N5404	MR754
600	1N4005	1N5397 MR5061	1N4723	MR506	1N5406	MR756
800	1N4006	1N5398	1N4724	MR508		MR758
1000	1N4007	1N5399	1N4725	MR510		MR760
I_{FSM} (Amps)	30	50	300	100	200	400
T_A @ Rated I_O (°C)	75	$T_L = 70$	75	95	$T_L = 105$	60
T_C @ Rated I_O (°C)						
T_J (Max) (°C)	175	175	175	175	175	175

	I_O, Average Rectified Forward Current (Amperes)										
	12	**20**	**24**	**25**	**30**		**40**	**50**	**25**	**35**	**40**
	245A-02 (DO-203AA) Metal	339-02 Plastic	193-04 Plastic		43-02 (DO-21) Metal		42A-01 (DO-203AB) Metal	43-04 Metal	309A-03	309A-02	
V_{RRM} (Volts)											
50	MR1120 1N1199,A,B	MR2000	MR2400	MR2500	1N3491	1N3659	1N1183A	MR5005	MDA2500	MDA3500	
100	MR1121 1N1200,A,B	MR2001	MR2401	MR2501	1N3492	1N3660	1N1184A	MR5010	MDA2501	MDA3501	
200	MR1122 1N1202,A,B	MR2002	MR2402	MR2502	1N3493	1N3661	1N1186A	MR5020	MDA2502	MDA3502	MDA4002
400	MR1124 1N1204,A,B	MR2004	MR2404	MR2504	1N3495	1N3663	1N1188A	MR5040	MDA2504	MDA3504	MDA4004
600	MR1126 1N1206,A,B	MR2006	MR2406	MR2506			1N1190A		MDA2506	MDA3506	MDA4006
800	MR1128	MR2008		MR2508					MDA2508	MDA3508	MDA4008
1000	MR1130	MR2010		MR2510					MDA2510	MDA3510	
I_{FSM} (Amps)	300	400	400	400	300	400	800	600	400	400	800
T_A @ Rated I_O (°C)											
T_C @ Rated I_O (°C)	150	150	125	150	130	100	150	150	55	55	35
T_J (Max) (°C)	190	175	175	175	175	175	190	195	175	175	175

FIGURE 2–65

A selection of rectifier diodes based on maximum ratings of I_O, I_{FSM}, and V_{RRM}.

2–9 REVIEW QUESTIONS

1. List the three rating categories typically given on all diode data sheets.
2. Define each of the following parameters: V_F, I_R, and I_O.
3. Define I_{FSM}, V_{RRM}, and V_{RSM}.
4. From Figure 2–65, select a diode to meet the following specifications: $I_O = 3$ A, $I_{FSM} = 300$ A, and $V_{RRM} = 100$ V.

2–10 ■ TROUBLESHOOTING

The backbone for nearly all electronic circuits is the power supply. Several types of failures can occur in power supplies. In this section, we will expand on our earlier coverage of troubleshooting by looking at specific power supply failures and the effects they would have on the supply's operation. Then we'll look at one more example of how you might troubleshoot a regulated power supply.

After completing this section, you should be able to

❑ Troubleshoot a power supply using accepted techniques
  ❑ Discuss the steps in forming a troubleshooting plan
  ❑ Explain fault analysis
  ❑ Describe symptoms that are likely for different failures

As discussed in Section 1–5, the first step in troubleshooting is to analyze the clues (symptoms) of a failure. These clues should lead to a logical plan for troubleshooting. The plan for a circuit that has never worked will be different than one that has been working. The history of past failures, or failures in a similar circuit, can also be a clue to a failure.

It is always useful to have a good understanding of the circuit you are troubleshooting and a schematic. There is no one plan that fits all situations; the one to use depends on the type and complexity of the circuit or system, the nature of the problem, and the preference of the individual technician.

A Troubleshooting Plan

Above all else, efficient troubleshooting requires logical thinking and a plan that will find the simple problems (such as a blown fuse) quickly. As an example, consider the following plan for troubleshooting a power supply that has failed in operation.

Step 1: Ask questions of the person reporting the trouble. When did it fail? (Right after it was plugged in? After running for 2 hours at maximum current out?) How do you know it failed? (Smoke? Low voltage?)

Step 2: Power check: Make sure the power cord is plugged in and the fuse is not burned out. Check that the controls are set for proper operation. Something this simple is often the cause of the problem. Perhaps the operator did not understand the correct settings required for controls.

Step 3: Sensory check: Beyond the power check, the simplest troubleshooting method relies on your senses of observation to check for obvious defects. Remove power, open the supply, and do a visual check for broken wires, poor solder connections, burned out fuses, and the like. Also, when certain types of components fail, you may be able to detect a smell of smoke if you happen to be there when it fails or shortly afterward. Since some failures are temperature dependent, you can sometimes use your sense of touch (carefully!) to detect an overheated component.

Step 4: Isolate the failure. Apply power to the supply while it is on the bench and trace the voltage. As described in Section 1–5, you can start in the middle of the circuit and do half-splitting or check the voltage at successive test points from the input side until you get an incorrect measurement. Some problems are more difficult than simply finding no voltage, but tracing should isolate the problem to an area or a component.

Having reviewed a plan for troubleshooting, let's turn to fault analysis. When you find a symptom, you next need to ask the question, *If component X fails in the circuit, what are the symptoms?* You will apply fault analysis when you find an incorrect voltage or waveform. For example, assume you observe high ripple at the input to a regulator. From your knowledge of circuit operation, you might reason that a defective or incorrect value capacitor could be the culprit. The following discussion describes four possible failures and gives more examples of fault analysis.

Open Fuse or Circuit Breaker

An overcurrent-protection device is essential to virtually all electronic equipment. These devices prevent damage to the equipment in case of a failure or an overload condition and reduce the probability of a violent failure. Overcurrent-protection devices include fuses, circuit breakers, solid-state current-limiting devices, and thermal overload devices. The circuit breakers for the ac line cannot be counted on to protect electronic devices as they only open when the current is 15 A or more in the ac line, far too high to offer protection to most electronic equipment.

If a single fuse is present, it is usually on the primary side of the transformer, and will be rated for 115 or 230 VAC at a current that is consistent with the maximum power rating of the supply. Frequently, protection devices may also be included on the secondary side, especially if a single transformer has multiple outputs. A fuse is designed to carry its rated current indefinitely (and will typically carry 120% of its rated current for several hours). Fuses are available in fast- and slow-blow versions. A fast-blow fuse opens in a few milliseconds when overloaded; a slow-blow fuse can survive transient overloads such as occurs when power is first applied. Most of the time, slow-blow fuses should be used on the ac side of a power supply circuit.

Testing for an open fuse is relatively simple. Glass fuses can be checked by inspection or checked with an ohmmeter (be certain power is disconnected from the circuit). If power is still applied, a blown fuse will have voltage across it (provided there are no other opens in the path, such as a switch). Usually an open fuse is symptomatic of a short circuit or overload; however, fuses can have fatigue failure and may be the only problem in the circuit.

Before replacing a blown fuse, the technician should check for the cause. If the fuse simply opened, it may be a fatigue failure. If the fuse has blown violently (as evidenced by complete vaporization of the wire inside), it is most likely that it opened due to another problem. Look for a short circuit with an ohmmeter; it can be the load, the filter capacitor, or other component that has shorted. Look for any visible signs of an overheated or damaged component. If the fuse is replaced, it is important to replace it with the identical type and current rating as specified by the manufacturer. The wrong fuse can cause further damage and may be a safety hazard.

Open Diode

Consider the full-wave, center-tapped rectifier in Figure 2–66. Assume that diode D_1 has failed open. This causes diode D_2 to conduct on only the negative cycle. With an oscilloscope connected to the output, as shown in part (a), you would observe a larger-than-normal ripple voltage at a frequency of 60 Hz rather than 120 Hz. Disconnecting the filter capacitor, you would observe a half-wave rectified voltage, as in part (b).

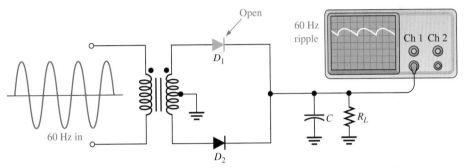

(a) Ripple should be less and have a frequency of 120 Hz. Instead it is greater in amplitude with a frequency of 60 Hz.

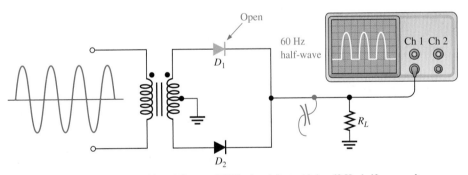

(b) With C removed, output should be a full-wave 120 Hz signal. Instead it is a 60 Hz, half-wave voltage.

FIGURE 2–66
Symptoms of an open diode in a full-wave, center-tapped rectifier.

With the filter capacitor in the circuit, the half-wave signal will allow it to discharge more than it would with a normal full-wave signal, resulting in a larger ripple voltage. Basically, the same observations would be made for an open failure of diode D_2.

An open diode in a bridge rectifier would create symptoms identical to those just discussed for the center-tapped rectifier. As illustrated in Figure 2–67, the open diode would prevent current through R_L during half of the input cycle (in this case, the negative half). As a result, there would be a half-wave output and an increased ripple voltage at 60 Hz, as discussed before.

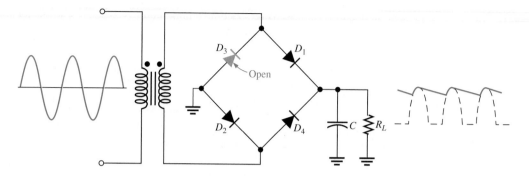

FIGURE 2–67
Effect of an open diode in a bridge rectifier. A half-wave operation results in an increased ripple at 60 Hz.

Generally, the easiest test for an open diode in a full-wave power supply is to measure the ripple frequency. If the ripple frequency is the same as the input ac frequency, look for an open diode or a connection problem with a diode (such as a cracked trace).

Shorted Diode

A shorted diode is one that has failed such that it has a very low resistance in both directions. If a diode suddenly became shorted in a bridge rectifier, normally a fuse would blow or other circuit protection would be activated. If the supply was not protected by a fuse, the shorted diode will most likely cause the transformer to be damaged or cause the other diode in series to open, as illustrated in Figure 2–68.

In part (a) of Figure 2–68, current is supplied to the load through the shorted diode during the first positive half-cycle, just as though it were forward-biased. During the negative half-cycle, the current is shorted through D_1 and D_4, as shown in part (b). It is this excessive current that leads to the second failure; hence, when a shorted diode is discovered, it is a good idea to check other components for a failure.

Shorted or Leaky Filter Capacitor

Electrolytic capacitors can appear shorted (or have high "leakage" current) when they fail. One cause of failure that produces symptoms of a short occurs when an electrolytic capacitor is put in backwards, an error that can happen with newly manufactured circuit boards. As in the case of a shorted diode, the normal symptom is a blown fuse due to excessive current. A leaky capacitor is a form of partial failure; it can be represented by a leakage resistance in parallel with the capacitor, as shown in Figure 2–69(a). The effect of the leakage resistance is to reduce the discharging time constant, causing an increase in ripple voltage on the output, as shown in Figure 2–69(b). A leaky capacitor may simply overheat; a capacitor should never show signs of overheating. For an unfused supply, a shorted capacitor would most likely cause one or more diodes or the transformer to open due to excessive current. In any event, there would be no dc voltage on the output.

When a defective capacitor is replaced, it is important to observe the working voltage specification as well as the size of the capacitor. If the working voltage specification is exceeded, the replacement is likely to fail again. In addition, it is vitally important to observe the polarity of the capacitor. An electrolytic capacitor installed backwards can literally explode.

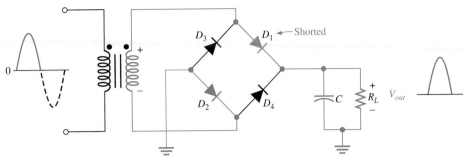

(a) Positive half-cycle: The shorted diode acts as a forward-biased diode, so the load current is normal. D_3 and D_4 are reverse-biased.

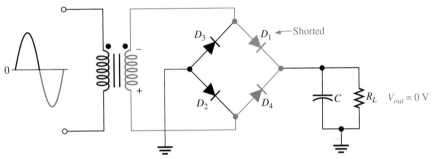

(b) Negative half-cycle: The shorted diode places forward-biased D_4 across the secondary. As a result D_1, D_4, or the transformer secondary will probably burn open, or a fuse (not shown) will open.

FIGURE 2–68
Effect of shorted diode in a bridge rectifier. Conducting paths are shown in color.

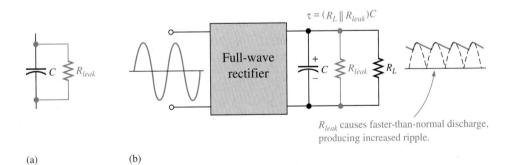

FIGURE 2–69
Effect of a leaky filter capacitor on the output of a full-wave rectifier. The capacitor has a dc path simulating a parallel resistance to the load.

Troubleshooting a Regulated Power Supply

As indicated at the beginning of this section, the plan for repairing any electronic equipment depends on the observed symptoms. Let's assume you have a supply like the one shown previously in Figure 2–28 that has blown a fuse immediately after it was connected to a printed circuit card. Your thinking might go like this: "The power supply was working fine until I added the card; perhaps I exceeded the current limit of the supply." Here you have considered the conditions and hypothesized a possible cause. The first step then is to remove the load and test the supply to see if this clears the problem. If so, then check the current requirement of the card that was added, or check to see if it is drawing too much current. If not, the problem is in the supply.

What if a power supply is completely dead but has a good fuse? In this case, start tracing the voltage to isolate the problem. For example, you could check for voltage on the primary of the transformer; if there is voltage, then test the secondary voltage. If the primary does not have voltage, check the path for the ac—the switch and connections to the transformer. If a single open is in series with the ac line, the full ac voltage will appear across that open.

Let's assume you have found that the transformer has ac on the primary but no voltage on the secondary. This indicates that the transformer is open, either the primary or the secondary winding. An ohmmeter should confirm this. Before replacing it, you should look why the failure occurred. Transformers seldom fail if they are operated properly. The likelihood is that another component shorted in the circuit. Look for a shorted diode or capacitor as an initial cause.

As previously stated, the exact strategy for troubleshooting depends on what is found at each step, the ease of making a particular test, and the likely cause of a failure. The key is that the technician use a series of logical tests to reduce the problem to the exact cause.

2–10 REVIEW QUESTIONS

1. What would you expect to see if R_1 of the power supply in Figure 2–29 were open?
2. You are checking a 60 Hz full-wave bridge rectifier and observe that the output has a 60 Hz ripple. What failure(s) do you suspect?
3. You observe that the output ripple of a full-wave rectifier is much greater than normal but its frequency is still 120 Hz. What component do you suspect?

2–11 ■ A SYSTEM APPLICATION

The dc power supply in any electronic system provides to all parts of the system a constant dc voltage and sufficient dc current from which all the circuits operate. In other words, the dc power supply energizes the system. In this section, you will be dealing with the dc power supply in a radio receiver system.

After completing this section, you should be able to

❏ Apply what you have learned in this chapter to a system application
   ❏ See how a diode rectifier is used in a system application
   ❏ See how the filter operates in a system application
   ❏ See how the 110 V ac voltage from a standard outlet is converted to the dc supply voltage

❏ Translate between a printed circuit (PC) board and a schematic
❏ Troubleshoot some common power supply failures

A dc power supply PC board with the components assembled on it represents a principal application of the diode and the full-wave bridge rectifier as well as the filter capacitor and an integrated circuit regulator. You should now be familiar with how all of these components work together to make a power supply.

Now, so that you can take a closer look at the power supply, let's take it out of the system and put it on the troubleshooter's bench.

TROUBLESHOOTER'S BENCH

Identifying the Components

On the PC board shown in Figure 2–70, the diodes are labeled D1, D2, D3, and D4. The banded end of the diode is the cathode, as indicated in Figure 2–71(a). The capacitors are labeled C1 and C2. These are cylindrical electrolytic capacitors as shown in Figure 2–71(b). Capacitor C1 filters the input to the regulator. Capacitor C2 removes any noise

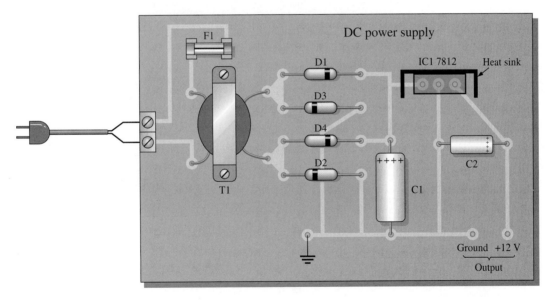

FIGURE 2–70

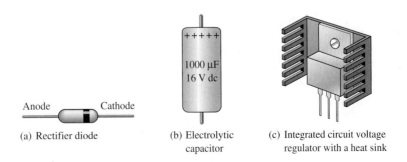

(a) Rectifier diode

(b) Electrolytic capacitor

(c) Integrated circuit voltage regulator with a heat sink

FIGURE 2–71

spikes that appear on the line and improves the transient response of the power supply. The 3-terminal voltage regulator is labeled IC1. The thick bracket-shaped object is a heat sink for conducting heat away from the device. A pictorial view is shown in Figure 2–71(c). Recall that IC voltage regulators were discussed in Section 2–6. The regulator in this circuit is a 7812. Also mounted in the power supply are the transformer (T1) and the fuse (F1).

■ ACTIVITY 1 Relate the PC Board to the Schematic

Carefully follow the conductive traces on the PC board to see how the components are interconnected. Compare each point (*A* through *J*) on the PC board with the corresponding point on the schematic in Figure 2–72. This exercise will develop your skill in going from a PC board to a schematic or vice versa—a very important skill for a technician. Notice that the IC regulator on the PC board has been drawn to let you see the connections and traces under it. For each point on the PC board, place the letter of the corresponding point or points on the schematic.

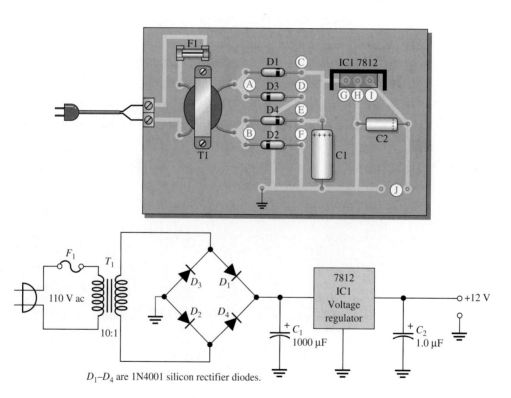

D_1–D_4 are 1N4001 silicon rectifier diodes.

FIGURE 2–72

■ ACTIVITY 2 Analyze the Power Supply Circuit

With 110 V rms applied to the input, determine what the voltages should be at each point indicated (1, 2, 3, and 4) in Figure 2–73 using the oscilloscope or digital multimeter as indicated.

FIGURE 2–73

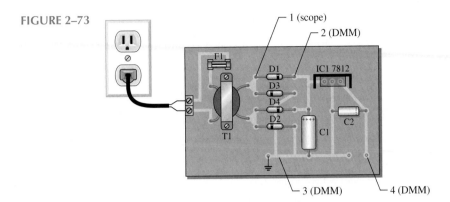

■ **ACTIVITY 3 Write a Technical Report**

Discuss the detailed operation from input to output including voltage values. Tell what each component does and why it is in the circuit.

■ **ACTIVITY 4 Troubleshoot the Power Supply for each of the Following Problems by Stating the Probable Cause or Causes**

Refer to Figure 2–73.

1. No voltage at point 1.
2. No voltage at point 2.
3. A full-wave 120 Hz voltage with a peak value of about 15 V at point 2.
4. A large 120 Hz ripple voltage with a peak of about 15 V at point 2.
5. A half-wave 60 Hz voltage with a peak of about 15 V at point 2.
6. 15 V dc at point 2 but no voltage at point 4.

COLOR INSERT

■ **ACTIVITY 5 Troubleshooter's Bench Special Assignment**

Go to Troubleshooter's Bench 1 in the color insert section and carry out the assignment stated there.

2–11 REVIEW QUESTIONS

1. List the major parts that make up a basic regulated dc power supply.
2. What is the purpose of a power supply in a system application?
3. What is the input voltage to a typical power supply?
4. From the data sheet for the 1N4001 diode in Appendix A, determine its PIV.
5. Are the diodes used in this power supply operating close to their PIV rating?
6. What is the purpose of capacitor C_1 in the power supply?
7. What is the purpose of capacitor C_2 in the power supply?
8. Why is the voltage regulator connected to a heat sink?
9. What do you think would happen if C_2 were open?

■ SUMMARY

- The Bohr model of an atom consists of a nucleus containing positively charged protons and uncharged neutrons orbited by negatively charged electrons.
- Atomic shells are energy bands. The outermost shell containing electrons is the valence shell.
- Silicon is the predominant semiconductive material.
- Atoms within a semiconductor crystal structure are held together with covalent bonds.
- Electron-hole pairs are thermally produced.
- A *p*-type semiconductor is made by adding trivalent impurity atoms to an intrinsic (pure) semiconductor.
- An *n*-type semiconductor is made by adding pentavalent impurity atoms to an intrinsic (pure) semiconductor.
- The depletion region is a region adjacent to the *pn* junction containing no majority carriers.
- Forward bias permits majority carrier current through the *pn* junction.
- Reverse bias prevents majority carrier current.
- A *pn* junction is called a diode.
- Reverse breakdown occurs when the reverse-biased voltage exceeds a specified value.
- Three types of rectifiers are the half-wave, the center-tapped full-wave, and the bridge. The center-tapped and the bridge are both types of full-wave rectifiers.
- The single diode in a half-wave rectifier conducts for half of the input cycle and has the entire output current in it. The output frequency equals the input frequency.
- Each diode in the center-tapped full-wave rectifier and the bridge rectifier conduct for one-half of the input cycle but share the total current. The output frequency of a full-wave rectifier is twice the input frequency.
- The PIV (peak inverse voltage) is the maximum voltage appearing across a reverse-biased diode.
- A capacitor-input filter provides a dc output approximately equal to the peak of the input.
- Ripple voltage is caused by the charging and discharging of the filter capacitor.
- Three-terminal integrated circuit regulators provide a nearly constant dc output from an unregulated dc input.
- Regulation of output voltage over a range of input voltages is called input or line regulation.
- Regulation of output voltage over a range of load currents is called load regulation.
- Diode limiters (clippers) cut off voltage above or below specified levels.
- Diode clampers add a dc level to an ac signal.
- The zener diode operates in reverse breakdown.
- A zener diode maintains an essentially constant voltage across its terminals over a specified range of zener currents.
- Zener diodes are used to establish a reference voltage and in basic regulator circuits.
- A varactor diode acts as a variable capacitor under reverse-biased conditions.
- The capacitance of a varactor varies inversely with reverse-biased voltage.
- Diode symbols are shown in Figure 2–74.

(a) Rectifier   (b) Zener   (c) Varactor   (d) LED   (e) Photodiode

FIGURE 2–74
Diode symbols.

■ GLOSSARY

Key terms are in color. All terms are included in the end-of-book glossary.

Anode (semiconductor diode definition) The terminal of a semiconductor diode that is more positive with respect to the other terminal when it is biased in the forward direction.

Barrier potential The inherent voltage across the depletion region of a *pn* junction.

Bias The application of dc voltage to a diode or other electronic device to produce a desired mode of operation.

Cathode (semiconductor diode definition) The terminal of a diode that is more negative with respect to the other terminal when it is biased in the forward direction.

Center tap A connection at the midpoint of the secondary of a transformer.

Clamper A circuit that adds a dc level to an ac signal; also called a *dc restorer*.

Conduction electron An electron that has broken away from the valence band of the parent atom and is free to move from atom to atom within the atomic structure of a material; also called a *free electron*.

Covalent bond A type of chemical bond in which atoms share electron pairs.

Crystal A solid in which the particles form a regular, repeating pattern.

Depletion region The area near a *pn* junction on both sides that has no majority carriers.

Diode An electronic device that permits current in only one direction.

Discrete device An individual electrical or electronic component that must be used in combination with other components to form a complete functional circuit.

Doping The process of imparting impurities to an intrinsic semiconductive material in order to control its conduction characteristics.

Electroluminescence The process of releasing light energy by the recombination of electrons in a semiconductor.

Electron The basic particle of negative electrical charge in matter.

Energy The ability to do work.

Filter A type of electrical circuit that passes certain frequencies and rejects all others.

Forward bias The condition in which a *pn* junction conducts current.

Full-wave rectifier A circuit that converts an alternating sine wave into a pulsating dc voltage consisting of both halves of a sine wave for each input cycle.

Germanium A semiconductive material.

Half-wave rectifier A circuit that converts an alternating sine wave into a pulsating dc voltage consisting of one-half of a sine wave for each input cycle.

Hole A mobile vacancy in the electronic valence structure of a semiconductor. A hole acts like a positively charged particle.

Integrated circuit (IC) A type of circuit in which all the components are constructed on a single tiny chip of silicon.

Intrinsic (pure) An intrinsic semiconductor is one in which the charge concentration is essentially the same as a pure crystal with relatively few free electrons.

Ion An atom that has gained or lost a valence electron resulting in a net positive or negative charge.

Light-emitting diode (LED) A type of diode that emits light when there is forward current.

Limiter A circuit that removes part of a waveform above or below a specified level; also called a *clipper*.

Line regulation The change in output voltage for a given change in line (input) voltage, normally expressed as a percentage.

Load regulation The change in output voltage for a given change in load current, normally expressed as a percentage.

Photodiode   A diode whose reverse resistance changes with incident light.

PN **junction**   The boundary between *n*-type and *p*-type materials.

Power supply   A device that converts ac or dc voltage into a voltage or current suitable for use in various applications to power electronic equipment. The most common form is to convert ac from the utility line to a constant dc voltage.

Recombination   The process of a free electron in the conduction band falling into a hole in the valence band of an atom.

Rectifier   An electronic circuit that converts ac into pulsating dc.

Regulator   An electronic circuit that is connected to the output of a rectifier and maintains an essentially constant output voltage despite changes in the input, the load current, or the temperature.

Reverse bias   The condition in which a *pn* junction blocks current.

Ripple voltage   The variation in the dc voltage on the output of a filtered rectifier caused by the slight charging and discharging action of the filter capacitor.

Semiconductor   A material that has a conductance value between that of a conductor and that of an insulator. Silicon and germanium are examples.

Shell   An energy level in which electrons orbit the nucleus of an atom.

Silicon   A semiconductive material used in diodes and transistors.

Terminal   An external contact point on an electronic device.

Valence electron   An electron in the outermost shell or orbit of an atom.

Varactor   A diode that is used as a voltage-variable capacitor.

Voltage regulation   The process of maintaining an essentially constant output voltage over variations in input voltage or load.

Zener diode   A type of diode that operates in reverse breakdown (called zener breakdown) to provide voltage regulation.

■ **KEY FORMULAS**

(2–1)	$V_{p(out)} = V_{p(in)} - 0.7 \text{ V}$	Half-wave and full-wave rectifier peak output voltage
(2–2)	$V_{out} = V_{sec} - 1.4 \text{ V}$	Bridge rectifier peak output voltage
(2–3)	$\text{Line regulation} = \left(\dfrac{\Delta V_{OUT}}{\Delta V_{IN}}\right)100\%$	Line regulation expressed as a percentage
(2–4)	$\text{Load regulation} = \left(\dfrac{V_{NL} - V_{FL}}{V_{FL}}\right)100\%$	Load regulation expressed as a percentage
(2–5)	$Z_Z = \dfrac{\Delta V_Z}{\Delta I_Z}$	Zener impedance

■ **SELF-TEST**   *Answers are at the end of the chapter.*

1. When a neutral atom loses or gains a valence electron, the atom becomes
   (a) covalent   (b) a metal   (c) a crystal   (d) an ion

2. Atoms within a semiconductor crystal are held together by
   (a) metallic bonds   (b) subatomic particles
   (c) covalent bonds   (d) the valence band

3. Free electrons exist in the
   (a) valence band   (b) conduction band
   (c) lowest band   (d) recombination band

4. A hole is a
   (a) vacancy in the valence band   (b) vacancy in the conduction band
   (c) positive electron   (d) conduction-band electron

5. The widest energy gap between the valence band and the conduction band occurs in
   (a) semiconductors   (b) insulators   (c) conductors   (d) a vacuum

6. The process of adding impurity atoms to a pure semiconductive material is called
   (a) recombination   (b) crystallization   (c) bonding   (d) doping

7. In a semiconductor diode, the region near the *pn* junction consisting of positive and negative ions
   is called the
   (a) neutral zone   (b) recombination region
   (c) depletion region   (d) diffusion area

8. In a semiconductor diode, the two bias conditions are
   (a) positive and negative   (b) blocking and nonblocking
   (c) open and closed   (d) forward and reverse

9. The voltage across a forward-biased silicon diode is approximately
   (a) 0.7 V   (b) 0.3 V   (c) 0 V   (d) dependent on the bias voltage

10. In Figure 2–75, identify the forward-biased diode(s).
    (a) D_1 (b) D_2 (c) D_3 (d) D_1 and D_3

FIGURE 2–75

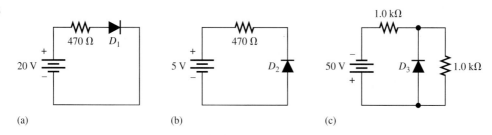

(a)   (b)   (c)

11. When the positive lead of an ohmmeter is connected to the cathode of a diode and the negative
    lead is connected to the anode, the meter reads
    (a) a very low resistance
    (b) an infinitely high resistance
    (c) a high resistance initially, decreasing to about 100 Ω
    (d) a gradually increasing resistance

12. The output frequency of a full-wave rectifier with a 60 Hz sinusoidal input is
    (a) 30 Hz   (b) 60 Hz   (c) 120 Hz   (d) 0 Hz

13. If a single diode in a center-tapped full-wave rectifier opens, the output is
    (a) 0 V   (b) half-wave rectified
    (c) reduced in amplitude   (d) unaffected

14. During the positive half-cycle of the input voltage in a bridge rectifier,
    (a) one diode is forward-biased   (b) all diodes are forward-biased
    (c) all diodes are reverse-biased   (d) two diodes are forward-biased

15. The process of changing a half-wave or a full-wave rectified voltage to a constant dc voltage is
    called
    (a) filtering   (b) ac to dc conversion
    (c) damping   (d) ripple suppression

16. Assume a particular IC regulator attenuates the input ripple by 60 dB. The output ripple will be attenuated by a factor of

(a) 60   (b) 600   (c) 1000   (d) 1,000,000

17. The purpose of a small capacitor placed across the output of an IC regulator is to

(a) improve transient response   (b) couple the output signal to the load
(c) filter the ac   (d) protect the IC regulator

18. A diode limiting circuit

(a) removes part of a waveform
(b) inserts a dc level
(c) produces an output equal to the average value of the input
(d) increases the peak value of the input

19. A clamping circuit is also known as

(a) an averaging circuit   (b) an inverter
(c) a dc restorer   (d) an ac restorer

20. The zener diode is designed for operation in

(a) zener breakdown   (b) forward bias
(c) reverse bias   (d) avalanche breakdown

21. Zener diodes are widely used as

(a) current limiters   (b) power distributors
(c) voltage regulators   (d) variable resistors

22. Varactor diodes are used as

(a) variable resistors   (b) variable current sources
(c) variable inductors   (d) variable capacitors

23. LEDs are based on the principle of

(a) forward bias   (b) electroluminescence
(c) photon sensitivity   (d) electron-hole recombination

24. In a photodiode, light produces

(a) reverse current   (b) forward current
(c) electroluminescence   (d) dark current

TROUBLESHOOTER'S QUIZ   *Answers are at the end of the chapter.*

Refer to Figure 2–79(a).

❑ If the power supply voltage is set for 10 V instead of 12 V,

1. The positive peak voltage of the output will

(a) increase   (b) decrease   (c) not change

2. The negative peak voltage of the output will

(a) increase   (b) decrease   (c) not change

3. The voltage across the 2.2 kΩ resistor will

(a) increase   (b) decrease   (c) not change

Refer to Figure 2–80(a).

❑ If the diode is open,

4. The peak-to-peak output voltage will

(a) increase   (b) decrease   (c) not change

5. The center of the output waveform will

(a) increase   (b) decrease   (c) not change

❏ If the capacitor is shorted,

   6. The peak-to-peak output voltage will

      (a) increase   **(b)** decrease   **(c)** not change

   7. The center of the output waveform will

      (a) increase   **(b)** decrease   **(c)** not change

Refer to Figure 2–88.

❏ If the capacitor is smaller than 1000 μF,

   8. The ripple frequency will

      (a) increase   **(b)** decrease   **(c)** not change

   9. The amplitude of the ripple voltage will

      (a) increase   **(b)** decrease   **(c)** not change

❏ If the zener diode is open,

   10. The output voltage will

      (a) increase   **(b)** decrease   **(c)** not change

■ PROBLEMS

Answers to odd-numbered problems are at the end of the book.

SECTION 2–5 Rectifiers

1. Sketch the waveforms for the load current and voltage for Figure 2–76. Show the peak values.

2. Determine the peak voltage and the peak power delivered to R_L in Figure 2–77.

FIGURE 2–76

FIGURE 2–77

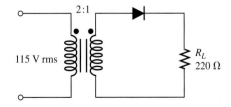

3. Consider the circuit in Figure 2–78.

   (a) What type of circuit is this?

   (b) What is the total peak secondary voltage?

   (c) Find the peak voltage across each half of the secondary.

   (d) Sketch the voltage waveform across R_L.

   (e) What is the peak current through each diode?

   (f) What is the PIV for each diode?

FIGURE 2–78

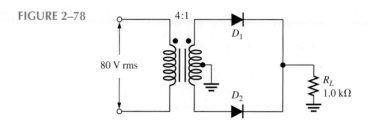

4. Show how to connect the diodes in a center-tapped rectifier in order to produce a negative-going full-wave voltage across the load resistor.

5. What PIV rating is required for the diodes in a bridge rectifier that produces an average output voltage of 50 V?

SECTION 2–6 Rectifier Filters and IC Regulators

6. The ideal dc output voltage of a capacitor-input filter is the (peak, average) value of the rectified input.

7. The minimum ripple rejection for a 7805 regulator is 68 dB. Compute the output ripple voltage for an input that has 150 mV of ripple.

8. If the no-load output voltage of a regulator is 15.5 V and the full-load output is 14.9 V, what is the percent load regulation?

9. Assume a regulator has a percent load regulation of 0.5%. What is output voltage at full-load if the unloaded output is 12.0 V?

10. For the variable output supply shown in Figure 2–29, what setting of R_2 would provide an output of 5.0 V?

11. For the variable output supply shown in Figure 2–29, what maximum output voltage could be obtained if R_2 were replaced with a 1.5 kΩ resistor?

SECTION 2–7 Diode Limiting and Clamping Circuits

12. Sketch the output waveforms for each circuit in Figure 2–79.

13. Describe the output waveform of each circuit in Figure 2–80. Assume that the RC time constant is much greater than the period of the input.

FIGURE 2–79

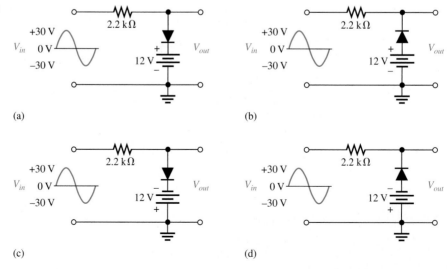

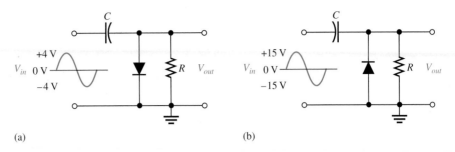

FIGURE 2–80

SECTION 2–8   Special-Purpose Diodes

14. Figure 2–81 shows a zener diode regulator designed to hold 5.0 V at the output. Assume the zener resistance is zero and the zener current ranges from 2 mA minimum (I_{ZK}) to 30 mA maximum (I_{ZM}). What are the minimum and maximum input source voltages for these currents?

15. A certain zener diode has a $V_Z = 7.5$ V and an $Z_Z = 5\ \Omega$ at a certain current. Sketch the equivalent circuit.

16. To what value must R be adjusted in Figure 2–82 to make $I_Z = 40$ mA? Assume that $V_Z = 12$ V at 30 mA and $Z_Z = 30\ \Omega$.

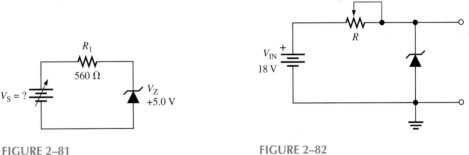

FIGURE 2–81   **FIGURE 2–82**

17. Assume the output of a zener regulator circuit drops from 8.0 V with no load to 7.8 V with a 500 Ω load. What is the percent load regulation?

18. Figure 2–83 is a curve of reverse voltage versus capacitance for a certain varactor. Determine the change in capacitance if V_R varies from 5 V to 20 V.

19. Refer to Figure 2–83 and determine the value of V_R that produces a capacitance of 25 pF.

FIGURE 2–83

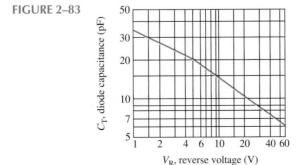

20. What capacitance value is required for each of the varactors in Figure 2–84 to produce a resonant frequency of 1 MHz?

21. At what value must the control voltage be set in Problem 20 if the varactors have the characteristic curve in Figure 2–83?

22. When the switch in Figure 2–85 is closed, will the microammeter reading increase or decrease? Assume that D_1 and D_2 are optically coupled.

23. With no incident light, there is a certain amount of reverse current in a photodiode. What is this current called?

FIGURE 2–84

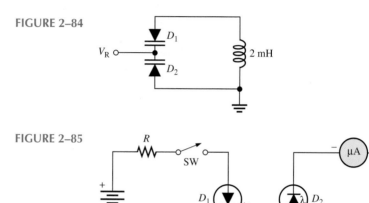

FIGURE 2–85

SECTION 2–9   The Diode Data Sheet

24. From the data sheet in Appendix A, determine how much peak inverse voltage that a 1N1183A diode can withstand.

25. Repeat Problem 24 for a 1N1188A.

26. If the peak output voltage of a full-wave bridge rectifier is 50 V, determine the minimum value of the surge-limiting resistor required when 1N1183A diodes are used.

SECTION 2–10   Troubleshooting

27. From the meter readings in Figure 2–86, determine the most likely problem. State how you could quickly isolate the exact location of the problem.

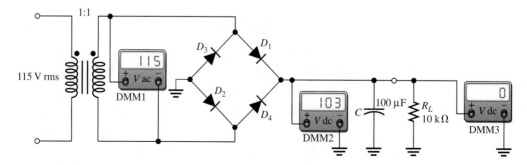

FIGURE 2–86

28. Each part of Figure 2–87 shows oscilloscope displays of rectifier output voltages. In each case, determine whether or not the rectifier is functioning properly and, if it is not, what is (are) the most likely failure(s). Assume all displays are set for the same time per division.

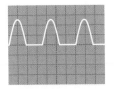 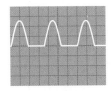 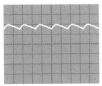 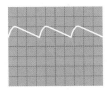

(a) Output of a half-wave unfiltered rectifier   (b) Output of a full-wave unfiltered rectifier   (c) Output of a full-wave filter   (d) Output of same full-wave filter as part (c)

FIGURE 2–87

29. For each set of measured voltages at the points (1, 2, and 3) indicated in Figure 2–88, determine if they are correct and if not identify the most likely fault(s). State what you would do to correct the problem once it is isolated.
   (a) $V_1 = 110$ V rms, $V_2 \cong 30$ V dc, $V_3 \cong 12$ V dc
   (b) $V_1 = 110$ V rms, $V_2 \cong 30$ V dc, $V_3 \cong 30$ V dc
   (c) $V_1 = 0$ V, $V_2 = 0$ V, $V_3 = 0$ V
   (d) $V_1 = 110$ V rms, $V_2 \cong 30$ V peak full-wave 120 Hz voltage, $V_3 \cong 12$ V, 120 Hz pulsating voltage
   (e) $V_1 = 110$ V rms, $V_2 = 0$ V, $V_3 = 0$ V

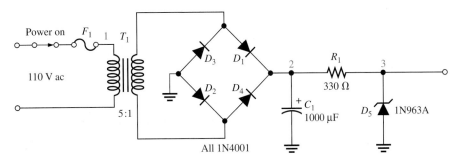

FIGURE 2–88

30. Determine the most likely failure in the circuit board of Figure 2–89 for each of the following symptoms. State the corrective action you would take in each case. The normal output of the transformer is 12.6 V ac. The regulator (IC1) is a 7812.
   (a) No voltage between points 1 and 6.
   (b) 110 V rms between points 1 and 6 but no voltage between points 2 and 5.
   (c) 110 V rms between points 1 and 6 but 11.5 V rms between points 2 and 5.
   (d) A pulsating full-wave rectified voltage with a peak of 19 V between point 3 and ground.
   (e) Excessive 120 Hz ripple voltage between point 3 and ground.
   (f) Ripple voltage has a frequency of 60 Hz between point 3 and ground.
   (g) 17 V dc with 120 Hz ripple at point 3 but no dc voltage at point 4.

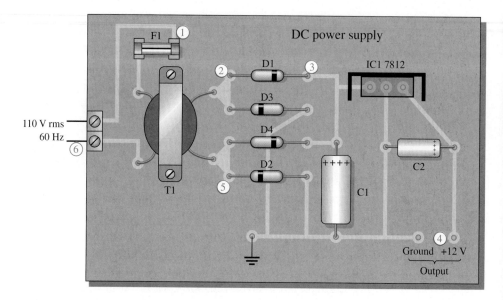

FIGURE 2–89

31. Draw the schematic for the circuit board in Figure 2–90 and determine what the correct output voltages should be.

32. The ac input of the circuit board in Figure 2–90 is connected to the secondary of a transformer with a turns ratio of 1 that is operating from the 110 V ac source. When you measure the output voltages, both are zero. What do you think has failed, and what is the reason for this failure?

FIGURE 2–90

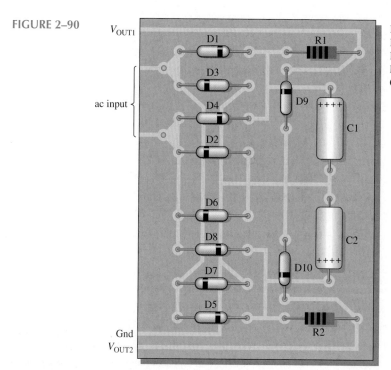

D1–D8: 1N4001
D9: 1N754
D10: 1N970 A
R1–R2: 220 Ω
C1–C2: 1000 μF

33. Select a transformer turns ratio that will provide the proper secondary voltage compatible with the circuit board in Figure 2–90.

34. In Figure 2–90, V_{OUT2} is zero and V_{OUT1} is correct. What are possible reasons for this?

■ **ANSWERS TO REVIEW QUESTIONS**

Section 2–1

1. Conduction band; valence band

2. An electron is thermally raised to the conduction band leaving a vacancy (hole) in the valence band.

3. The gap between the valence band and the conduction band is narrower in a semiconductor than in an insulator.

Section 2–2

1. By the addition of pentavalent impurities into the semiconductive material

2. By the addition of trivalent impurities into the semiconductive material

3. The boundary between p and n materials

4. 0.7 V

Section 2–3

1. Forward, reverse

2. Forward

3. Reverse

4. A rapid buildup of current when sufficient reverse bias is applied to the diode

Section 2–4

1. Forward-bias and reverse-bias

2. The reverse-breakdown region

3. As a switch

4. The barrier potential and the forward (bulk) resistance

Section 2–5

1. Bridge

2. Less

3. The peak of the negative alteration

4. 50% (with no filter)

Section 2–6

1. The charging and discharging of the capacitor

2. Increases ripple

3. Much better ripple rejection, line and load regulation, thermal protection

4. *Line regulation:* Constant output voltage for varying input voltage
Load regulation: Constant output voltage for varying load current

Section 2–7

1. Limiters clip off or remove portions of a waveform. Clampers insert a dc level.
2. Reversing the diode causes the limiter to clip the other side of the waveform.
3. The bias voltage must be 5 V − 0.7 V = 4.3 V.
4. The capacitor acts as a battery.

Section 2–8

1. In breakdown
2. The maximum current, above which the diode may be damaged
3. It is a variable capacitor.
4. The diode capacitance decreases.
5. Gallium arsenide, gallium arsenide phosphide, gallium phosphide
6. Dark current

Section 2–9

1. The three rating categories on a diode data sheet are maximum ratings, electrical characteristics, and mechanical data.
2. V_F is the maximum instantaneous forward voltage drop, I_R is reverse current, and I_O is peak average forward current.
3. I_{FSM} is maximum forward surge current, V_{RRM} is maximum reverse peak repetitive voltage, and V_{RSM} is maximum reverse peak nonrepetitive voltage.
4. The 1N4720 has an $I_O = 3.0$ A, $I_{FSM} = 300$ A, and $V_{RRM} = 100$ V.

Section 2–10

1. The output would not change as the potentiometer was varied; it would be slightly more than 1.25 V.
2. Open diode
3. The filter capacitor

Section 2–11

1. Transformer, rectifier, filter, and regulator
2. It provides all parts of the system with dc voltage and current to operate the circuits.
3. AC line voltage, 100 V rms (Portable power supplies use a battery.)
4. 50 V, designated V_{RRM}
5. No, they are experiencing approximately 16 V; they are rated for 50 V.
6. Filtering
7. Capacitor C_2 removes any noise spikes that appear on the line and improves the transient response of the power supply.
8. The voltage regulator must dissipate several watts of power. The heat sink helps remove heat produced, thus protecting the regulator.
9. If capacitor C_2 is open, noise spikes may appear on the output and the output voltage may have momentary changes during load changes.

■ **ANSWERS TO PRACTICE EXERCISES FOR EXAMPLES**	**2–1**	2.3 V, 3.0 V
	2–2	320 V
	2–3	39.8 μV
	2–4	3.7%
	2–5	The peak voltage drops to 8.7 V and −0.7 V.
	2–6	The output is clipped at +10.7 V and −10.7 V.
	2–7	The output would be a sinusoidal wave that goes from approximately −0.7 V to +47.3 V.
	2–8	8 Ω
	2–9	6.8 V and 28.9 V
	2–10	100 kHz to 318 kHz

■ **ANSWERS TO SELF-TEST**

1. (d)	**2.** (c)	**3.** (b)	**4.** (a)	**5.** (b)
6. (d)	**7.** (c)	**8.** (d)	**9.** (a)	**10.** (d)
11. (b)	**12.** (c)	**13.** (b)	**14.** (d)	**15.** (a)
16. (c)	**17.** (a)	**18.** (a)	**19.** (c)	**20.** (a)
21. (c)	**22.** (d)	**23.** (b)	**24.** (a)	

■ **ANSWERS TO TROUBLESHOOTER'S QUIZ**

1. decrease	**2.** not change	**3.** increase	**4.** not change
5. increase	**6.** not change	**7.** increase	**8.** not change
9. increase	**10.** increase		

3

BIPOLAR JUNCTION TRANSISTORS (BJTs)

Courtesy Hewlett-Packard Company

■ CHAPTER OBJECTIVES

❏ Describe the basic construction and operation of bipolar junction transistors (BJTs)
❏ Explain the operation of the four basic BJT bias circuits
❏ Discuss transistor parameters and characteristics and use them to analyze a transistor circuit
❏ Understand and analyze the operation of common-emitter amplifiers
❏ Understand and analyze the operation of common-collector amplifiers
❏ Understand and analyze the operation of common-base amplifiers
❏ Explain how a transistor can be used as a switch
❏ Identify various types of transistor package configurations

- ❏ Troubleshoot various faults in transistor circuits
- ❏ Apply what you have learned in this chapter to a system application

■ CHAPTER INTRODUCTION

Two basic types of **transistors** are the bipolar junction transistor (BJT) and the field-effect transistor (FET). In this chapter, you will be introduced to the first of these types—the bipolar junction transistor. The chapter begins with a discussion of dc operation and bias circuits. You will see how these various bias circuits operate and how basic types of discrete amplifiers work in linear and switching applications. You will also learn to read manufacturer's data sheets. The chapter ends with a system application.

■ A SYSTEM APPLICATION

Transistors are used in one form or another in just about every electronic system imaginable. A simple electronic security system is the focus of this chapter's system application section. You will see how transistors, operating in their switching modes, are used to perform a basic function in a system. Also, you will see how the transistor is used to operate nonelectronic devices such as, in this case, the electromechanical relay.

For the system application in Section 3–10, in addition to the other topics, be sure you understand

- ❏ How a bipolar junction transistor works
- ❏ How the currents in a transistor are calculated
- ❏ The meaning of saturation and cutoff
- ❏ The principles of biasing a transistor
- ❏ How to read a data sheet

www. VISIT THE COMPANION WEBSITE

Study Aids for This Chapter Are Available at

http://www.prenhall.com/floyd

3–1 ■ STRUCTURE OF BIPOLAR JUNCTION TRANSISTORS

The basic structure of the bipolar junction transistor (BJT) determines its operating characteristics. In this section, you will see how semiconductive materials form a BJT, and you will learn the standard transistor symbols. You will see the application of a load line to a basic transistor circuit for setting up proper dc currents and voltages (bias) in a transistor circuit.

After completing this section, you should be able to

❏ Describe the basic construction and operation of bipolar junction transistors (BJTs)
  ❏ Distinguish between *npn* and *pnp* transistors
  ❏ Define BJT currents and explain how they are related
  ❏ Interpret the characteristic curves for a BJT
  ❏ Explain how a dc load line is constructed for a transistor circuit
  ❏ Define the terms *cutoff* and *saturation*

The **bipolar junction transistor (BJT)** is constructed with three doped semiconductor regions called **emitter**, **base**, and **collector**. These three regions are separated by two *pn* junctions. The two types of bipolar transistors are shown in Figure 3–1. One type consists of two *n* regions separated by a thin *p* region (*npn*), and the other type consists of two *p* regions separated by a thin *n* region (*pnp*). Both types are widely used; however, because the *npn* type is more prevalent, it will be used for most of the discussion which follows.

FIGURE 3–1
Construction of bipolar junction transistors.

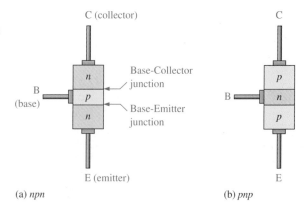

(a) *npn*   (b) *pnp*

The *pn* junction joining the base region and the emitter region is called the *base-emitter junction.* The *pn* junction joining the base region and the collector region is called the *base-collector junction,* as indicated in Figure 3–1(a). These junctions act just like the diode junctions discussed in Chapter 2 and are frequently referred to as the base-emitter diode and the base-collector diode. Each region is connected to a wire lead, labeled E, B, and C for emitter, base, and collector, respectively. Although the emitter and collector regions are made from the same type of material, the doping level and other characteristics are different.

Figure 3–2 shows the schematic symbols for the *npn* and *pnp* bipolar transistors. (Note that the arrow on an *npn* transistor is *not* pointing i*n.*) The term **bipolar** refers to the use of both holes and electrons as carriers in the transistor structure.

FIGURE 3–2
Standard bipolar junction transistor symbols.

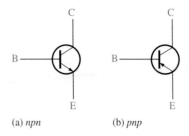

(a) *npn*   (b) *pnp*

Transistor Operation

In order for the transistor to operate properly, the two *pn* junctions must be supplied with external dc bias voltages to set the proper operating conditions. Figure 3–3 shows the proper bias arrangement for both *npn* and *pnp* transistors. In both cases, the base-emitter (BE) junction is forward-biased and the base-collector (BC) junction is reverse-biased. This is called *forward-reverse bias.* Both *npn* and *pnp* transistors normally use this forward-reverse bias, but the bias voltage polarities and the current directions are reversed between the two types.

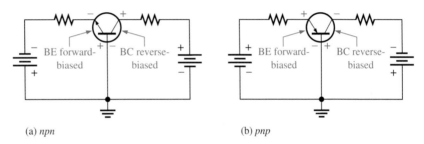

(a) *npn*   (b) *pnp*

FIGURE 3–3
Forward-reverse bias of a bipolar junction transistor.

To illustrate transistor action, let's examine what happens inside an *npn* transistor when the junctions are forward-reverse biased. (The same concepts can be applied to a *pnp* transistor by reversing the polarities.) The forward bias from base to emitter narrows the BE depletion region, and the reverse bias from base to collector widens the BC depletion region, as depicted in Figure 3–4. The heavily doped *n*-type emitter region is teeming with conduction-band (free) electrons that easily diffuse through the forward-biased BE junction into the *p*-type base region, just as in a forward-biased diode.

The base region is lightly doped and very narrow so that it has a very limited number of holes. Thus, only a small percentage of all the electrons flowing through the BE junction can combine with the available holes in the base. These relatively few recombined electrons flow out of the base lead as valence electrons, forming the small base current, as shown in Figure 3–4.

Most of the electrons flowing from the emitter into the narrow lightly doped base region do not recombine but diffuse into the BC depletion region. Once in the region, they are pulled through the reverse-biased BC junction by the electric field set up by the force of attraction between the positive and negative ions. Actually, you can think of the electrons as being pulled across the reverse-biased BC junction by the attraction of the collector supply voltage. The electrons now move through the collector region, out through the

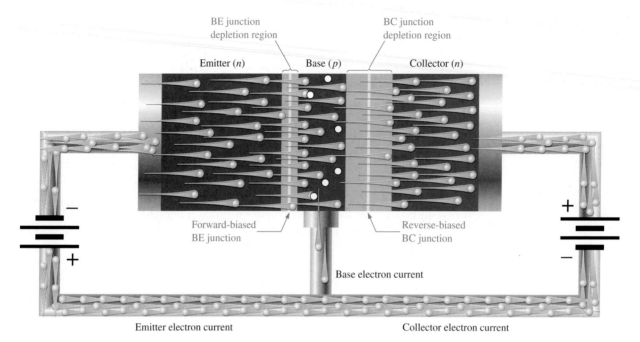

FIGURE 3–4
Illustration of BJT action. The base region is very narrow, but it is shown wider here for clarity.

collector lead, and into the positive terminal of the external dc source, thereby forming the collector current, as shown. The amount of collector current depends directly on the amount of base current and is essentially independent of the dc collector voltage.

The bottom line is this: *A small base current can control a larger collector current.* Because the controlling element is base current and it controls a larger collector current, the bipolar transistor is essentially a current amplifier. This concept of a small control element for a large current is analogous to deForest's control grid mentioned in Section 1–1.

Transistor Currents

Kirchhoff's current law (KCL) says the total current entering a junction must be equal to the total current leaving that junction. Applying this law to both the *npn* and the *pnp* transistors shows that the emitter current (I_E) is the sum of the collector current (I_C) and base current (I_B), expressed as follows:

$$I_E = I_C + I_B \qquad\qquad (3\text{–}1)$$

The base current, I_B, is very small compared to I_E or I_C, which leads to the approximation $I_E \cong I_C$, a useful assumption for analyzing transistor circuits. Examples of an *npn* and a *pnp* small-signal transistor with representative currents are shown on the meters in Figures 3–5(a) and 3–5(b), respectively. Notice the polarity of the ammeters and supply voltages are reversed between the *npn* and the *pnp* transistors. The capital-letter subscripts indicate dc values.

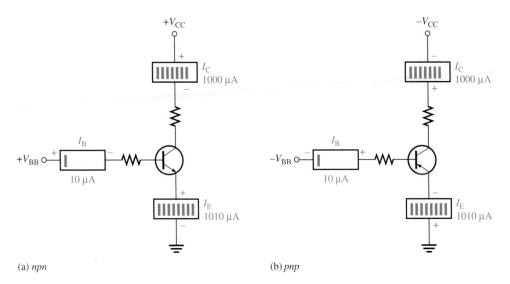

(a) *npn*   (b) *pnp*

FIGURE 3–5
Currents in small-signal transistors.

DC Beta (β_{DC})

When a transistor is operated within certain limits, the collector current is proportional to the base current. The **dc beta** (β_{DC}), the current gain of a transistor, is the ratio of the dc collector current to the dc base current.

$$\beta_{DC} = \frac{I_C}{I_B} \qquad\qquad\qquad \textbf{(3–2)}$$

The dc beta (β_{DC}) represents a constant of proportionality called the current gain and is usually designated as h_{FE} on transistor data sheets. It is valid as long as the transistor is operated within its linear range. For this case, the collector current is equal to β_{DC} multiplied by the base current. For the examples in Figure 3–5, $\beta_{DC} = 100$.

The values for β_{DC} vary widely and depend on the type of transistor. They are typically from 20 (power transistors) to 200 (small-signal transistors). Even two transistors of the same type can have current gains that are quite different. Although the current gain is necessary for a transistor to be useful as an amplifier, good designs do not rely on a particular value of β_{DC} to operate properly.

Transistor Voltages

The three dc voltages for the biased transistor in Figure 3–6 are the emitter voltage (V_E), the collector voltage (V_C), and the base voltage (V_B). These single-subscript voltages mean that they are referenced to ground. The collector power supply voltage, V_{CC}, is shown with repeated subscript letters. Because the emitter is grounded, the collector voltage is equal to the dc supply voltage, V_{CC}, less the drop across R_C.

$$V_C = V_{CC} - I_C R_C$$

Kirchhoff's voltage law (KVL) says the sum of the voltage drops and rises around a closed path is zero. The previous equation is an application of this law.

FIGURE 3–6
Bias voltages.

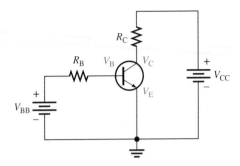

As mentioned earlier, the base-emitter diode is forward-biased when the transistor is operating normally. The forward-biased base-emitter diode drop, V_{BE}, is approximately 0.7 V. This means that the base voltage is one diode drop larger than the emitter voltage. In equation form,

$$V_B = V_E + V_{BE} = V_E + 0.7 \text{ V}$$

In the configuration of Figure 3–6, the emitter is the reference terminal, so $V_E = 0$ V and $V_B = 0.7$ V.

EXAMPLE 3–1

Determine I_B, I_C, I_E, V_B, and V_C in Figure 3–7, where β_{DC} is 50.

FIGURE 3–7

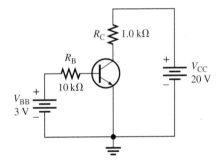

Solution Since V_E is ground, $V_B = \mathbf{0.7\ V}$. The drop across R_B is $V_{BB} - V_B$, so I_B is calculated as follows:

$$I_B = \frac{V_{BB} - V_B}{R_B} = \frac{3 \text{ V} - 0.7 \text{ V}}{10 \text{ k}\Omega} = \mathbf{0.23\ mA}$$

Now you can find I_C, I_E, and V_C.

$$I_C = \beta_{DC}I_B = 50(0.23 \text{ mA}) = \mathbf{11.5\ mA}$$
$$I_E = I_C + I_B = 11.5 \text{ mA} + 0.23 \text{ mA} = \mathbf{11.7\ mA}$$
$$V_C = V_{CC} - I_C R_C = 20 \text{ V} - (11.5 \text{ mA})(1.0 \text{ k}\Omega) = \mathbf{8.5\ V}$$

*Practice Exercise** Determine I_B, I_C, I_E, V_{CE}, and V_{CB} in Figure 3–7 for the following values: $R_B = 22$ kΩ, $R_C = 220$ Ω, $V_{BB} = 6$ V, $V_{CC} = 9$ V, and $\beta_{DC} = 90$.

---

* Answers are at the end of the chapter.

Characteristic Curves for a BJT

Base-Emitter Characteristic   The characteristic *IV* curve for the base-emitter junction is shown in Figure 3–8. As you can see, it is identical to that of an ordinary diode. You can model the base-emitter junction with any of the three diode models presented in Chapter 2. For most work, the offset model is sufficiently accurate. This means, if you are troubleshooting a bipolar transistor circuit, you can look for 0.7 V across the base-emitter junction (as in a forward-biased diode) to determine if the transistor is conducting. If the voltage is zero, the transistor is not conducting; if it is much larger than 0.7 V, it is likely that the transistor has an open base-emitter junction.

FIGURE 3–8
Base-emitter characteristic.

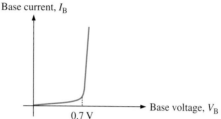

Collector Characteristic   Recall that the collector current is proportional to the base current ($I_C = \beta_{DC}I_B$). If there is no base current, the collector current is zero. In order to plot the collector characteristic, a base current must be selected and held constant. A circuit such as that in Figure 3–9(a) can be used to generate a set of collector *IV* curves to show how I_C varies with V_{CE} for a given base current. These curves are called the **collector characteristic curves**.

Notice in the circuit diagram that both dc supply voltages, V_{BB} and V_{CC}, are adjustable. If V_{BB} is set to produce a specific value of I_B and V_{CC} is zero, then $I_C = 0$ and $V_{CE} = 0$. Now, as V_{CC} is gradually increased, V_{CE} will increase and so will I_C, as indicated on the color-shaded portion of the curve between points *A* and *B* in Figure 3–9(b).

When V_{CE} reaches ≈ 0.7 V, the base-collector junction becomes reverse-biased and I_C reaches its full value determined by the relationship $I_C = \beta_{DC}I_B$. Ideally, I_C levels off to an almost constant value as V_{CE} continues to increase. This action appears to the right of point *B* on the curve. In practice, I_C increases slightly as V_{CE} increases due to the widening of the base-collector depletion region, which results in fewer holes for recombination in the base region. The steepness of this rise is determined by a parameter called the *forward Early voltage,* named after J. M. Early.

By setting I_B to other constant values, you can produce additional I_C versus V_{CE} curves, as shown in Figure 3–9(c). These curves constitute a "family" of collector curves for a given transistor. A family of curves allows you to visualize the complex situation when three variables interact. By holding one of them (I_B) constant, you can see the relation between the other two (I_C versus V_{CE}).

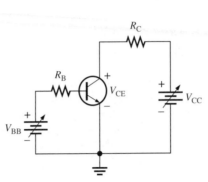

(a) Circuit

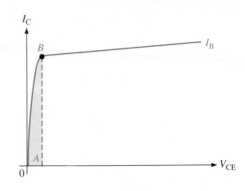

(b) I_C curve for one value of I_B with V_{CE} changing

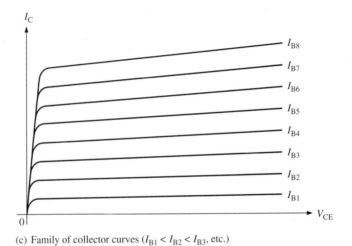

(c) Family of collector curves ($I_{B1} < I_{B2} < I_{B3}$, etc.)

FIGURE 3–9
Collector characteristic curves.

EXAMPLE 3–2

Sketch the family of collector curves for the circuit in Figure 3–10 for $I_B = 5$ μA to 25 μA in 5 μA increments. Assume that $\beta_{DC} = 100$.

FIGURE 3–10

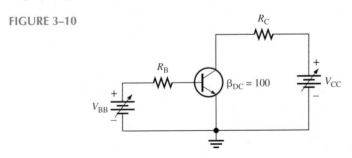

Solution Table 3–1 shows the calculations of I_C using the relationship $I_C = \beta_{DC}I_B$. The resulting curves are plotted in Figure 3–11. To account for the forward Early voltage, the resulting curves are shown with an arbitrary upward slope as previously discussed.

TABLE 3–1

I_B	I_C
5 μA	0.5 mA
10 μA	1.0 mA
15 μA	1.5 mA
20 μA	2.0 mA
25 μA	2.5 mA

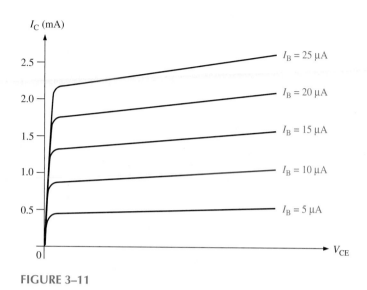

FIGURE 3–11

Practice Exercise Ideally, where would the curve for $I_B = 0$ appear on the graph?

Cutoff and Saturation

When $I_B = 0$, the transistor is in **cutoff** and there is essentially no collector current except for a very tiny amount of collector leakage current, I_{CEO}, which can usually be neglected. In cutoff, both the base-emitter and the base-collector junctions are reverse-biased. When you are troubleshooting a transistor that is in cutoff, you can assume the collector current is zero; therefore, there is no voltage drop across the collector resistor. As a result, the collector-to-emitter voltage will be very nearly equal to the supply voltage.

Now let's consider the opposite situation. When the base-emitter junction in Figure 3–9 becomes forward-biased and the base current is increased, the collector current also increases and V_{CE} decreases as a result of more drop across R_C. According to Kirchhoff's voltage law, if the voltage across R_C increases, the drop across the transistor must decrease.

Ideally, when the base current is high enough, the entire V_{CC} is dropped across R_C with no voltage between the collector and emitter. This condition is known as **saturation**. Saturation occurs when the supply voltage, V_{CC}, is across the total resistance of the collector circuit, R_C. The saturation current for this particular configuration is given by Ohm's law.

$$I_{C(sat)} = \frac{V_{CC}}{R_C}$$

Once the base current is high enough to produce saturation, further increases in base current have no effect on the collector current, and the relationship $I_C = \beta_{DC}I_B$ is no longer valid. When V_{CE} reaches its saturation value, $V_{CE(sat)}$, which is ideally zero, the base-collector junction becomes forward-biased.

When you are troubleshooting transistor circuits, a quick check for cutoff or saturation provides useful information. Remember a transistor in cutoff has nearly the entire supply voltage between collector and emitter; a saturated transistor, in practice, has a very small voltage drop between collector and emitter (typically 0.1 V).

DC Load Line

Recall from Section 1–3 that a Thevenin circuit is drawn as a voltage source in series with a resistor. Consider the circuit in Figure 3–12(a). The collector voltage source, V_{CC}, and the collector resistor, R_C, form a Thevenin source; the transistor is the load. The minimum and maximum current that can be provided from this source are zero and V_{CC}/R_C. These are, of course, the cutoff and saturation values as previously defined. Note that the saturation and cutoff points depend only on the Thevenin circuit; the transistor does not affect these points. A straight line drawn between cutoff and saturation defines the dc load line for this circuit, as shown in Figure 3–12(b). This line represents all possible dc operating points for the circuit.

The *IV* curve for any type of load can be added to the same plot as the dc load line to obtain a graphical picture of the circuit operation, as shown earlier in Section 1–3. Figure

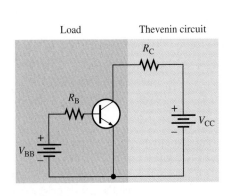

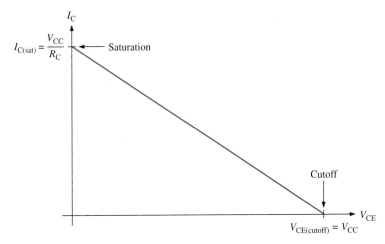

(a) The collector circuit, shown in the colored box, is a Thevenin circuit. The transistor circuit, shown in the gray box, is the load.

(b) DC load line for the circuit in (a)

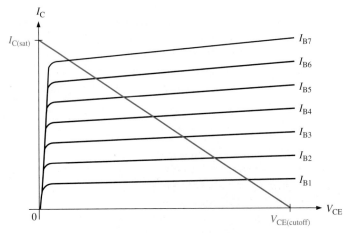

(c) Collector characteristic curves superimposed on the dc load line

FIGURE 3–12

3–12(c) shows a dc load line superimposed on a set of ideal collector characteristic curves. Any value of I_C and the corresponding V_{CE} will fall on this line, as long as dc operation is maintained.

Now let's see how the dc load line and the characteristic curves for a transistor can be used to illustrate transistor operation. Assume you have a transistor with the characteristic curves shown in Figure 3–13(a) and you install it in the dc test circuit shown in Figure 3–13(b). A graphical solution can be used to find currents and voltages by drawing a dc load line. First, the cutoff point on the load line is determined. When the transistor is cut off, there is essentially no collector current. Thus, the collector-emitter voltage and current are

$$V_{CE(cutoff)} = V_{CC} = 12 \text{ V}$$

and

$$I_{C(cutoff)} = 0 \text{ mA}$$

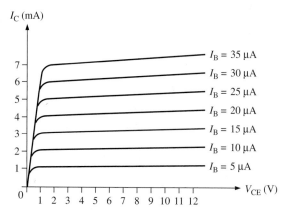

(a) Characteristic curves

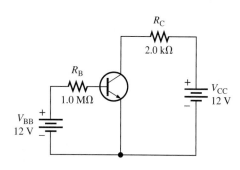

(b) DC test circuit

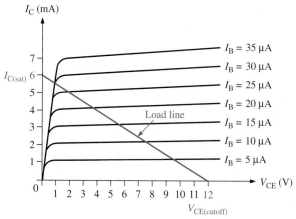

(c) Load line and characteristic curves

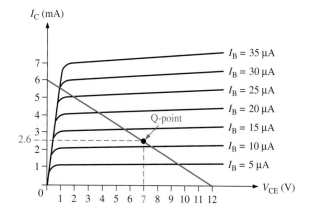

(d) Locating the Q-point

FIGURE 3–13

Next, the saturation point on the load line is determined. When the transistor is saturated, V_{CE} is nearly zero. Therefore, V_{CC} is dropped across R_C. The saturation value of the collector current, $I_{C(sat)}$, is found by applying Ohm's law to the collector resistor.

$$I_{C(sat)} = \frac{V_{CC}}{R_C} = \frac{12 \text{ V}}{2.0 \text{ k}\Omega} = 6.0 \text{ mA}$$

This value is the maximum value for I_C. It cannot possibly be increased without changing V_{CC} or R_C.

Next, the cutoff and saturation points are plotted on the same plot as the characteristic curves and a straight line, which is the load line, is drawn between them. This represents all possible operating points for the circuit. Figure 3–13(c) shows the load line and the characteristic curves for the transistor together on the same plot.

Q-Point

Before the actual collector current can be found, the value of the base current, I_B, needs to be established. Referring to the original circuit, it is apparent that the base power supply, V_{BB}, is across the series combination of the base resistor, R_B, and the forward-biased base-emitter junction. This means that the voltage across the base resistor is

$$V_{R_B} = V_{BB} - V_{BE} = 12 \text{ V} - 0.7 \text{ V} = 11.3 \text{ V}$$

By applying Ohm's law, you can find the base current.

$$I_B = \frac{V_{R_B}}{R_B} = \frac{11.3 \text{ V}}{1.0 \text{ M}\Omega} = 11.3 \text{ }\mu\text{A}$$

The point at which the actual base current line intersects the load line is the quiescent or Q-point for the circuit. The Q-point is found on the graph by interpolating between the 10 μA and 15 μA base current lines. The coordinates of the Q-point are the values for I_C and V_{CE} at that point, as illustrated in Figure 3–13(d). Reading these values from the plot, you find the value of I_C to be approximately 2.6 mA and V_{CE} to be approximately 7.0 V.

The plot in Figure 3–13(d) completely describes the dc operating conditions for the amplifier circuit. When troubleshooting, you won't take the time to draw load lines; rather you will learn to apply the basic math for circuits to obtain an idea of what should be occurring with a given circuit. However, the load line provides a useful mental picture for describing the dc conditions for the transistor.

3–1 REVIEW QUESTIONS*

1. What are the three BJT currents called?
2. Explain the difference between *saturation* and *cutoff.*
3. What is the definition of β_{DC}?

* Answers are at the end of the chapter.

3–2 ■ BJT BIAS CIRCUITS

In this section, methods for biasing a bipolar junction transistor are presented. Biasing is the application of the appropriate dc voltages to cause the transistor to operate properly. It can be accomplished with any of several basic circuits. The choice of biasing circuit depends largely on the application. You will learn about four biasing methods and see the advantages and disadvantages of each method.

After completing this section, you should be able to

❑ Explain the operation of the four basic BJT bias circuits
   ❑ Describe a base bias circuit
   ❑ Describe a collector-feedback bias circuit
   ❑ Describe a voltage-divider bias circuit
   ❑ Describe an emitter bias circuit

For linear amplifiers, the signal must swing in both the positive and negative directions. Transistors operate with current in one direction only. In order for a transistor to amplify an ac signal, the ac signal needs to be superimposed on a dc level that sets the operating point. Bias circuits set the dc level at a point that allows the ac signal to vary in both the positive and negative directions without driving the transistor into saturation or cutoff.

Base Bias

The simplest biasing circuit is **base bias**. For the *npn* transistor, shown in Figure 3–14(a), a resistor (R_B) is connected between the base and supply voltage. Note that this is essentially the same circuit that was introduced in Figure 3–9(a) and used to plot the characteristic curve. The only difference is that the base and the collector power supplies have been combined into a single supply (referred to as V_{CC}). Although this bias method is simple, it is not a good choice for linear amplifiers for reasons that will be discussed.

    The *pnp* transistor can be set up using a negative supply as shown in Figure 3–14(b) or it can be run with a positive supply by applying the positive supply voltage to the emitter as shown in Figure 3–14(c). Either of these arrangements provide a path for base current through the base-emitter junction. In turn, this base current causes a collector current

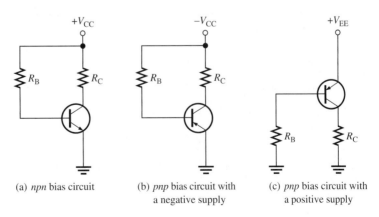

(a) *npn* bias circuit

(b) *pnp* bias circuit with a negative supply

(c) *pnp* bias circuit with a positive supply

FIGURE 3–14
Base bias circuits.

that is β_{DC} times larger than the base current (assuming linear operation). Thus, the collector current, for linear operation, is

$$I_C = \beta_{DC}I_B$$

The base resistor, R_B, has the base current, I_B, through it. From Ohm's law, you can substitute for I_B and obtain

$$I_C = \beta_{DC}\left(\frac{V_{R_B}}{R_B}\right)$$

$$I_C = \beta_{DC}\left(\frac{V_{CC} - V_{BE}}{R_B}\right) \tag{3–3}$$

This formula gives the collector current for base bias as long as the transistor is not in saturation. It is derived for the case with no emitter resistor, so this formula can only be applied to this configuration.

As mentioned previously, transistors can have very different current gains. Typical transistors of the same type can have β_{DC} values that vary by a factor of 3! In addition, current gain is a function of the temperature; as temperature increases, the base-emitter voltage decreases and β_{DC} increases. As a result, the collector current can vary widely between similar circuits with base bias. Circuits that depend on a particular β_{DC} cannot be expected to operate in a consistent manner. For this reason, base bias is rarely used for linear circuits.

Because it uses only a single resistor for bias, base bias is a good choice in switching applications, where the transistor is always operated in either saturation or cutoff. For switching amplifiers, Equation (3–3) does not apply.

EXAMPLE 3–3

The manufacturer's specification for a 2N3904 transistor shows that β_{DC} has a range from 100 to 300 (given in Fig. 3–25). Assume a 2N3904 is used in the base-biased circuit shown in Figure 3–15. Compute the minimum and maximum collector current based on this specification. (Note that this is effectively the same circuit that was solved with load line analysis in Figure 3–13 except it now is shown with a single power supply.)

Solution The base-emitter junction is forward-biased, causing a 0.7 V drop. The voltage across R_B is

$$V_{R_B} = V_{CC} - V_{BE} = 12\ \text{V} - 0.7\ \text{V} = 11.3\ \text{V}$$

FIGURE 3–15

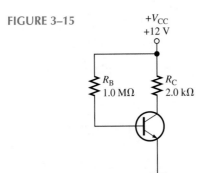

The base current can be found by applying Ohm's law to the base resistor.

$$I_B = \frac{V_{R_B}}{R_B} = \frac{11.3 \text{ V}}{1.0 \text{ M}\Omega} = 11.3 \text{ } \mu A$$

With linear operation, the collector current is β_{DC} times larger than the base current. Therefore, the minimum collector current is

$$I_{C(min)} = \beta_{DC}I_B = (100)(11.3 \text{ } \mu A) = \textbf{1.13 mA}$$

The maximum collector current is

$$I_{C(max)} = \beta_{DC}I_B = (300)(11.3 \text{ } \mu A) = \textbf{3.39 mA}$$

Notice that a 300% change in β_{DC} caused a 300% change in collector current.

Practice Exercise   If you measured 2.5 mA of collector current in the circuit of Figure 3–15, what is the β_{DC} for the transistor?

Collector-Feedback Bias

Another type of bias arrangement is the **collector-feedback bias** circuit shown in Figure 3–16 for an *npn* transistor. (A *pnp* transistor can be operated identically except for a negative supply voltage.) The base resistor, R_B, is connected to the collector rather than to V_{CC}, as the base bias arrangement discussed previously. The base resistor will be a smaller value than in base bias because the collector voltage is less than V_{CC} in normal operation.

FIGURE 3–16
Collector-feedback bias.

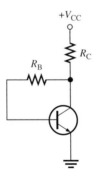

Collector feedback uses an important idea in electronics called **negative feedback** to achieve stability. Negative feedback returns a portion of the output back to the input in a manner to cancel changes that may occur. The negative feedback connection provides a relatively stable Q-point as a result.

Let's see how it works. In Figure 3–16, the collector voltage provides the bias to the base-emitter junction. The negative feedback creates a compensating effect that tends to keep the Q-point stable. Assume the β_{DC} increases due to a temperature increase. This causes I_C to increase and, in turn, more voltage drops across R_C. With more voltage dropped across R_C, V_C will decrease which, in turn, means it will supply *less* bias current. Thus, the original increase in β_{DC} has been compensated for, in part, by a smaller bias current. This compensating action is what is meant by negative feedback. Other applications for negative feedback will be described later.

The collector current for collector-feedback bias is derived from the application of Kirchhoff's voltage law (KVL). By writing a loop equation around the base circuit, the following equation for the collector current can be derived:

$$I_C = \frac{V_{CC} - V_{BE}}{R_C + R_B/\beta_{DC}} \tag{3-4}$$

This equation is valid for either an *npn* or a *pnp* transistor (be careful of signs). Equation (3–4) is applied in the next example to show how the effect of a different β_{DC} is compensated for by feedback.

EXAMPLE 3–4

As you saw earlier, a 2N3904 transistor has a β_{DC} range from 100 to 300. Assume a 2N3904 is used in the collector-feedback biased circuit shown in Figure 3–17. Compute the minimum and maximum collector current based on this specification.

FIGURE 3–17

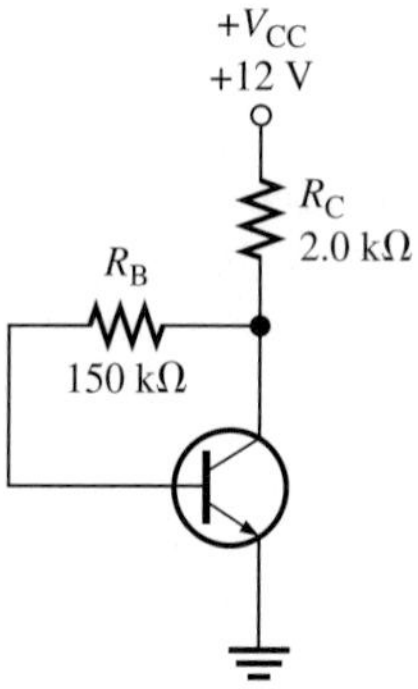

Solution   Substitute the values given for $\beta_{DC} = 100$ into Equation (3–4).

$$I_{C(min)} = \frac{V_{CC} - V_{BE}}{R_C + R_B/\beta_{DC}} = \frac{12\ V - 0.7\ V}{2.0\ k\Omega + 150\ k\Omega/100} = \textbf{3.2 mA}$$

Repeat the calculation for $\beta_{DC} = 300$.

$$I_{C(max)} = \frac{V_{CC} - V_{BE}}{R_C + R_B/\beta_{DC}} = \frac{12\ V - 0.7\ V}{2.0\ k\Omega + 150\ k\Omega/300} = \textbf{4.5 mA}$$

Note that a 300% change in β_{DC} resulted in only a 40% change in collector current for this case which is a considerable improvement over the base-bias case in Example 3–3.

Practice Exercise   Compute the minimum and maximum value of V_{CE} for the range of β_{DC} given in the problem. Notice that $V_E = 0$; therefore, $V_{CE} = V_C$.

As illustrated by Example 3–4, collector feedback bias offers greater stability than base bias with the same number of components. It still doesn't have the highest degree of stability required for many linear circuits. The highest stability for single supply operation is offered by voltage-divider bias.

Voltage-Divider Bias

As you have seen, the principal disadvantage to base bias is its dependency on β_{DC}. Collector-feedback bias offered greater stability than base bias, but an even higher degree of stability can be obtained with **voltage-divider bias**. Voltage-divider bias is the most widely used form of biasing because it uses only one supply voltage and produces bias that is essentially independent of β_{DC}. In fact, looking at the equations for voltage-divider bias reveals that neither β_{DC} nor any other transistor parameter is included. Essentially, good voltage-divider designs are independent of the transistor that is used.

The voltage-divider rule, one of the most useful rules from your basic dc/ac circuits course, allows you to compute the voltage across a series resistive branch in a circuit. Figure 3–18(a) illustrates a basic voltage divider. To find the output voltage, a ratio of the resistance to total resistance is set up and multiplied by the input voltage.

$$V_{OUT} = \left(\frac{R_2}{R_1 + R_2} \right) V_{IN}$$

When setting up the ratio for the voltage-divider rule, the resistor (in this case, R_2) that the output is taken across is the numerator and the sum of the resistances is the denominator.

FIGURE 3–18
Voltage dividers.

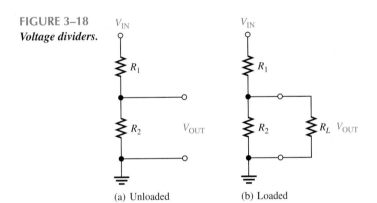

(a) Unloaded (b) Loaded

When a load resistor is placed across the output of a voltage divider, as in Figure 3–18(b), the output voltage decreases because of the loading effect. *As long as the load resistor is large compared to the divider resistors, the loading effect is small and can be ignored.*

Voltage-divider bias is shown in Figure 3–19. In this configuration, two resistors, R_1 and R_2, form a voltage divider that keeps the base voltage nearly the same for any load that requires a small current. This voltage forward-biases the base-emitter junction, resulting in a small base current. With voltage-divider bias, the transistor acts as a high resistance load on the divider. This will tend to make the base voltage slightly less than the unloaded value. In actual voltage-divider bias circuits, the effect is generally small, so the loading effects can be ignored. In any case, this loading effect can be minimized by the choice of R_1 and R_2. As a rule of thumb, these resistors should have a current that is at least ten times *larger* than the base current to avoid variations in the base voltage when a transistor with a different β_{DC} is used. This is called *stiff bias* because the base voltage is relatively independent of the base current.

The steps in computing the parameters for a voltage-divider bias circuit are straightforward applications of the voltage-divider rule and Ohm's law. Based on the assumption

FIGURE 3–19
Voltage-divider bias.

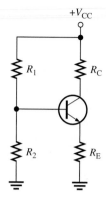

of no significant loading effect, you can use the voltage-divider rule described earlier to compute the base voltage. The voltage-divider rule applied to Figure 3–19 is

$$V_B = \left(\frac{R_2}{R_1 + R_2}\right)V_{CC}$$
(3–5)

The emitter voltage is one diode drop less than the base voltage. (For *pnp* transistors, it is one diode drop higher).

$$V_E = V_B - V_{BE}$$

$$V_E = V_B - 0.7 \text{ V}$$
(3–6)

With the emitter voltage known, the emitter current is found by Ohm's law.

$$I_E = \frac{V_E}{R_E}$$

The collector current is approximately the same as the emitter current.

$$I_C \cong I_E$$

The collector voltage can now be found. It is V_{CC} less the drop across the collector resistor, found by Ohm's law. Writing this as an equation,

$$V_C = V_{CC} - I_C R_C$$
(3–7)

To find the collector-emitter voltage, V_{CE}, subtract the emitter voltage, V_E, from the collector voltage, V_C.

$$V_{CE} = V_C - V_E$$

Example 3–5 illustrates this procedure for finding the dc parameters for a circuit.

EXAMPLE 3–5   Find V_B, V_E, I_E, I_C, and V_{CE} for the circuit in Figure 3–20.

FIGURE 3–20

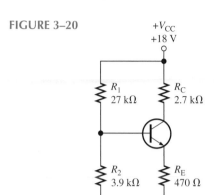

Solution   Begin by finding the base voltage using the voltage-divider rule.

$$V_B = \left(\frac{R_2}{R_1 + R_2}\right)V_{CC} = \left(\frac{3.9 \text{ k}\Omega}{27 \text{ k}\Omega + 3.9 \text{ k}\Omega}\right)18 \text{ V} = \mathbf{2.27 \text{ V}}$$

The emitter voltage is one diode drop less than the base voltage.

$$V_E = V_B - V_{BE} = 2.27 \text{ V} - 0.7 \text{ V} = \mathbf{1.57 \text{ V}}$$

Next, find the emitter current from Ohm's law.

$$I_E = \frac{V_E}{R_E} = \frac{1.57 \text{ V}}{470 \text{ }\Omega} = \mathbf{3.34 \text{ mA}}$$

Using the approximation $I_C \cong I_E$,

$$I_C = \mathbf{3.34 \text{ mA}}$$

Now find the collector voltage.

$$V_C = V_{CC} - I_C R_C = 18 \text{ V} - (3.34 \text{ mA})(2.7 \text{ k}\Omega) = 8.98 \text{ V}$$

The collector-emitter voltage is

$$V_{CE} = V_C - V_E = 8.98 \text{ V} - 1.57 \text{ V} = \mathbf{7.41 \text{ V}}$$

Practice Exercise   Find V_B, V_E, I_E, I_C, and V_{CE} for the circuit in Figure 3–20 if the power supply voltage was incorrectly set to +12 V.

Figure 3–21 shows two configurations for voltage-divider bias with a *pnp* transistor. As in the case of base bias, either negative or positive supply voltages can be used for bias. With a negative supply, shown in Figure 3–21(a), the voltage is applied to the collector. With a positive supply, shown in Figure 3–21(b), the voltage is applied to the emitter. The transistor is frequently drawn upside down to place the supply voltage on top; this means that the emitter resistor is also on top. The equations for *npn* transistors can be applied to *pnp* transistors, but you need to be careful of algebraic signs.

FIGURE 3–21
Voltage-divider bias for pnp transistors.

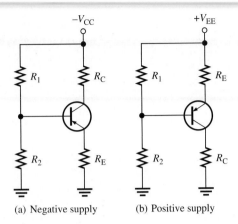

(a) Negative supply   (b) Positive supply

EXAMPLE 3–6   Find V_B, V_E, I_E, I_C, and V_{CE} for the *pnp* circuit in Figure 3–22.

FIGURE 3–22

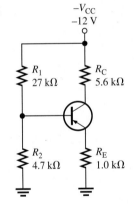

Solution   Begin by finding the base voltage using the voltage-divider rule.

$$V_B = \left(\frac{R_2}{R_1 + R_2}\right)V_{CC} = \left(\frac{4.7\ k\Omega}{27\ k\Omega + 4.7\ k\Omega}\right)(-12\ V) = -1.78\ V$$

The equation for V_E is the same one used for the *npn* transistor but note the signs. The emitter voltage is one diode drop *greater* than the base voltage for a forward-biased *pnp* transistor.

$$V_E = V_B - V_{BE} = -1.78 - (-0.7\ V) = -1.08\ V$$

Now find the emitter current using Ohm's law.

$$I_E = \frac{V_E}{R_E} = \frac{-1.08\ V}{1.0\ k\Omega} = -1.08\ mA$$

Using the approximation $I_C \cong I_E$,

$$I_C = -1.08\ mA$$

Now find the collector voltage.

$$V_C = V_{CC} - I_C R_C = -12\ V - (-1.08\ mA)(5.6\ k\Omega) = -5.96\ V$$

The collector-emitter voltage is

$$V_{CE} = V_C - V_E = -5.96 \text{ V} - (-1.08 \text{ V}) = \mathbf{-4.88 \text{ V}}$$

Notice that V_{CE} is negative for a *pnp* circuit.

Practice Exercise   Find V_B, V_E, I_E, I_C, and V_{CE} for the circuit in Figure 3–22 if R_E is changed to 1.2 kΩ.

Emitter Bias

Emitter bias is a very stable form of bias that uses both positive and negative power supplies and a single bias resistor that, in the usual configuration, puts the base voltage near ground potential. It is the type of bias used in most integrated circuit amplifiers.

Emitter bias circuits for *npn* and *pnp* configurations are shown in Figure 3–23. As in the other bias circuits, the key difference between the *npn* and *pnp* versions is that the polarity of the power supplies are opposite to each other.

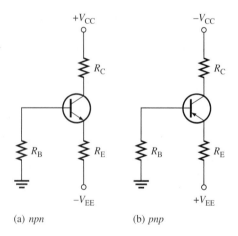

(a) *npn*   (b) *pnp*

FIGURE 3–23
Emitter bias circuits.

For stable bias, the base resistor is selected to drop only a few tenths of a volt. For the *npn* case, the emitter voltage is approximately −1 V due to the small drop across R_B and the forward-bias drop of the base-emitter junction of 0.7 V. For the *pnp* version, the emitter voltage is approximately +1 V. When troubleshooting, a quick check of the emitter voltage will reveal if the transistor is conducting and if the bias voltage is correct.

The emitter current is computed by applying Ohm's law to the emitter resistor. The approximation that $I_C \cong I_E$ is used to calculate the collector current and the collector voltage is again found by applying the following equation:

$$V_C = V_{CC} - I_C R_C$$

EXAMPLE 3–7   Find V_E, I_E, I_C, and V_{CE} for the emitter bias circuit in Figure 3–24.

FIGURE 3–24

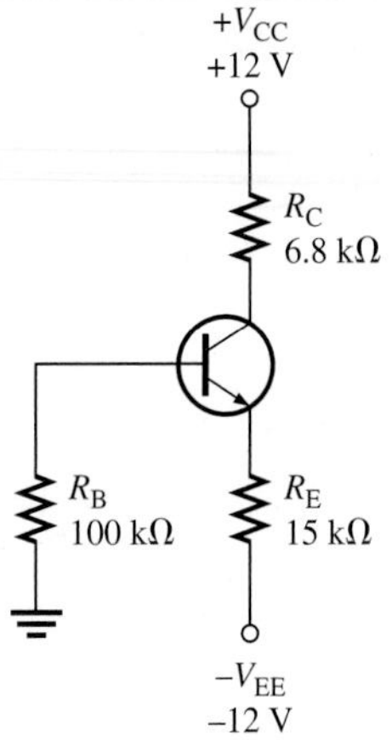

Solution   Start with the approximation $V_E \cong -1$ V. This implies the voltage across R_E is 11 V. Applying Ohm's law to the emitter resistor,

$$I_E = \frac{V_{R_E}}{R_E} = \frac{11 \text{ V}}{15 \text{ k}\Omega} = \textbf{0.73 mA}$$

The collector current is approximately equal to the emitter current.

$$I_C \cong \textbf{0.73 mA}$$

Now find the collector voltage.

$$V_C = V_{CC} - I_C R_C = 12 \text{ V} - (0.73 \text{ mA})(6.8 \text{ k}\Omega) = 7.0 \text{ V}$$

Find V_{CE} by subtracting V_E from V_C.

$$V_{CE} = V_C - V_E = 7.0 \text{ V} - (-1 \text{ V}) = \textbf{8.0 V}$$

Practice Exercise   Find V_E if the base of the transistor in Figure 3–24 were shorted to ground.

3–2 REVIEW QUESTIONS

1. Name the four types of bias circuits for BJTs.
2. What are the steps for finding V_{CE} with stiff voltage-divider bias?
3. What dc emitter voltage do you expect to find with emitter bias on a *pnp* transistor?

3–3 ■ DATA SHEET PARAMETERS AND AC CONSIDERATIONS

The backbone of analog electronics is the linear amplifier, a circuit that produces a larger signal that is a replica of a smaller one. In the last section, you saw how bias is used to provide the necessary dc conditions for the transistor to operate. In this section, we look at factors that affect the ac signal.

After completing this section, you should be able to

❏ Discuss transistor parameters and use them to analyze a transistor circuit
    ❏ Compare the notation used for dc and ac quantities
    ❏ Discuss principal characteristics given on manufacturer's data sheets for BJTs
    ❏ Explain the function of coupling and bypass capacitors
    ❏ Explain how an amplifier produces voltage gain

DC and AC Quantities

In the first part of this chapter, dc values were used to set up the operating conditions for transistors. These dc quantities of voltage and current used the standard italic capital letters with nonitalic capital-letter subscripts such as V_E, I_E, I_C, and V_{CE}. Lowercase italic subscripts are used to show ac quantities of rms, peak, and peak-to-peak voltages and currents such as V_e, I_e, I_c, and V_{ce}. Instantaneous quantities are indicated with both lowercase italic letters and subscripts such as v_e, i_e, i_c, and v_{ce}.

In addition to currents and voltages, resistances often have different values from an ac viewpoint compared to a dc viewpoint (see Section 1–1 for a review of dc versus ac resistance). Lowercase italic subscripts are used to identify ac resistance values. For example, R_C represents a dc collector resistance and R_c represents an ac collector resistance. You will see the need for this distinction as we discuss amplifiers. Internal resistances that are part of the transistor's equivalent circuit are written as lowercase italic letters (sometimes with a prime) and subscripts. For example, r'_e represents the internal ac emitter resistance and $R_{in(tot)}$ represents the total ac resistance that an amplifier presents to a signal source.

One parameter that is different for dc and ac circuits is β. The dc beta (β_{DC}) for a circuit was previously defined as the ratio of the collector current, I_C, to the base current, I_B. The **ac beta (β_{ac})** is defined as a small *change* in collector current divided by a corresponding *change* in base current. A changing quantity is written using ac notation and is a ratio of the collector current, I_c, to the base current, I_b (note the lowercase italic subscripts). On manufacturer's data sheets, β_{ac} is usually shown as h_{fe}. In equation form,

$$\beta_{ac} = \frac{I_c}{I_b} \tag{3–8}$$

The difference between β_{ac} and β_{DC} for a given transistor is normally quite small and is due to small nonlinearities in the characteristic curves. For almost all designs, these differences are not important but should be understood when reading data sheets.

Manufacturer's Data Sheet

Figure 3–25 shows a partial data sheet for the 2N3903 and 2N3904 *npn* transistors. Notice that the maximum collector-emitter voltage (V_{CEO}) is 40 V. The "O" in the subscript indicates that the voltage is measured from collector (C) to emitter (E) with the base open (O).

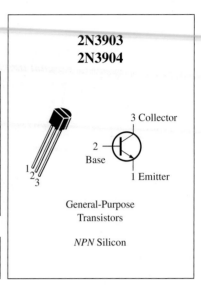

2N3903
2N3904

3 Collector

2
Base

1 Emitter

General-Purpose
Transistors

NPN Silicon

Maximum Ratings

Rating	Symbol	Value	Unit
Collector-Emitter voltage	V_{CEO}	40	V dc
Collector-Base voltage	V_{CBO}	60	V dc
Emitter-Base voltage	V_{EBO}	6.0	V dc
Collector current — continuous	I_C	200	mA dc
Total device dissipation @ $T_A = 25°C$ Derate above 25°C	P_D	625 5.0	mW mW/C°
Total device dissipation @ $T_C = 25°C$ Derate above 25°C	P_D	1.5 12	Watts mW/C°
Operating and storage junction Temperature range	T_J, T_{stg}	−55 to +150	°C

Thermal Characteristics

Characteristic	Symbol	Max	Unit
Thermal resistance, junction to case	$R_{\theta JC}$	83.3	C°/W
Thermal resistance, junction to ambient	$R_{\theta JA}$	200	C°/W

Electrical Characteristics ($T_A = 25°C$ unless otherwise noted.)

Characteristic		Symbol	Min	Max	Unit
OFF Characteristics					
Collector-Emitter breakdown voltage ($I_C = 1.0$ mA dc, $I_B = 0$)		$V_{(BR)CEO}$	40	–	V dc
Collector-Base breakdown voltage ($I_C = 10$ µA dc, $I_E = 0$)		$V_{(BR)CBO}$	60	–	V dc
Emitter-Base breakdown voltage ($I_E = 10$ µA dc, $I_C = 0$)		$V_{(BR)EBO}$	6.0	–	V dc
Base cutoff current ($V_{CE} = 30$ V dc, $V_{EB} = 3.0$ V dc)		I_{BL}	–	50	nA dc
Collector cutoff current ($V_{CE} = 30$ V dc, $V_{EB} = 3.0$ V dc)		I_{CEX}	–	50	nA dc
ON Characteristics					
DC current gain		h_{FE}			–
($I_C = 0.1$ mA dc, $V_{CE} = 1.0$ V dc)	2N3903 2N3904		20 40	– –	
($I_C = 1.0$ mA dc, $V_{CE} = 1.0$ V dc)	2N3903 2N3904		35 70	– –	
($I_C = 10$ mA dc, $V_{CE} = 1.0$ V dc)	2N3903 2N3904		50 100	150 300	
($I_C = 50$ mA dc, $V_{CE} = 1.0$ V dc)	2N3903 2N3904		30 60	– –	
($I_C = 100$ mA dc, $V_{CE} = 1.0$ V dc)	2N3903 2N3904		15 30	– –	

Collector-Emitter saturation voltage
    ($I_C = 10$ mA dc, $I_B = 1.0$ mA dc)
    ($I_C = 50$ mA dc, $I_B = 5.0$ mA dc)
		$V_{CE(sat)}$	–
			–
				0.2
				0.3
					V dc

Base-Emitter saturation voltage
    ($I_C = 10$ mA dc, $I_B = 1.0$ mA dc)
    ($I_C = 50$ mA dc, $I_B = 5.0$ mA dc)
		$V_{BE(sat)}$	0.65
			–
				0.85
				0.95
					V dc

FIGURE 3–25
Partial data sheet for 2N3903 and 2N3904 npn transistors.

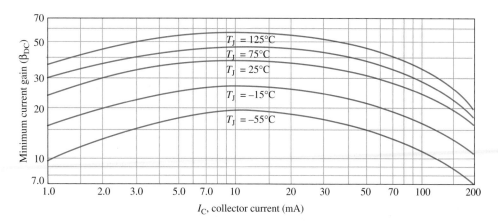

FIGURE 3–26
Variation of β_{DC} with I_C for several temperatures.

In this textbook, we use $V_{CE(max)}$ for clarity. Also notice that the maximum collector current is 200 mA.

On this data sheet, the dc current gain (β_{DC}) is given as h_{FE}. Minimum values of h_{FE} are listed on the data sheet under *ON Characteristics*. Note that the dc current gain is not really a constant but varies with the collector current. The current gain also changes with temperature as shown in Figure 3–26. The three variables, β_{DC}, I_C, and temperature, are plotted with a family of curves for a typical transistor. Keeping the junction temperature constant and increasing I_C causes β_{DC} to increase gradually to a maximum. A further increase in I_C beyond this maximum point causes β_{DC} to decrease. If I_C is held constant and the temperature is varied, β_{DC} changes directly with the temperature.

A transistor data sheet usually specifies β_{DC} at specific I_C values. Even at fixed values of I_C and temperature, β_{DC} varies from device to device for a given transistor. The β_{DC} specified at a certain value of I_C is usually the minimum value, $\beta_{DC(min)}$, although the maximum and typical values are sometimes specified.

The dc power dissipated in any component is the product of the current and voltage. For a transistor, the product of V_{CE} and I_C gives the power dissipated by the transistor. Like any other electronic device, the transistor has limitations on its operation. These limitations are stated in the form of maximum ratings and are normally specified on the manufacturer's data sheet. Typically, maximum ratings are given for collector-to-emitter voltage (V_{CE}), collector-to-base voltage (V_{CB}), emitter-to-base voltage (V_{EB}), collector current (I_C), and power dissipation (P_D). The product of V_{CE} and I_C must not exceed the maximum power dissipation specification. Both V_{CE} and I_C cannot be at their individual maximum values at the same time. If V_{CE} is maximum, I_C can be calculated as

$$I_C = \frac{P_{D(max)}}{V_{CE}}$$

If I_C is maximum, V_{CE} can be calculated as

$$V_{CE} = \frac{P_{D(max)}}{I_C}$$

For a given transistor, a maximum power dissipation curve can be plotted on the collector characteristic curves, as shown in Figure 3–27(a). These values are tabulated in Figure 3–27(b). For this transistor, $P_{D(max)}$ is 500 mW, $V_{CE(max)}$ is 20 V, and $I_{C(max)}$ is 50 mA.

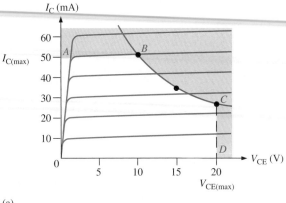

$P_{D(max)}$	V_{CE}	I_C
500 mW	10 V	50 mA
500 mW	15 V	33 mA
500 mW	20 V	25 mA

(a)

(b)

FIGURE 3–27
Maximum power dissipation curve.

The curve shows that this particular transistor cannot be operated in the shaded portion of the graph. $I_{C(max)}$ is the limiting rating between points *A* and *B*, $P_{D(max)}$ is the limiting rating between points *B* and *C*, and $V_{CE(max)}$ is the limiting rating between points *C* and *D*.

EXAMPLE 3–8

The transistor in Figure 3–28 has the following maximum ratings: $P_{D(max)} = 800$ mW, $V_{CE(max)} = 15$ V, $I_{C(max)} = 100$ mA, $V_{CB(max)} = 20$ V, and $V_{EB(max)} = 10$ V. Determine the maximum value to which V_{CC} can be adjusted without exceeding a rating. Which rating would be exceeded first?

FIGURE 3–28

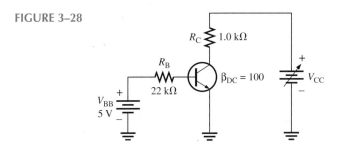

Solution First, find I_B so that you can determine I_C.

$$I_B = \frac{V_{BB} - V_{BE}}{R_B} = \frac{5 \text{ V} - 0.7 \text{ V}}{22 \text{ k}\Omega} = 195 \text{ }\mu\text{A}$$

$$I_C = \beta_{DC}I_B = (100)(195 \text{ }\mu\text{A}) = 19.5 \text{ mA}$$

I_C is much less than $I_{C(max)}$ and will not change with V_{CC}. It is determined only by I_B and β_{DC}.

The voltage drop across R_C is

$$V_{R_C} = I_C R_C = (19.5 \text{ mA})(1.0 \text{ k}\Omega) = 19.5 \text{ V}$$

Now you can determine the maximum value of V_{CC} when $V_{CE} = V_{CE(max)} = 15$ V.

$$V_{R_C} = V_{CC} - V_{CE}$$

Thus,

$$V_{CC(max)} = V_{CE(max)} + V_{R_C} = 15 \text{ V} + 19.5 \text{ V} = \mathbf{34.5 \text{ V}}$$

V_{CC} can be increased to 34.5 V, under the existing conditions, before $V_{CE(max)}$ is exceeded. However, at this point it is not known whether or not $P_{D(max)}$ has been exceeded.

$$P_D = V_{CE(max)}I_C = (15 \text{ V})(19.5 \text{ mA}) = 293 \text{ mW}$$

Since $P_{D(max)}$ is 800 mW, it is *not* exceeded when $V_{CC} = 34.5$ V. Thus, $V_{CE(max)} = 15$ V is the limiting rating in this case. If the base current is removed causing the transistor to turn off, $V_{CE(max)}$ will be exceeded because the entire supply voltage, V_{CC}, will be dropped across the transistor.

Practice Exercise Assume the transistor in Figure 3–28 has the following maximum ratings: $P_{D(max)} = 500$ mW, $V_{CE(max)} = 25$ V, $I_{C(max)} = 200$ mA, $V_{CB(max)} = 30$ V, $V_{EB(max)} = 15$ V. Determine the maximum value to which V_{CC} can be adjusted without exceeding a rating. Which rating would be exceeded first?

AC and DC Equivalent Circuits

In Section 3–2, you solved the dc bias conditions necessary to set the operating conditions for the transistor. The first step in analyzing or troubleshooting any transistor amplifier is to find the dc conditions. After checking that the dc voltages are correct, the next step is to check ac signals. The equivalent ac circuit is quite different from the dc circuit. For example, a capacitor prevents dc from passing; thus, it appears as an open circuit to dc but looks like a short circuit to most ac signals. For this reason, you need to be able to look at the dc and ac equivalent circuits quite differently.

Recall from your dc/ac circuits course that the superposition principle allows you to find the voltage or current anywhere in a linear circuit due to a single voltage or current source acting alone. This is done by reducing all other sources to zero. To compute ac parameters, reduce the dc power supply to zero by mentally replacing it with a short and computing the ac parameters as if they were acting alone. Replacing the power supply with a short means that V_{CC} is actually at ground potential for the ac signal. This is called an *ac ground*. The concept of a ground point that is an ac signal ground but not a dc ground may be new to you. Just remember that an ac ground is a common reference point for the ac signal.

Coupling and Bypass Capacitors

A basic BJT amplifier is shown in Figure 3–29. The difference between this circuit and the one in Figure 3–19 is the addition of an ac signal source, three capacitors, and a load resistor. In addition, the emitter resistor is divided into two resistors.

The ac signal is brought into and out of the amplifier through series capacitors (C_1 and C_3) called **coupling capacitors**. As mentioned previously, a capacitor normally appears as a short to the ac signal and an open to dc. This means that coupling capacitors can pass the ac signal while simultaneously blocking the dc voltage. The input coupling capacitor, C_1, passes the ac signal from the source to the base while isolating the source from the dc bias voltage. The output coupling capacitor, C_3, passes the signal on to the load while isolating the load from the power supply voltage. Notice that the coupling capacitors are in series with the signal path.

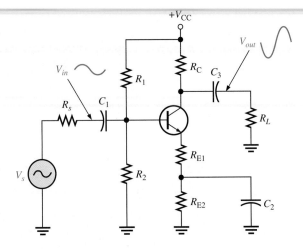

FIGURE 3–29
A basic transistor amplifier.

Capacitor C_2 is different; it is connected in parallel with one of the emitters resistors. This causes the ac signal to *bypass* the emitter resistor; thus, it is called a **bypass capacitor**. The purpose of the bypass capacitor is to increase the gain of the amplifier for reasons you will learn later. Since the bypass capacitor is an ac short, *both* ends of the capacitor are at ac ground. Whenever one side of a capacitor is connected to ground, the other side is also a ground to the ac signal. Remember this if you are troubleshooting; you shouldn't find an ac signal on either side of a bypass capacitor. If you do, the capacitor may be open.

Amplification

The signal source, V_s, shown in Figure 3–29 causes variations in base current which, in turn, cause the much larger emitter and collector currents to vary about the Q-point in phase with the base current. However, when the collector current *increases*, the collector voltage *decreases* and vice-versa. Thus, the sinusoidal collector-to-emitter voltage varies above and below its Q-point value 180° out of phase with the base voltage. A transistor always inverts the signal between the base and the collector. Amplification occurs because a small swing in base current produces a large variation in collector voltage.

3–3 REVIEW QUESTIONS

1. Does the β_{DC} of a transistor increase or decrease with temperature?
2. Generally, what effect does an increase in I_C have on the β_{DC}?
3. How can you find the dc power dissipated in a transistor?
4. What is the allowable collector current in a transistor with $P_{D(max)} = 320$ mW when $V_{CE} = 8$ V?
5. Explain the difference between a coupling capacitor and a bypass capacitor.

3–4 ■ COMMON-EMITTER AMPLIFIERS

The common-emitter (CE) is a type of BJT amplifier configuration in which the emitter is the reference for the input and output signals. In this section, a specific CE amplifier is introduced and used to illustrate certain ac parameters.

After completing this section, you should be able to

❑ Understand and analyze the operation of common-emitter amplifiers
  ❑ Draw the equivalent ac circuit for a CE amplifier
  ❑ Compute the voltage gain and the input and output resistances for a CE amplifier
  ❑ Discuss why the ac load line differs from the dc load line
  ❑ Draw the ac load line for a CE amplifier and find the Q-point

The **common-emitter (CE)** amplifier, the most widely used type of BJT amplifier, has the emitter as the reference terminal for the input and output signal. Figure 3–30(a) shows a CE amplifier that produces an amplified and inverted output signal at the load resistor. The input signal, V_{in}, is capacitively coupled to the base through C_1, causing the base current to vary above and below its dc bias value. This variation in base current produces a corresponding variation in collector current. The variation in collector current is much larger than the variation in base current because of the current gain through the transistor. This produces a larger variation in the collector voltage which is out of phase with the base signal voltage. This variation in collector voltage is then capacitively coupled to the load and appears as the output voltage, V_{out}.

Now let's look closely at the amplifier in Figure 3–30(a) and examine its dc and ac parameters. The dc parameters were worked out in Example 3–5 (Figure 3–20) and the method is briefly reviewed here. Notice that the original 470 Ω emitter resistor is now composed of two series resistors, R_{E1} and R_{E2}, that add to the same 470 Ω. This has no effect on the dc currents or voltages, but because of the presence of the bypass capacitor, C_2, the ac resistance of the emitter circuit is different.

There is voltage-divider bias, so first find the dc base voltage by applying the voltage-divider rule. The emitter voltage is 0.7 V less than the base voltage due to the base-emitter

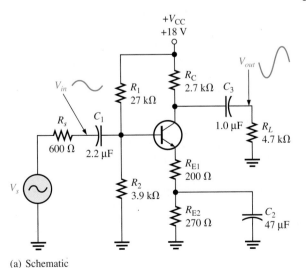

(a) Schematic

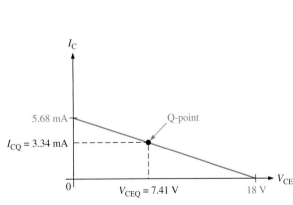

(b) DC load line

FIGURE 3–30
A basic common-emitter amplifier.

diode drop. Next, find the emitter current by applying Ohm's law to the emitter resistor. This calculates to 3.34 mA of emitter current, which is approximately the same as the collector current; therefore, the voltage drop across R_C can also be found by Ohm's law. The results from Example 3–5 showed that V_C is 8.98 V and that V_{CE} is 7.41 V. Recall that I_C and V_{CE} define the Q-point for the circuit. Because these are the values of I_C and V_{CE} at the Q-point, they are given special labels: I_{CQ} and V_{CEQ}, respectively.

A graphical picture of the parameters just reviewed may help you visualize the dc parameters. You can determine the load line by finding the saturation current and the collector-emitter cutoff voltage for the circuit. Recall that the saturation current is the current when the collector-to-emitter voltage is approximately zero. Thus,

$$I_{C(sat)} = \frac{V_{CC}}{R_C + R_{E1} + R_{E2}} = \frac{18 \text{ V}}{2.7 \text{ k}\Omega + 200 \text{ }\Omega + 270 \text{ }\Omega} = 5.68 \text{ mA}$$

At cutoff, there is no current, so the entire supply voltage, V_{CC}, is across the collector to emitter. These two points, saturation and cutoff, allow you to construct the dc load line, as shown in Figure 3–30(b). All possible operating points, with no ac signal, are shown. The Q-point is located on the load line using the earlier calculation.

The AC Equivalent Circuit

Recall that the ac signal "sees" a different circuit than does the dc source for several reasons. If you apply the superposition theorem to the circuit in Figure 3–30(a) and show the capacitors as shorts, you can redraw the CE amplifier from the perspective of the ac signal. This is shown in Figure 3–31. The power supply has been replaced with an ac ground, shown in color. The capacitors have been replaced with shorts, and R_{E2}, is eliminated because it is bypassed with C_2.

FIGURE 3–31
AC equivalent circuit for Figure 3–30(a).

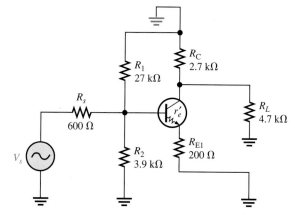

The ac equivalent circuit also shows an internal resistance in the base-emitter diode (using the offset-resistance model described in Section 2–4). This internal resistance, called r'_e, plays a role in the gain and input impedance of the amplifier so is generally included in the ac equivalent circuit. Because it is an ac resistance, it is sometimes called the **dynamic emitter resistance**. The value of this ac resistance is related to the dc emitter current as follows:

$$r'_e = \frac{25 \text{ mV}}{I_E} \tag{3–9}$$

The derivation of this formula is in Appendix B.

EXAMPLE 3–9

Find the dynamic emitter resistance, r'_e, for the circuit in Figure 3–30(a).

Solution   The emitter current was found to be 3.34 mA (see Example 3–5). Substituting into Equation (3–9),

$$r'_e = \frac{25 \text{ mV}}{I_E} = \frac{25 \text{ mV}}{3.34 \text{ mA}} = \textbf{7.5 } \boldsymbol{\Omega}$$

Practice Exercise   Compute r'_e for a transistor with an emitter current of 100 μA.

Voltage Gain

The voltage gain, A_v, of the CE amplifier is the ratio of the output signal voltage to the input signal voltage, V_{out}/V_{in}. The output voltage, V_{out}, is measured at the collector and the input voltage, V_{in}, is measured at the base. Because the base-emitter junction is forward-biased, the signal voltage at the emitter is approximately equal to the signal voltage at the base. Thus, since $V_b = V_e$, the voltage gain is

$$A_v = -\frac{V_c}{V_e} = -\frac{I_c R_c}{I_e R_e}$$

Since $I_c \cong I_e$, the voltage gain reduces to the ratio of ac collector resistance to ac emitter resistance.

$$A_v \cong -\frac{R_c}{R_e} \tag{3–10}$$

The negative sign in the gain formula is added to indicate inversion, meaning the output signal is out of phase with the input signal. Note that the gain is written as a ratio of two ac resistances; you will see this idea again with the other amplifier configurations.

The gain formula is a useful and simple way to quickly determine what the voltage gain of a CE amplifier should be. When you're troubleshooting, you need to know what signal to expect; remember that the collector and emitter resistances used in calculating gain are the *total* ac resistance. The following summarizes these ideas:

❑ *The emitter ac circuit*   In the emitter circuit, you need to include the internal base-emitter diode resistance (r'_e) and the fixed resistor that is not bypassed with a capacitor. The internal r'_e appears to be in series with the unbypassed emitter resistance in the ac emitter circuit. Incidentally, this unbypassed resistor, shown as R_{E1} in Figure 3–30(a), serves an important role in determining the gain and keeping the gain stable. It also increases the input resistance of the amplifier as you will see later. Sometimes it is called a *swamping resistor* because it tends to "swamp" out the uncertain value of r'_e.

❑ *The collector ac circuit*   From the vantage point of the collector, the collector resistor and the load resistor appear to be in parallel. Thus the ac resistance, R_c, of the collector is simply $R_C \parallel R_L$. An example should clarify this.

EXAMPLE 3–10

Find A_v for the circuit in Figure 3–30(a).

Solution The ac resistance in the emitter circuit, R_e, is composed of r'_e in series with the unbypassed R_{E1}. From Example 3–9, $r'_e = 7.5\ \Omega$. Therefore,

$$R_e = r'_e + R_{E1} = 7.5\ \Omega + 200\ \Omega = 207.5\ \Omega$$

Next, find the ac resistance as viewed from the transistor's collector.

$$R_c = R_C \parallel R_L = 2.7\ k\Omega \parallel 4.7\ k\Omega = 1.71\ k\Omega$$

Substituting into Equation (3–10),

$$A_v \cong -\frac{R_c}{R_e} = -\frac{1.71\ k\Omega}{207.5\ \Omega} = -8.3$$

Again, the negative sign is used to show that the amplifier inverts the signal.

Practice Exercise Assume the bypass capacitor in Figure 3–30(a) were open. What effect would this have on the gain?

Input Resistance

The input resistance for an amplifier, $R_{in(tot)}$, (called the input impedance when capacitive or inductive effects are included) was introduced in Section 1–4 and Figure 1–19. It is an ac parameter that acts like a load in series with the source resistance. As long as the input resistance is high compared to the source resistance, most of the voltage will appear at the input and the loading effect is small. If the input resistance is small compared to the source resistance, most of the source voltage will drop across its own series resistance, leaving little for the amplifier to amplify.

One of the problems with the CE amplifier is that the input resistance is dependent on β_{ac}. As you have seen, this parameter is highly variable, so you can't calculate input resistance *exactly* for a given amplifier without knowing the β_{ac}. Despite this, you can minimize the effect of β_{ac} and increase the total input resistance by adding a swamping resistor to the emitter circuit. You can then obtain a reasonable estimate of the input resistance, which for most purposes will enable you to determine if a given amplifier is appropriate for the job at hand.

The input circuit for the CE amplifier in Figure 3–30(a) has been redrawn in Figure 3–32 to eliminate the output circuit. R_C is not part of the input circuit because of the reverse-biased base-collector junction. For the input ac signal, there are three parallel paths to ground. Looking in from the source, the three paths consist of R_1, R_2, and a path through the transistor's base-emitter circuit. It is these three parallel paths that comprise the input resistance of the circuit. We define this resistance as $R_{in(tot)}$ because it represents the total input resistance including the bias resistors. The base-emitter path, however, has β_{ac} dependency because of the transistor's current gain. The equivalent resistances, R_{E1} and r'_e, appear to be larger in the base circuit than in the emitter circuit because of this current gain. The resistors in the emitter circuit must be multiplied by β_{ac} to obtain their equivalent resistance in the base circuit. Therefore, the formula for calculating total input resistance is

$$R_{in(tot)} = R_1 \parallel R_2 \parallel [\beta_{ac}(r'_e + R_{E1})] \tag{3–11}$$

FIGURE 3–32
Equivalent ac input circuit for the CE amplifier in Figure 3–30(a).

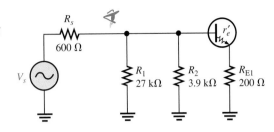

EXAMPLE 3–11

Find $R_{in(tot)}$ for the circuit in Figure 3–30(a). Assume the β_{ac} is 120.

Solution   The internal ac emitter resistance, r'_e, was found to be 7.5 Ω in Example 3–9. Substituting into Equation (3–11),

$$R_{in(tot)} = R_1 \parallel R_2 \parallel [\beta_{ac}(r'_e + R_{E1})]$$
$$= 27 \text{ k}\Omega \parallel 3.9 \text{ k}\Omega \parallel [120 (7.5 \ \Omega + 200 \ \Omega)] = \textbf{3.0 k}\boldsymbol{\Omega}$$

Practice Exercise   Compute $R_{in(tot)}$ for the circuit in Figure 3–30(a) if β_{ac} is 200.

Output Resistance

Recall that the model for an amplifier (Section 1–4) includes a Thevenin voltage source driving a series resistance or a Norton current source driving a parallel resistance. In both of these models, the resistance is the same. It is the equivalent output resistance of the amplifier.

To find the output resistance of any CE amplifier, look back from the output coupling capacitor as illustrated in Figure 3–33. The transistor appears as a current source in parallel with the collector resistor. This is the same as the equivalent Norton circuit in Figures 1–11 and 1–19(b).

FIGURE 3–33
Equivalent ac output circuit for the CE amplifier.

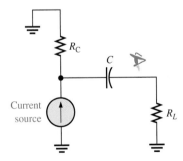

Recall that the internal resistance of an ideal current source is infinite. With this in mind, you can see that the output resistance for the CE amplifier is simply the collector resistance, R_C.

The AC Load Line

For most troubleshooting work, it is useful to be able to quickly estimate a circuit's voltage and current values. Although technicians seldom use them in their normal work, load lines are a useful tool for understanding a transistor's operation and may offer insight into a circuit's limitations, such as clipping levels.

As discussed in Section 3–1, a dc load line can be drawn for a basic transistor circuit that consists of a series collector resistor, R_C, and a voltage source, V_{CC}. As shown in Figure 3–12(a), this series combination formed a Thevenin circuit that was represented graphically by a dc load line that crossed the y-axis at saturation. Recall that the load line in Figure 3–12(b) was independent of the transistor, which served as the load.

For ac, the Thevenin resistance is more complicated because of the presence of capacitors and an internal emitter resistance, r_e'. In high-frequency circuits, inductors may also play a role. Even though r_e' is internal to the transistor, it is considered part of the Thevenin resistance. Capacitive coupling and bypass capacitors are also present in most practical ac circuits. Capacitors are normally treated as shorts for the ac signal, meaning that the ac resistance (R_{ac}) of the collector-emitter circuit is reduced. Example 3–12 will illustrate this concept.

A dc and an ac load line are shown together in Figure 3–34 for a capacitively coupled amplifier. The Q-point is the same for both load lines because when the ac signal is reduced to zero, operation must still occur at the Q-point. Notice that the ac saturation current is greater than the dc saturation current (because the ac resistance is smaller). In addition, the ac collector-emitter cutoff voltage is less than the dc collector-emitter cutoff voltage. The ac load line locates all possible operation points (collector current versus collector-emitter voltage) for the ac signal.

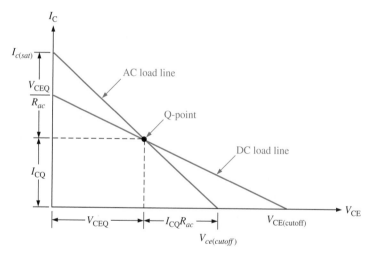

FIGURE 3–34
The dc and ac load lines.

The ac saturation and ac cutoff points can be computed for the ac load line. The ac load line crosses the y axis at $I_{c(sat)}$. This point is found by starting at the dc Q-point (I_{CQ}) and adding a term that includes the ac resistance of the collector-emitter circuit, R_{ac}, as shown on Figure 3–34. The equation for ac saturation is

$$I_{c(sat)} = I_{CQ} + \frac{V_{CEQ}}{R_{ac}}$$

The ac load line touches the x axis at $V_{ce(cutoff)}$. This point is also found by starting at the dc Q-point (V_{CEQ}) and adding a term that includes the ac resistance, R_{ac}. The equation for ac cutoff is

$$V_{ce(cutoff)} = V_{CEQ} + I_{CQ}R_{ac}$$

EXAMPLE 3–12   Draw the ac load line for the circuit in Figure 3–30(a).

Solution   The dc load line for this circuit was shown in Figure 3–30(b) and is shown in Figure 3–35 for reference. The Q-point coordinates are V_{CEQ} = 7.41 V and I_{CQ} = 3.34 mA.

Before locating the ac load line, it is necessary to find the ac resistance of the collector-emitter circuit. As you know, the emitter circuit has r'_e + R_{E1} in series. The collector circuit has the parallel combination of $R_C \parallel R_L$. The total ac resistance of the collector-emitter circuit is

$$R_{ac} = r'_e + R_{E1} + (R_C \parallel R_L)$$

In Example 3–9, r'_e was found to be 7.5 Ω. Substituting this value and the other fixed resistors into the previous equation results in

$$R_{ac} = 7.5\ \Omega + 200\ \Omega + (2.7\ k\Omega \parallel 4.7\ k\Omega) = 1.92\ k\Omega$$

Now, find the ac collector saturation current.

$$I_{c(sat)} = I_{CQ} + \frac{V_{CEQ}}{R_{ac}} = 3.34\ mA + \frac{7.41\ V}{1.92\ k\Omega} = 7.20\ mA$$

Next, find the ac collector-emitter cutoff voltage.

$$V_{ce(cutoff)} = V_{CEQ} + I_{CQ}R_{ac} = 7.41\ V + (3.34\ mA)(1.92\ k\Omega) = 13.8\ V$$

Together, the ac collector saturation current, the Q-point, and the ac collector-emitter cutoff voltage establish a straight line. The ac load line can now be drawn and is shown in Figure 3–35.

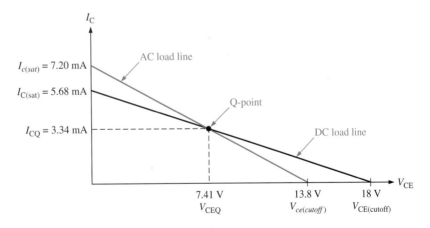

FIGURE 3–35
DC and ac load lines for the circuit in Figure 3–30(a).

Practice Exercise   What happens to the Q-point and the ac load line if the load resistor is changed from 4.7 kΩ to 2.7 kΩ?

One interesting way of viewing the operation of an amplifier is to superimpose a set of characteristic curves for the transistor on the ac load line. This is shown in Figure 3–36 for a typical transistor that could be used with the CE amplifier from Figure 3–30(a). Lines projected from the peaks of the base current across to the I_C axis and lines from the ac load line down to the V_{CE} axis indicate the peak-to-peak variations of the collector current and collector-to-emitter voltage, as shown. For the transistor in this example, if an input signal causes the base current to vary from approximately 13 μA to 18 μA, the output collector current will vary from approximately 2.9 mA to 3.9 mA. In addition, V_{CE} varies from approximately 6.3 V to 8.1 V for this same signal. The ac load line also gives a quick visual indication when the signal exceeds the linear range of the amplifier and shows the current and voltage range that a particular signal will encompass.

FIGURE 3–36

AC load line superimposed on a typical transistor characteristic.

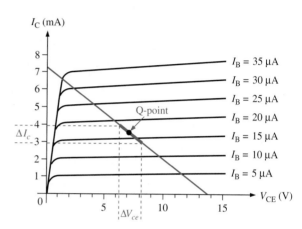

3–4 REVIEW QUESTIONS

1. Which terminal of a CE amplifier is the input terminal? Which is the output terminal?
2. What is the advantage of a high input resistance in an amplifier?
3. How is the gain determined in a CE amplifier?

3–5 ■ COMMON-COLLECTOR AMPLIFIERS

The common-collector (CC) amplifier, commonly referred to as an emitter-follower, is the second of the three basic amplifier configurations. The input is applied to the base and the output is at the emitter. The CC amplifier provides current gain; the voltage gain is approximately 1. It is frequently used as a buffer or driver because of its high input resistance.

After completing this section, you should be able to

❏ Understand and analyze the operation of common-collector amplifiers
  ❏ Draw the equivalent ac circuit for a CC amplifier
  ❏ Explain why the voltage gain for a CC amplifier is approximately 1

❑ Compute the current gain and the input and output resistances for a CC amplifier
❑ Explain why a darlington pair has a very high β

Figure 3–37(a) shows a **common-collector (CC)** circuit with a voltage-divider bias. The collector is connected directly to the dc power supply which is an ac ground. Notice that the input is applied to the base and the output is taken from the emitter. The output signal is in phase with the input signal. Looking from the input coupling capacitor to the base, the equivalent ac circuit has the bias resistors and the resistors in the emitter circuit as shown in Figure 3–37(b).

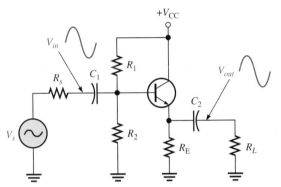

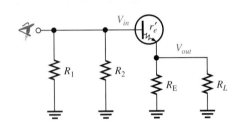

(a) Typical common-collector (CC) or emitter-follower amplifier

(b) Equivalent ac circuit

FIGURE 3–37

Voltage Gain

The ac circuit of Figure 3–37(b) can be simplified by combining the parallel emitter and load resistors into one equivalent resistor ($R_E \parallel R_L$), as shown in Figure 3–38. This circuit is used to analyze the voltage gain of the CC amplifier.

FIGURE 3–38
Equivalent ac input circuit for the CC amplifier.

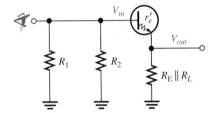

As in all linear amplifiers, the voltage gain in a CC amplifier is $A_v = V_{out}/V_{in}$. In the analysis of the gain, the bias resistors are not included because they do not directly affect the input signal (although they do cause a loading effect on the source). Notice in Figure 3–38 that the input voltage is across r'_e in series with $R_E \parallel R_L$. The output is across only $R_E \parallel R_L$. As long as r'_e is small compared to $R_E \parallel R_L$ (the usual case), you can ignore the small drop across r'_e. This means that the input and output voltages are nearly the same. Therefore,

$$A_v \cong 1$$

(3–12)

Because of the small drop across r'_e, the actual gain is slightly less than 1. For practical circuits, this difference is not important. If you are checking the input and output of a CC amplifier with an oscilloscope, expect to see nearly identical signals. Since the output voltage on the emitter follows the input voltage, the CC amplifier is often called an *emitter-follower*. There is no phase inversion in a CC amplifier.

You might wonder, if the CC amplifier has unity voltage gain, what is its value? The answer is that it has current gain. CC amplifiers are used when it is necessary to drive a low-impedance load, such as a speaker. In order to solve for current gain, you need to first analyze the input and output ac resistance.

Input Resistance

The emitter-follower is characterized by a high input resistance, which makes it a very useful circuit. Because of the high input resistance, the emitter-follower can be used as a buffer to minimize loading effects when one circuit is driving another. The derivation of the input resistance viewed from the base is similar to that for the CE amplifier. Looking in from the source, the CC amplifier has the same parallel paths as in the CE amplifier with voltage-divider bias, as shown in the equivalent circuit in Figure 3–38. In this case, however, there are no bypass capacitors in the emitter circuit. The total input resistance has a similar equation as in the CE case but with a different ac emitter resistance ($R_E \parallel R_L$).

$$R_{in(tot)} = R_1 \parallel R_2 \parallel [\beta_{ac}(r'_e + R_E \parallel R_L)] \qquad (3\text{–}13)$$

In most practical circuits, r'_e is much smaller than $R_E \parallel R_L$ and is ignored in the calculation. Further, the ac resistance of the transistor's emitter circuit is generally much larger than the paths through the bias resistors (because of β_{ac}). For a quick approximation of the total input resistance, you can just find the equivalent resistance of R_1 in parallel with R_2.

Output Resistance

The equivalent CC amplifier output circuit is shown in Figure 3–39 with the perspective of looking back from the output coupling capacitor. Resistor R_{base} represents the source and bias resistors in the base circuit. From the vantage point of the emitter circuit, these appear to be very small (their value is divided by β_{ac}). In practical circuits, they can be ignored; from the emitter's perspective, the base appears to be nearly at ac ground. This leaves only r'_e in parallel with R_E. Since R_E is much larger than r'_e, it also can be ignored.[1] For basic analysis purposes, the output resistance for the CC amplifier is quite simple—it's just r'_e!

FIGURE 3–39
Equivalent ac output circuit for the CC amplifier.

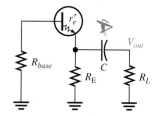

---

[1] Notice that the small equivalent series base resistor tends to be cancelled by the larger parallel emitter resistor, justifying the simplifying assumptions.

Current Gain

The signal current gain for the emitter-follower is I_{load}/I_s where I_{load} is the ac current in the load resistor and I_s is the ac signal current from the source. The current I_s is calculated using Ohm's law as $V_{in}/R_{in(tot)}$. Since the voltage gain is approximately 1, the input voltage is also across the load. Thus, the load current is V_{in}/R_L. Taking the ratio of the currents results in the current gain.

$$A_i = \frac{I_{load}}{I_s} = \frac{V_{in}/R_L}{V_{in}/R_{in(tot)}}$$

$$A_i = \frac{R_{in(tot)}}{R_L} \qquad \qquad (3\text{--}14)$$

This is a useful result and shows that the current gain, A_i, for the loaded CC amplifier is a ratio of the total input resistance to the load resistance. As noted with the earlier voltage gain equations, current gain can also be written as a ratio of resistances.

EXAMPLE 3–13   Determine the total input resistance, $R_{in(tot)}$, and the approximate voltage gain and current gain to the load of the emitter-follower in Figure 3–40. Assume the β_{ac} is 140.

FIGURE 3–40

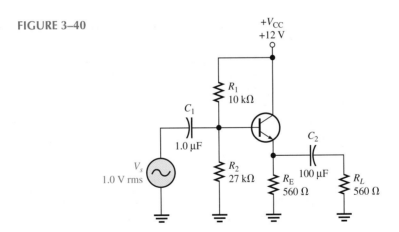

Solution   Although r'_e can be ignored for the calculation of the total input resistance, it is useful to review the method for finding r'_e. The value of r'_e is determined from I_E, so the first step is to find the dc conditions. The base voltage is found from the voltage-divider rule.

$$V_B = \left(\frac{R_2}{R_1 + R_2}\right)V_{CC} = \left(\frac{27 \text{ k}\Omega}{10 \text{ k}\Omega + 27 \text{ k}\Omega}\right)12 \text{ V} = 8.76 \text{ V}$$

The emitter voltage is approximately $V_B - V_{BE} = 8.06$ V. The emitter current is found from Ohm's law.

$$I_E = \frac{V_E}{R_E} = \frac{8.06 \text{ V}}{560 \text{ }\Omega} = 14.4 \text{ mA}$$

The value of r'_e is

$$r'_e = \frac{25 \text{ mV}}{I_E} = \frac{25 \text{ mV}}{14.4 \text{ mA}} = 1.7 \text{ }\Omega$$

Since this value is small compared to the emitter and load resistors, it can be ignored.
The total input resistance is

$$R_{in(tot)} = R_1 \parallel R_2 \parallel [\beta_{ac}(R_E \parallel R_L)]$$
$$= 10 \text{ k}\Omega \parallel 27 \text{ k}\Omega \parallel [140(560 \text{ }\Omega \parallel 560 \text{ }\Omega)] = \mathbf{6.15 \text{ k}\Omega}$$

Neglecting r'_e, the voltage gain is

$$A_v = \mathbf{1}$$

The current gain (to the load resistor) is

$$A_i = \frac{R_{in(tot)}}{R_L} = \frac{6.15 \text{ k}\Omega}{560 \text{ }\Omega} = \mathbf{11}$$

Practice Exercise Assume R_1 and R_2 were both doubled in value. How would this change affect the voltage gain? How would the change affect the current gain?

The Darlington Pair

One reason for using a CC amplifier is that it offers high input resistance. The input resistance of the CC amplifier is limited by the size of the bias resistors and the β_{ac} of the transistor. If β_{ac} could be made higher, larger-value bias resistors can still supply the necessary base current and the transistor's input resistance would look higher still.

One way to boost input resistance is to use a darlington pair, as shown in Figure 3–41. A darlington pair consists of two cascaded transistors with the collectors connected; the emitter of the first drives the base of the second. This configuration achieves β_{ac} multiplication. In effect, the darlington pair is a "super beta" transistor that looks like a single transistor with a beta equal to the product of the individual betas.

$$\beta_{ac} = \beta_{ac1}\beta_{ac2}$$

The main advantage of the darlington pair is that the circuit can achieve high input resistance and high current gain. Darlington pairs can be used in any circuit in which a very

FIGURE 3–41
Darlington pair.

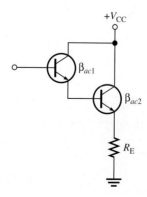

high β is desirable. Darlington transistors are available in a single package configuration that look like any other transistor. For example, the 2N6426 is a small-signal darlington transistor with a minimum β of 30,000.

3–5 REVIEW QUESTIONS

1. What is another name for a common-collector amplifier?
2. What is the ideal maximum voltage gain of a CC amplifier?
3. What are the most important characteristics of the CC amplifier?

3–6 ■ COMMON-BASE AMPLIFIERS

The third basic amplifier configuration is the common-base (CB). The CB amplifier provides high voltage gain but has low input resistance. For this reason, it is not as widely used as other types but is used in certain high-frequency applications and in a circuit called a differential amplifier that we will discuss in Chapter 6.

After completing this section, you should be able to

❏ Understand and analyze the operation of common-base amplifiers
  ❏ Draw the equivalent ac circuit for a CB amplifier
  ❏ Compute the voltage gain and the input and output resistances for a CB amplifier

A typical **common-base (CB)** amplifier using voltage-divider bias is shown in Figure 3–42(a). The base is at signal (ac) ground due to C_3, and the input signal is applied to the emitter. The output is coupled through C_2 from the collector to the load resistor. Figure 3–42(b) shows the equivalent ac circuit. Capacitors and the dc source have been replaced with shorts. This causes the bias resistors to be shorted in the equivalent circuit. The basic difference between this circuit and the common-emitter circuit is how the signal is fed to the amplifier.

FIGURE 3–42

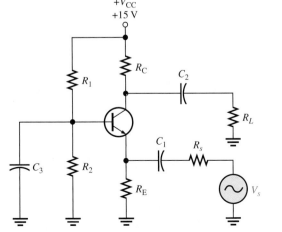

(a) Typical common-base (CB) amplifier

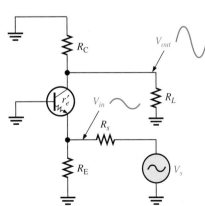

(b) Equivalent ac circuit

Voltage Gain

As in the CE and CC amplifiers, the voltage gain of the CB amplifier is the ratio of V_{out}/V_{in}. For the CB amplifier, V_{out} is the ac collector voltage, V_c, and V_{in} is the ac emitter voltage, V_e. With this in mind, the voltage gain formula is developed as follows:

$$A_v = \frac{V_c}{V_e} = \frac{I_c(R_C \parallel R_L)}{I_e(r'_e \parallel R_E)}$$

The ac collector and emitter currents are nearly the same, so they cancel. Since $R_E \gg r'_e$, you can approximate $r'_e \parallel R_E$ as just r'_e. Further, $R_C \parallel R_L$ represents the ac resistance of the collector, R_c. Thus, the voltage gain is again a ratio of resistances.

$$A_v = \frac{R_C \parallel R_L}{r'_e \parallel R_E}$$

or

$$A_v \cong \frac{R_c}{r'_e}$$

This equation says the voltage gain is approximately equal to the ratio of the ac collector resistance to the internal ac emitter resistance. In this case, the emitter resistance is composed only of r'_e. The more general case is when a swamping resistor is added in the emitter circuit, considered next.

Voltage Gain with a Swamping Resistor   One problem with the basic CB amplifier in Figure 3–42 is that it tends to distort larger signals because the only resistance on the input side is r'_e, which depends, to some extent, on the signal amplitude. A large signal causes changes in r'_e and therefore the gain. Figure 3–43 shows a modification of the basic amplifier with typical values for a small-signal transistor shown. The modification is the addition of a small-value swamping resistor, R_{E1}, in series with r'_e. As in the case of a CE amplifier, this additional fixed resistor produces greater gain stability and increases the input resistance but at the price of reduced gain. For the CB amplifier, it can also significantly improve linearity because the swamping resistor is a fixed quantity, independent of the signal amplitude.

FIGURE 3–43
CB amplifier with swamping resistor.

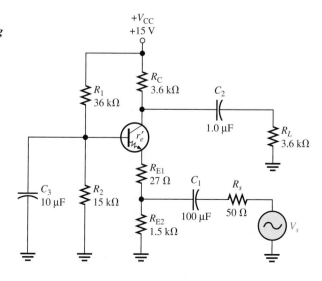

Since the swamping resistor is in series with r'_e, its value is added to r'_e to obtain the approximate ac emitter resistance (ignoring the much larger parallel R_{E2}). The voltage gain of the CB amplifier is still the ac collector resistance, R_c, divided by the ac emitter resistance, R_e, but now includes the swamping resistor.

$$A_v \cong \frac{R_C \| R_L}{r'_e + R_{E1}}$$

$$A_v \cong \frac{R_c}{R_e} \qquad \text{(3–15)}$$

Note that this gain equation is the same as that for the CE amplifier except that the CB amplifier does not invert the signal so the sign of the gain is positive.

Input Resistance

For the basic amplifier without a swamping resistor (Figure 3–42), R_E appears in parallel with r'_e when viewed from the source. However, r'_e is normally small compared to R_E, so you can generally ignore R_E when finding the input resistance. Therefore, the total input resistance of a CB amplifier without a swamping resistor is approximately r'_e. This is the main disadvantage to the CB amplifier. Although the input resistance is small compared to the CE and CC amplifiers, in certain high-frequency applications, this can be an advantage.

As in the case of the CE amplifier, a swamping resistor increases the input resistance. With a swamping resistor, the input resistance is approximately $r'_e + R_{E1}$. This approximation again ignores the contribution of resistor R_{E2}, which appears to be a large parallel path for the input signal. Thus,

$$R_{in(tot)} \cong r'_e + R_{E1} \qquad \text{(3–16)}$$

Output Resistance

The output circuit of the CB amplifier is identical to that of the CE amplifier; therefore, the output resistance is the same (see discussion in Section 3–4). When looking back from the output coupling capacitor, the output resistance for both the CB and CE amplifiers is simply the collector resistance, R_C.

$$R_{out} = R_C$$

EXAMPLE 3–14   Find the total input resistance and the voltage gain for the CB amplifier in Figure 3–43.

Solution   In order to determine r'_e, it is first necessary to find I_E. Find the dc voltage on the base using the voltage-divider rule.

$$V_B = \left(\frac{R_2}{R_1 + R_2}\right) V_{CC} = \left(\frac{15\ k\Omega}{36\ k\Omega + 15\ k\Omega}\right) 15\ V = 4.41\ V$$

The emitter voltage is one diode drop less than the base.

$$V_E = V_B - V_{BE} = 4.41\ V - 0.7\ V = 3.71\ V$$

From Ohm's law, the emitter current is

$$I_E = \frac{V_E}{R_E} = \frac{3.71 \text{ V}}{1.53 \text{ k}\Omega} = 2.43 \text{ mA}$$

The value of r'_e can now be found.

$$r'_e = \frac{25 \text{ mV}}{I_E} = \frac{25 \text{ mV}}{2.43 \text{ mA}} = 10.3 \text{ }\Omega$$

Looking from the input coupling capacitor, the total input resistance is the sum of the swamping resistor and r'_e.

$$R_{in(tot)} = r'_e + R_{E1} = 10.3 \text{ }\Omega + 27 \text{ }\Omega = \textbf{37.3 }\boldsymbol{\Omega}$$

The signal voltage gain is the ratio of the collector ac resistance to the emitter ac resistance. The collector ac resistance, R_c, is equal to $R_C \parallel R_L$. The emitter ac resistance is equal to $r'_e + R_{E1}$. Therefore, the voltage gain is

$$A_v \cong \frac{R_c}{R_e} = \frac{R_C \parallel R_L}{r'_e + R_{E1}} = \frac{3.6 \text{ k}\Omega \parallel 3.6 \text{ k}\Omega}{10.3 \text{ }\Omega + 27 \text{ }\Omega} = \textbf{48}$$

Practice Exercise   What is the output voltage if the input voltage at the coupling capacitor is a 50 mV peak-to-peak sinusoidal wave?

The bias methods introduced in Section 3–2 can be applied to CB amplifiers. Emitter bias requires fewer components but requires dual power supplies. With emitter bias on a CB amplifier, the base can be connected directly to ground, rather than through a capacitor and there are no bias resistors. Example 3–15 illustrates emitter bias with a CB amplifier using a *pnp* transistor. Try estimating the dc parameters and the gain before you look at the solution.

EXAMPLE 3–15   Find the total input resistance and voltage gain for the CB amplifier in Figure 3–44.

FIGURE 3–44

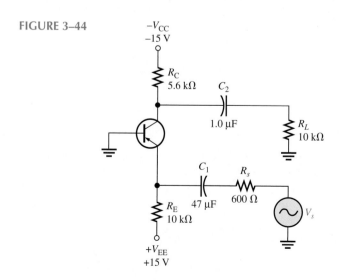

Solution   Since the base is grounded, the emitter voltage is 0.7 V above ground. It can be shown with an equation as

$$V_E = V_B - V_{BE} = 0 \text{ V} - (-0.7 \text{ V}) = +0.7 \text{ V}$$

Applying Ohm's law for the emitter current,

$$I_E = \frac{V_{R_E}}{R_E} = \frac{V_{EE} - V_E}{R_E} = \frac{15 \text{ V} - 0.7 \text{ V}}{10 \text{ k}\Omega} = 1.43 \text{ mA}$$

The value of r_e' is

$$r_e' = \frac{25 \text{ mV}}{I_E} = \frac{25 \text{ mV}}{1.43 \text{ mA}} = 17.5 \text{ }\Omega$$

Since there is no swamping resistor, the total input resistance (looking from the input coupling capacitor) is just r_e'. Therefore, the total input resistance is

$$R_{in(tot)} = r_e' = \textbf{17.5 }\Omega$$

The signal voltage gain, measured from the input coupling capacitor to the load resistor, is

$$A_v \cong \frac{R_c}{R_e} \cong \frac{R_C \| R_L}{r_e'} = \frac{5.6 \text{ k}\Omega \| 10 \text{ k}\Omega}{17.5 \text{ }\Omega} = \textbf{205}$$

Practice Exercise   What dc collector voltage should be observed for the circuit in Figure 3–44?

Summary of AC Parameters for CE, CC, and CB Amplifiers

Table 3–2 summarizes the important ac characteristics of each of the three basic voltage amplifier configurations. Also, relative values are indicated for general comparison of the amplifiers. The input resistance depends on the particular circuit including the type of bias. Voltage-divider bias is assumed for all amplifiers with an unbypassed emitter resistor (R_{E1}) in the CE and CB configurations. These configurations are the same as have been discussed in the previous sections.

TABLE 3–2

Comparison of amplifier ac parameters. Voltage-divider bias is assumed for all amplifiers with an unbypassed emitter resistor in the CE and CB configurations.

	CE	CC	CB
Voltage gain	$A_v \cong -\dfrac{R_c}{R_e}$ High	$A_v \cong 1$ Low	$A_v \cong \dfrac{R_c}{R_e}$ High
Input resistance	$R_{in(tot)} = R_1 \| R_2 \| [\beta_{ac}(r_e' + R_{E1})]$ Low	$R_{in(tot)} = R_1 \| R_2 [\beta_{ac}(r_e' + R_E \| R_L)]$ High	$R_{in(tot)} = r_e' + R_{E1}$ Very low
Output resistance	R_C High	$\cong r_e'$ Low	R_C High

3–6 REVIEW QUESTIONS

1. Can the same voltage gain be achieved with a CB as with a CE amplifier?
2. Is the input resistance of a CB amplifier very low or very high?
3. What is the advantage of using a swamping resistor with a CB amplifier?

3–7 ■ THE BIPOLAR TRANSISTOR AS A SWITCH

In the previous sections, we discussed the transistor as a linear amplifier. Another major application area is switching applications used in digital systems. The first large-scale use of digital circuits was in telephone systems. Today, computers form the most important application of switching circuits using integrated circuits (ICs). Discrete transistor switching circuits are used when it is necessary to supply higher currents or operate at a different voltage than can be obtained from ICs.

After completing this section, you should be able to

❑ Explain how a transistor can be used as a switch
  ❑ Compute the saturation current for a transistor switch
  ❑ Explain how a transistor switching circuit with hysteresis changes states

Figure 3–45 illustrates the basic operation of a transistor as a **switch**. A switch is a two-state device that is either open or closed. In part (a), the transistor is in cutoff because the base-emitter *pn* junction is not forward-biased. In this condition, there is, ideally, an open between collector and emitter, as indicated by the open switch equivalent. In part (b), the transistor is in saturation because the base-emitter *pn* junction is forward-biased and the base current is large enough to cause the collector current to reach its saturated value. In this condition, there is, ideally, a short between collector and emitter, as indicated by the closed-switch equivalent. Actually, a voltage drop of a few tenths of a volt normally occurs across the transistor when it is saturated.

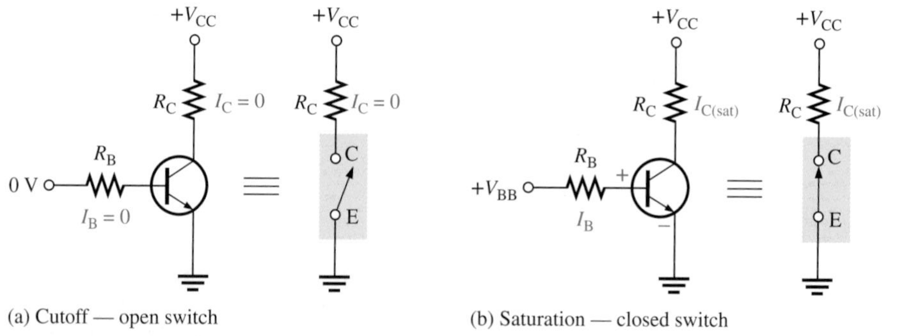

(a) Cutoff — open switch   (b) Saturation — closed switch

FIGURE 3–45
Ideal switching action of a transistor.

Conditions in Cutoff

As mentioned before, a transistor is in cutoff when the base-emitter *pn* junction is not forward-biased. Neglecting leakage current, all of the currents are zero, and V_{CE} is equal to V_{CC}.

$$V_{CE(cutoff)} = V_{CC}$$

Conditions in Saturation

When the emitter junction is forward-biased and there is enough base current to produce a maximum collector current, the transistor is saturated. Since V_{CE} is very small at saturation, the entire power supply voltage drops across the collector resistor. An approximation for the collector current is

$$I_{C(sat)} \cong \frac{V_{CC}}{R_C}$$

The minimum value of base current needed to produce saturation is

$$I_{B(min)} \cong \frac{I_{C(sat)}}{\beta_{DC}}$$

I_B should be significantly greater than $I_{B(min)}$ to keep the transistor well into saturation and to account for beta variations in different transistors.

EXAMPLE 3–16

(a) For the transistor switching circuit in Figure 3–46, what is V_{CE} when $V_{IN} = 0$ V?

(b) What minimum value of I_B is required to saturate this transistor if β_{DC} is 200? Assume $V_{CE(sat)} = 0$ V.

(c) Calculate the maximum value of R_B when $V_{IN} = 5$ V.

FIGURE 3–46

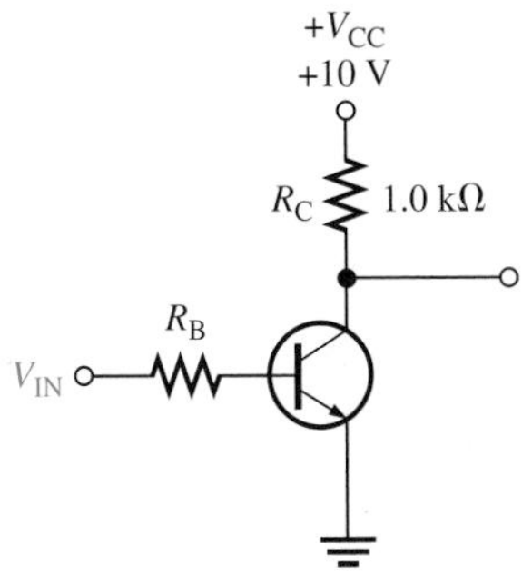

Solution

(a) When $V_{IN} = 0$ V, the transistor is in cutoff (acts like an open switch) and
$$V_{CE} = V_{CC} = \textbf{10 V.}$$

(b) Since $V_{CE(sat)} = 0$ V,

$$I_{C(sat)} \cong \frac{V_{CC}}{R_C} = \frac{10 \text{ V}}{1.0 \text{ k}\Omega} = 10 \text{ mA}$$

$$I_{B(min)} = \frac{I_{C(sat)}}{\beta_{DC}} = \frac{10 \text{ mA}}{200} = \textbf{0.05 mA}$$

This is the value of I_B necessary to drive the transistor to the point of saturation. Any further increase in I_B will drive the transistor deeper into saturation but will not increase I_C.

(c) When the transistor is saturated, $V_{BE} = 0.7$ V. The voltage across R_B is

$$V_{R_B} = V_{IN} - V_{BE} = 5 \text{ V} - 0.7 \text{ V} = 4.3 \text{ V}$$

The maximum value of R_B needed to allow a minimum I_B of 0.05 mA is calculated by Ohm's law as follows:

$$R_B = \frac{V_{R_B}}{I_B} = \frac{4.3 \text{ V}}{0.05 \text{ mA}} = \textbf{86 k}\boldsymbol{\Omega}$$

Practice Exercise   Determine the minimum value of I_B required to saturate the transistor in Figure 3–46 if β_{DC} is 125 and $V_{CE(sat)}$ is 0.2 V.

Improvements to the One-Transistor Switching Circuit

The basic switching circuit shown in Figure 3–45 has a threshold voltage at which it changes from *off to on* or *on to off*. Unfortunately, the threshold is not an absolute point because a transistor can operate between cutoff and saturation, a condition not desirable in a switching circuit. A second transistor can dramatically improve the switching action, providing a sharp threshold. The circuit is shown in Figure 3–47, designed with a light-emitting diode (LED) output so that you can construct it if you choose and observe the switching action. The circuit works as follows. When V_{IN} is very low, Q_1 is off since it does not have sufficient base current. Q_2 will be in saturation because it can obtain ample base current through R_2 and the LED is on. As the base voltage for Q_1 is increased, Q_1 begins to conduct. As Q_1 approaches saturation, the base voltage of Q_2 suddenly drops, causing it to go from a saturated to cutoff condition rapidly. The output voltage of Q_2 drops and the LED goes out.

Another improvement for basic switching circuits is the addition of hysteresis. For switching circuits, **hysteresis** means that there are two threshold voltages depending on whether the output is already high or already low. Figure 3–48 illustrates the situation. When the input voltage rises, it must cross the upper threshold before switching takes place. It does not switch at *A* or *B* because the lower threshold is inactive. When the signal crosses the upper threshold at point *C,* the output switches, and immediately the threshold

FIGURE 3–47
A two-transistor switching circuit with a sharp threshold.

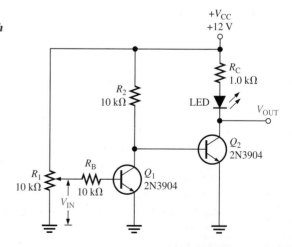

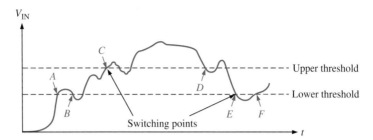

FIGURE 3–48
Hysteresis causes the circuit to switch at points C and E but not at the other points.

changes to a lower value. The output does not switch back at D but instead must cross the lower threshold at E before returning to the original state. Again, the threshold changes to the upper level, so switching does not take place at point F. The major advantage of hysteresis in a switching circuit is noise immunity. As you can see from this example, the output only changed twice despite a very noisy input.

The schematic for a transistor circuit with hysteresis is shown in Figure 3–49. As the potentiometer is turned in one direction, the output will switch once, even if the potentiometer is "noisy." When the output switches, the common-emitter resistor, R_E, causes the threshold voltage to change. This is caused by the different saturation currents for the two transistors; the threshold is different when the output is in cutoff than when the output is saturated.

FIGURE 3–49
A discrete transistor switching circuit with hysteresis.

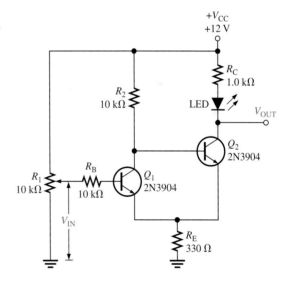

3–7 REVIEW QUESTIONS

1. When a transistor is used as a switching device, in what two states is it operated?
2. When does the collector current reach its maximum value?
3. When is the collector current approximately zero?
4. When is V_{CE} equal to V_{CC}?
5. What is meant by hysteresis in a switching circuit?

3–8 ■ TRANSISTOR PACKAGES AND TERMINAL IDENTIFICATION

Transistors are available in a wide range of package types for various applications. Those with mounting studs or heat sinks are usually power transistors. Low-power and medium-power transistors are usually found in smaller metal or plastic cases. Still another package classification is for high-frequency devices. You should be familiar with common transistor packages and be able to identify the emitter, base, and collector terminals. This section is about transistor packages and terminal identification.

After completing this section, you should be able to

❑ Identify various types of transistor package configurations
  ❑ List three broad categories of transistors
  ❑ Recognize various types of cases and identify the pin configurations

Transistor Categories

Manufacturers generally classify their bipolar junction transistors into three broad categories: general-purpose/small-signal devices, power devices, and RF (radio frequency/microwave) devices. Although each of these categories, to a large degree, has its own unique package types, you will find certain types of packages used in more than one device category. While keeping in mind there is some overlap, we will look at transistor packages for each of the three categories, so that you will be able to recognize a transistor when you see one on a circuit board and have a good idea of what general category it is in.

General-Purpose/Small-Signal Transistors   General-purpose/small-signal transistors are generally used for low- or medium-power amplifiers or switching circuits. The packages are either plastic or metal cases. Certain types of packages contain multiple transistors. Figure 3–50 illustrates common plastic cases, Figure 3–51 shows packages called *metal cans,* and Figure 3–52 shows multiple-transistor packages. Some of the multiple-transistor packages such as the dual-in-line (DIP) and the small-outline (SO) are the same as those used for many integrated circuits. Typical pin connections are shown so you can identify the emitter, base, and collector.

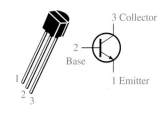

(a) TO-92 or TO-226AA

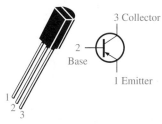

(b) TO-92 or TO-226AE

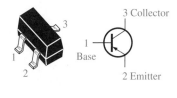

(c) SOT-23 or TO-236AB

FIGURE 3–50

Plastic cases for general-purpose/small-signal transistors. Both old and new JEDEC TO numbers are given. Pin configurations may vary. Always check the data sheet.

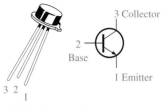

(a) TO-18 or TO-206AA

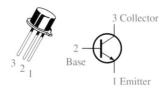

(b) TO-39 or TO-205AD

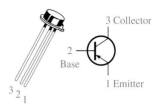

(c) TO-46 or TO-206AB

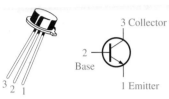

(d) TO-52 or TO-206AC

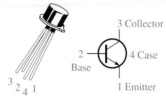

(e) TO-72 or TO-206AF

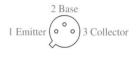

(f) Pin configuration (bottom view). Emitter is closest to tab.

FIGURE 3–51
Metal cans for general-purpose/small-signal transistors.

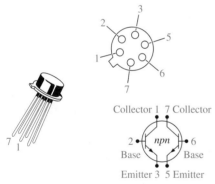

(a) Dual metal can. Tab indicates pin 1.

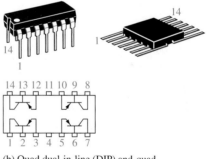

(b) Quad dual-in-line (DIP) and quad flat-pack. Dot indicates pin 1.

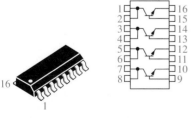

(c) Quad small outline (SO) package for surface-mount technology

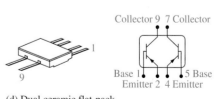

(d) Dual ceramic flat-pack

FIGURE 3–52
Typical multiple-transistor packages.

Power Transistors Power transistors are used to handle large currents (typically more than 1 A) and/or large voltages. For example, the final audio stage in a stereo system uses a power transistor amplifier to drive the speakers. Figure 3–53 shows some common package configurations. In most applications, the metal tab or the metal case is common to the collector and is thermally connected to a heat sink for heat dissipation. Notice in part (g) how the small transistor chip is mounted inside the much larger package.

RF Transistors RF transistors are designed to operate at extremely high frequencies and are commonly used for various purposes in communications systems and other high-frequency applications. Their unusual shapes and lead configurations are designed to optimize certain high-frequency parameters. Figure 3–54 shows some examples.

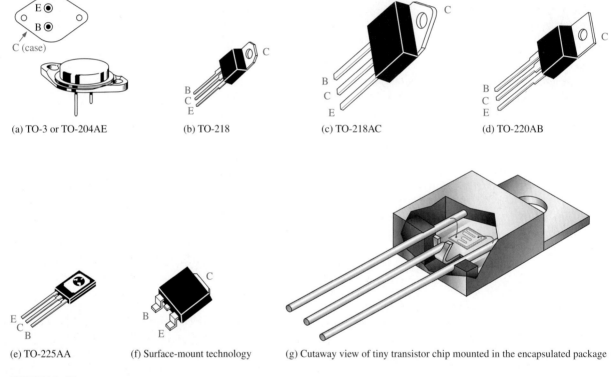

(a) TO-3 or TO-204AE (b) TO-218 (c) TO-218AC (d) TO-220AB

(e) TO-225AA (f) Surface-mount technology (g) Cutaway view of tiny transistor chip mounted in the encapsulated package

FIGURE 3–53
Typical power transistors.

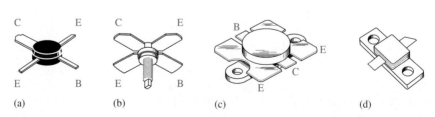

(a) (b) (c) (d)

FIGURE 3–54
Examples of RF transistors.

3–8 REVIEW QUESTIONS

1. List the three broad categories of bipolar junction transistors.

2. In a single-transistor metal case, how do you identify the leads?

3. In power transistors, the metal mounting tab or case is connected to which transistor region?

3–9 ■ TROUBLESHOOTING

As you already know, a critical skill in electronics is the ability to identify a circuit malfunction and to isolate the failure to a single component if possible. In this section, the basics of troubleshooting transistor bias circuits and testing individual transistors are covered.

After completing this section, you should be able to

❑ Troubleshoot various faults in transistor circuits
  ❑ Define *floating point*
  ❑ Use voltage measurements to identify a fault in a transistor circuit
  ❑ Use a DMM to test a transistor
  ❑ Explain how a transistor can be viewed in terms of a diode equivalent
  ❑ Discuss in-circuit and out-of-circuit testing
  ❑ Discuss point-of-measurement in troubleshooting
  ❑ Discuss leakage and gain measurements

Troubleshooting a Biased Transistor

Several faults can occur in a simple transistor bias circuit. Possible faults are open bias resistors, open or resistive connections, shorted connections, and opens or shorts internal to the transistor itself. Figure 3–55 is a basic transistor bias circuit with all voltages referenced to ground. The two bias voltages are $V_{BB} = 3$ V and $V_{CC} = 9$ V. The correct voltages at the base and collector are shown. Analytically, these voltages are determined as follows. A $\beta_{DC} = 200$ is taken as midway between the minimum and maximum values of h_{FE} given on the data sheet for the 2N3904 in Figure 3–25. A different h_{FE} (β_{DC}), of course, will produce different results for the given circuit.

$$V_B = V_{BE} = 0.7 \text{ V}$$

$$I_B = \frac{V_{BB} - V_{BE}}{R_B} = \frac{3 \text{ V} - 0.7 \text{ V}}{56 \text{ k}\Omega} = \frac{2.3 \text{ V}}{56 \text{ k}\Omega} = 41.1 \text{ }\mu\text{A}$$

$$I_C = \beta_{DC}I_B = 200(41.1 \text{ }\mu\text{A}) = 8.2 \text{ mA}$$

$$V_C = V_{CC} - I_C R_C = 9 \text{ V} - (8.2 \text{ mA})(560 \text{ }\Omega) = 4.4 \text{ V}$$

Several faults that can occur in the circuit and the accompanying symptoms are illustrated in Figure 3–56. Symptoms are shown in terms of measured voltages that are incorrect. The term **floating point** refers to a point in the circuit that is not electrically connected to ground or a "solid" voltage. Normally, very small and sometimes fluctuating voltages in the μV to low mV range are generally observed at floating points. The faults in Figure 3–56 are typical but do not represent all possible faults that may occur.

FIGURE 3–55
A basic transistor bias circuit.

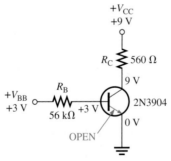

FIGURE 3–55
A basic transistor bias circuit.

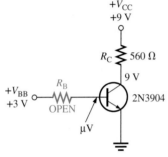

(a) *Fault:* Open base resistor.
Symptoms: Readings from μV to a few mV at base due to floating point. 9 V at collector because transistor is in cutoff.

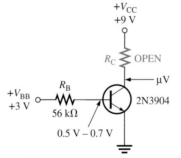

(b) *Fault:* Open collector resistor.
Symptoms: Readings from μV to a few mV at collector due to floating point. 0.5 V – 0.7 V at base due to forward voltage drop across the base-emitter junction.

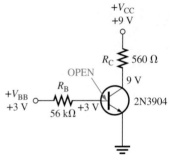

(c) *Fault:* Base internally open.
Symptoms: 3 V at base lead. 9 V at collector because transistor is in cutoff.

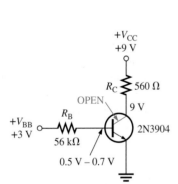

(d) *Fault:* Collector internally open.
Symptoms: 0.5 V – 0.7 V at base lead due to forward voltage drop across base-emitter junction. 9 V at collector because the open prevents collector current.

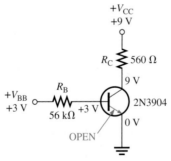

(e) *Fault:* Emitter internally open.
Symptoms: 3 V at base lead. 9 V at collector because there is no collector current. 0 V at the emitter as normal.

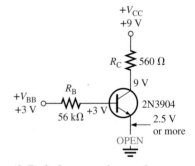

(f) *Fault:* Open ground connection.
Symptoms: 3 V at base lead. 9 V at collector because there is no collector current. 2.5 V or more at the emitter due to the forward voltage drop across the base-emitter junction. The measuring voltmeter provides a forward current path through its internal resistance.

FIGURE 3–56
Typical faults and symptoms in the basic transistor bias circuit.

Testing a Transistor with a DMM

A digital multimeter can be used as a fast and simple way to check a transistor for open or shorted junctions. For this test, you can view the transistor as two diodes connected as shown in Figure 3–57 for both *npn* and *pnp* transistors. The base-collector junction is one diode and the base-emitter junction is the other.

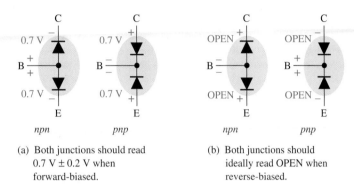

(a) Both junctions should read 0.7 V ± 0.2 V when forward-biased.

(b) Both junctions should ideally read OPEN when reverse-biased.

FIGURE 3–57
A transistor viewed as two diodes.

Recall that a good diode will show an extremely high resistance (or open) with reverse bias and a very low resistance with forward bias. A defective open diode will show an extremely high resistance (or open) for both forward and reverse bias. A defective shorted or resistive diode will show zero or a very low resistance for both forward and reverse bias. An open diode is the most common type of failure. Since the transistor *pn* junctions are, in effect diodes, the same basic characteristics apply.

The DMM Diode Test Position   Many digital multimeters (DMMs) have a *diode test* position that provides a convenient way to test a transistor. A typical DMM, as shown in Figure 3–58, has a small diode symbol to mark the position of the function switch. When set to diode test, the meter provides an internal voltage sufficient to forward-bias and reverse-bias a transistor junction. This internal voltage may vary among different makes of DMM, but 2.5 V to 3.5 V is a typical range of values. The meter provides a voltage reading to indicate the condition of the transistor junction under test.

When the Transistor Is Not Defective   In Figure 3–58(a), the V Ω (positive) lead of the meter is connected to the base of an *npn* transistor and the COM (negative) lead is connected to the emitter to forward-bias the base-emitter junction. If the junction is good, you will get a reading of between 0.5 V and 0.9 V, with 0.7 V being typical for forward bias.

In Figure 3–58(b), the leads are switched to reverse-bias the base-emitter junction, as shown. If the transistor is working properly, you will get a voltage reading based on the meter's internal voltage source. The 2.6 V shown in the figure represents a typical value and indicates that the junction has an extremely high reverse resistance with essentially all of the internal voltage appearing across it.

The process just described is repeated for the base-collector junction as shown in Figure 3–58(c) and (d). For a *pnp* transistor, the polarity of the meter leads are reversed for each test.

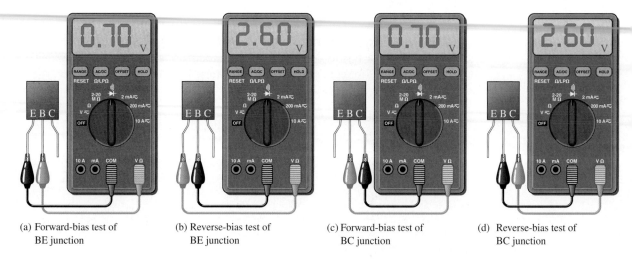

(a) Forward-bias test of
    BE junction

(b) Reverse-bias test of
    BE junction

(c) Forward-bias test of
    BC junction

(d) Reverse-bias test of
    BC junction

FIGURE 3–58
Typical DMM test of a properly functioning npn transistor. Leads are reversed for a pnp test.

When the Transistor Is Defective When a transistor has failed with an open junction or internal connection, you get an open circuit voltage reading (2.6 V is typical for many DMMs) for both the forward-bias and the reverse-bias conditions for that junction, as illustrated in Figure 3–59(a). If a junction is shorted, the meter reads 0 V in both forward- and reverse-bias tests, as indicated in part (b). Sometimes, a failed junction may exhibit a small resistance for both bias conditions rather than a pure short. In this case, the meter will show a small voltage much less than the correct open voltage. For example, a resistive junction may result in a reading of 1.1 V in both directions rather than the correct readings of 0.7 V forward and 2.6 V reverse.

    Some DMMs provide a test socket on their front panel for testing a transistor for the h_{FE} (β_{DC}) value. If the transistor is inserted improperly in the socket or if it is not functioning

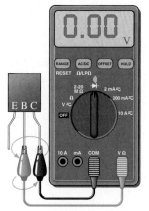

(a) Forward-bias test and reverse-
    bias test give the same reading
    (2.60 V is typical) for an open
    BC junction.

(b) Forward- and reverse-bias tests
    for a shorted junction give the
    same 0 V reading. If the junction
    is resistive, the reading is less
    than 2.6 V.

FIGURE 3–59
Testing a defective npn transistor. Leads are reversed for a pnp.

properly due to a faulty junction or internal connection, a typical meter will flash a 1 or display a 0. If a value of β_{DC} within the normal range for the specific transistor is displayed, the device is functioning properly. The normal range of β_{DC} can be determined from the data sheet.

Checking a Transistor with the OHMs Function   DMMs that do not have a diode test position or an h_{FE} socket can be used to test a transistor for open or shorted junctions by setting the function switch to an OHMs range. For the forward-bias check of a good transistor *pn* junction, you will get a resistance reading that can vary depending on the meter's internal battery. Many DMMs do not have sufficient voltage on the OHMs range to fully forward-bias a junction, and you may get a reading of from several hundred to several thousand ohms.

For the reverse-bias check of a good transistor, you will get an out-of-range indication on most DMMs because the reverse resistance is too high to measure. An out-of-range indication may be a flashing 1 or a display of dashes, depending on the particular DMM.

Even though you may not get accurate forward and reverse resistance readings on a DMM, the relative readings are sufficient to indicate a properly functioning transistor *pn* junction. The out-of-range indication shows that the reverse resistance is very high, as you expect. The reading of a few hundred to a few thousand ohms for forward bias indicates that the forward resistance is small compared to the reverse resistance, as you expect.

Transistor Testers

A comprehensive test of a transistor can be performed with a transistor curve tracer, shown in Figure 3–60(a). The curve tracer can show the characteristic curve for all types of transistors, as well as other solid-state devices. It can measure most of the important parameters for these devices. Some advanced curve tracers can perform these measurements automatically and include automated setup and sequencing through a variety of measurements, data storage, and direct hard copy output of measurements for documentation.

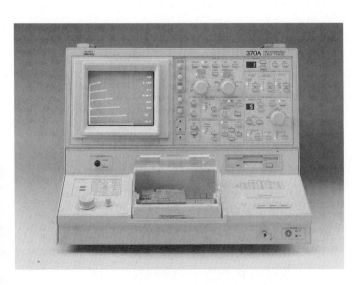

(a) Curve tracer

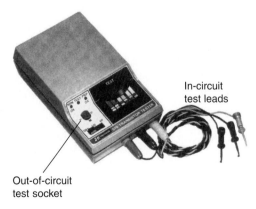

In-circuit test leads

Out-of-circuit test socket

(b) Transistor tester

FIGURE 3–60

Transistor curve tracer and tester. Part (a) copyright 2000, Tektronix, Inc. Reprinted with permission. All rights reserved. Part (b) courtesy of BK Precision, Maxtec International Corp.

There are various reasons for measuring the characteristics of a device. In engineering work, it is useful to know certain parameters to completely understand a circuit's performance. Component manufacturers measure characteristics in order to develop better products and to characterize production runs. Sometimes a curve tracer is used for incoming test, quality control, or to sort components. Finally, of course, there are educational reasons to study the parameters of various active devices.

Although a curve tracer is the ultimate transistor tester, a simpler device is often useful for checking bipolar transistors. It is handy to be able to test a transistor in the circuit, particularly if it is soldered in place. Good troubleshooting practice dictates that you do not unsolder a component unless you are reasonably sure it is bad or you simply cannot isolate the trouble any other way. The basic tester in Figure 3–60(b) allows you to test a bipolar transistor while it is still in the circuit board. The three clip-leads are connected to the transistor and give a positive indication if the transistor is good. A circuit that is not working may have a good transistor or a bad one as illustrated in the simplified circuit in the following two cases.

Case 1 If the transistor tests defective, it should be carefully removed and replaced with a known good one. An out-of-circuit check of the replacement device is usually a good idea, just to make sure it is OK. The transistor is plugged into the socket on the transistor tester for out-of-circuit tests.

Case 2 If the transistor tests good in-circuit but the circuit is not working properly, examine the circuit board for a poor connection at the collector pad or for a break in the connecting trace. A poor solder joint often results in an open or a highly resistive contact. A troubleshooter can isolate the problem with voltage measurements. The physical point at which you actually measure the voltage is very important in this case. For example, if you check the collector lead when there is an external open at the collector pad, you will get a floating point. If you measure on the connecting trace or on the R_C lead, you will read V_{CC}. This situation is illustrated in Figure 3–61.

Importance of Point-of-Measurement in Troubleshooting In case 2, if you had taken the initial measurement on the transistor lead itself and the open were *internal* to the transistor as shown in Figure 3–62, you would have measured V_{CC}. This would have

FIGURE 3–61
The indication of an open, when it is in the external circuit, depends on where you measure.

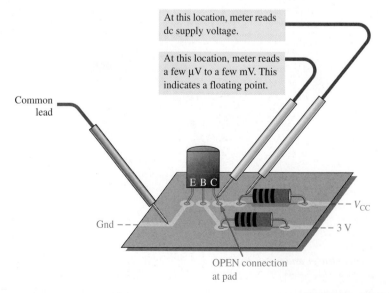

At this location, meter reads dc supply voltage.

At this location, meter reads a few μV to a few mV. This indicates a floating point.

Common lead

E B C

Gnd --

-- V_{CC}

-- 3 V

OPEN connection at pad

FIGURE 3–62

Illustration of an internal open. Compare with Figure 3–61.

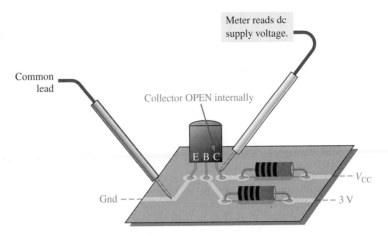

indicated a defective transistor even before the tester was used. This simple concept emphasizes the importance of point-of-measurement in certain troubleshooting situations.

EXAMPLE 3–17   What fault do the measurements in Figure 3–63 indicate? (The probe is shown in three different locations.)

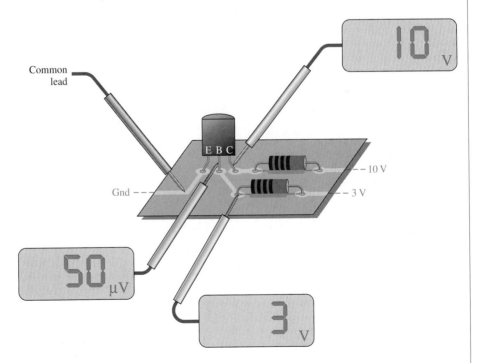

FIGURE 3–63

Solution   The transistor is in cutoff, as indicated by the 10 V measurement on the collector lead. The base bias voltage of 3 V appears on the PC board contact but not on the transistor lead as indicated by the floating point measurement. This shows that there is an open external to the transistor between the two measured base points. Check the solder joint at the base contact on the PC board. If the open were internal, there would be 3 V on the base lead.

Practice Exercise   If the meter in Figure 3–63 that now reads 3 V indicates a floating point when touching the circuit board pad, what is the most likely fault?

Leakage Measurement

Very small leakage currents exist in all transistors and, in most cases, are small enough to neglect (usually nA). When a transistor is connected as shown in Figure 3–64(a) with the base open ($I_B = 0$), it is in cutoff. Ideally $I_C = 0$; but actually there is a small current from collector to emitter, as mentioned earlier, called I_{CEO} (collector-to-emitter current with base open). This leakage current is usually in the nA range for silicon. A faulty transistor will often have excessive leakage current and can be checked in a transistor tester, which connects an ammeter as shown in part (a). Another leakage current in transistors is the reverse collector-to-base current, I_{CBO}. This is measured with the emitter open, as shown in Figure 3–64(b). If it is excessive, a shorted collector-base junction is likely.

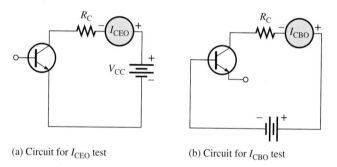

(a) Circuit for I_{CEO} test   (b) Circuit for I_{CBO} test

FIGURE 3–64
Leakage current test circuits.

Gain Measurement

In addition to leakage tests, the typical transistor tester also checks the β_{DC}. A known value of I_B is applied and the resulting I_C is measured. The reading will indicate the value of the I_C/I_B ratio, although in some units only a relative indication is given. Most testers provide for an in-circuit β_{DC} check, so that a suspected device does not have to be removed from the circuit for testing.

3–9 REVIEW QUESTIONS

1. If a transistor on a circuit board is suspected of being faulty, what should you do?
2. In a transistor bias circuit, such as the one in Figure 3–55, what happens if R_B opens?
3. In a circuit such as the one in Figure 3–55, what are the base and collector voltages if there is an external open between the emitter and ground?

3–10 ■ A SYSTEM APPLICATION

A basic electronic security system has several parts, but in this section we are concerned with the transistor circuits that detect an open in the loops containing remote sensors for windows and doors. In this particular application, the transistors are used as switching devices.

After completing the section, you should be able to

❏ Apply what you have learned in this chapter to a system application
   ❏ See how transistors are used in a switching application
   ❏ See how a transistor is used to activate a relay
   ❏ See how one transistor is used to drive another transistor
   ❏ Translate between a printed circuit board and a schematic
   ❏ Troubleshoot some common transistor circuit failures

A Brief Description of the System

The circuit board shown in the system diagram in Figure 3–65 contains the transistor circuits for detecting when one of the remote switch loops is open. There are three remote zones in this particular system. Zone 1 protects all window/doors in one area of the structure, and Zone 2 protects all windows/doors in a second area. Zone 3 protects the main entry door.

When an intrusion occurs, a switch sensor at the point of intrusion breaks contact and opens the zone loop. This causes the input to the monitoring circuit for that zone to go

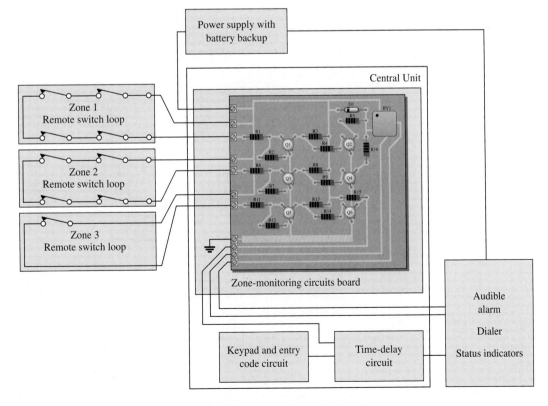

FIGURE 3–65

to a 0 V level and activate the circuit which, in turn, energizes the relay. The closure of the relay sets off an audible alarm and/or initiates an automatic telephone dialing sequence. Either of the monitoring circuits for zone 1 or zone 2 can energize the common relay.

For zone 3, which is the main entry, the output of the monitoring circuit goes to a time-delay circuit to allow time for keying in an entry code that will disarm the system. If, after a sufficient time, no code or an incorrect code has been entered, the time-delay circuit will set off an alarm and/or the telephone dialing sequence.

There are many aspects to a security system and a system can range from very simple to very complex. However, in this application we are concentrating only on the zone-monitoring circuits board.

Now, so you can take a closer look at the monitoring circuits, let's take the board out of the system and put it on the troubleshooter's bench.

TROUBLESHOOTER'S BENCH

Identifying the Components

There are three separate monitoring circuits on the board in Figure 3–66, each consisting of two transistors and their associated resistors. The transistors are labeled Q1, Q2, and so forth. There is one relay labeled RY1 and a diode D1 for suppression of negative tran-

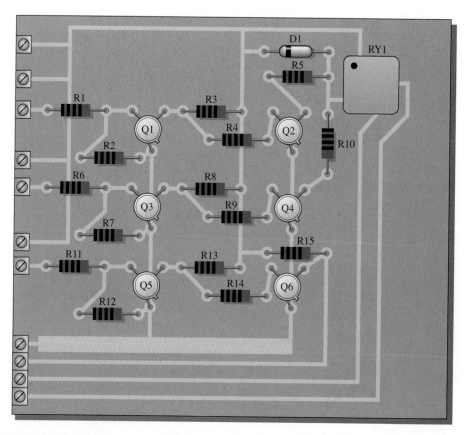

FIGURE 3–66

sients across the relay coil. As shown in Figure 3–67(a), the transistors are housed in TO-18 (TO-206AA) metal cans (remember, the emitter is closest to the tab). A detail of the relay is shown in Figure 3–67(b). You are already familiar with resistors and diodes.

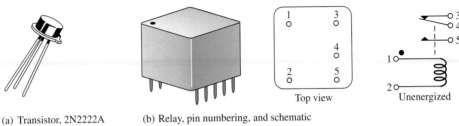

(a) Transistor, 2N2222A   (b) Relay, pin numbering, and schematic

FIGURE 3–67

■ ACTIVITY 1   Relate the PC Board to the Schematic

Carefully follow the traces on the PC board to see how the components are interconnected. Compare each point (A through X) on the circuit board with the corresponding point on the schematic in Figure 3–68. For each point on the PC board, place the letter of the corresponding point or points on the schematic.

■ ACTIVITY 2   Analyze the Circuits

Calculate the amount of base current and collector current for each transistor when it is saturated. Refer to the β_{DC} (h_{FE}) information on the data sheet for the 2N2222 in Appendix A to make sure there is sufficient base current to keep the transistor well in saturation. Assume that the relay has a coil resistance of 24 Ω and requires a minimum of 400 mA to operate.

■ ACTIVITY 3   Write a Technical Report

Discuss the detailed operation of the zone-monitoring circuits on the PC board. Also, describe the inputs and outputs indicating what other areas of the system they connect to and describe the purpose served by each one.

■ ACTIVITY 4   Develop a Complete Test Procedure

Someone with no knowledge of how this circuitry works should be able to use your test procedure and verify that all of the circuits operate properly. The circuit board test is to be done on the test bench and not in the functional system, so you must simulate operational conditions.

Step 1: Begin with the most basic and necessary thing—power.

Step 2: Develop a detailed step-by-step procedure for simulating the external remote loops.

Step 3: For each loop-monitoring circuit, specify the points at which to measure voltage, the value of the voltage, and under what conditions it is to be measured. Detail each step required to fully check out all the circuits.

Step 4: Specify in detail any other tests in addition to voltage measurements.

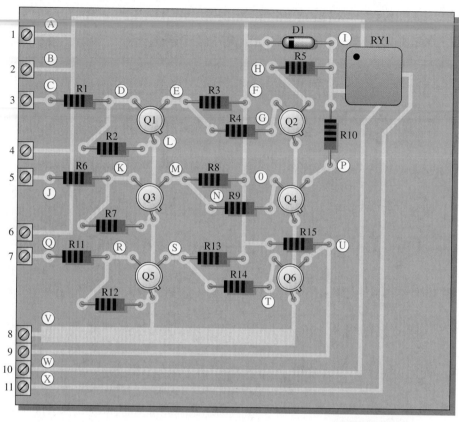

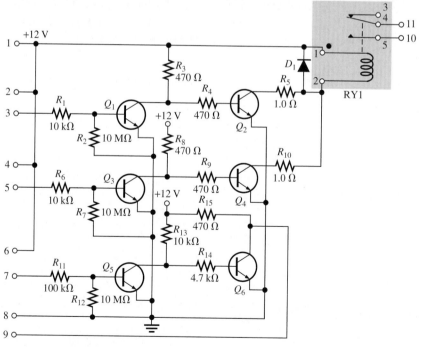

FIGURE 3–68

■ **ACTIVITY 5   Troubleshoot the Circuit Board for Each of the Following Problems by Stating the Probable Cause or Causes in Each Case**

1. Approximately 5.65 V at the collector of Q_1 with the remote loop switch closed.
2. Relay is energized (contact from pin 10 to pin 11) continuously and independent of remote switches.
3. Collector lead of Q_4 floating.
4. Pin 9 is always at 12 V. Q_6 collector lead is also at 12 V.
5. When 12 V is applied to pin 5 with the closed-loop switch, the base of Q_3 is 0 V.
6. Approximately 3.6 V at the collector of Q_5 when 12 V is applied to pin 7.

3–10 REVIEW QUESTIONS

1. State the basic purpose of the zone-monitoring circuits board in the security system.
2. How many of the circuits on the board are identical?
3. When is there a contact closure between pins 10 and 11 on the PC board?
4. Based on your knowledge of coils and diodes, why do you think D_1 is in the circuit?

■ SUMMARY

- A bipolar junction transistor (BJT) consists of three regions: emitter, base, and collector. The term *bipolar* refers to two types of current: electron current and hole current.
- The three regions of a BJT are separated by two *pn* junctions.
- The two types of bipolar transistor are the *npn* and the *pnp*.
- In normal operation, the base-emitter (BE) junction is forward-biased and the base-collector (BC) junction is reverse-biased.
- The three currents in a BJT are base current, emitter current, and collector current. They are related to each other by this formula: $I_E = I_C + I_B$.
- Characteristic collector curves for a BJT are a family of curves showing I_C versus V_{CE} for a given set of base currents.
- When a BJT is in cutoff, there is essentially no collector current except for a very tiny amount of collector leakage current, I_{CEO}. V_{CE} is a maximum.
- When a BJT is saturated, there is maximum collector current as determined by the external circuit.
- A load line represents all possible operating points for a circuit, including cutoff and saturation. The point at which the actual base current line intersects the load line is the quiescent or Q-point for the circuit.
- Base bias uses a single resistor between the power supply and the base terminal.
- Collector-feedback bias uses a single resistor between the collector and base terminals.
- Voltage-divider bias is a very stable form of bias that uses two resistors to form a voltage divider in the base circuit.
- Emitter bias is a very stable form of bias that uses both positive and negative power supplies and a single resistor between the base terminal and ground.
- DC values are identified with capital-letter nonitalic subscripts; ac values are identified with lowercase italic subscripts.

- Manufacturer's data sheets typically show maximum voltage, current, and power ratings for various parameters.

- Coupling capacitors are connected in series with the ac signal to bring it into and out of the amplifier.

- Bypass capacitors are connected in parallel with a resistor to provide an ac path around the resistor.

- Common-emitter (CE), common-collector (CC), and common-base (CB) designations refer to the common terminal for the ac signal.

- Voltage gain for CE, CC, and CB amplifiers can be found using a ratio of ac resistances.

- A darlington pair is a two-transistor configuration that is equivalent to a single very high-beta transistor.

- In switching circuits, transistors are designed to operate at either cutoff or saturation, the equivalent of an open or closed switch.

■ GLOSSARY

Key terms are in color. All terms are included in the end-of-book glossary.

ac beta (β_{ac})   The ratio of a change in collector current to a corresponding change in base current in a bipolar junction transistor.

Base   One of the semiconductor regions in a BJT.

Base bias   A form of bias in which a single resistor is connected between a BJT's base and V_{CC}.

Bipolar   Characterized by two *pn* junctions.

Bipolar junction transistor (BJT)   A transistor constructed with three doped semiconductor regions separated by two *pn* junctions.

Bypass capacitor   A capacitor connected in parallel with a resistor to provide the ac signal with a low impedance path.

Collector   One of the semiconductor regions in a BJT.

Collector characteristic curves   A set of collector *IV* curves that show how I_C varies with V_{CE} for a given base current.

Collector feedback bias   A form of bias, used in CE and CB amplifiers, in which a single resistor is connected between a BJT's base and its collector.

Common-base (CB)   A BJT amplifier configuration in which the base is the common terminal to an ac signal or ground.

Common-collector (CC)   A BJT amplifier configuration in which the collector is the common terminal to an ac signal or ground.

Common-emitter (CE)   A BJT amplifier configuration in which the emitter is the common terminal to an ac signal or ground.

Coupling capacitor   A capacitor connected in series with the ac signal and used to block dc voltages.

Cutoff   The nonconducting state of a transistor.

dc beta (β_{DC})   The ratio of collector current to base current in a bipolar junction transistor.

Dynamic emitter resistance (r'_e)   The ac resistance of the emitter; it is determined by the dc emitter current.

Emitter   One of the three semiconductor regions in a BJT.

Emitter bias   A very stable form of bias requiring two power supplies. The emitter is connected through a resistor to one supply; another resistor is connected between a BJT's base and ground.

Floating point   A point in the circuit that is not electrically connected to ground or a "solid" voltage.

Hysteresis   Characteristic of a circuit in which two different trigger levels create an offset or lag in the switching action.

Negative feedback   The process of returning a portion of the output back to the input in a manner to cancel changes that may occur at the input.

Saturation   The state of a BJT in which the collector current has reached a maximum and is independent of the base current.

Switch   An electrical or electronic device for opening and closing a current path.

Transistor   A semiconductor device used for amplification and switching applications in electronic circuits.

Voltage-divider bias   A very stable form of bias in which a voltage divider is connected between V_{CC} and ground; the output of the divider supplies bias current to the base of a BJT.

■ **KEY FORMULAS**

(3–1)	$I_E = I_C + I_B$	Relationship of key transistor currents
(3–2)	$\beta_{DC} = \dfrac{I_C}{I_B}$	Definition of β_{DC}
(3–3)	$I_C = \beta_{DC}\left(\dfrac{V_{CC} - V_{BE}}{R_B}\right)$	Collector current for base bias
(3–4)	$I_C = \dfrac{V_{CC} - V_{BE}}{R_C + R_B/\beta_{DC}}$	Collector current for collector-feedback bias
(3–5)	$V_B = \left(\dfrac{R_2}{R_1 + R_2}\right)V_{CC}$	Base voltage for voltage-divider bias
(3–6)	$V_E = V_B - 0.7\ \text{V}$	Emitter voltage for voltage-divider bias
(3–7)	$V_C = V_{CC} - I_C R_C$	Collector voltage for CE and CB amplifiers
(3–8)	$\beta_{ac} = \dfrac{I_c}{I_b}$	Definition of β_{ac}
(3–9)	$r'_e = \dfrac{25\ \text{mV}}{I_E}$	AC emitter resistance
(3–10)	$A_v \cong -\dfrac{R_c}{R_e}$	Voltage gain for CE amplifier
(3–11)	$R_{in(tot)} = R_1 \parallel R_2 \parallel [\beta_{ac}(r'_e + R_{E1})]$	Input resistance for CE amplifier with voltage-divider bias (R_{E1} is not bypassed)
(3–12)	$A_v \cong 1$	Voltage gain for CC amplifier
(3–13)	$R_{in(tot)} = R_1 \parallel R_2 \parallel [\beta_{ac}(r'_e + R_E \parallel R_L)]$	Input resistance for CC amplifier with voltage-divider bias and load resistor
(3–14)	$A_i = \dfrac{R_{in(tot)}}{R_L}$	Current gain for CC amplifier
(3–15)	$A_v \cong \dfrac{R_c}{R_e}$	Voltage gain for a CB amplifier
(3–16)	$R_{in(tot)} \cong r'_e + R_{E1}$	Input resistance for CB amplifier with swamping resistor

■ **SELF-TEST**   *Answers are at the end of the chapter.*

1. The *n*-type regions in an *npn* bipolar junction transistor are
   (a) collector and base   (b) collector and emitter
   (c) base and emitter   (d) collector, base, and emitter

2. The *n*-region in a *pnp* transistor is the
   (a) base   (b) collector   (c) emitter   (d) case

3. For normal operation of an *npn* transistor, the base must be
   (a) disconnected
   (b) negative with respect to the emitter
   (c) positive with respect to the emitter
   (d) positive with respect to the collector

4. Beta (β) is the ratio of
   (a) collector current to emitter current
   (b) collector current to base current
   (c) emitter current to base current
   (d) output voltage to input voltage

5. Two currents that are nearly the same in normal operation are
   (a) collector and base   (b) collector and emitter
   (c) base and emitter   (d) input and output

6. If the base current for a transistor operating below saturation is increased, the collector current
   (a) increases and the emitter current decreases
   (b) decreases and the emitter current decreases
   (c) increases and the emitter current does not change
   (d) increases and the emitter current increases

7. A saturated bipolar transistor can be recognized by
   (a) a very small voltage between the collector and emitter
   (b) V_{CC} between collector and emitter
   (c) a base emitter drop of 0.7 V
   (d) no base current

8. The voltage gain for a common-emitter (CE) amplifier can be expressed as a ratio of
   (a) ac collector resistance to ac input resistance
   (b) ac emitter resistance to ac input resistance
   (c) dc collector resistance to dc emitter resistance
   (d) none of the above

9. The voltage gain for a common-collector (CC) amplifier
   (a) depends on the input signal
   (b) depends on the transistor's β
   (c) is approximately 1
   (d) none of the above

10. In a CE amplifier, the capacitor from emitter to ground is called the
    (a) coupling capacitor   (b) decoupling capacitor
    (c) bypass capacitor   (d) tuning capacitor

11. If the capacitor from emitter to ground in a CE amplifier is removed, the voltage gain
    (a) increases   (b) decreases
    (c) is not affected   (d) becomes erratic

12. When the collector resistor in a CE amplifier is increased in value, the voltage gain
    (a) increases   (b) decreases
    (c) is not affected   (d) becomes erratic

13. The input resistance of a CE amplifier is affected by
    (a) the bias resistors   **(b)** the collector resistor
    (c) answers (a) and (b)   **(d)** neither (a) nor (b)

14. The output signal of a CB amplifier is always
    (a) in phase with the input signal   **(b)** out of phase with the input signal
    (c) larger than the input signal   **(d)** equal to the amplitude of the input signal

15. The output signal of a CC amplifier is always
    (a) in phase with the input signal   **(b)** out of phase with the input signal
    (c) larger than the input signal   **(d)** exactly equal to the input signal

16. A darlington pair is two transistors connected to give
    (a) very high voltage gain   **(b)** very high β
    (c) very low input resistance   **(d)** very low output resistance

17. Compared to CE and CC amplifiers, the common-base (CB) amplifier has
    (a) a lower input resistance   **(b)** a much larger voltage gain
    (c) a larger current gain   **(d)** a higher input resistance

18. Compared to a normal transistor switch, a transistor switch with hysteresis has
    (a) high input impedance   **(b)** faster switching time
    (c) higher output current   **(d)** two switching thresholds

TROUBLESHOOTER'S QUIZ   *Answers are at the end of the chapter.*

Refer to Figure 3–75.

❏ If R_2 is open,

    1. V_B will
        (a) increase   **(b)** decrease   **(c)** not change

    2. V_C will
        (a) increase   **(b)** decrease   **(c)** not change

❏ If R_C is open,

    3. V_B will
        (a) increase   **(b)** decrease   **(c)** not change

    4. V_C will
        (a) increase   **(b)** decrease   **(c)** not change

Refer to Figure 3–78.

❏ If R_E is 560 Ω instead of 390 Ω,

    5. The dc collector voltage will
        (a) increase   **(b)** decrease   **(c)** not change

    6. The voltage gain will
        (a) increase   **(b)** decrease   **(c)** not change

❏ If C_2 is open,

    7. The dc emitter voltage will
        (a) increase   **(b)** decrease   **(c)** not change

    8. The voltage gain will
        (a) increase   **(b)** decrease   **(c)** not change

9. The input impedance will

    (a) increase   (b) decrease   (c) not change

Refer to Figure 3–80.

❑ If V_{CC} is 15 V,

10. The voltage gain will

    (a) increase   (b) decrease   (c) not change

11. The input impedance will

    (a) increase   (b) decrease   (c) not change

■ PROBLEMS

Answers to odd-numbered problems are at the end of the book.

SECTION 3–1   Structure of Bipolar Junction Transistors

1. What is the exact value of I_C for $I_E = 5.34$ mA and $I_B = 47.5$ μA?

2. A certain transistor has an $I_C = 25$ mA and an $I_B = 200$ μA. Determine the β_{DC}.

3. In a certain transistor circuit, the base current is 2% of the 30 mA emitter current. Determine the approximate collector current.

4. Find V_E and I_C in Figure 3–69.

5. Determine the I_B, I_C, and V_C for the transistor circuit in Figure 3–70. Assume $\beta_{DC} = 75$.

6. Draw the dc load line for the transistor circuit in Figure 3–71.

7. Determine I_B, I_C, and V_C in Figure 3–71.

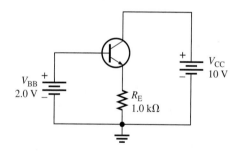

FIGURE 3–69

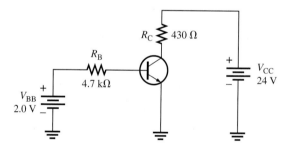

FIGURE 3–70

FIGURE 3–71

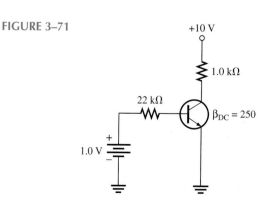

SECTION 3–2 BJT Bias Circuits

8. For the base-biased *npn* transistor in Figure 3–72, assume $\beta_{DC} = 100$. Find I_C and V_{CE}.

9. Repeat Problem 8 for $\beta_{DC} = 300$. (*Hint:* The transistor is now saturated!).

10. For the base-biased *pnp* transistor in Figure 3–73, assume $\beta_{DC} = 200$. Find I_C and V_{CE}.

11. For each of the following conditions in the circuit of Figure 3–73, determine if the collector current will increase, decrease, or remain the same:
   (a) the base is shorted to ground   **(b)** R_C is smaller
   (c) the transistor has a higher β **(d)** the temperature increases
   (e) R_B is smaller

12. For the collector-feedback bias circuit in Figure 3–74, determine I_C and V_{CE}. Assume $\beta_{DC} = 100$.

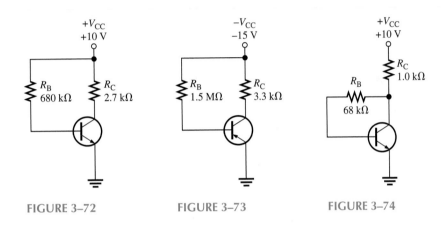

FIGURE 3–72   FIGURE 3–73   FIGURE 3–74

13. For the voltage-divider biased circuit in Figure 3–75, determine I_C and V_{CE}.

14. For the voltage-divider biased (*pnp*) circuit in Figure 3–76, determine I_C and V_{CE}.

15. Determine the end points for the dc load line, $I_{C(sat)}$ and $V_{CE(cutoff)}$ for Figure 3–76.

16. For the emitter-bias circuit in Figure 3–77, determine I_C and V_{CE}.

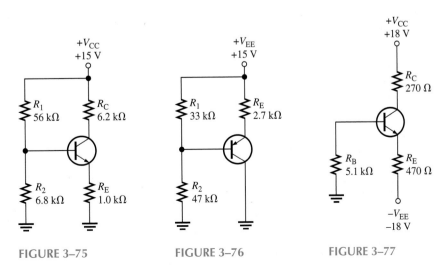

FIGURE 3–75   FIGURE 3–76   FIGURE 3–77

17. For the circuit in Figure 3–77, determine dc power dissipated in the transistor.

18. Assume the transistor in Figure 3–77 is a 2N3904. Can the power supplies be increased to 24 V without exceeding $P_{D(max)}$? (Data sheet is Figure 3–25).

19. Assume that R_C in Figure 3–77 was replaced with a 330 Ω resistor.
    (a) What is the new value of I_C and V_{CE}?
    (b) What is the power dissipated in R_C as a result of this change?
    (c) What is the power dissipated in the transistor as a result of this change?

20. A certain transistor is to be operated at a collector current of 50 mA. How high can V_{CE} go without exceeding a $P_{D(max)}$ of 1.2 W?

SECTION 3–4 Common-Emitter Amplifiers

21. Determine the dc voltages, V_B, V_E, and V_C, with respect to ground in Figure 3–78.

22. Determine the voltage gain for the CE amplifier in Figure 3–78.

23. The amplifier in Figure 3–79 has a variable gain control, using a 100 Ω potentiometer for R_E with the wiper ac grounded. As the potentiometer is adjusted, more or less of R_E is bypassed to ground, thus varying the gain. The total R_E remains constant to dc, keeping the bias fixed. Determine the maximum and minimum gains for this amplifier.

24. If a load resistance of 600 Ω is placed on the output of the amplifier in Figure 3–79, what is the maximum gain?

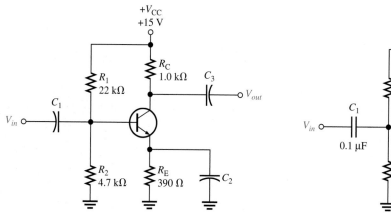

FIGURE 3–78

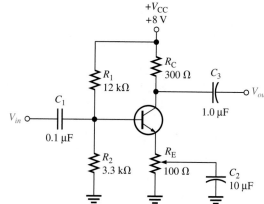

FIGURE 3–79

SECTION 3–5 Common-Collector Amplifiers

25. For the CC amplifier in Figure 3–80, compute the total ac input resistance and the current gain to the load. Assume $\beta_{ac} = 150$.

26. Draw the ac equivalent circuit for the CC amplifier in Figure 3–80.

27. Compute the ac saturation current ($I_{c(sat)}$) and ac cutoff voltage ($V_{ce(cutoff)}$) for the CC amplifier in Figure 3–80.

28. For the *pnp* CC amplifier in Figure 3–76, show where the input and output signals should be connected.

FIGURE 3–80

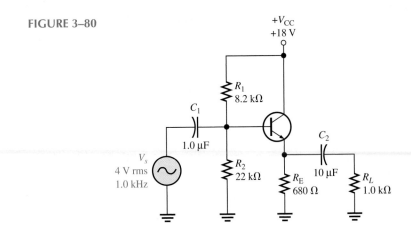

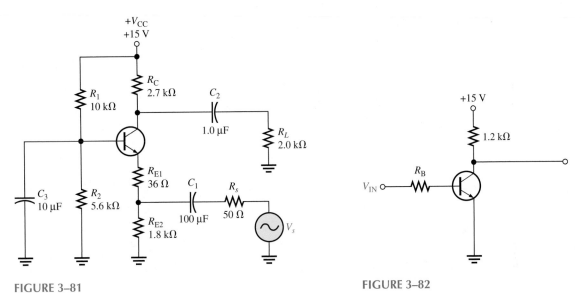

FIGURE 3–81

FIGURE 3–82

SECTION 3–6 Common-Base Amplifiers

29. What is the main disadvantage of the CB amplifier compared to the CE and the emitter-follower?

30. For the CB amplifier in Figure 3–81, compute the following: V_B, V_E, V_C, V_{CE}, r'_e, A_v.

31. For the CB amplifier in Figure 3–81, compute the input resistance, $R_{in(tot)}$.

32. For the CB amplifier in Figure 3–81, what is the purpose of R_{E1}?

SECTION 3–7 The Bipolar Transistor as a Switch

33. Determine $I_{C(sat)}$ for Q_1 and Q_2 in Figure 3–47.

34. The transistor in Figure 3–82 has $\beta_{DC} = 100$. Determine the maximum value of R_B that will ensure saturation when V_{IN} is 5 V.

SECTION 3–8   Transistor Packages and Terminal Identification

35. Identify the leads on the transistors in Figure 3–83. Bottom views are shown.

36. What is the most probable category of each transistor in Figure 3–84?

FIGURE 3–83

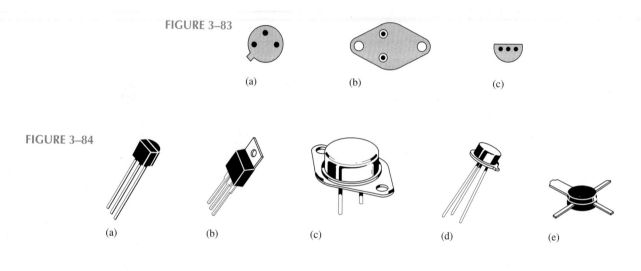

(a)   (b)   (c)

FIGURE 3–84

(a)   (b)   (c)   (d)   (e)

SECTION 3–9   Troubleshooting

37. In an out-of-circuit test of a good *npn* transistor, what should an analog ohmmeter indicate when its positive probe is touching the emitter and the negative probe is touching the base? When its positive probe is touching the base and the negative probe is touching the emitter?

38. What is the most likely problem, if any, in each circuit of Figure 3–85? Assume a $\beta_{DC} = 75$.

39. What is the value of the dc beta of each transistor in Figure 3–86?

SECTION 3–10   A System Application

40. This problem relates to the circuit board and schematic in Figure 3–68. A remote switch loop is connected between pins 2 and 3. When the remote switches are closed, the relay (RY1) contacts between pin 10 and pin 11 are normally open. When a remote switch is opened, the relay contacts do not close. Determine the possible causes of this malfunction.

41. This problem relates to Figure 3–68. The relay contacts remain closed between pins 10 and 11 no matter what any of the inputs are. This means that the relay is energized continuously. What are the possible faults?

42. This problem relates to Figure 3–68. Pin 9 stays at approximately 0.1 V, regardless of the input at pin 7. What do you think is wrong? What would you check first?

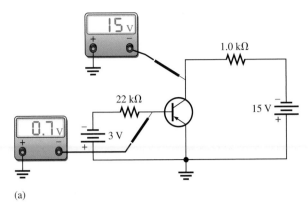

(a)

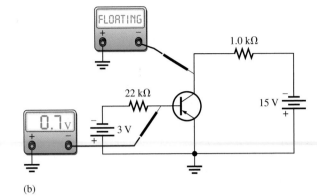

(b)

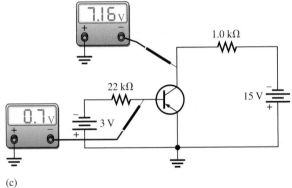

(c)

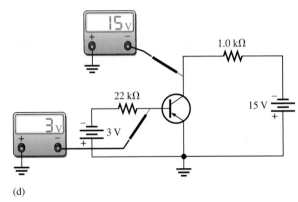

(d)

FIGURE 3–85

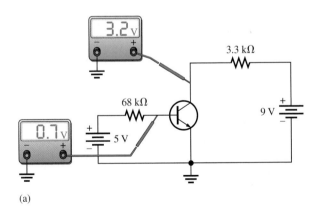

(a)

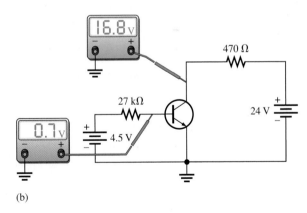

(b)

FIGURE 3–86

■ **ANSWERS TO REVIEW QUESTIONS**

Section 3–1

1. Emitter, base, collector

2. Saturation means there is maximum conduction and the voltage from collector to emitter is close to zero. Cutoff is essentially when there is no collector current and the power supply voltage appears between the collector and the emitter.

3. The ratio of collector current to base current in a bipolar junction transistor

Section 3–2

1. Base, collector-feedback, voltage-divider, and emitter

2. The following steps are for an *npn* transistor with a positive supply voltage:
   (a) Compute the base voltage using the voltage-divider rule.
   (b) Subtract 0.7 V to obtain the emitter voltage.
   (c) Apply Ohm's law to the emitter resistor to obtain the approximate collector current.
   (d) Using the collector current, find the voltage across the collector resistor by Ohm's law.
   (e) Subtract the drop across the collector resistor from the power supply to obtain the collector voltage.
   (f) Subtract the emitter voltage from the collector voltage to obtain V_{CE}.

3. Approximately +1 V. (This result assumes the base resistor is connected to ground and drops a few tenths of a volt.)

Section 3–3

1. Increases

2. β_{DC} increases with I_C to a certain value and then decreases.

3. Multiply I_C by V_{CE}.

4. 40 mA

5. A coupling capacitor is in series with the signal and passes it to or from a transistor. A bypass capacitor is in parallel with the signal and provides an ac path around a resistor.

Section 3–4

1. The input terminal is the base; the output terminal is the collector.

2. High input resistance reduces the loading effect on a source.

3. The ac collector resistance is divided by the ac emitter resistance.

Section 3–5

1. Emitter-follower

2. 1.0

3. Current gain, high input resistance, low output resistance

Section 3–6

1. Yes

2. Very low

3. Improved linearity, gain stability, increased input impedance

Section 3–7

1. Saturation (on) and cutoff (off)

2. At saturation

3. At cutoff

4. At cutoff

5. Two different switching thresholds

Section 3–8

1. Three categories of BJT are small signal/general purpose, power, and RF.

2. Going clockwise from tab: emitter, base, and collector (bottom view).

3. The metal mounting tab or case in power transistors is connected to the collector terminal.

Section 3–9

1. First, test it in-circuit.
2. If R_B opens, the transistor is in cutoff.
3. The base voltage is $+3$ V and the collector voltage is $+9$ V.

Section 3–10

1. To detect when there is an open in one of the remote loops
2. Two, zone 1 and zone 2 are identical.
3. When the relay is activated as a result of an open switch in zone 1 or zone 2 loops
4. To clip off any negative voltage induced in the coil to prevent possible damage to a transistor

■ **ANSWERS TO PRACTICE EXERCISES FOR EXAMPLES**

3–1 $I_B = 0.241$ mA, $I_C = 21.7$ mA, $I_E = 21.9$ mA, $V_{CE} = 4.23$ V, $V_{CB} = 3.53$ V

3–2   Along the x-axis

3–3   221

3–4   When β_{DC} is 100, V_{CE} is 5.6 V; when β_{DC} is 300, V_{CE} is 3.0 V.

3–5 $V_B = 1.51$ V, $V_E = 0.81$ V, $I_E = 1.73$ mA, $I_C = 1.73$ mA, $V_{CE} = 6.51$ V

3–6 $V_B = -1.78$ V, $V_E = -1.08$ V, $I_E = 0.90$ mA, $I_C = 0.90$ mA, $V_{CE} = -5.88$ V

3–7 $V_E = -0.7$ V

3–8 $V_{CC(max)} = 44.5$ V; $V_{CE(max)}$ is exceeded first.

3–9   250 Ω

3–10   Gain is reduced to -3.65.

3–11   3.15 kΩ

3–12   No effect on Q-point but the ac load line is steeper; $I_{c(sat)} = 8.1$ mA and $V_{ce(cutoff)} = 12.6$ V.

3–13   The voltage gain is still 1.0. The current gain rises to 19.

3–14   2.4 V pp

3–15 -6.99 V

3–16   78.4 μA

3–17 R_B is open (this can be due to a broken trace or pad at the contact).

■ **ANSWERS TO SELF-TEST**

1. (b)	2. (a)	3. (c)	4. (b)	5. (b)
6. (d)	7. (a)	8. (d)	9. (c)	10. (c)
11. (b)	12. (a)	13. (a)	14. (a)	15. (a)
16. (b)	17. (a)	18. (d)		

■ **ANSWERS TO TROUBLE-SHOOTER'S QUIZ**

1. increase	2. decrease	3. decrease	4. decrease
5. increase	6. decrease	7. not change	8. decrease
9. increase	10. not change	11. not change	

4

FIELD-EFFECT TRANSISTORS (FETs)

Courtesy Yuba College

■ CHAPTER OBJECTIVES

❑ Describe the basic classifications for field-effect transistors (FETs)

❑ Describe the construction and operation of junction field-effect transistors (JFETs)

❑ Describe three bias methods for JFETs and explain how each method works

❑ Explain the operation of metal-oxide semiconductor field-effect transistors (MOSFETs)

❑ Discuss and analyze MOSFET bias circuits

❑ Describe the operation of FET linear amplifiers

❑ Discuss two switching applications of FETs

❑ Apply what you have learned in this chapter to a system application

■ CHAPTER INTRODUCTION

This chapter introduces the field-effect transistor (FET), a transistor that works on an entirely different principle than the bipolar junction transistor (BJT). Although the idea for the FET precedes the invention of the BJT by decades, it wasn't until the 1960s that commercial production of FETs was possible. For certain applications, the FET is superior to the BJT. In other applications, a mix of the two types produces a circuit with optimum characteristics. A circuit that takes advantage of the best characteristics of both types is presented in the system application in Section 4–8.

■ A SYSTEM APPLICATION

The system for this chapter is a small preamplifier that uses both field-effect and bipolar transistors to take advantage of the best characteristics of both types. The input circuit uses a field-effect transistor because of the extremely high input resistance it offers; it has current-source biasing provided by another FET. The signal is then amplified by two BJTs to provide reasonably high gain with good linearity. The amplifier in this system application is dc coupled; for this reason, it needs no bias resistors or coupling capacitors, thus reducing the parts count and cost. A preamp such as this can serve as an "active antenna" for a radio receiver in outlying areas.

For the system application in Section 4–8, in addition to the other topics, be sure you understand

❑ How a JFET works
❑ What a common-drain amplifier with current-source biasing is
❑ How to estimate the dc and ac voltages in an amplifier
❑ How to relate a schematic to a printed circuit (PC) board layout

www.   VISIT THE COMPANION WEBSITE

Study Aids for This Chapter Are Available at

http://www.prenhall.com/floyd

4–1 ■ STRUCTURE OF FIELD-EFFECT TRANSISTORS (FETs)

Recall that the bipolar junction transistor (BJT) is a current-controlled device; that is, the base current controls the amount of collector current. The field-effect transistor (FET) is a voltage-controlled device in which the voltage at the gate terminal controls the amount of current through the device. Both the BJT and the FET can be used as an amplifier and in switching applications.

After completing this section, you should be able to

❑ Describe the basic classifications for field-effect transistors (FETs)
  ❑ Discuss principal differences between FETs and BJTs

The FET Family

Field-effect transistors (FETs) are a class of semiconductors that operate on an entirely different principle than bipolar transistors. In a FET, a narrow conducting channel is connected to two leads called the **source** and the **drain**. This channel is made from either an *n*-type or *p*-type material. As the name *field-effect* implies, conduction in the channel is controlled by an electric field, established by a voltage applied to a third lead called the **gate**. In a *junction* FET (or JFET), the gate forms a *pn* junction with the channel. The other type of FET, called the *MOSFET* (for *Metal Oxide Semiconductor FET*), uses an insulated gate to control conduction in the channel. (The terms *insulated gate* and *MOSFET* refer to the same type of device.) The insulation is an extremely thin layer (<1 μm) of glass (typically SiO_2). Figure 4–1 is an overview of the FET family, showing the various types available. All of these types are discussed in this chapter.

FETs were actually thought of long before bipolar junction transistors (BJTs). J. E. Lilienfeld applied for a patent for a FET in 1925 (granted in 1930), but it wasn't until the 1960s that FETs became commercially available. Today, FETs are used in most computer

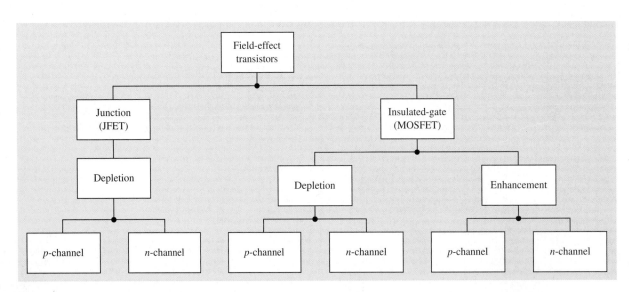

FIGURE 4–1
Classification of field-effect transistors.

integrated circuits (ICs) because of several important advantages they have over BJTs, particularly with respect to manufacturing of large-scale ICs. For digital circuits, MOSFETs have become the dominant type of transistor for several reasons. They can be fabricated in much smaller areas than BJTs, they are relatively easy to manufacture on ICs, and they produce simpler circuits with no resistors or diodes. Most microprocessors and computer memories use FET technology. A brief look at how FETs are used in certain ICs is included in Section 4–7.

Compared to the BJT, the FET family is more diverse. A characteristic that differs between various types of FETs is their dc behavior. For example, JFETs are biased differently than E-MOSFETs. For this reason, the dc bias characteristics for each type are discussed in this chapter separately. Fortunately, bias circuits are fairly easy to understand. Before proceeding to bias circuits, the characteristics of the transistors that make up the FET family will be discussed.

Common to all FETs is very high input resistance and low electrical noise. In addition, both JFETs and MOSFETs respond the same way to ac signals and have similar ac equivalent circuits. JFETs achieve their high input resistance because the input *pn* junction is always operated with reverse bias; MOSFETs achieve their high input resistance because of the insulated gate. Although all FETs have high input resistance, they do not have the high gain of bipolar junction transistors. BJTs are also inherently more linear than FETs. For certain applications, FETs are superior; for other applications, BJTs are superior. Many designs take advantage of both types and include a mix of FETs and BJTs. You need to understand both types of transistors.

4–1 REVIEW QUESTIONS*

1. What are the three terminals of a FET called?
2. What is another name for an insulated-gate FET?
3. Why are MOSFETs the dominant type of transistor used in ICs?
4. What are some important differences between BJTs and FETs?

_____

* Answers are at the end of the chapter.

4–2 ■ JFET CHARACTERISTICS

In this section, you will see how the JFET operates as a voltage-controlled, constant-current device and study the drain characteristic curve and the transconductance curve. You will also learn about cutoff and pinch-off as well as JFET input resistance and capacitance.

After completing this section, you should be able to

❑ Describe the construction and operation of junction field-effect transistors (JFETs)
  ❑ Draw the symbol for an *n*-channel or a *p*-channel JFET
  ❑ Interpret the drain characteristic curve for a JFET including the ohmic and constant-current regions
  ❑ Explain the parameters g_m, I_{DSS}, I_{GSS}, C_{iss}, $V_{GS(off)}$, and V_P
  ❑ Describe the transconductance curve for a JFET and explain how it relates to the drain characteristic curve

JFET Operation

Figure 4–2(a) shows the basic structure of an *n*-channel **junction field-effect transistor** (**JFET**). Wire leads are connected to each end of the *n* channel; the drain is shown at the upper end, and the source is at the lower end. This channel is a conductor: for an *n*-channel JFET, electrons are the carrier; for a *p*-channel JFET, holes are the carrier. With no external voltages, the channel can conduct current in either direction.

FIGURE 4–2
Basic structure of the two types of JFET.

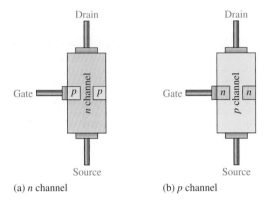

(a) *n* channel   (b) *p* channel

In an *n*-channel device, a *p* material is diffused into the *n*-channel to form a *pn* junction and is connected to the gate lead. The diagram in Figure 4–2(a) shows *p*-material diffused into two regions that are normally connected internally by the manufacturer to form a single gate. (A special-purpose JFET, called a dual-gate JFET, has a separate lead to each of these regions.) In the structure diagrams, the interconnection of both *p* regions is omitted for simplicity, with a connection to only one shown. A *p*-channel JFET is shown in Figure 4–2(b).

As previously stated, the channel in a JFET is a narrow conduction path between the gate and the source. The width of the channel, and therefore its ability to conduct current, is controlled by the gate voltage. With no gate voltage, the channel conducts the maximum current. When reverse bias is applied to the gate, the channel width narrows, and the conductivity drops.

To illustrate this operation, Figure 4–3(a) shows normal operating voltages applied to an *n*-channel device. V_{DD} provides a positive drain-to-source voltage, causing electrons to flow from the source to the drain. For an *n*-channel JFET, reverse-biasing of the gate-source junction is done with a negative gate voltage. V_{GG} sets the reverse-biased voltage between the gate and the source, as shown. Notice that there should *never* be any forward-biased junctions in a FET; this is one of the principal differences between FETs and BJTs.

The channel width, and thus the channel resistance, is controlled by varying the gate voltage, thereby controlling the amount of drain current, I_D. This concept is illustrated in Figure 4–3(b) and (c). The white areas represent the depletion region created by the reverse bias. It is wider toward the drain end of the channel because the reverse-biased voltage between the gate and the drain is greater than that between the gate and the source.

JFET Symbols

The schematic symbols for both *n*-channel and *p*-channel JFETs are shown in Figure 4–4. Notice that the arrow on the gate points "in" for *n*-channel and "out" for *p*-channel.

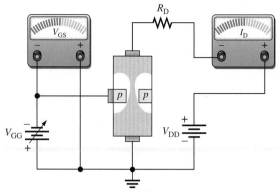

(a) JFET biased for conduction

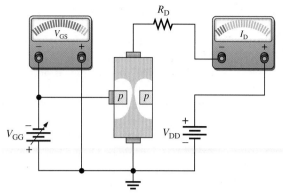

(b) Greater V_{GG} narrows the channel which increases the resistance of the channel and decreases I_D.

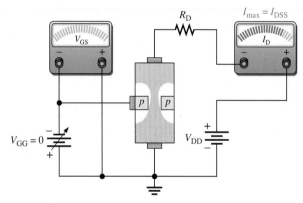

(c) Less V_{GG} widens the channel which decreases the resistance of the channel and increases I_D.

FIGURE 4–3

Effects of V_{GG} on channel width, resistance, and drain current ($V_{GG} = V_{GS}$).

FIGURE 4–4
JFET schematic symbols.

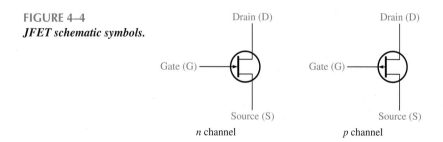

n channel

p channel

Drain Characteristic Curve

The drain characteristic curve is a plot of the drain current, I_D, versus the drain-to-source voltage, V_{DS}, which corresponds to a BJT's collector current, I_C, versus collector-to-emitter voltage, V_{CE}. There are, however, some significant differences between BJT characteristics and FET characteristics. Since the FET is a voltage-controlled device, the third variable on the FET characteristic (V_{GS}) has units of voltage instead of current (I_B) in the case of the BJT.

The characteristics for *n*-channel devices are introduced in this section. *P*-channel devices operate in the same way but with opposite polarities. Generally, *n*-channel JFETs have better specifications than their *p*-channel counterparts, so they are more popular.

Consider an *n*-channel JFET where the gate-to-source voltage is zero ($V_{GS} = 0$ V). This zero voltage is produced by shorting the gate to the source, as in Figure 4–5(a) where both are grounded. As V_{DD} (and thus V_{DS}) is increased from 0 V, I_D will increase proportionally, as shown in the graph of Figure 4–5(b) between points *A* and *B*. In this region, the channel resistance is essentially constant because the depletion region is not large enough to have a significant effect. This region is called the **ohmic region** because V_{DS} and I_D are related by Ohm's law. The value of the resistance can be changed by the gate voltage; thus, it is possible to use a JFET as a voltage-controlled resistor. An application will be shown later in Figure 10–11 (Wien bridge). Further applications are found in Experiments 14,15, and 27 in the accompanying Laboratory Exercises Manual.

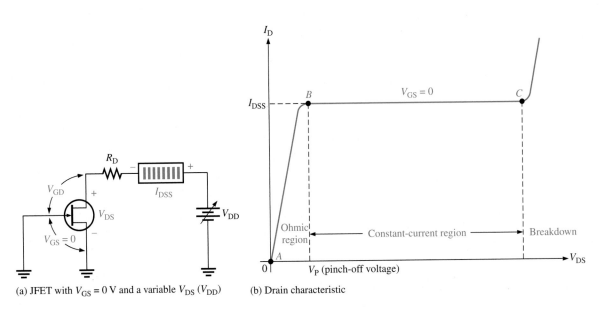

(a) JFET with $V_{GS} = 0$ V and a variable V_{DS} (V_{DD})   (b) Drain characteristic

FIGURE 4–5
The drain characteristic curve of a JFET for $V_{GS} = 0$ V showing pinch-off.

At point *B* in Figure 4–5(b), the curve levels off and I_D becomes essentially constant. As V_{DS} increases from point *B* to point *C*, the reverse-bias voltage from gate to drain (V_{GD}) produces a depletion region large enough to offset the increase in V_{DS}, thus keeping I_D relatively constant. This region is called the **constant-current region**.

Pinch-Off Voltage

For $V_{GS} = 0$ V, the value of V_{DS} at which I_D becomes essentially constant (point *B* on the curve in Figure 4–5 (b)) is the **pinch-off voltage**, V_P. Notice that the pinch-off voltage is a positive value for an *n*-channel JFET. For a given JFET, V_P has a fixed value. As you can see, a continued increase in V_{DS} above the pinch-off voltage produces an almost constant drain current. This value of drain current is I_{DSS} (*D*rain to *S*ource current with gate *S*horted) and is always specified on JFET data sheets. I_{DSS} is the maximum drain current that a specific JFET can produce regardless of the external circuit, and it is always specified for the condition, $V_{GS} = 0$ V.

Continuing along the graph in Figure 4–5(b), breakdown occurs at point C when I_D begins to increase very rapidly with any further increase in V_{DS}. Breakdown can result in irreversible damage to the device, so JFETs are always operated below breakdown and within the constant-current region (between points B and C on the graph).

V_{GS} Controls I_D

Let's connect a bias voltage, V_{GG}, from gate to source as shown in Figure 4–6(a). As V_{GS} is set to increasingly more negative values by adjusting V_{GG}, a family of drain characteristic curves is produced as shown in Figure 4–6(b). Notice that I_D decreases as the magnitude of V_{GS} is increased to larger negative values because of the narrowing of the channel. Also notice that, for each increase in V_{GS}, the JFET reaches pinch-off (where constant current begins) at values of V_{DS} less than V_P. So, the amount of drain current is controlled by V_{GS}.

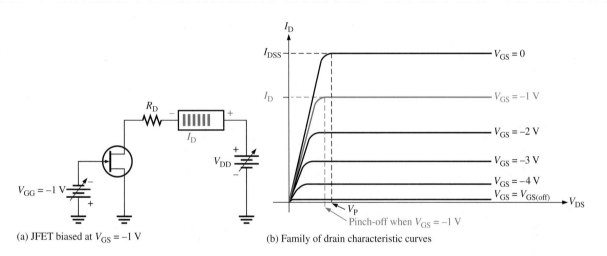

(a) JFET biased at $V_{GS} = -1$ V

(b) Family of drain characteristic curves

FIGURE 4–6

Pinch-off occurs at a lower V_{DS} as V_{GS} is increased to more negative values.

Cutoff Voltage

The value of V_{GS} that makes I_D approximately zero is the cutoff voltage, $V_{GS(off)}$. The JFET must be operated between $V_{GS} = 0$ V and $V_{GS(off)}$. For this range of gate-to-source voltages, I_D will vary from a maximum of I_{DSS} to a minimum of almost zero.

As you have seen, for an n-channel JFET, the more negative V_{GS} is, the smaller I_D becomes in the constant-current region. When V_{GS} has a sufficiently large negative value, I_D is reduced to zero. This cutoff effect is caused by the widening of the depletion region to a point where it completely closes the channel as shown in Figure 4–7.

Comparison of Pinch-Off and Cutoff

The pinch-off voltage is measured on the drain characteristic. For an n-channel device, it is the positive voltage at which the drain current becomes constant when $V_{GS} = 0$ V. Cutoff can also be measured on the drain characteristic and represents the negative gate-to-source voltage that reduces the drain current to zero.

FIGURE 4–7
JFET at cutoff.

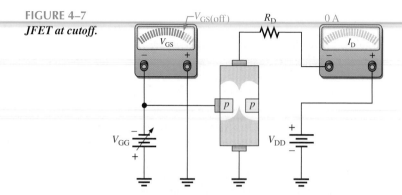

$V_{GS(off)}$ and V_P are always equal in magnitude but opposite in sign. A data sheet usually will give either $V_{GS(off)}$ or V_P, but not both. However, when you know one, you have the other. For example, if $V_{GS(off)} = -5$ V, then $V_P = +5$ V.

EXAMPLE 4–1

For the *n*-channel JFET in Figure 4–8, $V_{GS(off)} = -4$ V and $I_{DSS} = 12$ mA. Determine the *minimum* value of V_{DD} required to put the device in the constant-current region of operation.

FIGURE 4–8

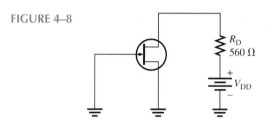

Solution Since $V_{GS(off)} = -4$ V, $V_P = 4$ V. The minimum value of V_{DS} for the JFET to be in its constant-current region is

$$V_{DS} = V_P = 4 \text{ V}$$

In the constant-current region with $V_{GS} = 0$ V,

$$I_D = I_{DSS} = 12 \text{ mA}$$

The drop across the drain resistor is

$$V_{R_D} = (12 \text{ mA})(560 \text{ }\Omega) = 6.7 \text{ V}$$

Applying Kirchhoff's law around the drain circuit gives

$$V_{DD} = V_{DS} + V_{R_D} = 4 \text{ V} + 6.7 \text{ V} = \textbf{10.7 V}$$

This is the minimum value of V_{DD} to make $V_{DS} = V_P$ and to put the device in the constant-current region.

*Practice Exercise** If V_{DD} is increased to 15 V, what is the drain current?

---

* Answers are at the end of the chapter.

JFET Transconductance Curves

A useful way of looking at any circuit is to show the output for a given input, as done earlier in Section 1–4 for an amplifier. This characteristic is called a *transfer curve.*

Since the JFET is controlled by a negative voltage on the input (gate) and the output is drain current, the transfer curve is a plot of I_D, plotted on the y-axis, as a function of V_{GS}, plotted on the x-axis. When the output unit (mA) is divided by the input unit (V), the result is the unit of conductance (mS). You can think of a voltage at the input being transferred to the output as a current; thus, the prefix "trans" is added to the word *conductance* forming the word **transconductance**. The transconductance curve is a plot of the transfer characteristic (I_D versus V_{GS}) of a FET. Transconductance is listed in data sheets as g_m or y_{fs}.

A representative curve for an n-channel JFET is shown in Figure 4–9(a). Generally, all types of FETs have a transconductance curve with this same basic shape. The curves shown are typical for the MPF102[1], a general-purpose n-channel JFET.

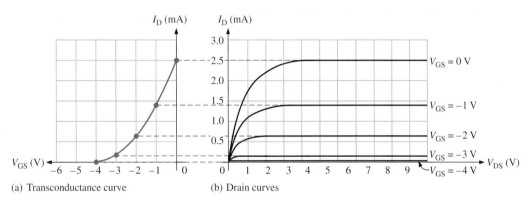

(a) Transconductance curve   (b) Drain curves

FIGURE 4–9
Representative characteristic curves for an MPF102 n-channel JFET.

The transconductance characteristic is directly related to the drain characteristic as shown in Figure 4–9(b). Notice that both plots have the same vertical axis, representing I_D. Transconductance is an ac parameter so its value is found at any point on the curve by dividing a small change in drain current by a small change in gate-to-source voltage.

$$g_m = \frac{\Delta I_D}{\Delta V_{GS}}$$

This equation can be written with ac notation as simply

$$g_m = \frac{I_d}{V_{gs}} \qquad \textbf{(4–1)}$$

The transconductance curve is not a straight line, implying that the relation between the output current and the input voltage is nonlinear. This is an important point: FETs have a nonlinear transconductance curve. This means that they tend to add distortion to an input signal. Distortion is not always a bad thing; for example, in radio frequency mixers, JFETs have an advantage over BJTs because of this characteristic. However, some JFETs (such as

---

[1] Data sheet for MPF102 available at http://www.onsemi.com

the 2N4339) are designed with physical geometries that minimize distortion for audio applications. In addition, the designer can minimize distortion by keeping the signal level low (below about 100 mV). Other design techniques (such as the biasing method used in the system application in Section 4–8) have been used to minimize distortion.

EXAMPLE 4–2

For the curve in Figure 4–10, determine the transconductance at $I_D = 1.0$ mA.

FIGURE 4–10

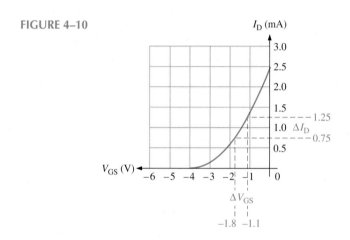

Solution Select a small change in I_D and divide it by the corresponding change in V_{GS} at 1.0 mA. The graphical method is shown in Figure 4–10. From the graph, the transconductance is

$$g_m = \frac{\Delta I_D}{\Delta V_{GS}} = \frac{1.25 \text{ mA} - 0.75 \text{ mA}}{-1.1 \text{ V} - (-1.8 \text{ V})} = \textbf{0.714 mS}$$

Practice Exercise Find the transconductance at $I_D = 1.5$ mA.

JFET Input Resistance and Capacitance

As you know, a *pn* junction has a very high resistance when it is reverse-biased. A JFET operates with its gate-source junction reverse-biased; therefore, the input resistance at the gate is very high. This very high input resistance is a major advantage of the JFET over the bipolar junction transistor with its forward-biased base-emitter junction.

JFET data sheets often specify the input resistance by giving a value for the gate reverse current, I_{GSS}, at a certain gate-to-source voltage. The input resistance can then be determined using the following equation. The vertical lines indicate an absolute value (an unsigned value).

$$R_{IN} = \left| \frac{V_{GS}}{I_{GSS}} \right| \tag{4–2}$$

For example, the 2N5457[2] data sheet lists a maximum I_{GSS} of -1 nA for $V_{GS} = -15$ V at 25°C. Using these values, you find that the input resistance is

$$R_{IN} = \left| \frac{V_{GS}}{I_{GSS}} \right| = \frac{15 \text{ V}}{1 \text{ nA}} = 15 \text{ G}\Omega$$

As you can see from this result, the input resistance of this JFET is incredibly high. However, R_{IN} drops considerably as temperature increases (as shown in Example 4–3).

The reverse-biased *pn* junction at the input provides the high input resistance associated with a reverse-biased diode, but it also means that a JFET will generally have higher input capacitance than a bipolar junction transistor. Recall that a reverse-biased *pn* junction acts as a capacitor whose capacitance depends on the amount of reverse voltage (see Section 2–8). The input capacitance, C_{iss}, of a JFET is greater than that of a BJT because of this reverse-biased *pn* junction. For example, the 2N5457 has a maximum C_{iss} of 7 pF for $V_{GS} = 0$ V.

EXAMPLE 4–3

The data sheet for an MPF3821 *n*-channel JFET shows a maximum I_{GSS} of -0.1 nA at 25°C for $V_{GS} = -30$ V and a maximum I_{GSS} of -100 nA at 150°C for $V_{GS} = -30$ V. Determine the minimum input resistance at 25°C.

Solution
$$R_{IN} = \left| \frac{V_{GS}}{I_{GSS}} \right| = \frac{30 \text{ V}}{0.1 \text{ nA}} = \textbf{300 G}\boldsymbol{\Omega}$$

Practice Exercise   Determine the minimum input resistance for the MPF3821 JFET at a temperature of 150°C.

4–2 REVIEW QUESTIONS

1. What is another name for the transfer curve for a JFET?
2. Does a *p*-channel JFET require a positive or a negative voltage for V_{GS}?
3. How is the drain current controlled in a JFET?
4. The drain-to-source voltage at the pinch-off point of a particular JFET is 7 V. If the gate-to-source voltage is zero, what is V_P?
5. The V_{GS} of a certain *n*-channel JFET is increased negatively. Does the drain current increase or decrease?
6. What value must V_{GS} have to produce cutoff in a *p*-channel JFET with a $V_P = -3$ V?

4–3 ■ JFET BIASING

Using some of the JFET characteristics discussed in the previous section, we will now see how to dc bias JFETs. The purpose of biasing is to select a proper dc gate-to-source voltage to establish a desired value of drain current. Because the gate is reverse-biased, the methods for applying bias with a bipolar junction transistor do not work for JFETs.

---

[2] Data sheet for 2N5457 available at http://www.onsemi.com

After completing this section, you should be able to

❑ Describe three bias methods for JFETs and explain how each method works
   ❑ Use a transconductance curve to choose a reasonable value for a self-bias resistor
   ❑ Explain how voltage-divider bias or current-source bias produces a more stable Q-point than self-bias

Self-Biasing a JFET

Biasing a FET is relatively easy. An *n*-channel JFET is shown for the following examples. Keep in mind that a *p*-channel JFET just reverses the polarities. To set up reverse bias requires a negative V_{GS} for an *n*-channel JFET. This can be achieved using the self-bias arrangement shown in Figure 4–11. Notice that the gate is biased at 0 V by resistor R_G connected to ground. Although reverse leakage current, I_{GSS}, does produce a very tiny voltage across R_G, it is neglected in most cases; it can be assumed that R_G has no current and no voltage drop across it. The purpose of R_G is to tie the gate to a solid 0 V without affecting any ac signal that will be added later. Since the gate current is negligible, R_G can be large (typically 1.0 MΩ or more), resulting in very high input resistance to low frequency ac signals.[3]

FIGURE 4–11

Self-biased n-channel JFET.

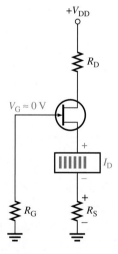

If the gate is at zero volts, how do you obtain the required negative bias on the gate-source junction? The answer is that you make the source *positive* with respect to the gate, producing the required reverse bias. For the *n*-channel JFET in Figure 4–11, I_D produces a voltage drop across R_S with the polarity shown, making the source terminal positive with respect to ground. Since $V_G = 0$ V, and $V_S = I_D R_S$, the gate-to-source voltage is

$$V_{GS} = V_G - V_S = 0 - I_D R_S$$

Thus,

$$V_{GS} = -I_D R_S$$

This result shows that the gate-to-source voltage is negative, producing the required reverse bias. In this analysis, an *n*-channel JFET was used for illustration. Again, the *p*-channel JFET also requires reverse bias, but the polarity of all voltages is opposite those of the *n*-channel JFET.

---

[3] The capacitance effect at high frequencies can significantly reduce the effective input impedance.

The drain voltage with respect to ground is determined as follows:

$$V_D = V_{DD} - I_D R_D \qquad \text{(4–3)}$$

Since $V_S = I_D R_S$, the drain-to-source voltage is

$$V_{DS} = V_D - V_S$$

$$V_{DS} = V_{DD} - I_D(R_D + R_S) \qquad \text{(4–4)}$$

EXAMPLE 4–4

Find V_{DS} and V_{GS} in Figure 4–12. For the particular JFET in this circuit, the internal parameter values such as g_m, $V_{GS(off)}$, and I_{DSS} are such that a drain current (I_D) of approximately 5.0 mA is produced. Another JFET, even of the same type, may not produce the same results when connected in this circuit due to variations in parameter values.

FIGURE 4–12

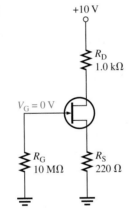

Solution $V_S = I_D R_S = (5.0 \text{ mA})(220 \text{ } \Omega) = 1.10 \text{ V}$

$V_D = V_{DD} - I_D R_D = 10 \text{ V} - (5.0 \text{ mA})(1.0 \text{ k}\Omega) = 5.00 \text{ V}$

Therefore,

$$V_{DS} = V_D - V_S = 5.00 \text{ V} - 1.10 \text{ V} = \textbf{3.90 V}$$

and

$$V_{GS} = V_G - V_S = 0 \text{ V} - 1.10 \text{ V} = \textbf{–1.10 V}$$

Practice Exercise   Determine V_{DS} and V_{GS} in Figure 4–12 when $I_D = 3.0$ mA.

Graphical Methods

Recall that the *IV* curve for a resistor, *R,* is a straight line with a slope of $1/R$. To compare the plot of the self-bias resistor with the transconductance curve, both lines are plotted in the second quadrant; the resistance is plotted with a slope of $-1/R$.

The transconductance curve for the MPF3821 can be used to illustrate how a reasonable value of a self-bias resistor (R_S) is selected. Assume you have an MPF3821 with the transconductance curve shown in Figure 4–13. Draw a straight line from the origin to the

FIGURE 4–13

Graphical analysis of self-bias.

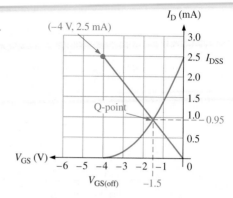

point where $V_{GS(off)}$ (-4 V) intersects I_{DSS} (2.5 mA). The reciprocal of the slope of this line represents a reasonable choice for R_S.

$$R_S = \frac{|V_{GS(off)}|}{I_{DSS}} = \frac{4 \text{ V}}{2.5 \text{ mA}} = 1.6 \text{ k}\Omega$$

The absolute (unsigned) value of $V_{GS(off)}$ is used. The resulting 1.6 kΩ resistor is available as a standard 5% value, or you could select a 1.5 kΩ standard 10% resistor instead. The point where the two lines cross represents the Q-point.[4] This Q-point represents V_{GS} and I_D for this particular case; it shows that $V_{GS} = -1.5$ V at $I_D = 0.95$ mA.

Self-bias produces a form of negative feedback to help compensate for different device characteristics between various JFETs. For instance, assume the transistor is replaced with one with a lower transconductance. As a result, the new drain current will be less, causing a smaller voltage drop across R_S. This reduced voltage tends to turn the JFET on more, compensating for the lower transconductance of the new transistor. The effect of a range of transconductance curves is best illustrated by two examples.

EXAMPLE 4–5

A 2N5457 general-purpose JFET has the following specifications: $I_{DSS(min)} = 1$ mA, $I_{DSS(max)} = 5$ mA, $V_{GS(off)(min)} = -0.5$ V, $V_{GS(off)(max)} = -6$ V. Select a self-bias resistor for this JFET.

Solution Typical of small-signal JFETs, the range of I_{DSS} and $V_{GS(off)}$ is very large. To select the best resistor, check the extremes of the specified values $V_{GS(off)}$ and I_{DSS}.

$$R_S = \frac{|V_{GS(off)(min)}|}{I_{DSS(min)}} = \frac{0.5 \text{ V}}{1.0 \text{ mA}} = 500 \text{ }\Omega$$

$$R_S = \frac{|V_{GS(off)(max)}|}{I_{DSS(max)}} = \frac{6 \text{ V}}{5.0 \text{ mA}} = 1.2 \text{ k}\Omega$$

A good choice is **820 Ω,** a standard value between these extremes. To see what this looks like on the transconductance curve, sketch the curves with this resistor and plot the maximum and minimum Q-points. This is done in Figure 4–14. Despite the ex-

---

[4] This is a form of load line analysis. This type of load line is called the *bias load line.*

FIGURE 4–14

Effect of a wide range of transconductance curves on the Q-point.

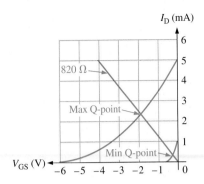

treme variation between the minimum and maximum specification, the 820 Ω resistor represents a good choice for either.

Practice Exercise Estimate the largest and smallest I_D expected for a 2N5457 that is self-biased with an 820 Ω resistor.

EXAMPLE 4–6

The MPF3821 JFET shown in the circuit of Figure 4–15(a) has the transconductance curve shown in Figure 4–15(b). From the transconductance curve, determine V_S and I_D. Using this result, determine the value of V_{DS}.

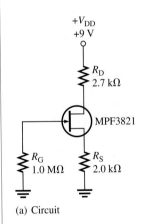

(a) Circuit

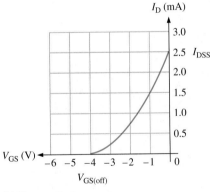

(b) Transconductance curve

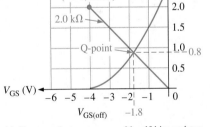

(c) Transconductance curve with self-bias resistor plotted

FIGURE 4–15

Solution Plot the line representing a 2.0 kΩ resistor by selecting the origin and any point on the resistor's line. A convenient point is $V_{GS} = -4$ V, $I_D = 2$ mA. The line between the origin and this point represents a 2.0 kΩ self-bias resistor as shown in Figure 4–15(c).

The intersection of the 2.0 kΩ resistor and the transconductance curve represents V_{GS} and I_D for this case. Reading the plot, you can see that $V_{GS} = -1.8$ V and I_D is **0.8 mA.** Since $V_G = 0$ V, $V_S = +1.8$ V.

The voltage across the drain resistor is found by Ohm's law.

$$V_{R_D} = I_D R_D = (0.8 \text{ mA})(2.7 \text{ k}\Omega) = 2.16 \text{ V}$$

To obtain the drain voltage, subtract the previous result from V_{DD}.

$$V_D = V_{DD} - V_{R_D} = 9.0 \text{ V} - 2.16 \text{ V} = 6.84 \text{ V}$$
$$V_{DS} = V_D - V_S = 6.84 \text{ V} - 1.8 \text{ V} = \textbf{5.04 V}$$

Interestingly, this result can be obtained graphically by load line analysis. The load line for the circuit is superimposed on the drain curves (from Figure 4–9(b)) as shown in Figure 4–16.

FIGURE 4–16

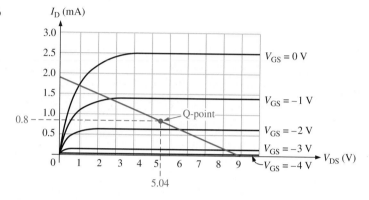

Practice Exercise   Confirm that the load line, drawn in Figure 4–16 represents the circuit in Figure 4–15(a).

Voltage-Divider Bias

Although self-bias is satisfactory for many applications, the operating point is dependent on the transconductance curve as you have seen. The bias can be made more stable with the addition of a voltage divider on the gate circuit, forcing the gate to a positive voltage. Since the JFET must still operate with a negative gate-source bias, a larger source resistor is used than in normal self-bias. The circuit is shown in Figure 4–17.

The gate voltage is found by applying the voltage-divider rule to R_1 and R_2.

$$V_G = \left(\frac{R_2}{R_1 + R_2}\right)V_{DD} \tag{4–5}$$

FIGURE 4–17
Voltage-divider bias and self-bias.

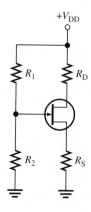

Remember, if you are troubleshooting any JFET circuit, the source voltage has to be equal or larger than the gate voltage. The drain current is in both R_D and R_S. Since I_D is dependent on the transconductance of the JFET, the precise value of V_D and V_S cannot be determined from the circuit values alone because of the manufacturing spread of FETs. In general, a JFET linear amplifier should be designed such that V_{DS} is in the range from about 25% to 50% of V_{DD}. Even without knowing the parameters for the transistor, you can verify that the bias is set up correctly by checking V_{DS}.

EXAMPLE 4–7

Assume you are troubleshooting the 2N5458[5] JFET shown in Figure 4–18. You do not know the transconductance of the transistor, but you need to find out if the circuit is working properly.

(a) Estimate the expected V_G and V_S.

(b) Assume you measured the source voltage and found it was +5.4 V. Is the circuit functioning as expected? Based on this measurement, what is the drain voltage?

(c) Assume you replace the transistor. The measured source voltage for the new transistor is +4.0 V. From this measurement, what is the expected drain voltage?

FIGURE 4–18

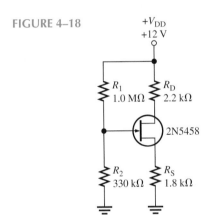

Solution

(a) Start with V_G because this value can be computed accurately and quickly with the voltage-divider rule. The gate voltage is about one-fourth of V_{DD} as shown in the following equation:

$$V_G = \left(\frac{R_2}{R_1 + R_2}\right)V_{DD} = \left(\frac{330\ k\Omega}{1.0\ M\Omega + 330\ k\Omega}\right)12\ V = \textbf{2.98 V}$$

You know immediately that if the circuit is operating properly, the source voltage must be more positive than this value. Your estimate of the source voltage should be about **+4 V.**

(b) The measured value of +5.4 V may indicate a problem. This is larger than the expected 4 V and is nearly half of V_{DD}. Since R_D is even larger than R_S, it should drop even more of the total voltage. A quick check of V_{DS} shows that it is 0 V! This confirms a problem with the circuit; there is probably a drain-to-source short in the transistor, causing V_D to also be **5.4 V.**

---

[5] Data sheet for 2N5458 available at http://www.fairchildsemi.com

(c) The drain current for the new transistor can be found by applying Ohm's law to the source resistor.

$$I_D = \frac{4.0 \text{ V}}{1.8 \text{ k}\Omega} = 2.2 \text{ mA}$$

Subtracting V_{R_D} from V_{DD} gives V_D.

$$V_D = V_{DD} - I_D R_D = 12 \text{ V} - (2.2 \text{ mA})(2.2 \text{ k}\Omega) = \textbf{7.16 V}$$

Practice Exercise   Assume the drain resistor were open in the circuit of Figure 4–18. What voltage do you expect in this case for V_G, V_S, and V_D?

Current Sources

Before discussing current-source biasing, let's review current sources. An ideal current source is a device that provides a fixed current that is independent of any load connected to it. The *IV* curve for an ideal current source is shown with a horizontal line in Figure 4–19(a). Recall that the slope of an *IV* curve is inversely proportional to resistance. A horizontal line implies that the internal resistance of the ideal source is infinite. A circuit model of an ideal current source is shown in Figure 4–19(b).

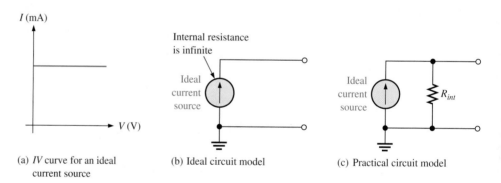

(a) *IV* curve for an ideal current source

(b) Ideal circuit model

(c) Practical circuit model

FIGURE 4–19
Current sources. The arrow in the current source symbol always points to the positive side of the source.

As you have seen, both FETs and BJTs have a region called the constant-current region on their characteristic curves. This region is depicted with a nearly horizontal line, representing the internal resistance of the source, which is very high, indeed. For most applications, a FET or a BJT can be assumed to be an ideal current source. In those cases where the internal resistance is taken into account, the Norton model, discussed in Section 1–3, is used. The Norton model for a practical current source is shown in Figure 4–19(c) with the Norton resistance representing the internal resistance of the current source.

Current-Source Biasing

This form of bias is widely used in ICs but requires an extra transistor. One transistor acts as a current source to force I_D to stay constant, creating a very stable form of bias. Current-source biasing also can improve the gain, as you will see later.

Two examples of current-source biasing are shown in Figure 4–20. In Figure 4–20(a), Q_2 is a FET constant-current source that provides a current to Q_1. The amount of current is determined by the I_{DSS} of Q_2 and the value of R_S. Since I_{DSS} varies between transistors, the amount of current depends on the particular transistor that is selected. The current source must *not* provide more current than the I_{DSS} of Q_1 to ensure that the V_{GS} of Q_1 is negative.

FIGURE 4–20
Current-source biasing.

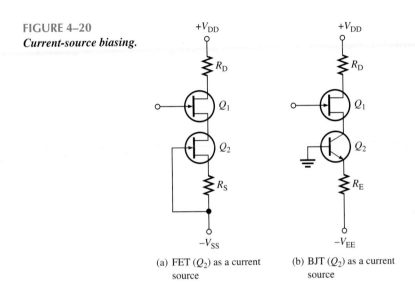

(a) FET (Q_2) as a current
source

(b) BJT (Q_2) as a current
source

For consistency between transistors, the arrangement in Figure 4–20(b) is better. Here the current is provided by a BJT. Since the base is grounded, the emitter voltage will be -0.7 V due to the forward-biased base-emitter junction. This means there is a constant voltage across R_E; thus, a constant current in the FET. Again, it is important that this current is less than the I_{DSS} of Q_1.

EXAMPLE 4–8

Figure 4–21 shows a current-source biasing circuit. What is I_D?

FIGURE 4–21

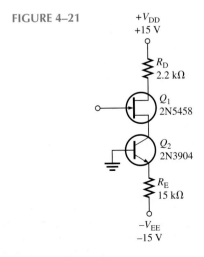

Solution You should recognize emitter bias for the bipolar junction transistor but with no base resistor. Since the base is connected directly to ground, the emitter voltage is -0.7 V due to the forward-biased base-emitter *pn* junction. This means that the voltage across R_E is 14.3 V and the current in R_E is constant. From Ohm's law, find the emitter current as

$$I_E = \frac{V_{R_E}}{R_E} = \frac{14.3 \text{ V}}{15 \text{ k}\Omega} = 0.95 \text{ mA}$$

This current is provided to the FET. Therefore, I_D = **0.95 mA**.

Practice Exercise What is the minimum I_{DSS} for the FET (for proper operation) in Figure 4–21?

4–3 REVIEW QUESTIONS

1. What two parameters for a JFET could you use to choose a reasonable value of self-bias resistor?
2. Why can't the bias circuits for BJTs be used for JFETs?
3. In a certain self-biased *n*-channel JFET circuit, I_D = 8 mA and R_S = 1.0 kΩ. What is V_{GS}?
4. For a JFET with current-source biasing, what parameter must not be exceeded by the current source?

4–4 ■ MOSFET CHARACTERISTICS

The metal-oxide semiconductor field-effect transistor (MOSFET) is the other major category of field-effect transistors. The MOSFET differs from the JFET in that it has no pn junction structure; instead, the gate of the MOSFET is insulated from the channel by a very thin silicon dioxide (SiO_2) layer. The two basic types of MOSFETs are depletion (D) and enhancement (E).

After completing this section, you should be able to

❑ Explain the operation of metal-oxide semiconductor field-effect transistors (MOSFETs)
  ❑ Describe the differences in the construction of MOSFETs
  ❑ Draw the symbol for an *n*-channel or a *p*-channel D-MOSFET or E-MOSFET
  ❑ Explain how a MOSFET functions in the depletion and the enhancement modes
  ❑ Interpret the drain characteristic curve for a MOSFET
  ❑ Describe the transconductance curve for a MOSFET and explain how it relates to the drain characteristic curve
  ❑ Discuss specific handling precautions for MOSFET devices

Depletion MOSFET (D-MOSFET)

One type of **MOSFET** is the depletion MOSFET (D-MOSFET) and Figure 4–22 illustrates its basic structure. The drain and source are diffused into the substrate material and then connected by a narrow channel adjacent to the insulated gate. Both *n*-channel and

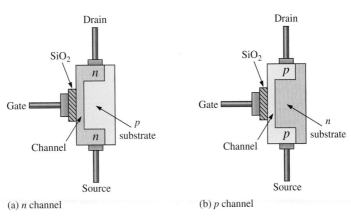

FIGURE 4–22
Basic structure of D-MOSFETs.

p-channel devices are shown in the figure; however, *p*-channel D-MOSFETs are not widely used. The basic operation is the same for both types, except the voltage polarities for the *p*-channel device are opposite those of the *n*-channel. For simplicity, *n*-channel devices are described in this section.

The D-MOSFET can be operated in either of two modes—the depletion mode or the enhancement mode—and is sometimes called a *depletion-enhancement MOSFET*. Since the gate is insulated from the channel, either a positive or a negative gate voltage can be applied. The *n*-channel D-MOSFET operates in the **depletion mode** when a negative gate-to-source voltage is applied and in the **enhancement mode** when a positive gate-to-source voltage is applied. These devices are generally operated in the depletion mode.

Depletion Mode   Visualize the gate as one plate of a parallel-plate capacitor and the channel as the other plate. The silicon dioxide insulating layer is the dielectric. With a negative gate voltage, the negative charges on the gate repel conduction electrons from the channel, leaving positive ions in their place. Thereby, the *n*-channel is depleted of some of its electrons, thus decreasing the channel conductivity. The greater the negative voltage on the gate, the greater the depletion of *n*-channel electrons. At a sufficiently negative gate-to-source voltage, $V_{GS(off)}$, the channel is totally depleted and the drain current is zero. This depletion mode is illustrated in Figure 4–23(a). Like the *n*-channel JFET, the *n*-channel D-MOSFET conducts drain current for gate-to-source voltages between $V_{GS(off)}$ and 0 V. In addition, the D-MOSFET conducts for values of V_{GS} above 0 V.

Enhancement Mode   With an *n*-channel device and with a positive gate voltage, more conduction electrons are attracted into the channel, thus increasing (enhancing) the channel conductivity, as illustrated in Figure 4–23(b).

D-MOSFET Symbols   Figure 4–24 shows the schematic symbols for both the *n*-channel and the *p*-channel D-MOSFETs. The substrate, indicated by the arrow, is normally (but not always) connected internally to the source. Sometimes the substrate is brought out as another lead. An inward substrate arrow is for *n*-channel, and an outward arrow is for *p*-channel.

Because the MOSFET is a field-effect device like the JFET, you might expect it to have similar characteristics as the JFET. This is indeed the case. The transfer characteristic (I_D versus V_{GS}) for the 2N3631, an *n*-channel D-MOSFET, is shown in Figure 4–25. It has the same shape as the one given earlier for the *n*-channel JFET (Figure 4–9(a)), but note that both negative and positive values of V_{GS} are shown on the transfer characteristic representing operation in the depletion region and the enhancement region respectively. This

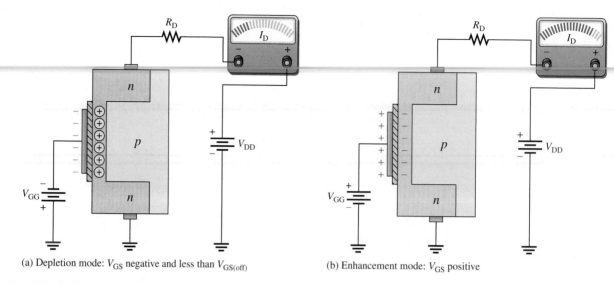

(a) Depletion mode: V_{GS} negative and less than $V_{GS(off)}$ (b) Enhancement mode: V_{GS} positive

FIGURE 4–23
Operation of n-channel D-MOSFETs.

FIGURE 4–24
D-MOSFET schematic symbols.

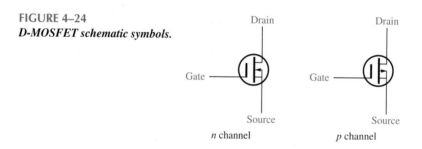

n channel   *p* channel

FIGURE 4–25
Transfer characteristic for a 2N3631 D-MOSFET (typical values shown).

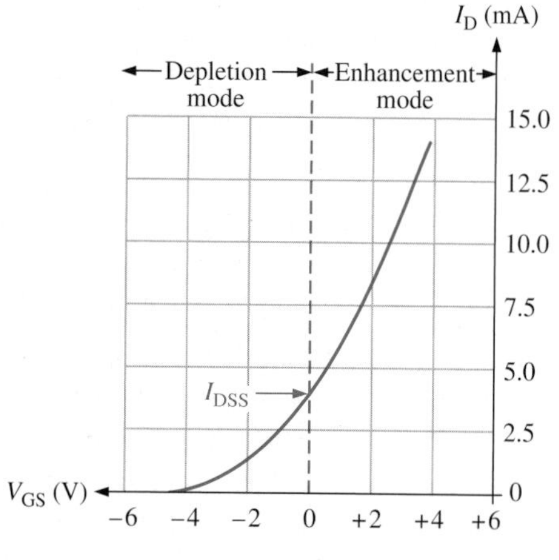

particular curve indicates the I_D is approximately 4.0 mA when V_{GS} is 0 V. Since V_{GS} is 0 V, the point is I_{DSS}. Notice that operation with currents higher than I_{DSS} is permissible with a D-MOSFET but not with a JFET.

Enhancement MOSFET (E-MOSFET)

This type of MOSFET operates only in the enhancement mode and has no depletion mode. It differs in construction from the D-MOSFET in that it has no physical channel. Notice in Figure 4–26(a) that the substrate extends completely to the SiO$_2$ layer.

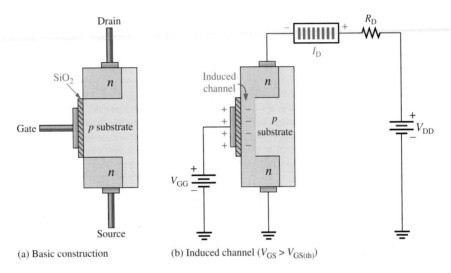

FIGURE 4–26
E-MOSFET construction and operation (n-channel).

For an *n*-channel device, a positive gate voltage above a threshold value, $V_{GS(th)}$, induces a channel by creating a thin layer of negative charges in the substrate region adjacent to the SiO$_2$ layer, as shown in Figure 4–26(b). The conductivity of the channel is enhanced by increasing the gate-to-source voltage, thus pulling more electrons into the channel. For any gate voltage below the threshold value, there is no channel.

The schematic symbols for the *n*-channel and *p*-channel E-MOSFETs are shown in Figure 4–27. The broken lines symbolize the absence of a physical channel.

Because the channel is closed unless a voltage is applied to the gate, an E-MOSFET can be thought of as a normally off device. Again the transfer characteristic has the same shape as the JFET and D-MOSFET, but now the gate of an *n*-channel device must be made positive in order to cause conduction. This means that the $V_{GS(off)}$ specification will be a

FIGURE 4–27
E-MOSFET schematic symbols.

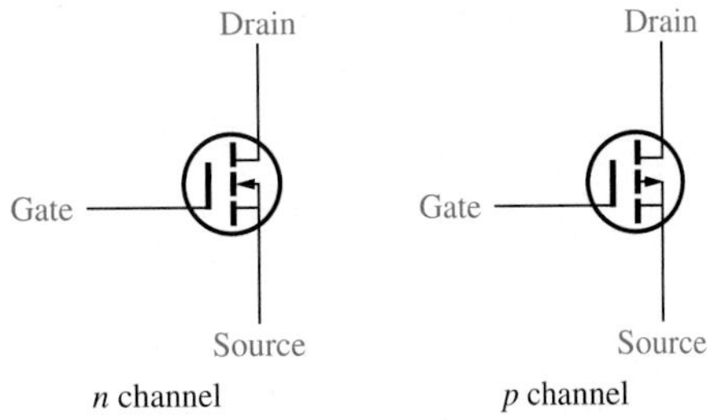

n channel

p channel

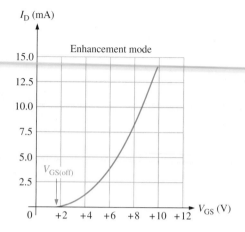

FIGURE 4–28

Transfer characteristic for a typical E-MOSFET.

positive voltage for an *n*-channel E-MOSFET. A typical characteristic is shown in Figure 4–28. Compare it to the D-MOSFET characteristic in Figure 4–25.

Handling Precautions

Because the gate of a MOSFET is insulated from the channel, the input resistance is extremely high (ideally infinite). The gate leakage current, I_{GSS}, for a typical MOSFET is in the pA range, whereas the gate reverse current for a typical JFET is in the nA range. The input capacitance, of course, results from the insulated gate structure. Excess static charge can accumulate because the input capacitance combines with the very high input resistance and can result in damage to the device as a result of electrostatic discharge (ESD). In fact, ESD is the single largest cause of failure with MOSFET devices. To avoid ESD and possible damage, the following precautions should be taken:

1. Metal-oxide semiconductor (MOS) devices should be shipped and stored in conductive foam.

2. All instruments and metal benches used in assembly or testing should be connected to earth ground (round prong of wall outlets).

3. The assembler's or handler's wrist should be connected to earth ground with a length of wire and a high-value series resistor.

4. Never remove a MOS device (or any other device, for that matter) from the circuit while the power is on.

5. Do not apply signals to a MOS device while the dc power supply is off.

4–4 REVIEW QUESTIONS

1. Name two types of MOSFETs, and describe the major difference in construction.

2. If the gate-to-source voltage in a D-MOSFET is zero, is there current from drain to source?

3. If the gate-to-source voltage in an E-MOSFET is zero, is there current from drain to source?

4. Can a D-MOSFET have a higher current than I_{DSS} and remain within the specified drain current?

4–5 ■ MOSFET BIASING

As with BJTs and JFETs, bias establishes the appropriate dc operating conditions that provide a stable operating point for centering the ac signal. MOSFET biasing circuits are similar to those you have already seen for BJTs and JFETs. The particular bias circuit depends on whether one or two supplies are used and the type of MOSFET (depletion or enhancement).

After completing this section, you should be able to

❏ Discuss and analyze MOSFET bias circuits
    ❏ Explain why a D-MOSFET has more bias options than any other type of transistor
    ❏ Explain zero bias
    ❏ Discuss three methods for biasing an E-MOSFET

D-MOSFET Bias

As you know, D-MOSFETs can be operated with either positive or negative values of V_{GS}. When V_{GS} is negative, operation is in the depletion mode; when it is positive, operation is in the enhancement mode. A D-MOSFET has the advantage of being able to operate in both modes; it is the only type of transistor that can do this.

Zero Bias The most basic bias method is to set $V_{GS} = 0$ V so that an ac signal at the gate varies the gate-to-source voltage above and below this bias point. Figure 4–29 shows the circuit. Because it is effective and simple, it is the preferred method for biasing a D-MOSFET. The operating point is set between depletion and enhancement operation. Since $V_{GS} = 0$ V, $I_D = I_{DSS}$, as indicated. The drain-to-source voltage is expressed as follows:

$$V_{DS} = V_{DD} - I_{DSS}R_D$$

FIGURE 4–29
A zero-biased D-MOSFET.

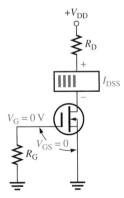

EXAMPLE 4–9   Determine the drain-to-source voltage in the circuit of Figure 4–30. The MOSFET data sheet gives $I_{DSS} = 12$ mA.

FIGURE 4–30

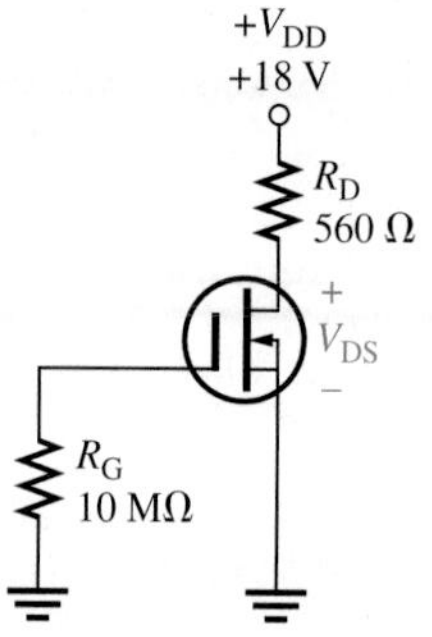

+V_{DD}
+18 V

R_D
560 Ω

+
V_{DS}
−

R_G
10 MΩ

Solution   Since $I_D = I_{DSS} = 12$ mA, the drain-to-source voltage is

$$V_{DS} = V_{DD} - I_{DSS}R_D = 18 \text{ V} - (12 \text{ mA})(560 \text{ Ω}) = \mathbf{11.28 \text{ V}}$$

Practice Exercise   Find V_{DS} in Figure 4–30 when $I_{DSS} = 20$ mA.

Other Bias Arrangements   As you know, the D-MOSFET can operate in the depletion or enhancement mode. Because of this versatility, any of the bias circuits you have studied for the BJT and the JFET can also be applied to D-MOSFETs. Figure 4–31 illustrates three popular methods for biasing, but you may see other methods in practice.

The bias circuit in Figure 4–31(a) uses a combination of voltage divider and self-bias as seen earlier with JFETs. The voltage at the gate is computed by the voltage-divider formula, which is quite accurate for any FET device because of the negligible loading effect. The gate voltage is the same as given for JFETs (see Equation (4–5)):

$$V_G = \left(\frac{R_2}{R_1 + R_2}\right)V_{DD}$$

The resistors that form the voltage divider are usually quite large (in the megohm range) because of the high input resistance of the gate terminal. The voltages at the other terminals depend on specific device parameters.

FIGURE 4–31
Other D-MOSFET bias circuits.

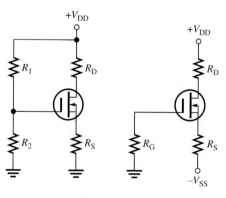

(a) Voltage-divider with self-bias

(b) Source bias

(c) Current-source bias

When positive and negative supplies are used, the source-bias arrangement in Figure 4–31(b) is frequently used. This is similar to emitter bias seen with BJTs. Ideally, the gate circuit looks like an open circuit, so you would expect the gate voltage to be at ground potential.

Current-source biasing is a form of bias that is common in operational amplifiers and is shown with a BJT current source in Figure 4–31(c). Other current sources, including FETs can be used. The current source sets the value of the source and drain current. By analyzing the current source (as in Example 4–8), the expected drop across the drain resistor can be computed from Ohm's law.

E-MOSFET Bias

E-MOSFETs must have a V_{GS} greater than the threshold value, $V_{GS(th)}$. Any of the bias circuits developed for BJTs (except base bias) could be used with appropriate values for E-MOSFETs. Figure 4–32 shows two common ways to bias an n-channel E-MOSFET. (D-MOSFETs can also be biased using these methods.) In either the drain-feedback or the voltage-divider bias arrangement, the purpose is to make the gate voltage more positive than the source by an amount exceeding $V_{GS(th)}$.

FIGURE 4–32
E-MOSFET biasing arrangements.

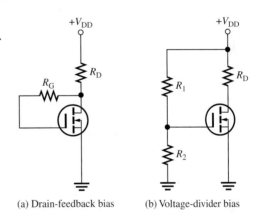

(a) Drain-feedback bias   (b) Voltage-divider bias

In the drain-feedback bias circuit in Figure 4–32(a), there is negligible gate current and, therefore, no voltage drop across R_G. As a result, $V_{GS} = V_{DS}$.

The voltage-divider bias is a straight-forward application of the voltage-divider rule you have already seen. Again, the voltage divider appears to be unloaded because of the high input resistance, so you can compute the gate voltage accurately using Equation (4–5).

EXAMPLE 4–10   Determine the amount of drain current in Figure 4–33. The E-MOSFET has a $V_{GS(th)}$ of 3 V.

FIGURE 4–33

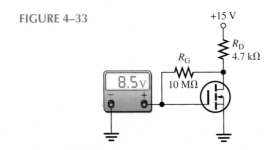

Solution The meter indicates that $V_{GS} = 8.5$ V. Since this is a drain-feedback configuration, $V_{DS} = V_{GS} = 8.5$ V.

$$I_D = \left(\frac{V_{DD} - V_{DS}}{R_D}\right) = \frac{15\ V - 8.5\ V}{4.7\ k\Omega} = \textbf{1.38 mA}$$

Practice Exercise Determine I_D if the meter in Figure 4–33 reads 5 V.

4–5 REVIEW QUESTIONS

1. For a D-MOSFET biased at $V_{GS} = 0$ V, is the drain current equal to 0, I_{GSS}, or I_{DSS}?
2. Why can't an E-MOSFET use zero bias?
3. For an *n*-channel E-MOSFET with $V_{GS(th)} = 2$ V, V_{GS} must be in excess of what value in order to conduct?

4–6 ■ FET LINEAR AMPLIFIERS

Field-effect transistors, both JFETs and MOSFETs, can be used as linear amplifiers in any of three circuit configurations similar to the bipolar junction transistor's CE, CC, and CB amplifiers you studied earlier. The FET configurations are common-source (CS), common-drain (CD), and common-gate (CG). The CS and CD amplifiers are characterized by high input impedance and low noise, making them excellent choices as the first stage of an amplifier. The common-gate amplifier has few applications, so it is discussed only briefly here.

After completing this section, you should be able to

❑ Describe the operation of FET linear amplifiers
   ❑ Describe the three FET configurations for linear amplifiers: common-source (CS), common-drain (CD), and common-gate (CG)
   ❑ Given the transconductance, compute the gain for any FET amplifier
   ❑ Explain why a CD amplifier with current-source biasing is significantly better than a single-stage CD amplifier

Transconductance of FETs

The transfer characteristic for a FET is the transconductance curve as was shown in Figure 4–9(a). FETs are fundamentally different than BJTs because they are voltage-controlled devices. The output drain *current* is controlled by the input gate *voltage*. As an ac parameter, transconductance was earlier defined as

$$g_m = \frac{I_d}{V_{gs}}$$

Considered in terms of output current (I_d) divided by input voltage (V_{gs}), transconductance is essentially the gain of the FET by itself. Unlike β_{ac}, a pure number, transcon-

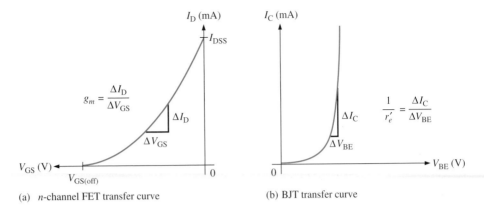

FIGURE 4–34

Comparison of the transfer curve for an n-channel FET with a BJT.

ductance (g_m) has units of the siemen (the reciprocal of resistance). Many data sheets continue to use the older unit, the mho (ohm spelled backwards). The transconductance of a particular FET can be measured directly as shown in Figure 4–34(a). Notice that the transconductance is the slope of the transfer curve and it is *not* a constant, but depends on the drain current.

Figure 4–34(b) shows an analogous situation for the input to a BJT. The base voltage, applied across the base-emitter *pn* junction, "sees" an ac resistance that depends on the dc emitter current. This small ac resistance plays an important role in determining the gain of a BJT amplifier as you saw in Section 3–4.

The *reciprocal* of g_m is analogous to r'_e for BJTs. Most ac models for a FET use g_m as one of the key parameters; however, to make the transition from BJT amplifiers to FET amplifiers, it is useful to define a parameter representing the ac source resistance of the FET.

$$r'_s = \frac{1}{g_m} \qquad (4\text{–}6)$$

The concept of r'_s leads to voltage gain equations that are analogous to those developed in Chapter 3 for BJTs. A mental picture of r'_s for a JFET is shown in Figure 4–35. The gate is shown with a dotted line to remind you that, from the gate's perspective, the input resistance is nearly infinite (because of the input's reverse-biased diode). Although the gate voltage controls the drain current, it does it with negligible current. Unfortunately, r'_s for a FET is not as predictable as r'_e is for a BJT and it is generally larger than r'_e. Data sheets don't show this parameter, but they do show a range of values for g_m (also shown as y_{fs}), so you can obtain an approximate value of r'_s by taking the reciprocal of the typical value of g_m. For example, if y_{gs} is shown as 2000 μS on a data sheet, $r'_s = 500 \ \Omega$.

FIGURE 4–35

The internal source resistance r'_s is analogous to r'_e for a BJT. The dotted line is a reminder that gate current is negligible because of extremely high input resistance.

Common-Source Amplifiers

JFET   Figure 4–36 shows a common-source (CS) amplifier with a self-biased *n*-channel JFET. An ac source is capacitively coupled to the gate. The resistor, R_G, serves two purposes: (a) It keeps the gate at approximately 0 V dc (Because I_{GSS} is extremely small), and (b) its large value (usually several megohms) prevents loading of the ac signal source. The bias voltage is created by the drop across R_S. The bypass capacitor, C_2, keeps the source of the FET effectively at ac ground.

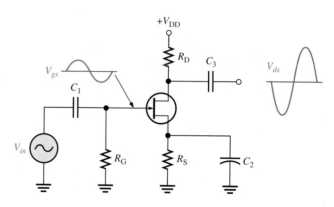

FIGURE 4–36
JFET common-source amplifier.

The signal voltage causes the gate-to-source voltage to swing above and below its Q-point value, causing a swing in drain current. As the drain current increases, the voltage drop across R_D also increases, causing the drain voltage (with respect to ground) to decrease.

The drain current swings above and below its Q-point value in phase with the gate-to-source voltage. The drain-to-source voltage swings above and below its Q-point value 180° out of phase with the gate-to-source voltage, as illustrated in Figure 4–36.

D-MOSFET   Figure 4–37 shows a zero-biased *n*-channel D-MOSFET with an ac source capacitively coupled to the gate. The gate is at approximately 0 V dc and the source terminal is at ground, thus making $V_{GS} = 0$ V.

FIGURE 4–37
Zero-biased D-MOSFET common-source amplifier.

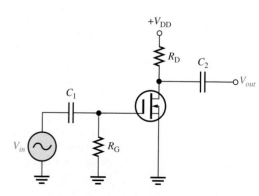

The signal voltage causes V_{gs} to swing above and below its 0 value, producing a swing in I_d. The negative swing in V_{gs} produces the depletion mode, and I_d decreases. The positive swing in V_{gs} produces the enhancement mode, and I_d increases.

E-MOSFET   Figure 4–38 shows a voltage-divider biased, n-channel E-MOSFET with an ac signal source capacitively coupled to the gate. The gate is biased with a positive voltage such that $V_{GS} > V_{GS(th)}$. As with the JFET and D-MOSFET, the signal voltage produces a swing in V_{gs} above and below its Q-point value. This swing, in turn, causes a swing in I_d. Operation is entirely in the enhancement mode.

FIGURE 4–38
Common-source E-MOSFET amplifier with voltage-divider bias.

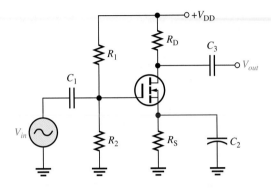

Voltage Gain   Voltage gain, A_v, of an amplifier always equals V_{out}/V_{in}. In the case of the CS amplifier, V_{in} is equal to V_{gs} (due to the bypass capacitor) and V_{out} is equal to the signal voltage developed across R_d, the ac drain resistance. In a CS amplifier with no load, the ac and dc drain resistances are equal: $R_d = R_D$. Thus, $V_{out} = I_d R_d$.

$$A_v = \frac{V_{out}}{V_{in}} = \frac{I_d R_d}{V_{gs}}$$

Since $g_m = I_d / V_{gs}$, the common-source voltage gain is

$$A_v = -g_m R_d \qquad \textbf{(4–7)}$$

This is the traditional voltage gain equation for the CS amplifier. The negative sign is added to Equation (4–7) to indicate that it is an inverting amplifier. The gain for the CS amplifier can be expressed in a similar form to the common-emitter (CE) amplifier as a ratio of ac resistances. By substituting $1/r_s'$ for g_m, the voltage gain can be written as

$$A_v = -\frac{R_d}{r_s'} \qquad \textbf{(4–8)}$$

Compare this result with Equation (3–10) that gives the voltage gain for a CE amplifier: $A_v = -R_c/R_e$. Both equations show voltage gain as a ratio of ac resistances.

Input Resistance   Because the input to a CS amplifier is at the gate, the input resistance to the transistor is extremely high. As you know, this extremely high resistance is produced by the reverse-biased pn junction in a JFET and by the insulated gate structure in a MOSFET. For practical work, the transistor's input circuit looks open.

When the transistor's internal resistance is ignored, the input resistance seen by the signal source is determined only by the bias resistor (or resistors). With self-bias, it is

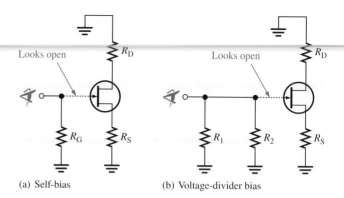

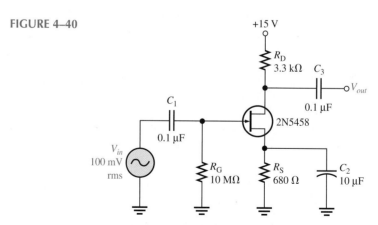

(a) Self-bias   (b) Voltage-divider bias

FIGURE 4–39
Input resistance is determined by the bias resistors.

simply the gate resistor, R_G, as shown in the equivalent ac circuit looking into the gate in Figure 4–39.

With voltage-divider bias, the power supply is at ac ground and the gate again appears as an open. The two voltage-divider resistors are seen by the ac source in parallel. The input resistance is the parallel combination of R_1 and R_2.

$$R_{in} \cong R_1 \parallel R_2$$

EXAMPLE 4–11

(a) What is the total output voltage (dc + ac) of the amplifier in Figure 4–40? The g_m is 1500 μS, I_D is 2.0 mA, and $V_{GS(off)}$ is 3 V.

(b) What is the input resistance seen by the signal source?

FIGURE 4–40

Solution

(a) First, find the dc output voltage.

$$V_D = V_{DD} - I_D R_D = 15 \text{ V} - (2 \text{ mA})(3.3 \text{ k}\Omega) = 8.4 \text{ V}$$

Next, find the voltage gain.

$$A_v = -g_m R_d = -(1500 \text{ μS})(3.3 \text{ k}\Omega) = -5.0$$

Alternatively, the voltage gain could be found by computing r'_s and using the ratio of ac drain resistance to ac source resistance.

$$r'_s = \frac{1}{g_m} = \frac{1}{1500 \ \mu S} = 667 \ \Omega$$

$$A_v = -\frac{R_d}{r'_s} = -\frac{3.3 \ k\Omega}{667 \ \Omega} = -5.0$$

The output voltage is the gain times the input voltage.

$$V_{out} = A_v V_{in} = (-5.0)(100 \ mV) = -0.5 \ V \ rms$$

The negative sign indicates the output waveform is inverted.

The total output voltage is an ac signal with a peak-to-peak value of $0.5 \ V \times 2.828 = \mathbf{1.4 \ V}$, riding on a dc level of **8.4 V**.

(b) The input resistance is

$$R_{in} \cong R_G = \mathbf{10 \ M\Omega}$$

Practice Exercise What happens to the g_m if the source resistor is made larger? Does this affect the gain?

Common-Drain (CD) Amplifier

A **common-drain (CD)** JFET amplifier is shown in Figure 4–41 with voltages indicated. Self-biasing is used in this circuit. The input signal is applied to the gate through a coupling capacitor, and the output is at the source terminal. There is no drain resistor. This circuit, of course, is analogous to the BJT emitter-follower and is sometimes called a *source-follower*. It is a widely used FET circuit because of its very high input impedance.

FIGURE 4–41
JFET common-drain amplifier (source-follower).

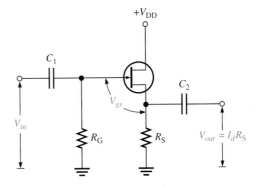

Voltage Gain As in all amplifiers, the voltage gain is $A_v = V_{out}/V_{in}$. Like the emitter-follower, the source-follower has an ideal voltage gain of 1, but in practice it is less (typically between 0.5 and 1.0). To compute the voltage gain, the voltage-divider rule can be applied to the circuit shown in Figure 4–42(a). First, the circuit is simplified to the ac equivalent shown in Figure 4–42(b). The gate resistor does not affect the ac, so it is not shown. The gate input is shown with a dotted line to remind you that it looks like an open to the ac input signal. The load and source resistors are in parallel and can be combined to form an equivalent ac source resistance, R_s, that is in series with the internal resistance r'_s

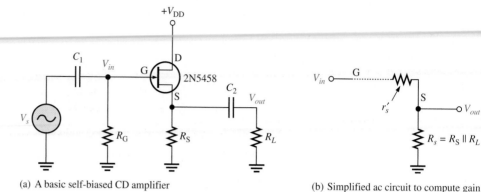

(a) A basic self-biased CD amplifier (b) Simplified ac circuit to compute gain

FIGURE 4–42

(or $1/g_m$). The input is across both R_s and r'_s, but the output is taken across R_s only. Therefore, the output voltage is

$$V_{out} = V_{in}\left(\frac{R_s}{r'_s + R_s}\right)$$

Dividing by V_{in} results in the equation for voltage gain.

$$A_v = \frac{R_s}{r'_s + R_s} \qquad \textbf{(4–9)}$$

Again, note that gain can be written as a ratio of ac resistances. This equation is easy to recall if you keep in mind that it is based on the voltage-divider rule.

An alternate voltage-gain equation, derived in Appendix B, is as follows:

$$A_v = \frac{g_m R_s}{1 + g_m R_s} \qquad \textbf{(4–10)}$$

This formula yields the same result as Equation (4–9).

Input Resistance   Because the input signal is applied to the gate, the input resistance seen by the input signal source is the same as the CS-amplifier configuration discussed previously. For practical work, you can ignore the very high resistance of the transistor's input. The input resistance is determined by the bias resistor or resistors as done with the CS amplifier. For self-bias, the input resistance is equal to the gate resistor R_G.

$$R_{in} \cong R_G$$

With voltage-divider bias, the voltage-divider resistors are seen by the source as a parallel path to ground. Thus for voltage-divider bias, the input resistance is

$$R_{in} \cong R_1 \parallel R_2$$

EXAMPLE 4–12   Determine the minimum and maximum voltage gain of the amplifier in Figure 4–43(a) based on the data sheet information in Figure 4–43(b).

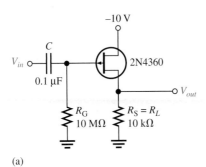

(a)

Electrical Characteristics ($T_A = 25°C$ unless otherwise noted)

Characteristic	Symbol	Min	Max	Unit
OFF Characteristics				
Gate-Source breakdown voltage ($I_G = 10$ μA dc, $V_{DS} = 0$)	$V_{(BR)GSS}$	20	–	V dc
Gate-Source cutoff voltage ($V_{DS} = -10$ V dc, $I_D = 1.0$ μA dc)	$V_{GS(off)}$	0.7	10	V dc
Gate-reverse current ($V_{GS} = 15$ V dc, $V_{DS} = 0$) ($V_{GS} = 15$ V dc, $V_{DS} = 0$, $T_A = 65°C$)	I_{GSS}	– –	10 0.5	nA dc μA dc
ON Characteristics				
Zero-Gate voltage drain current* ($V_{DS} = -10$ V dc, $V_{GS} = 0$)	I_{DSS}	3.0	30	mA dc
Gate-source voltage ($V_{DS} = -10$ V dc, $I_D = 0.3$ mA dc)	V_{GS}	0.4	9.0	V dc
Small-Signal Characteristics				
Drain-Source "ON" resistance ($V_{GS} = 0$, $I_D = 0$, $f = 1.0$ kHz)	$r_{ds(on)}$	–	700	Ohms
Forward-transadmittance* ($V_{DS} = -10$ V dc, $V_{GS} = 0$, $f = 1.0$ kHz)	$|y_{fs}|$	2000	8000	μmhos
Forward-transconductance ($V_{DS} = -10$ V dc, $V_{GS} = 0$, $f = 1.0$ MHz)	$Re(y_{fs})$	1500	–	μmhos
Output admittance ($V_{DS} = -10$ V dc, $V_{GS} = 0$, $f = 1.0$ kHz)	$|y_{os}|$	–	100	μmhos
Input capacitance ($V_{DS} = -10$ V dc, $V_{GS} = 0$, $f = 1.0$ MHz)	C_{iss}	–	20	pF
Reverse transfer capacitance ($V_{DS} = -10$ V dc, $V_{GS} = 0$, $f = 1.0$ MHz)	C_{rss}	–	5.0	pF
Common-Source noise figure ($V_{DS} = -10$ V dc, $I_D = 1.0$ mA dc, $R_G = 1.0$ Megohm, $f = 100$ Hz)	NF	–	5.0	dB
Equivalent short-circuit input noise voltage ($V_{DS} = -10$ V dc, $I_D = 1.0$ mA dc, $f = 100$ Hz, BW = 15 Hz)	E_n	–	0.19	μV/$\sqrt{Hz}$

*Pulse test: Pulse width ≤ 630 ms, Duty cycle ≤ 10%

(b)

FIGURE 4–43

Solution   On the data sheet, g_m is shown as y_{fs}. The range is 2000 μS to 8000 μS (shown as 2000 μmhos on the data sheet). The maximum value of r'_s is

$$r'_s = \frac{1}{g_m} = \frac{1}{2000\ \mu S} = 500\ \Omega$$

The ac source resistance, R_s, is simply the load resistor, R_L which is R_S. Substituting into Equation (4–9), the minimum voltage gain is

$$A_{v(min)} = \frac{R_s}{r'_s + R_s} = \frac{10 \text{ k}\Omega}{500 \text{ }\Omega + 10 \text{ k}\Omega} = \mathbf{0.95}$$

The minimum value of r'_s is

$$r'_s = \frac{1}{g_m} = \frac{1}{8000 \text{ }\mu S} = 125 \text{ }\Omega$$

The maximum voltage gain is then

$$A_{v(max)} = \frac{R_s}{r'_s + R_s} = \frac{10 \text{ k}\Omega}{125 \text{ }\Omega + 10 \text{ k}\Omega} = \mathbf{0.99}$$

Notice that the gain is slightly less than 1. When r'_s is small compared to the ac source resistance, then a good approximation is $A_v = 1$. Since the output voltage is at the source, it is in phase with the gate (input) voltage.

Practice Exercise   Determine the approximate input resistance seen by the source for the amplifier in Figure 4–43(a).

CD Amplifier with Current-Source Biasing   The CD amplifier can be improved significantly by the addition of a current source as shown in Figure 4–44. The current source not only provides bias (as described in Section 4–3) but also acts as the load for the CD amplifier. As you know, a current source looks like a very high resistance to the ac signal, so the voltage gain is very close to the ideal value of 1.0.

The current source load offers other significant advantages including higher input resistance, lower distortion, and the ability to dc couple the signal at both the input and output (no coupling capacitors). The output voltage from a regular source-follower (such as given in the previous example) is riding on a dc level that is equal to the magnitude of V_{GS} (since the gate is at 0 V). For a p-channel device, the dc offset is negative; for an n-channel device, it is positive. Ideally, current-source biasing does not add *any* dc offset to the out-

FIGURE 4–44

CD amplifier with current-source bias.

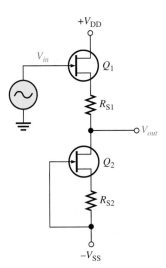

put. This is particularly useful in applications such as the preamp to an oscilloscope that must pass any dc component of the signal to the rest of the vertical amplifier.

For optimum results, the two FETs and two resistors in Figure 4–44 should match. This means that both transistors will have identical transfer and output characteristics. Both transistors will have same V_{GS} (since they have the same drain current). This drain current drops the same voltage (V_{GS} again) across both resistors, compensating for the bias. This ensures that the output will be close to 0 V when the input is 0 V. One way to ensure matching transistors is to use a dual device (two matching transistors in one package) such as type U401.[6]

---

[6] Data sheet for U401 available at http://www.vishay.com

EXAMPLE 4–13

Determine the drain current, I_D, and the source voltage, V_S, of Q_1 for the CD amplifier with current-source bias shown in Figure 4–45(a). Assume the FETs are matched and each has a transconductance curve as shown in Figure 4–45(b).

FIGURE 4–45

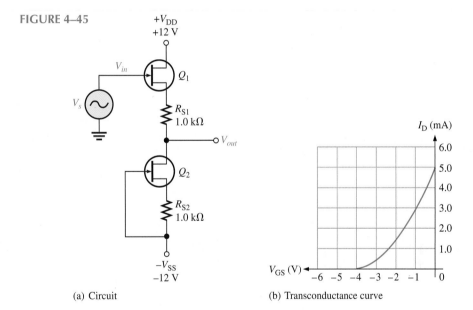

(a) Circuit (b) Transconductance curve

Solution On the transconductance curve draw a line representing the 1.0 kΩ bias resistor for the current source (Q_2). This is shown in Figure 4–46. The crossing point in-

FIGURE 4–46

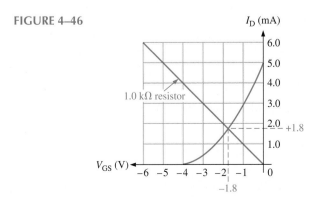

dicates that I_D is approximately **1.8 mA** at V_{GS} of -1.8 V. This current in R_{S1} causes the source of Q_1 to be at **+1.8 V.**

Practice Exercise   Determine I_D and $V_{S(Q1)}$ if both resistors are changed to 2.0 kΩ.

Common-Gate Amplifier

As mentioned in the section introduction, the **common-gate (CG)** amplifier has limited use by itself, but it is used in the second stage of a FET differential amplifier (discussed in Chapter 6). It also has application at high frequencies. Although it has a voltage gain comparable to the CD amplifier, the input resistance is low, cancelling one of the major advantages of FETs. A basic CG amplifier is shown in Figure 4–47. The input signal is applied to the source terminal through C_1 and taken from the drain terminal through C_2. The voltage gain is the same as that of a CS amplifier but with no phase inversion.

$$A_v = \frac{R_d}{r'_s}$$

An alternate gain formula is

$$A_v = g_m R_d$$

FIGURE 4–47
JFET common-gate amplifier.

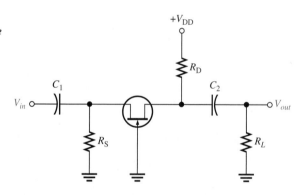

The principal advantage of FETs for linear applications is their high input resistance. Looking into the CG amplifier, you see the source resistor in parallel with r'_s. Usually the source resistor is large enough to ignore; therefore, the input resistance is approximately

$$R_{in} \cong r'_s$$

Alternatively, the input resistance is approximately

$$R_{in} \cong \frac{1}{g_m}$$

This result shows why the CG amplifier has low input resistance.

4–6 REVIEW QUESTIONS

1. How is the gain computed for a CS amplifier?
2. What are the principal advantages of a CD amplifier with current-source biasing over a single-stage CD amplifier?
3. Which of the three configurations (CS, CD, or CG) do not invert the input signal?
4. What is the principal disadvantage of a CG amplifier?
5. Which characteristics of FETs make them excellent choices for the first stage of an amplifier?

4–7 ■ FET SWITCHING CIRCUITS

There are two basic switching circuits that use FETs: analog switches and digital switches. Switching circuits are often found as the interface between analog and digital circuits, so both types are presented in this section. For most (but not all) switching applications, FETs are superior to BJTs because they require almost no drive current. Switching applications are frequently found in industry where the need for automatic control of high currents is common.

After completing this section, you should be able to

❑ Discuss two switching applications of FETs
   ❑ Explain the difference between an analog and a digital switch
   ❑ Compute the on resistance, $r_{DS(on)}$, given the $V_{GS(off)}$ and I_{DSS} parameters for a FET
   ❑ Explain the operating points on a load line for a digital switch
   ❑ Explain how to interface a logic circuit to an ac load using a FET switch and a relay
   ❑ Describe the switching process in CMOS logic

Types of Solid-State Switches

There are two distinctly different switching applications for FETs. The first is the *analog switch* (sometimes called an ac switch) which takes advantage of the FET's low internal drain-source resistance when it is on and its high drain-source resistance when it is off to either pass or block an ac signal. Analog switches are connected in series with a signal; and when the switch is closed, the signal should be passed to the load without regard to the waveform polarity and with no significant change. Ideally, when the switch is open, the signal should be completely blocked and the switch should appear as an open circuit. While FETs cannot quite match ideal behavior, they offer the major advantage of very fast electronic control. Usually, analog switches are used in low-power low-voltage applications; for example, to turn on only one of several inputs to an analog-to-digital converter.

    The second type of switching circuit is a *digital switch* (sometimes called a dc or logic switch) which turns on or turns off current to a device (such as a motor or relay). This type of switch was introduced earlier (in Section 3–7) with BJTs. FET digital switches are frequently designed for power switching where current and voltages may be large. A control voltage causes the switch to appear open or closed. The control signal is frequently generated by a computer or a logic circuit that does not have sufficient drive capability by itself to perform the needed function. FET digital switches are controlled by a voltage applied to the gate, allowing drain current to turn on or turn off the device.

The Analog Switch

JFET Analog Switch   An *n*-channel JFET analog switch (transmission gate) is shown in Figure 4–48. The JFET is switched between the on and off state with a control voltage applied to the gate. To turn the JFET on, V_{GS} is made equal to zero volts. This can create a potential problem because the input (source) voltage varies and could cause the gate-source *pn* junction to be forward-biased. The solution is to connect a resistor between the source and gate terminals and add a diode in the gate circuit to prevent V_{GS} from becoming positive. To turn the JFET off, the channel must be pinched off. This is done by making the control voltage more negative than the lowest value of the signal plus the pinch-off voltage, V_P. (Recall that the magnitude of V_P is the same as $V_{GS(off)}$.)

FIGURE 4–48
A JFET analog switch.

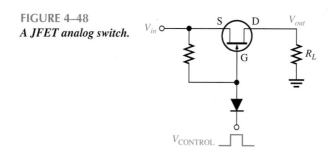

JFETs are particularly well suited as analog switches. Because the gate draws essentially no current, they have an extremely high off resistance which provides a high degree of isolation from other signals. When the control signal is applied, the channel resistance ($r_{DS(on)}$) between the drain and source is relatively small but constant. The typical JFET switch shown in Figure 4–49 is in series with a load resistance. As long as the load is much larger than $r_{DS(on)}$, the output voltage is approximately the same as the input voltage.

FIGURE 4–49
Simplified equivalent circuit for a JFET analog switch.

The on-state channel resistance is a function of the $V_{GS(off)}$ and I_{DSS} parameters, which can be found on manufacturer's specification sheets. The equation for $r_{DS(on)}$ is

$$r_{DS(on)} = -\frac{V_{GS(off)}}{2I_{DSS}}$$

(4–11)

EXAMPLE 4–14   Compute $r_{DS(on)}$ for a JFET with the transconductance curve shown in Figure 4–45(b).

Solution   Reading from the graph, $V_{GS(off)} = -4$ V and $I_{DSS} = 5.0$ mA. Substituting into the equation for $r_{DS(on)}$ gives

$$r_{DS(on)} = -\frac{V_{GS(off)}}{2I_{DSS}} = -\frac{-4 \text{ V}}{2(5.0 \text{ mA})} = 400 \text{ }\Omega$$

Related Exercise   What fraction of the input voltage appears at the output if the JFET switch in this example is connected to a 10 kΩ load resistor?

JFET switches are fast and switching is accomplished from control voltages that can be obtained from standard logic families. A disadvantage of JFET switches is that they are also more prone to switching transients appearing on the signal lines when control signals change states. All in all, the advantages far outweigh the disadvantages and JFET switches are widely used in switching applications for instrumentation systems.

Specially designed FET analog switches are available in IC form with switching speeds of less than 1 ns and capacitances of under 3 pF. They are useful in circuits such as fast ADCs (analog-to-digital converters), and DACs (digital-to-analog converters), choppers, and sample-and-hold circuits.

MOSFET Analog Switch   MOSFETs are also widely used for analog switching applications. They offer simpler circuits than JFETs and both positive and negative control. A basic MOSFET switch, using a *p*-channel E-MOSFET, is shown in Figure 4–50. A disadvantage to MOSFET switches is that the on resistance tends to be higher than that of JFETs. In the special case of switching a high current analog signal, power MOSFETs are available that can switch as high as 10 A with a control circuit that supplies essentially no current to the gate. There are no high current JFETs available for this application.

FIGURE 4–50
A p-channel MOSFET analog switch.

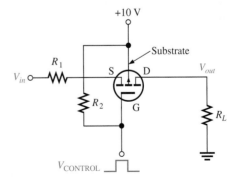

Solid-State Relays   A related analog switch is called a *solid-state relay (SSR)*. SSRs are packaged with an optical isolator on the input circuit and power MOSFET output. In the past, mechanical relays have been necessary for applications such as low-level multiplexing of analog signals. However, because of extremely high off-state resistance, SSRs are now commonly used in these applications.

The Digital Switch

Discrete MOSFET Switches   Ideally, any digital switch is either open or closed. In practice, the operation of a FET switch can be described by a load line drawn on the transistor's characteristic curve. A circuit example, shown in Figure 4–51(a) uses a BS107A switching transistor, designed for small-power applications. The typical characteristic curve for the BS107A[7], from the manufacturer's specification sheet, and the load line for the circuit are shown in Figure 4–51(b). The drain resistor represents a load (such as a light bulb) that requires approximately 200 mA of current at 12 V. The power supply is set higher than +12 V to allow for the drop across the transistor. Notice that when the transistor is on, there is a small drop (about 1 V) across it as indicated by $V_{DS(on)}$. The series resistor in the gate circuit is optional; it is often added by designers to protect the MOSFET. In addition, the schematic symbol shows a reverse-biased diode between the drain and source; this is because the body of the transistor is connected internally to the source, forming a diode. You cannot use a power MOSFET as an analog switch with bipolar signals.

Figure 4–51(c) shows how a small transistor like the BS107A can drive a 12 V relay coil, which in turn controls the ac line voltage for a small motor or appliance. The coil and

---

[7] Data sheet for BS107A available at http://www.onsemi.com

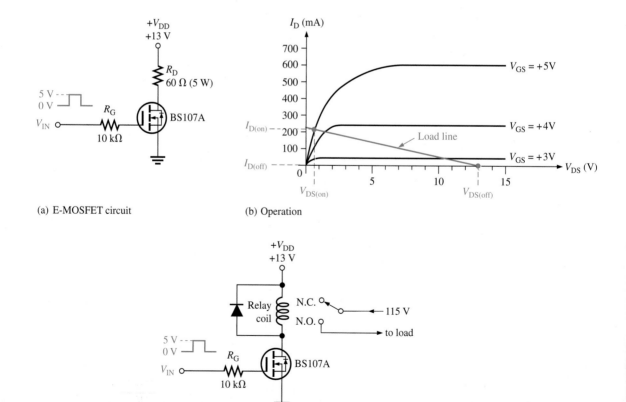

(a) E-MOSFET circuit

(b) Operation

(c) Controlling line voltage with a relay

FIGURE 4–51

a normally reverse-biased protection diode are inserted in the drain circuit, in place of the drain resistor. The diode prevents the MOSFET from being destroyed from a highly induced negative voltage when the current is rapidly switched off.

A partial manufacturer's specification sheet is shown in Figure 4–52 for the BS107A. Notice that the typical on resistance at 250 mA is 4.8 Ω with a gate-source voltage of 10 V. The static drain-to-source resistance when the device is on is listed on the data

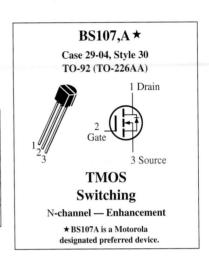

BS107,A ★

Case 29-04, Style 30
TO-92 (TO-226AA)

1 Drain

2 Gate

3 Source

TMOS
Switching

N-channel — Enhancement

★ BS107A is a Motorola
designated preferred device.

Maximum Ratings

Rating	Symbol	Value	Unit
Drain-Source voltage	V_{DS}	200	V dc
Gate-Source voltage	V_{GS}	±20	V dc
Drain current Continuous(1) Pulsed(2)	I_D I_{DM}	250 500	mA dc
Total device dissipation @ $T_A = 25°C$ Derate above 25°C	P_D	350	mW
Operating and storage junction temperature range	T_J, T_{stg}	−55 to +150	°C

(1) The Power Dissipation of the package may result in a lower continuous drain current.
(2) Pulse width ≤ 300 μs, Duty cycle ≤ 2.0%.

Electrical Characteristics ($T_A = 25°C$ unless otherwise noted.)

Characteristic	Symbol	Min	Typ	Max	Unit
OFF Characteristics					
Zero gate voltage drain current ($V_{DS} = 130$ V, $V_{GS} = 0$)	I_{DSS}	–	–	30	nA dc
Drain-Source breakdown voltage ($V_{GS} = 0, I_D = 100$ μA)	$V_{(BR)DSX}$	200	–	–	V dc
Gate-reverse current ($V_{GS} = 15$ V dc, $V_{DS} = 0$)	I_{GSS}	–	0.01	10	nA dc
ON Characteristics*					
Gate threshold voltage ($I_D = 1.0$ mA, $V_{DS} = V_{GS}$)	$V_{GS(th)}$	1.0	–	3.0	V dc
Static drain-source on-resistance BS107 ($V_{GS} = 2.6$ V, $I_D = 20$ mA) ($V_{GS} = 10$ V, $I_D = 200$ mA) BS107A ($V_{GS} = 10$ V_{DC}) ($I_D = 100$ mA) ($I_D = 250$ mA)	$r_{DS(on)}$	 – – – –	 – – 4.5 4.8	 28 14 6.0 6.4	Ohms
Small-Signal Characteristics					
Input capacitance ($V_{DS} = 25$ V, $V_{GS} = 0, f = 1.0$ MHz)	C_{iss}	–	60	–	pF
Reverse transfer capacitance ($V_{DS} = 25$ V, $V_{GS} = 0, f = 1.0$ MHz)	C_{rss}	–	6.0	–	pF
Output capacitance ($V_{DS} = 25$ V, $V_{GS} = 0, f = 1.0$ MHz)	C_{oss}	–	30	–	pF
Forward transconductance ($V_{DS} = 25$ V, $I_D = 250$ mA)	g_{fs}	200	400	–	mmhos
Switching Characteristics					
Turn-On time	t_{on}	–	6.0	15	ns
Turn-Off time	t_{off}	–	12	15	ns

*Pulse test: Pulse width ≤ 300 μs, Duty cycle ≤ 2.0%.

FIGURE 4–52
Partial manufacturer's data sheet for a BS107A MOSFET transistor.

sheet as $r_{DS(on)}$. The BS107A requires 10 V to turn on with the specified drain current and resistance as shown on the specification sheet.

If you compute the voltage across a 4.8 Ω resistor, in place of the MOSFET in Figure 4–51, you will find that the drop across the MOSFET is approximately 1 V as shown graphically in Figure 4–51(b). In the on state, the transistor should normally be saturated to ensure the smallest resistance between the drain and source. The gate voltage must be large enough to ensure this saturation. Even though this is a low value, the FET is not a perfect switch; the on-state resistance of an ideal switch is 0 Ω.

As with any device, a MOSFET switch must be operated within certain limits specified by the manufacturer. In the off state, the MOSFET must be able to withstand the power supply voltage which will appear between the drain and source terminals. On a data sheet, this maximum voltage is listed as $V_{(BR)DSX}$ which is the maximum drain-source voltage with the gate shorted to the source. For the BS107A, this value is a minimum of 200 V (see Figure 4–52).

MOSFET switches have several advantages over BJT switches in power applications. One advantage is that they are easier to drive because the input is a voltage (rather than a current in the case of the BJT). For small-power applications, this difference isn't too important, but when the load current exceeds a few amps, the BJT requires a hefty drive current compared to the MOSFET's simple voltage control. Another advantage is that MOSFETs in general have immunity to an effect called *thermal runaway*. Thermal runaway can happen to BJTs. As they get hot, they tend to conduct more current, which causes them to get hotter yet. With a power MOSFET, when the current reaches some level, it starts to conduct *less,* helping avoid the problem of thermal runaway.

IC Switching Circuits   FET switching circuits are widely used in ICs. Let's look at how MOSFETs are used in logic circuits for switching. IC logic circuits are a form of digital switch, in which the output is at one extreme of the load line. One of the most popular types of logic, called CMOS (for *Complementary Metal-Oxide Semiconductors*) uses both *p*-type and *n*-type E-MOSFETs. CMOS is used in computer memories and in many other IC logic circuits.[8] The two types of transistors are typically connected in an arrangement called a *half-bridge* similar to that shown in Figure 4–53(a), a basic CMOS inverter. The

---

[8] CMOS technology is also used in IC analog switches.

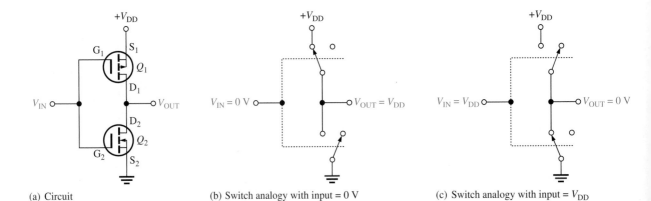

(a) Circuit   (b) Switch analogy with input = 0 V   (c) Switch analogy with input = V_{DD}

FIGURE 4–53
CMOS inverter circuit.

function of the inverter is to cause the output to be either at a high or low voltage (called a *logic level*) which is just opposite from the input.

Except during the very brief switching time, the two transistors do not conduct at the same time. The positive power supply, marked V_{DD}, is connected to Q_1's source (a *p*-channel device), and ground is connected to Q_2's source (an *n*-channel device). The positive supply is usually marked as a "drain" voltage, V_{DD}, (as if an *n*-type device were used). The gates of both transistors are connected together at the input, and the drains of both transistors are connected together at the output. Remember that a positive gate-to-source voltage turns on an *n*-channel E-MOSFET but turns off a *p*-channel E-MOSFET.

If the input voltage is near ground, Q_2 is off (it requires a positive gate voltage to turn on) and Q_1 is on (since the gate is negative with respect to the source). This causes the output to be nearly equal to the power-supply voltage, V_{DD}. A switch analogy of this condition is shown in Figure 4–53(b). If the input voltage is increased past the threshold voltage for Q_2, it conducts; and the increasing input voltage causes Q_1 to reduce conduction. This rapid switching action occurs at about one-half the supply voltage. At this point, the lower transistor appears as a short and the upper transistor appears as an open, as shown in the switch analogy in Figure 4–53(c). As a result, the output approaches 0 V.

The two transistors in the CMOS circuit described are available in a single package that uses medium power FETs in a half-bridge (the MMDF2C05E, rated for 2.0 A continuously). Applications for this device include various small motor-control circuits. One interesting application solves the problem associated with reversing a dc motor, a common problem for devices such as a chart recorder in which an analog input is plotted as a function of time. The direction of the motor is controlled by the input voltage. With proper bias, a very small analog voltage at the input can change the direction of the motor. Two half-bridges are set up as shown in Figure 4–54. Assume V_{IN} is a low voltage, causing Q_1 to be biased on. The inverter causes a high voltage to appear on the right side, turning on Q_4. The other transistors are off. Current through Q_1 and Q_4 passes through the motor. When the input voltage is raised past a threshold, all transistors switch states, causing current to be in the opposite direction through Q_2, Q_3, and the motor. The input voltage can come from an analog source or a digital source.

FIGURE 4–54

A MOSFET switching circuit that can reverse the polarity to a motor.

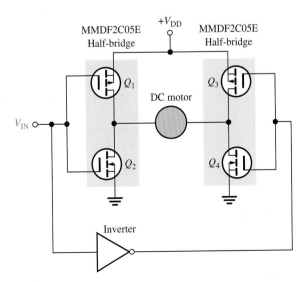

4–7 REVIEW QUESTIONS

1. What is the difference between an analog and a digital switch?
2. What are the attributes of an ideal analog switch?
3. What advantage do MOSFETs have for digital switches?
4. How does a CMOS inverter work?
5. When V_{IN} is at 0 V in Figure 4–54, which side of the motor is the most positive?

4–8 ■ A SYSTEM APPLICATION

The preamplifier you saw at the opening of this chapter is designed for applications where a high input resistance, low noise amplifier is required. This preamp could serve to boost signals that are much less than 1 mV from a nonideal, short antenna where a full-blown antenna system would take up too much room. The FET's high input resistance allows the antenna to work over a broad range of frequencies.

After completing this section, you should be able to

❏ Apply what you have learned in this chapter to a system application
   ❏ See how a FET is used in small system
   ❏ See how one transistor can be used to drive another
   ❏ See how an amplifier can be dc coupled, eliminating bias resistors
   ❏ Relate the schematic to the PC board layout

A Brief Description of the System

The preamplifier shown in Figure 4–55 uses a common-drain amplifier (Q_1) with current-source biasing as described in Section 4–6. If the two transistors are matched, the dc voltage at Q_3 is 0 V. Q_3 is a common-emitter amplifier with emitter bias. Normally, a base resistor is used to force the base to approximately 0 V, but it is not necessary because of the current-source biasing on the input stage. This also eliminates the need for a coupling capacitor. The collector of Q_3 is directly connected to the base of a *pnp* transistor (Q_4), again eliminating the need for a coupling capacitor and bias resistors. The dc voltage at this point is determined by the emitter current and can be adjusted by R_5. The purpose of Q_3 is to provide additional gain and move the dc level back to 0 V at the output. Overall, the preamplifier provides very high input resistance, low noise, reasonable gain, with economy of components.

COLOR
INSERT

Now, so you can take a closer look at the preamplifier, let's take it out of the system and put it on the troubleshooter's bench. (The PC board is also shown at the end of the color insert section.)

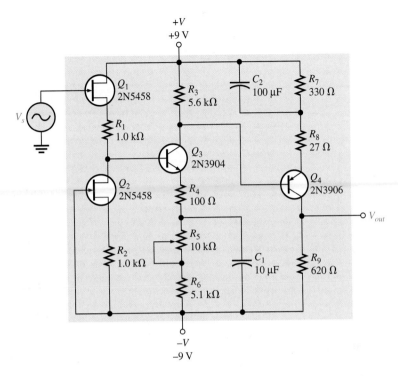

FIGURE 4–55
Schematic of preamplifier.

TROUBLESHOOTER'S BENCH

Identifying the Components

There are four transistors, eight fixed resistors, two capacitors, and a potentiometer on the board. These components are shown in the schematic in Figure 4–55 and on the board in Figure 4–56. Transistor numbers, however, were not added to the PC board. The physical appearance and pinout of the transistors are shown in Figure 4–57.

■ ACTIVITY 1 Relate the PC Board to the Schematic

Carefully follow the traces on the PC board to see how the components are interconnected. Compare each point *(A through F)* on the circuit board with the corresponding point on the schematic in Figure 4–55. For each point on the PC board, place the letter of the corresponding point or points on the schematic. Label the transistors on the PC board with the proper number (Q_1 through Q_4) from the schematic. There are four jacks shown on the PC board. Determine what point on the schematic each jack represents.

■ ACTIVITY 2 Compute the DC Voltages in the Circuit

To determine if a circuit is operating properly, it is useful to start with the expected dc parameters. From your knowledge of transistors gained in this and the previous chapter, compute the following dc parameters: $V_{B(Q3)}$, $V_{E(Q3)}$, $V_{C(Q3)}$, $V_{E(Q4)}$, $V_{C(Q4)}$. Assume the potentiometer is set in the center.

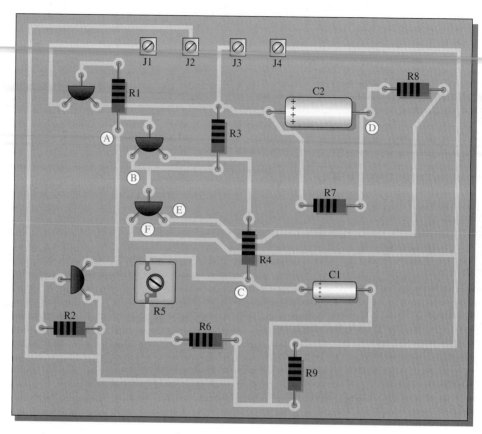

FIGURE 4–56
Preamplifier PC board.

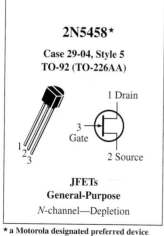

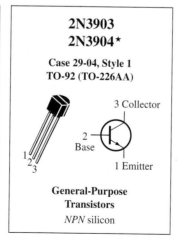

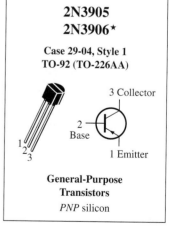

2N5458★

Case 29-04, Style 5
TO-92 (TO-226AA)

1 Drain

3
Gate

2 Source

**JFETs
General-Purpose**

N-channel—Depletion

**2N3903
2N3904★**

Case 29-04, Style 1
TO-92 (TO-226AA)

3 Collector

2
Base

1 Emitter

**General-Purpose
Transistors**

NPN silicon

**2N3905
2N3906★**

Case 29-04, Style 1
TO-92 (TO-226AA)

3 Collector

2
Base

1 Emitter

**General-Purpose
Transistors**

PNP silicon

★ a Motorola designated preferred device

FIGURE 4–57
Transistor pinouts.

Your calculation should indicate that the output voltage is less than 1.0 V. This can be adjusted to exactly 0 V with the potentiometer. The required potentiometer setting depends on the specific transistors used in the circuit, but will be about 2.5 kΩ.

■ **ACTIVITY 3   Compute the AC Voltages in the Circuit**

To check the ac conditions in the circuit, it is useful to inject a small signal at the input. Assume your troubleshooter's bench has a signal generator that can be set to 1 mV at 10 kHz. (If you cannot obtain 1 mV, you could add a voltage divider at the input and reduce the signal generator's voltage by some known amount.) The gain calculation is complicated by loading effects of one stage on another, (which include knowing the BJT's β_{ac}). However, you should be able to obtain a reasonable estimate of the expected gain. Compute the following ac parameters: $V_{b(Q3)}$, $V_{c(Q3)}$, $V_{c(Q4)}$.

■ **ACTIVITY 4   Write a Technical Report**

In your report, summarize the operation of the circuit, and describe the purpose of each transistor. Describe the dc and ac parameters that you determined in Activity 2 and Activity 3.

■ **ACTIVITY 5   Troubleshoot the Circuit for Each of the Following Problems by Stating the Probable Cause or Causes in Each Case**

Assume the input is connected to a 1.0 mV, 10 kHz source for each of the following problems. The power supplies are on and verified to be correct.

1. The output ac voltage is higher than expected (approximately 700 mV). The dc voltages are nearly the same as the computed normal parameters. The ac signal at the collector of Q_3 is normal.

2. There is no output ac signal. A check of the dc conditions reveals that the collector voltage of Q_3 is 9.0 V and the base voltage and emitter voltage are both 0 V.

3. The output ac voltage is much smaller than expected (approximately 8 mV). The dc voltages are normal. The ac signal at the collector of Q_3 is less than 1 mV.

4. The output ac voltage is smaller than expected (approximately 200 mV). The dc voltage at the collector of Q_3 is 3.3 V. The ac signal at the collector of Q_3 is only 8 mV.

4–8 REVIEW QUESTIONS

1. What is the purpose of Q_2?
2. Assume Q_1 and Q_2 match and their drain currents are 1.7 mA. What is the source voltage for Q_1?
3. Does the potentiometer affect the gain? Explain your answer.
4. What type of amplifier (CE, CB, CC) is Q_4?

■ **SUMMARY**

- FETs can be broadly classified into JFETs and MOSFETs. JFETs have a reverse-biased gate-source *pn* junction at the input; MOSFETs have an insulated-gate input.
- MOSFETs are classified as either depletion mode or enhancement mode. The D-MOSFET has a physical channel between the drain and the source; the E-MOSFET does not.
- All FETs are either *n*-channel or *p*-channel.

■ The three terminals on a FET are the source, drain, and gate that correspond to the emitter, collector, and base of a BJT.

■ JFETs have very high input resistance due to the reverse-biased gate-source *pn* junction. MOSFETs have a very high input resistance due to the insulated gate input.

■ JFETs are normally on devices. Drain current is controlled by the amount of reverse bias on the gate-source *pn* junction.

■ D-MOSFETs are normally on devices. Drain current is controlled by the amount of bias on the gate-source *pn* junction. A D-MOSFET can have either forward bias or reverse bias on the gate-source *pn* junction.

■ E-MOSFETs are normally off devices. Drain current is controlled by the amount of forward bias on the gate-source *pn* junction.

■ The drain characteristic curve for FETs is divided between an ohmic region and a constant-current region.

■ The transconductance curve is a plot of drain current versus gate-source voltage.

■ MOSFET devices need special handling procedures to avoid destructive static electricity.

■ JFETs can be biased by self-bias, a combination of self-bias and voltage-divider bias, or current-source bias.

■ A D-MOSFET can operate with a positive, negative, or zero gate-to-source voltage so it can be biased by several different methods.

■ An E-MOSFET can be biased by the same methods as a BJT (except base bias).

■ A common-source (CS) amplifier has high voltage gain and high input resistance.

■ A common-drain (CD) amplifier has unity (or less) voltage gain and high input resistance.

■ A CD amplifier can be improved significantly with current-source biasing.

■ A common-gate (CG) amplifier has high voltage gain but low input resistance.

■ The voltage gain of various amplifiers can be computed by a ratio of resistances (including internal resistance).

■ Analog switches pass or block a signal.

■ Digital switches turn on or off a device.

■ Digital switches are designed to operate in either saturation or cutoff.

■ MOSFETs have important advantages as digital switches, particularly for high-current applications.

■ GLOSSARY

Key terms are in color. All terms are included in the end-of-book glossary.

C_{iss} The common-source input capacitance of a FET as seen looking into the gate.

Common-drain (CD)   A FET amplifier configuration in which the drain is the grounded terminal.

Common-gate (CG)   A FET amplifier configuration in which the gate is the grounded terminal.

Common-source (CS)   A FET amplifier configuration in which the source is the grounded terminal.

Constant-current region   The region on the drain characteristic of a FET in which the drain current is independent of the drain-to-source voltage.

Depletion mode   A class of FETs that is on with zero-gate voltage and is turned off by gate voltage. All JFETs and some MOSFETs are depletion-mode devices.

Drain   One of the three terminals of a field-effect transistor; it is one end of the channel.

Enhancement mode   A MOSFET in which the channel is formed (or enhanced) by the application of a gate voltage.

Field-effect transistor (FET)   A voltage-controlled device in which the voltage at the gate terminal controls the amount of current through the device.

Gate One of the three terminals of a field-effect transistor. A voltage applied to the gate controls drain current.

I_{DSS} The drain current in a FET when the gate is shorted to the source. For JFETs, this is the maximum allowed current.

I_{GSS} The gate-reverse current in a FET. The value is based on a specified gate-to-source voltage.

Junction field-effect transistor (JFET) A type of FET that operates with a reverse-biased *pn* junction to control current in a channel. It is a depletion-mode device.

MOSFET Metal-oxide semiconductor field-effect transistor; one of two major types of FET. It uses a SiO_2 layer to insulate the gate lead from the channel. MOSFETs can be either depletion mode or enhancement mode.

Ohmic region The region on the drain characteristic of a FET with low values of V_{DS} in which the channel resistance can be changed by the gate voltage; in this region the FET can be operated as a voltage-controlled resistor.

Pinch-off voltage The value of the drain-to-source voltage of a FET at which the drain current becomes constant when the gate-to-source voltage is zero.

$r_{DS(on)}$ The resistance of the channel of a FET measured between the drain and the source when the FET is fully on and only a small voltage is between the drain and the source.

Source One of the three terminals of a field-effect transistor; it is one end of the channel.

Transconductance The gain of a FET; it is determined by a small change in drain current divided by a corresponding change in gate-to-source voltage. It is measured in siemens or mhos.

$V_{GS(off)}$ The voltage applied between the gate and the source that is just sufficient to turn off a FET. The exact point is arbitrary; some manufacturers use a specific very small current to determine it.

■ KEY FORMULAS

(4–1)	$g_m = \dfrac{I_d}{V_{gs}}$	Transconductance of a FET
(4–2)	$R_{IN} = \left\lvert \dfrac{V_{GS}}{I_{GSS}} \right\rvert$	Input resistance. It is the gate-source voltage divided by the gate-reverse current.
(4–3)	$V_D = V_{DD} - I_D R_D$	DC drain voltage for a FET
(4–4)	$V_{DS} = V_{DD} - I_D(R_D + R_S)$	DC drain-to-source voltage for a FET
(4–5)	$V_G = \left(\dfrac{R_2}{R_1 + R_2}\right)V_{DD}$	Gate voltage with voltage-divider bias
(4–6)	$r_s' = \dfrac{1}{g_m}$	Equivalent internal ac source resistance for computing voltage gain
(4–7)	$A_v = -g_m R_d$	Voltage gain for a CS amplifier
(4–8)	$A_v = -\dfrac{R_d}{r_s'}$	Alternate voltage gain for a CS amplifier
(4–9)	$A_v = \dfrac{R_s}{r_s' + R_s}$	Voltage gain for a CD amplifier
(4–10)	$A_v = \dfrac{g_m R_s}{1 + g_m R_s}$	Alternate voltage gain for a CD amplifier
(4–11)	$r_{DS(on)} = -\dfrac{V_{GS(off)}}{2I_{DSS}}$	Channel resistance

Answers are at the end of the chapter.

1. A type of transistor that is normally on when the gate-to-source voltage is zero is
   (a) JFET   (b) D-MOSFET   (c) E-MOSFET
   (d) answers (a) and (b)   (e) answers (a) and (c)

2. A bias method that can be used for D-MOSFETs is
   (a) voltage divider   (b) drain feedback   (c) current source
   (d) self   (e) either (a), (b), (c), or (d)

3. In normal operation, the gate-source *pn* junction for a JFET is
   (a) reverse-biased   (b) forward-biased   (c) either (a) or (b)   (d) neither (a) nor (b)

4. When the voltage between the gate and source of a JFET is zero, the drain current will be
   (a) zero   (b) I_{DSS} (c) I_{GSS} (d) none of these answers

5. One reason an *n*-channel D-MOSFET can have zero bias is it
   (a) can operate in either depletion mode or enhancement mode
   (b) does not have an insulated gate
   (c) does not have a channel
   (d) will not have drain current when operated with zero bias

6. A feature of FETs that is superior to BJTs is their
   (a) high gain   (b) low distortion   (c) high input resistance   (d) all of these answers

7. An amplifier with high voltage gain and high input resistance is a common-
   (a) gate   (b) source   (c) drain
   (d) answers (a), (b), and (c)   (e) neither (a), (b), nor (c)

8. An amplifier that inverts the signal between input and output is a common-
   (a) gate   (b) source   (c) drain
   (d) answers (a), (b), and (c)   (e) neither (a), (b), nor (c)

9. A transistor that has a closed channel unless a voltage is applied to the gate is
   (a) a JFET   (b) a D-MOSFET   (c) an E-MOSFET
   (d) all of these answers   (e) none of these answers

10. The value of the drain-to-source voltage of a FET at which the drain current becomes constant when the gate-to-source voltage is zero is called the
    (a) bias voltage   (b) pinch-off voltage   (c) saturation voltage   (d) cutoff voltage

11. The voltage gain of a common-drain amplifier cannot exceed
    (a) 1.0   (b) 2.0   (c) 10   (d) 100

12. A type of electronic switching circuit which can be used to connect a given signal to the input of an analog-to-digital converter (ADC) is a(n)
    (a) analog switch   (b) digital switch
    (c) logic switch   (d) bipolar switch

13. Refer to Figure 4–58. The schematic symbol for a *p*-channel E-MOSFET is
    (a) a   (b) b   (c) c   (d) d   (e) e   (f) f

FIGURE 4–58

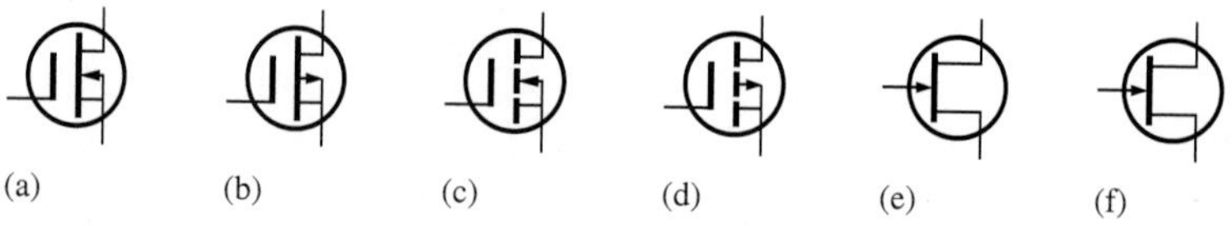

     (a)   (b)   (c)   (d)   (e)   (f)

14. Refer to Figure 4–58. The schematic symbol for a *n*-channel D-MOSFET is
    (a) a   (b) b   (c) c   (d) d   (e) e   (f) f

15. The type of device used in a CMOS switching circuit is
    (a) *n*-channel D-MOSFET   (b) *p*-channel D-MOSFET
    (c) both (a) and (b)   (d) neither (a) nor (b)

TROUBLESHOOTER'S QUIZ *Answers are at the end of the chapter.*

Refer to Figure 4–62(a).

❏ If R_G is 1.0 MΩ instead of 10 MΩ,

1. The gate voltage will

    (a) increase   **(b)** decrease   **(c)** not change

2. The drain current will

    (a) increase   **(b)** decrease   **(c)** not change

3. The input resistance will

    (a) increase   **(b)** decrease   **(c)** not change

Refer to Figure 4–63.

❏ If the positive power supply falls to +9.0 V, but the negative supply voltage is unchanged,

4. The drain current will

    (a) increase   **(b)** decrease   **(c)** not change

5. The voltage across R_D will

    (a) increase   **(b)** decrease   **(c)** not change

Refer to Figure 4–70.

❏ If capacitor C_2 is open,

6. The dc voltage at the source terminal of the MOSFET will

    (a) increase   **(b)** decrease   **(c)** not change

7. The ac voltage at the source terminal of the MOSFET will

    (a) increase   **(b)** decrease   **(c)** not change

8. The gain will

    (a) increase   **(b)** decrease   **(c)** not change

❏ If resistor R_2 is open,

9. The dc gate voltage will

    (a) increase   **(b)** decrease   **(c)** not change

10. The drain current will

    (a) increase   **(b)** decrease   **(c)** not change

■ PROBLEMS

Answers to odd-numbered problems are at the end of the book.

SECTION 4–1   Structure of Field-Effect Transistors

1. What type of transistor conducts with a reverse-biased *pn* junction at the input?

2. What type of transistor has an insulated gate?

SECTION 4–2   JFET Characteristics

3. The V_{GS} of a *p*-channel JFET is increased from +1 V to +3 V.

    (a) Does the depletion region narrow or widen?

    (b) Does the resistance of the channel increase or decrease?

    (c) Does the transistor conduct more current or less?

4. Why must the gate-to-source voltage of an *n*-channel JFET always be either zero or negative?

5. A JFET has a specified pinch-off voltage of -5 V. When $V_{GS} = 0$, what is V_{DS} at the point where I_D becomes constant?

6. An *n*-channel JFET is biased such that $V_{GS} = -2$ V using self-bias. The gate resistor is connected to ground.
   (a) What is V_S?
   (b) What is the value of $V_{GS(off)}$ if V_P is specified to be 6 V?

7. A certain JFET data sheet gives $V_{GS(off)} = -8$ V, $I_{DSS} = 10$ mA, and $I_{GSS} = 1.0$ nA at 25°C.
   (a) When $V_{GS} = 0$, what is I_D for values of V_{DS} above pinch-off?
   (b) When $V_{GS} = -4$ V, what is R_{IN} at 25°C?
   (c) What happens to R_{IN} if the temperature increases?

8. A certain *p*-channel JFET has a $V_{GS(off)} = +6$ V. What is I_D when $V_{GS} = +8$ V?

9. The JFET in Figure 4–59 has a $V_{GS(off)} = -4$ V and an $I_{DSS} = 2.5$ mA. Assume that you increase the supply voltage, V_{DD}, beginning at 0 until the ammeter reaches a steady value. At this point,
   (a) What does the voltmeter read?
   (b) What does the ammeter read?
   (c) What is V_{DD}?

10. Assume a JFET has the transconductance curve shown in Figure 4–60.
    (a) What is I_{DSS}?
    (b) What is $V_{GS(off)}$?
    (c) What is the transconductance at a drain current of 2.0 mA?

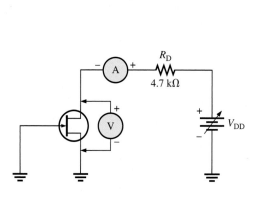

FIGURE 4–59

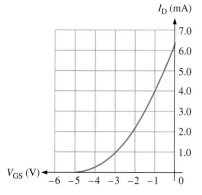

FIGURE 4–60

SECTION 4–3   JFET Biasing

11. Assume the JFET with the transconductance curve shown in Figure 4–60 is connected in the circuit shown in Figure 4–61.
    (a) What is V_S?
    (b) What is I_D?
    (c) What is V_{DS}?

12. Assume the JFET in Figure 4–61 is replaced with one with a lower transconductance.
    (a) What will happen to V_{GS}?
    (b) What will happen to V_{DS}?

13. For each circuit in Figure 4–62, determine V_{DS} and V_{GS}.

14. Assume the circuit shown in Example 4–6 has $R_S = 1.0$ kΩ and $R_D = 3.7$ kΩ. (The load line does not change as a result of these changes, but the Q-point moves). Find the new I_D and V_{GS}.

15. Find I_D and V_D for the current-source biased JFET in Figure 4–63.

FIGURE 4–61

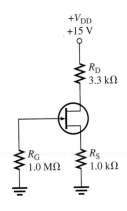

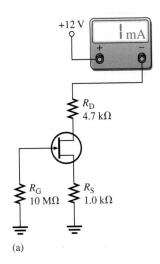

(a)

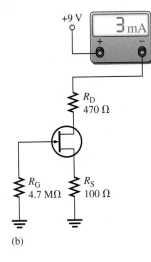

(b)

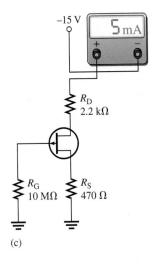

(c)

FIGURE 4–62

FIGURE 4–63

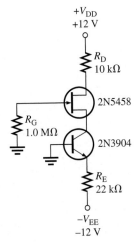

SECTION 4–4 MOSFET Characteristics

16. Sketch the schematic symbols for *n*-channel and *p*-channel D-MOSFETs and E-MOSFETs. Label the terminals.

17. Explain why MOSFETs have an extremely high input resistance at the gate.

18. An *n*-channel D-MOSFET with a positive V_{GS} is operating in what mode?

19. A certain E-MOSFET has a $V_{GS(th)} = 3$ V. What is the minimum V_{GS} for the device to turn on?

SECTION 4–5 MOSFET Biasing

20. Determine in which mode (depletion or enhancement) each D-MOSFET in Figure 4–64 is biased.

21. Each E-MOSFET in Figure 4–65 has a $V_{GS(th)}$ of +5 V or −5 V, depending on whether it is an *n*-channel or a *p*-channel device. Determine whether each MOSFET is on or off.

22. The drain current for the E-MOSFET shown in Figure 4–66 is 3.0 mA.
    (a) What type of bias is this?
    (b) Can a JFET use this type of bias?
    (c) Compute the value of V_D.
    (d) Compute the value of V_G.

23. Sketch the load line for the circuit of Figure 4–66 and indicate the Q-point. ($I_D = 3.0$ mA).

FIGURE 4–64

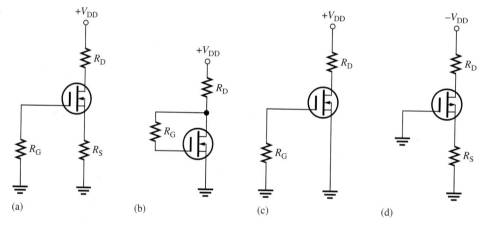

(a)   (b)   (c)   (d)

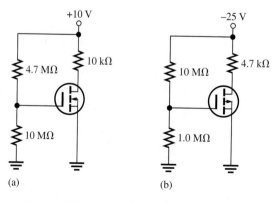

(a)   (b)

FIGURE 4–65

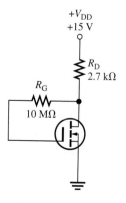

FIGURE 4–66

FIGURE 4–67

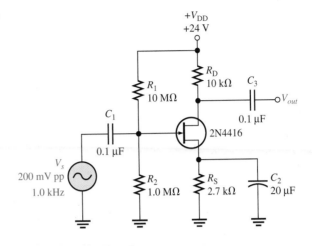

SECTION 4–6   FET Linear Amplifiers

24. (a) Assume $V_{GS} = -2.0$ V for the circuit in Figure 4–67. Determine V_G, V_S, and V_D.
   (b) If $g_m = 3000$ μmho, what is the voltage gain?
   (c) What is V_{out}?

25. Determine the gain of the amplifier in Figure 4–67 when a 27 kΩ load is connected from the output to ground. $g_m = 3000$ μmhos

26. For the circuit in Figure 4–67, what effect would an open R_1 have on
   (a) V_G (b) A_v (c) I_D

27. The minimum specified value of g_m for a 2N5457 is 1000 μmhos and the maximum specified value is 5000 μmhos. From these values, find the minimum and maximum gain for the CD amplifier in Figure 4–68.

28. Assume the g_m of Q_1 is 1500 μmhos for the amplifier in Figure 4–69.
   (a) Compute I_D.
   (b) If $g_m = 1500$ μmho, what is the voltage gain?
   (c) What is V_{out}?
   (d) What is the purpose of C_2? What happens if it is open?

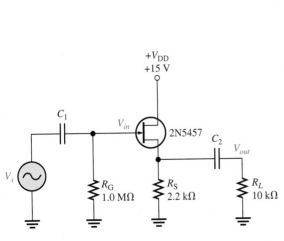

FIGURE 4–68

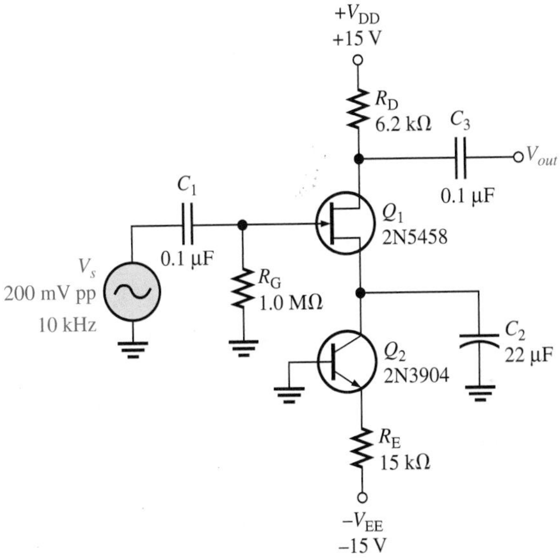

FIGURE 4–69

29. Assume the amplifier in Figure 4–69 has no output voltage. A check of the dc conditions reveals that the drain voltage is 15 V. Name at least three failures that can account for this.

30. Refer to Figure 4–69. Assume the dc voltages and the ac input voltage are correct, but V_{out} is very small. What failure can account for this?

31. Refer to Figure 4–69. What is the minimum value of I_{DSS} that Q_1 can have before the gate-source *pn* junction is forward-biased?

32. Assume the source voltage for the D-MOSFET in Figure 4–70 is measured and found to be 1.6 V.
    (a) Compute I_D and V_{DS}.
    (b) If $g_m = 2000$ µmho, what is the voltage gain?
    (c) Compute the input resistance of the amplifier.
    (d) Is the D-MOSFET operating in the depletion or the enhancement mode?

FIGURE 4–70

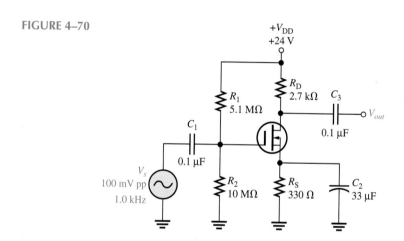

33. Repeat Problem 32 (a) and (b) if a 5.1 kΩ load is connected between V_{out} and ground.

34. Assume Q_1 and Q_2 in Figure 4–71 have matching characteristics and I_{DSS} is 1.5 mA.
    (a) What is I_D?
    (b) What is the approximate gain?
    (c) If Q_2 is replaced with a transistor with an I_{DSS} of 1.0 mA, what problem will occur?

SECTION 4–7   FET Switching Circuits

35. Explain why $r_{DS(on)}$ is one of the most important specifications for an analog switch.

36. The 2N5555 is an *n*-channel JFET switching transistor. The specification sheet shows $r_{DS(on)(max)} = 150$ Ω and $I_{DSS(min)} = 15$ mA. It also shows a maximum drain current of 2.0 µA when $V_{GS} = -10$ V at a temperature of 100°C. Assume it is used in the analog switch circuit of Figure 4–72 for the worst case conditions given here.
    (a) What is the output voltage when $V_{GS} = 0$ V?
    (b) What is the output voltage when $V_{GS} = -10$ V?

37. What are the two operating conditions for a transistor in a digital switch called?

38. Could a D-MOSFET be substituted for the E-MOSFET in Figure 4–51(c)? Explain your answer.

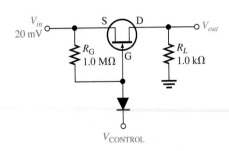

FIGURE 4–71   FIGURE 4–72

■ **ANSWERS**
TO REVIEW
QUESTIONS

Section 4–1

1. Drain, source and gate

2. MOSFET

3. They require smaller areas than BJTs, are easy to manufacture in ICs, and produce simpler circuits.

4. BJTs are controlled by a current, FETs are controlled by a voltage. BJT circuits have higher gain but lower input resistance.

Section 4–2

1. Transconductance curve

2. Positive

3. By the gate-to-source voltage

4. 7 V

5. Decrease

6. +3 V

Section 4–3

1. $V_{GS(off)}$ and I_{DSS}.

2. The base-emitter *pn* junction in a BJT is forward biased; the gate-source *pn* junction in a JFET is reverse biased.

3. −8 V

4. I_{DSS}

Section 4–4

1. Depletion MOSFET and enhancement-only MOSFET. The D-MOSFET has a physical channel; the E-MOSFET does not.

2. Yes; the current is I_{DSS}.

3. No

4. Yes

Section 4–5

1. I_{DSS}
2. It is a normally off device. It must have forward bias to turn it on.
3. $+2$ V

Section 4–6

1. It is the transconductance (g_m) times the ac drain resistance (R_d) *or* the ratio of the ac drain resistance (R_d) to the internal ac source resistance (r'_s).
2. A CD amplifier with current-source biasing has higher input resistance, no bias resistors, and can be dc coupled with 0 V dc output.
3. CD and CG
4. Low input resistance
5. High input resistance and low noise

Section 4–7

1. An analog switch passes or blocks an ac signal; a digital switch turns on or off a device.
2. When closed, it has no resistance to the signal; when open, it is an infinite resistance.
3. They are voltage controlled and draw no drive current. They can control a large current to a device and they are immune to thermal runaway.
4. N-channel and p-channel E-MOSFETs are connected with common gates and drains, and the output is connected to the drains. The n-channel source is connected to ground and the p-channel source is connected to a positive supply voltage. When the input is greater than one-half the power supply voltage, the n-channel device is on, causing the output to be near ground; when the input is less than one-half the power supply voltage, the p-channel MOSFET is on causing the output to be near the power supply voltage.
5. The left side

Section 4–8

1. It is the current source for Q_1.
2. 1.7 V
3. No, the ac signal bypasses the potentiometer through C_1.
4. CE

■ **ANSWERS TO PRACTICE EXERCISES FOR EXAMPLES**

4–1 I_D remains at approximately 12 mA.

4–2 ≈ 1.0 mS

4–3 300 MΩ

4–4 $V_{DS} = 6.34$ V, $V_{GS} = -0.66$ V

4–5 $I_{D(min)} \cong 0.3$ mA; $I_{D(max)} \cong 2.3$ mA

4–6 The dc cutoff is $+9$ V; the dc saturation current is 1.91 mA. These represent the end points on the x-axis and the y-axis for the load line.

4–7 The open on the drain leaves the gate forward-biased through the voltage divider on the base. About 11 μA is in the path that includes the source resistor and the 1.0 MΩ bias resistor. The source voltage is therefore approximately 20 mV: the drain voltage is the same as the source voltage because of the common channel, and the gate voltage is about 500 mV since the gate source *pn* junction has only a very small forward current.

4–8 I_{DSS} cannot be less than 0.95 mA.

4–9 6.8 V

4–10 2.13 mA

4–11 A larger source resistor reduces (slightly) the transconductance. As a result, the gain will be reduced.

4–12 10 MΩ

4–13 I_D is approximately 1.1 mA; $V_{S(Q1)}$ is approximately 2.2 V.

4–14 0.96

■ **ANSWERS TO SELF-TEST**

1. (a)	**2.** (e)	**3.** (a)	**4.** (b)	**5.** (a)
6. (c)	**7.** (b)	**8.** (b)	**9.** (c)	**10.** (b)
11. (a)	**12.** (a)	**13.** (d)	**14.** (a)	**15.** (d)

■ **ANSWERS TO TROUBLE-SHOOTER'S QUIZ**

1. not change	**2.** not change	**3.** decrease	**4.** not change
5. decrease	**6.** not change	**7.** increase	**8.** decrease
9. increase	**10.** increase		

5

MULTISTAGE, RF, AND POWER AMPLIFIERS

■ **CHAPTER OBJECTIVES**

❑ Determine the ac parameters for a
   capacitively coupled multistage amplifier
❑ Describe the characteristics of high-frequency
   amplifiers and give practical considerations
   for implementing high-frequency circuits
❑ Describe the characteristics of transformer-
   coupled amplifiers, tuned amplifiers, and
   mixers
❑ Determine basic dc and ac parameters for
   direct-coupled amplifiers and describe how
   negative feedback can stabilize the gain of an
   amplifier
❑ Compute key ac and dc parameters for class
   A power amplifiers and discuss operation
   along the ac load line
❑ Compute key ac and dc parameters for class
   B power amplifiers including bipolar and FET
   types
❑ Give principal features and describe
   applications for IC power amplifiers
❑ Apply what you have learned in this chapter
   to a system application

■ CHAPTER INTRODUCTION

The previous two chapters have introduced single-stage amplifiers whose primary function was to increase the voltage of a signal. You should be familiar with the biasing and ac parameters for both BJTs and FETs.

When very small signals must be amplified, such as from an antenna, variations about the Q-point are relatively small. Amplifiers designed to amplify these signals are called small-signal amplifiers. They may also be designed specifically for high frequencies. Frequently, it is useful to have additional stages of gain; this is particularly true in high-frequency applications, where the frequencies of interest are restricted to a certain bandwidth.

In this chapter, you will study various types of multistage amplifiers, with particular emphasis on high-frequency considerations including noise, cabling, and eliminating unwanted oscillations. Then, the emphasis shifts to the important requirement of delivering power to a load. For these applications, power amplifiers are needed. The chapter ends with an introduction to integrated circuit (IC) power amplifiers.

■ A SYSTEM APPLICATION

The system application in this chapter is a small power amplifier that is part of a public address system. The input is from a preamp such as the one you studied in Chapter 4. You will apply the knowledge you gain in this chapter to understanding its operation so that you can troubleshoot any fault in the circuit.

For the system application in Section 5–8, be sure you understand
❑ How a class AB amplifier works
❑ How to compute the maximum power delivered to a load from a power amplifier

5–1 ■ CAPACITIVELY COUPLED AMPLIFIERS

Two or more transistors can be connected together to form an amplifier called a multistage amplifier. In Section 1–4, a simplified amplifier model was introduced. You are now ready to apply this simplified model to actual amplifier circuits to determine their overall performance. In this section, you will learn about capacitively coupled amplifiers, also called RC coupled amplifiers. Capacitive coupling is the most widely used method for passing the ac signal to the next stage.

After completing this section, you should be able to

❑ Determine the ac parameters for a capacitively coupled multistage amplifier
  ❑ Compute the overall gain, the input resistance, and the output resistance of a two-stage capacitively coupled amplifier
  ❑ Discuss how oscillation and noise problems can be alleviated in multistage amplifiers

Two or more transistors can be connected together to enhance the performance of an amplifier. Each transistor that amplifies the signal is considered a **stage**. Frequently, the first stage of an amplifier must have very high input resistance to avoid loading the source. In addition, the first stage needs to be designed for low noise operation because the very small signal voltage can easily be obscured by noise. Succeeding stages are designed to increase the amplitude of the signal without adding distortion.

Probably the simplest way to add gain to an amplifier is to capacitively couple two stages together as shown in Figure 5–1. In this case, both stages are identical CE amplifiers with the output of the first connected to the input of the second stage. Capacitive coupling prevents the dc bias of one stage from affecting the dc bias of another stage because capacitors block dc. Although the dc path is open, the coupling capacitor provides almost no opposition to the ac signal and the signal passes to the next stage.

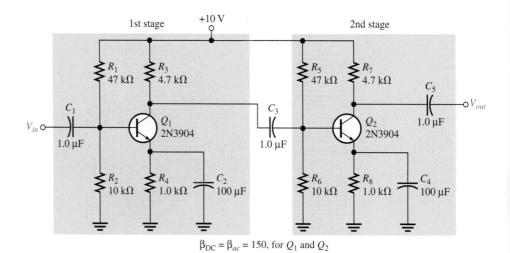

$\beta_{DC} = \beta_{ac} = 150$, for Q_1 and Q_2

FIGURE 5–1
A two-stage CE amplifier.

The analysis of the circuit starts with the dc conditions, as explained in Section 3–2. To compute the base voltage of either stage, use the voltage-divider rule.

$$V_B \cong \left(\frac{R_2}{R_1 + R_2}\right) V_{CC} = \left(\frac{10 \text{ k}\Omega}{47 \text{ k}\Omega + 10 \text{ k}\Omega}\right) 10 \text{ V} = 1.7 \text{ V}$$

This estimate is slightly high because it is made for an unloaded voltage divider. After subtracting 0.7 V for the base-emitter diode, the emitter voltage is 1.0 V, which results in an emitter current of

$$I_E = \frac{V_E}{R_E} = \frac{1.0 \text{ V}}{1.0 \text{ k}\Omega} = 1.0 \text{ mA}$$

The emitter current is also approximately equal to the collector current.

Loading Effects

Recall from Section 1–4 that amplifiers were shown as a block diagram with essential parameters only. The ac model is simply a dependent voltage source with a series resistance (a Thevenin circuit). To compute the overall gain of the amplifier, each transistor stage in Figure 5–2 can be modeled in a similar manner. Only three parameters need to be known: the unloaded (No-Load) voltage gain ($A_{v(NL)}$), the total input resistance, ($R_{in(tot)}$), and the output resistance (R_{out}). Notice that the unloaded output voltage is the input voltage times the unloaded gain.

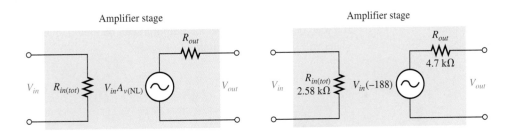

(a) One-stage amplifier model

(b) Values for one stage of the amplifier in Figure 5–1

FIGURE 5–2

Start by finding the unloaded gain of one stage. Because the two stages are identical, the unloaded gain is the same for both. The input resistance of the second stage acts as a load on the first stage. Thus, the loaded gain of the first stage can be found by assuming it has a load resistor equal to $R_{in(tot)}$ of stage 2. This lowers the gain of the first stage but can be considered separately from the unloaded gain calculation. An illustration of this idea should clarify how the basic amplifier model can simplify determining the overall gain.

As you know, the unloaded gain of a CE amplifier is the ratio of the ac collector resistance to the ac emitter resistance. This unloaded gain is dependent on r'_e, which in turn depends on I_E, so the calculation should be considered approximate.

Since the unloaded gain is being computed, the ac collector resistance, R_c, is the same as the actual collector resistor, R_C, which is 4.7 kΩ. The ac emitter resistance is approximately

$$r'_e \cong \frac{25 \text{ mV}}{I_E} = \frac{25 \text{ mV}}{1.0 \text{ mA}} = 25 \text{ }\Omega$$

The unloaded gain, $A_{v(NL)}$, is approximately

$$A_{v(NL)} = -\frac{R_c}{R_e} = -\frac{R_C}{r'_e} = -\frac{4.7 \text{ k}\Omega}{25 \text{ }\Omega} = -188$$

The input resistance of the CE amplifier was discussed in Section 3–4. The equation for input resistance with voltage-divider bias and no swamping resistor is

$$R_{in(tot)} = R_1 \parallel R_2 \parallel (\beta_{ac} r'_e)$$

By substitution, the input resistance of the amplifier, assuming a β_{ac} of 150, in Figure 5–1 is

$$R_{in(tot)} \cong 47 \text{ k}\Omega \parallel 10 \text{ k}\Omega \parallel [150(25 \text{ }\Omega)] \cong 2.58 \text{ k}\Omega$$

The output resistance is the resistance looking back to the collector circuit and is simply the collector resistor.

$$R_{out} = R_C = 4.7 \text{ k}\Omega$$

These values can be entered onto the model as shown in Figure 5–2(b).

The two stages that comprise the amplifier are now connected in Figure 5–3. In this drawing, the unloaded gain for each stage is shown below the Thevenin source, and the model is used to find the overall gain. The overall gain is the product of three terms:

1. The unloaded voltage gain of the first stage
2. The gain of the voltage divider consisting of the input resistance of the second stage with the output resistance of the first stage
3. The unloaded gain of the second stage

If a load resistor is added to the output, it can be included as another voltage-divider term. (This will be shown in Example 5–1.)

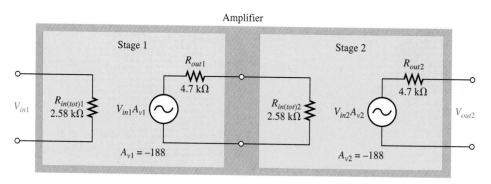

FIGURE 5–3
AC model of the complete two-stage amplifier.

The unloaded gain of each stage is -188, as previously calculated. The voltage divider between the stages accounts for loading effects. It consists of $R_{in(tot)2}$ for stage 2 and R_{out1} for stage 1. The gain (attenuation) of this voltage divider is

$$A_{v(divider)} = \frac{R_{in(tot)2}}{R_{out1} + R_{in(tot)2}} = \frac{2.58 \text{ k}\Omega}{4.7 \text{ k}\Omega + 2.58 \text{ k}\Omega} = 0.35$$

The overall voltage gain is the product of the three gains.

$$A_{v(tot)} = A_{v1}A_{v(divider)}A_{v2} = (-188)(0.35)(-188) \cong 12,400$$

This product indicates the voltage gain is fairly large. If an input signal of 100 μV, for example, is applied to the first stage and the attenuation of the input base circuit is neglected, an output from the second stage of $(100 \mu V)(12,400) = 1.24$ V will result. Again, a factor that must be kept in mind is that this answer is approximate because the gain is very dependent on the value of r_e' and the specific transistors used. At the price of reduced gain, greater stability can be achieved by adding a swamping resistor in the emitter circuit. This will tend to make the circuit produce consistent gain that is independent of the specific transistor.

The gain for amplifiers is frequently expressed as a decibel voltage gain. For the amplifier just considered, the unloaded decibel voltage gain of each stage is

$$A_v' = 20 \log|A_v| = 20 \log(188) = 45.5 \text{ dB}$$

The gain (attenuation) of the voltage divider between the stages is

$$A_{v(divider)}' = 20 \log(0.35) = -9.1 \text{ dB}$$

The overall decibel voltage gain is the sum of the individual decibel voltage gains.

$$A_{v(tot)}' = A_{v1}' + A_{v(divider)}' + A_{v2}' = 45.5 \text{ dB} - 9.1 \text{ dB} + 45.5 \text{ dB} = 81.9 \text{ dB}$$

EXAMPLE 5–1

Draw the simplified ac model and compute the overall gain for the two-stage preamplifier in Figure 5–4. Assume the g_m of Q_1 is 1500 μmhos (typical for a 2N5458) and the β_{ac} is 150 (typical of the 2N3904). The first stage provides a very high input resistance circuit with low noise. The second stage provides voltage gain.

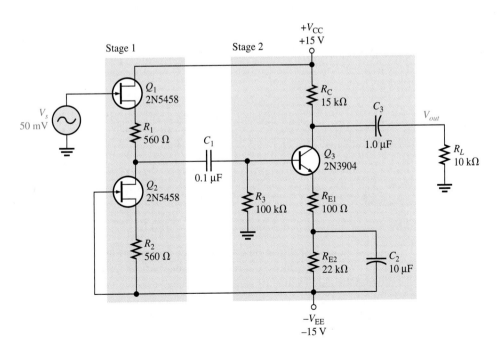

FIGURE 5–4

Solution Start with the dc parameters. The input stage shown in Figure 5–4 is composed of a FET (Q_1) that uses current-source biasing from Q_2. The drain of Q_2 should be approximately 0 V. The second stage is composed of the BJT, Q_3, with emitter biasing. Because of emitter biasing, the emitter voltage will be close to -1 V. Applying Ohm's law, the emitter current is approximately

$$I_E = \frac{V_E - (-V_{EE})}{R_{E1} + R_{E2}} = \frac{-1 \text{ V} - (-15 \text{ V})}{100 \text{ }\Omega + 22 \text{ k}\Omega} = 0.63 \text{ mA}$$

The collector dc voltage is

$$V_C = V_{CC} - I_E R_C = 15 \text{ V} - (0.63 \text{ mA})(15 \text{ k}\Omega) = 5.6 \text{ V}$$

Now determine the ac parameters. The input resistance of the first stage, $R_{in(tot)1}$, is that of a reverse-biased *pn* junction. It is very high; a precise value depends on I_{GSS}. It is sufficient to estimate it as >1 MΩ. The output resistance of the first stage is the 560 Ω source resistor in series with r'_s. Therefore,

$$R_{out1} = R_1 + r'_s = 560 \text{ }\Omega + \frac{1}{g_m} = 560 \text{ }\Omega + \frac{1}{1500 \text{ }\mu\text{mhos}} = 1.23 \text{ k}\Omega$$

The value of r'_e is

$$r'_e = \frac{25 \text{ mV}}{I_E} \cong \frac{25 \text{ mV}}{0.63 \text{ mA}} \cong 40 \text{ }\Omega$$

The unloaded voltage gain of the first stage is 1.0 (because of the current-source biasing). The unloaded voltage gain of the second stage is

$$A_{v2} = -\frac{R_c}{R_e} = -\frac{R_C}{r'_e + R_{E1}} \cong -\frac{15 \text{ k}\Omega}{40 \text{ }\Omega + 100 \text{ }\Omega} \cong -107$$

The input resistance of the second stage acts as a load on the first stage. To find the input resistance, note that the emitter-bias resistor and the ac resistance of the base form a parallel combination given by

$$R_{in(tot)2} = R_3 \| [\beta_{ac}(r'_e + R_{E1})]$$

Assuming a β_{ac} of 150,

$$R_{in(tot)2} \cong 100 \text{ k}\Omega \| [150(40 \text{ }\Omega + 100 \text{ }\Omega)] \cong 17.3 \text{ k}\Omega$$

The output resistance is just the collector resistor, R_C.

$$R_{out2} = R_C = 15 \text{ k}\Omega$$

These values are shown on the simplified circuit in Figure 5–5; the unloaded gain values are also shown. The output resistance and the load resistor form a voltage divider that reduces the gain of the last stage. Thus, the overall voltage gain is

$$A_{v(tot)} = (A_{v1})\left(\frac{R_{in(tot)2}}{R_{out1} + R_{in(tot)2}}\right)(A_{v2})\left(\frac{R_L}{R_{out2} + R_L}\right)$$

$$= (1)\left(\frac{17.3 \text{ k}\Omega}{1.23 \text{ k}\Omega + 17.3 \text{ k}\Omega}\right)(-107)\left(\frac{10 \text{ k}\Omega}{15 \text{ k}\Omega + 10 \text{ k}\Omega}\right) = \mathbf{-40}$$

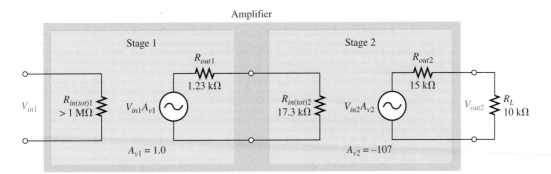

FIGURE 5–5

Practice Exercise*
(a) If the input voltage in Figure 5–4 is 10 mV, what is the output voltage?
(b) What is the voltage gain of the amplifier in Figure 5–4 if the load resistor is removed?

---

* Answers are at the end of the chapter.

Unwanted Oscillation and Noise

Multistage amplifiers require careful design to avoid unwanted oscillations. When large signals are present in the same circuit as small signals, the large signal can have an adverse effect on the small signal due to unwanted feedback paths. This problem is compounded in high-frequency amplifiers because feedback paths tend to have lower reactance, causing more unwanted feedback. For example, protoboards have stray capacitances between rows that can lead to feedback and noise problems when constructing multistage amplifiers on them. It is usually helpful to isolate the various stages by connecting a capacitor between V_{CC} and ground at each stage; this technique is seen frequently in commercial printed circuit (PC) boards. Capacitors should be connected very close to the point that V_{CC} is applied to a stage and should have short lead lengths.

In addition to unwanted oscillations, noise voltages (unwanted electrical disturbances) can be a problem for multistage amplifiers. The ratio of signal to noise determines if the noise is sufficient to disrupt the signal. When the signal is small, a little noise voltage has a greater effect than when the signal is large. This means the first stage of an amplifier is the most important stage because of the very small signal level. FETs have the advantage for high-impedance sources; but when the source impedance is lower, (<1 MΩ), bipolar transistors can provide excellent low-noise performance.

Much has been written on the subject of noise in circuits. The "cure" for noise problems depends on the source of the noise, the path into the circuit, the type of noise, and other details. Many times, there is no *one* solution to a noise problem. Noise can enter a circuit from sources external to the circuit (such as fluorescent lighting) by capacitive or inductive coupling, through the power supply, or from within the circuit (thermal noise). The following are suggestions for avoiding noise problems:

1. Keep wiring short to avoid "antennas" in circuits (particularly low-level input lines) and make signal return loops as small as possible.

2. Use capacitors between power supply and ground at each stage and make sure the power supply is properly filtered.

3. Reduce noise sources, if possible, and separate or shield the noise source and the circuit. Use shielded wiring, twisted pair, or shielded twisted pair wiring for low-level signals.

4. Ground circuits at a single point, and isolate grounds that have high currents from those with low currents by running separate ground lines back to the single point. Ground current from a high current ground can generate noise in another part of a circuit because of IR drops in the conductive paths.

5. Keep the bandwidth of amplifiers no larger than necessary to amplify only the desired signal, not extra noise.

5–1 REVIEW QUESTIONS*

1. What three parameters are needed for each stage of a multistage amplifier to determine the gain?
2. What is the decibel gain of the amplifier in Example 5–1?
3. What advantages do FETs have as the input stage to a multistage amplifier?
4. Why is the first stage of a multistage amplifier the most important for reducing noise?

---

* Answers are at the end of the chapter.

5–2 ■ RF AMPLIFIERS

In general, radio frequencies (RF) are those frequencies that are useful for radio transmission—from a practical low frequency of about 10 kHz to above 300 GHz. Above approximately 100 kHz, amplifiers often use tuned resonant circuits at the input, output, or load, so many people prefer to consider amplifiers that use frequencies above 100 kHz as RF amplifiers. At these higher frequencies, amplifiers are designed to provide gain only for those frequencies within a certain band. In this section, practical considerations for high-frequency amplifiers are given. In the next section, you will see how high-frequency signals can be coupled with transformers from one stage to another.

After completing this section, you should be able to

❏ Describe the characteristics of high-frequency amplifiers and give practical considerations for implementing high-frequency circuits
  ❏ Explain the need for transmission lines when working with high frequencies
  ❏ Find the characteristic impedance of a cable, given the inductance and capacitance per unit length
  ❏ Explain the proper way to terminate a cable to avoid reflections
  ❏ Describe important ac considerations for RF amplifiers
  ❏ Explain what is meant by neutralization
  ❏ Explain how AGC works

Transmission Lines

When high-frequency signals or fast-rising digital signals are transferred from one point to another, the wiring can have adverse effects such as attenuation of the signal, a decrease in high-frequency response, and noise pickup. These effects are important for signal path lengths of a few inches when the frequency is above approximately 100 MHz or the rise time of a digital signal is less than approximately 4 ns.

Consider two wires that form a transmission line to send a high-frequency signal form one point to another. The wires have an inductance, *L*, that appears to be in series along the wires and a capacitance, *C*, that appears to be in parallel between the conductors. (Two conductors separated by an insulator form a capacitor.) At high frequencies, the reactance of the series inductance increases and the reactance of the parallel capacitance decreases. The *L* and *C* of the wiring is not "lumped" in one place but is distributed over the entire length of the line.

Figure 5–6(a) illustrates an equivalent circuit for a very short section of transmission line in which the inductance and capacitance are drawn as discrete components, but keep in mind they are distributed evenly throughout the wire. The inductance is shown as four inductors, each with an inductance of *L*/4. The capacitance of the short section is shown as *C*. Resistance is also present but forms a minor part of the impedance at high frequencies, so it can be ignored.

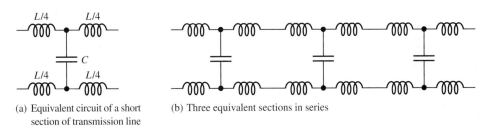

(a) Equivalent circuit of a short section of transmission line

(b) Three equivalent sections in series

FIGURE 5–6
Equivalent circuit for a transmission line at high frequencies.

To help you understand a transmission line, the equivalent circuit for a very short section is extended into a series of short sections of discrete inductors and capacitors connected together as shown in Figure 5–6(b). If additional sections are added to the equivalent circuit, you would observe an interesting effect; after about ten sections, the impedance of the line hardly changes as more and more sections are added. *The impedance does not depend on the length of the line!* The constant value of impedance is known as the *characteristic impedance* of the line. At high frequencies, the value of the characteristic impedance of a line is given by the following equation:

$$Z_0 = \sqrt{\frac{L}{C}}$$

(5–1)

where Z_0 = characteristic impedance of the line, in ohms (Ω)
L = inductance per unit length of line, in henries (H)
C = capacitance per unit length of line, in farads (F)

Note that *L* and *C* have to be measured for the same length of line. Since the equation uses a ratio, the length used does not matter, only that the lengths used are the same. The impedance of a cable is a function of its geometry and the type of dielectric that is used to construct the cable. There are various types of cables for high-frequency applications. They should all have a relatively high bandwidth and have a constant impedance independent of the length.

One common type of high-frequency transmission line is called **coax**. With coax, a center conductor is surrounded by a tubular conducting shield. At high frequencies, this outer conductor acts as a shield and helps avoid radiating signals from the cable or adding unwanted external interference noise to the signal. Various types of coax are available with a range of characteristics, including power handling, high-frequency characteristics, and the characteristic impedance.

It is important to install the type of cable that was designed for use in a given system. For example, video systems are standardized on 75 Ω coaxial cable. Depending on the type of line, transmission lines have characteristic impedances that are typically between 50 Ω and 100 Ω for coax cable and up to several hundred ohms for parallel conductors.

Because of its high bandwidth, coax is applied to some communication systems where many different voice channels can be put on the same cable. Frequency separation filters allow transmission in both directions at the same time.

EXAMPLE 5–2

Determine the characteristic impedance of RG58C coaxial cable, which has a capacitance of 28.5 pF per foot and an inductance of 7.12 μH per 100 feet. (RG means *R*adio *G*rade.)

Solution   To determine the characteristic impedance, the capacitance and inductance must be specified for the same length of cable. To convert the capacitance to pF per 100 feet, multiply the capacitance per foot by 100. Therefore, C is 2850 pF per 100 feet. Substituting into Equation (5–1),

$$Z_0 = \sqrt{\frac{L}{C}} = \sqrt{\frac{7.12\ \mu H}{2850\ pF}} = \mathbf{50\ \Omega}$$

Practice Exercise   Compute the characteristic impedance of RG59B coaxial cable which has a capacitance of 21.0 pF per foot and and inductance of 11.8 μH per 100 feet.

Terminating Transmission Lines

At high frequencies, even a short transmission line may be long compared to the wavelength of the signal. When the signal from the source (the incident wave) reaches the end of the line, it can be reflected back toward the source (the reflected wave). The incident and reflected waves interact along the length of the line, creating a **standing wave** on the line. A standing wave is a stationary wave formed by the interaction of an incident and reflected wave.

Standing waves create undesired effects such as "ghosts" in a television signal and increased noise. To prevent standing waves, transmission lines are terminated with a resistive load equal to the characteristic impedance of the line. The entire transmission line looks resistive to the source when it is terminated in this manner. When properly terminated, all of the signal power is dissipated in the terminating resistor. Improper termination can result in reflections and the wrong signal level.

High-Frequency Considerations

Inductive Effects   At high frequencies, (above approximately 10 MHz), wire is no longer a simple conductive path but becomes an effective inductor. This happens because of an effect known as **skin effect**, which causes current to move to the outside surface of conductors. The addition of inductance in a signal path is generally undesirable because it increases the reactance in the signal line and can increase noise in the circuit. To avoid adverse inductive effects, wiring in high-frequency circuits should be kept as short as possible.

Capacitance Effects   At high frequencies, transistor amplifiers are less effective because of the increased effect of capacitance. All active devices have internal capacitances between their terminals. These internal capacitances appear as a low-impedance path for high-frequency analog signals, reducing the effectiveness of the device. In digital circuits, internal capacitances limit the speed at which a pulse can change from one level to another. Transistors for high frequencies are specially designed to minimize the internal capacitances.

The adverse effects of capacitance are magnified by inverting amplifiers (such as a common-source or common-emitter amplifier) due to a form of positive feedback called the *Miller effect.* As a rule, it is best to try to minimize capacitance in high-frequency circuits by keeping wiring as short as possible and by avoiding high-gain inverting type amplifiers.

Another effect of capacitance is that undesired oscillations can be produced in high-frequency amplifiers. Oscillations are eliminated by a technique called *neutralization,* discussed later in this section.

Tuned Amplifiers

Amplifiers with resonant circuits are common in communication applications because communication systems use high frequencies. Frequencies greater than 100 kHz are commonly referred to as RF for *R*adio *F*requency. Amplifiers that are designed for these high frequencies are called RF amplifiers. The techniques you learned for finding dc bias conditions with low-frequency amplifiers are the same for RF amplifiers but need to be modified for ac considerations. Low-frequency amplifiers are untuned; they are designed to amplify a wide band of frequencies.

Tuned amplifiers are different; they are designed to amplify a specific band of frequencies while eliminating any signals outside the band. They use a parallel *LC* resonant circuit for a load to provide a high impedance to the ac signal and thus produce a high gain at the resonant frequency. The center frequency (assuming *Q* is large) of the tuned circuit can be computed from the basic resonant frequency equation:

$$f_r = \frac{1}{2\pi\sqrt{LC}} \tag{5–2}$$

The bandwidth of a tuned amplifier is determined by the *Q* (quality factor) of the resonant circuit. The **quality factor (*Q*)** is a dimensionless number that is the ratio of the maximum energy stored in a cycle to the energy lost in a cycle. From a practical view, the inductor almost always determines the *Q;* consequently *Q* is often expressed as the ratio of the inductive reactance, X_L, to the resistance, *R*. It is also the ratio of the center (resonant) frequency, f_r, to the bandwidth, *BW*.

$$Q = \frac{X_L}{R} = \frac{f_r}{BW} \tag{5–3}$$

The response of a parallel resonant circuit depends on the *Q* of the circuit, as illustrated in Figure 5–7. The *Q* for an RF circuit depends on the type of inductor; it can range from 50 to 250 for ferrite-core inductors and even higher for air-core inductors.

FIGURE 5–7

Impedance of a parallel resonant circuit as a function of frequency.

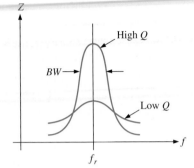

A basic tuned RF amplifier, using a JFET, is shown in Figure 5–8. The gate and drain circuits each have a parallel resonant circuit using a transformer coil and a capacitor to form the resonant circuit. (Transformer coupling is discussed in Section 5–3.) The dashed-line capacitor between the drain and gate represents an internal capacitance of the transistor, a capacitance of only a few pF. The drain circuit presents a high impedance to the ac signal but easily passes the dc quiescent current through the transformer's primary coil, which looks like a small resistance to dc.

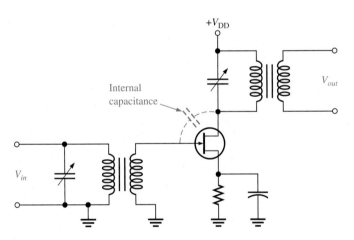

FIGURE 5–8

A tuned RF amplifier.

Despite the fact that the internal capacitance between the drain and gate is very small, at high frequencies it may produce sufficient positive feedback (in phase) from the output back to the input to cause the amplifier to break into undesired oscillation. To prevent this, a neutralization circuit is sometimes necessary, particularly for high-impedance circuits.

Neutralization is the process of adding the same amount of negative feedback (out of phase) to just cancel the positive (in phase) feedback due to the internal capacitances of an amplifier. Figure 5–9 shows a popular neutralization circuit called *Hazeltine neutralization*. The neutralization is accomplished by sending a negative feedback signal from the output back to the input through a small neutralization capacitor, C_n, which is adjusted to just cancel the undesired positive-feedback signal. Notice that the drain power supply is now connected through a center-tapped transformer.

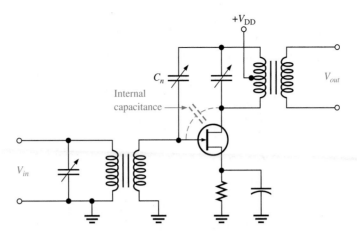

FIGURE 5–9
Hazeltine neutralization. C_n cancels the internal capacitance.

Figure 5–10 shows another popular circuit for an RF amplifier that uses a dual-gate D-MOSFET to amplify antenna signals. The dual-gate configuration simplifies adding **automatic gain control (AGC)** to the circuit because the signals are combined within the MOSFET. The AGC reduces the gain when a larger signal is received and increases the gain for smaller signals. The RF signal is connected to the lower gate while the upper gate is used to control the gain. The AGC signal is a negative dc voltage (for an *n*-channel device) that is developed from a later stage in the amplifier. The AGC voltage is proportional to the input signal strength. A large input signal produces a larger AGC voltage, tending to pinch off the channel and thus reducing the gain.

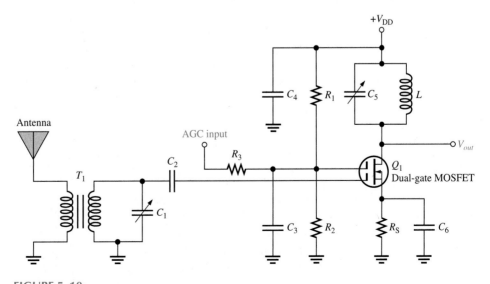

FIGURE 5–10
RF amplifier using a dual-gate depletion-mode MOSFET.

5–2 REVIEW QUESTIONS

1. What are the advantages of coax cable for passing an RF signal?
2. What is meant by neutralization?
3. What is AGC?
4. What is the definition of *Q?*

5–3 ■ TRANSFORMER-COUPLED AMPLIFIERS

Transformers can be used to couple a signal from one stage to another. Although principally used in high-frequency designs, they are also found in low-frequency power amplifiers. When the signal frequency is in the RF range (>100 kHz), stages within an amplifier are frequently coupled with tuned transformers, which form a resonant circuit. In this section, you will see examples of transformer-coupled amplifiers—both low-frequency and high-frequency tuned amplifiers.

After completing this section, you should be able to

❑ Describe the characteristics of transformer-coupled amplifiers, tuned amplifiers, and mixers
  ❑ Describe in general how a transformer-coupled amplifier operates
  ❑ Determine the ac and dc load line for a transformer-coupled amplifier
  ❑ Explain how a high frequency is converted to a lower frequency with a mixer
  ❑ Give advantages for using an IF amplifier in a high-frequency application

Low-Frequency Applications

Most amplifiers require that the dc signal be isolated from the ac signal. In Section 5–1, you learned how a capacitor could be used to pass the ac signal while blocking the dc signal. Transformers also block dc (because they provide no direct path) and pass ac.

In addition, transformers provide a useful means of matching the impedance of one part of a circuit to another. From basic dc/ac circuits courses, recall that a load on the secondary side of a transformer is changed by the transformer when looking from the primary side. A step-down transformer causes the load to look larger on the primary side as expressed by

$$R'_L = \left(\frac{N_{pri}}{N_{sec}}\right)^2 R_L \qquad (5\text{–}4)$$

where $\quad R'_L$ = reflected resistance on the primary side
N_{pri}/N_{sec} = ratio of primary turns to secondary turns
R_L = load resistance on the secondary side

Transformers can be used at the input, the output, or between stages to couple the ac signal from one part of a circuit to another. By matching impedances in a power transformer, maximum power can be transferred (discussed in Section 5–4). Transformers can also be used for matching the impedance of a source to a line. Line-matching transformers are used primarily for low-impedance circuits (<200 Ω). For voltage amplifiers, a transformer can also step up the voltage to the next stage (but never the power).

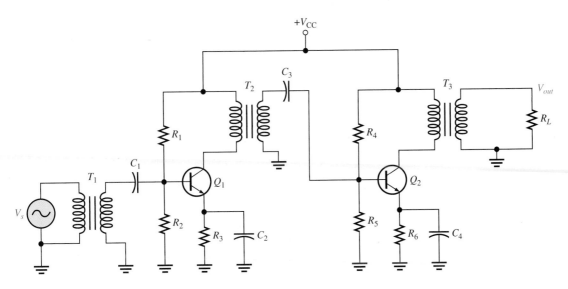

FIGURE 5–11
A basic transformer-coupled amplifier showing input, coupling, and output transformers.

Figure 5–11 shows examples of transformer coupling in a two-stage amplifier. Small low-frequency transformers are occasionally used in certain microphones or other transducers to couple a signal to an amplifier.

Although transformer coupling can give higher efficiency than *RC* coupling, transformer coupling is not widely applied to low-frequency designs because of two major drawbacks. First, transformers are more expensive and are much bulkier than capacitors. Second, they tend to have poorer response at high frequencies due to the reactance of the coils. For these reasons, low-frequency transformer coupling is not commonly used except in certain class A power amplifiers.

EXAMPLE 5–3

Assume the component values for the second stage of Figure 5–11 are as follows: $R_4 = 5.1$ kΩ, $R_5 = 2.7$ kΩ, $R_6 = R_E = 680$ Ω, and $R_L = 50$ Ω; transformer T_3 is a 5:1 step-down transformer and $V_{CC} = 12$ V. Draw the dc and ac load lines for the stage.

Solution Start with the dc conditions. Find the base voltage by applying the voltage-divider rule to the bias resistors.

$$V_B = \left(\frac{R_5}{R_4 + R_5}\right)V_{CC} = \left(\frac{2.7 \text{ k}\Omega}{5.1 \text{ k}\Omega + 2.7 \text{ k}\Omega}\right)12 \text{ V} = 4.2 \text{ V}$$

Next, calculate the emitter voltage and current.

$$V_E = V_B - V_{BE} = 4.2 \text{ V} - 0.7 \text{ V} = 3.5 \text{ V}$$

$$I_E = \frac{V_E}{R_E} = \frac{3.5 \text{ V}}{680 \text{ }\Omega} = 5.15 \text{ mA}$$

The emitter current is approximately equal to the collector current and represents the current at the Q-point, I_{CQ}. The dc resistance of the transformer primary is small and can be ignored. With this assumption, the transformer has no effect on the dc load line. V_{CE} is the difference between V_{CC} and the drop across the emitter resistor.

$$V_{CEQ} \cong V_{CC} - V_E = 12 \text{ V} - 3.5 \text{ V} = 8.5 \text{ V}$$

You can now draw the dc load line because you know two of the points on the line (the Q-point and $V_{CE(cutoff)}$). As confirmation, determine the dc saturation current by finding the current if the transistor's collector is shorted to the emitter.

$$I_{C(sat)} = \frac{V_{CC}}{R_E} = \frac{12 \text{ V}}{680 \text{ }\Omega} = 17.6 \text{ mA}$$

Now proceed to find the ac load line by starting with the ac resistance in the collector circuit. To find the equivalent reflected resistance of the load resistor on the primary side, apply Equation (5–4).

$$R'_L = \left(\frac{N_{pri}}{N_{sec}}\right)^2 R_L = \left(\frac{5}{1}\right)^2 50 \text{ }\Omega = 1.25 \text{ k}\Omega$$

This represents the entire ac resistance (ignoring r'_e) for the collector-emitter circuit as the emitter resistor is bypassed by C_4. Recall from Section 3–4 that the ac saturation current is found from

$$I_{c(sat)} = I_{CQ} + \frac{V_{CEQ}}{R_{ac}}$$

Substituting the values of I_E and R'_L for I_{CQ} and R_{ac}, the ac saturation current is

$$I_{c(sat)} = 5.15 \text{ mA} + \frac{8.5 \text{ V}}{1.25 \text{ k}\Omega} = 11.95 \text{ mA}$$

From the two points found (the Q-point and $I_{c(sat)}$), you can construct the ac load line. As confirmation, determine the ac cutoff voltage, $V_{ce(cutoff)}$.

$$V_{ce(cutoff)} = V_{CEQ} + I_{CQ}R_{ac} = 8.5 \text{ V} + (5.15 \text{ mA})(1.25 \text{ k}\Omega) = 14.9 \text{ V}$$

The result is shown in Figure 5–12.

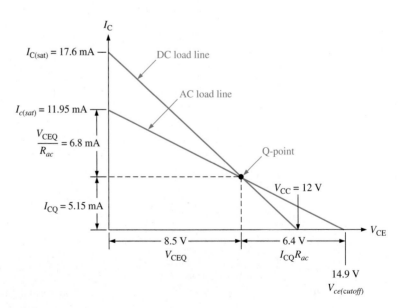

FIGURE 5–12

Practice Exercise   If you replaced the transformer in this example with one with a turns ratio of 6:1, what, if anything, happens to the Q-point? What happens to the ac load line?

Example 5–3 illustrates an important point about transformer-coupled amplifiers. Unlike previous examples that showed capacitively coupled amplifiers, the ac load line is not as steep as the dc load line. The ac saturation current is *lower* than the dc saturation current and the ac cutoff voltage is *larger* than the dc cutoff voltage (V_{CC}).

High-Frequency Applications

At higher frequencies, transformers are much smaller, less expensive, and offer important advantages for coupling signals over a limited bandwidth. As you saw in the last section, at high frequencies a transformer primary can be connected with a parallel capacitor to form a high-Q resonant circuit. Frequently, the secondary winding, with an appropriate capacitor across it, is also connected as a resonant circuit.

From basic dc/ac courses, you learned that a parallel resonant circuit is an LC combination that has an impedance maximum at the resonant frequency. This high impedance at the resonant frequency means that the gain of the amplifier can be very high at frequencies near the resonant frequency while offering little opposition to dc. This forms a very efficient narrow bandwidth amplifier (typically 10 kHz) with gains as high as 1000 or so. Furthermore, the amplifier is tailored to amplify a very narrow band of frequencies containing the signal of interest and not amplify other frequencies.

During signal processing, a radio frequency is usually converted to a lower frequency by mixing the RF with an oscillator. The new lower frequency that is produced is called an *Intermediate Frequency* or IF. Tuned transformer coupling is important in both RF and IF amplifiers.

The principal advantage to using IF is that it is a fixed frequency and requires no changes in the tuned circuit for any given RF signal (within design limits). This is accomplished by causing the oscillator to "track" the RF. Since the IF is fixed, it is easy to amplify with a fixed-resonant circuit, without need for the user to adjust any controls. This idea, first developed by Major Edwin Armstrong during World War I, is found in most communication equipment and is also used in the spectrum analyzer, an important piece of high-frequency test equipment.

Figure 5–13 shows an example of a two-stage tuned amplifier that uses resonant circuits at both the input to the first stage and the output from the second stage. Transformer coupling is used between the stages. A circuit similar to this is part of most communication equipment and consists of an RF amplifier and a mixer. The RF amplifier tunes and amplifies the high-frequency signal from a station. The **mixer** is a nonlinear circuit that combines this signal with a sine wave generated from an oscillator.

The oscillator's frequency is set to a fixed difference from the RF. When the RF and oscillator signals are mixed in a nonlinear circuit, they produce two new frequencies: the sum of the input signals and the difference of the input signals. The second resonant circuit is tuned to the difference frequency, while rejecting all other frequencies. This difference frequency is the IF signal that is amplified further by the IF amplifier section. The advantage of an IF section is that it is specifically designed to process a single frequency.

Let's examine the circuit in Figure 5–13 further. The first tuned circuit consists of the primary of T_1 which resonates with C_1 to tune a station. Stations not at the resonant

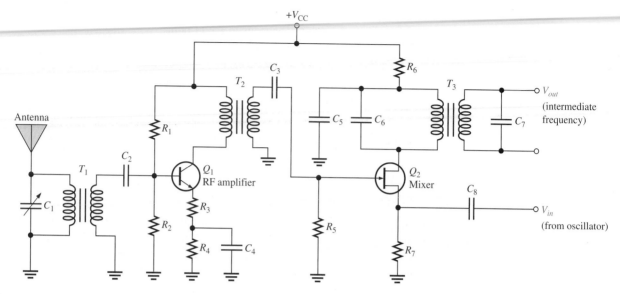

FIGURE 5–13
A tuned amplifier consisting of an RF stage and a mixer.

frequency are rejected by the resonant circuit. Notice that Q_1 is biased with stable voltage-divider bias. There is no collector resistor, but instead, the ac signal "sees" the primary of transformer T_2 as a load. The gain for this stage is determined by the reactance in the collector circuit divided by the ac emitter resistance consisting of R_3 and r'_e.

The RF signal is passed to the gate of Q_2 by transformer T_2 where it is combined with the signal from the oscillator (not shown). Note that Q_2 is a CS amplifier for the RF signal, but a CG amplifier for the oscillator signal. The resonant circuit in the output of Q_2 is tuned to the desired difference frequency. Thus, the output of Q_2 is the intermediate frequency (IF), which is sent to the next stage for further amplification. In order to generate the intermediate frequency, Q_2 must operate as a nonlinear amplifier. FETs fulfill this role nicely. Often, the mixer is combined with the RF amplifier using a two-gate MOSFET such as introduced earlier.

Notice in Figure 5–13 that a resistor, R_6, is in series with the voltage from the power supply. This resistor and C_5 form a low-pass filter called a **decoupling network** that helps isolate the circuit from other amplifiers and helps prevent unwanted oscillations. The resistor is a small value (typically 100 Ω) and the capacitor is selected to have a reactance that is <10% of this value at the operating frequency. (For example, a 100 Ω resistor can be bypassed with a capacitor that has a reactance of approximately 10 Ω.)

An IF amplifier is shown in Figure 5–14. The IF transformer is designed for the specific intermediate frequency selected. The IF amplifier is in all respects an RF amplifier; the only difference between an IF and an RF amplifier is the function it serves in a given circuit. An IF amplifier uses a tuned input circuit and tuned output circuit to selectively amplify the intermediate frequency. The capacitor that forms the primary resonant circuit and the transformer are inside a metal enclosure that provides shielding. The exact intermediate frequency is adjusted with a tuning slug that is moved in and out of the core. Again, a decoupling network is included (R_3 and C_3). When tuning the IF circuit, it is important to use a high-impedance, low-capacitance test instrument to avoid changing the circuit response due to instrument loading.

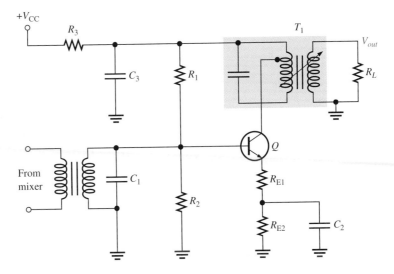

FIGURE 5–14
An IF amplifier.

The dc parameters for the circuit in Figure 5–14 are found the same way as in any CE amplifier. However, the tuned circuit affects the ac parameters differently than a circuit with just a collector resistor. With the parallel resonant circuit in the collector, the voltage gain of the amplifier is the ratio of the impedance of the resonant circuit to the ac resistance of the emitter circuit.

$$A_v = \frac{Z_c}{R_e}$$

where Z_c = impedance of the collector circuit
R_e = ac resistance of the emitter circuit

The impedance of the resonant circuit depends on the frequency and the Q of the resonant circuit. You can find the impedance, Z_c, at the resonance if you know X_L and Q.

$$Z_c = QX_L$$

Because of transformer action, the load resistor is part of the equivalent primary circuit and affects the frequency response of the tuned circuit; a smaller load resistor produces a lower Q and a broader response. Example 5–4 illustrates these ideas.

EXAMPLE 5–4

Assume the IF amplifier in Figure 5–15 has an IF transformer tuned to 455 kHz (a standard intermediate frequency). The primary has an inductance of 99.5 μH and a resistance of 5.6 Ω. Internally, there is a 1250 pF capacitor connected in parallel with the primary.

(a) Find the Q of the resonant circuit, the unloaded voltage gain, $A_{v(NL)}$, and the bandwidth, *BW*.

(b) Find the voltage gain and new bandwidth if a 1.0 kΩ load resistor is connected across the secondary, causing the Q of the resonant circuit to be reduced to 20.

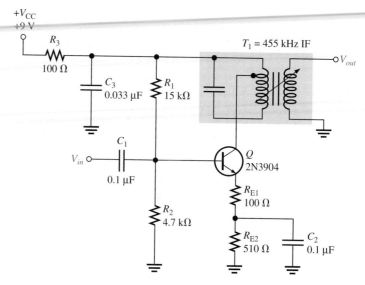

FIGURE 5–15

Solution

(a) Compute the dc parameters first. The dc drop across R_3 is very small and can be ignored.

$$V_B = \left(\frac{R_2}{R_1 + R_2}\right)V_{CC} = \left(\frac{4.7\ \text{k}\Omega}{15\ \text{k}\Omega + 4.7\ \text{k}\Omega}\right)9\ \text{V} = 2.15\ \text{V}$$

$$V_E = V_B - V_{BE} = 2.15\ \text{V} - 0.7\ \text{V} = 1.45\ \text{V}$$

$$I_E = \frac{V_E}{R_{E1} + R_{E2}} = \frac{1.45\ \text{V}}{100\ \Omega + 510\ \Omega} = 2.38\ \text{mA}$$

The ac parameters are

$$r'_e \cong \frac{25\ \text{mV}}{I_E} = \frac{25\ \text{mV}}{2.38\ \text{mA}} = 10.5\ \Omega$$

$$X_L = 2\pi f L = 2\pi (455\ \text{kHz})(99.5\ \mu\text{H}) = 284\ \Omega$$

$$Q = \frac{X_L}{R} = \frac{284\ \Omega}{5.6\ \Omega} = \textbf{50.7}$$

$$Z_c \cong QX_L = (50.7)(284\ \Omega) = 14.4\ \text{k}\Omega$$

$$A_{v(NL)} = \frac{Z_c}{R_e} = \frac{Z_c}{r'_e + R_{E1}} = \frac{14.4\ \text{k}\Omega}{10.5\ \Omega + 100\ \Omega} = \textbf{130}$$

$$BW = \frac{f_r}{Q} = \frac{455\ \text{kHz}}{50.7} = \textbf{9.0 kHz}$$

(b) The addition of a load has no effect on the dc parameters, r'_e, or X_L. The reflected resistance of the secondary reduces the Q and therefore the impedance of the resonant circuit. The new impedance of the resonant circuit is

$$Z_c = QX_L = (20)(284\ \Omega) = 5.68\ \text{k}\Omega$$

The voltage gain and bandwidth are

$$A_v = \frac{Z_c}{R_e} = \frac{Z_c}{r'_e + R_{E1}} = \frac{5.68 \text{ k}\Omega}{10.5 \ \Omega + 100 \ \Omega} = \mathbf{51}$$

$$BW = \frac{f_r}{Q} = \frac{455 \text{ kHz}}{20} = \mathbf{23 \text{ kHz}}$$

Practice Exercise   What effect, if any, would an open C_2 have on the voltage gain? What effect on the bandwidth?

5–3 REVIEW QUESTIONS

1. What is the difference between an RF and an IF signal?
2. What is the function of a mixer?
3. What are the two signals mixed in a mixer?
4. What effect does a load resistor on the secondary of a tuned transformer have on the Q of the tuned circuit?
5. Why should a high-impedance, low-capacitance instrument be used to test an IF stage?

5–4 ■ DIRECT-COUPLED AMPLIFIERS

Another important method for coupling signals is called direct coupling. With direct coupling, there are no coupling capacitors or transformers between stages. Depending on how the input and output signals are coupled, some amplifiers can operate with frequencies all the way down to dc. In this section, a direct-coupled amplifier is introduced, then negative feedback is added to stabilize the bias and the gain. Direct coupling will be applied again in Section 5–5 to power amplifiers.

After completing this section, you should be able to

❏ Determine basic dc and ac parameters for direct-coupled amplifiers and describe how negative feedback can stabilize the gain of an amplifier
  ❏ Describe how direct-coupled stages obtain bias
  ❏ Compute dc and ac parameters for a direct-coupled amplifier
  ❏ Explain how negative feedback can stabilize bias and gain

Figure 5–16 shows a direct-coupled amplifier. Direct coupling is from the collector of Q_1 to the base of Q_2. Since the stages are direct coupled, bias current for Q_2 is supplied by Q_1, eliminating the need for any bias resistors for Q_2 and eliminating a coupling capacitor between the stages. Although the stages are direct coupled, it is necessary in this particular amplifier to ac couple the input and output signals (through capacitors) to prevent the external signal source and the load from disturbing the dc voltages.

Bias for Q_2 is supplied through R_{C1}, the collector resistor for Q_1. Transistor Q_1 is relatively independent of β because it has voltage-divider bias, but Q_2 uses base bias, a method not desired in linear amplifiers because of the variation in β. In addition, thermal

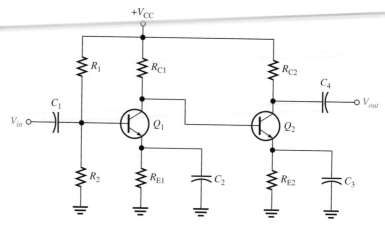

FIGURE 5–16
A direct-coupled amplifier without feedback.

changes will cause the circuit to drift. Although this particular amplifier has fewer components than comparable capacitively coupled amplifiers, the drawbacks mentioned may outweigh the advantages. A relatively simple change—the addition of negative feedback—can cure the problem of β dependency and drift.

Negative Feedback for Bias Stability

The circuit in Figure 5–17 is a modification that greatly improves the bias stability of the amplifier in Figure 5–16 with an added bonus of reducing the parts count. Again, the input and output signals are capacitively coupled to avoid disturbing the bias voltages. Because there are two transistors, the feedback network, shown in color, takes advantage of the extra gain over a single transistor and produces excellent stability for variations in β and for variations in temperature. This is similar to the negative feedback introduced in Section 3–2 for collector-feedback bias.

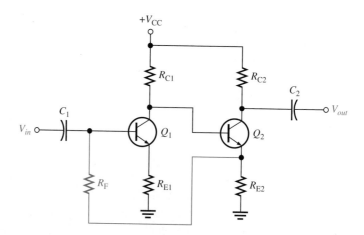

FIGURE 5–17
A direct coupled amplifier with negative feedback that produces bias stability.

Let's look at how the feedback works in Figure 5–17. Start with Q_2 and note that its base is forward-biased by R_{C1}, causing a collector current in Q_2 or $I_{C(Q2)}$. This current causes the emitter voltage of Q_2 to rise, turning on Q_1. As Q_1 conducts more, its collector voltage drops, reducing the bias on Q_2. This action reduces the bias on Q_2 to a stable point determined by the particular design values.

The collector current for Q_1 can be computed by writing Kirchhoff's voltage law (KVL) around the path that includes V_{CC}, R_{C1}, $V_{BE(Q2)}$, R_F, $V_{BE(Q1)}$, and R_{E1}. From this, the collector current in Q_1 is approximately

$$I_{C(Q1)} = \frac{V_{CC} - 2V_{BE}}{R_{C1} + \dfrac{R_F}{\beta} + R_{E1}}$$

The circuit is designed such that R_{C1} is much larger than R_F/β or R_{E1}. Thus, $I_{C(Q1)}$ is almost completely independent of the value of β, producing a stable value of collector voltage in Q_1 and a stable base voltage in Q_2. Thus, the β dependency problem associated with base bias is no longer a factor.

The collector current in Q_1 causes the collector voltage of Q_1 to be

$$V_{C(Q1)} = V_{CC} - I_{C(Q1)}R_{C1}$$

which is also the base voltage of Q_2. The emitter voltage of Q_2 is $V_{C(Q1)} - 0.7$ V and the emitter current is found from Ohm's law.

$$I_{E(Q2)} = \frac{V_{C(Q1)} - 0.7 \text{ V}}{R_{E2}}$$

The feedback resistor, R_F, is not included in the calculation of $I_{E(Q2)}$ because it is much larger than R_{E2}. The approximation that $I_C \cong I_E$ enables you to find the collector current in Q_2. Subtracting the voltage drop across R_{C2} from V_{CC} allows you to find $V_{C(Q2)}$. The following example illustrates a typical set of parameters for this circuit.

EXAMPLE 5–5

(a) For the circuit in Figure 5–18, find the dc parameters, $I_{C(Q1)}$, $V_{C(Q1)}$, $I_{E(Q2)}$, and $V_{C(Q2)}$ if the β of each transistor is 200.

(b) Compute the voltage gain and the output voltage if a 5 mV input signal is applied to the amplifier.

FIGURE 5–18

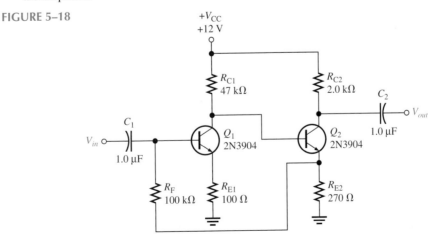

Solution

(a) Begin with the dc parameters. The collector current in Q_1 is approximately

$$I_{C(Q1)} = \frac{V_{CC} - 2V_{BE}}{R_{C1} + \dfrac{R_F}{\beta} + R_{E1}} = \frac{12\ \text{V} - 2(0.7\ \text{V})}{47\ \text{k}\Omega + \dfrac{100\ \text{k}\Omega}{200} + 100\ \Omega} = \mathbf{0.223\ mA}$$

This current is in R_{C1} (plus the small base current of Q_2 that is not included). Find $V_{C(Q1)}$ as follows:

$$V_{C(Q1)} = V_{CC} - I_{C(Q1)}R_{C1} = 12\ \text{V} - (0.223\ \text{mA})(47\ \text{k}\Omega) = \mathbf{1.52\ V}$$

The emitter current in Q_2 is

$$I_{E(Q2)} = \frac{V_{C(Q1)} - 0.7\ \text{V}}{R_{E2}} = \frac{1.52\ \text{V} - 0.7\ \text{V}}{270\ \Omega} = \mathbf{3.03\ mA}$$

Since $I_{C(Q2)} \cong I_{E(Q2)}$, this implies that the collector voltage of Q_2 is

$$V_{C(Q2)} = V_{CC} - I_{C(Q2)}R_{C2} = 12\ \text{V} - (3.03\ \text{mA})(2.0\ \text{k}\Omega) = \mathbf{5.94\ V}$$

Although this is a very stable arrangement, the calculation is sensitive to the exact collector current in Q_1 and the V_{BE} drop across Q_2. Measured dc parameters will vary somewhat from the computed values depending on the particular transistors used.

(b) To compute the voltage gain, it is first necessary to find r'_e for each transistor. Assume $I_E = I_C$.

$$r'_{e(Q1)} = \frac{25\ \text{mV}}{I_{E(Q1)}} = \frac{25\ \text{mV}}{0.223\ \text{mA}} = 112\ \Omega$$

$$r'_{e(Q2)} = \frac{25\ \text{mV}}{I_{E(Q2)}} = \frac{25\ \text{mV}}{3.03\ \text{mA}} = 9\ \Omega$$

Then the unloaded voltage gain for each amplifier can be computed.

$$A_{v1(NL)} = -\frac{R_c}{R_e} = -\frac{R_{C1}}{R_{E1} + r'_{e(Q1)}} = -\frac{47\ \text{k}\Omega}{100\ \Omega + 112\ \Omega} = -222$$

$$A_{v2(NL)} = -\frac{R_c}{R_e} = -\frac{R_{C2}}{R_{E2} + r'_{e(Q2)}} = -\frac{2.0\ \text{k}\Omega}{270\ \Omega + 9\ \Omega} = -7.2$$

Next, compute the input and output resistance for each amplifier stage. Notice that the input resistance shows β dependency.

$$R_{in(tot)1} = [\beta_{ac}(R_{E1} + r'_{e(Q1)})] \parallel R_F$$
$$= [200(100\ \Omega + 112\ \Omega)] \parallel 100\ \text{k}\Omega = 29.8\ \text{k}\Omega$$
$$R_{out1} = R_{C1} = 47\ \text{k}\Omega$$
$$R_{in(tot)2} = \beta_{ac}(R_{E2} + r'_{e(Q2)}) = 200(270\ \Omega + 9\ \Omega) = 55.8\ \text{k}\Omega$$
$$R_{out2} = R_{C2} = 2.0\ \text{k}\Omega$$

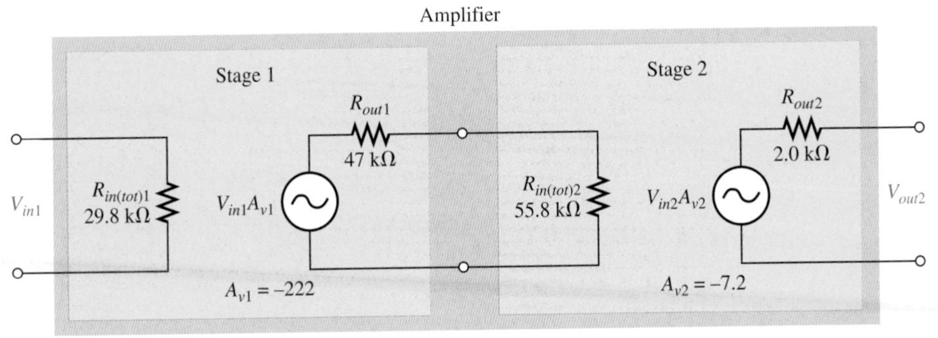

Amplifier

FIGURE 5–19

You can now compute the overall gain. Use the model of a multistage amplifier, introduced in Chapter 1 and shown in Figure 5–19. The overall voltage gain is

$$A_{v(overall)} = (A_{v1(NL)})\left(\frac{R_{in(tot)2}}{R_{out1} + R_{in(tot)2}}\right)(A_{v2(NL)})$$

$$= (-222)\left(\frac{55.8 \text{ k}\Omega}{47 \text{ k}\Omega + 55.8 \text{ k}\Omega}\right)(-7.2) = \mathbf{867}$$

Although this is a relatively high gain, measured values are likely to vary somewhat due to simplifying assumptions and β dependency. The variation is mainly due to the second term which is dependent on β.

With the gain calculation completed, the output voltage can be found.

$$V_{out} = A_{v(overall)}V_{in} = (867)(5 \text{ mV}) = \mathbf{4.34 \text{ V}}$$

Practice Exercise   Determine the overall voltage gain of the amplifier if the β is 100 for each transistor.

Negative Feedback for Gain Stability

The amplifier illustrated in Example 5–5 has high gain but the gain is somewhat dependent on the β. Negative feedback provided excellent bias stability that was not dependent on a particular β. It's also possible to provide excellent gain stability that is independent of β by using negative feedback with the ac signal. As you will see, negative feedback produces a self-correcting action that stabilizes the voltage gain. The modification in Figure 5–20 shows how this is achieved.

First, a bypass capacitor, C_2, is connected in parallel with R_{E2} in order to boost the voltage gain even higher; this will produce greater gain stability when feedback is added. The gain without feedback is called **open-loop voltage gain**, which will be described further when you study operational amplifiers in Chapter 6. For the amplifier in Figure 5–20, the addition of the emitter capacitor boosts the open-loop voltage gain by a factor of approximately two.[1] Then a new path is added, consisting of C_3 and R_{F2}, to return a fraction of the output ac signal back to Q_1. The fraction that is returned is determined by the voltage

---

[1] The gain would be even higher except for the adverse loading effect on Q_1 due to lower input resistance of Q_2.

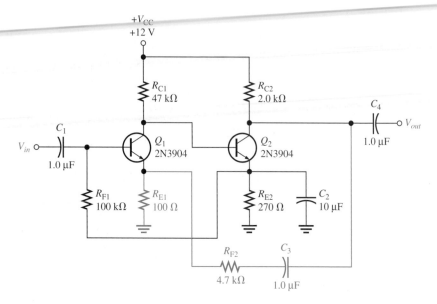

FIGURE 5–20

Modification of the circuit in Example 5–5 to improve gain stability.

divider consisting of R_{F2} and R_{E1}. For the amplifier in Figure 5–20, the feedback voltage, V_f, is equal to the output voltage multiplied by the feedback fraction.

$$V_f = \left(\frac{R_{E1}}{R_{E1} + R_{F2}} \right) V_{out}$$

This feedback voltage tends to cancel the original input signal. The signal that is amplified by the open-loop voltage gain is the small *difference* in the input and negative feedback signals. As a result, the net voltage gain of the amplifier is controlled by the amount of feedback. This net gain with feedback is called the **closed-loop voltage gain**. As mentioned, the closed-loop voltage gain is determined by the amount of output signal that is returned.

An implication of a very large open-loop voltage gain is that the difference between the feedback and input signals is very small at the input to Q_1. For the amplifier in Figure 5–20, the ac signal on the base and emitter of Q_1 will have nearly the same amplitude.

Here's how the negative feedback works to achieve gain stability. Suppose the voltage gain increases due to heating (causing r'_e to be smaller). The increased open-loop gain causes the output voltage to increase and, in turn, increases the negative feedback voltage. This reduces the difference voltage at Q_1. Thus, the original change in gain is almost completely canceled by the self-correcting action of negative feedback.

Now assume a technician replaces one of the transistors with one with a lower β than in the original circuit. This causes a decrease in the open-loop gain of the amplifier. Now there will be a smaller feedback voltage that causes the difference voltage to be larger. Since there is a larger difference voltage, the original effect of a lower β has little net effect on the output voltage and, again, gain stability is achieved.

The net voltage gain of the amplifier is approximately equal to the reciprocal of the feedback fraction. For the amplifier in Figure 5–20, the net gain is

$$A_v = \left(\frac{R_{E1} + R_{F2}}{R_{E1}} \right) = \left(\frac{100 \ \Omega + 4.7 \ \text{k}\Omega}{100 \ \Omega} \right) = 48$$

As you can see, it is easy to change the gain by simply changing the value of R_{F2}. In fact, a gain control can be easily added by using a variable resistor in place of R_{F2}.

Another dc coupled amplifier is the circuit introduced in Chapter 4 as the system application. It is shown again in Figure 5–21 for reference. An important advantage of this circuit is the ability to amplify low frequencies all the way down to dc. The potentiometer (R_5) is adjusted so that the output voltage is zero when the input voltage is zero. For this particular design, zero output is obtained by using a *pnp* transistor for the last stage.

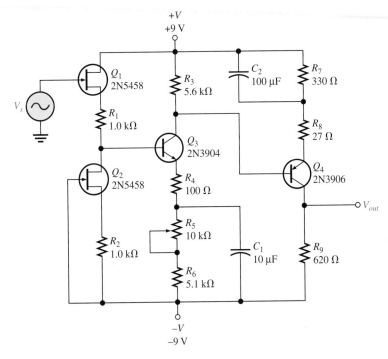

FIGURE 5–21
A direct-coupled preamp.

The amplifier in Figure 5–21 uses unbypassed emitter resistors (R_4 and R_8) to achieve gain stability. This is a form of negative feedback that produces gain stability (at the price of reduced gain) that you saw earlier in Chapters 3 and 4. The FET input stage gave the added advantage of extremely high input resistance.

5–4 REVIEW QUESTIONS

1. What are the main advantages of a direct-coupled amplifier?
2. How does negative feedback produce bias or gain stability?
3. Why does the addition of a bypass capacitor in the emitter circuit of a CE amplifier improve the gain stability but not the bias stability?
4. How is gain determined for the circuit in Figure 5–20?

5–5 ■ CLASS A POWER AMPLIFIERS

When an amplifier is biased such that it always operates in the linear region where the output signal is an amplified replica of the input signal, it is a class A amplifier. The discussion and formulas in the previous sections apply to class A operation. Power amplifiers are those amplifiers that have the objective of delivering power to a load. This means that components must be considered in terms of their ability to dissipate heat.

After completing this section, you should be able to

❏ Compute key ac and dc parameters for class A power amplifiers and discuss operation along the ac load line
  ❏ Explain why a centered Q-point is important for a class A power amplifier
  ❏ Determine the voltage gain and power gain for a multistage amplifier
  ❏ Determine the efficiency of a class A power amplifier

In a small-signal amplifier, the ac signal moves over a small percentage of the total ac load line. When the output signal is larger and approaches the limits of the ac load line, the amplifier is a **large-signal** type. Both large-signal and small-signal amplifiers are considered to be **class A** if they operate in the active region at all times. Class A power amplifiers are large-signal amplifiers with the objective of providing power (rather than voltage) to a load. As a rule of thumb, an amplifier may be considered to be a power amplifier if it is necessary to consider the problem of heat dissipation in components (>1/4 W).

Heat Dissipation

Power transistors (and other power devices) must dissipate excessive internally generated heat. For bipolar power transistors, the collector terminal is the critical junction; for this reason, the transistor's case is always connected to the collector terminal. The case of all power transistors is designed to provide a large contact area between it and an external heat sink. Heat from the transistor flows through the case to the heat sink and then dissipates in the surrounding air. Heat sinks vary in size, number of fins, and type of material. Their size depends on the heat dissipation requirement and the maximum ambient temperature in which the transistor is to operate. In high-power applications (a few hundred watts), a cooling fan may be necessary.

Centered Q-Point

Recall (from Section 3–4) that the dc and ac load lines cross at the Q-point. When the Q-point is at the center of the ac load line, a maximum class A signal can be obtained. You can see this concept by examining the graph of the load line for a given amplifier in Figure 5–22(a). This graph shows the ac load line with the Q-point at its center. The collector current can vary from its Q-point value, I_{CQ}, up to its saturation value, $I_{c(sat)}$, and down to its cutoff value of zero. Likewise, the collector-to-emitter voltage can swing from its Q-point value, V_{CEQ}, up to its cutoff value, $V_{ce(cutoff)}$, and down to its saturation value of near zero. This operation is indicated in Figure 5–22(b). The peak value of the collector current equals I_{CQ}, and the peak value of the collector-to-emitter voltage equals V_{CEQ} in this case. This signal is the maximum that can be obtained from the class A amplifier. Actually, the output cannot quite reach saturation or cutoff, so the practical maximum is slightly less.

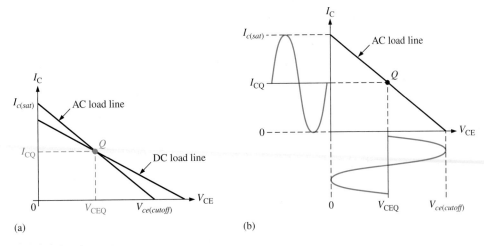

FIGURE 5–22

Maximum class A output occurs when the Q-point is centered on the ac load line.

If the Q-point is not centered on the ac load line, the output signal is limited. Figure 5–23 shows a load line with the Q-point moved away from center toward cutoff. The output variation is limited by cutoff in this case. The collector current can only swing down to near zero and an equal amount above I_{CQ}. The collector-to-emitter voltage can only swing up to its cutoff value and an equal amount below V_{CEQ}. This situation is illustrated in Figure 5–23(a). If the amplifier is driven any further than this, it will "clip" at cutoff, as shown in Figure 5–23(b).

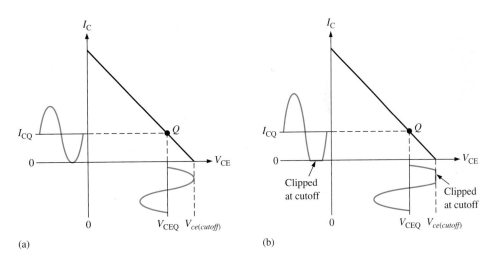

FIGURE 5–23

Q-point closer to cutoff.

Figure 5–24 shows a load line with the Q-point moved away from center toward saturation. In this case, the output variation is limited by saturation. The collector current can only swing up to near saturation and an equal amount below I_{CQ}. The collector-to-emitter voltage can only swing down to its saturation value and an equal amount above V_{CEQ}. This situation is illustrated in Figure 5–24(a). If the amplifier is driven any further, it will "clip" at saturation, as shown in Figure 5–24(b).

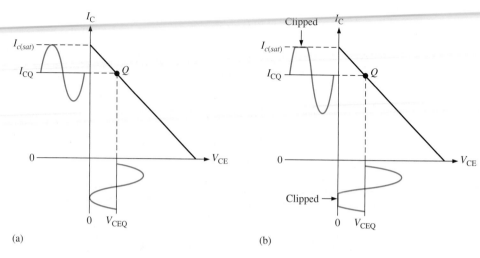

(a)

(b)

FIGURE 5–24
Q-point closer to saturation.

Power Gain

A power amplifier delivers power to a load. The **power gain** of an amplifier is the ratio of the power delivered to the load to the input power. In general, power gain is

$$A_p = \frac{P_L}{P_{in}} \tag{5–5}$$

where A_p = power gain
P_L = signal power delivered to the load
P_{in} = signal power delivered to the amplifier

The power gain can be computed by any of several formulas, depending on what is known. Frequently, the easiest way to obtain power gain is from input resistance, load resistance, and voltage gain. To see how this is done, recall that power can be expressed in terms of voltage and resistance as

$$P = \frac{V^2}{R}$$

For ac power, the voltage is expressed as rms. The output power delivered to the load is

$$P_L = \frac{V_L^2}{R_L}$$

The input power delivered to the amplifier is

$$P_{in} = \frac{V_{in}^2}{R_{in}}$$

By substituting into Equation (5–5), the following useful relationship can be found:

$$A_p = \frac{V_L^2}{V_{in}^2}\left(\frac{R_{in}}{R_L}\right)$$

$$A_p = A_v^2\left(\frac{R_{in}}{R_L}\right) \tag{5–6}$$

Equation (5–6) says that the power gain to an amplifier is the voltage gain squared times the ratio of the input resistance to the output load resistance. It can be applied to any amplifier. For example, assume a common-collector (CC) amplifier has an input resistance of 10 kΩ and a load resistance of 100 Ω. Since a CC amplifier has a voltage gain of approximately 1, the power gain is

$$A_p = A_v^2\left(\frac{R_{in}}{R_L}\right) = 1^2\left(\frac{10\ k\Omega}{100\ \Omega}\right) = 100$$

For a CC amplifier, A_p is approximately equal to the ratio of the input resistance to the output load resistance.

DC Quiescent Power

The power dissipation of a transistor with no signal input is the product of its Q-point current and voltage.

$$P_{DQ} = I_{CQ}V_{CEQ} \tag{5–7}$$

The only way a class A power amplifier can supply power to a load is to maintain a quiescent current that is at least as large as the peak current requirement for the load current. A signal will not increase the power dissipated by the transistor but actually causes less total power to be dissipated. The quiescent power, given in Equation (5–3), is the maximum power that a class A amplifier must handle. The transistor's power rating will normally exceed this value.

Output Power

In general, the output signal power is the product of the rms load current and the rms load voltage. The maximum unclipped ac signal occurs when the Q-point is centered on the ac load line. For a CE amplifier with a centered Q-point, the maximum peak voltage swing is

$$V_{c(max)} = I_{CQ}R_c$$

The rms value is $0.707V_{c(max)}$.
   The maximum peak current swing is

$$I_{c(max)} = \frac{V_{CEQ}}{R_c}$$

The rms value is $0.707I_{c(max)}$.
   To find the maximum power output of the signal, you use the rms values of maximum current and voltage. The maximum power out from a class A amplifier is

$$P_{out(max)} = (0.707I_c)(0.707V_c)$$

$$P_{out(max)} = 0.5I_{CQ}V_{CEQ} \tag{5–8}$$

EXAMPLE 5–6   Determine the ac model for the class A power amplifier in Figure 5–25. Use the ac model of a two-stage amplifier to compute the voltage gain and power gain.

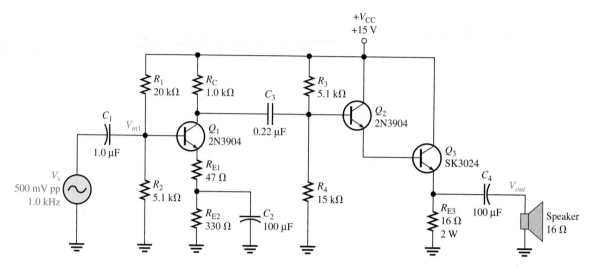

FIGURE 5–25

Solution   Begin by finding the basic parameters for each amplifier stage: $A_{v(NL)}$, R_{in}, and R_{out}. (Q_2 and Q_3 will be treated as a single transistor, named $Q_{2,3}$, in the second stage.)

Stage 1 parameters (Q_1):

The unloaded voltage gain of the first stage is the collector resistance, R_C, divided by the ac emitter resistance, which is the sum of R_{E1} and $r'_{e(Q1)}$. To estimate $r'_{e(Q1)}$, you first need to find I_E. The base voltage is approximately 2.7 V due to the loading effects on the input voltage divider. The emitter voltage is approximately one diode drop less, or 2.0 V. By Ohm's law, the emitter current is

$$I_{E(Q1)} = \frac{V_{E(Q1)}}{R_{E1} + R_{E2}} = \frac{2.0 \text{ V}}{47 \text{ }\Omega + 330 \text{ }\Omega} = 5.3 \text{ mA}$$

and $r'_{e(Q1)}$ is approximately 25 mV/5.3 mA = 5 Ω. The unloaded voltage gain is

$$A_{v1(NL)} = -\frac{R_c}{R_e} = -\frac{R_C}{R_{E1} + r'_{e(Q1)}} = -\frac{1000 \text{ }\Omega}{47 \text{ }\Omega + 5 \text{ }\Omega} = -19.2$$

The input resistance of the first stage is composed of three parallel paths (as discussed in Section 3–4). These include the two bias resistors and the ac resistance of the emitter circuit multiplied by β_{ac} of Q_1. The path through Q_1 has a small effect on R_{in} and is also dependent on β_{ac}. A reasonable estimate is to assume a β_{ac} for Q_1 of 200.

$$R_{in(tot)1} = [(R_{E1} + r'_{e(Q1)})\beta_{ac(Q1)}] \| R_1 \| R_2$$
$$= [(47 \text{ }\Omega + 5 \text{ }\Omega)200] \| 20 \text{ k}\Omega \| 5.1 \text{ k}\Omega = 2.9 \text{ k}\Omega$$

The output resistance of the first stage is just the collector resistor, R_C, which is 1.0 kΩ.

Stage 2 parameters ($Q_{2,3}$):

$Q_{2,3}$ is a darlington pair that is configured as a CC amplifier. The voltage gain of the second stage is approximately 1 for a CC amplifier. Therefore,

$$A_{v2} = 1.0$$

Find the input resistance of the second stage by the same method as that used for the first stage. There are three parallel paths to ac ground. Looking into the base of Q_2 from the coupling capacitor (C_3), the three paths are one path through R_3, a separate path through R_4, and a path through $Q_{2,3}$.

Only the bias resistors are important in this calculation because the path through the darlington transistors has a much higher resistance. You can obtain a reasonable estimate of the input resistance of the second stage by ignoring the path through the transistors and computing only the parallel combination of R_3 and R_4.

$$R_{in(tot)2} \cong R_4 \parallel R_3 = 15 \text{ k}\Omega \parallel 5.1 \text{ k}\Omega = 3.8 \text{ k}\Omega$$

A more precise calculation includes the path through $Q_{2,3}$.

$$R_{in(tot)2} \cong [(R_L \parallel R_{E3})\beta_{ac(Q3)}\beta_{ac(Q2)}] \parallel R_4 \parallel R_3$$

Generally, the β_{ac} of a power transistor is smaller than for signal transistors. A nominal value of 50 for the power transistor (Q_3) and 200 for the signal transistor (Q_2) is reasonable. Therefore, substituting values, the input resistance of the second stage is

$$R_{in(tot)2} = [(16 \ \Omega \parallel 16 \ \Omega)50 \times 200] \parallel 15 \text{ k}\Omega \parallel 5.1 \text{ k}\Omega = 3.6 \text{ k}\Omega$$

Notice that the more precise calculation changes the answer by only 6%.

The output resistance of the second stage is very small (less than 1 Ω) and can be ignored.

$$R_{out2} \cong 0 \ \Omega$$

Overall result:

Using the computed parameters, the ac model of the amplifier can be modeled as shown in Figure 5–26.

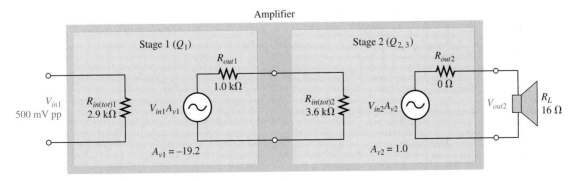

FIGURE 5–26
Amplifier model. (V_{in} is shown as V_{in1} for the first stage).

The overall voltage gain is computed by the method introduced in Section 1–4. The last voltage divider consisting of the speaker and output resistance is not included in the calculation because the output resistance is negligible.

$$A_{v(tot)} = A_{v1}\left(\frac{R_{in(tot)2}}{R_{out1} + R_{in(tot)2}}\right)A_{v2} = -19.2\left(\frac{3.6 \text{ k}\Omega}{1.0 \text{ k}\Omega + 3.6 \text{ k}\Omega}\right)1.0 = \mathbf{-15}$$

The power gain can be computed using Equation (5–6).

$$A_p = A_{v(tot)}^2\left(\frac{R_{in(tot)1}}{R_L}\right) = (-15)^2\left(\frac{2.9 \text{ k}\Omega}{16 \text{ }\Omega}\right) = \mathbf{41,000}$$

Practice Exercise   Express the power gain as a decibel power gain. (Review Section 1–4 if necessary).

Efficiency

The **efficiency** of any amplifier is the ratio of the signal power supplied to the load to the power from the dc supply. The maximum signal power that can be obtained is given by Equation (5–8). The average power supply current, I_{CC}, is equal to I_{CQ} and the supply voltage is at least $2V_{CEQ}$. Therefore, the dc power is

$$P_{DC} = I_{CC}V_{CC} = 2I_{CQ}V_{CEQ}$$

The maximum efficiency of a capacitively coupled load is

$$eff_{max} = \frac{P_{out}}{P_{DC}} = \frac{0.5I_{CQ}V_{CEQ}}{2I_{CQ}V_{CEQ}} = 0.25$$

The maximum efficiency of a capacitively coupled class A amplifier cannot be higher than 0.25, or 25%, and, in practice, is usually considerably less (about 10%). Although the efficiency can be made higher by transformer coupling the signal to the load, there are drawbacks to transformer coupling. These drawbacks include the size and cost of transformers as well as potential distortion problems when the transformer core begins to saturate. In general, the low efficiency of class A power amplifiers limits their usefulness to small power applications that require only a few watts of load power.

EXAMPLE 5–7

Determine the efficiency of the power amplifier in Figure 5–25.

Solution   The efficiency is the ratio of the signal power in the load to the power supplied by the dc source. The input voltage is 500 mV peak-to-peak which is 176 mV rms. The input power is, therefore,

$$P_{in} = \frac{V_{in}^2}{R_{in}} = \frac{(176 \text{ mV})^2}{2.9 \text{ k}\Omega} = 10.7 \text{ }\mu\text{W}$$

The output power is

$$P_{out} = P_{in}A_p = (10.7 \text{ }\mu\text{W})(41,000) = 0.44 \text{ W}$$

Most of the power from the dc source is supplied to the output stage. The current in the output stage can be computed from the emitter voltage of Q_3 which is approximately 9.5 V, producing a current of 0.60 A. Neglecting the other transistors and bias circuits, the total dc supply current is about 0.6 A. The power from the dc source is

$$P_{DC} = I_{CC}V_{CC} = (0.6 \text{ A})(15 \text{ V}) = 9 \text{ W}$$

Therefore, the efficiency of the amplifier for this input is

$$eff = \frac{P_{out}}{P_{DC}} = \frac{0.44 \text{ W}}{9 \text{ W}} \cong \textbf{0.05}$$

This represents an efficiency of 5%.

Practice Exercise Explain what happens to the efficiency if R_{E3} were replaced with the speaker? What disadvantage does this have?

5–5 REVIEW QUESTIONS

1. What is the purpose of a heat sink?
2. Which lead of a bipolar transistor is connected to the case?
3. What are the two types of clipping with a class A amplifier?
4. What is the maximum theoretical efficiency for a class A amplifier?
5. How can the power gain of a CC amplifier be expressed in terms of a ratio of resistances?

5–6 ■ CLASS B POWER AMPLIFIERS

When an amplifier is biased such that it operates in the linear region for 180° of the input cycle and is in cutoff for 180°, it is a class B amplifier. The primary advantage of a class B amplifier over a class A amplifier is that the class B is much more efficient; you can get more output power for a given amount of input power. Class B amplifiers are generally configured with at least two active devices that alternately amplify the positive and negative part of the input waveform. This arrangement is called push-pull.

After completing this section, you should be able to

❑ Compute key ac and dc parameters for class B power amplifiers including bipolar and FET types
  ❑ Describe two configurations for push-pull amplifiers
  ❑ Describe crossover distortion and how to overcome it
  ❑ Explain how class AB operation differs from class B operation
  ❑ Describe how to avoid thermal problems with bipolar class AB amplifiers
  ❑ Discuss characteristics of MOSFET class B amplifiers

Class B operation refers to operation when the Q-point is located at cutoff, causing the output current to vary only during one-half of the input cycle. In a linear amplifier, two devices are required for a complete cycle; one amplifies the positive cycle and the other amplifies the negative cycle. As you will see, this arrangement has a great advantage for power amplifiers as it greatly increases the efficiency. For this reason, they are widely used as power amplifiers.

The Q-point Is at Cutoff

The class B amplifier is biased at cutoff so that $I_{CQ} = 0$ and $V_{CEQ} = V_{CE(cutoff)}$. Thus, there is *no dc current or power dissipated* when there is no signal as in the case of the class A amplifier. When a signal drives a class B amplifier into conduction, it then operates in its linear region. This is illustrated in Figure 5–27 with an emitter-follower circuit.

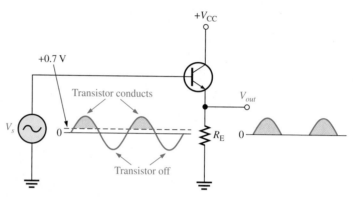

FIGURE 5–27
Common-collector class B amplifier.

Push-Pull Operation

As you can see, the circuit in Figure 5–27 only conducts for the positive half of the cycle. To amplify the entire cycle, it is necessary to add a second class B amplifier that operates on the negative half. The combination of two class B amplifiers working together is called **push-pull** operation.

There are two common approaches for using push-pull amplifiers to reproduce the entire waveform. The first approach uses transformer coupling. The second uses two **complementary symmetry transistors**; these are a matching pair of *npn/pnp* BJTs or a matching pair of *n*-channel/*p*-channel FETs.

Transformer Coupling Transformer coupling is illustrated in Figure 5–28. The input transformer has a center-tapped secondary that is connected to ground, producing phase inversion of one side with respect to the other. The input transformer thus converts the input signal to two out-of-phase signals for the transistors. Notice that both transistors are *npn* types. Because of the signal inversion, Q_1 will conduct on the positive part of the cycle and Q_2 will conduct on the negative part. The output transformer combines the signals by permitting current in both directions, even though one transistor is always cut off. The positive power supply signal is connected to the center tap of the output transformer.

Complementary Symmetry Transistors Figure 5–29 shows one of the most popular types of push-pull class B amplifiers using two emitter-followers and both positive and negative power supplies. This is a complementary amplifier because one emitter-follower uses an *npn* transistor and the other a *pnp*, which conduct on opposite alternations of the input cycle. Notice that there is no dc base bias voltage ($V_B = 0$). Thus, only the sig-

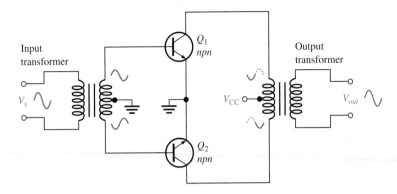

FIGURE 5–28
Transformer coupled push-pull amplifiers. Q_1 conducts during the positive half-cycle; Q_2 conducts during the negative half-cycle. The two halves are combined by the output transformer.

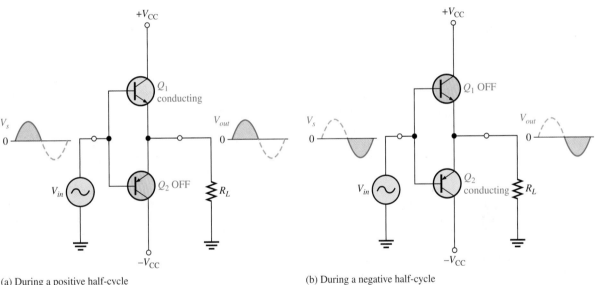

(a) During a positive half-cycle

(b) During a negative half-cycle

FIGURE 5–29
Class B push-pull operation.

nal voltage drives the transistors into conduction. Transistor Q_1 conducts during the positive half of the input cycle, and Q_2 conducts during the negative half.

Crossover Distortion

When the dc base voltage is zero, the input signal voltage must exceed V_{BE} before a transistor conducts. As a result, there is a time interval between the positive and negative alternations of the input when neither transistor is conducting, as shown in Figure 5–30. The resulting distortion in the output waveform is called **crossover distortion**.

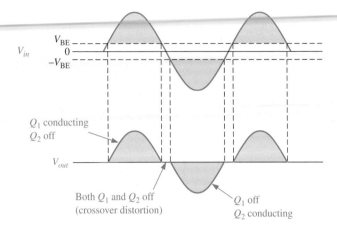

Q_1 conducting
Q_2 off

Both Q_1 and Q_2 off
(crossover distortion)

Q_1 off
Q_2 conducting

FIGURE 5–30
Crossover distortion in a class B amplifier.

Biasing the Push-Pull Amplifier

To overcome crossover distortion, the biasing is adjusted to just overcome the V_{BE} of the transistors; this results in a modified form of operation called **class AB**. In class AB operation, the push-pull stages are biased into slight conduction, even when no input signal is present. This can be done with a voltage divider and diode arrangement, as shown in Figure 5–31. When the diode characteristics of D_1 and D_2 are closely matched to the characteristics of the transistor base-emitter junctions, the current in the diodes and the current in the transistors is the same; this is called a **current mirror**. This current mirror produces the desired class AB operation and eliminates crossover distortion.

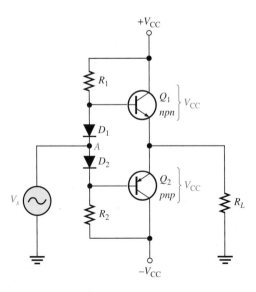

FIGURE 5–31
Biasing the push-pull amplifier to eliminate crossover distortion.

In the bias path, R_1 and R_2 are of equal value, as are the positive and negative supply voltages. This forces the voltage at point A to equal 0 V and eliminates the need for an input coupling capacitor. The dc voltage on the output is also 0 V. Assuming that both diodes and both transistors are identical, the drop across D_1 equals the V_{BE} of Q_1, and the drop across D_2 equals the V_{BE} of Q_2. Since they are matched, the diode current will be the same as I_{CQ}. The diode current and I_{CQ} can be found by applying Ohm's law to either R_1 or R_2 as follows:

$$I_{CQ} = \frac{V_{CC} - 0.7 \text{ V}}{R_1}$$

This small current required of class AB operation eliminates the crossover distortion but has the potential for thermal instability if the transistor's V_{BE} drops are not matched to the diode drops or if the diodes are not in thermal equilibrium with the transistors. Heat in the power transistors decreases the base-emitter voltage and tends to increase current. If the diodes are warmed the same amount, the current is stabilized; but if the diodes are in a cooler environment, they cause I_{CQ} to increase even more. More heat is produced in an unrestrained cycle known as *thermal runaway*.[2] To keep this from happening, the diodes should have the same thermal environment as the transistors. In stringent cases, a small resistor in the emitter of each transistor can alleviate thermal runaway.

Crossover distortion also occurs in transformer-coupled amplifiers such as that shown in Figure 5–28. To eliminate it in this case, a 0.7 V is applied to the input transformer's secondary that just biases both transistors into conduction. The bias voltage to produce this drop can be derived from the power supply using a single diode as shown in Figure 5–32.

AC Operation

Consider the ac load line for Q_1 of the class AB amplifier in Figure 5–31. The Q-point is slightly above cutoff. (In a true class B amplifier, the Q-point is at cutoff.) The ac cutoff

---

[2] The base-emitter voltage in a BJT drops about 2 mV/°C.

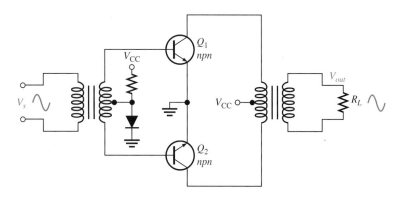

FIGURE 5–32
Eliminating crossover distortion in a transformer-coupled push-pull amplifier. The diode compensates for the base-emitter drop of the transistors and produces class AB operation.

voltage for a two-supply operation is at V_{CC} with an I_{CQ} as given earlier. The ac saturation current for a two-supply operation with a push-pull amplifier is

$$I_{c(sat)} = \frac{V_{CC}}{R_L} \qquad (5\text{--}9)$$

The ac load line for the *npn* transistor is as shown in Figure 5–33. The dc load line can be found by drawing a line that passes through V_{CEQ} and the dc saturation current, $I_{C(sat)}$. But the saturation current for dc is the current if the collector to emitter is shorted on both transistors! This assumed short across the power supplies obviously would cause maximum current from the supplies and implies the dc load line passes almost vertically through the cutoff as shown. Operation along the dc load line, such as caused by thermal runaway, could produce such a high current that the transistors are destroyed.

Figure 5–34(a) illustrates the ac load line for Q_1 of the class B amplifier in Figure 5–34(b). In the case illustrated, a signal is applied that swings over the region of the ac load line shown in bold. At the upper end of the ac load line, the voltage across the transistor (V_{ce}) is a minimum, and the output voltage is maximum.

FIGURE 5–33

Load lines for a complementary symmetry push-pull amplifier. Only the load lines for the npn transistor are shown.

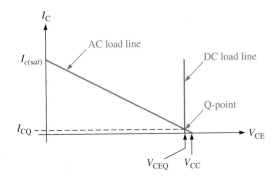

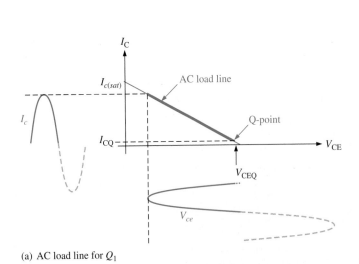

(a) AC load line for Q_1

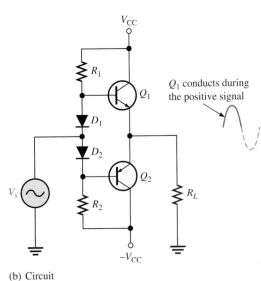

(b) Circuit

FIGURE 5–34

Under maximum conditions, transistors Q_1 and Q_2 are alternately driven from near cutoff to near saturation. During the positive alternation of the input signal, the Q_1 emitter is driven from its Q-point value of 0 to nearly V_{CC}, producing a positive peak voltage a little less than V_{CC}. Likewise, during the negative alternation of the input signal, the Q_2 emitter is driven from its Q-point value of 0 V, to near $-V_{CC}$, producing a negative peak voltage almost equal to $-V_{CC}$. Although it is possible to operate close to the saturation current, this type of operation results in increased distortion of the signal.

The ac saturation current given in Equation (5–9) is also the peak output current. Each transistor can essentially operate over its entire load line. Recall that in class A operation, the transistor can also operate over the entire load line but with a significant difference. In class A operation, the Q-point is near the middle and there is significant current in the transistors even with no signal. In class B operation, when there is no signal, the transistors have only a very small current and therefore dissipate very little power. Thus, the efficiency of a class B amplifier can be much higher than a class A amplifier. It can be shown that the maximum theoretical efficiency of a class B amplifier is 79%.

Single-Supply Operation

Push-pull amplifiers using complementary symmetry transistors can be operated from a single voltage source as shown in Figure 5–35. The circuit operation is the same as that described previously, except the bias is set to force the output emitter voltage to be $V_{CC}/2$ instead of zero volts used with two supplies. Because the output is not biased at zero volts, capacitive coupling for the input and output is necessary to block the bias voltage from the source and the load resistor. Ideally, the output voltage can swing from zero to V_{CC}, but in practice it does not quite reach these ideal values.

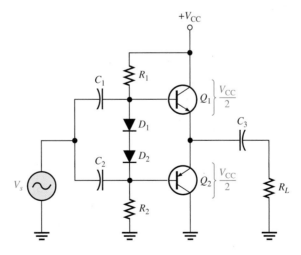

FIGURE 5–35
Single ended push-pull amplifier.

EXAMPLE 5–8

Determine the ideal maximum peak output voltage and current for the circuit shown in Figure 5–36.

FIGURE 5–36

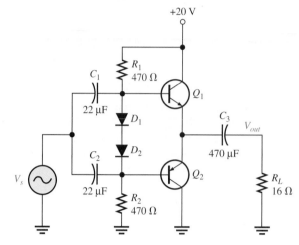

Solution The ideal maximum peak output voltage is

$$V_{p(out)} \cong V_{CEQ} \cong \frac{V_{CC}}{2} = \frac{20 \text{ V}}{2} = \textbf{10 V}$$

The ideal maximum peak current is

$$I_{p(out)} \cong I_{c(sat)} \cong \frac{V_{CEQ}}{R_L} = \frac{10 \text{ V}}{16 \text{ }\Omega} = \textbf{0.63 A}$$

The actual maximum values of voltage and current are slightly smaller.

Practice Exercise Compute the maximum peak output voltage and current if the supply voltage is raised to +30 V.

MOSFET Push-Pull Amplifiers

When MOSFETs were first introduced to the commercial market, they were unable to handle the large currents required of power devices. In recent years, the advances in MOSFET technology have made high-power MOSFETs available and offer some important advantages in the design of power amplifiers for both digital and analog circuits. MOSFETs are very reliable, providing their specified voltage, current, and temperature ratings are not exceeded.

Comparing MOSFETs to BJTs, there are several important advantages but also some disadvantages to MOSFETs. The principal advantages of MOSFETs over BJTs is that their biasing networks are simpler, their drive requirements are simpler, and they can be connected in parallel for added drive capability. In addition, MOSFETs are not generally prone to thermal instability; as they get hotter, they tend to have less current (just the opposite of bipolar transistors)[3]. In switching applications (discussed in Section 4–7), they can switch faster than BJTs. MOSFET switches are widely used in both digital logic and in high power switching circuits.

---

[3] An exception to this rule is with high voltages and low currents; the temperature characteristics are reversed, and MOSFETs *can* experience thermal runaway problems.

The BJT has the edge when the voltage drop across the transistor is important and, as a result, may be more efficient than a MOSFET in certain cases. In addition, bipolar transistors are not as prone to **electrostatic discharge (ESD)** that can kill a MOSFET. Most MOSFETs are shipped with the pins shorted together with a ring; they should be soldered into a circuit before the shorting ring is removed.

A simplified class B amplifier using complementary symmetry E-MOSFETs and two-supply operation is shown in Figure 5–37(a). Recall that an E-MOSFET is normally off but can be turned on when the input exceeds the threshold voltage. For logic devices, the on voltage is typically between 1 V and 2 V; for standard devices the threshold is higher. When the signal exceeds the positive threshold voltage of Q_1, it conducts; likewise, when the signal is below the negative threshold voltage of Q_2, it conducts. Thus, the n-channel device conducts on the positive cycle; the p-channel device conducts on the negative cycle.

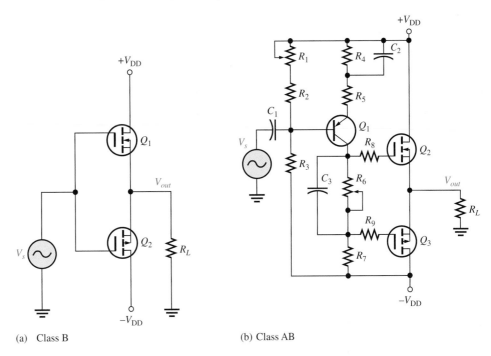

(a) Class B   (b) Class AB

FIGURE 5–37
MOSFET push-pull amplifiers.

As in the case of the BJT push-pull amplifier, the transistor does not conduct just above zero signal voltage, which causes crossover distortion. If each transistor is biased just at the threshold voltage, the MOSFETs will operate in class AB, as shown in the circuit in Figure 5–37(b). This amplifier includes a bipolar transistor amplifier as a driver and other components to assure a reasonably linear output from an E-MOSFET push-pull stage. Of course, there are other features to this basic design in commercial amplifiers.

The basic class AB push-pull amplifier shown in Figure 5–37(b) includes a common-emitter stage that amplifies the input signal and couples the signal to the gates of the push-pull stage, consisting of Q_2 and Q_3. Notice that R_6 is bypassed with capacitor C_3 to allow identical ac signals to be applied to the push-pull stage. Potentiometer R_6 is used to develop the proper dc voltage to set the bias to the threshold voltages of Q_2 and Q_3. It is adjusted to minimize cross-over distortion. Potentiometer R_1 is adjusted to zero the output dc output voltage with no input signal.

This type of amplifier can give increased power out by simply adding another pair of MOSFETs in parallel; however, this can sometimes cause unwanted oscillations. To prevent this, gate resistors can be used to isolate the MOSFETs from each other. Although not strictly required in this simplified amplifier, they are shown as R_8 and R_9. Power amplifiers with parallel E-MOSFET transistors can deliver over 100 W of power.

EXAMPLE 5–9

The n-channel E-MOSFET shown in Figure 5–38 has a threshold voltage of $+2.0$ V and the p-channel E-MOSFET has a threshold voltage of -2.0 V. What resistance setting for R_6 will bias the transistors to class AB operation? At this setting, what power is delivered to the load if the input signal is 100 mV? Assume that potentiometer R_1 is set to 440 Ω.

FIGURE 5–38

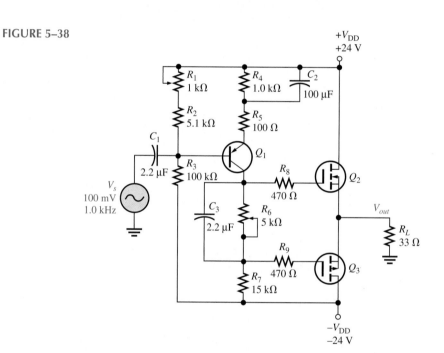

Solution Start by computing the dc parameter for the CE amplifier. The base voltage is determined by the voltage divider composed of R_1, R_2, and R_3. The standard voltage-divider equation is modified to account for the fact that the divider is not referenced to ground.

$$V_B = V_{R3} - V_{DD} = \left(V_{DD} - (-V_{DD})\right)\left(\frac{R_3}{R_1 + R_2 + R_3}\right) - V_{DD}$$

$$= \left(24 \text{ V} - (-24 \text{ V})\right)\left(\frac{100 \text{ k}\Omega}{440 \text{ }\Omega + 5.1 \text{ k}\Omega + 100 \text{ k}\Omega}\right) - 24 \text{ V} = 21.5 \text{ V}$$

The emitter voltage is one diode drop higher than the base voltage (because the transistor is a *pnp* type).

$$V_E = V_B + 0.7 \text{ V} = 21.5 \text{ V} + 0.7 \text{ V} = 22.2 \text{ V}$$

Calculate the emitter current from Ohm's law.

$$I_E = \frac{V_{DD} - V_E}{R_4 + R_5} = \frac{24 \text{ V} - 22.2 \text{ V}}{1.1 \text{ k}\Omega} = 1.64 \text{ mA}$$

The required drop across R_6 is the difference in the threshold voltages.

$$V_{R6} = V_{TH(Q1)} - V_{TH(Q2)} = 2.0 \text{ V} - (-2.0 \text{ V}) = 4.0 \text{ V}$$

Use Ohm's law to determine the required setting for R_6.

$$R_6 = \frac{V_{R6}}{I_{R6}} = \frac{4.0 \text{ V}}{1.64 \text{ mA}} = \textbf{2.4 k}\boldsymbol{\Omega}$$

This setting produces class AB operation, so the output voltage replicates the input of the MOSFET (less a small drop across the internal MOSFET resistance). Determine the gain of the CE amplifier using the ratio of unbypassed collector resistance (R_7) to the unbypassed emitter resistance (R_5) and r'_e.

$$r'_e = \frac{25 \text{ mV}}{I_E} = \frac{25 \text{ mV}}{1.64 \text{ mA}} = 15.2 \text{ }\Omega$$

and

$$A_v = \frac{R_7}{R_5 + r'_e} = \frac{15 \text{ k}\Omega}{100 \text{ }\Omega + 15.2 \text{ }\Omega} = 130$$

Assuming no internal drop in the MOSFETs, the output voltage is

$$V_{out} = A_v V_{in} = (130)(100 \text{ mV}) = 13 \text{ V}$$

The power out is

$$P_L = \frac{V_{out}^2}{R_L} = \frac{13 \text{ V}^2}{33 \text{ }\Omega} = \textbf{5.1 W}$$

Practice Exercise   Compute the setting of R_6 if the threshold voltages for the MOSFETs are $+1.5$ V and -1.5 V.

5–6 REVIEW QUESTIONS

1. What is the advantage to two-supply operation with a class B complementary symmetry amplifier?
2. What is crossover distortion and how is it avoided?
3. What is the maximum theoretical efficiency for a class B amplifier?
4. Where should an E-MOSFET, operating as a class AB amplifier, be biased?

5–7 ■ IC POWER AMPLIFIERS

An integrated circuit (IC) is a network of interconnected circuit elements (transistors, diodes, resistors) in a single piece of silicon that forms a functioning circuit. For analog electronics, the operational amplifier, introduced in Chapter 6, is the most common type of IC. In this section, you will learn about IC circuits specifically designed to provide power to a load. Two specific IC audio amplifiers, the National Semiconductor LM384 and the Motorola MC34119, are introduced.

After completing this section, you should be able to

❑ Give principal features and describe applications for IC power amplifiers
   ❑ Describe principal specifications of an integrated circuit power amplifier
   ❑ Show how to configure an LM384 audio power amplifier as a basic amplifier
   ❑ Explain how an LM384 can be used as the amplifier for an intercom system

Originally, small integrated circuit (IC) power amplifiers were designed for audio applications; they were designed to directly connect a speaker to the output. As applications were expanded, so was the proliferation of device types. Today, there are many IC amplifiers specifically designed as power amplifiers. Their principal advantages over small discrete power amplifiers are high reliability and low cost.

IC power amplifiers are used in applications ranging from small consumer appliances to power supplies, industrial motor controllers, and regulator designs. Most contain a class A or AB power amplifier stage along with associated driver stages and frequently include voltage gain.

The typical output power from a power IC is a few watts although some ICs can provide 100 W or more to a load with external power transistors. The maximum output power from any power amplifier is very dependent on proper heat sinking. Manufacturers' data sheets provide information on heat sinks required for IC power amplifiers.

The National Semiconductor LM384

An example of a typical small-power audio amplifier is the LM384[4]. It is available in a standard 14 pin dual-in-line package that contains a metal heat sink tab as shown in Figure 5–39. A copper frame is connected to the center three pins on either side (pins 3, 4, 5, and 10, 11, 12) to form a heat sink that is connected to ground.

---

[4] Data sheet for LM384 available at http://www.national.com

FIGURE 5–39
LM384 dual-in-line packages.

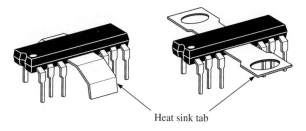

Heat sink tab

The LM384 has an internally fixed gain of 50 and operates on a single supply voltage in the range from 9 V to 24 V. The output voltage is centered at one-half the supply voltage. Choice of supply voltage depends on the required power out and the load. In addition, like many IC power amplifiers, it features short-circuit protection and thermal shutdown circuitry. It can provide up to 5 W of power to a load with appropriate heat sinking. Without an external heat sink, the maximum power output is reduced to 1.5 W.

Internally, the circuit has an emitter-follower and a differential voltage amplifier (discussed in Chapter 6). This is followed by a CE driver stage and a single-ended push-pull output stage. All stages are dc coupled. The LM384 has two inputs, one is an inverting input (labeled $-$), and the other is a noninverting input (labeled $+$).

Operation of the LM384 as a basic power amplifier requires only the addition of a few external components, as shown in Figure 5–40. In this illustration, the load consists of a small speaker, capacitively coupled through C_3 to the output. The inverting input of the LM384 is connected to the variable leg of R_1, a potentiometer that serves as a volume control. The noninverting input is shown connected to ground. Depending on the application, either input can be used by itself or the signal can be connected between the inputs. The *RC* network, composed of R_2 and C_2, suppresses by low-pass filtering action the very high-frequency oscillation that can interfere with any nearby sensitive RF circuits.

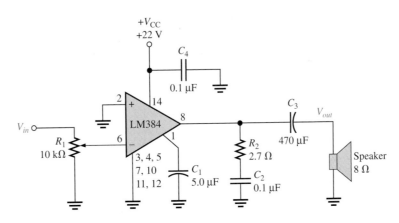

FIGURE 5–40
A basic amplifier using the LM384 audio power amplifier.

With a few modifications, the basic amplifier can become the heart of a small system, such as the intercom system shown in Figure 5–41. In this system, a small step-up transformer of 1:25 multiplies the basic gain of 50 to 1250. One speaker acts as a microphone while the second serves the traditional role of speaker. The DPDT switch controls which speaker is the talker and which is the listener. In the talk position, speaker 1 is the microphone while speaker 2 is the speaker; in the listen position, the roles are reversed.

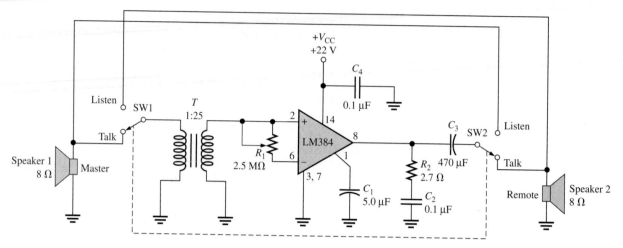

FIGURE 5–41
A basic intercom using an LM384 as the amplifier.

Power Output The data sheet for the LM384 gives a curve of the output power to a load as a function of device dissipation, as shown in Figure 5–42. The curves are different for various load resistors; an 8 Ω load resistor is shown. The power supply voltage can be selected based on the required output power and minimum distortion required. The heat sink must be capable of dissipating the device power. For example, if the heat sink can handle 3 W of power from the device, a 22 V power supply can be used to deliver nearly 5 W to the load at 3% total harmonic distortion (THD).

The Motorola MC34119 Low-Power Audio Amplifier

The Motorola MC34119 is a low-power IC audio amplifier that is designed to run on battery power for telephone applications such as telephone speaker phones. It is contained in an 8-pin DIP package (and is available in a surface-mount package). It can drive a small

FIGURE 5–42
LM384 power dissipation as a function of output power in an 8 Ω load.

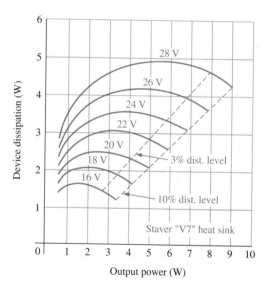

load (8 Ω to 100 Ω) with more than 250 mW for a 6 V supply voltage and can deliver more power using higher supply voltages. A chip-disable input allows for powering down or muting the input and reduces the current drain from the battery to 65 μA (typical).

Figure 5–43 illustrates a basic amplifier with the MC34119. The voltage gain can be controlled by two external resistors, (R_f and R_i) that can set the gain from less than 1 to 200 and is equal to the ratio of R_f to R_i. (Gain control is discussed in Chapter 6.) For the circuit shown, the voltage gain is 5.0. The output is coupled directly to the load (in this case, a speaker) with no coupling capacitor.

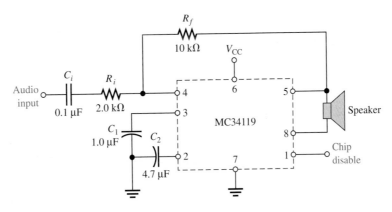

FIGURE 5–43
A small IC amplifier designed to operate from a battery.

5–7 REVIEW QUESTIONS

1. What are some applications for IC power amplifiers?
2. What advantage do IC power amplifiers have over discrete designs?
3. What is the voltage gain of an LM384?
4. What are the two inputs to an LM384 called?

5–8 ■ A SYSTEM APPLICATION

In this system application, you will work on a power amplifier that is part of a public address system. The input for the power amplifier is from a preamp such as the one in Chapter 4.

After completing this section, you should be able to

❏ Apply what you have learned in this chapter to a system application
  ❏ Compute the overall power gain for a power amplifier
  ❏ Troubleshoot the amplifier for common problems

A Brief Description of the System

A power amplifier and a horn speaker are used to form the output block of a public address system as shown in the diagram of Figure 5–44. The +12 V power supply from Chapter 2 and the preamp in Chapter 4 could be used for the other blocks in the system. The focus of this application is the power amplifier circuit board.

FIGURE 5–44
Basic public address/paging system block diagram.

The schematic of the push-pull power amplifier is shown in Figure 5–45. The circuit is a class AB amplifier implemented with darlington power transistors. The base-emitter junctions of two additional darlington transistors of the same type are used in the bias circuit to assure matched thermal characteristics. The output is delivered to a speaker with a nominal 16 Ω impedance.

FIGURE 5–45
Schematic of the power amplifier circuit.

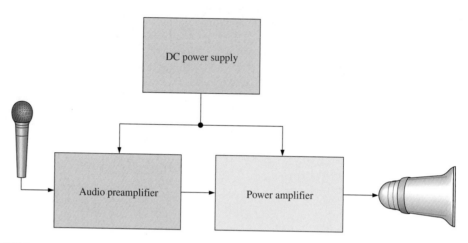

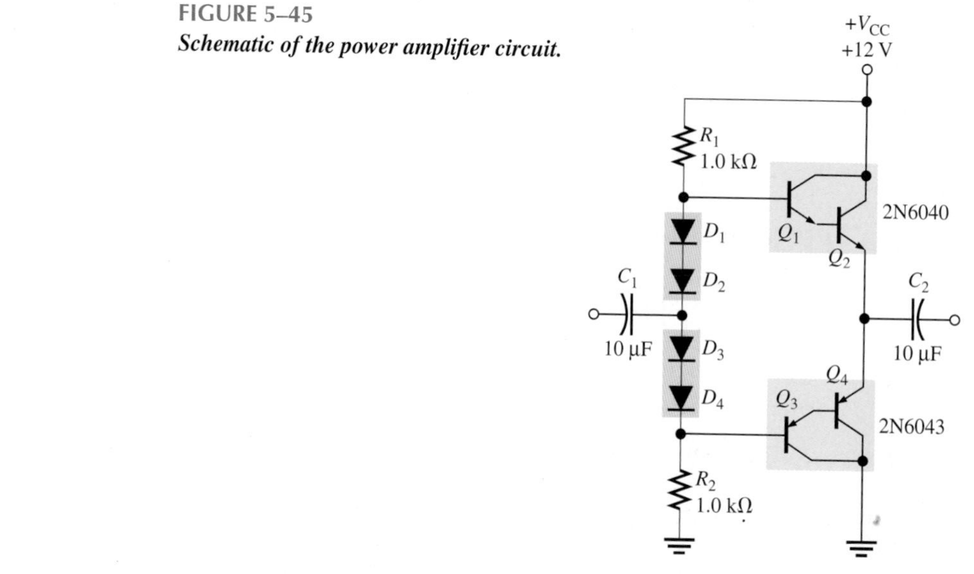

Rather than applying the signal to each base using two input coupling capacitors, this design capacitively couples the output of the preamplifier to the midpoint between the bias diodes. Since the diodes are forward-biased, the signal voltage is equally developed at the bases of Q_1 and Q_3. Darlington transistors are used in this application because of their high betas to prevent excessive loading of the preamplifier and a resulting loss of voltage gain.

Each darlington pair is packaged in a single case which looks like one transistor. The darlington transistors are mounted to a heat sink to prevent overheating and reduction of maximum power dissipation. Heat sinks come in many forms. In this application, a metal bracket with cooling fins is attached to the printed circuit board. The transistors are mounted with their collectors thermally connected but electrically isolated from the heat sink using a mica insulator.

Now, so you can take a closer look at a power amplifier, let's take it out of the system and put it on the troubleshooter's bench.

TROUBLESHOOTER'S BENCH

■ ACTIVITY 1   Analysis of the Power Amplifier Circuit

Refer to the partial data sheet in Figure 5–46.

❏ Determine the minimum input resistance of the push-pull amplifier.

❏ Verify that the power rating of each transistor in the push-pull amplifier is not exceeded under maximum signal conditions.

❏ Compute the maximum power delivered to the speaker (assuming 16 Ω impedance).

■ ACTIVITY 2   Relate the PC Board to the Schematic

❏ Check out the printed circuit board in Figure 5–47 to verify that it is correct according to the schematic. One interconnection, shown as a slightly darker trace, is on the back side of the board between the two blank pads. Each blank pad provides a feedthrough.

❏ Label a copy of the board with component and input/output designations in agreement with the schematic.

■ ACTIVITY 3   Develop a Test Procedure

❏ Develop a step-by-step set of instructions on how to check the power amplifier circuit board for proper operation at a test frequency of 5 kHz using the test points (circled numbers) indicated in the troubleshooter's bench setup of Figure 5–48. A function generator is the signal source.

❏ Specify voltage values for all the measurements to be made. Provide a fault analysis for all possible component failures.

■ ACTIVITY 4   Troubleshoot Two Circuit Boards

Problems have developed in two prototype boards. Based on the sequence of troubleshooter's bench measurements for each board indicated in Figure 5–49, determine the most likely fault in each case. The circled numbers indicate test point connections to the circuit board.

<table>
<tr><td colspan="2">

Plastic Medium-Power
Complementary Silicon Transistors

... designed for general-purpose amplifier and low-speed switching applications.
- High DC current gain —
  h_{FE} = 2500 (Typ) @ I_C = 4.0 A dc
- Collector-Emitter sustaining voltage — @ 100 mA dc
  $V_{CEO(sus)}$ = 60 V dc (Min) — 2N6040, 2N6043
         = 80 V dc (Min) — 2N6041, 2N6044
         = 100 V dc (Min) — 2N6042, 2N6045
- Low collector-emitter saturation voltage
  $V_{CE(sat)}$ = 2.0 V dc (Max) @ I_C = 4.0 A dc — 2N6040,41,2N6043,44
        = 2.0 V dc (Max) @ I_C = 3.0 A dc — 2N6042, 2N6045
- Monolithic construction with built-in base-emitter
  shunt resistors

</td><td>

Darlington
8 ampere

Complementary Silicon
Power Transistors

60-80-100 VOLTS
75 WATTS

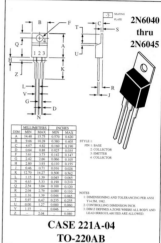

2N6040
thru
2N6045

CASE 221A-04
TO-220AB

</td></tr>
</table>

Maximum Ratings

Rating	Symbol	2N6040 2N6043 MJE6040 MJE6043	2N6041 2N6044 MJE6041 MJE6044	2N6042 2N6045 MJE6045	Unit
Collector-Emitter voltage	V_{CEO}	60	80	100	V dc
Collector-Base voltage	V_{CB}	60	80	100	V dc
Emitter-Base voltage	V_{EB}	←	5.0	→	V dc
Collector Current—Continuous Peak	I_C	←	8.0 16	→	A dc
Base current	I_B	←	120	→	mA dc
Total power dissipation @ T_C = 25°C Derate above 25°C	P_D	←	75 0.60	→	Watts W/C°
Total power dissipation @ T_A = 25°C Derate above 25°C	P_D	←	2.2 0.0175	→	Watts W/C°
Operating and storage junction, Temperature range	T_J, T_{stg}	←	−65 to +150	→	°C

Electrical Characteristics (T_C = 25°C unless otherwise noted)

Characteristic	Symbol	Min	Max	Unit
DC current gain (I_C = 4.0 A dc, V_{CE} = 4.0 V dc) 2N6040,41,2N6043,44,MJE6040,41, MJE6043,44 (I_C = 3.0 A dc, V_{CE} = 4.0 V dc) 2N6042, 2N6045, MJE6045 (I_C = 8.0 A dc, V_{CE} = 4.0 V dc) All Types	h_{FE}	1000 1000 100	20,000 20,000 —	–
Collector-Emitter saturation voltage (I_C = 4.0 A dc, I_B = 16 mA dc) 2N6040,41,2N6043,44,MJE6040,41,MJE6043,44 (I_C = 3.0 A dc, I_B = 12 mA dc) 2N6042,2N6045,MJE6045 (I_C = 8.0 A dc, I_B = 80 mA dc) All Types	$V_{CE(sat)}$	— — —	2.0 2.0 4.0	V dc
Base-Emitter saturation voltage (I_C = 8.0 A dc, I_B = 80 mA dc)	$V_{CE(sat)}$	—	4.5	V dc
Base-Emitter on voltage (I_C = 4.0 A dc, V_{CE} = 4.0 V dc)	$V_{BE(on)}$	—	2.8	V dc

Dynamic Characteristics

Small-signal current gain (I_C = 3.0 A dc, V_{CE} = 4.0 V dc, f = 1.0 MHz)	$\|h_{fe}\|$	4.0	—	
Output capacitance (V_{CB} = 10 V dc, I_E = 0, f = 1.0 MHz) 2N6040/2N6042, MJE6040 2N6043/2N6045, MJE6043/MJE6045	C_{ob}	— —	300 200	pF
Small-signal current gain (I_C = 3.0 A dc, V_{CE} = 4.0 V dc, f = 1.0 kHz)	h_{fe}	300	—	–

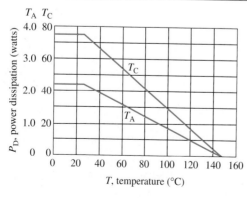

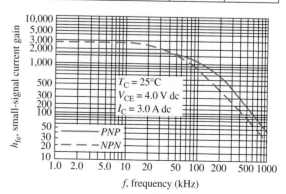

FIGURE 5–46

Partial data sheet for complementary darlington transistors 2N6040 (npn) and 2N6043 (pnp).

FIGURE 5–47
Power amplifier circuit
board and transistor pin
configuration.

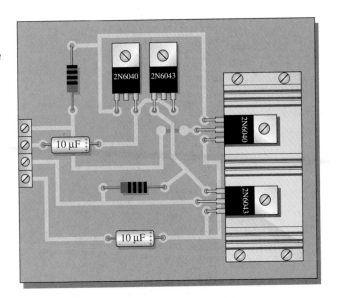

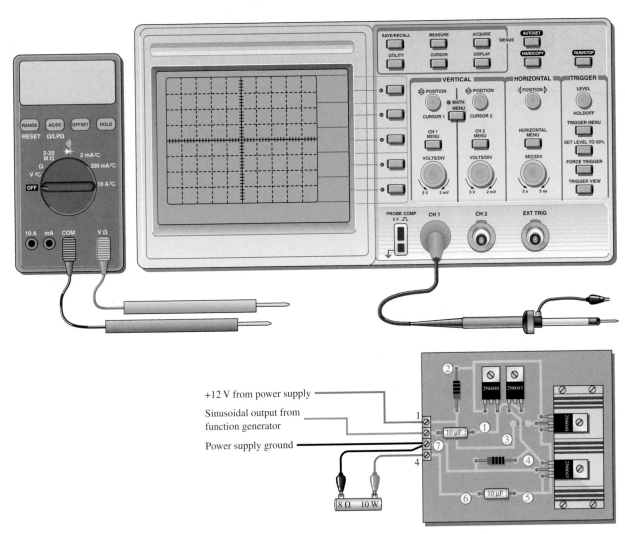

+12 V from power supply

Sinusoidal output from
function generator

Power supply ground

FIGURE 5–48
Troubleshooter's bench setup for the power amplifier board.

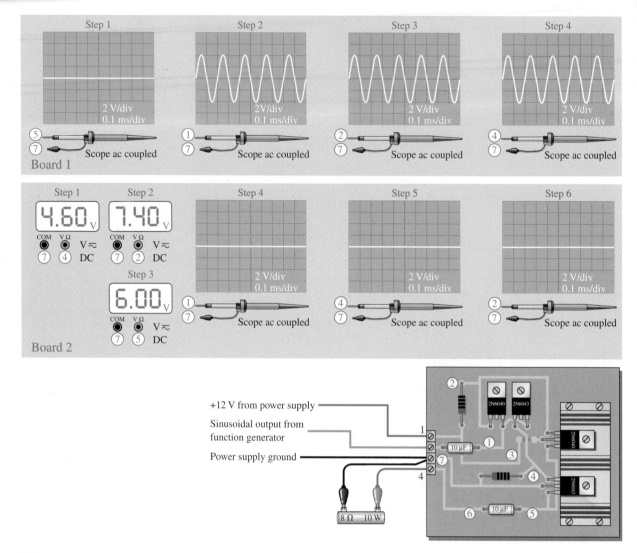

FIGURE 5–49
Test results for two faulty circuit boards.

■ ACTIVITY 5   Write a Technical Report

Submit a technical report on the power amplifier circuit board using an organized format that includes the following:

1. A physical description of the circuits.

2. A discussion of the operation of the circuits.

3. A list of the specifications.

4. A list of parts with part numbers if available.

5. A list of the types of problems on the two faulty circuit boards.

6. A complete description of how you determined the problem on each of the faulty circuit boards.

5–8 REVIEW QUESTIONS

1. Explain the advantage of a push-pull amplifier for the public address system over a class A amplifier.
2. Why is it necessary to include input and output coupling capacitors in the power amplifier circuit?
3. Why were darlington transistors used in the design of the push-pull amplifier?

■ SUMMARY

- Three ways to couple amplifier stages together are capacitive coupling, transformer coupling, and direct coupling.
- Capacitive coupling and transformer coupling provide a low-impedance ac path while blocking dc. Direct coupling requires that the dc conditions from one stage are compatible with the requirements of the next stage.
- General points to alleviate noise problems in amplifiers are

  1. Keep wiring short and make signal return loops as small as possible.
  2. Use bypass capacitors between power supply and ground.
  3. Reduce noise sources and separate or shield the noise source and the circuit.
  4. Ground circuits at a single point, and isolate grounds that have high currents from those with low currents.
  5. Keep the bandwidth of amplifiers no larger than necessary.

- One common type of high-frequency cable is coax, a form of shielded wiring. Coax typically has a characteristic impedance between 50 Ω and 100 Ω. It should be terminated in its characteristic impedance to prevent reflections.
- Tuned amplifiers use one or more resonant circuits to select a band of frequencies.
- A mixer combines a radio frequency (RF) signal with a sine wave generated from a local oscillator to produce an intermediate frequency (IF) that is amplified by an amplifier tuned to the IF.
- Negative feedback produces a self-correcting action that can produce excellent bias stability and gain stability in amplifiers.
- The voltage gain of an amplifier without feedback is called open-loop voltage gain. The voltage gain of an amplifier with feedback is called closed-loop voltage gain.
- A class A amplifier operates entirely in the linear region of the transistor's characteristic curves. The transistor conducts during the full 360° of the input cycle.
- The Q-point must be centered on the ac load line for maximum class A output signal swing.
- The maximum efficiency of a class A amplifier is 25%.
- A class B amplifier operates in the linear region for half of the input cycle (180°), and it is in cutoff for the other half.
- The Q-point is at cutoff for class B operation.
- Class B amplifiers are normally operated in a push-pull configuration in order to produce an output that is a replica of the input.
- The maximum efficiency of a class B amplifier is 79%.
- A class AB amplifier is biased slightly above cutoff and operates in the linear region for slightly more than 180° of the input cycle.
- Class AB eliminates the crossover distortion found in pure class B.

■ GLOSSARY

Key terms are in color. All terms are included in the end-of-book glossary.

Automatic gain control (AGC)   A feedback system that reduces the gain for larger signals and increases the gain for smaller signals.

Class A   An amplifier that operates in the active region at all times.

Class AB   An amplifier that is biased into slight conduction; the Q-point is slightly above cutoff.

Class B   An amplifier that has the Q-point located at cutoff, causing the output current to vary only during one-half of the input cycle.

Closed-loop voltage gain   The net voltage gain of an amplifier when negative feedback is included.

Coax   A transmission line, principally used for high frequencies, in which a center conductor is surrounded by a tubular conducting shield.

Complementary symmetry transistors   These are a matching pair of *npn/pnp* BJTs or a matching pair of *n*-channel/*p*-channel FETs.

Crossover distortion   Distortion in the output of a class B push-pull amplifier at the point where each transistor changes from the cutoff state to the on state.

Current mirror   A circuit that uses matching diode junctions to form a current source. The current in a diode is reflected as a matching current in the other junction (which is typically the base-emitter junction of a transistor). Current mirrors are commonly used to bias a push-pull amplifier.

Decoupling network   A low-pass filter that provides a low impedance path to ground for high-frequency signals.

Efficiency (power)   The ratio of the signal power supplied to the load to the power from the dc supply.

Electrostatic discharge (ESD)   The discharge of a high voltage through an insulating path that frequently destroys a device.

Intermediate frequency   A fixed frequency that is lower than the RF, produced by beating an RF signal with an oscillator frequency.

Large-signal   A signal that operates an amplifier over a significant portion of its load line.

Mixer   A nonlinear circuit that combines two signals and produces the sum and difference frequencies.

Neutralization   A method of preventing unwanted oscillations by adding negative feedback to just cancel the positive feedback caused by internal capacitances of an amplifier.

Open-loop voltage gain   The voltage gain of an amplifier without external feedback.

Power gain   The ratio of the power delivered to the load to the input power of an amplifier.

Push-Pull   A type of class B amplifier with two transistors in which one transistor conducts for one half-cycle and the other conducts for the other half-cycle.

Quality factor (Q)   A dimensionless number that is the ratio of the maximum energy stored in a cycle to the energy lost in a cycle.

Skin effect   The phenomenon at high frequencies which causes current to move to the outside surface of conductors.

Stage   Each transistor in a multistage amplifier that amplifies a signal.

Standing wave   A stationary wave on a transmission line formed by the interaction of an incident and reflected wave.

■ **KEY FORMULAS**

(5–1) $Z_0 = \sqrt{\dfrac{L}{C}}$ Characteristic impedance of a transmission line

(5–2) $f_r = \dfrac{1}{2\pi\sqrt{LC}}$ Resonant frequency (high Q resonant circuit)

(5–3) $Q = \dfrac{X_L}{R} = \dfrac{f_r}{BW}$ Quality factor of a resonant circuit

(5–4) $R'_L = \left(\dfrac{N_{pri}}{N_{sec}}\right)^2 R_L$ Reflected resistance of a load resistor by a transformer

(5–5) $A_p = \dfrac{P_L}{P_{in}}$ Amplifier power gain

(5–6) $A_p = A_v^2\left(\dfrac{R_{in}}{R_L}\right)$ Alternate amplifier power gain

(5–7) $P_{DQ} = I_{CQ}V_{CEQ}$ Power dissipation of a transistor

(5–8) $P_{out(max)} = 0.5I_{CQ}V_{CEQ}$ Maximum power from a class A amplifier

(5–9) $I_{c(sat)} = \dfrac{V_{CC}}{R_L}$ AC saturation current for a two-supply operation with a push-pull amplifier

■ **SELF-TEST**

Answers are at the end of the chapter.

1. If an amplifier has a decibel voltage gain of 60 dB, the actual gain is
   (a) 600   (b) 1000   (c) 1200   (d) 1,000,000

2. If two identical stages with a gain of 25 dB each are connected together and the equivalent voltage divider has an attenuation of 5 dB, the overall gain of the amplifier is
   (a) 20 dB   (b) 45 dB   (c) 55 dB   (d) 70 dB

3. Noise can enter a circuit
   (a) by capacitive or inductive coupling   (b) through the power supply
   (c) from within the circuit   (d) all of these answers

4. The impedance of coax cable is typically
   (a) less than 10 Ω (b) 50 Ω to 100 Ω
   (c) 200 Ω to 500 Ω (d) answer depends on the length of the line

5. Neutralization is a technique used with high-frequency amplifiers to eliminate
   (a) oscillations   (b) noise
   (c) distortion   (d) all of these answers

6. The quality factor, Q, is a pure number that is the ratio of
   (a) X_L to X_C (b) X_L to R
   (c) X_C to R (d) none of these answers

7. In a tuned circuit, if the Q is high, the
   (a) resistance is high   (b) bandwidth is small
   (c) frequency is low   (d) power is high

8. Negative feedback can provide excellent
   (a) bias stability   (b) gain stability
   (c) both (a) and (b)   (d) neither (a) nor (b)

9. The peak current a class A amplifier can deliver to a load depends on the
   (a) maximum rating of the power supply   (b) quiescent current
   (c) current in the bias resistors   (d) size of the heat sink

10. An amplifier that operates in the linear region at all times is
    (a) Class A   (b) Class AB
    (c) Class B   (d) all of these answers

11. The efficiency of a power amplifier is the ratio of the power delivered to the load to the
    (a) input signal power   (b) power dissipated in the last stage
    (c) power from the power supply   (d) none of these answers

12. Crossover distortion is a problem for
    (a) Class A amplifiers   (b) Class AB amplifiers
    (c) Class B amplifiers   (d) all of these answers

13. A current mirror in a push-pull amplifier should give an I_{CQ} that is
    (a) equal to the current in the bias resistors and diodes
    (b) twice the current in the bias resistors and diodes
    (c) half the current in the bias resistors and diodes
    (d) zero

14. To avoid crossover distortion with an E-MOSFET push-pull amplifier, you should bias the MOSFETs with
    (a) a current mirror   (b) self-bias
    (c) a voltage divider   (d) a separate power supply

15. Typically, an IC power amplifier
    (a) does not need a heat sink   (b) costs more than a discrete circuit
    (c) has high reliability   (d) all of these answers

TROUBLESHOOTER'S QUIZ   *Answers are at the end of the chapter.*

Refer to Figure 5–50.

❑ If C_3 is open,

1. The dc drain voltage of Q_1 will

   (a) increase   (b) decrease   (c) not change

2. The ac drain voltage of Q_1 will

   (a) increase   (b) decrease   (c) not change

❑ If R_{E1} is 0 Ω because of a direct short across it,

3. The dc emitter current will

   (a) increase   (b) decrease   (c) not change

4. The overall voltage gain of the amplifier will

   (a) increase   (b) decrease   (c) not change

5. The input resistance of Q_2 will

   (a) increase   (b) decrease   (c) not change

Refer to Figure 5–57.

❑ If Q_2 has an open emitter,

6. The positive side of the ac output voltage will

   (a) increase   (b) decrease   (c) not change

7. The negative side of the ac output voltage will

    (a) increase   **(b)** decrease   **(c)** not change

❏ If D_1 is shorted,

8. The bias current in R_1 will

    (a) increase   **(b)** decrease   **(c)** not change

9. The ac output voltage will

    (a) increase   **(b)** decrease   **(c)** not change

Refer to Figure 5–59.

❏ If C_2 is open,

10. The gain will

    (a) increase   **(b)** decrease   **(c)** not change

❏ If C_3 is open,

11. The distortion will

    (a) increase   **(b)** decrease   **(c)** not change

❏ If R_8 is shorted,

12. The gain will

    (a) increase   **(b)** decrease   **(c)** not change

■ PROBLEMS

Answers to odd-numbered problems are at the end of the book.

SECTION 5–1 Capacitively Coupled Amplifiers

1. For the two-stage capacitively coupled amplifier shown in Figure 5–50, compute the overall voltage gain, the input resistance, and the output resistance. Assume the g_m of the JFET is 2700 µS and β_{ac} of the BJT is 150.

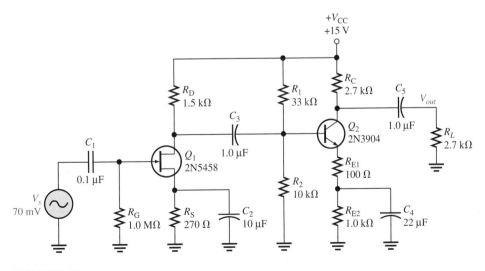

FIGURE 5–50

2. Using the results of Problem 1, draw the ac amplifier model for the two-stage amplifier in Figure 5–50. (See Figure 5–5 for an example).

3. For the two-stage amplifier modeled in Figure 5–51, determine the ordinary gain and the decibel voltage gain.

4. Assume a 1.0 kΩ load is connected to the amplifier modeled in Figure 5–51. What is the new gain?

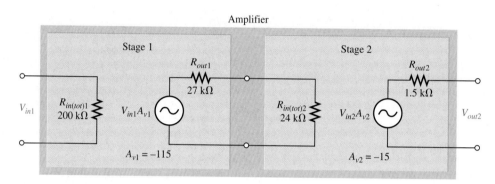

FIGURE 5–51

5. Assume a two-stage amplifier is constructed from two identical amplifiers with the following specifications: $R_{in} = 30$ kΩ, $R_{out} = 2$ kΩ, $A_{v(NL)} = 80$.
   (a) Draw the ac model of the amplifier.
   (b) What is the overall gain when the two stages are connected together?
   (c) If a 3 kΩ load resistor is connected to the amplifier, what is the overall gain?

6. Determine the decibel gain for the amplifier in Problem 5(b).

7. Assume Q_3 in Figure 5–4 is replaced with a darlington transistor with a β_{ac} of 10,000. What effect will this have on the amplifier's gain?

SECTION 5–2   RF Amplifiers

8. RG180B/U is a coaxial cable with a nominal impedance of 95 Ω and a capacitance per foot of 15.5 pF/foot. Determine the inductance per foot for this cable.

9. Why is it important to terminate a high-frequency cable in its characteristic impedance?

10. For the circuit in Figure 5–9, assume the capacitor in the drain circuit is 68 pF and the coil has an inductance of 300 μH with a resistance of 15.2 Ω. What is the center frequency for the amplifier?

11. For the circuit in Figure 5–9, what happens to the gain if the input signal is larger? Explain your answer.

12. Assume a parallel resonant circuit is constructed from a 200 μH inductor with 9.5 Ω of resistance and a 1000 pF capacitor.
    (a) What is the resonant frequency?
    (b) What is the Q?
    (c) What is the bandwidth?

SECTION 5–3   Transformer-Coupled Amplifiers

13. Assume a 10:1 step-down transformer has a load of 100 Ω connected across the secondary. What is the reflected resistance in the primary circuit?

14. The audio frequency power amplifier shown in Figure 5–52 has a 3:1 step-down transformer in the collector circuit with a 16 Ω load resistor connected to the secondary. Determine the gain of the circuit. (Since r'_e is small compared to R_{E1}, it can be ignored.)

FIGURE 5–52

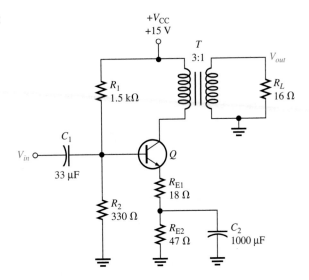

15. Draw the dc and ac load lines for the circuit in Figure 5–52. (Assume the dc resistance of the transformer is very small and can be ignored.)

16. The amplifier in Figure 5–53 is a low-power audio amplifier with collector-feedback bias. The transformer is a step-down impedance-matching transformer designed to give a reflected resistance in the primary of 1000 Ω when the load is 8 Ω (such as the speaker). The dc resistance of the primary winding is 66 Ω.
    (a) Compute V_{CE} and the I_E for the transistor assuming $\beta_{ac} = \beta_{DC} = 150$.
    (b) Compute A_v, A_p, and power delivered to the load when the input is 500 mV pp.

FIGURE 5–53

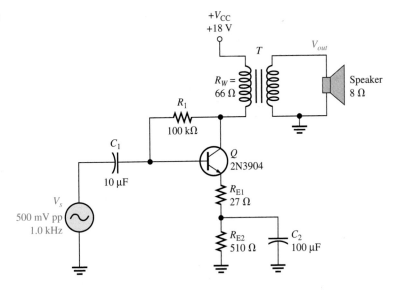

17. Assume the IF amplifier in Figure 5–54 has an IF transformer with a primary inductance of 180 μH and a resistance of 6.5 Ω. Internally, there is a 680 pF capacitor connected in parallel with the primary. Find the Q of the resonant circuit, the unloaded voltage gain, $A_{v(NL)}$, and the bandwidth, BW.

18. What is the purpose of R_3 and C_3 in the circuit of Figure 5–54?

FIGURE 5–54

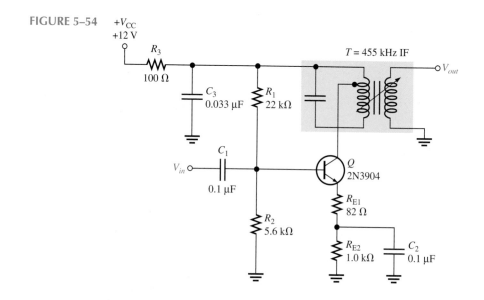

SECTION 5–4 **Direct-Coupled Amplifiers**

19. Figure 5–55 shows two dc coupled CC amplifiers (Q_2 and Q_3) with no coupling capacitors required at the input or output. Q_1 is a current source for Q_2 and produces very high input resistance for the amplifier.

    (a) Assuming the base of Q_2 is at zero volts, determine the following dc parameters: $I_{C(Q2)}$, $V_{B(Q3)}$, $I_{C(Q3)}$, $V_{E(Q3)}$.

    (b) Assuming a 5 V rms input signal, what power is delivered to the load resistor?

20. What are advantages of a dc coupled CC amplifier such as shown in Figure 5–55?

21. Draw the dc load line for Q_3 in Figure 5–55.

22. Assume the emitter resistor for Q_3 is open. Will the base-emitter junction of Q_3 still be forward-biased? What happens to the collector current in Q_3?

23. For the circuit in Figure 5–20, assume a 10 kΩ resistor is substituted for R_{F2}. What effect, if any, does this change have on

    (a) the dc emitter voltage of Q_1?

    (b) the voltage gain?

    (c) the input resistance of the amplifier?

24. Each of three cascaded amplifiers has a decibel voltage gain of 15. What is the overall decibel voltage gain? What is the overall linear voltage gain?

FIGURE 5–55

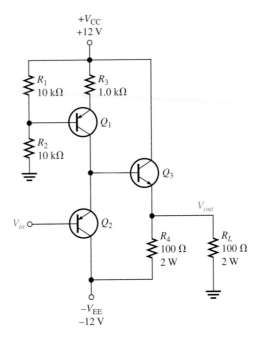

SECTION 5–5 Class A Power Amplifiers

25. Figure 5–56 shows a CE power amplifier in which the collector resistor serves also as the load resistor. Assume $\beta_{DC} = \beta_{ac} = 100$.
    (a) Determine the dc Q-point (I_{CQ} and V_{CEQ}).
    (b) Determine the voltage gain and the power gain.

26. For the circuit in Figure 5–56, determine the following:
    (a) the power dissipated in the transistor with no load.
    (b) the total power from the power supply with no load.
    (c) the signal power in the load with a 500 mV input.

27. Refer to the circuit in Figure 5–56. What changes would be necessary to convert the circuit to a *pnp* transistor with a positive supply? What advantage would this have?

FIGURE 5–56

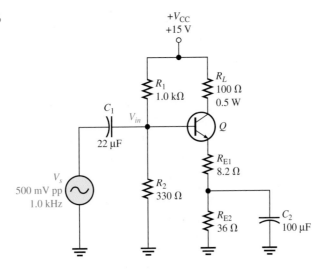

28. Assume a CC amplifier has an input resistance of 2.2 kΩ and drives an output load of 50 Ω. What is the power gain?

SECTION 5–6 Class B Power Amplifiers

29. Refer to the Class AB amplifier in Figure 5–57.
(a) Determine the dc parameters $V_{B(Q1)}$, $V_{B(Q2)}$, V_E, I_{CQ}, $V_{CEQ(Q1)}$, $V_{CEQ(Q2)}$.
(b) For the 5 V rms input, determine the power delivered to the load resistor.

30. Draw the load line for the *npn* transistor in Figure 5–57. Label the saturation current, $I_{c(sat)}$, and show the Q-point.

FIGURE 5–57

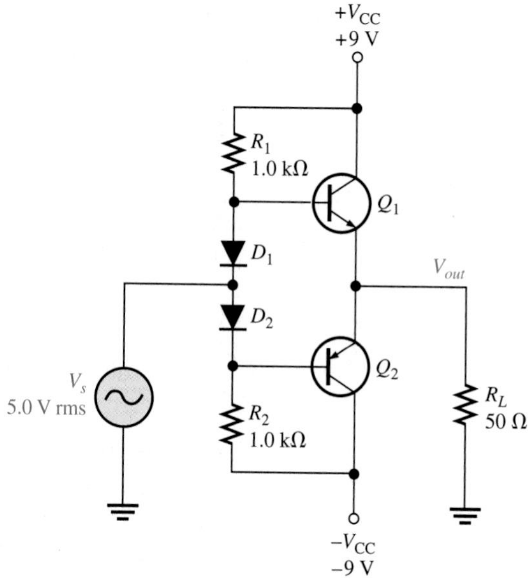

31. Refer to the Class AB amplifier in Figure 5–58 operating with a single power supply.
(a) Determine the dc parameters $V_{B(Q1)}$, $V_{B(Q2)}$, V_E, I_{CQ}, $V_{CEQ(Q1)}$, $V_{CEQ(Q2)}$.
(b) Assuming the input voltage is 10 V pp, determine the power delivered to the load resistor.

32. Refer to the Class AB amplifier in Figure 5–58.
(a) What is the maximum power that could be delivered to the load resistor?
(b) Assume the power supply voltage is raised to 24 V. What is the new maximum power that could be delivered to the load resistor?

33. Refer to the Class AB amplifier in Figure 5–58. What fault or faults could account for each of the following troubles?
(a) a positive half-wave output signal
(b) zero volts on both bases and the emitters
(c) no output; emitter voltage = $+15$ V
(d) crossover distortion observed on the output waveform

34. Assume the *n*-channel E-MOSFET shown in Figure 5–59 has a threshold voltage of 2.75 V and the *p*-channel E-MOSFET has a threshold voltage of -2.75 V.
(a) What resistance setting for R_5 will bias the output transistors to class AB operation?
(b) Assuming the input voltage is 150 mV rms, what is the rms voltage delivered to the load?
(c) What is the power delivered to the load with this setting?

FIGURE 5–58

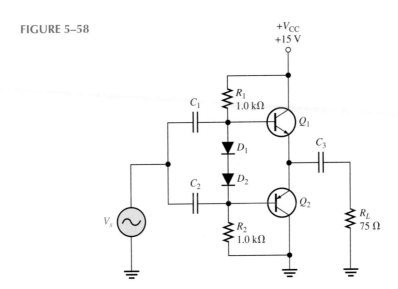

FIGURE 5–59

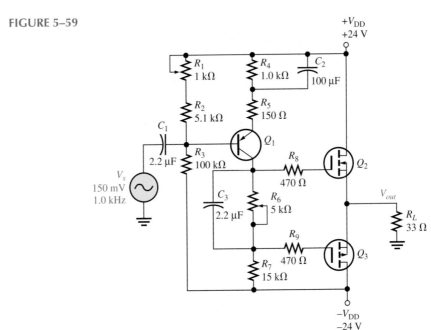

SECTION 5–7   IC Power Amplifiers

35. Refer to Figure 5–42. Assume you are operating an LM384 from a +18 V power supply and are dissipating 3 W in the load. How much power must the LM384 and its heat sink dissipate?

36. Assume you want to increase the voltage gain of the MC34119 in Figure 5–43 by changing R_f. What value should you specify if the gain required is 15?

SECTION 5–8   A System Application

37. Describe the output if the base-emitter junction of Q_2 opened (Figure 5–45).

38. Why is it necessary to isolate the transistor's metal tab from the heat sink? What would happen if they were not isolated?

■ ANSWERS TO REVIEW QUESTIONS

Section 5–1

1. R_{in}, R_{out}, and $A_{v(NL)}$
2. 32 dB
3. Low noise and high input resistance
4. The signal levels are small and can easily be obscured by noise.

Section 5–2

1. Coax is a high-frequency cable with constant impedance independent of length. The outer conductor helps shield the signal and prevent interference.
2. Neutralization is a method of preventing unwanted oscillations by adding *negative* feedback to just cancel the *positive* feedback caused by internal capacitances of an amplifier.
3. *Automatic Gain Control*
4. Q is a dimensionless number that is the ratio of the maximum energy stored in a cycle to the energy lost in a cycle. It can be expressed as $Q = X_L/R = f_r/BW$.

Section 5–3

1. RF indicates *Radio Frequency* and is any frequency useful for radio transmission. IF means *Intermediate Frequency*; it represents a frequency that has been shifted for processing.
2. A mixer combines two signals in a nonlinear circuit, producing the sum and difference frequencies as a result. The difference frequency becomes the IF.
3. The RF signal and a signal from the local oscillator
4. A load resistor affects the Q by reflecting a resistance into the primary side. Since Q is the ratio of X_L to R, the increase in R decreases Q.
5. Any instrument connected to the circuit can change the Q of the circuit because of resistance loading and can change the frequency because of capacitance loading.

Section 5–4

1. Direct coupling reduces the parts count and allows for frequencies down to dc.
2. Negative feedback returns a portion of the output in a way that tends to cancel changes in the bias circuit or the gain.
3. The capacitor has no effect on the dc circuit, but it increases the open-loop gain. A higher open-loop gain means a small change in a circuit parameters will have less effect.
4. The gain is determined by the reciprocal of the feedback fraction.

Section 5–5

1. To dissipate excessive heat
2. The collector
3. Cutoff and saturation clipping
4. 25%
5. It is the ratio of the input resistance to the output resistance

Section 5–6

1. The signal can be direct coupled at the input and output; parts count can be reduced.
2. Crossover distortion occurs when the input signal is less than the base-emitter drop of the push-pull amplifier. It can be avoided by biasing the class B amplifier on slightly, producing class AB operation.

3. 79%

4. At the threshold voltage

Section 5–7

1. Small amplifiers for consumer products (TVs, radios, telephone speaker phones), alarms, intercoms, small motor controls, and the like

2. Simplified application, reliability, and cost

3. 50

4. Inverting and noninverting

Section 5–8

1. The efficiency is much higher for a push-pull amplifier than a class A amplifier.

2. For single-supply operation, the coupling capacitors isolate the bias voltage from the input and output.

3. Darlington transistors reduce the loading effect of the push-pull amplifier on the driving circuit.

■ **ANSWERS TO PRACTICE EXERCISES FOR EXAMPLES**

5–1 **(a)** 400 mV   **(b)** 99.9

5–2 75 Ω

5–3 The Q-point is unchanged; the ac load line is flatter due to increased resistance.

5–4 An open C_2 reduces the gain. The BW is not affected.

5–5 The β affects the input resistance of both stages. The gain is affected only by the input resistance of Q_2 and reduces the gain to 595.

5–6 46 dB

5–7 The efficiency goes up because no power is wasted in R_{E3}. The disadvantage is that the speaker has a dc current (the emitter current) in the coil.

5–8 15 V, 0.94 A

5–9 1.83 kΩ

■ **ANSWERS TO SELF-TEST**

1. (b)	**2.** (b)	**3.** (d)	**4.** (b)	**5.** (a)
6. (b)	**7.** (b)	**8.** (c)	**9.** (b)	**10.** (a)
11. (c)	**12.** (c)	**13.** (a)	**14.** (c)	**15.** (c)

■ **ANSWERS TO TROUBLE-SHOOTER'S QUIZ**

1. not change	**2.** increase	**3.** increase	**4.** increase
5. decrease	**6.** not change	**7.** decrease	**8.** increase
9. decrease	**10.** decrease	**11.** increase	**12.** not change

6

OPERATIONAL AMPLIFIERS

Copyright 2000, Tektronix, Inc. Reprinted with permission. All rights reserved.

■ CHAPTER OBJECTIVES

❑ Describe the basic op-amp and its characteristics
❑ Discuss the differential amplifier and its operation
❑ Discuss several op-amp parameters
❑ Explain negative feedback in op-amp circuits
❑ Analyze three op-amp configurations
❑ Describe impedances of the three op-amp configurations
❑ Troubleshoot op-amp circuits
❑ Apply what you have learned in this chapter to a system application

■ KEY TERMS

❑ Operational amplifier
❑ Differential amplifier
❑ Single-ended mode
❑ Differential-mode
❑ Common-mode
❑ Common-mode rejection ratio (CMRR)
❑ Open-loop voltage gain
❑ Slew rate
❑ Negative feedback
❑ Closed-loop voltage gain
❑ Noninverting amplifier
❑ Voltage-follower
❑ Inverting amplifier

■ CHAPTER INTRODUCTION

So far, you have studied a number of important electronic devices. These devices, such as the diode and the transistor, are separate devices that are individually packaged and interconnected in a circuit with other devices to form a complete, functional unit. Such devices are referred to as *discrete components*.

Now you will learn more about analog (linear) integrated circuits where many transistors, diodes, resistors, and capacitors are fabricated on a single tiny chip of semiconductive material and packaged in a single case to form a functional circuit. You were introduced to a specific integrated circuit (IC) in Chapter 5 that was designed for the specific purpose of audio amplification.

In this chapter, you are introduced to a general-purpose IC, the operational amplifier (op-amp), which is the most versatile and widely used of all linear integrated circuits. Although the op-amp is made up of many resistors, diodes, and transistors, it is treated as a single device. This means that you will be concerned with what the circuit does more from an external viewpoint than from an internal, component-level viewpoint.

■ A SYSTEM APPLICATION

In medical laboratories, an instrument known as a spectrophotometer is used to analyze chemicals in solutions by determining how much absorption of light occurs over a range of wavelengths. A basic system is shown in Figure 6–35. Light is passed through a prism; and as the light source and prism are pivoted, different wavelengths of visible light pass through the slit. The wavelength coming through the slit at a given pivot angle passes through the solution and is detected by a photocell. The op-amp circuit is used to amplify the output of the photocell and send the signal to a processor and display instrument. Since every chemical and compound absorbs light in a different way, the output of the spectro-photometer can be used to accurately identify the contents of the solution.

For the system application in Section 6–8, in addition to the other topics, be sure you understand
❑ The functions of the inputs and outputs of an op-amp
❑ How an op-amp works
❑ The pin configurations of an op-amp

www.   VISIT THE COMPANION WEBSITE

Study Aids for This Chapter Are Available at

http://www.prenhall.com/floyd

6–1 ■ INTRODUCTION TO OPERATIONAL AMPLIFIERS

Early operational amplifiers (op-amps) were used primarily to perform mathematical operations such as addition, subtraction, integration, and differentiation, thus the term operational. These early devices were constructed with vacuum tubes and worked with high voltages. Today, op-amps are linear integrated circuits that use relatively low supply voltages and are reliable and inexpensive.

After completing this section, you should be able to

❑ Describe the basic op-amp and its characteristics
   ❑ Recognize the op-amp symbol
   ❑ Identify the terminals on op-amp packages
   ❑ Describe the ideal op-amp
   ❑ Describe the practical op-amp

Symbol and Terminals

The standard **operational amplifier** (op-amp) symbol is shown in Figure 6–1(a). It has two input terminals, the inverting (−) input and the noninverting (+) input, and one output terminal. The typical op-amp operates with two dc supply voltages, one positive and the other negative, as shown in Figure 6–1(b). Usually these dc voltage terminals are left off the schematic symbol for simplicity but are always understood to be there. Some typical op-amp IC packages are shown in Figure 6–1(c).

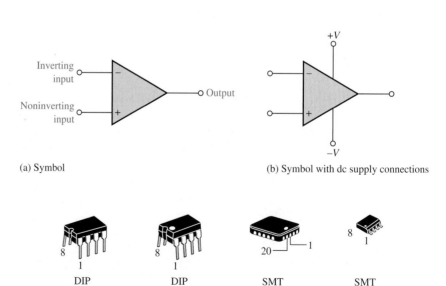

(a) Symbol

(b) Symbol with dc supply connections

DIP   DIP   SMT   SMT

(c) Typical packages. Pin 1 is indicated by a notch or dot on dual-in-line (DIP) and surface-mount technology (SMT) packages, as shown.

FIGURE 6–1
Op-amp symbols and packages.

The Ideal Op-Amp

To illustrate what an op-amp is, let's consider its *ideal* characteristics. A practical op-amp, of course, falls short of these ideal standards, but it is much easier to understand and analyze the device from an ideal point of view.

First, the ideal op-amp has *infinite voltage gain* and an *infinite input impedance* (open), so that it does not load the driving source. Finally, it has a *zero output impedance*. These characteristics are illustrated in Figure 6–2. The input voltage V_{in} appears between the two input terminals, and the output voltage is A_vV_{in}, as indicated by the internal voltage source symbol. The concept of infinite input impedance is a particularly valuable analysis tool for the various op-amp configurations, which will be discussed in Section 6–5.

FIGURE 6–2
Ideal op-amp representation.

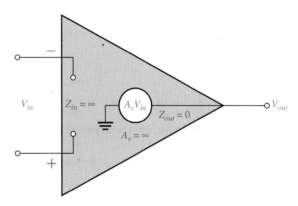

The Practical Op-Amp

Although modern integrated circuit (IC) op-amps approach parameter values that can be treated as ideal in many cases, no practical op-amp can be ideal. Any device has limitations, and the IC op-amp is no exception. Op-amps have both voltage and current limitations. Peak-to-peak output voltage, for example, is usually limited to slightly less than the difference between the two supply voltages. Output current is also limited by internal restrictions such as power dissipation and component ratings.

Characteristics of a practical op-amp are *high voltage gain, high input impedance, low output impedance,* and *wide bandwidth*. Some of these are illustrated in Figure 6–3.

FIGURE 6–3
Practical op-amp representation.

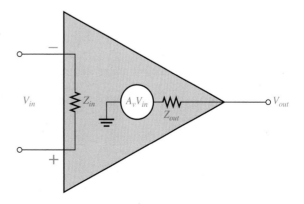

6–1 REVIEW QUESTIONS*

1. What are the connections to a basic op-amp?
2. Describe some of the characteristics of a practical op-amp.

_____

* Answers are at the end of the chapter.

6–2 ■ THE DIFFERENTIAL AMPLIFIER

The op-amp typically consists of at least one differential amplifier stage. Because the differential amplifier (diff-amp) is the input stage of an op-amp, it is fundamental to the op-amp's internal operation. Therefore, it is useful to have a basic understanding of this type of circuit.

After completing this section, you should be able to

❑ Discuss the differential amplifier and its operation
  ❑ Explain single-ended input operation
  ❑ Explain differential-input operation
  ❑ Explain common-mode operation
  ❑ Define *common-mode rejection ratio*
  ❑ Discuss the use of differential amplifiers in op-amps

A basic **differential amplifier** (diff-amp) circuit and its symbol are shown in Figure 6–4. The diff-amp stage that makes up part of an op-amp provides high voltage gain and common-mode rejection (defined later in this section).

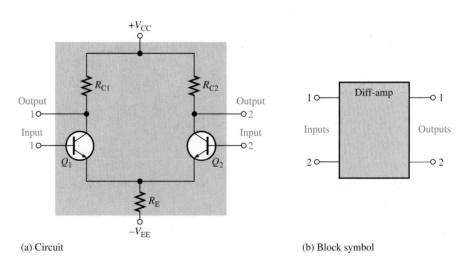

(a) Circuit   (b) Block symbol

FIGURE 6–4
Basic differential amplifier.

Basic Operation

The following discussion is in relation to Figure 6–5 and consists of a basic dc analysis of the diff-amp's operation.

First, when both inputs are grounded (0 V), the emitters are at −0.7 V, as indicated in Figure 6–5(a). It is assumed that the transistors, Q_1 and Q_2, are identically matched by

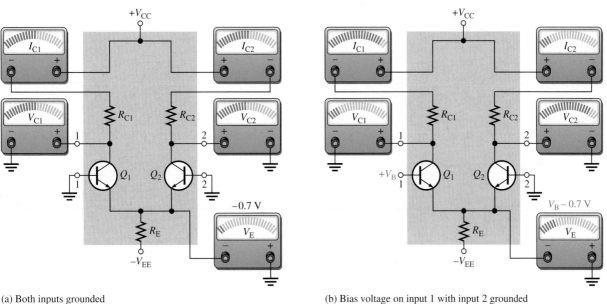

(a) Both inputs grounded

(b) Bias voltage on input 1 with input 2 grounded

(c) Bias voltage on input 2 with input 1 grounded

FIGURE 6–5

Basic operation of a differential amplifier (ground is zero volts) showing relative changes in currents and voltages.

careful process control during manufacturing so that their dc emitter currents are the same when there is no input signal. Thus,

$$I_{E1} = I_{E2}$$

Since both emitter currents combine through R_E,

$$I_{E1} = I_{E2} = \frac{I_{R_E}}{2}$$

where

$$I_{R_E} = \frac{V_E - V_{EE}}{R_E}$$

Based on the approximation that $I_C \cong I_E$, it can be stated that

$$I_{C1} = I_{C2} \cong \frac{I_{R_E}}{2}$$

Since both collector currents and both collector resistors are equal (when the input voltage is zero),

$$V_{C1} = V_{C2} = V_{CC} - I_{C1}R_{C1}$$

This condition is illustrated in Figure 6–5(a).

Next, input 2 remains grounded, and a positive bias voltage is applied to input 1, as shown in Figure 6–5(b). The positive voltage on the base of Q_1 increases I_{C1} and raises the emitter voltage to

$$V_E = V_B - 0.7 \text{ V}$$

This action reduces the forward bias (V_{BE}) of Q_2 because its base is held at 0 V (ground), thus causing I_{C2} to decrease as indicated in part (b) of the diagram. The net result is that the increase in I_{C1} causes a decrease in V_{C1}, and the decrease in I_{C2} causes an increase in V_{C2}, as shown.

Finally, input 1 is grounded and a positive bias voltage is applied to input 2, as shown in Figure 6–5(c). The positive bias voltage causes Q_2 to conduct more, thus increasing I_{C2}. Also, the emitter voltage is raised. This reduces the forward bias of Q_1, since its base is held at ground, and causes I_{C1} to decrease. The result is that the increase in I_{C2} produces a decrease in V_{C2}, and the decrease in I_{C1} causes V_{C1} to increase, as shown.

Modes of Signal Operation

Single-Ended Input In the **single-ended mode**, one input is grounded and the signal voltage is applied only to the other input, as shown in Figure 6–6. In the case where the signal voltage is applied to input 1 as in part (a), an inverted, amplified signal voltage appears at output 1 as shown. Also, a signal voltage appears in phase at the emitter of Q_1. Since the emitters of Q_1 and Q_2 are common, the emitter signal becomes an input to Q_2, which functions as a common-base amplifier. The signal is amplified by Q_2 and appears, noninverted, at output 2. This action is illustrated in part (a).

In the case where the signal is applied to input 2 with input 1 grounded, as in Figure 6–6(b), an inverted, amplified signal voltage appears at output 2. In this situation, Q_1 acts as a common-base amplifier, and a noninverted, amplified signal appears at output 1. This action is illustrated in part (b) of the figure.

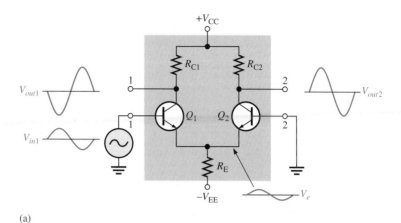

(a)

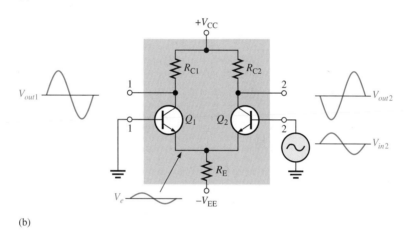

(b)

FIGURE 6–6
Single-ended operation of a differential amplifier.

Differential Input   In the **differential mode**, two opposite-polarity (out-of-phase) signals are applied to the inputs, as shown in Figure 6–7(a). This type of operation is also referred to as *double-ended*. As you will see, each input affects the outputs.

Figure 6–7(b) shows the output signals due to the signal on input 1 acting alone as a single-ended input. Figure 6–7(c) shows the output signals due to the signal on input 2 acting alone as a single-ended input. Notice in parts (b) and (c) that the signals on output 1 are of the same polarity. The same is also true for output 2. By superimposing both output 1 signals and both output 2 signals, you get the total differential operation, as shown in Figure 6–7(d).

Common-Mode Input   One of the most important aspects of the operation of a differential amplifier can be seen by considering the **common-mode** condition where signal voltages of the same phase, frequency, and amplitude are applied to the two inputs, as shown in Figure 6–8(a). Again, by considering each input signal as acting alone, you can understand the basic operation.

Figure 6–8(b) shows the output signals due to the signal on only input 1, and Figure 6–8(c) shows the output signals due to the signal on only input 2. Notice that the corresponding signals on output 1 are of the opposite polarity, and so are the ones on output 2. When the input signals are applied to both inputs, the outputs are superimposed and they cancel, resulting in a zero output voltage, as shown in Figure 6–8(d).

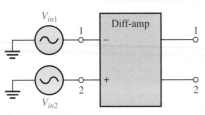

(a) Differential inputs

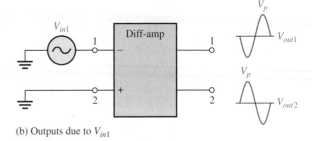

(b) Outputs due to V_{in1}

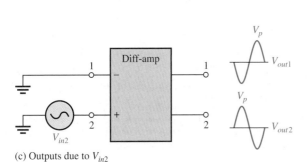

(c) Outputs due to V_{in2}

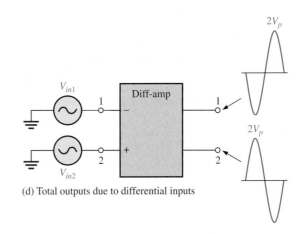

(d) Total outputs due to differential inputs

FIGURE 6–7
Differential operation of a differential amplifier.

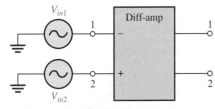

(a) Common-mode inputs

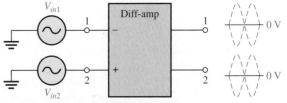

(b) Outputs due to V_{in1}

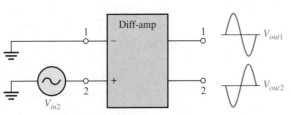

(c) Outputs due to V_{in2}

(d) Outputs cancel when common-mode signals are applied. Output
signals of equal amplitude but opposite phase cancel, producing
0 V on each output.

FIGURE 6–8
Common-mode operation of a differential amplifier.

This action is called *common-mode rejection*. Its importance lies in the situation where an unwanted signal appears commonly on both diff-amp inputs. Common-mode rejection means that this unwanted signal will not appear on the outputs to distort the desired signal. Common-mode signals (noise) generally are the result of the pick-up of radiated energy on the input lines, from adjacent lines, or the 60 Hz power line, or other sources.

Common-Mode Rejection Ratio

Wanted signals appear on only one input or with opposite polarities on both input lines. These wanted signals are amplified and appear on the outputs as previously discussed. Unwanted signals (noise) appearing with the same polarity on both input lines are essentially cancelled by the diff-amp and do not appear on the outputs. The measure of an amplifier's ability to reject common-mode signals is a parameter called the **common-mode rejection ratio (CMRR)**.

Ideally, a differential amplifier provides a very high gain for desired signals (single-ended or differential) and zero gain for common-mode signals. Practical diff-amps, however, do exhibit a very small common-mode gain (usually much less than 1), while providing a high differential voltage gain (usually several thousand). The higher the differential gain with respect to the common-mode gain, the better the performance of the diff-amp in terms of rejection of common-mode signals. This suggests that a good measure of the diff-amp's performance in rejecting unwanted common-mode signals is the ratio of the differential gain $A_{v(d)}$ to the common-mode gain, A_{cm}. This ratio is the common-mode rejection ratio, CMRR.

$$CMRR = \frac{A_{v(d)}}{A_{cm}} \qquad \textbf{(6–1)}$$

The higher the CMRR, the better. A very high value of CMRR means that the differential gain $A_{v(d)}$ is high and the common-mode gain A_{cm} is low.

The CMRR is often expressed in decibels (dB) as

$$CMRR' = 20 \log\left(\frac{A_{v(d)}}{A_{cm}}\right) \qquad \textbf{(6–2)}$$

EXAMPLE 6–1

A certain diff-amp has a differential voltage gain of 2000 and a common-mode gain of 0.2. Determine the CMRR and express it in decibels.

Solution $A_{v(d)} = 2000$, and $A_{cm} = 0.2$. Therefore,

$$CMRR = \frac{A_{v(d)}}{A_{cm}} = \frac{2000}{0.2} = \textbf{10,000}$$

Expressed in dB,

$$CMRR' = 20 \log(10,000) = \textbf{80 dB}$$

*Practice Exercise** Determine the CMRR and express it in dB for an amplifier with a differential voltage gain of 8500 and a common-mode gain of 0.25.

---

* Answers are at the end of the chapter.

A CMRR of 10,000, for example, means that the desired input signal (differential) is amplified 10,000 times more than the unwanted noise (common-mode). So, as an example, if the amplitudes of the differential input signal and the common-mode noise are equal, the desired signal will appear on the output 10,000 times greater in amplitude than the noise. Thus, the noise or interference has been essentially eliminated.

Example 6–2 illustrates further the idea of common-mode rejection and the general signal operation of the differential amplifier.

EXAMPLE 6–2

The differential amplifier shown in Figure 6–9 has a differential voltage gain of 2500 and a CMRR of 30,000. In part (a), a single-ended input signal of 500 μV rms is applied. At the same time, a 1 V, 60 Hz common-mode interference signal appears on both inputs as a result of radiated pick-up from the ac power system. In part (b), differential input signals of 500 μV rms each are applied to the inputs. The common-mode interference is the same as in part (a).

(a) Determine the common-mode gain.
(b) Express the CMRR in dB.
(c) Determine the rms output signal for Figure 6–9(a) and (b).
(d) Determine the rms interference voltage on the output.

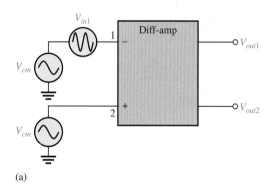

(a)

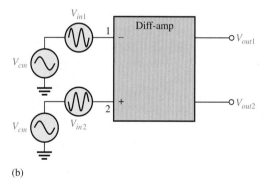

(b)

FIGURE 6–9

Solution

(a) $\text{CMRR} = \dfrac{A_{v(d)}}{A_{cm}}$. Therefore,

$$A_{cm} = \frac{A_{v(d)}}{\text{CMRR}} = \frac{2500}{30,000} = \mathbf{0.083}$$

(b) $\text{CMRR}' = 20 \log(30,000) = \mathbf{89.5\ dB}$

(c) In Figure 6–9(a), the differential input voltage is the difference between the voltage on input 1 and that on input 2. Since input 2 is grounded, its voltage is zero. Therefore,

$$V_{in(d)} = V_{in1} - V_{in2} = 500\ \mu\text{V} - 0\ \text{V} = 500\ \mu\text{V}$$

The output signal voltage in this case is taken at output 1.

$$V_{out1} = A_{v(d)}V_{in(d)} = (2500)(500\ \mu\text{V}) = \mathbf{1.25\ V\ rms}$$

In Figure 6–9(b), the differential input voltage is the difference between the two opposite-polarity, 500 μV signals.

$$V_{in(d)} = V_{in1} - V_{in2} = 500 \ \mu V - (-500 \ \mu V) = 1000 \ \mu V = 1 \ mV$$

The output voltage signal is

$$V_{out1} = A_{v(d)}V_{in(d)} = (2500)(1 \ mV) = \textbf{2.5 V rms}$$

This shows that a differential input (two opposite-polarity signals) results in a gain that is double that for a single-ended input.

(d) The common-mode input is 1 V rms. The common-mode gain A_{cm} is 0.083. The interference (common-mode) voltage on the output is, therefore,

$$A_{cm} = \frac{V_{out(cm)}}{V_{in(cm)}}$$

$$V_{out(cm)} = A_{cm}V_{in(cm)} = (0.083)(1 \ V) = \textbf{0.083 V}$$

Practice Exercise   The amplifier in Figure 6–9 has a differential voltage gain of 4200 and a CMRR of 25,000. For the same single-ended and differential input signals as described in the example: **(a)** Find A_{cm}. **(b)** Express the CMRR in dB. **(c)** Determine the rms output signal for parts (a) and (b) of the figure. **(d)** Determine the rms interference (common-mode) voltage appearing on the output.

Internal Block Diagram of an Op-Amp

A typical op-amp is made up of three types of amplifier circuits: a *differential amplifier,* a *voltage amplifier,* and a *push-pull amplifier,* as shown in Figure 6–10.

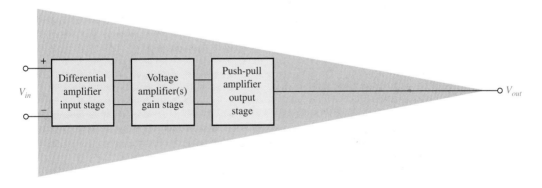

FIGURE 6–10
Basic internal arrangement of an op-amp.

A differential amplifier is the input stage for the op-amp; it has two inputs and provides amplification of the difference voltage between the two inputs. The voltage amplifier is usually a class A amplifier that provides additional op-amp gain. Some op-amps may have more than one voltage amplifier stage. A push-pull class B amplifier is used for the output stage.

6–2 REVIEW QUESTIONS

1. Distinguish between differential and single-ended inputs.

2. Define *common-mode rejection?*

3. For a given value of differential gain, does a higher CMRR result in a higher or lower common-mode gain?

6–3 ■ OP-AMP DATA SHEET PARAMETERS

In this section, several important op-amp parameters are defined. (These are listed in the objectives that follow.) Also several IC op-amps are compared in terms of these parameters.

After completing this section, you should be able to

❑ Discuss several op-amp parameters
   ❑ Define *input offset voltage*
   ❑ Discuss input offset voltage drift with temperature
   ❑ Define *input bias current*
   ❑ Define *input impedance*
   ❑ Define *input offset current*
   ❑ Define *output impedance*
   ❑ Discuss common-mode input voltage range
   ❑ Discuss open-loop voltage gain
   ❑ Define *common-mode rejection ratio*
   ❑ Define *slew rate*
   ❑ Discuss frequency response
   ❑ Compare the parameters of several types of IC op-amps

Input Offset Voltage

The ideal op-amp produces zero volts out for zero volts in. In a practical op-amp, however, a small dc voltage, $V_{OUT(error)}$, appears at the output when no differential input voltage is applied. Its primary cause is a slight mismatch of the base-emitter voltages of the differential input stage of an op-amp, as illustrated in Figure 6–11(a).

The output voltage of the differential input stage is expressed as

$$V_{OUT(error)} = I_{C2}R_C - I_{C1}R_C$$

A small difference in the base-emitter voltages of Q_1 and Q_2 causes a small difference in the collector currents. This results in a nonzero value of V_{OUT}, which is the error voltage. (The collector resistors are equal.)

As specified on an op-amp data sheet, the **input offset voltage** (V_{OS}) is the differential dc voltage required between the inputs to force the differential output to zero volts. V_{OS} is demonstrated in Figure 6–11(b). Typical values of input offset voltage are in the range of 2 mV or less. In the ideal case, it is 0 V.

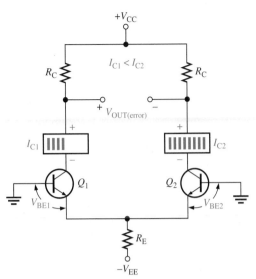

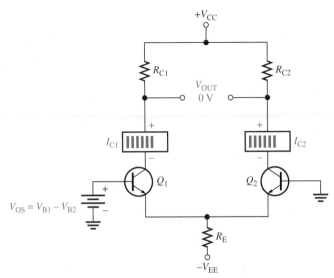

(a) A V_{BE} mismatch (V_{BE1} different than V_{BE2}) causes a small output error voltage.

(b) The input offset voltage is the difference in the voltage between the inputs that is necessary to eliminate the output error voltage (makes $V_{OUT} = 0$).

FIGURE 6–11
Illustration of input offset voltage, V_{OS}.

Input Offset Voltage Drift with Temperature

The **input offset voltage drift** is a parameter related to V_{OS} that specifies how much change occurs in the input offset voltage for each degree change in temperature. Typical values range anywhere from about 5 μV per degree Celsius to about 50 μV per degree Celsius. Usually, an op-amp with a higher nominal value of input offset voltage exhibits a higher drift.

Input Bias Current

You have seen that the input terminals of a bipolar differential amplifier are the transistor bases and, therefore, the input currents are the base currents.

The **input bias current** is the dc current required by the inputs of the amplifier to properly operate the first stage. By definition, the input bias current is the *average* of both input currents and is calculated as follows:

$$I_{BIAS} = \frac{I_1 + I_2}{2}$$

The concept of input bias current is illustrated in Figure 6–12.

FIGURE 6–12
Input bias current is the average of the two op-amp input currents.

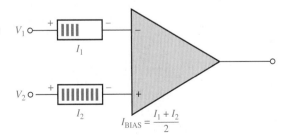

Input Impedance

Two basic ways of specifying the input impedance of an op-amp are the differential and the common mode. The **differential input impedance** is the total resistance between the inverting and the noninverting inputs, as illustrated in Figure 6–13(a). Differential input impedance is measured by determining the change in bias current for a given change in differential input voltage. The **common-mode input impedance** is the resistance between each input and ground and is measured by determining the change in bias current for a given change in common-mode input voltage. It is depicted in Figure 6–13(b).

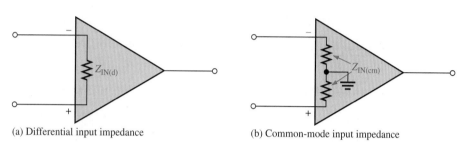

(a) Differential input impedance   (b) Common-mode input impedance

FIGURE 6–13
Op-amp input impedance.

Input Offset Current

Ideally, the two input bias currents are equal, and thus their difference is zero. In a practical op-amp, however, the bias currents are not exactly equal.

The **input offset current,** I_{OS}, is the difference of the input bias currents, expressed as an absolute value.

$$I_{OS} = |I_1 - I_2|$$

Actual magnitudes of offset current are usually at least an order of magnitude (ten times) less than the bias current. In many applications, the offset current can be neglected. However, high-gain, high-input impedance amplifiers should have as little I_{OS} as possible because the difference in currents through large input resistances develops a substantial offset voltage, as shown in Figure 6–14.

The offset voltage developed by the input offset current is

$$V_{OS} = I_1 R_{in} - I_2 R_{in} = (I_1 - I_2)R_{in} = I_{OS}R_{in}$$

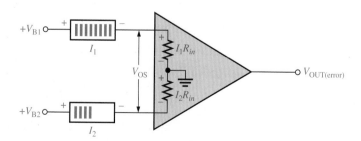

FIGURE 6–14
Effect of input offset current.

The error created by I_{OS} is amplified by the gain A_v of the op-amp and appears in the output as

$$V_{OUT(error)} = A_v I_{OS} R_{in}$$

A change in offset current with temperature affects the error voltage. Values of temperature coefficient for the offset current in the range of 0.5 nA per degree Celsius are common.

Output Impedance

Output impedance is the resistance viewed from the output terminal of the op-amp, as indicated in Figure 6–15.

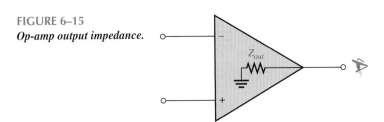

FIGURE 6–15
Op-amp output impedance.

Common-Mode Input Voltage Range

All op-amps have limitations on the range of voltages over which they will operate. The **common-mode input voltage range** is the range of input voltages which, when applied to both inputs, will not cause clipping or other output distortion. Many op-amps have common-mode ranges of no more than ± 10 V with dc supply voltages of ± 15 V, while in others the output can go as high as the supply voltages (this is called rail-to-rail).

Open-Loop Voltage Gain

The **open-loop voltage gain**, A_{ol}, of an op-amp is the internal voltage gain of the device and represents the ratio of output voltage to input voltage when there are no external components. The open-loop voltage gain is set entirely by the internal design. Open-loop voltage gain can range up to 200,000 and is *not a well-controlled parameter*. Data sheets often refer to the open-loop voltage gain as the *large-signal voltage gain*.

Common-Mode Rejection Ratio for an Op-Amp

The *common-mode rejection ratio* (CMRR) was discussed in conjunction with the diff-amp. Similarly, for an op-amp, CMRR is a measure of an op-amp's ability to reject common-mode signals. An infinite value of CMRR means that the output is zero when the same signal is applied to both inputs (common-mode).

An infinite CMRR is never achieved in practice, but a good op-amp does have a very high value of CMRR. As previously mentioned, common-mode signals are undesired interference voltages such as 60 Hz power-supply ripple and noise voltages due to pick-up of radiated energy. A high CMRR enables the op-amp to virtually eliminate these interference signals from the output.

The accepted definition of CMRR for an op-amp is the open-loop voltage gain (A_{ol}) divided by the common-mode gain.

$$CMRR = \frac{A_{ol}}{A_{cm}} \tag{6–3}$$

It is commonly expressed in decibels as follows:

$$CMRR' = 20 \log\left(\frac{A_{ol}}{A_{cm}}\right) \tag{6–4}$$

EXAMPLE 6–3

A certain op-amp has an open-loop voltage gain of 100,000 and a common-mode gain of 0.25. Determine the CMRR and express it in decibels.

Solution

$$CMRR = \frac{A_{ol}}{A_{cm}} = \frac{100,000}{0.25} = \textbf{400,000}$$

$$CMRR' = 20 \log(400,000) = \textbf{112 dB}$$

Practice Exercise If a particular op-amp has a CMRR' of 90 dB and a common-mode gain of 0.4, what is the open-loop voltage gain?

Slew Rate

The maximum rate of change of the output voltage in response to a step input voltage is the **slew rate** of an op-amp. The slew rate is dependent upon the high-frequency response of the amplifier stages within the op-amp.

Slew rate is measured with an op-amp connected as shown in Figure 6–16(a). This particular op-amp connection is a unity-gain, noninverting configuration which will be discussed later. It gives a worst-case (slowest) slew rate. Recall that the high-frequency components of a voltage step are contained in the rising edge and that the upper critical fre-

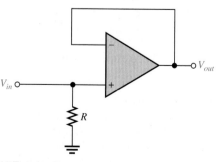

(a) Test circuit

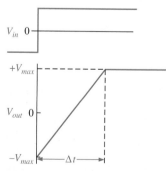

(b) Step input voltage and the resulting output voltage

FIGURE 6–16
Slew rate measurement.

quency of an amplifier limits its response to a step input. The lower the upper critical frequency is, the more gradual the slope on the output for a step input.

A pulse is applied to the input as shown in Figure 6–16(b), and the ideal output voltage is measured as indicated. The width of the input pulse must be sufficient to allow the output to "slew" from its lower limit to its upper limit, as shown. As you can see, a certain time interval, Δt, is required for the output voltage to go from its lower limit $-V_{max}$ to its upper limit $+V_{max}$, once the input step is applied. The slew rate is expressed as

$$\text{Slew rate} = \frac{\Delta V_{out}}{\Delta t} \tag{6–5}$$

where $\Delta V_{out} = +V_{max} - (-V_{max})$. The unit of slew rate is volts per microsecond (V/μs).

EXAMPLE 6–4

The output voltage of a certain op-amp appears as shown in Figure 6–17 in response to a step input. Determine the slew rate.

FIGURE 6–17

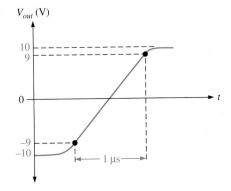

Solution   The output goes from the lower to the upper limit in 1 μs. Since this response is not ideal, the limits are taken at the 90% points, as indicated, So, the upper limit is +9 V and the lower limit is −9 V. The slew rate is

$$\text{Slew rate} = \frac{\Delta V}{\Delta t} = \frac{+9\text{ V} - (-9\text{ V})}{1\text{ μs}} = \textbf{18 V/μs}$$

Practice Exercise   When a pulse is applied to an op-amp, the output voltage goes from −8 V to +7 V in 0.75 μs. What is the slew rate?

Frequency Response

The internal amplifier stages that make up an op-amp have voltage gains limited by junction capacitances. Although the differential amplifiers used in op-amps are somewhat different from the basic amplifiers discussed, the same principles apply. An op-amp has no internal coupling capacitors, however; therefore, the low-frequency response extends down to dc (0 Hz). Frequency-related characteristics of op-amps will be discussed in the next chapter.

Comparison of Op-Amp Parameters

Table 6–1 provides a comparison of values of some of the parameters just described for several common IC op-amps. Any values not listed were not given on the manufacturer's data sheet.

TABLE 6–1

Op-amp	Input offset voltage (mV) (max)	Input bias current (nA) (max)	Input impedance (MΩ) (min)	Open-loop gain (typ)	Slew rate (V/μs) (typ)	CMRR′ (dB) (min)	Comment
LM741[1]	6	500	0.3	200,000	0.5	70	Industry standard
LM101A[1]	7.5	250	1.5	160,000	—	80	General-purpose
OP113[2]	0.075	600	—	2,400,000	1.2	100	Low noise, low drift
OP177[2]	0.01	1.5	26	12,000,000	0.3	130	Ultra precision
OP184[2]	0.065	350	—	240,000	2.4	60	Precision, rail-to-rail[3]
AD8009[2]	5	150	—	—	5500	50	BW=700 MHz, ultra fast, low distortion, current feedback
AD8041[2]	7	2000	0.16	56,000	160	74	BW=160 MHz, rail-to-rail
AD8055[2]	5	1200	10	3500	1400	82	Very fast voltage feedback

[1]Data sheet available at http://www.national.com

[2]Data sheet available at http://www.analogdevices.com

[3]Rail-to-rail means that the output voltage can go as high as the supply voltages.

Other Features

Most available op-amps have three important features: short-circuit protection, no latch-up, and input offset nulling. Short-circuit protection keeps the circuit from being damaged if the output becomes shorted, and the no latch-up feature prevents the op-amp from hanging up in one output state (high or low voltage level) under certain input conditions. Input offset nulling is achieved by an external potentiometer that sets the output voltage at precisely zero with zero input.

6–3 REVIEW QUESTIONS

1. List ten or more op-amp parameters.
2. Which two parameters, not including frequency response, are frequency dependent?

6–4 ■ NEGATIVE FEEDBACK

Negative feedback is one of the most useful concepts in electronics, particularly in op-amp applications. Negative feedback is the process whereby a portion of the output voltage of an amplifier is returned to the input with a phase angle that opposes (or subtracts from) the input signal.

After completing this section, you should be able to

❏ Explain negative feedback in op-amp circuits
   ❏ Describe the effects of negative feedback
   ❏ Discuss why negative feedback is used

Negative feedback is illustrated in Figure 6–18. The inverting $(-)$ input effectively makes the feedback signal 180° out of phase with the input signal. The op-amp has extremely high gain and amplifies the *difference* in the signals applied to the inverting and noninverting inputs. A very tiny difference in these two signals is all the op-amp needs to produce the required output. *When negative feedback is present, the noninverting and inverting inputs are nearly identical.* This concept can help you figure out what signal to expect in many op-amp circuits.

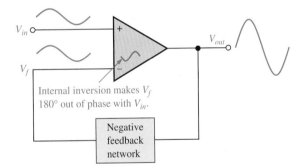

FIGURE 6–18
Illustration of negative feedback.

V_{in}

V_f

Internal inversion makes V_f 180° out of phase with V_{in}.

V_{out}

Negative feedback network

Now let's review how negative feedback works and why the signals at the inverting and noninverting terminals are nearly identical when negative feedback is used. Assume a 1.0 V input signal is applied to the noninverting terminal and the open-loop gain of the op-amp is 100,000. The amplifier responds to the voltage at its noninverting input terminal and moves the output toward saturation. Immediately, a fraction of this output is returned to the inverting terminal through the feedback path. But if the feedback signal ever reaches 1.0 V, there is nothing left for the op-amp to amplify! Thus, the feedback signal tries (but never quite succeeds) in matching the input signal. The gain is controlled by the amount of feedback used. When you are troubleshooting an op-amp circuit with negative feedback present, remember that the two inputs will look identical on a scope but in fact are very slightly different.

Now suppose something happens that reduces the internal gain of the op-amp. This causes the output signal to drop a small amount, returning a smaller signal to the inverting input via the feedback path. This means the difference between the signals is larger than it was. The output increases, compensating for the original drop in gain. The net change in the output is so small, it can hardly be measured. The main point is that any variation in the amplifier is immediately compensated for by the negative feedback, resulting in a very stable, predictable output.

Why Use Negative Feedback?

As you have seen, the inherent open-loop gain of a typical op-amp is very high (usually greater than 100,000). Therefore, an extremely small difference in the two input voltages drives the op-amp into its saturated output states. In fact, even the input offset voltage of the op-amp can drive it into saturation. For example, assume $V_{in} = 1$ mV and $A_{ol} = 100,000$. Then,

$$V_{in}A_{ol} = (1 \text{ mV})(100,000) = 100 \text{ V}$$

Since the output level of an op-amp can never reach 100 V, it is driven into saturation and the output is limited to its maximum output levels, as illustrated in Figure 6–19 for both a positive and a negative input voltage of 1 mV.

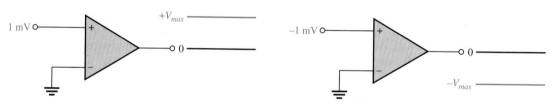

FIGURE 6–19

Without negative feedback, an extremely small difference in the two input voltages drives the op-amp to its output limits and it becomes nonlinear.

The usefulness of an op-amp operated in this manner is severely restricted and is generally limited to comparator applications (to be studied in Chapter 8). With negative feedback, the overall closed-loop voltage gain (A_{cl}) can be reduced and controlled so that the op-amp can function as a linear amplifier. In addition to providing a controlled, stable voltage gain, negative feedback also provides for control of the input and output impedances and amplifier bandwidth. Table 6–2 summarizes the general effects of negative feedback on op-amp performance.

TABLE 6–2

	Voltage Gain	Input Z	Output Z	Bandwidth
Without negative feedback	A_{ol} is too high for linear amplifier applications	Relatively high (see Table 6–1)	Relatively low	Relatively narrow (because the gain is so high)
With negative feedback	A_{cl} is set to desired value by the feedback network	Can be increased or reduced to a desired value depending on type of circuit	Can be reduced to a desired value	Significantly wider

6–4 REVIEW QUESTIONS

1. What are the benefits of negative feedback in an op-amp circuit?
2. Why is it necessary to reduce the gain of an op-amp from its open-loop value?
3. When troubleshooting an op-amp circuit in which negative feedback is present, what do you expect to observe on the input terminals?

6–5 ■ OP-AMP CONFIGURATIONS WITH NEGATIVE FEEDBACK

In this section, we will discuss three basic ways in which an op-amp can be connected using negative feedback to stabilize the gain and increase frequency response. As mentioned, the extremely high open-loop gain of an op-amp creates an unstable situation because a small noise voltage on the input can be amplified to a point where the amplifier is driven out of its linear region. Also, unwanted oscillations can occur. In addition, the open-loop gain parameter of an op-amp can vary greatly from one device to the next. Negative feedback takes a portion of the output and applies it back out of phase with the input, creating an effective reduction in gain. This closed-loop gain is usually much less than the open-loop gain and independent of it.

After completing this section, you should be able to

❏ Analyze three op-amp configurations
  ❏ Identify the noninverting amplifier configuration
  ❏ Determine the voltage gain of a noninverting amplifier
  ❏ Identify the voltage-follower configuration
  ❏ Identify the inverting amplifier configuration
  ❏ Determine the voltage gain of an inverting amplifier

Closed-Loop Voltage Gain, A_{cl}

The **closed-loop voltage gain** is the voltage gain of an op-amp with negative feedback. The amplifier configuration consists of the op-amp and an external feedback network that connects the output to the inverting input. The closed-loop voltage gain is then determined by the component values in the feedback network and can be precisely controlled by them.

Noninverting Amplifier

An op-amp connected in a closed-loop configuration as a **noninverting amplifier** is shown in Figure 6–20. The input signal is applied to the noninverting (+) input. A portion of the output is applied back to the inverting (−) input through the feedback network. This constitutes negative feedback. The feedback fraction, B, is the portion of the output returned to the inverting input and determines the gain of the amplifier as you will see. This smaller feedback voltage, V_f, can be written

$$V_f = BV_{out}$$

FIGURE 6–20
Noninverting amplifier.

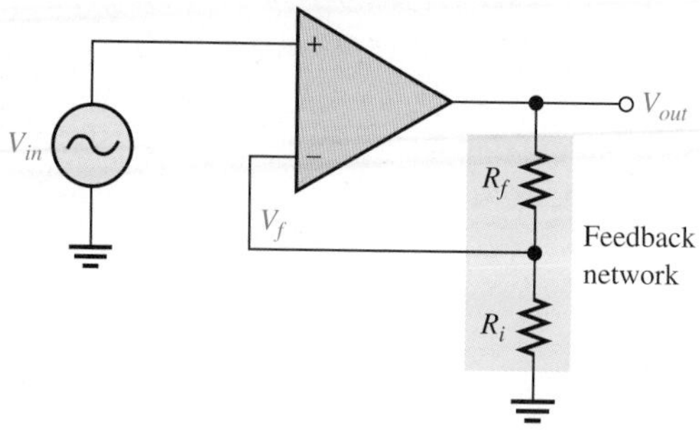

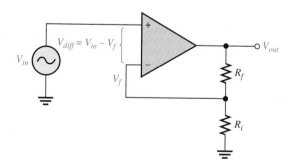

The differential voltage, V_{diff}, between the op-amp's input terminals is illustrated in Figure 6–21 and can be expressed as

$$V_{diff} = V_{in} - V_f$$

This input differential voltage is forced to be very small as a result of the negative feedback and the high open-loop gain, A_{ol}. Therefore, a close approximation is

$$V_{in} \cong V_f$$

By substitution,

$$V_{in} \cong BV_{out}$$

Rearranging,

$$\frac{V_{out}}{V_{in}} \cong \frac{1}{B}$$

The ratio of the output voltage to the input voltage is the closed-loop gain. This result shows that the closed-loop gain for the noninverting amplifier, $A_{cl(NI)}$, is approximately

$$A_{cl(NI)} = \frac{V_{out}}{V_{in}} \cong \frac{1}{B}$$

FIGURE 6–21
Differential input, $V_{in} - V_f$.

The feedback fraction is determined by R_i and R_f, which form a voltage-divider network. The fraction of the output voltage, V_{out}, that is returned to the inverting input is found by applying the voltage-divider rule to the feedback network.

$$V_{in} \cong BV_{out} \cong \left(\frac{R_i}{R_i + R_f}\right)V_{out}$$

Rearranging,

$$\frac{V_{out}}{V_{in}} = \left(\frac{R_i + R_f}{R_i}\right)$$

which can be expressed as follows:

$$A_{cl(NI)} = \frac{R_f}{R_i} + 1 \qquad \textbf{(6–6)}$$

Equation (6–6) shows that the closed-loop voltage gain, $A_{cl(NI)}$, of the noninverting (NI) amplifier is not dependent on the op-amp's open-loop gain but can be set by selecting values of R_i and R_f. This equation is based on the assumption that the open-loop gain is very high compared to the ratio of the feedback resistors, causing the input differential voltage, V_{diff}, to be very small. In nearly all practical circuits, this is an excellent assumption.

For those rare cases where a more exact equation is necessary, the output voltage can be expressed as

$$V_{out} = V_{in}\left(\frac{A_{ol}}{1 + A_{ol}B}\right)$$

The following formula gives the exact solution of the closed-loop gain:

$$A_{cl(NI)} = \frac{V_{out}}{V_{in}} = \left(\frac{A_{ol}}{1 + A_{ol}B}\right)$$

EXAMPLE 6–5

Determine the closed-loop voltage gain of the amplifier in Figure 6–22.

FIGURE 6–22

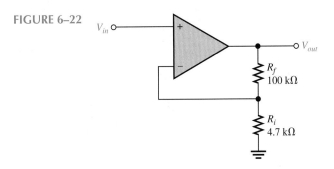

Solution This is a noninverting op-amp configuration. Therefore, the closed-loop voltage gain is

$$A_{cl(NI)} = \frac{R_f}{R_i} + 1 = \frac{100\,\text{k}\Omega}{4.7\,\text{k}\Omega} + 1 = \textbf{22.3}$$

Practice Exercise If R_f in Figure 6–22 is increased to 150 kΩ, determine the closed-loop gain.

Voltage-Follower The **voltage-follower** configuration is a special case of the noninverting amplifier where all of the output voltage is fed back to the inverting input by a straight connection, as shown in Figure 6–23. As you can see, the straight feedback connection has a voltage gain of approximately 1. The closed-loop voltage gain of a noninverting amplifier is $1/B$ as previously derived. Since $B = 1$, the closed-loop gain of the voltage-follower is

$$A_{cl(VF)} = 1 \qquad\qquad (6\text{--}7)$$

The most important features of the voltage-follower configuration are its very high input impedance and its very low output impedance. These features make it a nearly ideal buffer amplifier for interfacing high-impedance sources and low-impedance loads. This is discussed further in Section 6–6.

FIGURE 6–23
Op-amp voltage-follower.

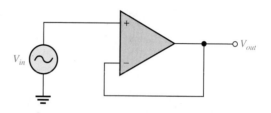

Inverting Amplifier

An op-amp connected as an **inverting amplifier** with a controlled amount of voltage gain is shown in Figure 6–24. The input signal is applied through a series input resistor (R_i) to the inverting input. Also, the output is fed back through R_f to the inverting input. The noninverting input is grounded.

FIGURE 6–24
Inverting amplifier.

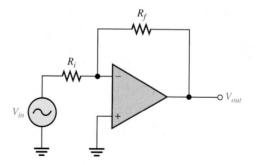

At this point, the ideal op-amp parameters mentioned earlier are useful in simplifying the analysis of this circuit. In particular, the concept of infinite input impedance is of great value. An infinite input impedance implies that there is *no* current out of the inverting input. If there is no current through the input impedance, then there must be *no* voltage drop between the inverting and noninverting inputs. This means that the voltage at the inverting ($-$) input is zero because the noninverting ($+$) input is grounded. This zero voltage at the inverting input terminal is referred to as *virtual ground*. This condition is illustrated in Figure 6–25(a).

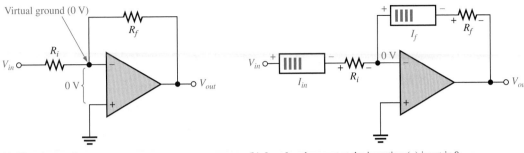

(a) Virtual ground

(b) $I_{in} = I_f$ and current at the inverting (−) input is 0.

FIGURE 6–25
Virtual ground concept and closed-loop voltage gain development for the inverting amplifier.

Since there is no current at the inverting input, the current through R_i and the current through R_f are equal, as shown in Figure 6–25(b).

$$I_{in} = I_f$$

The voltage across R_i equals V_{in} because of virtual ground on the other side of the resistor. Therefore,

$$I_{in} = \frac{V_{in}}{R_i}$$

Also, the voltage across R_f equals $-V_{out}$ because of virtual ground, and therefore,

$$I_f = \frac{-V_{out}}{R_f}$$

Since $I_f = I_{in}$,

$$\frac{-V_{out}}{R_f} = \frac{V_{in}}{R_i}$$

Rearranging the terms,

$$\frac{V_{out}}{V_{in}} = -\frac{R_f}{R_i}$$

Of course, V_{out}/V_{in} is the overall gain of the inverting amplifier.

$$A_{cl(I)} = -\frac{R_f}{R_i} \qquad \textbf{(6–8)}$$

Equation (6–8) shows that the closed-loop voltage gain $A_{cl(I)}$ of the inverting amplifier is the ratio of the feedback resistance R_f to the input resistance R_i. *The closed-loop gain is independent of the op-amp's internal open-loop gain.* Thus, the negative feedback stabilizes the voltage gain. The negative sign indicates inversion.

EXAMPLE 6–6

Given the op-amp configuration in Figure 6–26, determine the value of R_f required to produce a closed-loop voltage gain of -100.

FIGURE 6–26

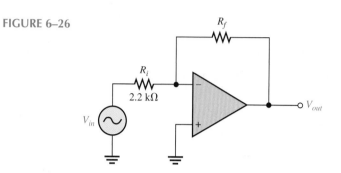

Solution Knowing that $R_i = 2.2 \text{ k}\Omega$ and $A_{cl(I)} = -100$, calculate R_f as follows:

$$A_{cl(I)} = -\frac{R_f}{R_i}$$

$$R_f = -A_{cl(I)}R_i = -(-100)(2.2 \text{ k}\Omega) = \mathbf{220 \text{ k}\Omega}$$

Practice Exercise

(a) If R_i is changed to 2.7 kΩ in Figure 6–26, what value of R_f is required to produce a closed-loop gain of -25?

(b) If R_f failed open, what would you expect to see at the output?

6–5 REVIEW QUESTIONS

1. What is the main purpose of negative feedback?

2. The closed-loop voltage gain of each of the op-amp configurations discussed is dependent on the internal open-loop voltage gain of the op-amp. (True or False)

3. The attenuation of the negative feedback network of a noninverting op-amp configuration is 0.02. What is the closed-loop gain of the amplifier?

6–6 ■ OP-AMP IMPEDANCES

In this section, you will see how a negative feedback connection affects the input and output impedances of an op-amp. The effects on both inverting and noninverting amplifiers are examined.

After completing this section, you should be able to

❑ Describe impedances of the three op-amp configurations
  ❑ Determine input and output impedances of a noninverting amplifier
  ❑ Determine input and output impedances of a voltage-follower
  ❑ Determine input and output impedances of an inverting amplifier

Input Impedance of the Noninverting Amplifier

Recall that negative feedback causes the feedback voltage, V_f, to nearly equal the input voltage, V_{in}. The difference between the input and feedback voltage, V_{diff}, is approximately zero, and ideally, can be assumed to have this value. This assumption implies that the input signal current to the op-amp is also zero. Since the input impedance is the ratio of input voltage to input current, the input impedance of a noninverting amplifier is

$$Z_{in} = \frac{V_{in}}{I_{in}} \cong \frac{V_{in}}{0} = \text{infinity } (\infty)$$

For many practical circuits, this assumption is good for obtaining a basic idea of the operation. A more exact analysis takes into account the fact that the input signal current is not zero.

The exact input impedance of this op-amp configuration is developed with the aid of Figure 6–27. For this analysis, a small differential voltage, V_{diff}, is assumed to exist between the two inputs, as indicated. This means that you cannot assume the op-amp's input impedance to be infinite or the input current to be zero. The input voltage can be expressed as

$$V_{in} = V_{diff} + V_f$$

FIGURE 6–27

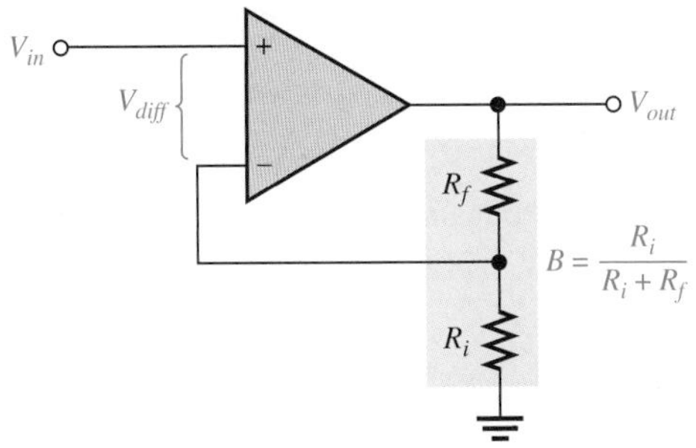

Substituting BV_{out} for V_f,

$$V_{in} = V_{diff} + BV_{out}$$

Since $V_{out} \cong A_{ol}V_{diff}$ (A_{ol} is the open-loop gain of the op-amp),

$$V_{in} = V_{diff} + A_{ol}BV_{diff} = (1 + A_{ol}B)V_{diff}$$

Because $V_{diff} = I_{in}Z_{in}$,

$$V_{in} = (1 + A_{ol}B)I_{in}Z_{in}$$

where Z_{in} is the open-loop input impedance of the op-amp (without feedback connections).

$$\frac{V_{in}}{I_{in}} = (1 + A_{ol}B)Z_{in}$$

V_{in}/I_{in} is the overall input impedance of the closed-loop noninverting configuration.

$$Z_{in(NI)} = (1 + A_{ol}B)Z_{in} \tag{6–9}$$

This equation shows that the input impedance of this amplifier configuration with negative feedback is much greater than the internal input impedance of the op-amp itself (without feedback).

Output Impedance of the Noninverting Amplifier

In addition to the input impedance, negative feedback also produces an advantage for the output impedance of an op-amp. The output impedance of an amplifier without feedback is relatively low. With feedback, the output impedance is even lower. For many applications, the assumption that the output impedance with feedback is zero will produce sufficient accuracy. That is,

$$Z_{out(\text{NI})} \cong 0$$

An exact analysis to find the output impedance with feedback is developed with the aid of Figure 6–28. By applying Kirchhoff's law to the output circuit,

$$V_{out} = A_{ol}V_{diff} - Z_{out}I_{out}$$

The differential input voltage is $V_{in} - V_f$; so, by assuming that $A_{ol}V_{diff} \gg Z_{out}I_{out}$, the output voltage can be expressed as

$$V_{out} \cong A_{ol}(V_{in} - V_f)$$

Substituting BV_{out} for V_f,

$$V_{out} \cong A_{ol}(V_{in} - BV_{out})$$

Remember, B is the attenuation of the negative feedback network. Expanding, factoring and rearranging terms,

$$A_{ol}V_{in} \cong V_{out} + A_{ol}BV_{out} = (1 + A_{ol}B)V_{out}$$

Since the output impedance of the noninverting configuration is $Z_{out(\text{NI})} = V_{out}/I_{out}$, you can substitute $I_{out}Z_{out(\text{NI})}$ for V_{out}; therefore,

$$A_{ol}V_{in} = (1 + A_{ol}B)I_{out}Z_{out(\text{NI})}$$

Dividing both sides of the above expression by I_{out} yields

$$\frac{A_{ol}V_{in}}{I_{out}} = (1 + A_{ol}B)Z_{out(\text{NI})}$$

The term on the left is the internal output impedance of the op-amp (Z_{out}) because, without feedback, $A_{ol}V_{in} = V_{out}$. Therefore,

$$Z_{out} = (1 + A_{ol}B)Z_{out(\text{NI})}$$

FIGURE 6–28

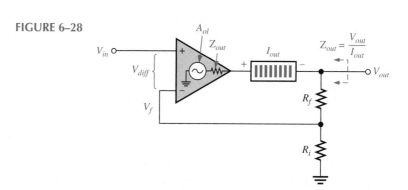

Thus,

$$Z_{out(NI)} = \frac{Z_{out}}{1 + A_{ol}B} \qquad (6\text{--}10)$$

This equation shows that the output impedance of this amplifier configuration with negative feedback is much less than the internal output impedance of the op-amp itself (without feedback) because it is divided by the factor $1 + A_{ol}B$.

EXAMPLE 6–7

(a) Determine the input and output impedances of the amplifier in Figure 6–29. The op-amp data sheet gives $Z_{in} = 2$ MΩ, $Z_{out} = 75$ Ω, and $A_{ol} = 200{,}000$.
(b) Find the closed-loop voltage gain.

FIGURE 6–29

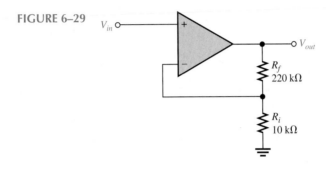

Solution
(a) The attenuation, B, of the feedback network is

$$B = \frac{R_i}{R_i + R_f} = \frac{10 \text{ k}\Omega}{230 \text{ k}\Omega} = 0.0435$$

$$Z_{in(NI)} = (1 + A_{ol}B)Z_{in} = [1 + (200{,}000)(0.0435)](2 \text{ M}\Omega)$$
$$= (1 + 8700)(2 \text{ M}\Omega) = 17{,}402 \text{ M}\Omega = \mathbf{17.4 \text{ G}\Omega}$$

$$Z_{out(NI)} = \frac{Z_{out}}{1 + A_{ol}B} = \frac{75 \text{ }\Omega}{1 + 8700} = 0.0086 \text{ }\Omega = \mathbf{8.6 \text{ m}\Omega}$$

(b) $A_{cl(NI)} = \dfrac{1}{B} = \dfrac{1}{0.0435} \cong \mathbf{23}$

Practice Exercise
(a) Determine the input and output impedances in Figure 6–29 for op-amp data sheet values of $Z_{in} = 3.5$ MΩ, $Z_{out} = 82$ Ω, and $A_{ol} = 135{,}000$.
(b) Find A_{cl}.

Voltage-Follower Impedances

Since the voltage-follower is a special case of the noninverting configuration, the same impedance formulas are used with $B = 1$.

$$Z_{in(VF)} = (1 + A_{ol})Z_{in} \tag{6-11}$$

$$Z_{out(VF)} = \frac{Z_{out}}{1 + A_{ol}} \tag{6-12}$$

As you can see, the voltage-follower input impedance is greater for a given A_{ol} and Z_{in} than for the noninverting configuration with the voltage-divider feedback network. Also, its output impedance is much smaller because B is normally much smaller than 1 for a noninverting configuration.

EXAMPLE 6–8

The same op-amp as in Example 6–7 is used in a voltage-follower configuration. Determine the input and output impedances.

Solution Since $B = 1$,

$$Z_{in(VF)} = (1 + A_{ol})Z_{in} = (1 + 200,000)(2 \text{ M}\Omega) = \textbf{400 G}\boldsymbol{\Omega}$$

$$Z_{out(VF)} = \frac{Z_{out}}{1 + A_{ol}} = \frac{75 \ \Omega}{1 + 200,000} = \textbf{375 } \boldsymbol{\mu\Omega}$$

Notice that $Z_{in(VF)}$ is much greater than $Z_{in(NI)}$, and $Z_{out(VF)}$ is much less than $Z_{out(NI)}$ from Example 6–7.

Practice Exercise If the op-amp in this example is replaced with one having a higher open-loop gain, how are the input and output impedances affected?

Impedances of the Inverting Amplifier

The input and output impedances of the inverting amplifier are developed with the aid of Figure 6–30. Both the input signal and the negative feedback are applied, through resistors, to the inverting terminal as shown.

FIGURE 6–30
Inverting amplifier.

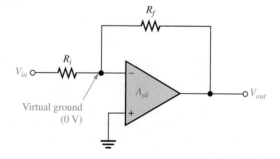

Input Impedance The input impedance for the inverting amplifier is

$$Z_{in(I)} \cong R_i \tag{6-13}$$

This is because the inverting input of the op-amp is at virtual ground (0 V) and the input source simply sees R_i to ground, as shown in Figure 6–31.

FIGURE 6–31

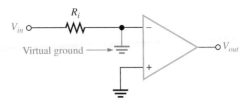

Output Impedance   As with the noninverting amplifier, the output impedance of the inverting amplifier is decreased by the negative feedback. In fact, the expression is the same as for the noninverting case.

$$Z_{out(I)} \cong \frac{Z_{out}}{1 + A_{ol}B} \qquad \textbf{(6–14)}$$

The output impedance of both the noninverting and the inverting configurations is very low; in fact, it is almost zero in practical cases. Because of this near zero output impedance, any load impedance connected to the op-amp output can vary greatly and not change the output voltage at all.

EXAMPLE 6–9

Find the values of the input and output impedances in Figure 6–32. Also, determine the closed-loop voltage gain. The op-amp has the following parameters: $A_{ol} = 50,000$; $Z_{in} = 4 \text{ M}\Omega$; and $Z_{out} = 50 \text{ } \Omega$.

FIGURE 6–32

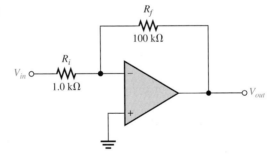

Solution $Z_{in(I)} \cong R_i = \textbf{1.0 k}\boldsymbol{\Omega}$

The feedback attenuation, B, is

$$B = \frac{R_i}{R_i + R_f} = \frac{1.0 \text{ k}\Omega}{101 \text{ k}\Omega} = 0.0099$$

Then

$$Z_{out(I)} = \frac{Z_{out}}{1 + A_{ol}B} = \frac{50 \text{ } \Omega}{1 + (50,000)(0.0099)} = \textbf{101 m}\boldsymbol{\Omega} \qquad \begin{matrix}\text{(zero for all} \\ \text{practical} \\ \text{purposes)}\end{matrix}$$

$$A_{cl(I)} = -\frac{R_f}{R_i} = -\frac{100 \text{ k}\Omega}{1.0 \text{ k}\Omega} = \textbf{–100}$$

Practice Exercise   Determine the input and output impedances and the closed-loop voltage gain in Figure 6–32. The op-amp parameters and circuit values are as follows: $A_{ol} = 100,000$; $Z_{in} = 5$ MΩ; $Z_{out} = 75$ Ω; $R_i = 560$ Ω; and $R_f = 82$ kΩ.

6–6 REVIEW QUESTIONS

1. How does the input impedance of a noninverting amplifier configuration compare to the input impedance of the op-amp itself?

2. When an op-amp is connected in a voltage-follower configuration, does the input impedance increase or decrease?

3. Given that $R_f = 100$ kΩ; $R_i = 2.0$ kΩ; $A_{ol} = 120,000$; $Z_{in} = 2$ MΩ; and $Z_{out} = 60$ Ω, what are $Z_{in(I)}$ and $Z_{out(I)}$ for an inverting amplifier configuration?

6–7 ■ TROUBLESHOOTING

As a technician, you will no doubt encounter situations in which an op-amp or its associated circuitry has malfunctioned. The op-amp is a complex integrated circuit with many types of internal failures possible. However, since you cannot troubleshoot the op-amp internally, you treat it as a single device with only a few connections to it. If it fails, you replace it just as you would a resistor, capacitor, or transistor.

After completing this section, you should be able to

❑ Troubleshoot op-amp circuits
   ❑ Analyze faults in a noninverting amplifier
   ❑ Analyze faults in a voltage-follower
   ❑ Analyze faults in an inverting amplifier

In op-amp configurations, there are only a few components that can fail. Both inverting and noninverting amplifiers have a feedback resistor, R_f, and an input resistor, R_i. Depending on the circuit, a load resistor, bypass capacitors, or a voltage compensation resistor may also be present. Any of these components can appear to be open or appear to be shorted. An open is not always due to the component itself, but may be due to a poor solder connection or a bent pin on the op-amp. Likewise, a short circuit may be due to a solder bridge. Of course, the op-amp itself can fail. Let's examine the basic configurations, considering only the feedback and input resistor failure modes and associated symptoms.

Faults in the Noninverting Amplifier

The first thing to do when you suspect a faulty circuit is to check for the proper power supply voltage. *The positive and negative supply voltages should be measured on the op-amp's pins* with respect to a nearby circuit ground. If either voltage is missing or incorrect, trace the power connections back toward the supply before making other checks. Check that the ground path is not open, giving a misleading power supply reading. If you have verified the supply voltages and ground path, possible faults with the basic amplifier are as follows.

Open Feedback Resistor If the feedback resistor, R_f, in Figure 6–33 opens, the op-amp is operating with its very high open-loop gain, which causes the input signal to drive the device into nonlinear operation and results in a severely clipped output signal as shown in part (a).

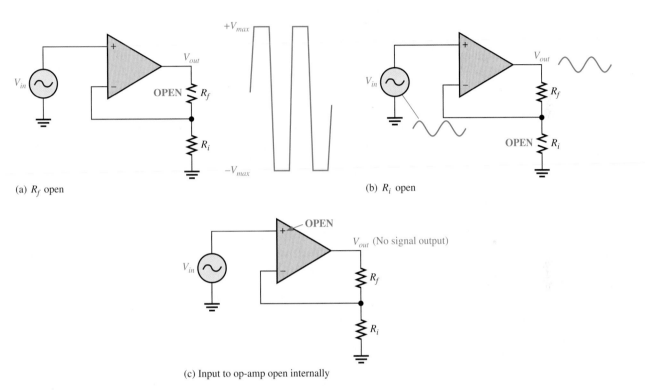

(a) R_f open

(b) R_i open

(c) Input to op-amp open internally

FIGURE 6–33
Faults in the noninverting amplifier.

Open Input Resistor In this case, you still have a closed-loop configuration. But, since R_i is open and effectively equal to infinity, ∞, the closed-loop gain from Equation (6–6) is

$$A_{cl(\text{NI})} = \frac{R_f}{R_i} + 1 = \frac{R_f}{\infty} + 1 = 0 + 1 = 1$$

This shows that the amplifier acts like a voltage-follower. You would observe an output signal that is the same as the input, as indicated in Figure 6–33(b).

Internally Open Noninverting Op-Amp Input In this situation, because the input voltage is not applied to the op-amp, the output is zero. This is indicated in Figure 6–33(c).

Other Op-Amp Faults In general, an internal failure will result in a loss or distortion of the output signal. The best approach is to first make sure that there are no external failures or faulty conditions. If everything else is good, then the op-amp must be bad.

Faults in the Voltage-Follower

The voltage-follower is a special case of the noninverting amplifier. Except for a bad power supply, a bad op-amp, or an open or short at a connection, about the only thing that can happen in a voltage-follower circuit is an open feedback loop. This would have the same effect as an open feedback resistor as previously discussed.

Faults in the Inverting Amplifier

Power Supply As in the case of the noninverting amplifier, the power supply voltages should be checked first. Power supply voltages should be checked on the op-amp's pins with respect to a nearby ground.

Open Feedback Resistor If R_f opens as indicated in Figure 6–34(a), the input signal still feeds through the input resistor and is amplified by the high open-loop gain of the op-amp. This forces the device to be driven into nonlinear operation, and you will see an output something like that shown. This is the same result as in the noninverting configuration.

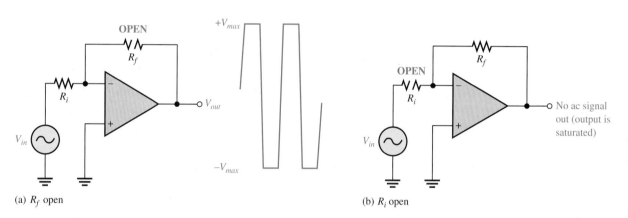

(a) R_f open (b) R_i open

FIGURE 6–34
Faults in the inverting amplifier.

Open Input Resistor This prevents the input signal from getting to the op-amp input, so there will be no output signal, as indicated in Figure 6–34(b).

Failures in the op-amp itself have the same effects as previously discussed for the noninverting amplifier.

6–7 REVIEW QUESTIONS

1. If you notice that the op-amp output is saturated, what should you check first?
2. If there is no op-amp output signal when there is a verified input signal, what should you check first?

6–8 ■ A SYSTEM APPLICATION

The spectrophotometer system presented at the beginning of this chapter combines light optics with electronics to analyze the chemical makeup of various solutions. This type of system is common in medical laboratories as well as many other areas. It is another example of a mixed system in which electronic circuits interface with other types of systems, such as mechanical and optical, to accomplish a specific function. When you are a technician or technologist in industry, you will probably be working with different types of mixed systems from time to time.

After completing this section, you should be able to

❏ Apply what you have learned in this chapter to a system application
   ❏ Understand the role of electronics in a mixed system
   ❏ See how an op-amp is used in the system
   ❏ See how an electronic circuit interfaces with an optical device
   ❏ Translate between a printed circuit board and a schematic
   ❏ Troubleshoot some common system problems

A Brief Description of the System

The light source shown in Figure 6–35 produces a beam of visible light containing a wide spectrum of wavelengths. Each component wavelength in the beam of light is refracted at a different angle by the prism as indicated. Depending on the angle of the platform as set by the pivot angle controller, a certain wavelength passes through the narrow slit and is transmitted through the solution under analysis. By precisely pivoting the light source and prism, a selected wavelength can be transmitted. Every chemical and compound absorbs different wavelengths of light in different ways, so the resulting light coming through the solution has a unique "signature" that can be used to define the chemicals in the solution.

The photocell on the circuit board produces a voltage that is proportional to the amount of light and wavelength. The op-amp circuit amplifies the photocell output and

FIGURE 6–35

sends the resulting signal to the processing and display unit where the type of chemical(s) in the solution is identified. The focus of this system application is the photocell/amplifier circuit board.

Now, so that you can take a closer look at the photocell/amplifier circuit board, let's take it out of the system and put it on the troubleshooter's bench.

TROUBLESHOOTER'S BENCH

■ ACTIVITY 1   Relate the PC Board to the Schematic

Develop a complete schematic by carefully following the conductive traces on the PC board shown in Figure 6–36 to see how the components are interconnected. Two interconnecting traces are on the reverse side of the board and are indicated as slightly darker traces. Refer to the chapter material or the 741 data sheet for the pin layout. This op-amp is housed in a surface-mount SO-8 package. A pad to which no component lead is connected represents a feedthrough to the other side.

FIGURE 6–36

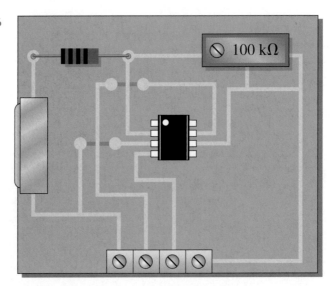

■ ACTIVITY 2   Analyze the Circuit

Step 1: Determine the resistance value to which the feedback rheostat must be adjusted for a voltage gain of 50.

Step 2: Assume the maximum linear output of the op-amp is 1 V less than the dc supply voltage. Determine the voltage gain required and the value to which the feedback resistor must be set to achieve the maximum linear output. The maximum voltage from the photocell is 0.5 V.

Step 3: The system light source produces wavelengths ranging from 400 nm to 700 nm, which is approximately the full range of visible light from violet to red. Determine the op-amp output voltage over this range of wavelengths in 50 nm intervals and plot a graph of the results. Refer to the photocell response characteristic in Figure 6–37.

FIGURE 6–37
Photocell response curve.

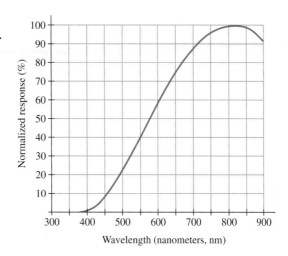

Wavelength (nanometers, nm)

■ ACTIVITY 3   Write a Technical Report

Describe the circuit operation. Be sure to identify the type of op-amp circuit configuration and explain the purpose of the potentiometer. Use the results of Activity 2 to specify the performance of the circuit.

■ ACTIVITY 4   Troubleshoot the Photocell/Amplifier Circuit for Each of the Following Symptoms By Stating the Probable Cause or Causes

1. No voltage at the op-amp output
2. Output of op-amp stays at approximately -9 V.
3. A small dc voltage on the op-amp output under no-light conditions
4. Zero output voltage as light source is pivoted with verified photocell output voltage

6–8 REVIEW QUESTIONS

1. What is the purpose of the 100 kΩ potentiometer on the circuit board?
2. Explain why the light source and prism must be pivoted.

■ SUMMARY

- The basic op-amp has three terminals not including power and ground: inverting ($-$) input, noninverting ($+$) input, and output.
- Most op-amps require both a positive and a negative dc supply voltage.
- The ideal (perfect) op-amp has infinite input impedance, zero output impedance, infinite open-loop voltage gain, infinite bandwidth, and infinite CMRR.
- A good practical op-amp has high input impedance, low output impedance, high open-loop voltage gain, and a wide bandwidth.
- A differential amplifier is normally used for the input stage of an op-amp.
- A differential input voltage appears between the inverting and noninverting inputs of a differential amplifier.

- A single-ended input voltage appears between one input and ground (with the other input grounded).

- A differential output voltage appears between two output terminals of a diff-amp.

- A single-ended output voltage appears between the output and ground of a diff-amp.

- Common mode occurs when equal in-phase voltages are applied to both input terminals.

- Input offset voltage produces an output error voltage (with no input voltage).

- Input bias current also produces an output error voltage (with no input voltage).

- Input offset current is the difference between the two bias currents.

- Open-loop voltage gain is the gain of an op-amp with no external feedback connections.

- Closed-loop voltage gain is the gain of an op-amp with external feedback.

- The common-mode rejection ratio (CMRR) is a measure of an op-amp's ability to reject common-mode inputs.

- Slew rate is the rate in volts per microsecond at which the output voltage of an op-amp can change in response to a step input.

- Figure 6–38 shows the op-amp symbol and the three basic op-amp configurations.

- All op-amp configurations listed use negative feedback. Negative feedback occurs when a portion of the output voltage is connected back to the inverting input such that it subtracts from the input voltage, thus reducing the voltage gain but increasing the stability and bandwidth.

- A noninverting amplifier configuration has a higher input impedance and a lower output impedance than the op-amp itself (without feedback).

- An inverting amplifier configuration has an input impedance approximately equal to the input resistor R_i and an output impedance approximately equal to the output impedance of the op-amp itself.

- The voltage-follower has the highest input impedance and the lowest output impedance of the three configurations.

FIGURE 6–38

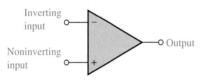

(a) Basic op-amp symbol

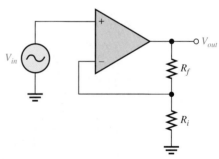

(b) Noninverting amplifier

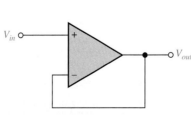

(c) Voltage-follower

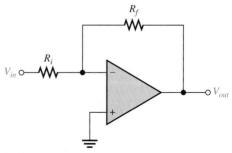

(d) Inverting amplifier

■ GLOSSARY

Key terms are in color. All terms are included in the end-of-book glossary.

Closed-loop voltage gain   The net voltage gain of an amplifier when negative feedback is included.

Common mode   A condition characterized by the presence of the same signal on both op-amp inputs.

Common-mode input impedance   The resistance between each input and ground.

Common-mode rejection ratio (CMRR)   The ratio of open-loop gain to common-mode gain; a measure of an op-amp's ability to reject common-mode signals.

Common-mode input voltage range   The range of input voltage which, when applied to both inputs, will not cause clipping or other output distortion.

Differential amplifier (diff-amp)   An amplifier that produces an output voltage proportional to the difference of the two input voltages.

Differential input impedance   The total resistance between the inverting and the noninverting inputs.

Differential mode   The input condition of an op-amp in which opposite polarity signals are applied to the two inputs.

Input bias current   The average dc current required by the inputs of an op-amp to properly operate the device.

Input offset current (I_{OS})   The difference of the input bias currents.

Input offset voltage (V_{OS})   The differential dc voltage required between the op-amp inputs to force the differential output to zero volts.

Input offset voltage drift   A parameter that specifies how much change occurs in the input offset voltage for each degree change in temperature.

Inverting amplifier   An op-amp closed-loop configuration in which the input signal is applied to the inverting input.

Negative feedback   The process of returning a portion of the output signal to the input of an amplifier such that it is out of phase with the input signal.

Noninverting amplifier   An op-amp closed-loop configuration in which the input signal is applied to the noninverting input.

Open-loop voltage gain   The internal gain of an op-amp without any external feedback.

Operational amplifier (op-amp)   A type of amplifier that has very high voltage gain, very high input impedance, very low output impedance, and good rejection of common-mode signals.

Output impedance   The resistance viewed from the output terminal of an op-amp.

Single-ended mode   The input condition of an op-amp in which one input is grounded and the signal voltage is applied only to the other input.

Slew rate   The rate of change of the output voltage of an op-amp in response to a step input.

Voltage-follower   A closed-loop, noninverting op-amp with a voltage gain of 1.

■ KEY FORMULAS

Differential Amplifiers

(6–1) $\quad$ CMRR $= \dfrac{A_{v(d)}}{A_{cm}}$ $\qquad$ Common-mode rejection ratio (diff-amp)

(6–2) $\quad$ CMRR$' = 20 \log\left(\dfrac{A_{v(d)}}{A_{cm}}\right)$ $\qquad$ Common-mode rejection ratio (dB) (diff-amp)

Op-Amp Parameters

(6–3) $\quad$ CMRR $= \dfrac{A_{ol}}{A_{cm}}$ $\qquad$ Common-mode rejection ratio (op-amp)

(6–4) $\quad$ CMRR$' = 20 \log\left(\dfrac{A_{ol}}{A_{cm}}\right)$ $\qquad$ Common-mode rejection ratio (dB) (op-amp)

(6–5) $\quad$ Slew rate $= \dfrac{\Delta V_{out}}{\Delta t}$ $\qquad$ Slew rate

Op-Amp Configurations

(6–6) $\quad A_{cl(NI)} = \dfrac{R_f}{R_i} + 1$ $\qquad$ Voltage gain (noninverting)

(6–7) $\quad A_{cl(VF)} = 1$ $\qquad$ Voltage gain (voltage-follower)

(6–8) $\quad A_{cl(I)} = -\dfrac{R_f}{R_i}$ $\qquad$ Voltage gain (inverting)

Op-Amp Impedances

(6–9) $\quad Z_{in(NI)} = (1 + A_{ol}B)Z_{in}$ $\qquad$ Input impedance (noninverting)

(6–10) $\quad Z_{out(NI)} = \dfrac{Z_{out}}{1 + A_{ol}B}$ $\qquad$ Output impedance (noninverting)

(6–11) $\quad Z_{in(VF)} = (1 + A_{ol})Z_{in}$ $\qquad$ Input impedance (voltage-follower)

(6–12) $\quad Z_{out(VF)} = \dfrac{Z_{out}}{1 + A_{ol}}$ $\qquad$ Output impedance (voltage-follower)

(6–13) $\quad Z_{in(I)} \cong R_i$ $\qquad$ Input impedance (inverting)

(6–14) $\quad Z_{out(I)} \cong \dfrac{Z_{out}}{1 + A_{ol}B}$ $\qquad$ Output impedance (inverting)

■ **SELF-TEST** $\qquad$ *Answers are at the end of the chapter.*

1. An integrated circuit (IC) op-amp has
   (a) two inputs and two outputs $\qquad$ (b) one input and one output
   (c) two inputs and one output

2. Which of the following characteristics does not necessarily apply to an op-amp?
   (a) High gain $\qquad\qquad\qquad$ (b) Low power
   (c) High input impedance $\qquad$ (d) Low output impedance

3. A differential amplifier
   (a) is part of an op-amp $\qquad$ (b) has one input and one output
   (c) has two outputs $\qquad\qquad$ (d) answers (a) and (c)

4. When a differential amplifier is operated single-ended,
   (a) the output is grounded
   (b) one input is grounded and a signal is applied to the other
   (c) both inputs are connected together
   (d) the output is not inverted

5. In the differential mode,
   (a) opposite polarity signals are applied to the inputs
   (b) the gain is 1
   (c) the outputs are different amplitudes
   (d) only one supply voltage is used

6. In the common mode,
   (a) both inputs are grounded
   (b) the outputs are connected together
   (c) an identical signal appears on both inputs
   (d) the output signals are in phase

7. Common-mode gain is
   (a) very high   **(b)** very low   **(c)** always unity   **(d)** unpredictable

8. Differential gain is
   (a) very high   **(b)** very low
   (c) dependent on the input voltage   **(d)** about 100

9. If $A_{v(d)} = 3500$ and $A_{cm} = 0.35$, the CMRR is
   (a) 1225   **(b)** 10,000   **(c)** 80 dB   **(d)** answers (b) and (c)

10. With zero volts on both inputs, an op-amp ideally should have an output
   (a) equal to the positive supply voltage
   (b) equal to the negative supply voltage
   (c) equal to zero
   (d) equal to the CMRR

11. Of the values listed, the most realistic value for open-loop gain of an op-amp is
   (a) 1   **(b)** 2000   **(c)** 80 dB   **(d)** 100,000

12. A certain op-amp has bias currents of 50 μA and 49.3 μA. The input offset current is
   (a) 700 nA   **(b)** 99.3 μA   **(c)** 49.65 μA   **(d)** none of these

13. The output of a particular op-amp increases 8 V in 12 μs. The slew rate is
   (a) 96 V/μs   **(b)** 0.67 V/μs   **(c)** 1.5 V/μs   **(d)** none of these

14. For an op-amp with negative feedback, the output is
   (a) equal to the input   **(b)** increased
   (c) fed back to the inverting input   **(d)** fed back to the noninverting input

15. The use of negative feedback
   (a) reduces the voltage gain of an op-amp   **(b)** makes the op-amp oscillate
   (c) makes linear operation possible   **(d)** answers (a) and (c)

16. Negative feedback
   (a) increases the input and output impedances
   (b) increases the input impedance and the bandwidth
   (c) decreases the output impedance and the bandwidth
   (d) does not affect impedances or bandwidth

17. A certain noninverting amplifier has an R_i of 1.0 kΩ and an R_f of 100 kΩ. The closed-loop gain is
   (a) 100,000   **(b)** 1000   **(c)** 101   **(d)** 100

18. If the feedback resistor in Question 17 is open, the voltage gain
   (a) increases   **(b)** decreases   **(c)** is not affected   **(d)** depends on R_i

19. A certain inverting amplifier has a closed-loop gain of 25. The op-amp has an open-loop gain of 100,000. If another op-amp with an open-loop gain of 200,000 is substituted in the configuration, the closed-loop gain
   (a) doubles   **(b)** drops to 12.5   **(c)** remains at 25   **(d)** increases slightly

20. A voltage-follower
   (a) has a gain of one   **(b)** is noninverting
   (c) has no feedback resistor   **(d)** answers (a), (b), and (c)

TROUBLESHOOTER'S QUIZ   *Answers are at the end of the chapter.*

Refer to Figure 6–40.

❑ If the collector of Q_1 opens,

1. The dc output voltage will

(a) increase   (b) decrease   (c) not change

2. The current through R_3 will

(a) increase   (b) decrease   (c) not change

Refer to Figure 6–44.

❑ If R_i is open,

3. The closed-loop gain will

(a) increase   (b) decrease   (c) not change

4. For a given input signal, the output signal will

(a) increase   (b) decrease   (c) not change

❑ If R_f is open,

5. The output voltage will

(a) increase   (b) decrease   (c) not change

6. The open-loop gain will

(a) increase   (b) decrease   (c) not change

7. The closed-loop gain will

(a) increase   (b) decrease   (c) not change

Refer to Figure 6–48.

❑ If R_i is shorted,

8. The closed-loop gain will

(a) increase   (b) decrease   (c) not change

9. The input impedance will

(a) increase   (b) decrease   (c) not change

❑ If R_f is open,

10. The open-loop gain will

(a) increase   (b) decrease   (c) not change

❑ If R_f is smaller than the specified value,

11. The closed-loop gain will

(a) increase   (b) decrease   (c) not change

12. The open-loop gain will

(a) increase   (b) decrease   (c) not change

■ PROBLEMS

Answers to odd-numbered problems are at the end of the book.

SECTION 6–1 Introduction to Operational Amplifiers

1. Compare a practical op-amp to the ideal.

2. Two IC op-amps are available to you. Their characteristics are listed below. Choose the one you think is more desirable.
   Op-amp 1: $Z_{in} = 5\ M\Omega$, $Z_{out} = 100\ \Omega$, $A_{ol} = 50{,}000$
   Op-amp 2: $Z_{in} = 10\ M\Omega$, $Z_{out} = 75\ \Omega$, $A_{ol} = 150{,}000$

SECTION 6–2 The Differential Amplifier

3. Identify the type of input and output configuration for each basic differential amplifier in Figure 6–39.

4. The dc base voltages in Figure 6–40 are zero. Using your knowledge of transistor analysis, determine the dc differential output voltage. Assume that for Q_1, $I_C/I_E = 0.98$ and for Q_2, $I_C/I_E = 0.975$.

5. Identify the quantity being measured by each meter in Figure 6–41.

6. A differential amplifier stage has collector resistors of 5.1 kΩ each. If $I_{C1} = 1.35$ mA and $I_{C2} = 1.29$ mA, what is the differential output voltage?

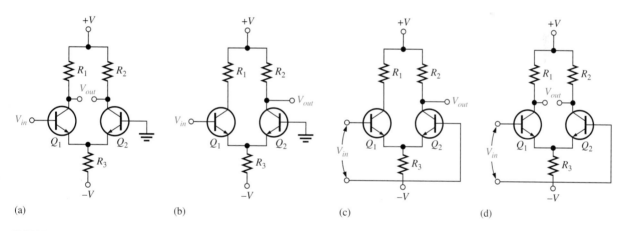

(a)   (b)   (c)   (d)

FIGURE 6–39

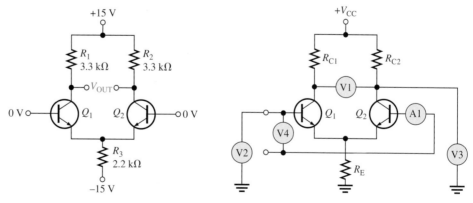

FIGURE 6–40   **FIGURE 6–41**

SECTION 6–3 Op-Amp Data Sheet Parameters

7. Determine the bias current, I_{BIAS}, given that the input currents to an op-amp are 8.3 μA and 7.9 μA.

8. Distinguish between input bias current and input offset current, and then calculate the input offset current in Problem 7.

9. A certain op-amp has a CMRR of 250,000. Convert this to dB.

10. The open-loop gain of a certain op-amp is 175,000. Its common-mode gain is 0.18. Determine the CMRR in dB.

11. An op-amp data sheet specifies a CMRR of 300,000 and an A_{ol} of 90,000. What is the common-mode gain?

12. Figure 6–42 shows the output voltage of an op-amp in response to a step input. What is the slew rate?

FIGURE 6–42

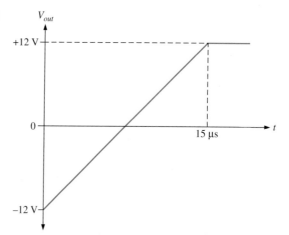

13. How long does it take the output voltage of an op-amp to go from -10 V to $+10$ V, if the slew rate is 0.5 V/μs?

SECTION 6–5 Op-Amp Configurations with Negative Feedback

14. Identify each of the op-amp configurations in Figure 6–43.

15. A noninverting amplifier has an R_i of 1.0 kΩ and an R_f of 100 kΩ. Determine V_f and B if $V_{out} = 5$ V.

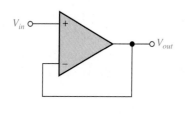

(a)

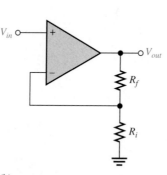

(b)

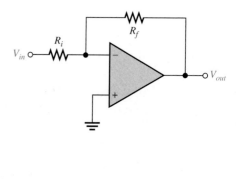

(c)

FIGURE 6–43

16. For the amplifier in Figure 6–44, determine the following:
  (a) $A_{cl(NI)}$ **(b)** V_{out} **(c)** V_f

17. Determine the closed-loop gain of each amplifier in Figure 6–45.

18. Find the value of R_f that will produce the indicated closed-loop gain in each amplifier in Figure 6–46.

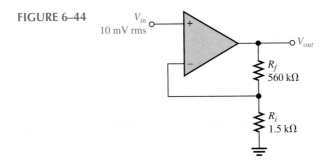

FIGURE 6–44

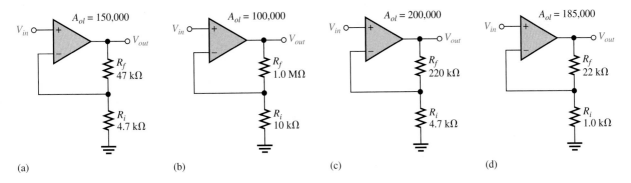

(a)   (b)   (c)   (d)

FIGURE 6–45

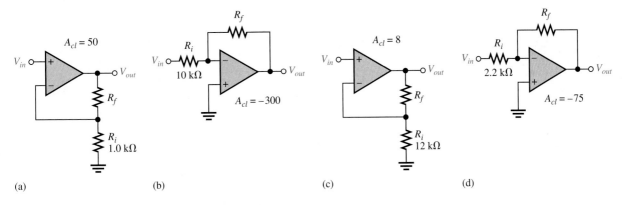

(a)   (b)   (c)   (d)

FIGURE 6–46

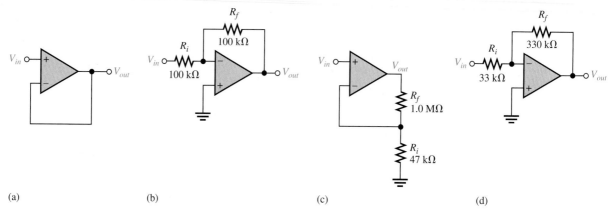

(a)   (b)   (c)   (d)

FIGURE 6–47

19. Find the gain of each amplifier in Figure 6–47.

20. If a signal voltage of 10 mV rms is applied to each amplifier in Figure 6–47, what are the output voltages and what is their phase relationship with inputs?

21. Determine the approximate values for each of the following quantities in Figure 6–48.
    (a) I_{in} **(b)** I_f **(c)** V_{out} **(d)** Closed-loop gain

FIGURE 6–48

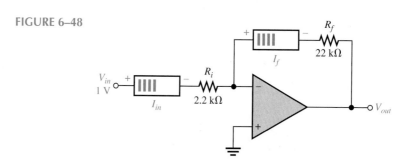

SECTION 6–6 Op-Amp Impedances

22. Determine the input and output impedances for each amplifier configuration in Figure 6–49.

23. Repeat Problem 22 for each circuit in Figure 6–50.

24. Repeat Problem 22 for each circuit in Figure 6–51.

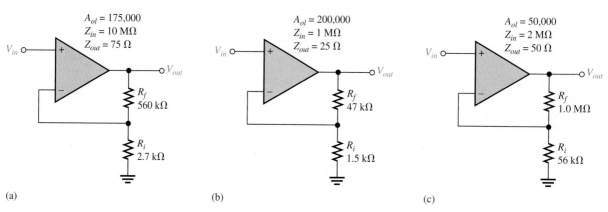

(a)   (b)   (c)

FIGURE 6–49

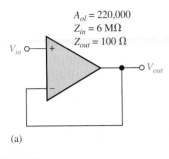

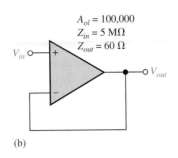

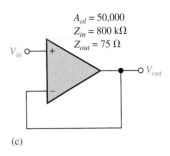

(a)   (b)   (c)

FIGURE 6–50

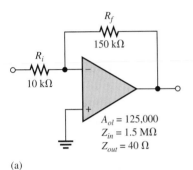

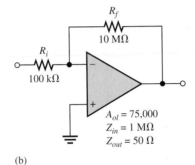

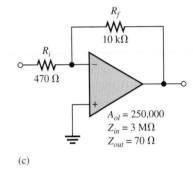

(a)   (b)   (c)

FIGURE 6–51

SECTION 6–7   Troubleshooting

25. Determine the most likely fault(s) for each of the following symptoms in Figure 6–52 with a 100 mV signal applied.
   (a) No output signal.
   (b) Output severely clipped on both positive and negative swings.

26. Determine the effect on the output if the circuit in Figure 6–52 has the following fault (one fault at a time).
   (a) Output pin is shorted to the inverting input.
   (b) R_3 is open.
   (c) R_3 is 10 kΩ instead of 910 Ω.
   (d) R_1 and R_2 are swapped.

FIGURE 6–52

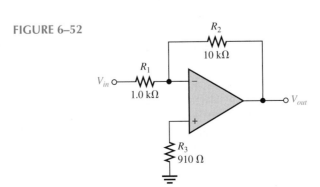

FIGURE 6–53

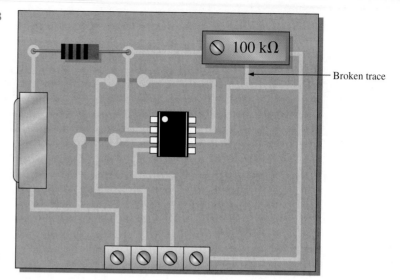

Broken trace

27. On the circuit board in Figure 6–53, what happens if the middle lead (wiper) of the 100 kΩ potentiometer is broken?

■ **ANSWERS TO REVIEW QUESTIONS**

Section 6–1

1. Inverting input, noninverting input, output, positive and negative supply voltages

2. A practical op-amp has high input impedance, low output impedance, high voltage gain, and wide bandwidth.

Section 6–2

1. Differential input is between two input terminals. Single-ended input is from one input terminal to ground (with other input grounded).

2. Common-mode rejection is the ability of an op-amp to produce very little output when the same signal is applied to both inputs.

3. A higher CMRR results in a lower common-mode gain.

Section 6–3

1. Input bias current, input offset voltage, drift, input offset current, input impedance, output impedance, common-mode input voltage range, CMRR, open-loop voltage gain, slew rate, frequency response

2. Slew rate and voltage gain are both frequency dependent.

Section 6–4

1. Negative feedback provides a stable controlled voltage gain, control of input and output impedances, and wider bandwidth.

2. The open-loop gain is so high that a very small signal on the input will drive the op-amp into saturation.

3. Both inputs will be the same.

Section 6–5

1. The main purpose of negative feedback is to stabilize the gain.
2. False
3. $A_{cl} = 1/0.02 = 50$

Section 6–6

1. The noninverting configuration has a higher Z_{in} than the op-amp alone.
2. Z_{in} increases in a voltage-follower.
3. $Z_{in(I)} \cong R_i = 2.0 \text{ k}\Omega$, $Z_{out(I)} \cong Z_{out} = 26 \text{ m}\Omega$.

Section 6–7

1. Check power supply voltages with respect to ground. Verify ground connections. Check for an open feedback resistor.
2. Verify power supply voltages and ground leads. For inverting amplifiers, check for open R_i. For noninverting amplifiers, check that V_{in} is actually on (+) pin; if so, check (−) pin for identical signal.

Section 6–8

1. The 100 kΩ potentiometer is the feedback resistor.
2. The light source and prism must be pivoted to allow different wavelengths of light to pass through the slit.

■ **ANSWERS TO PRACTICE EXERCISES FOR EXAMPLES**

6–1 34,000; 90.6 dB

6–2 (a) 0.168   (b) 87.96 dB   (c) 2.1 V rms, 4.2 V rms   (d) 0.168V

6–3 12,649

6–4 20 V/μs

6–5 32.9

6–6 (a) 67.5 kΩ   (b) The amplifier would have an open-loop gain producing a square wave.

6–7 (a) 20.6 GΩ, 14 mΩ   (b) 23

6–8 Input Z increases, output Z decreases.

6–9 $Z_{in(I)} = 560 \text{ Ω}$, $Z_{out(I)} = 110 \text{ mΩ}$, $A_{cl} = -146$

■ **ANSWERS TO SELF-TEST**

1. (c)	2. (b)	3. (d)	4. (b)	5. (a)
6. (c)	7. (b)	8. (a)	9. (d)	10. (c)
11. (d)	12. (a)	13. (b)	14. (c)	15. (d)
16. (b)	17. (c)	18. (a)	19. (c)	20. (d)

■ **ANSWERS TO TROUBLE-SHOOTER'S QUIZ**

1. increase	2. not change	3. decrease	4. decrease
5. increase	6. not change	7. increase	8. increase
9. decrease	10. not change	11. decrease	12. not change

7

OP-AMP RESPONSES

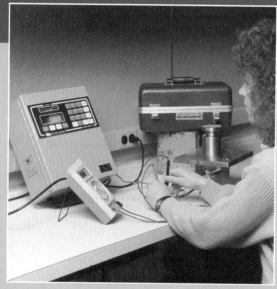

Courtesy Yuba College

■ CHAPTER OBJECTIVES

❑ Discuss the basic areas of op-amp responses
❑ Understand the open-loop response of an op-amp
❑ Understand the closed-loop response of an op-amp
❑ Discuss positive feedback and stability in op-amp circuits
❑ Explain op-amp phase compensation
❑ Apply what you have learned in this chapter to a system application

■ KEY TERMS

❑ Bandwidth
❑ Phase shift
❑ Positive feedback
❑ Loop gain
❑ Phase margin
❑ Stability

■ CHAPTER INTRODUCTION

In this chapter, you will learn about frequency response, bandwidth, phase shift, and other frequency-related parameters in op-amps. The effects of negative feedback will be further examined, and you will learn about stability requirements and how to compensate op-amp circuits to ensure stable operation.

■ A SYSTEM APPLICATION

Stereo systems use two separate frequency-modulated (FM) signals to reproduce sound as, for example, from the left and right sides of the stage in a concert performance. When the signal is processed by a stereo receiver, the sound comes out of both the left and right speakers, and you get the original sound effects in terms of direction and distribution. When a stereo broadcast is received by a single-speaker (monophonic) system, the sound from the speaker is actually the composite or sum of the left and right channel sounds so you get the original sound without separation. Op-amps can be used for several purposes in stereo systems such as this, but we will focus on the identical left and right channel audio amplifiers for this system application.

For the system application in Section 7–6, in addition to the other topics, be sure you understand

❏ The noninverting op-amp configuration
❏ How capacitors can be connected for frequency compensation
❏ Discrete transistor push-pull amplifier operation (from Chapter 5)

7–1 ■ BASIC CONCEPTS

In Chapter 6 you learned how closed-loop voltage gains of the basic op-amp configurations are determined, and the distinction between open-loop voltage gain and closed-loop voltage gain was established. Because of the importance of these two different types of voltage gain, the definitions are restated in this section.

After completing this section, you should be able to

❏ Discuss the basic areas of op-amp responses
    ❏ Explain open-loop gain
    ❏ Explain closed-loop gain
    ❏ Discuss the frequency dependency of gain
    ❏ Explain the open-loop bandwidth
    ❏ Explain the unity-gain bandwidth
    ❏ Determine phase shift

Open-Loop Gain

The *open-loop gain* (A_{ol}) of an op-amp is the internal voltage gain of the device and represents the ratio of output voltage to input voltage, as indicated in Figure 7–1(a). Notice that there are no external components, so the open-loop gain is set entirely by the internal design. Open-loop voltage gain varies widely for different op-amps. Table 6–1 listed the open-loop gain for some representative op-amps. Data sheets often refer to the open-loop gain as the *large-signal voltage gain.*

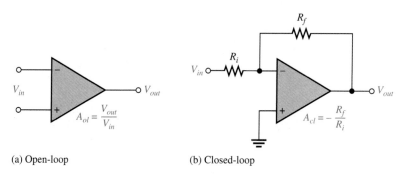

(a) Open-loop   (b) Closed-loop

FIGURE 7–1
Open-loop and closed-loop op-amp configurations.

Closed-Loop Gain

The *closed-loop gain* (A_{cl}) is the voltage gain of an op-amp with external feedback. The amplifier configuration consists of the op-amp and an external negative feedback network that connects the output to the inverting ($-$) input. The closed-loop gain is determined by the external component values, as illustrated in Figure 7–1(b) for an inverting amplifier configuration. The closed-loop gain can be precisely controlled by external component values.

The Gain Is Frequency Dependent

In Chapter 6, all of the gain expressions applied to the midrange gain and were considered independent of the frequency. The midrange open-loop gain of an op-amp extends from zero frequency (dc) up to a critical frequency at which the gain is 3 dB less than the midrange value. The difference here is that op-amps are dc amplifiers (no capacitive coupling between stages), and therefore, there is no lower critical frequency. This means that the midrange gain extends down to zero frequency (dc), and dc voltages are amplified the same as midrange signal frequencies.

An open-loop response curve (Bode plot) for a certain op-amp is shown in Figure 7–2. Most op-amp data sheets show this type of curve or specify the midrange open-loop gain. Notice that the curve rolls off (decreases) at −20 dB per decade (−6 dB per octave). The midrange gain is 200,000, which is 106 dB, and the critical (cutoff) frequency is approximately 10 Hz.

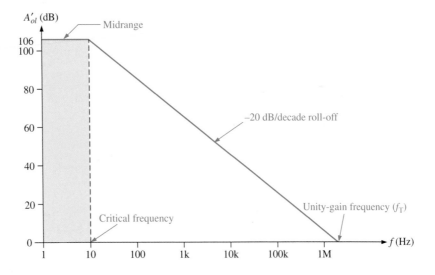

FIGURE 7–2
Ideal plot of open-loop voltage gain versus frequency for a typical op-amp. The frequency scale is logarithmic.

3 dB Open-Loop Bandwidth

The **bandwidth** of an ac amplifier is the frequency range between the points where the gain is 3 dB less than the midrange gain. In general, the bandwidth equals the upper critical frequency (f_{cu}) minus the lower critical frequency (f_{cl}).

$$BW = f_{cu} - f_{cl}$$

Since f_{cl} for an op-amp is zero, the bandwidth is simply equal to the upper critical frequency.

$$BW = f_{cu} \qquad\qquad\qquad \textbf{(7–1)}$$

From now on, we will refer to f_{cu} as simply f_c; and we will use open-loop (*ol*) or closed-loop (*cl*) subscript designators. For example, $f_{c(ol)}$ is the open-loop upper critical frequency and $f_{c(cl)}$ is the closed-loop upper critical frequency.

Unity-Gain Bandwidth

Notice in Figure 7–2 that the gain steadily decreases to a point where it is equal to one (0 dB). The value of the frequency at which this unity gain occurs is the *unity-gain bandwidth*.

Gain-Versus-Frequency Analysis

The *RC* lag (low-pass) networks within an op-amp are responsible for the roll-off in gain as the frequency increases. From basic ac circuit theory, the attenuation of an *RC* lag network, such as in Figure 7–3, is expressed as

$$\frac{V_{out}}{V_{in}} = \frac{X_C}{\sqrt{R^2 + X_C^2}}$$

Dividing both the numerator and denominator to the right of the equal sign by X_C,

$$\frac{V_{out}}{V_{in}} = \frac{1}{\sqrt{1 + R^2/X_C^2}}$$

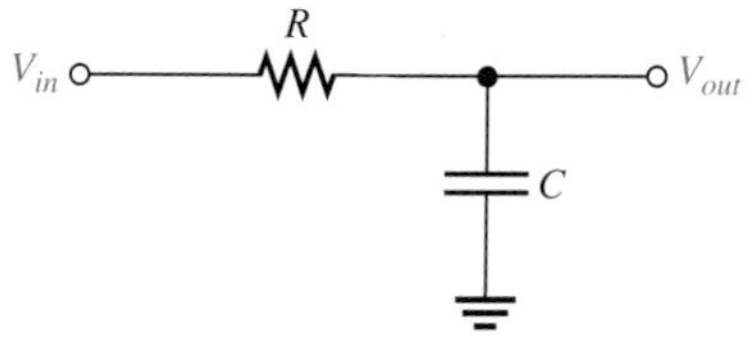

FIGURE 7–3
***RC* lag network.**

The critical frequency of an *RC* network is

$$f_c = \frac{1}{2\pi RC}$$

Dividing both sides by f gives

$$\frac{f_c}{f} = \frac{1}{2\pi RCf} = \frac{1}{(2\pi fC)R}$$

Since $X_C = 1/(2\pi fC)$, the above expression can be written as

$$\frac{f_c}{f} = \frac{X_C}{R}$$

Substituting this result into the second equation produces the following expression for the attenuation of an *RC* lag network:

$$\frac{V_{out}}{V_{in}} = \frac{1}{\sqrt{1 + f^2/f_c^2}}$$

FIGURE 7–4

Op-amp represented by gain element and internal RC network.

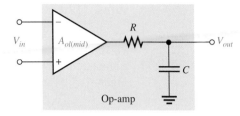

Op-amp

If an op-amp is represented by a voltage gain element with a gain of $A_{ol(mid)}$ and a single RC lag network, as shown in Figure 7–4, then the total open-loop gain of the op-amp is the product of the midrange open-loop gain $A_{ol(mid)}$ and the attenuation of the RC network.

$$A_{ol} = \frac{A_{ol(mid)}}{\sqrt{1 + f^2/f_c^2}} \qquad (7\text{--}2)$$

As you can see from Equation (7–2), the open-loop gain equals the midrange value when the signal frequency f is much less than the critical frequency f_c and drops off as the frequency increases. Since f_c is part of the open-loop response of an op-amp, we will refer to it as $f_{c(ol)}$.

The following example demonstrates how the open-loop gain decreases as the frequency increases above $f_{c(ol)}$.

EXAMPLE 7–1

Determine A_{ol} for the following values of f. Assume $f_{c(ol)} = 100$ Hz and $A_{ol(mid)} = 100,000$.

(a) $f = 0$ Hz   (b) $f = 10$ Hz   (c) $f = 100$ Hz   (d) $f = 1000$ Hz

Solution

(a) $A_{ol} = \dfrac{A_{ol(mid)}}{\sqrt{1 + f^2/f_{c(ol)}^2}} = \dfrac{100,000}{\sqrt{1 + 0}} = \mathbf{100,000}$

(b) $A_{ol} = \dfrac{100,000}{\sqrt{1 + (0.1)^2}} = \mathbf{99,500}$

(c) $A_{ol} = \dfrac{100,000}{\sqrt{1 + (1)^2}} = \dfrac{100,000}{\sqrt{2}} = \mathbf{70,700}$

(d) $A_{ol} = \dfrac{100,000}{\sqrt{1 + (10)^2}} = \mathbf{9950}$

*Practice Exercise** Find A_{ol} for the following frequencies. Assume $f_{c(ol)} = 200$ Hz and $A_{ol(mid)} = 80,000$.

(a) $f = 2$ Hz   (b) $f = 10$ Hz   (c) $f = 2500$ Hz

---

* Answers are at the end of the chapter.

Phase Shift

As you know, an *RC* network causes a propagation delay from input to output, thus creating a **phase shift** between the input signal and the output signal. An *RC* lag network such as found in an op-amp stage causes the output signal voltage to lag the input, as shown in Figure 7–5. From basic ac circuit theory, the phase shift, φ, is

$$\phi = -\tan^{-1}\left(\frac{R}{X_C}\right)$$

Since $R/X_C = f/f_c$,

$$\phi = -\tan^{-1}\left(\frac{f}{f_c}\right) \tag{7–3}$$

The negative sign indicates that the output lags the input. This equation shows that the phase shift increases with frequency and approaches $-90°$ as f becomes much greater than f_c.

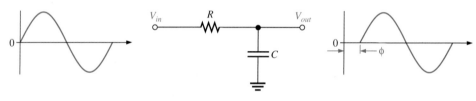

FIGURE 7–5
Output voltage lags input voltage.

EXAMPLE 7–2

Calculate the phase shift for an *RC* lag network for each of the following frequencies, and then plot the curve of phase shift versus frequency. Assume $f_c = 100$ Hz.

(a) $f = 1$ Hz   (b) $f = 10$ Hz   (c) $f = 100$ Hz
(d) $f = 1000$ Hz   (e) $f = 10$ kHz

Solution

(a) $\phi = -\tan^{-1}\left(\dfrac{f}{f_c}\right) = -\tan^{-1}\left(\dfrac{1\ \text{Hz}}{100\ \text{Hz}}\right) = \mathbf{-0.6°}$

(b) $\phi = -\tan^{-1}\left(\dfrac{10\ \text{Hz}}{100\ \text{Hz}}\right) = \mathbf{-5.7°}$

(c) $\phi = -\tan^{-1}\left(\dfrac{100\ \text{Hz}}{100\ \text{Hz}}\right) = \mathbf{-45.0°}$

(d) $\phi = -\tan^{-1}\left(\dfrac{1000\ \text{Hz}}{100\ \text{Hz}}\right) = \mathbf{-84.3°}$

(e) $\phi = -\tan^{-1}\left(\dfrac{10\ \text{kHz}}{100\ \text{Hz}}\right) = \mathbf{-89.4°}$

The phase shift versus frequency curve is plotted in Figure 7–6. Note that the frequency axis is logarithmic.

FIGURE 7–6

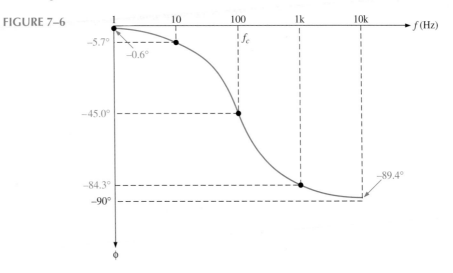

Practice Exercise   At what frequency, in this example, is the phase shift −60°?

7–1 REVIEW QUESTIONS*

1. How do the open-loop gain and the closed-loop gain of an op-amp differ?
2. The upper critical frequency of a particular op-amp is 100 Hz. What is its open-loop 3 dB bandwidth?
3. Does the open-loop gain increase or decrease with frequency above the critical frequency?

_____

* Answers are at the end of the chapter.

7–2 ■ OP-AMP OPEN-LOOP RESPONSE

In this section, you will learn about the open-loop frequency response and the open-loop phase response of an op-amp. Open-loop responses relate to an op-amp with no external feedback. The frequency response indicates how the voltage gain changes with frequency, and the phase response indicates how the phase shift between the input and output signal changes with frequency. The open-loop gain, like the β of a transistor, varies greatly from one device to the next of the same type.

After completing this section, you should be able to

❏ Understand the open-loop response of an op-amp
  ❏ Discuss how internal stages affect the overall response
  ❏ Discuss critical frequencies and roll-off rates
  ❏ Determine overall phase response

Frequency Response

In Section 7–1, an op-amp was assumed to have a constant roll-off of -20 dB/decade above its critical frequency. For a large number of op-amps, this is indeed the case. Op-amps that have a constant -20 dB/decade roll-off from f_c to unity gain are called *compensated op-amps*. A compensated op-amp has only one RC network that determines its frequency characteristic. Thus, the roll-off rate is the same as that of a basic RC network.

For some op-amp circuits, the situation is more complicated. The frequency response may be determined by several internal stages, where each stage has its own critical frequency. As a result, the overall response is affected by more than one cascaded stage and the overall response is a composite of the individual responses. An op-amp that has more than one critical frequency is called an **uncompensated op-amp**.

Uncompensated op-amps require careful attention to the feedback network to avoid oscillation. As an example, a three-stage op-amp is represented in Figure 7–7(a), and the frequency response of each stage is shown in Figure 7–7(b). As you have learned, dB gains are added so that the total op-amp frequency response is as shown in Figure 7–7(c). Since the roll-off rates are additive, the total roll-off rate increases by -20 dB/decade (-6 dB/octave) as each critical frequency is reached.

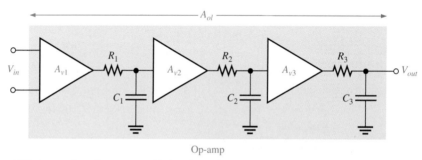

(a) Representation of an op-amp with three internal stages

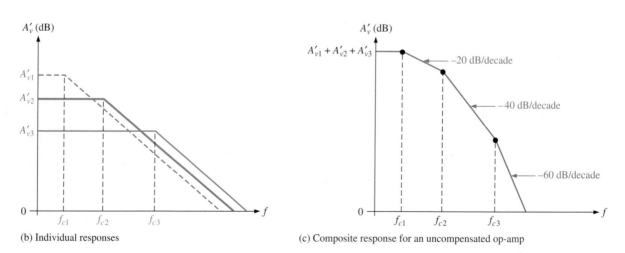

(b) Individual responses

(c) Composite response for an uncompensated op-amp

FIGURE 7–7

Op-amp open-loop frequency response.

Phase Response

In a multistage amplifier, each stage contributes to the total phase lag. As you have seen, each RC lag network can produce up to a $-90°$ phase shift. Since each stage in an op-amp includes an RC lag network, a three-stage op-amp, for example, can have a maximum phase lag of $-270°$. Also, the phase lag of each stage is less than $-45°$ when the frequency is below the critical frequency, equal to $-45°$ at the critical frequency, and greater than $-45°$ when the frequency is above the critical frequency. The phase lags of the stages of an op-amp are added to produce a total phase lag, according to the following formula for three stages:

$$\phi_{tot} = -\tan^{-1}\left(\frac{f}{f_{c1}}\right) - \tan^{-1}\left(\frac{f}{f_{c2}}\right) - \tan^{-1}\left(\frac{f}{f_{c3}}\right)$$

EXAMPLE 7–3

A certain op-amp has three internal amplifier stages with the following gains and critical frequencies:

Stage 1: $A'_{v1} = 40$ dB, $f_{c1} = 2000$ Hz
Stage 2: $A'_{v2} = 32$ dB, $f_{c2} = 40$ kHz
Stage 3: $A'_{v3} = 20$ dB, $f_{c3} = 150$ kHz

Determine the open-loop midrange dB gain and the total phase lag when $f = f_{c1}$.

Solution

$$A'_{ol(mid)} = A'_{v1} + A'_{v2} + A'_{v3} = 40 \text{ dB} + 32 \text{ dB} + 20 \text{ dB} = \textbf{92 dB}$$

$$\phi_{tot} = -\tan^{-1}\left(\frac{f}{f_{c1}}\right) - \tan^{-1}\left(\frac{f}{f_{c2}}\right) - \tan^{-1}\left(\frac{f}{f_{c3}}\right)$$

$$= -\tan^{-1}(1) - \tan^{-1}\left(\frac{2}{40}\right) - \tan^{-1}\left(\frac{2}{150}\right)$$

$$= -45° - 2.86° - 0.76° = \textbf{−48.6°}$$

Practice Exercise   The internal stages of a two-stage amplifier have the following characteristics: $A'_{v1} = 50$ dB, $A'_{v2} = 25$ dB, $f_{c1} = 1500$ Hz, and $f_{c2} = 3000$ Hz. Determine the open-loop midrange gain in dB and the total phase lag when $f = f_{c1}$.

7–2 REVIEW QUESTIONS

1. If the individual stage gains of an op-amp are 20 dB and 30 dB, what is the total gain in dB?

2. If the individual phase lags are $-49°$ and $-5.2°$, what is the total phase lag?

7–3 ■ OP-AMP CLOSED-LOOP RESPONSE

Op-amps are normally used in a closed-loop configuration with negative feedback in order to achieve precise control of the gain and bandwidth. In this section, you will see how feedback affects the gain and frequency response of an op-amp.

After completing this section, you should be able to

❑ Understand the closed-loop response of an op-amp
  ❑ Determine the closed-loop gain
  ❑ Explain the effect of negative feedback on bandwidth
  ❑ Explain gain-bandwidth product

Recall from Chapter 6 that midrange gain is reduced by negative feedback, as indicated by the following closed-loop gain expressions for the three configurations previously covered. For the noninverting amplifier,

$$A_{cl(NI)} = \frac{R_f}{R_i} + 1$$

For the voltage-follower,

$$A_{cl(VF)} \cong 1$$

For the inverting amplifier,

$$A_{cl(I)} \cong -\frac{R_f}{R_i}$$

Effect of Negative Feedback on Bandwidth

You have learned how negative feedback affects the gain; now you will learn how it affects the amplifier's bandwidth. The closed-loop critical frequency of an op-amp is

$$f_{c(cl)} = f_{c(ol)}(1 + BA_{ol(mid)}) \tag{7–4}$$

This expression shows that the closed-loop critical frequency, $f_{c(cl)}$, is higher than the open-loop critical frequency $f_{c(ol)}$ by the factor $1 + BA_{ol(mid)}$. Recall that B is the feedback attenuation, $R_i/(R_i + R_f)$. A derivation of Equation (7–4) can be found in Appendix B.

Since $f_{c(cl)}$ equals the bandwidth for the closed-loop amplifier, the bandwidth is also increased by the same factor.

$$BW_{cl} = BW_{ol}(1 + BA_{ol(mid)}) \tag{7–5}$$

EXAMPLE 7–4

A certain amplifier has an open-loop midrange gain of 150,000 and an open-loop 3 dB bandwidth of 200 Hz. The attenuation of the feedback loop is 0.002. What is the closed-loop bandwidth?

Solution

$$BW_{cl} = BW_{ol}(1 + BA_{ol(mid)}) = 200 \text{ Hz}[1 + (0.002)(150,000)] = \textbf{60.2 kHz}$$

Practice Exercise If $A_{ol(mid)} = 200{,}000$ and $B = 0.05$, what is the closed loop bandwidth?

Figure 7–8 graphically illustrates the concept of closed-loop response for a compensated op-amp. When the open-loop gain of an op-amp is reduced by negative feedback, the bandwidth is increased. The closed-loop gain is independent of the open-loop gain up to the point of intersection of the two gain curves. This point of intersection is the critical frequency, $f_{c(cl)}$, for the closed-loop response. Notice that beyond the closed-loop critical frequency the closed-loop gain has the same roll-off rate as the open-loop gain.

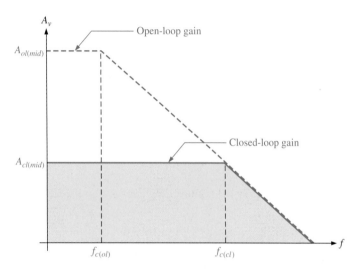

FIGURE 7–8
Closed-loop gain compared to open-loop gain.

Gain-Bandwidth Product

An increase in closed-loop gain causes a decrease in the bandwidth and vice versa, such that *the product of gain and bandwidth is a constant.* This is true as long as the roll-off rate is a fixed -20 dB/decade. If A_{cl} represents the gain of any of the closed-loop configurations and $f_{c(cl)}$ represents the closed-loop critical frequency (same as the bandwidth), then

$$A_{cl}f_{c(cl)} = A_{ol}f_{c(ol)}$$

The gain-bandwidth product is always equal to the frequency at which the op-amp's open-loop gain is unity (unity-gain bandwidth).[1]

$$A_{cl}f_{c(cl)} = \text{unity-gain bandwidth} \tag{7–6}$$

---

[1] Technically speaking, this equation is true only for noninverting configurations. Other cases are discussed in the lab manual.

EXAMPLE 7–5

Determine the bandwidth of each of the amplifiers in Figure 7–9. Both op-amps have an open-loop gain of 100 dB and a unity-gain bandwidth of 3 MHz.

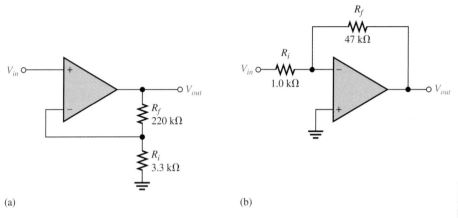

(a)

(b)

FIGURE 7–9

Solution

(a) For the noninverting amplifier in Figure 7–9(a), the closed-loop gain is

$$A_{cl(NI)} = \frac{R_f}{R_i} + 1 = \frac{220 \text{ k}\Omega}{3.3 \text{ k}\Omega} + 1 = 67.7$$

Use Equation (7–6) and solve for $f_{c(cl)}$ (where $f_{c(cl)} = BW_{cl}$).

$$f_{c(cl)} = BW_{cl} = \frac{\text{unity-gain } BW}{A_{cl}}$$

$$BW_{cl} = \frac{3 \text{ MHz}}{67.7} = \textbf{44.3 kHz}$$

(b) For the inverting amplifier in Figure 7–9(b), the closed-loop gain is

$$A_{cl(I)} = -\frac{R_f}{R_i} = -\frac{47 \text{ k}\Omega}{1.0 \text{ k}\Omega} = -47$$

Using the absolute value of $A_{cl(I)}$, the closed-loop bandwidth is

$$BW_{cl} = \frac{3 \text{ MHz}}{47} = \textbf{63.8 kHz}$$

Practice Exercise   Determine the bandwidth of each of the amplifiers in Figure 7–9. Both op-amps have an A'_{ol} of 90 dB and a unity-gain bandwidth of 2 MHz.

7–3 REVIEW QUESTIONS

1. Is the closed-loop gain always less than the open-loop gain?
2. A certain op-amp is used in a feedback configuration having a gain of 30 and a bandwidth of 100 kHz. If the external resistor values are changed to increase the gain to 60, what is the new bandwidth?
3. What is the unity-gain bandwidth of the op-amp in Question 2?

7–4 ■ POSITIVE FEEDBACK AND STABILITY

Stability is a consideration when using op-amps. Stable operation means that the op-amp does not oscillate under any condition. Instability produces oscillations, which are unwanted voltage swings on the output when there is no signal present on the input, or in response to noise or transient voltages on the input. This section may be treated as optional.

After completing this section, you should be able to

❏ Discuss positive feedback and stability in op-amp circuits
  ❏ Define *positive feedback*
  ❏ Define *loop gain*
  ❏ Define *phase margin* and discuss its importance
  ❏ Determine if an op-amp circuit is stable
  ❏ Summarize the criteria for stability

Positive Feedback

To understand stability, you must first examine instability and its causes. As you know, with negative feedback, the signal fed back to the input of an amplifier is out of phase with the input signal, thus subtracting from it and effectively reducing the voltage gain. As long as the feedback is negative, the amplifier is stable.

When the signal fed back from output to input is in phase with the input signal, a **positive feedback** condition exists and the amplifier can oscillate. That is, positive feedback occurs when the total phase shift through the op-amp and feedback network is 360°, which is equivalent to no phase shift (0°).

Loop Gain

For instability to occur, (a) there must be positive feedback, and (b) the loop gain of the closed-loop amplifier must be greater than 1. The **loop gain** of a closed-loop amplifier is defined to be the op-amp's open-loop gain times the attenuation of the feedback network.

$$\text{Loop gain} = A_{ol}B \qquad\qquad (7\text{–}7)$$

Phase Margin

Notice that for each amplifier configuration in Figure 7–10, the feedback loop is connected to the inverting input. There is an inherent phase shift of 180° because of the *inversion* between input and output. Additional phase shift (ϕ_{tot}) is produced by the *RC* lag networks (not shown) within the amplifier. So, the total phase shift around the loop is $180° + \phi_{tot}$.

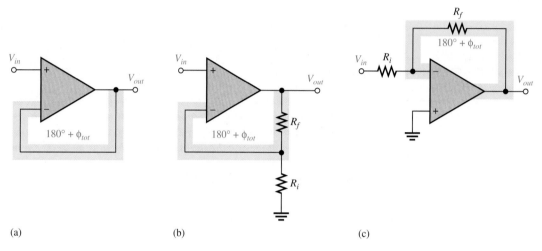

(a)   (b)   (c)

FIGURE 7–10
Feedback-loop phase shift.

The **phase margin**, ϕ_{pm}, is the amount of additional phase shift required to make the total phase shift around the loop 360°. (360° is equivalent to 0°.)

$$180° + \phi_{tot} + \phi_{pm} = 360°$$

$$\phi_{pm} = 180° - |\phi_{tot}| \tag{7–8}$$

If the phase margin is positive, the total phase shift is less than 360° and the amplifier is stable. If the phase margin is zero or negative, then the amplifier is potentially unstable because the signal fed back can be in phase with the input. As you can see from Equation (7–8), when the total lag network phase shift (ϕ_{tot}) equals or exceeds 180°, then the phase margin is 0° or negative and an unstable condition exists, which would cause the amplifier to oscillate.

Stability Analysis

Since most op-amp configurations use a loop gain greater than 1 ($A_{ol}B > 1$), the criteria for stability are based on the phase angle of the internal lag networks. As previously mentioned, operational amplifiers are composed of multiple stages, each of which has a critical frequency. For compensated op-amps, only one critical frequency is dominant, and stability due to the feedback is not a problem. Stability problems generally manifest themselves as unwanted oscillations. Feedback stability occurs near the unity-gain frequency for the op-amp.

To illustrate the concept of feedback **stability**, we will use an uncompensated three-stage op-amp with an open-loop response as shown in the Bode plot of Figure 7–11. For this case, there are three different critical frequencies, which indicate three internal *RC* lag networks. At the first critical frequency, f_{c1}, the gain begins rolling off at −20 dB/decade; when the second critical frequency, f_{c2}, is reached, the gain decreases at −40 dB/decade; and when the third critical frequency, f_{c3}, is reached, the gain drops at −60 dB/decade.

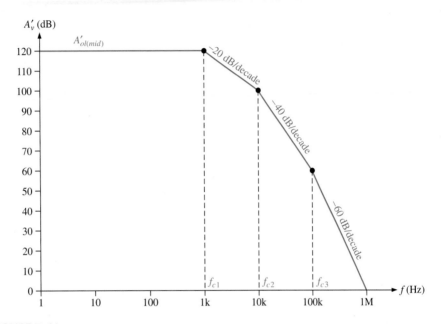

FIGURE 7–11
Bode plot of example of three-stage op-amp response.

To analyze an uncompensated closed-loop amplifier for stability, the phase margin must be determined. A positive phase margin will indicate that the amplifier is stable for a given value of closed-loop gain. Three example cases will be considered in order to demonstrate the conditions for instability.

Case 1   The closed-loop gain intersects the open-loop response on the −20 dB/decade slope, as shown in Figure 7–12. The midrange closed-loop gain is 106 dB, and the closed-loop critical frequency is 5 kHz. If we assume that the amplifier is not operated out of its midrange, the maximum phase shift for the 106 dB amplifier occurs at the highest midrange frequency (in this case, 5 kHz). The total phase shift at this frequency due to the three lag networks is calculated as follows:

$$\phi_{tot} = -\tan^{-1}\left(\frac{f}{f_{c1}}\right) - \tan^{-1}\left(\frac{f}{f_{c2}}\right) - \tan^{-1}\left(\frac{f}{f_{c3}}\right)$$

where $f = 5$ kHz, $f_{c1} = 1$ kHz, $f_{c2} = 10$ kHz, and $f_{c3} = 100$ kHz. Therefore,

$$\phi_{tot} = -\tan^{-1}\left(\frac{5 \text{ kHz}}{1 \text{ kHz}}\right) - \tan^{-1}\left(\frac{5 \text{ kHz}}{10 \text{ kHz}}\right) - \tan^{-1}\left(\frac{5 \text{ kHz}}{100 \text{ kHz}}\right)$$

$$= -78.7° - 26.6° - 2.9° = -108.1°$$

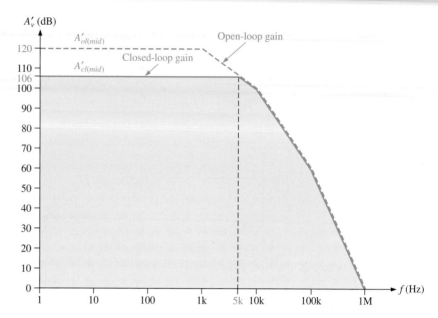

FIGURE 7–12

Case where closed-loop gain intersects open-loop gain on −20 dB/decade slope (stable operation).

The phase margin, ϕ_{pm}, is

$$\phi_{pm} = 180° - |\phi_{tot}| = 180° - 108.1° = +71.9°$$

The phase margin is positive, so the amplifier is stable for all frequencies in its midrange. In general, an amplifier is stable for all midrange frequencies if its closed-loop gain intersects the open-loop response curve on a −20 dB/decade slope.

Case 2   The closed-loop gain is lowered to where it intersects the open-loop response on the −40 dB/decade slope, as shown in Figure 7–13. The midrange closed-loop gain in this case is 80 dB, and the closed-loop critical frequency is approximately 30 kHz. The total phase shift at $f = 30$ kHz due to the three lag networks is calculated as follows:

$$\phi_{tot} = -\tan^{-1}\left(\frac{30 \text{ kHz}}{1 \text{ kHz}}\right) - \tan^{-1}\left(\frac{30 \text{ kHz}}{10 \text{ kHz}}\right) - \tan^{-1}\left(\frac{30 \text{ kHz}}{100 \text{ kHz}}\right)$$

$$= -88.1° - 71.6° - 16.7° = -176.4°$$

The phase margin is

$$\phi_{pm} = 180° - 176.4° = +3.6°$$

The phase margin is positive, so the amplifier is still stable for frequencies in its midrange, but a very slight increase in frequency above f_c would cause it to oscillate. Therefore, it is marginally stable and may oscillate due to other paths. It is very close to instability because instability occurs where $\phi_{pm} = 0°$. As a general rule, a minimum 45° phase margin is recommended to avoid marginal conditions.

Case 3   The closed-loop gain is further decreased until it intersects the open-loop response on the −60 dB/decade slope, as shown in Figure 7–14. The midrange closed-loop

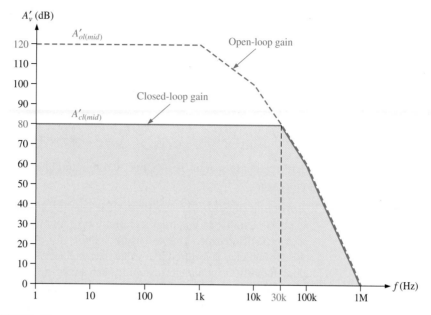

FIGURE 7–13

Case where closed-loop gain intersects open-loop gain on −40 dB/decade slope (marginally stable operation).

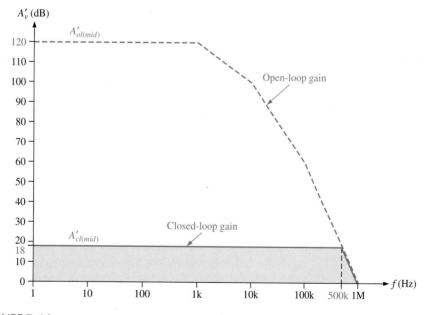

FIGURE 7–14

Case where closed-loop gain intersects open-loop gain on −60 dB/decade slope (unstable operation).

gain in this case is 18 dB, and the closed-loop critical frequency is 500 kHz. The total phase shift at $f = 500$ kHz due to the three lag networks is

$$\phi_{tot} = -\tan^{-1}\left(\frac{500 \text{ kHz}}{1 \text{ kHz}}\right) - \tan^{-1}\left(\frac{500 \text{ kHz}}{10 \text{ kHz}}\right) - \tan^{-1}\left(\frac{500 \text{ kHz}}{100 \text{ kHz}}\right)$$

$$= -89.9° - 88.9° - 78.7° = -257.5°$$

The phase margin is

$$\phi_{pm} = 180° - 257.5° = -77.5°$$

Here the phase margin is negative and the amplifier is unstable at the upper end of its midrange.

Summary of Stability Criteria The stability analysis of the three example cases has demonstrated that an amplifier's closed-loop gain must intersect the open-loop gain curve on a -20 dB/decade slope to ensure stability for all of its midrange frequencies. If the closed-loop gain is lowered to a value that intersects on a -40 dB/decade slope, then marginal stability or complete instability can occur. In the previous situations (Cases 1, 2, and 3), the closed-loop gain should be greater than 72 dB.

If the closed-loop gain intersects the open-loop response on a -60 dB/decade slope, instability will definitely occur at some frequency within the amplifier's midrange unless a specially designed feedback network is used. Therefore, to ensure stability for all of the midrange frequencies, an op-amp must be operated at a closed-loop gain such that the roll-off rate beginning at its dominant critical frequency does not exceed -20 dB/decade.

Troubleshooting Unwanted Oscillations

The stability problems mentioned in this section can be brought under control, even in the case of a negative phase margin (Case 3), by specially designed feedback networks. A lead network in the feedback path can be used to increase the phase margin and thus increase the stability. In some cases, a complicated feedback network with an amplifier or other active element is added to a design to increase stability.

Not all stability problems are due to the feedback network. If oscillations are not near the unity-gain frequency of the op-amp, the feedback loop is probably not the culprit. Causes of oscillations can include the presence of an external feedback path, a grounding problem, or an extraneous noise signal coupled into the power supply lines. When oscillations are a problem, a simple test is to increase the gain and see if they disappear. (This means the closed-loop gain will intersect the open-loop gain at a higher point.) If the oscillations persist, the problem may be something other than a negative phase margin.

To eliminate unwanted oscillations, check ground paths (try to use single-point grounding), add bypass capacitors to the supply voltages, and try to eliminate extraneous capacitive coupling paths to the input. A coupling path may not be obvious but can be due to a protoboard, especially if it has no ground plane, or may be caused by long leads in the circuit (remember that wires have capacitance). Power supply noise can produce feedback in the amplifier that can result in oscillations. At low frequencies, a simple bypass capacitor (1 μF to 10 μF tantalum) may be all that is necessary to solve the problem. At high frequencies, a single bypass capacitor may have a self-resonance, requiring the addition of a secondary bypass capacitor.

Occasionally, oscillations are due to interference from nearby sources and may require shielding. It is also possible to induce oscillations when a low-level signal shares a common ground path with a high-level signal or because of long leads in the circuit layout.

Try reconstructing the circuit with shorter leads, paying attention to ground paths and making sure a ground plane is present, if possible.

7–4 REVIEW QUESTIONS

1. Under what feedback condition can an amplifier oscillate?
2. How much can the phase shift of an amplifier's internal RC network be before instability occurs? What is the phase margin at the point where instability begins?
3. What is the maximum roll-off rate of the open-loop gain of an op-amp for which the device will still be stable?

7–5 ■ OP-AMP COMPENSATION

The last section demonstrated that instability can occur when an op-amp's response has roll-off rates exceeding −20 dB/decade and the op-amp is operated in a closed-loop configuration having a gain curve that intersects a higher roll-off rate portion of the open-loop response. In situations like those examined in the last section, the closed-loop voltage gain is restricted to very high values. In many applications, lower values of closed-loop gain are necessary or desirable. To allow op-amps to be operated at low closed-loop gain, phase lag compensation is required. This section may be treated as optional.

After completing this section, you should be able to

❑ Explain op-amp phase compensation
   ❑ Describe phase-lag compensation
   ❑ Explain a compensating circuit
   ❑ Apply single-capacitor compensation
   ❑ Apply feedforward compensation

Phase Lag Compensation

As you have seen, the cause of instability is excessive phase shift through an op-amp's internal lag networks. When these phase shifts equal or exceed 180°, the amplifier can oscillate. **Compensation** is used to either eliminate open-loop roll-off rates greater than −20 dB/decade or extend the −20 dB/decade rate to a lower gain. These concepts are illustrated in Figure 7–15.

Compensating Network

There are two basic methods of compensation for integrated circuit op-amps: internal and external. In either case an RC network is added. The basic compensating action is as follows. Consider first the RC network shown in Figure 7–16(a). At low frequencies where the reactance of the compensating capacitor, X_{C_c}, is extremely large, the output voltage approximately equals the input voltage. When the frequency reaches its critical value, $f_c = 1/[2\pi(R_1 + R_2)C_c]$, the output voltage decreases at −20 dB/decade. This roll-off rate continues until $X_{C_c} \cong 0$, at which point the output voltage levels off to a value determined

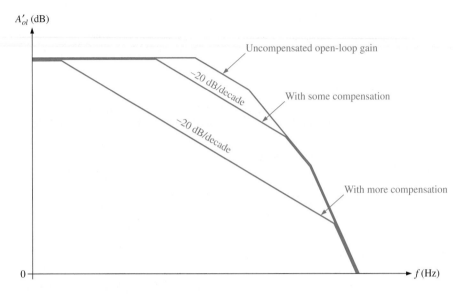

FIGURE 7–15

Bode plot illustrating effect of phase compensation on open-loop gain of typical op-amp.

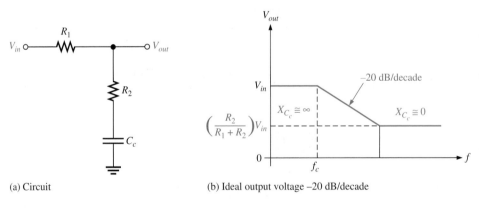

(a) Circuit

(b) Ideal output voltage –20 dB/decade

FIGURE 7–16

Basic compensating network action.

by R_1 and R_2, as indicated in Figure 7–16(b). This is the principle used in the phase compensation of an op-amp.

To see how a compensating network changes the open-loop response of an op-amp, refer to Figure 7–17. This diagram represents a two-stage op-amp. The individual stages are within the color-shaded blocks along with the associated lag networks. A compensating network is shown connected at point *A* on the output of stage 1.

The critical frequency of the compensating network is set to a value less than the dominant (lowest) critical frequency of the internal lag networks. This causes the −20 dB/decade roll-off to begin at the compensating network's critical frequency. The roll-off of the compensating network continues up to the critical frequency of the dominant lag network. At this point, the response of the compensating network levels off, and the

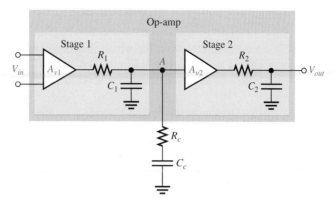

FIGURE 7–17
Representation of op-amp with compensation.

−20 dB/decade roll-off of the dominant lag network takes over. The net result is a shift of the open-loop response to the left, thus reducing the bandwidth, as shown in Figure 7–18. The response curve of the compensating network is shown in proper relation to the overall open-loop response.

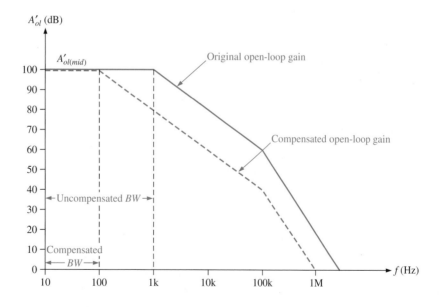

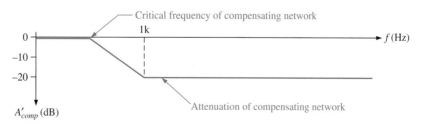

FIGURE 7–18
Example of compensated op-amp frequency response.

EXAMPLE 7–6   A certain op-amp has the open-loop response in Figure 7–19. As you can see, the lowest closed-loop gain for which stability is assured is approximately 40 dB (where the closed-loop gain line still intersects the −20 dB/decade slope). In a particular application, a 20 dB closed-loop gain is required.

(a) Determine the critical frequency for the compensating network.
(b) Sketch the ideal response curve for the compensating network.
(c) Sketch the total ideal compensated open-loop response.

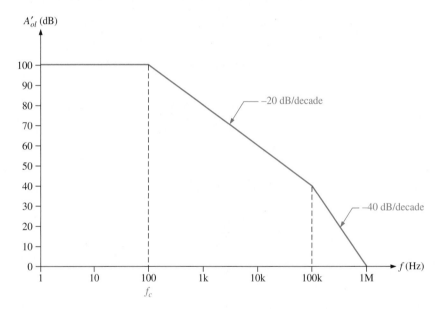

FIGURE 7–19
Original open-loop response.

Solution

(a) The gain must be dropped so that the −20 dB/decade roll-off extends down to 20 dB rather than to 40 dB. To achieve this, the midrange open-loop gain must be made to roll off a decade sooner. Therefore, the critical frequency of the compensating network must be 10 Hz.

(b) The roll-off of the compensating network must end at 100 Hz, as shown in Figure 7–20(a).

(c) The total open-loop response resulting from compensation is shown in Figure 7–20(b).

FIGURE 7–20

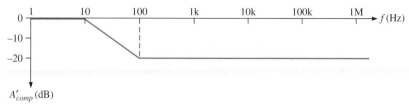

(a) Compensating network response

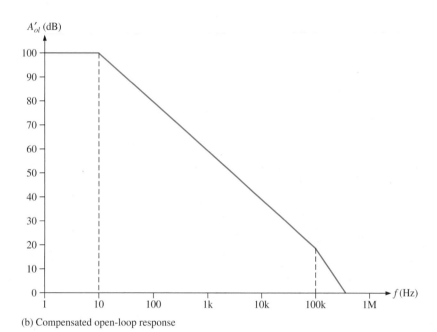

(b) Compensated open-loop response

Practice Exercise   In this example, what is the uncompensated bandwidth? What is the compensated bandwidth?

Extent of Compensation

A larger compensating capacitor will cause the open-loop roll-off to begin at a lower frequency and thus extend the −20 dB/decade roll-off to lower gain levels, as shown in Figure 7–21(a). With a sufficiently large compensating capacitor, an op-amp can be made unconditionally stable, as illustrated in Figure 7–21(b), where the −20 dB/decade slope is extended all the way down to unity gain. This is normally the case when internal compensation is provided by the manufacturer. An internally, fully compensated op-amp can be used for any value of closed-loop gain and remain stable. The 741 is an example of an internally fully compensated device.

A disadvantage of internally fully compensated op-amps is that bandwidth is sacrificed; thus the slew rate is decreased. Therefore, many IC op-amps have provisions for external compensation. Figure 7–22 shows typical package layouts of an LM101A op-amp with pins available for external compensation with a small capacitor. With provisions for external connections, just enough compensation can be used for a given application without sacrificing more performance than necessary.

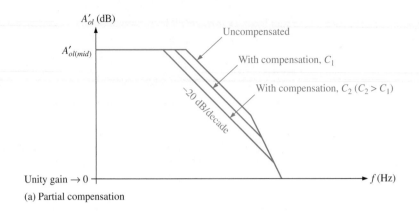

(a) Partial compensation

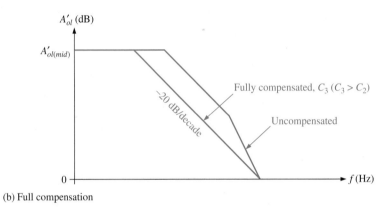

(b) Full compensation

FIGURE 7–21
Extent of compensation.

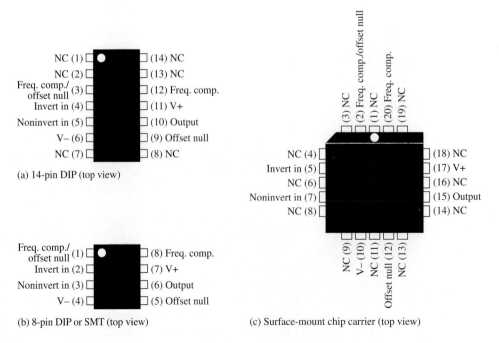

(a) 14-pin DIP (top view)

(b) 8-pin DIP or SMT (top view)

(c) Surface-mount chip carrier (top view)

FIGURE 7–22
Typical op-amp packages.

Single-Capacitor Compensation

As an example of compensating an IC op-amp, a capacitor C_1 is connected to pins 1 and 8 of an LM101A in an inverting amplifier configuration, as shown in Figure 7–23(a). Part (b) of the figure shows the open-loop frequency response curves for two values of C_1. The 3 pF compensating capacitor produces a unity-gain bandwidth approaching 10 MHz. Notice that the -20 dB/decade slope extends to a very low gain value. When C_1 is increased ten times to 30 pF, the bandwidth is reduced by a factor of ten. Notice that the -20 dB/decade slope now extends through unity gain.

When the op-amp is used in a closed-loop configuration, as in Figure 7–23(c), the useful frequency range depends on the compensating capacitor. For example, with a closed-loop gain of 40 dB as shown in part (c), the bandwidth is approximately 10 kHz for $C_1 = 30$ pF and increases to approximately 100 kHz when C_1 is decreased to 3 pF.

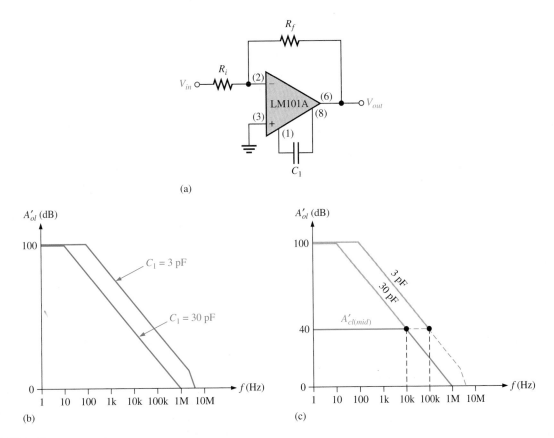

(a)

(b)

(c)

FIGURE 7–23

Example of single-capacitor compensation of an LM101A op-amp.

Feedforward Compensation

Another method of phase compensation is called **feedforward**. This type of compensation results in less bandwidth reduction than the method previously discussed. The basic concept is to bypass the internal input stage of the op-amp at high frequencies and drive the higher-frequency second stage, as shown in Figure 7–24.

FIGURE 7–24
Feedforward compensation showing high-frequency bypassing of first stage.

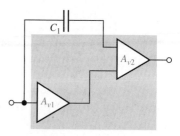

Feedforward compensation of an LM101A is shown in Figure 7–25(a). The feedforward capacitor C_1 is connected from the inverting input to the compensating terminal. A small capacitor is needed across R_f to ensure stability. The Bode plot in Figure 7–25(b) shows the feedforward compensated response and the standard compensated response that was discussed previously. The use of feedforward compensation is restricted to the inverting amplifier configuration. Other compensation methods are also used. Often, recommendations are provided by the manufacturer on the data sheet.

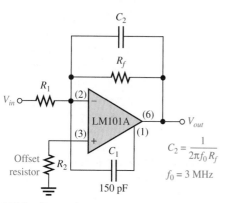

$$C_2 = \frac{1}{2\pi f_0 R_f}$$

$$f_0 = 3 \text{ MHz}$$

(a) Manufacturers' recommended configuration

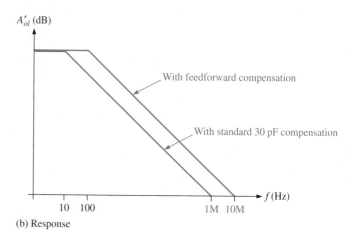

(b) Response

FIGURE 7–25
Feedforward compensation of an LM101A op-amp and the response curves.

7–5 REVIEW QUESTIONS

1. What is the purpose of phase compensation?
2. What is the main difference between internal and external compensation?
3. When you compensate an amplifier, does the bandwidth increase or decrease?

7–6 ■ A SYSTEM APPLICATION

In this system application, we are focusing on the two audio amplifier boards in an FM stereo receiver. Both boards are identical except one is for the left channel sound and the other is for the right channel sound. This circuit is a good example of a mixed use of an integrated circuit and discrete components.

After completing this section, you should be able to

❑ Apply what you have learned in this chapter to a system application
  ❑ See how an op-amp is used as an audio amplifier
  ❑ Identify the functions of various components on the board
  ❑ Analyze the circuit's operation
  ❑ Translate between a printed circuit board and a schematic
  ❑ Troubleshoot some common amplifier failures

A Brief Description of the System

Some general information about the stereo system might be helpful before you concentrate on the audio amplifiers. When an FM stereo broadcast is received by a standard single-speaker system, the output to the speaker is equal to the sum of the left plus the right channel audio, so you get the original sound without separation. When a stereo receiver is used, the full stereo effect is reproduced by the two speakers. Stereo FM signals are transmitted on a carrier frequency of 88 MHz to 108 MHz. The complete stereo signal consists of three modulating signals. These are the sum of the left and right channel audio, the difference of the left and right channel audio, and a pilot subcarrier. These three signals are detected and are used to separate out the left and right channel audio by special circuits. The channel audio amplifiers then amplify each signal equally and drive the speakers. It is not necessary for you to understand this process for the purposes of this system application, although you may be interested in doing further study in this area on your own.

The two channel audio amplifiers are identical, so we will look at only one. The op-amp serves basically as a preamplifier that drives the power amplifier stage.

Now, so that you can take a closer look at one of the audio amplifier boards, let's take one out of the system and put it on the troubleshooter's bench.

TROUBLESHOOTER'S BENCH

■ ACTIVITY 1   Relate the PC Board to the Schematic

The schematic for the audio amplifier board in Figure 7–26 is shown in Figure 7–27. Using this schematic, locate and label each component on the PC board. The board has several feedthrough pads for connections that are on the back side. Backside traces are shown as darker lines. Compare the board to the schematic.

■ ACTIVITY 2   Analyze the Circuit

Step 1: Determine the midrange voltage gain.

Step 2: Determine the lower critical frequency. Given that the upper critical frequency is 15 kHz, what is the bandwidth?

Step 3: Determine the maximum peak-to-peak input voltage that can be applied without producing a distorted output signal. Assume that the maximum output peaks are 1 V less than the supply voltages.

■ ACTIVITY 3   Write a Technical Report

Describe the overall operation of the circuit and the function of each component. In discussing the general operation and basic purpose of each component, make sure you identify the negative feedback loop, the type of op-amp configuration, which components

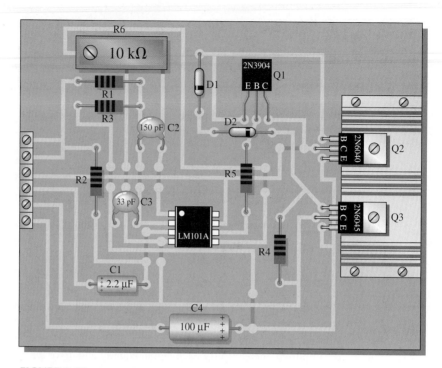

FIGURE 7–26

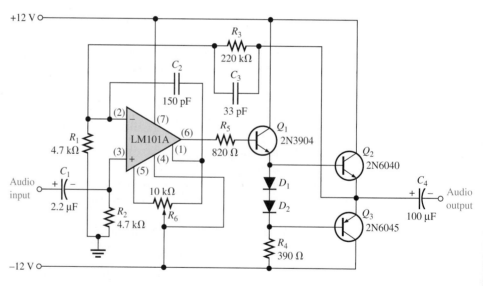

FIGURE 7–27

determine the voltage gain, which components set the lower critical frequency, and the purpose of each of the capacitors. Use the results of Activity 2 when appropriate.

■ **ACTIVITY 4 Troubleshoot the Audio Amplifier PC Boards for Each of the Following Problems by Stating the Probable Cause or Causes in Each Case**

1. No final output signal when there is a verified input signal.
2. The positive half-cycle of the output voltage is severely distorted or missing.
3. Output severely clipped on both positive and negative cycles.

COLOR
INSERT

■ **ACTIVITY 5 Troubleshooter's Bench Special Assignment**

Go to Troubleshooter's Bench 2 in the color insert section and carry out the assignment that is stated there.

7–6 REVIEW QUESTIONS

1. How can the lower critical frequency of the amplifier be reduced?
2. Which transistors form the class B power amplifier?
3. What is the purpose of Q_1 and what type of circuit is it?

■ SUMMARY

- Open-loop gain is the voltage gain of an op-amp without feedback.
- Closed-loop gain is the voltage gain of an op-amp with negative feedback.
- The closed-loop gain is always less than the open-loop gain.
- The midrange gain of an op-amp extends down to dc.
- Above the critical frequency, the gain of an op-amp decreases.
- The internal RC lag networks that are inherently part of the amplifier stages cause the gain to roll off as frequency goes up.
- The internal RC lag networks also cause a phase shift between input and output signals.
- Negative feedback lowers the gain and increases the bandwidth.
- The product of gain and bandwidth is constant for a compensated op-amp.
- The gain-bandwidth product equals the frequency at which unity voltage gain occurs.
- Positive feedback occurs when the total phase shift through the op-amp (including 180° inversion) and feedback network is 0° (equivalent to 360°) or more.
- The phase margin is the amount of additional phase shift required to make the total phase shift around the loop 360°.
- When the closed-loop gain of an op-amp intersects the open-loop response curve on a −20 dB/decade (−6 dB/octave) slope, the amplifier is stable.
- When the closed-loop gain intersects the open-loop response curve on a slope greater than −20 dB/decade, the amplifier can be either marginally stable or unstable.
- A minimum phase margin of 45° is recommended to provide a sufficient safety factor for stable operation.

- A fully compensated op-amp has a -20 dB/decade roll-off all the way down to unity gain.
- Compensation reduces bandwidth and increases slew rate.
- Internally compensated op-amps such as the 741 are available. These are usually fully compensated with a large sacrifice in bandwidth.
- Externally compensated op-amps such as LM101A are available. External compensating networks can be connected to specified pins, and the compensation can be tailored to a specific application. In this way, bandwidth and slew rate are not degraded more than necessary.

■ GLOSSARY

Key terms are in color. All terms are included in the end-of-book glossary.

Bandwidth   The range of frequencies between the lower critical frequency and the upper critical frequency.

Compensation   The process of modifying the roll-off rate of an amplifier to ensure stability.

Feedforward   A method of frequency compensation in op-amp circuits.

Loop gain   An op-amp's open-loop voltage gain times the attenuation of the feedback network.

Phase shift   The relative angular displacement of a time-varying function relative to a reference.

Phase margin   The difference between the total phase shift through an amplifier and 180°; the additional amount of phase shift that can be allowed before instability occurs.

Positive feedback   The return of a portion of the output signal to the input such that it reinforces the output. This output signal is in phase with the input signal.

Stability   A condition in which an amplifier circuit does not oscillate.

Uncompensated op-amp   An op-amp with more than one critical frequency.

■ KEY FORMULAS

(7–1) $BW = f_{cu}$ Op-amp bandwidth

(7–2) $A_{ol} = \dfrac{A_{ol(mid)}}{\sqrt{1 + f^2/f_c^2}}$ Open-loop gain

(7–3) $\phi = -\tan^{-1}\left(\dfrac{f}{f_c}\right)$ RC phase shift

(7–4) $f_{c(cl)} = f_{c(ol)}(1 + BA_{ol(mid)})$ Closed-loop critical frequency

(7–5) $BW_{cl} = BW_{ol}(1 + BA_{ol(mid)})$ Closed-loop bandwidth

(7–6) $A_{cl}f_{c(cl)} =$ unity-gain bandwidth

(7–7)   Loop gain $= A_{ol}B$

(7–8) $\phi_{pm} = 180° - |\phi_{tot}|$ Phase margin

■ SELF-TEST

Answers are at the end of the chapter.

1. The open-loop gain of an op-amp is always
   (a) less than the closed-loop gain
   (b) equal to the closed-loop gain
   (c) greater than the closed-loop gain
   (d) a very stable and constant quantity for a given type of op-amp

2. The bandwidth of an ac amplifier having a lower critical frequency of 1 kHz and an upper critical frequency of 10 kHz is
   (a) 1 kHz   (b) 9 kHz   (c) 10 kHz   (d) 11 kHz

3. The bandwidth of a dc amplifier having an upper critical frequency of 100 kHz is
    (a) 100 kHz   (b) unknown   (c) infinity   (d) 0 kHz

4. The midrange open-loop gain of an op-amp
    (a) extends from the lower critical frequency to the upper critical frequency
    (b) extends from 0 Hz to the upper critical frequency
    (c) rolls off at −20 dB/decade beginning at 0 Hz
    (d) answers (b) and (c)

5. The frequency at which the open-loop gain is equal to one is called
    (a) the upper critical frequency   (b) the cutoff frequency
    (c) the notch frequency   (d) the unity-gain frequency

6. Phase shift through an op-amp is caused by
    (a) the internal RC networks   (b) the external RC networks
    (c) the gain roll-off   (d) negative feedback

7. Each RC network in an op-amp
    (a) causes the gain to roll off at −6 dB/octave
    (b) causes the gain to roll off at −20 dB/decade
    (c) reduces the midrange gain by 3 dB
    (d) answers (a) and (b)

8. When negative feedback is used, the gain-bandwidth product of an op-amp
    (a) increases   (b) decreases   (c) stays the same   (d) fluctuates

9. If a certain noninverting op-amp has a midrange open-loop gain of 200,000 and a unity-gain fre-
    quency of 5 MHz, the gain-bandwidth product is
    (a) 200,000 Hz   (b) 5,000,000 Hz
    (c) 1×10^{12} Hz   (d) not determinable from the information given

10. If a certain noninverting op-amp has a closed-loop gain of 20 and an upper critical frequency of
    10 MHz, the gain-bandwidth product is
    (a) 200 MHz   (b) 10 MHz   (c) the unity-gain frequency   (d) answers (a) and (c)

11. Positive feedback occurs when
    (a) the output signal is fed back to the input in-phase with the input signal
    (b) the output signal is fed back to the input out-of-phase with the input signal
    (c) the total phase shift through the op-amp and feedback network is 360°
    (d) answers (a) and (c)

12. For a closed-loop op-amp circuit to be unstable,
    (a) there must be positive feedback
    (b) the loop gain must be greater than 1
    (c) the loop gain must be less than 1
    (d) answers (a) and (b)

13. The amount of additional phase shift required to make the total phase shift around a closed loop
    equal to zero is called
    (a) the unity-gain phase shift   (b) phase margin
    (c) phase lag   (d) phase bandwidth

14. For a given value of closed-loop gain, a positive phase margin indicates
    (a) an unstable condition   (b) too much phase shift
    (c) a stable condition   (d) nothing

15. The purpose of phase-lag compensation is to
    (a) make the op-amp stable at very high values of gain
    (b) make the op-amp stable at low values of gain
    (c) reduce the unity-gain frequency
    (d) increase the bandwidth

TROUBLESHOOTER'S QUIZ   *Answers are at the end of the chapter.*

Refer to Figure 7–31(a).

❑ If R_f is 100 kΩ instead of the specified 150 kΩ,

    1. For a low-frequency input signal, the gain will

        (a) increase   **(b)** decrease   **(c)** not change

    2. The bandwidth will

        (a) increase   **(b)** decrease   **(c)** not change

    3. The gain-bandwidth product will

        (a) increase   **(b)** decrease   **(c)** not change

❑ If the op-amp has an $f_{c(ol)}$ of 200 Hz instead of the specified 150 Hz,

    4. The bandwidth will

        (a) increase   **(b)** decrease   **(c)** not change

    5. For a low-frequency input signal, the gain will

        (a) increase   **(b)** decrease   **(c)** not change

Refer to Figure 7–27.

❑ If C_1 is 0.22 μF instead of the specified 2.2 μF,

    6. The lower cutoff frequency will

        (a) increase   **(b)** decrease   **(c)** not change

    7. The upper cutoff frequency will

        (a) increase   **(b)** decrease   **(c)** not change

❑ If C_3 is open,

    8. The stability will

        (a) increase   **(b)** decrease   **(c)** not change

❑ If R_3 is larger than the specified value,

    9. The bandwidth will

        (a) increase   **(b)** decrease   **(c)** not change

    10. The gain-bandwidth product will

        (a) increase   **(b)** decrease   **(c)** not change

❑ If the compensating capacitor C_2 is open,

    11. The bandwidth will

        (a) increase   **(b)** decrease   **(c)** not change

    12. The stability will

        (a) increase   **(b)** decrease   **(c)** not change

■ PROBLEMS

Answers to odd-numbered problems are at the end of the book.

SECTION 7–1 Basic Concepts

1. The midrange open-loop gain of a certain op-amp is 120 dB. Negative feedback reduces this gain by 50 dB. What is the closed-loop gain?

2. The upper critical frequency of an op-amp's open-loop response is 200 Hz. If the midrange gain is 175,000, what is the ideal gain at 200 Hz? What is the actual gain? What is the op-amp's open-loop bandwidth?

3. An RC lag network has a critical frequency of 5 kHz. If the resistance value is 1.0 kΩ, what is X_C when $f = 3$ kHz?

4. Determine the attenuation of an RC lag network with $f_c = 12$ kHz for each of the following frequencies.
   (a) 1 kHz (b) 5 kHz (c) 12 kHz (d) 20 kHz (e) 100 kHz

5. The midrange open-loop gain of a certain op-amp is 80,000. If the open-loop critical frequency is 1 kHz, what is the open-loop gain at each of the following frequencies?
   (a) 100 Hz (b) 1 kHz (c) 10 kHz (d) 1 MHz

6. Determine the phase shift through each network in Figure 7–28 at a frequency of 2 kHz.

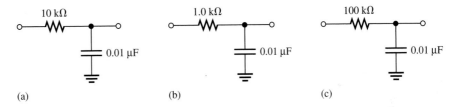

(a)　　　　　　　　　(b)　　　　　　　　　(c)

FIGURE 7–28

7. An RC lag network has a critical frequency of 8.5 kHz. Determine the phase for each frequency and plot a graph of its phase angle versus frequency.
   (a) 100 Hz (b) 400 Hz (c) 850 Hz
   (d) 8.5 kHz (e) 25 kHz (f) 85 kHz

SECTION 7–2 Op-Amp Open-Loop Response

8. A certain op-amp has three internal amplifier stages with midrange gains of 30 dB, 40 dB, and 20 dB. Each stage also has a critical frequency associated with it as follows: $f_{c1} = 600$ Hz, $f_{c2} = 50$ kHz, and $f_{c3} = 200$ kHz.
   (a) What is the midrange open-loop gain of the op-amp, expressed in dB?
   (b) What is the total phase shift through the amplifier, including inversion, when the signal frequency is 10 kHz?

9. What is the gain roll-off rate in Problem 8 between the following frequencies?
   (a) 0 Hz and 600 Hz (b) 600 Hz and 50 kHz
   (c) 50 kHz and 200 kHz (d) 200 kHz and 1 MHz

SECTION 7–3 Op-Amp Closed-Loop Response

10. A certain amplifier has an open-loop gain in midrange of 180,000 and an open-loop critical frequency of 1500 Hz. If the attenuation of the feedback path is 0.015, what is the closed-loop bandwidth?

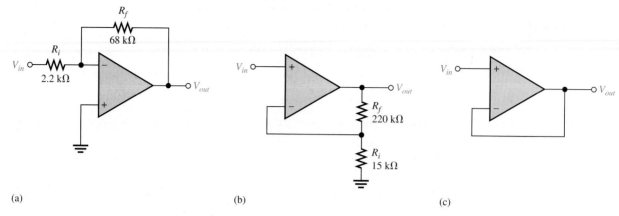

(a)   (b)   (c)

FIGURE 7–29

11. Determine the midrange gain in dB of each amplifier in Figure 7–29. Are these open-loop or closed-loop gains?

12. Given that $f_{c(ol)}$ = 750 Hz, A'_{ol} = 89 dB, and $f_{c(cl)}$ = 5.5 kHz, determine the closed-loop gain in dB.

13. What is the unity-gain bandwidth in Problem 12?

14. For each amplifier in Figure 7–30, determine the closed-loop gain and bandwidth. The op-amps in each circuit exhibit an open-loop gain of 125 dB and a unity-gain bandwidth of 2.8 MHz.

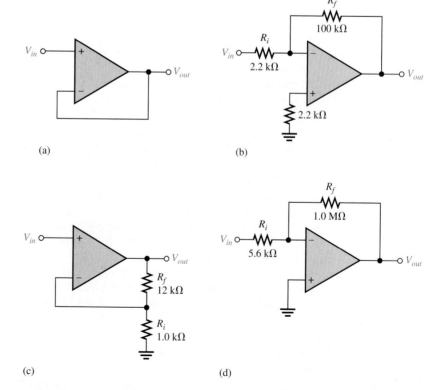

(a)   (b)

(c)   (d)

FIGURE 7–30

FIGURE 7–31

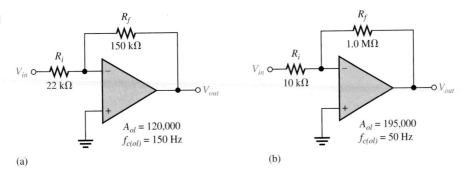

(a)

(b)

15. Which of the amplifiers in Figure 7–31 has the smaller bandwidth?

SECTION 7–4   Positive Feedback and Stability

16. It has been determined that the op-amp circuit in Figure 7–32 has three internal critical frequencies as follows: 1.2 kHz, 50 kHz, 250 kHz. If the midrange open-loop gain is 100 dB, is the amplifier configuration stable, marginally stable, or unstable?

17. Determine the phase margin for each value of phase lag.
   (a) 30°   (b) 60°   (c) 120°   (d) 180°   (e) 210°

18. A certain op-amp has the following internal critical frequencies in its open-loop response: 125 Hz, 25 kHz, and 180 kHz. What is the total phase shift through the amplifier when the signal frequency is 50 kHz?

19. Each graph in Figure 7–33 shows both the open-loop and the closed-loop response of a particular op-amp configuration. Analyze each case for stability.

FIGURE 7–32

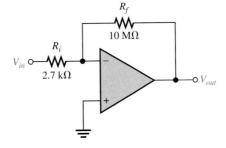

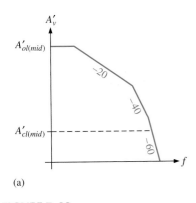

(a)

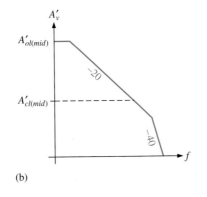

(b)

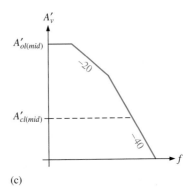

(c)

FIGURE 7–33

SECTION 7–5 Op-Amp Compensation

20. A certain operational amplifier has an open-loop response curve as shown in Figure 7–34. A particular application requires a 30 dB closed-loop midrange gain. In order to achieve a 30 dB gain, compensation must be added because the 30 dB line intersects the uncompensated open-loop gain on the -40 dB/decade slope and, therefore, stability is not assured.
  (a) Find the critical frequency of the compensating network such that the -20 dB/decade slope is lowered to a point where it intersects the 30 dB gain line.
  (b) Sketch the ideal response curve for the compensating network.
  (c) Sketch the total ideal compensated open-loop response.

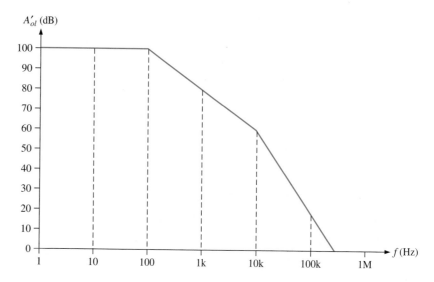

FIGURE 7–34

21. The open-loop gain of a certain op-amp rolls off at -20 dB/decade, beginning at $f = 250$ Hz. This roll-off rate extends down to a gain of 60 dB. If a 40 dB closed-loop gain is required, what is the critical frequency for the compensating network?

22. Repeat Problem 21 for a closed-loop gain of 20 dB.

■ **ANSWERS TO REVIEW QUESTIONS**

Section 7–1

1. Open-loop gain is without feedback, and closed-loop gain is with negative feedback. Open-loop gain is larger.

2. $BW = 100$ Hz

3. A_{ol} decreases.

Section 7–2

1. $A'_{v(tot)} = 20$ dB $+ 30$ dB $= 50$ dB

2. $\phi_{tot} = -49° + (-5.2°) = -54.2°$

Section 7–3

1. Yes, A_{cl} is always less than A_{ol}.
2. $BW = 3,000 \text{ kHz}/60 = 50 \text{ kHz}$
3. unity-gain $BW = 3,000 \text{ kHz}/1 = 3 \text{ MHz}$

Section 7–4

1. Positive feedback
2. $180°$, $0°$
3. -20 dB/decade (-6 dB/octave)

Section 7–5

1. Phase compensation increases the phase margin at a given frequency.
2. Internal compensation is full compensation; external compensation can be tailored to maximize bandwidth.
3. Bandwidth decreases.

Section 7–6

1. f_{cl} can be reduced by increasing C_1 or R_2.
2. Q_2 and Q_3
3. Q_1 is an emitter-follower buffer stage.

■ **ANSWERS TO PRACTICE EXERCISES FOR EXAMPLES**

7–1 (a) 80,000   (b) 79,900   (c) 6400
7–2 173 Hz
7–3 75 dB; $-71.6°$
7–4 2.00 MHz
7–5 (a) 29.6 kHz   (b) 42.6 kHz
7–6 100 Hz; 10 Hz

■ **ANSWERS TO SELF-TEST**

1. (c)	**2.** (b)	**3.** (a)	**4.** (b)	**5.** (d)
6. (a)	**7.** (d)	**8.** (c)	**9.** (b)	**10.** (d)
11. (d)	**12.** (d)	**13.** (b)	**14.** (c)	**15.** (b)

■ **ANSWERS TO TROUBLESHOOTER'S QUIZ**

1. decrease	**2.** increase	**3.** not change	**4.** increase
5. not change	**6.** increase	**7.** not change	**8.** decrease
9. decrease	**10.** not change	**11.** increase	**12.** decrease

8

BASIC OP-AMP CIRCUITS

Courtesy Hewlett-Packard Company

■ CHAPTER OBJECTIVES

❑ Understand the operation of several basic
  comparator circuits
❑ Understand the operation of several types of
  summing amplifiers
❑ Understand the operation of integrators and
  differentiators
❑ Understand the operation of several special
  op-amp circuits
❑ Troubleshoot basic op-amp circuits
❑ Apply what you have learned in this chapter
  to a system application

■ KEY TERMS

❑ Comparator
❑ Hysteresis
❑ Schmitt trigger
❑ Bounding
❑ Summing amplifier
❑ Integrator
❑ Differentiator
❑ Constant-current source
❑ Current-to-voltage converter
❑ Voltage-to-current converter
❑ Peak detector

■ CHAPTER INTRODUCTION

In the last two chapters, you learned about the principles, operation, and characteristics of the operational amplifier. Op-amps are used in such a wide variety of applications that it is impossible to cover all of them in one chapter, or even in one book. Therefore, in this chapter, we will examine some of the more fundamental applications to illustrate how versatile the op-amp is and to give you a foundation in basic op-amp circuits.

■ A SYSTEM APPLICATION

This system application illustrates a very interesting application of three types of op-amp circuits that will be studied in this chapter: the summing amplifier, integrator, and comparator. The system diagram in Figure 8–47 shows one basic type of analog-to-digital converter that takes an audio input, such as voice or music, and converts it to binary codes that can be recorded digitally. Analog-to-digital converters are covered thoroughly in Chapter 14.

Op-amps are a key part of this system, and we will be focusing on the analog-to-digital converter board to see how these circuits are used in a representative application. The digital circuits are discussed just enough to allow you to understand what the overall system does. You do not need to have a background in digital circuits for our purposes here. However, this particular system application points out the fact, again, that many systems include combinations of both analog and digital circuits.

For the system application in Section 8–6, in addition to the other topics, be sure you understand
❑ How a summing amplifier works
❑ How an integrator works
❑ How a comparator works

www.   VISIT THE COMPANION WEBSITE

Study Aids for This Chapter Are Available at

http://www.prenhall.com/floyd

8–1 ◼ COMPARATORS

Operational amplifiers are often used as nonlinear devices to compare the amplitude of one voltage with another. In this application, the op-amp is used in the open-loop configuration, with the input voltage on one input and a reference voltage on the other.

After completing this section, you should be able to

❑ Understand the operation of several basic comparator circuits
    ❑ Describe the operation of a zero-level detector
    ❑ Describe the operation of a nonzero-level detector
    ❑ Discuss how input noise affects comparator operation
    ❑ Define *hysteresis*
    ❑ Explain how hysteresis reduces noise effects
    ❑ Describe a Schmitt trigger circuit
    ❑ Describe the operation of bounded comparators
    ❑ Describe the operation of a window comparator
    ❑ Discuss two comparator applications including analog-to-digital conversion

Zero-Level Detection

One application of an op-amp used as a **comparator** is to determine when an input voltage exceeds a certain level. Figure 8–1(a) shows a zero-level detector. Notice that the inverting (−) input is grounded to produce a zero level and that the input signal voltage is applied to the noninverting (+) input. Because of the high open-loop voltage gain, a very small difference voltage between the two inputs drives the amplifier into saturation, causing the output voltage to go to its limit.

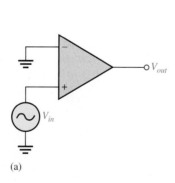
(a)

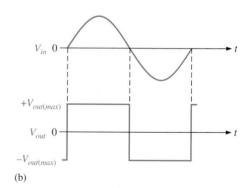
(b)

FIGURE 8–1
The op-amp as a zero-level detector.

For example, consider an op-amp having A_{ol} = 100,000. A voltage difference of only 0.25 mV between the inputs could produce an output voltage of (0.25 mV)(100,000) = 25 V *if* the op-amp were capable. However, since most op-amps have output voltage limitations of ±15 V or less, the device would be driven into saturation. For many comparison applications, special op-amp comparators are selected. These ICs are generally uncompensated to maximize speed. In less stringent applications, a general-purpose op-amp works nicely as a comparator.

Figure 8–1(b) shows the result of a sinusoidal input voltage applied to the noninverting input of the zero-level detector. When the sine wave is negative, the output is at its maximum negative level. When the sine wave crosses 0, the amplifier is driven to its opposite state and the output goes to its maximum positive level, as shown. As you can see, the zero-level detector can be used as a squaring circuit to produce a square wave from a sine wave.

Nonzero-Level Detection

The zero-level detector in Figure 8–1 can be modified to detect positive and negative voltages by connecting a fixed reference voltage to the inverting (−) input, as shown in Figure 8–2(a). A more practical arrangement is shown in Figure 8–2(b) using a voltage divider to set the reference voltage as follows:

$$V_{REF} = \frac{R_2}{R_1 + R_2}(+V) \tag{8–1}$$

where $+V$ is the positive op-amp supply voltage. The circuit in Figure 8–2(c) uses a zener diode to set the reference voltage ($V_{REF} = V_Z$). As long as the input voltage V_{in} is less than

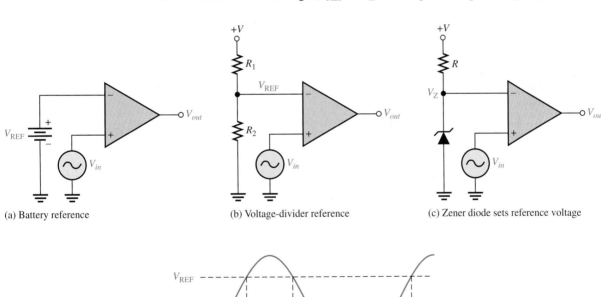

(a) Battery reference   (b) Voltage-divider reference   (c) Zener diode sets reference voltage

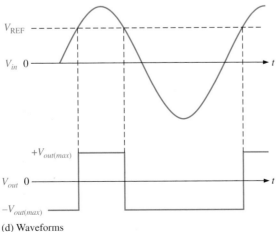

(d) Waveforms

FIGURE 8–2
Nonzero-level detectors.

V_{REF}, the output remains at the maximum negative level. When the input voltage exceeds the reference voltage, the output goes to its maximum positive state, as shown in Figure 8–2(d) with a sinusoidal input voltage.

EXAMPLE 8–1

The input signal in Figure 8–3(a) is applied to the comparator circuit in Figure 8–3(b). Make a sketch of the output showing its proper relationship to the input signal. Assume the maximum output levels of the op-amp are ± 12 V.

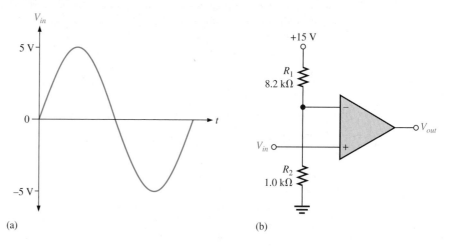

(a)

(b)

FIGURE 8–3

Solution   The reference voltage is set by R_1 and R_2 as follows:

$$V_{REF} = \frac{R_2}{R_1 + R_2} (+V) = \frac{1.0 \text{ k}\Omega}{8.2 \text{ k}\Omega + 1.0 \text{ k}\Omega} (+15 \text{ V}) = 1.63 \text{ V}$$

As shown in Figure 8–4, each time the input exceeds $+1.63$ V, the output voltage switches to its $+12$ V level, and each time the input goes below $+1.63$ V, the output switches back to its -12 V level.

FIGURE 8–4

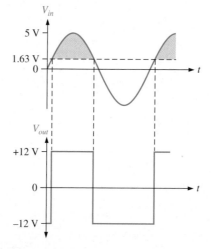

Practice Exercise* Determine the reference voltage in Figure 8–3 if $R_1 = 22$ kΩ and $R_2 = 3.3$ kΩ.

* Answers are at the end of the chapter.

Effects of Input Noise on Comparator Operation

In many practical situations, **noise** (unwanted voltage or current fluctuations) may appear on the input line. This noise voltage becomes superimposed on the input voltage, as shown in Figure 8–5, and can cause a comparator to erratically switch output states.

FIGURE 8–5
Sine wave with superimposed noise.

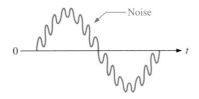

In order to understand the potential effects of noise voltage, consider a low-frequency sinusoidal voltage applied to the noninverting (+) input of an op-amp comparator used as a zero-level detector, as shown in Figure 8–6(a). Part (b) of the figure shows the input sine wave plus noise and the resulting output. As you can see, when the sine wave

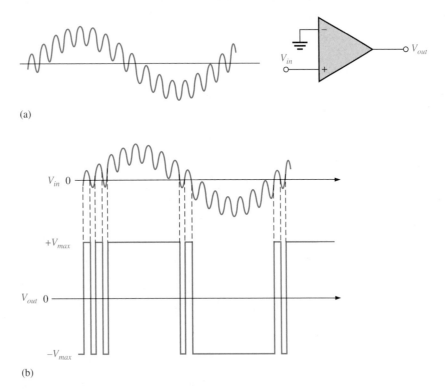

FIGURE 8–6
Effects of noise on comparator circuit.

approaches 0, the fluctuations due to noise cause the total input to vary above and below 0 several times, thus producing an erratic output voltage.

Reducing Noise Effects with Hysteresis

An erratic output voltage caused by noise on the input occurs because the op-amp comparator switches from its negative output state to its positive output state at the same input voltage level that causes it to switch in the opposite direction, from positive to negative. This unstable condition occurs when the input voltage hovers around the reference voltage, and any small noise fluctuations cause the comparator to switch first one way and then the other.

In order to make the comparator less sensitive to noise, a technique incorporating positive feedback, called **hysteresis**, can be used. Basically, hysteresis means that there is a higher reference level when the input voltage goes from a lower to higher value than when it goes from a higher to a lower value. A good example of hysteresis is a common household thermostat that turns the furnace on at one temperature and off at another.

The two reference levels are referred to as the upper trigger point (UTP) and the lower trigger point (LTP). This two-level hysteresis is established with a positive feedback arrangement, as shown in Figure 8–7. Notice that the noninverting (+) input is connected to a resistive voltage divider such that a portion of the output voltage is fed back to the input. The input signal is applied to the inverting (−) input in this case.

FIGURE 8–7
Comparator with positive feedback for hysteresis.

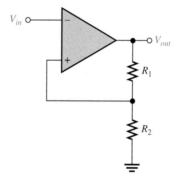

The basic operation of the comparator with hysteresis is as follows and is illustrated in Figure 8–8. Assume that the output voltage is at its positive maximum, $+V_{out(max)}$. The voltage fed back to the noninverting input is V_{UTP} and is expressed as

$$V_{UTP} = \frac{R_2}{R_1 + R_2}(+V_{out(max)})$$

When the input voltage V_{in} exceeds V_{UTP}, the output voltage drops to its negative maximum, $-V_{out(max)}$. Now the voltage fed back to the noninverting input is V_{LTP} and is expressed as

$$V_{LTP} = \frac{R_2}{R_1 + R_2}(-V_{out(max)})$$

The input voltage must now fall below V_{LTP} before the device will switch back to its other voltage level. This means that a small amount of noise voltage has no effect on the output, as illustrated by Figure 8–8.

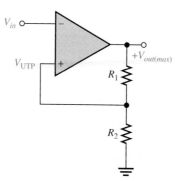

(a) Output at the maximum positive voltage

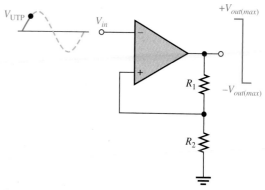

(b) Input exceeds UTP; output switches from the maximum positive voltage to the maximum negative voltage.

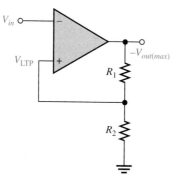

(c) Output at the maximum negative voltage

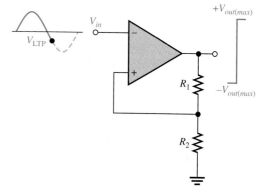

(d) Input goes below LTP; output switches from maximum negative voltage back to maximum positive voltage.

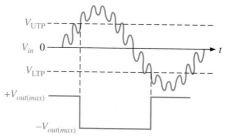

(e) Device triggers only once when UTP or LTP is reached; thus, there is immunity to noise that is riding on the input signal.

FIGURE 8–8

Operation of a comparator with hysteresis.

A comparator with hysteresis is sometimes known as a Schmitt trigger. The amount of hysteresis is defined by the difference of the two trigger levels.

$$V_{HYS} = V_{UTP} - V_{LTP} \tag{8-2}$$

EXAMPLE 8–2

Determine the upper and lower trigger points and the hysteresis for the comparator circuit in Figure 8–9. Assume that $+V_{out(max)} = +5 \text{ V}$ and $-V_{out(max)} = -5 \text{ V}$.

FIGURE 8–9

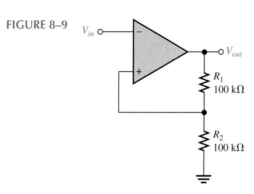

Solution

$$V_{UTP} = \frac{R_2}{R_1 + R_2}[+V_{out(max)}] = 0.5(5 \text{ V}) = \textbf{+2.5 V}$$

$$V_{LTP} = \frac{R_2}{R_1 + R_2}[-V_{out(max)}] = 0.5(-5 \text{ V}) = \textbf{-2.5 V}$$

$$V_{HYS} = V_{UTP} - V_{LTP} = 2.5 \text{ V} - (-2.5 \text{ V}) = \textbf{5 V}$$

Practice Exercise Determine the upper and lower trigger points and the hysteresis in Figure 8–9 for $R_1 = 68 \text{ k}\Omega$ and $R_2 = 82 \text{ k}\Omega$. The maximum output voltage levels are ±7 V.

Output Bounding

In some applications, it is necessary to limit the output voltage levels of a comparator to a value less than that provided by the saturated op-amp. A single zener diode can be used as shown in Figure 8–10 to limit the output voltage to the zener voltage in one direction and to the forward diode drop in the other. This process of limiting the output range is called **bounding**.

FIGURE 8–10
Comparator with output bounding.

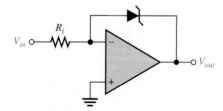

The operation is as follows. Since the anode of the zener is connected to the inverting (−) input, it is at virtual ground (≅0 V). Therefore, when the output voltage reaches a positive value equal to the zener voltage, it limits at that value, as illustrated in Figure 8–11. When the output switches negative, the zener acts as a regular diode and becomes forward-biased at 0.7 V, limiting the negative output voltage to this value, as shown. Turning the zener around limits the output voltage in the opposite direction.

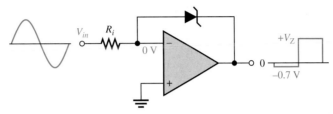

(a) Bounded at a positive value

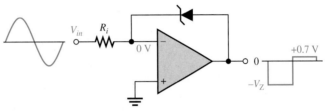

(b) Bounded at a negative value

FIGURE 8–11
Operation of a bounded comparator.

Two zener diodes arranged as in Figure 8–12 limit the output voltage to the zener voltage plus the forward voltage drop (0.7 V) of the forward-biased zener, both positively and negatively, as shown in Figure 8–12.

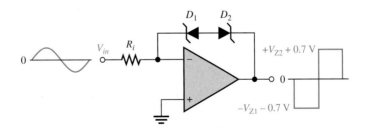

FIGURE 8–12
Double-bounded comparator.

EXAMPLE 8–3

Determine the output voltage waveform for Figure 8–13.

Solution   This comparator has both hysteresis and zener bounding.
    The voltage across D_1 and D_2 in either direction is 4.7 V + 0.7 V = 5.4 V. This is because one zener is always forward-biased with a drop of 0.7 V when the other one is in breakdown.

FIGURE 8–13

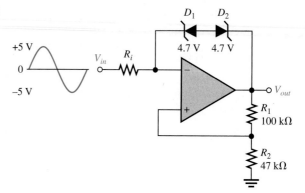

The voltage at the inverting (−) op-amp input is $V_{out} \pm 5.4$ V. Since the differential voltage is negligible, the voltage at the noninverting (+) op-amp input is also approximately $V_{out} \pm 5.4$ V. Thus,

$$V_{R1} = V_{out} - (V_{out} \pm 5.4 \text{ V}) = \pm 5.4 \text{ V}$$

$$I_{R1} = \frac{V_{R1}}{R_1} = \frac{\pm 5.4 \text{ V}}{100 \text{ k}\Omega} = \pm 54 \text{ }\mu\text{A}$$

Since the current at the noninverting input is negligible,

$$I_{R2} = I_{R1} = \pm 54 \text{ }\mu\text{A}$$

$$V_{R2} = R_2 I_{R2} = (47 \text{ k}\Omega)(\pm 54 \text{ }\mu\text{A}) = \pm 2.54 \text{ V}$$

$$V_{out} = V_{R1} + V_{R2} = \pm 5.4 \text{ V} \pm 2.54 \text{ V} = \pm 7.94 \text{ V}$$

The upper trigger point (UTP) and the lower trigger point (LTP) are as follows:

$$V_{UTP} = \left(\frac{R_2}{R_1 + R_2}\right)(+V_{out}) = \left(\frac{47 \text{ k}\Omega}{147 \text{ k}\Omega}\right)(+7.94 \text{ V}) = +2.54 \text{ V}$$

$$V_{LTP} = \left(\frac{R_2}{R_1 + R_2}\right)(-V_{out}) = \left(\frac{47 \text{ k}\Omega}{147 \text{ k}\Omega}\right)(-7.94 \text{ V}) = -2.54 \text{ V}$$

The output waveform for the given input voltage is shown in Figure 8–14.

FIGURE 8–14

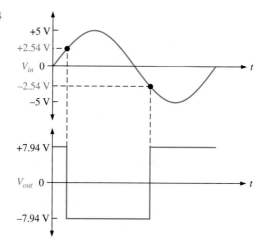

Practice Exercise   Determine the upper and lower trigger points for Figure 8–13 if $R_1 = 150$ kΩ, $R_2 = 68$ kΩ, and the zener diodes are 3.3 V devices.

Window Comparator

Two individual op-amp comparators arranged as in Figure 8–15 form what is known as a *window comparator.* This circuit detects when an input voltage is between two limits, an upper and a lower, called the "window."

FIGURE 8–15
A basic window comparator.

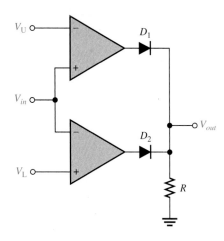

The upper and lower limits are set by reference voltages designated V_U and V_L. These voltages can be established with voltage dividers, zener diodes, or any type of voltage source. As long as V_{in} is within the window (less than V_U and greater than V_L), the output of each comparator is at its low saturated level. Under this condition, both diodes are reverse-biased and V_{out} is held at zero by the resistor to ground. When V_{in} goes above V_U or below V_L, the output of the associated comparator goes to its high saturated level. This action forward-biases the diode and produces a high-level V_{out}. This is illustrated in Figure 8–16 with V_{in} varying arbitrarily.

FIGURE 8–16
Example of window comparator operation.

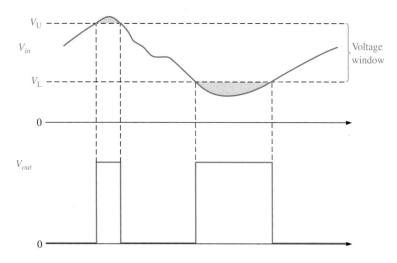

A Comparator Application: Over-Temperature Sensing Circuit

Figure 8–17 shows an op-amp comparator used in a precision over-temperature sensing circuit to determine when the temperature reaches a certain critical value. The circuit consists of a Wheatstone bridge with the op-amp used to detect when the bridge is balanced. One leg of the bridge contains a thermistor (R_1), which is a temperature-sensing resistor with a negative temperature coefficient (its resistance decreases as temperature increases and vice versa). The potentiometer (R_2) is set at a value equal to the resistance of the thermistor at the critical temperature. At normal temperatures (below critical), R_1 is greater than R_2, thus creating an unbalanced condition that drives the op-amp to its low saturated output level and keeps transistor Q_1 off.

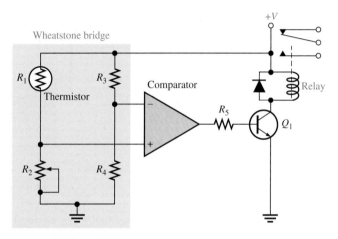

FIGURE 8–17
An over-temperature sensing circuit.

As the temperature increases, the resistance of the thermistor decreases. When the temperature reaches the critical value, R_1 becomes equal to R_2, and the bridge becomes balanced (since $R_3 = R_4$). At this point the op-amp switches to its high saturated output level, turning Q_1 on. This energizes the relay, which can be used to activate an alarm or initiate an appropriate response to the over-temperature condition.

A Comparator Application: Analog-to-Digital (A/D) Conversion

A/D conversion is a common interfacing process often used when a linear *analog* system must provide inputs to a *digital* system. Many methods for A/D conversion are available and some of these will be covered thoroughly in Chapter 14. However, in this discussion, only one type is used to demonstrate the concept.

The *simultaneous,* or *flash,* method of A/D conversion uses parallel comparators to compare the linear input signal with various reference voltages developed by a voltage divider. When the input voltage exceeds the reference voltage for a given comparator, a high level is produced on that comparator's output. Figure 8–18 shows an analog-to-digital converter (ADC) that produces three-digit binary numbers on its output, which represent the values of the analog input voltage as it changes. This converter requires seven comparators.

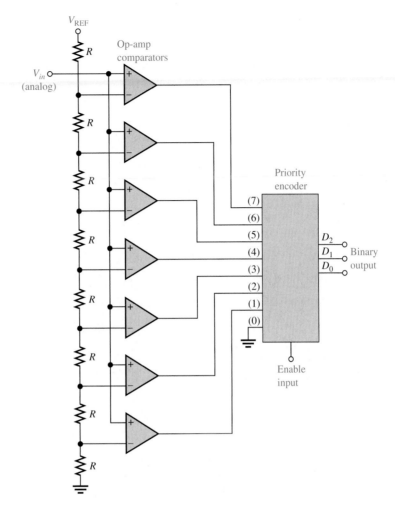

FIGURE 8–18

A simultaneous (flash) analog-to-digital converter (ADC) using op-amps as comparators.

In general, $2^n - 1$ comparators are required for conversion to an n-digit binary number. The large number of comparators necessary for a reasonably sized binary number is one of the drawbacks of this type of ADC. Its chief advantage is that it provides a fast conversion time.

The reference voltage for each comparator is set by the resistive voltage-divider network and V_{REF}. The output of each comparator is connected to an input of the priority encoder. The *priority encoder* is a digital device that produces a binary number on its outputs representing the highest-value input.

The encoder *samples* its input when a pulse occurs on the enable line (sampling pulse), and a three-digit binary number proportional to the value of the analog input signal appears on the encoder's outputs. The sampling rate determines the accuracy with which the sequence of binary numbers represents the changing input signal. The more samples taken in a given unit of time, the more accurately the analog signal is represented in digital form.

8–1 REVIEW QUESTIONS*

1. What is the reference voltage for each comparator in Figure 8–19?

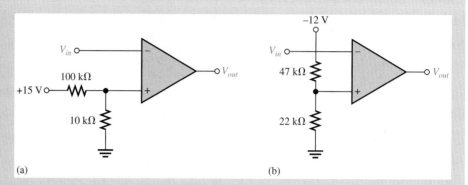

FIGURE 8–19

2. What is the purpose of hysteresis in a comparator?
3. Define the term *bounding* in relation to a comparator's output.

* Answers are at the end of the chapter.

8–2 ■ SUMMING AMPLIFIERS

The summing amplifier is a variation of the inverting op-amp configuration covered in Chapter 6. The summing amplifier has two or more inputs, and its output voltage is proportional to the negative of the algebraic sum of its input voltages. In this section, you will see how a summing amplifier works, and you will learn about the averaging amplifier and the scaling amplifier, which are variations of the basic summing amplifier.

After completing this section, you should be able to

❑ Understand the operation of several types of summing amplifiers
  ❑ Describe the operation of a unity-gain summing amplifier
  ❑ Discuss how to achieve any specified gain greater than unity
  ❑ Describe the operation of an averaging amplifier
  ❑ Describe the operation of a scaling adder
  ❑ Discuss a scaling adder used as a digital-to-analog converter

Summing Amplifier with Unity Gain

A two-input **summing amplifier** is shown in Figure 8–20, but any number of inputs can be used.

The operation of the circuit and derivation of the output expression are as follows. Two voltages, V_{IN1} and V_{IN2}, are applied to the inputs and produce currents I_1 and I_2, as shown. From the concepts of infinite input impedance and virtual ground, the voltage at the inverting ($-$) input of the op-amp is approximately 0 V, and therefore there is no current at

FIGURE 8–20
Two-input inverting summing amplifier.

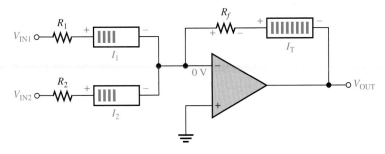

the input. This means that both input currents I_1 and I_2 combine at this summing point and form the total current, which is through R_f, as indicated ($I_T = I_1 + I_2$). Since $V_{OUT} = -I_T R_f$, the following steps apply.

$$V_{OUT} = -(I_1 + I_2)R_f = -\left(\frac{V_{IN1}}{R_1} + \frac{V_{IN2}}{R_2}\right)R_f$$

If all three of the resistors are equal in value ($R_1 = R_2 = R_f = R$), then

$$V_{OUT} = -\left(\frac{V_{IN1}}{R} + \frac{V_{IN2}}{R}\right)R = -(V_{IN1} + V_{IN2})$$

The previous equation shows that the output voltage has the same magnitude as the sum of the two input voltages but with a negative sign. A general expression is given in Equation (8–3) for a summing amplifier with n inputs, as shown in Figure 8–21 where all resistors are equal in value.

$$V_{OUT} = -(V_{IN1} + V_{IN2} + \cdots + V_{INn}) \qquad \textbf{(8–3)}$$

FIGURE 8–21
Summing amplifier with n inputs.

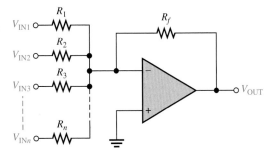

EXAMPLE 8–4   Determine the output voltage in Figure 8–22.

FIGURE 8–22

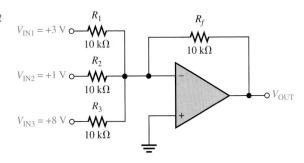

Solution

$$V_{OUT} = -(V_{IN1} + V_{IN2} + V_{IN3}) = -(3\text{ V} + 1\text{ V} + 8\text{ V}) = \mathbf{-12\text{ V}}$$

Practice Exercise   If a fourth input of +0.5 V is added to Figure 8–22 with a 10 kΩ resistor, what is the output voltage?

Summing Amplifier with Gain Greater Than Unity

When R_f is larger than the input resistors, the amplifier has a gain of R_f/R, where R is the value of each input resistor. The general expression for the output is

$$V_{OUT} = -\frac{R_f}{R}(V_{IN1} + V_{IN2} + \cdots + V_{INn}) \tag{8–4}$$

As you can see, the output has the same magnitude as the sum of all the input voltages multiplied by a constant determined by the ratio $-R_f/R$.

EXAMPLE 8–5

Determine the output voltage for the summing amplifier in Figure 8–23.

FIGURE 8–23

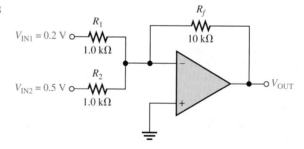

Solution $R_f = 10$ kΩ, and $R = R_1 = R_2 = 1.0$ kΩ. Therefore,

$$V_{OUT} = -\frac{R_f}{R}(V_{IN1} + V_{IN2}) = -\frac{10\text{ k}\Omega}{1.0\text{ k}\Omega}(0.2\text{ V} + 0.5\text{ V}) = -10(0.7\text{ V}) = \mathbf{-7\text{ V}}$$

Practice Exercise   Determine the output voltage in Figure 8–23 if the two input resistors are 2.2 kΩ and the feedback resistor is 18 kΩ.

Averaging Amplifier

A summing amplifier can be made to produce the mathematical average of the input voltages. This is done by setting the ratio R_f/R equal to the reciprocal of the number of inputs (n); that is, $R_f/R = 1/n$.

You obtain the average of several numbers by first adding the numbers and then dividing by the quantity of numbers you have. Examination of Equation (8–4) and a little thought will convince you that a summing amplifier will do this. The next example illustrates this idea.

EXAMPLE 8–6

Show that the amplifier in Figure 8–24 produces an output whose magnitude is the mathematical average of the input voltages.

FIGURE 8–24

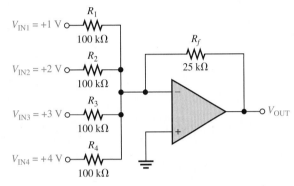

Solution Since the input resistors are equal, $R = 100\ k\Omega$. The output voltage is

$$V_{OUT} = -\frac{R_f}{R}(V_{IN1} + V_{IN2} + V_{IN3} + V_{IN4})$$

$$= -\frac{25\ k\Omega}{100\ k\Omega}(1\ V + 2\ V + 3\ V + 4\ V) = -\frac{1}{4}(10\ V) = \mathbf{-2.5\ V}$$

A simple calculation shows that the average of the input values is the same magnitude as V_{OUT} but of opposite sign.

$$V_{IN(avg)} = \frac{1\ V + 2\ V + 3\ V + 4\ V}{4} = \frac{10\ V}{4} = 2.5\ V$$

Practice Exercise Specify the changes required in the averaging amplifier in Figure 8–24 in order to handle five inputs.

Scaling Adder

A different weight can be assigned to each input of a summing amplifier by simply adjusting the values of the input resistors. As you have seen, the output voltage can be expressed as

$$V_{OUT} = -\left(\frac{R_f}{R_1}V_{IN1} + \frac{R_f}{R_2}V_{IN2} + \cdots + \frac{R_f}{R_n}V_{INn}\right) \qquad \textbf{(8–5)}$$

The weight of a particular input is set by the ratio of R_f to the resistance for that input. For example, if an input voltage is to have a weight of 1, then $R = R_f$. Or, if a weight of 0.5 is required, $R = 2R_f$. The smaller the value of R, the greater the weight, and vice versa.

EXAMPLE 8–7 Determine the weight of each input voltage for the scaling adder in Figure 8–25 and find the output voltage.

FIGURE 8–25

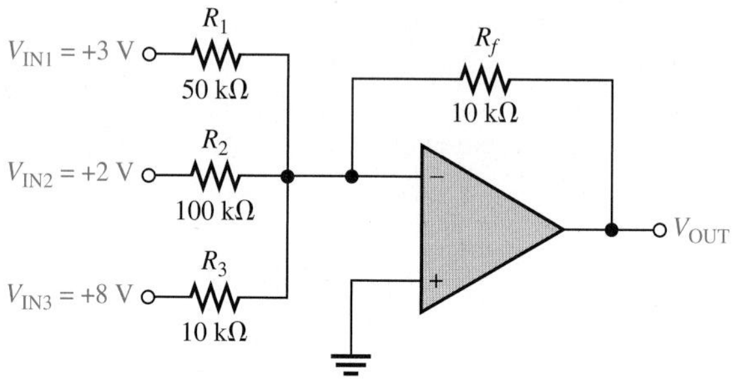

Solution

Weight of input 1: $\dfrac{R_f}{R_1} = \dfrac{10\ \text{k}\Omega}{50\ \text{k}\Omega} = \mathbf{0.2}$

Weight of input 2: $\dfrac{R_f}{R_2} = \dfrac{10\ \text{k}\Omega}{100\ \text{k}\Omega} = \mathbf{0.1}$

Weight of input 3: $\dfrac{R_f}{R_3} = \dfrac{10\ \text{k}\Omega}{10\ \text{k}\Omega} = \mathbf{1}$

The output voltage is

$$V_{\text{OUT}} = -\left(\frac{R_f}{R_1}V_{\text{IN1}} + \frac{R_f}{R_2}V_{\text{IN2}} + \frac{R_f}{R_3}V_{\text{IN3}}\right)$$

$$= -[0.2(3\ \text{V}) + 0.1(2\ \text{V}) + 1(8\ \text{V})] = -(0.6\ \text{V} + 0.2\ \text{V} + 8\ \text{V}) = \mathbf{-8.8\ V}$$

Practice Exercise Determine the weight of each input voltage in Figure 8–25 if $R_1 = 22\ \text{k}\Omega$, $R_2 = 82\ \text{k}\Omega$, $R_3 = 56\ \text{k}\Omega$, and $R_f = 10\ \text{k}\Omega$. Also find V_{OUT}.

A Scaling Adder Application: Digital-to-Analog (D/A) Conversion

D/A conversion is an important interface process for converting digital signals to analog (linear) signals. An example is a voice signal that is digitized for storage, processing, or transmission and must be changed back into an approximation of the original audio signal in order to drive a speaker. Digital-to-analog converters will be covered thoroughly in Chapter 14.

One method of D/A conversion uses a scaling adder with input resistor values that represent the binary weights of the digital input code. Figure 8–26 shows a four-digit digital-to-analog converter (DAC) of this type (called a *binary-weighted resistor DAC*). The switch symbols represent transistor switches for applying each of the four binary digits to the inputs.

The inverting (−) input is at virtual ground, so that the output voltage is proportional to the current through the feedback resistor R_f (sum of input currents). The lowest-value resistor R corresponds to the highest weighted binary input (2^3). All of the other resistors are multiples of R and correspond to the binary weights 2^2, 2^1, and 2^0.

FIGURE 8–26

A scaling adder as a four-digit digital-to-analog converter (DAC).

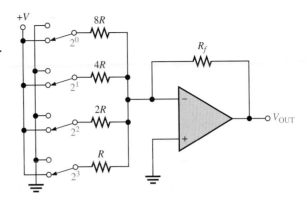

8–2 REVIEW QUESTIONS

1. Define *summing point.*
2. What is the value of R_f/R for a five-input averaging amplifier?
3. A certain scaling adder has two inputs, one having twice the weight of the other. If the resistor value for the lower weighted input is 10 kΩ, what is the value of the other input resistor?

8–3 ■ INTEGRATORS AND DIFFERENTIATORS

An op-amp integrator simulates mathematical integration, which is basically a summing process that determines the total area under the curve of a function. An op-amp differentiator simulates mathematical differentiation, which is a process of determining the instantaneous rate of change of a function. The integrators and differentiators shown in this section are idealized to show basic principles. Practical integrators often have an additional resistor or other circuitry in parallel with the feedback capacitor to prevent saturation. Practical differentiators may include a series resistor to reduce high frequency noise.

After completing this section, you should be able to

❑ Understand the operation of integrators and differentiators
   ❑ Identify an integrator
   ❑ Discuss how a capacitor charges
   ❑ Determine the rate of change of an integrator's output
   ❑ Identify a differentiator
   ❑ Determine the output voltage of a differentiator

The Op-Amp Integrator

An ideal **integrator** is shown in Figure 8–27. Notice that the feedback element is a capacitor that forms an *RC* circuit with the input resistor.

FIGURE 8–27
An ideal op-amp integrator.

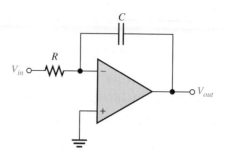

How a Capacitor Charges

To understand how the integrator works, it is important to review how a capacitor charges. Recall that the charge Q on a capacitor is proportional to the charging current (I_C) and the time (t).

$$Q = I_C t$$

Also, in terms of the voltage, the charge on a capacitor is

$$Q = CV_C$$

From these two relationships, the capacitor voltage can be expressed as

$$V_C = \left(\frac{I_C}{C}\right)t$$

This expression is an equation for a straight line which begins at zero with a constant slope of I_C/C. (Remember from algebra that the general formula for a straight line is $y = mx + b$. In this case, $y = V_C$, $m = I_C/C$, $x = t$, and $b = 0$.)

Recall that the capacitor voltage in a simple RC network is not linear but is exponential. This is because the charging current continuously decreases as the capacitor charges and causes the rate of change of the voltage to continuously decrease. The key thing about using an op-amp with an RC network to form an integrator is that the capacitor's charging current is made constant, thus producing a straight-line (linear) voltage rather than an exponential voltage. Now let's see why this is true.

In Figure 8–28, the inverting input of the op-amp is at virtual ground (0 V), so the voltage across R_i equals V_{in}. Therefore, the input current is

$$I_{in} = \frac{V_{in}}{R_i}$$

If V_{in} is a constant voltage, then I_{in} is also a constant because the inverting input always remains at 0 V, keeping a constant voltage across R_i. Because of the very high input

FIGURE 8–28
Currents in an integrator.

impedance of the op-amp, there is negligible current at the inverting input. This makes all of the input current charge the capacitor, so

$$I_C = I_{in}$$

The Capacitor Voltage   Since I_{in} is constant, so is I_C. The constant I_C charges the capacitor linearly and produces a linear voltage across C. The positive side of the capacitor is held at 0 V by the virtual ground of the op-amp. The voltage on the negative side of the capacitor decreases linearly from zero as the capacitor charges, as shown in Figure 8–29. This voltage is called a *negative ramp* and is the consequence of a constant positive input.

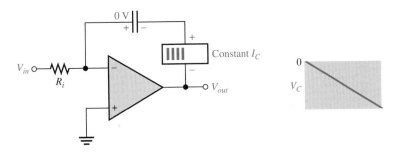

FIGURE 8–29
A linear ramp voltage is produced across C by the constant charging current.

The Output Voltage V_{out} is the same as the voltage on the negative side of the capacitor. When a constant positive input voltage in the form of a step or pulse (a pulse has a constant amplitude when high) is applied, the output ramp decreases negatively until the op-amp saturates at its maximum negative level. This is indicated in Figure 8–30.

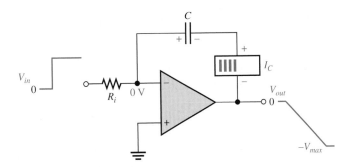

FIGURE 8–30
A constant input voltage produces a ramp on the output of the integrator.

Rate of Change of the Output   The rate at which the capacitor charges, and therefore the slope of the output ramp, is set by the ratio I_C/C, as you have seen. Since $I_C = V_{in}/R_i$, the rate of change or slope of the integrator's output voltage is

$$\frac{\Delta V_{out}}{\Delta t} = -\frac{V_{in}}{R_i C} \tag{8–6}$$

Integrators are especially useful in triangular-wave generators as you will see in Chapter 10.

EXAMPLE 8–8

(a) Determine the rate of change of the output voltage in response to the first input pulse in a pulse waveform, as shown for the integrator in Figure 8–31(a). The output voltage is initially zero.

(b) Describe the output after the first pulse. Draw the output waveform.

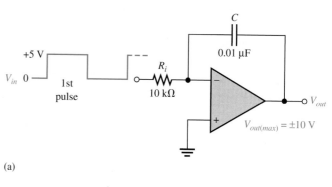

(a)

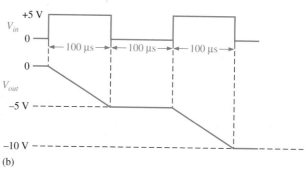

(b)

FIGURE 8–31

Solution

(a) The rate of change of the output voltage during the time that the input pulse is high is

$$\frac{\Delta V_{out}}{\Delta t} = -\frac{V_{in}}{R_i C} = -\frac{5 \text{ V}}{(10 \text{ k}\Omega)(0.01 \text{ }\mu\text{F})} = -50 \text{ kV/s} = \mathbf{-50 \text{ mV/}\mu\text{s}}$$

(b) The rate of change was found to be −50 mV/μs in part (a). When the input is at +5 V, the output is a negative-going ramp. When the input is at 0 V, the output is a constant level. In 100 μs, the voltage decreases.

$$\Delta V_{out} = (-50 \text{ mV/}\mu\text{s})(100 \text{ }\mu\text{s}) = \mathbf{-5 \text{ V}}$$

Therefore, the negative-going ramp reaches −5 V at the end of the pulse. The output voltage then remains constant at −5 V for the time that the input is zero. On the next pulse, the output again is a negative-going ramp that reaches −10 V. Since this is the maximum limit, the output remains at −10 V as long as pulses are applied. The waveforms are shown in Figure 8–31(b).

Practice Exercise Modify the integrator in Figure 8–31 to make the output change from 0 to −5 V in 50 μs with the same input.

The Op-Amp Differentiator

An ideal **differentiator** is shown in Figure 8–32. Notice how the placement of the capacitor and resistor differ from that in the integrator. The capacitor is now the input element. A differentiator produces an output that is proportional to the rate of change of the input voltage. Although a small-value resistor is normally used in series with the capacitor to limit the gain, it does not affect the basic operation and is not shown for purposes of this analysis.

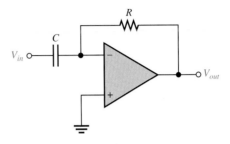

FIGURE 8–32
An ideal op-amp differentiator.

To see how the differentiator works, let's apply a positive-going ramp voltage to the input as indicated in Figure 8–33. In this case, $I_C = I_{in}$ and the voltage across the capacitor is equal to V_{in} at all times ($V_C = V_{in}$) because of virtual ground on the inverting input.

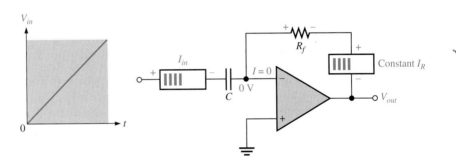

FIGURE 8–33
A differentiator with a ramp input.

From the basic formula, which is $V_C = (I_C/C)t$,

$$I_C = \left(\frac{V_C}{t}\right)C$$

Since the current at the inverting input is negligible, $I_R = I_C$. Both currents are constant because the slope of the capacitor voltage (V_C/t) is constant. The output voltage is also constant and equal to the voltage across R_f because one side of the feedback resistor is always 0 V (virtual ground).

$$V_{out} = I_R R_f = I_C R_f$$

$$V_{out} = -\left(\frac{V_C}{t}\right)R_f C \tag{8–7}$$

The output is negative when the input is a positive-going ramp and positive when the input is a negative-going ramp, as illustrated in Figure 8–34. During the positive slope of the

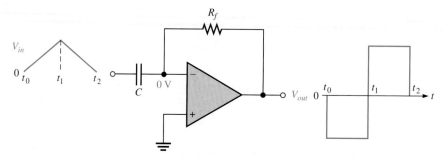

FIGURE 8–34
Output of a differentiator with a series of positive and negative ramps (triangle wave) on the input.

input, the capacitor is charging from the input source with constant current through the feedback resistor. During the negative slope of the input, the constant current is in the opposite direction because the capacitor is discharging.

Notice in Equation (8–7) that the term V_C/t is the slope of the input. If the slope increases, V_{out} becomes more negative. If the slope decreases, V_{out} becomes more positive. So, the output voltage is proportional to the negative slope (rate of change) of the input. The constant of proportionality is the time constant, R_fC.

EXAMPLE 8–9

Determine the output voltage of the op-amp differentiator in Figure 8–35 for the triangular-wave input shown.

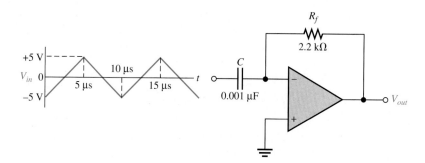

FIGURE 8–35

Solution Starting at $t = 0$, the input voltage is a positive-going ramp ranging from -5 V to $+5$ V (a $+10$ V change) in 5 μs. Then it changes to a negative-going ramp ranging from $+5$ V to -5 V (a -10 V change) in 5 μs.

Substituting into Equation (8–7), the output voltage for the positive-going ramp is

$$V_{out} = -\left(\frac{V_C}{t}\right)R_fC = -\left(\frac{10\ \text{V}}{5\ \mu\text{s}}\right)(2.2\ \text{k}\Omega)(0.001\ \mu\text{F}) = -4.4\ \text{V}$$

The output voltage for the negative-going ramp is calculated the same way.

$$V_{out} = -\left(\frac{V_C}{t}\right)R_fC = -\left(\frac{-10\ \text{V}}{5\ \mu\text{s}}\right)(2.2\ \text{k}\Omega)(0.001\ \mu\text{F}) = +4.4\ \text{V}$$

Finally, the output voltage waveform is graphed relative to the input as shown in Figure 8–36.

FIGURE 8–36

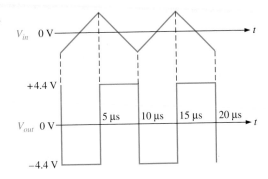

Practice Exercise   What would the output voltage be if the feedback resistor in Figure 8–35 is changed to 3.3 kΩ?

8–3 REVIEW QUESTIONS

1. What is the feedback element in an op-amp integrator?
2. For a constant input voltage to an integrator, why is the voltage across the capacitor linear?
3. What is the feedback element in an op-amp differentiator?
4. How is the output of a differentiator related to the input?

8–4 ■ CONVERTERS AND OTHER OP-AMP CIRCUITS

This section introduces a few more op-amp circuits that represent basic applications of the op-amp. You will learn about the constant-current source, the current-to-voltage converter, the voltage-to-current converter, and the peak detector. This is, of course, not a comprehensive coverage of all possible op-amp circuits but is intended only to introduce you to some common and basic uses.

After completing this section, you should be able to

❑ Understand the operation of several special op-amp circuits
  ❑ Identify and explain the operation of an op-amp constant-current source
  ❑ Identify and explain the operation of an op-amp current-to-voltage converter
  ❑ Identify and explain the operation of an op-amp voltage-to-current converter
  ❑ Explain how an op-amp can be used as a peak detector

Constant-Current Source

A **constant-current source** delivers a load current that remains constant when the load resistance changes. Figure 8–37 shows a basic circuit in which a stable voltage source (V_{in}) provides a constant current (I_i) through the input resistor (R_i). Since the inverting

FIGURE 8–37
A basic constant-current source.

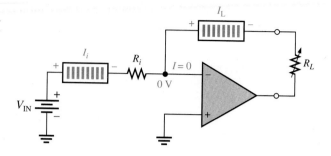

input of the op-amp is at virtual ground (0 V), the value of I_i is determined by V_{IN} and R_i as

$$I_i = \frac{V_{IN}}{R_i}$$

Now, since the internal input impedance of the op-amp is extremely high (ideally infinite), practically all of I_i is through R_L, which is connected in the feedback path. Since $I_i = I_L$,

$$I_L = \frac{V_{IN}}{R_i} \tag{8–8}$$

If R_L changes, I_L remains constant as long as V_{IN} and R_i are held constant.

Current-to-Voltage Converter

A **current-to-voltage converter** converts a variable input current to a proportional output voltage. A basic circuit that accomplishes this is shown in Figure 8–38(a). Since practically all of I_i is through the feedback path, the voltage dropped across R_f is $I_i R_f$. Because the left side of R_f is at virtual ground (0 V), the output voltage equals the voltage across R_f, which is proportional to I_i.

$$V_{OUT} = I_i R_f \tag{8–9}$$

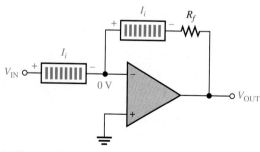

(a) Basic circuit

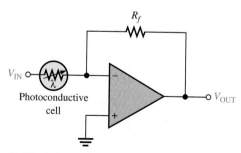

(b) Circuit for sensing light level and converting it to a proportional output voltage

FIGURE 8–38
Current-to-voltage converter.

A specific application of this circuit is illustrated in Figure 8–38(b), where a photo-conductive cell is used to sense changes in light level. As the amount of light changes, the current through the photoconductive cell varies because of the cell's change in resistance. This change in resistance produces a proportional change in the output voltage ($\Delta V_{OUT} = \Delta I_i R_f$).

Voltage-to-Current Converter

A basic **voltage-to-current converter** is shown in Figure 8–39. This circuit is used in applications where it is necessary to have an output (load) current that is controlled by an input voltage.

Neglecting the input offset voltage, both inverting and noninverting input terminals of the op-amp are at the same voltage, V_{IN}. Therefore, the voltage across R_1 equals V_{IN}. Since there is negligible current for the inverting input, the current through R_1 is the same as the current through R_L; thus,

$$I_L = \frac{V_{IN}}{R_1} \tag{8–10}$$

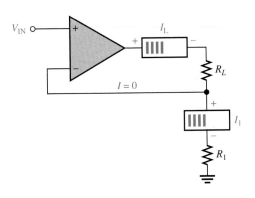

FIGURE 8–39

Voltage-to-current converter.

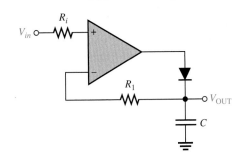

FIGURE 8–40

A basic peak detector.

Peak Detector

An interesting application of the op-amp is in a **peak detector** circuit such as the one shown in Figure 8–40. In this case the op-amp is used as a comparator. The purpose of this circuit is to detect the peak of the input voltage and store that peak voltage on a capacitor. For example, this circuit can be used to detect and store the maximum value of a voltage surge; this value can then be measured at the output with a voltmeter or recording device. The basic operation is as follows. When a positive voltage is applied to the noninverting input of the op-amp through R_i, the high-level output voltage of the op-amp forward-biases the diode and charges the capacitor. The capacitor continues to charge until its voltage reaches a value equal to the input voltage and thus both op-amp inputs are at the same voltage. At this point, the op-amp comparator switches, and its output goes to the low level. The diode is now reverse-biased, and the capacitor stops charging. It has reached a voltage equal to the peak of V_{in} and will hold this voltage until the charge eventually leaks off. If a greater input peak occurs, the capacitor charges to the new peak.

8–4 REVIEW QUESTIONS

1. For the constant-current source in Figure 8–37, the input reference voltage is 6.8 V and R_i is 10 kΩ. What value of constant current does the circuit supply to a 1 kΩ load? To a 5 kΩ load?

2. What element determines the constant of proportionality that relates input current to output voltage in the current-to-voltage converter?

8–5 ■ TROUBLESHOOTING

Although integrated circuit op-amps are extremely reliable and trouble-free, failures do occur from time to time. One type of internal failure mode is a condition where the op-amp output is "stuck" in a saturated state resulting in a constant high or constant low level, regardless of the input. Also, external component failures will produce various types of failure modes in op-amp circuits. Some examples are presented in this section.

After completing this section, you should be able to

❏ Troubleshoot basic op-amp circuits
  ❏ Identify failures in comparator circuits
  ❏ Identify failures in summing amplifiers

Figure 8–41 illustrates an internal failure of a comparator circuit that results in a "stuck" output.

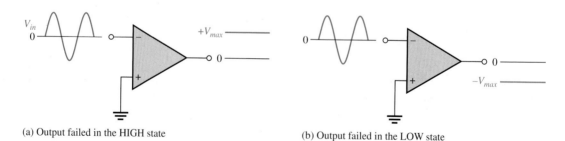

(a) Output failed in the HIGH state   (b) Output failed in the LOW state

FIGURE 8–41

Internal comparator failures typically result in the output being "stuck" in the HIGH or LOW state.

Symptoms of External Component Failures in Comparator Circuits

A comparator with zener-bounding and hysteresis is shown in Figure 8–42. In addition to a failure of the op-amp itself, a zener diode or one of the resistors could be faulty. For example, suppose one of the zener diodes opens. This effectively eliminates both zeners, and the circuit operates as an unbounded comparator, as indicated in Figure 8–43(a). With a shorted diode, the output is limited to the zener voltage (bounded) only in one direction depending on which diode remains operational, as illustrated in Figure 8–43(b). In the other direction, the output is held at the forward diode voltage.

FIGURE 8–42

A bounded comparator.

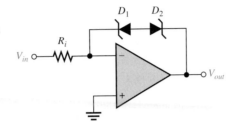

FIGURE 8–43

Examples of comparator circuit failures and their effects.

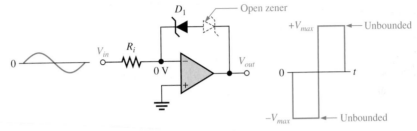

(a) The effect of an open zener

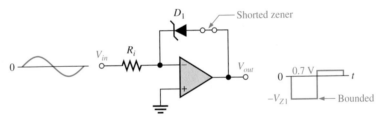

(b) The effect of a shorted zener

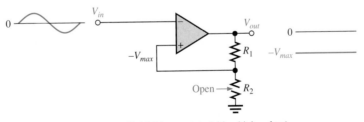

(c) Open R_2 causes output to "stick" in one state (either high or low)

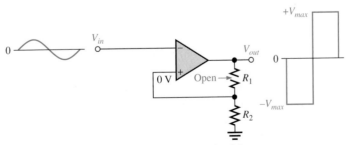

(d) Open R_1 forces the circuit to operate as a zero-level detector

Recall that R_1 and R_2 set the UTP and LTP for the hysteresis comparator. Now, suppose that R_2 opens. Essentially all of the output voltage is fed back to the noninverting input, and, since the input voltage will never exceed the output, the device will remain in one of its saturated states. This symptom can also indicate a faulty op-amp, as mentioned before. Now, assume that R_1 opens. This leaves the noninverting input near ground potential and causes the circuit to operate as a zero-level detector. These conditions are shown in parts (c) and (d) of Figure 8–43.

EXAMPLE 8–10

One channel of a dual-trace oscilloscope is connected to the comparator output and the other channel to the input signal, as shown in Figure 8–44. From the observed waveforms, determine if the circuit is operating properly, and if not, what the most likely failure is.

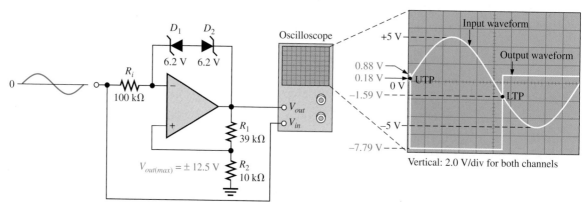

FIGURE 8–44

Solution   The output should be limited to ±8.67 V. However, the positive maximum is +0.88 V and the negative maximum is −7.79 V. This indicates that D_2 is shorted. Refer to Example 8–3 for analysis of the bounded comparator.

Practice Exercise   What would the output voltage look like if D_1 shorted rather than D_2?

Symptoms of Component Failures in Summing Amplifiers

If one of the input resistors in a unity-gain summing amplifier opens, the output will be less than the normal value by the amount of the voltage applied to the open input. Stated another way, the output will be the sum of the remaining input voltages.

If the summing amplifier has a nonunity gain, an open input resistor causes the output to be less than normal by an amount equal to the gain times the voltage at the open input.

EXAMPLE 8–11

(a) What is the normal output voltage in Figure 8–45?
(b) What is the output voltage if R_2 opens?
(c) What happens if R_5 opens?

FIGURE 8–45

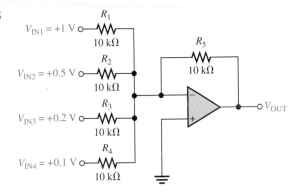

Solution

(a) $V_{OUT} = -(V_{IN1} + V_{IN2} + \cdots + V_{INn}) = -(1\,V + 0.5\,V + 0.2\,V + 0.1\,V) = \mathbf{-1.8\,V}$
(b) $V_{OUT} = -(1\,V + 0.2\,V + 0.1\,V) = \mathbf{-1.3\,V}$
(c) If R_5 opens, the circuit becomes a comparator and the output goes to $-V_{max}$.

Practice Exercise In Figure 8–45, assume $R_5 = 47\,k\Omega$. What is the output voltage if R_1 opens?

As another example, let's look at an averaging amplifier. An open input resistor will result in an output voltage that is the average of all the inputs with the open input averaged in as a zero.

EXAMPLE 8–12

(a) What is the normal output voltage for the averaging amplifier in Figure 8–46?
(b) If R_4 opens, what is the output voltage? What does the output voltage represent?

FIGURE 8–46

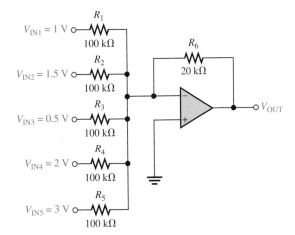

Solution   Since the input resistors are equal, $R = 100 \text{ k}\Omega$. $R_f = R_6$.

(a) $V_{OUT} = -\dfrac{R_f}{R}(V_{IN1} + V_{IN2} + \cdots + V_{INn})$

$= -\dfrac{20 \text{ k}\Omega}{100 \text{ k}\Omega}(1 \text{ V} + 1.5 \text{ V} + 0.5 \text{ V} + 2 \text{ V} + 3 \text{ V}) = -0.2(8 \text{ V}) = \mathbf{-1.6 \text{ V}}$

(b) $V_{OUT} = -\dfrac{20 \text{ k}\Omega}{100 \text{ k}\Omega}(1 \text{ V} + 1.5 \text{ V} + 0.5 \text{ V} + 3 \text{ V}) = -0.2(6 \text{ V}) = \mathbf{-1.2 \text{ V}}$

The 1.2 V result is the average of five voltages with the 2 V input replaced by 0 V. Notice that the output is not the average of the four remaining input voltages.

Practice Exercise   If R_4 is open, as was the case in this example, what would you have to do to make the output equal to the average of the remaining four input voltages?

8–5 REVIEW QUESTIONS

1. Describe one type of internal op-amp failure.
2. If a certain malfunction is attributable to more than one possible component failure, what would you do to isolate the problem?

8–6 ■ A SYSTEM APPLICATION

The system presented at the beginning of this chapter is a dual-slope analog-to-digital converter (ADC). This is one of several methods for A/D conversion. You were introduced to another type, called a simultaneous (or flash) ADC as an example in the chapter. The topic of data conversion including both ADCs and DACs is covered thoroughly in Chapter 14. Although A/D conversion is used for many purposes, in this particular application the converter is used to change an audio signal into digital form for recording. Although many parts of this system are digital, you will focus on the analog-to-digital converter board, which includes op-amps used in several types of circuits that you have learned about in this chapter.

After completing this section, you should be able to

❑ Apply what you have learned in this chapter to a system application
  ❑ Describe one way in which a summing amplifier is used
  ❑ State how the integrator is a key element in A/D conversion
  ❑ Explain how a comparator is used
  ❑ Translate between a printed circuit board and a schematic
  ❑ Troubleshoot some common system problems

A Brief Description of the System

The dual-slope ADC in Figure 8–47 accepts an audio signal and converts it to a series of digital codes for the purpose of recording. The audio signal voltage is applied to the sample-and-hold circuit. (Sample-and-hold circuits are covered in detail in Chapter 14.) At fixed intervals, sample pulses cause the amplitude at that point on the audio waveform to be converted to proportional dc levels that are then processed by the rest of the circuits and represented by a series of digital codes.

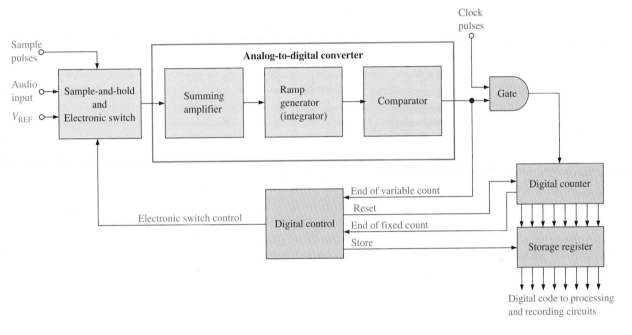

FIGURE 8–47
Basic dual-slope analog-to-digital converter.

The sample pulses occur at a much higher rate than the audio frequency so that a sufficient number of points on the audio waveform are sampled and converted to obtain an accurate digital representation of the audio signal. A rough approximation of the sampling process is illustrated in Figure 8–48. As the frequency of the sample pulses increases relative to the audio frequency, an increasingly accurate representation is achieved.

The summing amplifier has only one input active at a time. For example, when the audio input is switched in, the reference voltage input is zero and vice versa.

During the time between each sample pulse, the dc level from the sample-and-hold circuit is switched electronically into the summing amplifier on the ADC board. The output of the summing amplifier goes to the ramp generator which is an integrator circuit. At the same time, the digital counter starts counting up from zero. During the fixed time interval of the counting sequence, the integrator (ramp generator) produces a positive-going ramp voltage whose slope depends on the level of the sampled audio voltage. At the end of the fixed time interval, the ramp voltage at the output of the integrator has reached a voltage that is proportional to the sampled audio voltage. At this time, the digital control logic switches from the sample-and-hold input to the negative dc reference voltage input and resets the digital counter to zero.

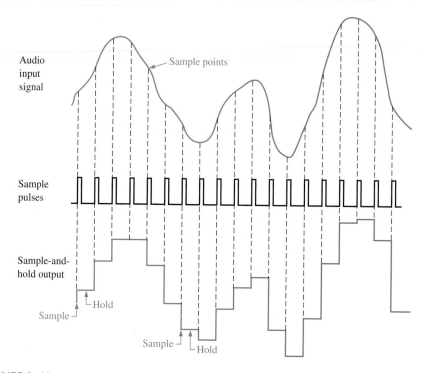

FIGURE 8–48

Sample-and-hold process. The sample-and-hold output is a rough approximation of the audio voltage for purposes of illustration. As the frequency of the sample pulses is increased, an increasingly accurate representation is achieved.

The summing amplifier applies this negative dc reference to the integrator input, which starts a negative-going ramp on the output. This ramp voltage has a slope that is fixed by the value V_{REF}. At the same time, the digital counter begins to count up again and will continue to count up until the negative-going ramp output of the integrator reaches zero volts.

At this point, the comparator switches to its negative saturated output voltage and disables the gate so that there are no additional clock pulses to the counter. At this time, the digital code in the counter is proportional to the time that it took for the negative-going ramp at the integrator output to reach zero and it will vary for each different sampled value.

Recall that the negative-going ramp started at a positive voltage that was dependent on the sampled value of the audio signal. Therefore, the digital code in the counter is also proportional to, and represents, the amplitude of the sampled audio voltage. This code is then shifted out to the register and then processed and recorded.

This process is repeated many times during a typical audio cycle. The result is a sequence of digital codes that represent the audio voltage amplitude as it varies with time. Figure 8–49 illustrates this for several sampled values. As mentioned, you will focus on the ADC board, which contains the summing amplifier, integrator, and comparator.

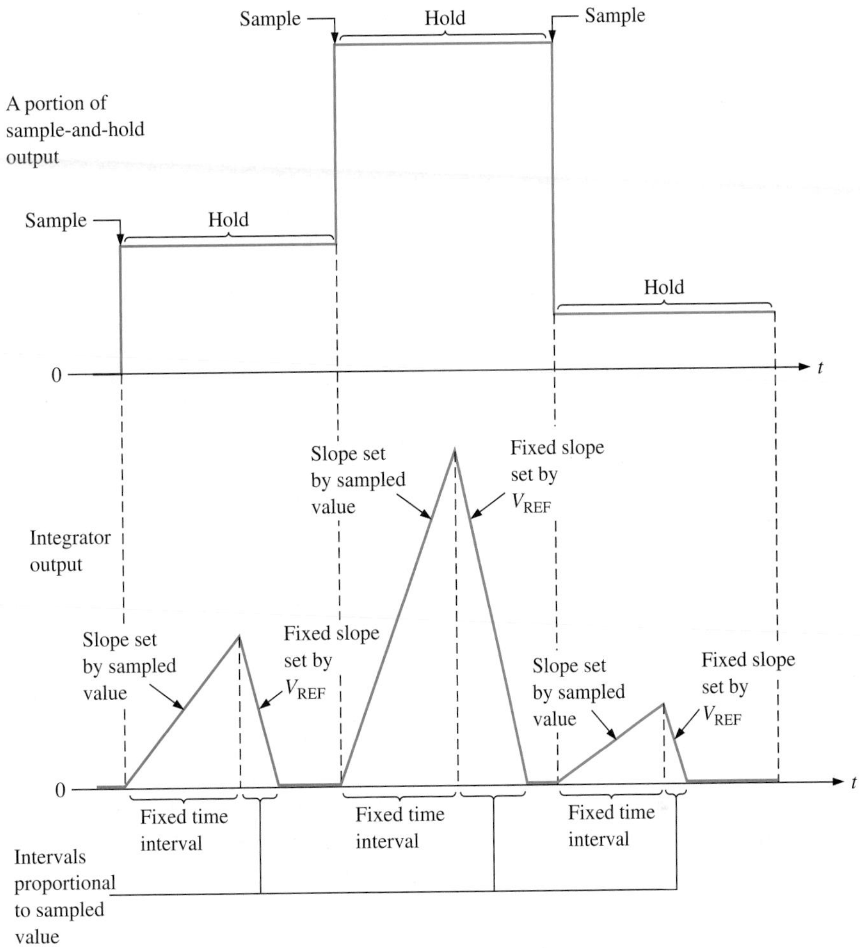

FIGURE 8–49

During the fixed-time interval of the positive-going ramp, the sampled audio input is applied to the integrator. During the variable-time interval of the negative-going ramp, the reference voltage is applied to the integrator. The counter sets the fixed-time interval and is then reset. Another count begins during the variable interval and the code in the counter at the end of this interval represents the sampled value.

Now, so you can take a closer look at the ADC board, let's take it out of the system and put it on the troubleshooter's bench.

TROUBLESHOOTER'S BENCH

■ ACTIVITY 1   Relate the PC Board to the Schematic

Identify each component on the circuit board in Figure 8–50 using the schematic in Figure 8–51. Also, identify each input and output on the board. There are several backside connections with corresponding feedthrough pads. Locate and identify the components associated with the summing amplifier, the integrator, and the comparator. Note that the two 10 kΩ potentiometers are for nulling the output offset voltage.

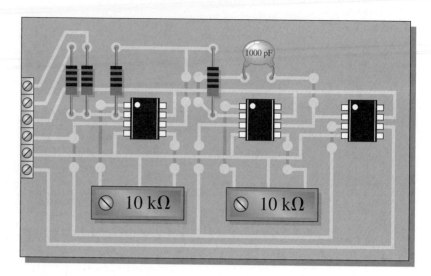

FIGURE 8–50
Analog-to-digital (ADC) board.

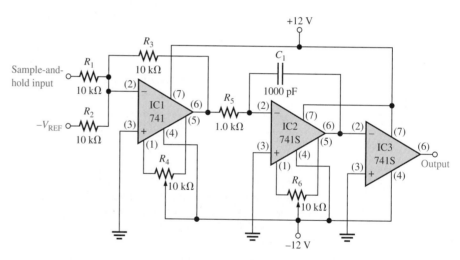

FIGURE 8–51
Schematic of the ADC.

■ **ACTIVITY 2 Analyze the Circuit**

Step 1: Determine the gain of the summing amplifier.

Step 2: Determine the slope of the integrator ramp in volts per microsecond when a sampled audio voltage of $+2$ V is applied.

Step 3: Determine the slope of the integrator ramp in volts per microsecond when the reference voltage of -8 V is applied.

Step 4: Given that the reference voltage is -8 V and the fixed-time interval of the negative-going slope is 1 μs, sketch the dual-slope output of the integrator when an instantaneous audio voltage of $+3$ V is sampled.

Step 5: Assuming that the maximum audio voltage to be sampled is +6 V, determine the maximum audio frequency that can be sampled by this particular system if there are to be 100 samples per cycle. What is the sample pulse rate in this case?

■ **ACTIVITY 3   Write a Technical Report**

Discuss the detailed operation of the ADC board circuitry and explain how it interfaces with the overall system. Discuss the purpose of each component on the circuit board.

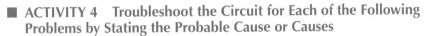

■ **ACTIVITY 4   Troubleshoot the Circuit for Each of the Following Problems by Stating the Probable Cause or Causes**

1. Zero volts on the output of IC1 when there are voltages on the sampled audio input and on the reference voltage input.
2. IC1 goes back and forth between its saturated states as the positive audio voltage and the negative reference voltage are alternately switched in.
3. The inverting input of IC1 never goes negative.
4. The output of IC2 stays at zero volts under normal operating conditions.

8–6 REVIEW QUESTIONS

1. Identify the summing amplifier, the integrator, and the comparator by IC number.
2. The 741S op-amps are high slew-rate devices with a minimum slew rate of 10 V/μs. Why are these used in this application?
3. What is the purpose of R_4 and R_6 in the circuit of Figure 8–51?
4. What type of output voltage does the ADC board produce and how is it used in the system?
5. If a sample pulse rate of 500 kHz is used, how long does each sampled audio voltage remain on the input to the ADC board?
6. Although IC1 is connected in the form of a summing amplifier, it does not actually perform a summing operation in this application. Why?

■ **SUMMARY**

- In an op-amp comparator, when the input voltage exceeds a specified reference voltage, the output changes state.
- Hysteresis gives an op-amp noise immunity.
- A comparator switches to one state when the input reaches the upper trigger point (UTP) and back to the other state when the input drops below the lower trigger point (LTP).
- The difference between the UTP and the LTP is the hysteresis voltage.
- Bounding limits the output amplitude of a comparator.
- The output voltage of a summing amplifier is proportional to the sum of the input voltages.
- An averaging amplifier is a summing amplifier with a closed-loop gain equal to the reciprocal of the number of inputs.
- In a scaling adder, a different weight can be assigned to each input, thus making the input contribute more or contribute less to the output.
- Integration is a mathematical process for determining the area under a curve.

- Integration of a step produces a ramp with a slope proportional to the amplitude.
- Differentiation is a mathematical process for determining the rate of change of a function.
- Differentiation of a ramp produces a step with an amplitude proportional to the slope.

■ GLOSSARY

Key terms are in color. All terms are included in the end-of-book glossary.

A/D conversion   A process whereby information in analog form is converted into digital form.

Bounding   The process of limiting the output range of an amplifier or other circuit.

Comparator   A circuit which compares two input voltages and produces an output in either of two states indicating the greater than or less than relationship of the inputs.

Constant-current source   A circuit that delivers a load current that remains constant when the load resistance changes.

Current-to-voltage converter   A circuit that converts a variable input current to a proportional output voltage.

D/A conversion   The process of converting a sequence of digital codes to an analog form.

Differentiator   A circuit that produces an inverted output which approximates the rate of change of the input function.

Hysteresis   The property that permits a circuit to switch from one state to the other at one voltage level and switch back to the original state at another lower voltage level.

Integrator   A circuit that produces an inverted output which approximates the area under the curve of the input function.

Noise   An unwanted voltage or current fluctuation.

Peak detector   A circuit used to detect the peak of the input voltage and store that peak value on a capacitor.

Schmitt trigger   A comparator with hysteresis.

Summing amplifier   A variation of a basic comparator circuit that is characterized by two or more inputs and an output voltage that is proportional to the magnitude of the algebraic sum of the input voltages.

Voltage-to-current converter   A circuit that converts a variable input voltage to a proportional output current.

■ KEY FORMULAS

Comparators

$$(8\text{--}1) \qquad V_{\text{REF}} = \frac{R_2}{R_1 + R_2}(+V) \qquad\qquad \text{Comparator reference}$$

$$(8\text{--}2) \qquad V_{\text{HYS}} = V_{\text{UTP}} - V_{\text{LTP}} \qquad\qquad \text{Hysteresis voltage}$$

Summing Amplifier

$$(8\text{--}3) \qquad V_{\text{OUT}} = -(V_{\text{IN1}} + V_{\text{IN2}} + \cdots + V_{\text{IN}n}) \qquad\qquad n\text{-input adder}$$

$$(8\text{--}4) \qquad V_{\text{OUT}} = -\frac{R_f}{R}(V_{\text{IN1}} + V_{\text{IN2}} + \cdots + V_{\text{IN}n}) \qquad\qquad \text{Scaling adder with gain}$$

$$(8\text{--}5) \qquad V_{\text{OUT}} = -\left(\frac{R_f}{R_1}V_{\text{IN1}} + \frac{R_f}{R_2}V_{\text{IN2}} + \cdots + \frac{R_f}{R_n}V_{\text{IN}n}\right) \qquad\qquad \text{Scaling adder}$$

Integrator and Differentiator

(8–6) $\dfrac{\Delta V_{out}}{\Delta t} = -\dfrac{V_{in}}{R_i C}$ Integrator output rate of change

(8–7) $V_{out} = -\left(\dfrac{V_C}{t}\right) R_f C$ Differentiator output voltage with ramp input

Miscellaneous

(8–8) $I_L = \dfrac{V_{IN}}{R_i}$ Constant-current source

(8–9) $V_{OUT} = I_i R_f$ Current-to-voltage converter

(8–10) $I_L = \dfrac{V_{in}}{R_1}$ Voltage-to-current converter

---

■ SELF-TEST

Answers are at the end of the chapter.

1. In a zero-level detector, the output changes state when the input
   (a) is positive   (b) is negative
   (c) crosses zero   (d) has a zero rate of change

2. The zero-level detector is one application of a
   (a) comparator   (b) differentiator
   (c) summing amplifier   (d) diode

3. Noise on the input of a comparator can cause the output to
   (a) hang up in one state
   (b) go to zero
   (c) change back and forth erratically between two states
   (d) produce the amplified noise signal

4. The effects of noise can be reduced by
   (a) lowering the supply voltage   (b) using positive feedback
   (c) using negative feedback   (d) using hysteresis
   (e) answers (b) and (d)

5. A comparator with hysteresis
   (a) has one trigger point   (b) has two trigger points
   (c) has a variable trigger point   (d) is like a magnetic circuit

6. In a comparator with hysteresis,
   (a) a bias voltage is applied between the two inputs
   (b) only one supply voltage is used
   (c) a portion of the output is fed back to the inverting input
   (d) a portion of the output is fed back to the noninverting input

7. Using output bounding in a comparator
   (a) makes it faster   (b) keeps the output positive
   (c) limits the output levels   (d) stabilizes the output

8. A window comparator detects when
   (a) the input is between two specified limits
   (b) the input is not changing
   (c) the input is changing too fast
   (d) the amount of light exceeds a certain value

9. A summing amplifier can have
   (a) only one input   (b) only two inputs   (c) any number of inputs

10. If the voltage gain for each input of a summing amplifier with a 4.7 kΩ feedback resistor is unity, the input resistors must have a value of
   (a) 4.7 kΩ
   (b) 4.7 kΩ divided by the number of inputs
   (c) 4.7 kΩ times the number of inputs

11. An averaging amplifier has five inputs. The ratio R_f/R_{in} must be
   (a) 5   (b) 0.2   (c) 1

12. In a scaling adder, the input resistors are
   (a) all the same value   (b) all of different values
   (c) each proportional to the weight of its inputs   (d) related by a factor of two

13. In an integrator, the feedback element is a
   (a) resistor   (b) capacitor   (c) zener diode   (d) voltage divider

14. For a step input, the output of an integrator is a
   (a) pulse   (b) triangular waveform   (c) spike   (d) ramp

15. The rate of change of an integrator's output voltage in response to a step input is set by
   (a) the RC time constant   (b) the amplitude of the step input
   (c) the current through the capacitor   (d) all of these

16. In a differentiator, the feedback element is a
   (a) resistor   (b) capacitor
   (c) zener diode   (d) voltage divider

17. The output of a differentiator is proportional to
   (a) the RC time constant   (b) the rate at which the input is changing
   (c) the amplitude of the input   (d) answers (a) and (b)

18. When you apply a triangular waveform to the input of a differentiator, the output is
   (a) a dc level   (b) an inverted triangular waveform
   (c) a square waveform   (d) the first harmonic of the triangular waveform

TROUBLESHOOTER'S QUIZ   *Answers are at the end of the chapter.*

Refer to Figure 8–53.

❑ If the value of R_1 is larger than specified,

1. The hysteresis voltage will

   (a) increase   (b) decrease   (c) not change

2. The sensitivity to noise voltage on the input will

   (a) increase   (b) decrease   (c) not change

Refer to Figure 8–56.

❑ If D_2 is open,

3. The output voltage will

   (a) increase   (b) decrease   (c) not change

4. The upper trigger point voltage will

   (a) increase   (b) decrease   (c) not change

5. The hysteresis voltage will

   (a) increase   (b) decrease   (c) not change

Refer to Figure 8–58.

❑ If the value of R_1 is less than specified,

6. The output voltage will

  (a) increase   **(b)** decrease   **(c)** not change

7. The voltage at the inverting input will

  (a) increase   **(b)** decrease   **(c)** not change

Refer to Figure 8–60.

❑ If C is open,

8. The rate of change of the output voltage in response to the step input will

  (a) increase   **(b)** decrease   **(c)** not change

9. The maximum voltage at the output will

  (a) increase   **(b)** decrease   **(c)** not change

Refer to Figure 8–61.

❑ If C is open,

10. The output signal voltage in response to a periodic triangular input waveform will

  (a) increase   **(b)** decrease   **(c)** not change

❑ If the value of R is larger than specified,

11. The output voltage in response to the periodic triangular input waveform will

  (a) increase   **(b)** decrease   **(c)** not change

12. The period of the output will

  (a) increase   **(b)** decrease   **(c)** not change

■ **PROBLEMS**   *Answers to odd-numbered problems are at the end of the book.*

SECTION 8–1   Comparators

1. A certain op-amp has an open-loop gain of 80,000. The maximum saturated output levels of this particular device are ±12 V when the dc supply voltages are ±15 V. If a differential voltage of 0.15 mV rms is applied between the inputs, what is the peak-to-peak value of the output?

2. Determine the output level (maximum positive or maximum negative) for each comparator in Figure 8–52.

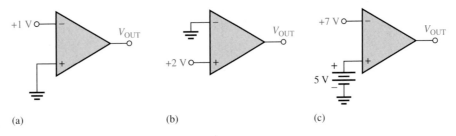

(a)   (b)   (c)

FIGURE 8–52

3. Calculate the V_{UTP} and V_{LTP} in Figure 8–53. $V_{out(max)} = -10$ V.

4. What is the hysteresis voltage in Figure 8–53?

FIGURE 8–53

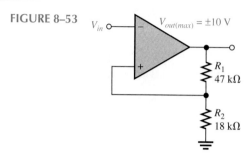

5. Sketch the output voltage waveform for each circuit in Figure 8–54 with respect to the input. Show voltage levels.

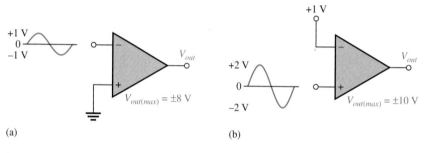

(a)   (b)

FIGURE 8–54

6. Determine the hysteresis voltage for each comparator in Figure 8–55. The maximum output levels are ±11 V.

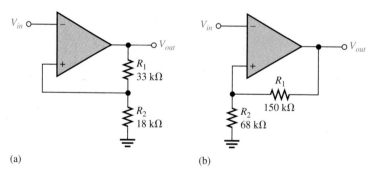

(a)   (b)

FIGURE 8–55

7. A 6.2 V zener diode is connected from the output to the inverting input in Figure 8–53 with the cathode at the output. What are the positive and negative output levels?

8. Determine the output voltage waveform in Figure 8–56.

FIGURE 8–56

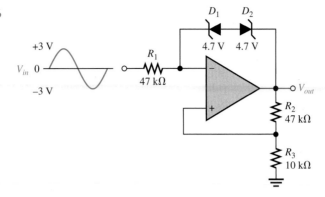

SECTION 8–2 Summing Amplifiers

9. Determine the output voltage for each circuit in Figure 8–57.

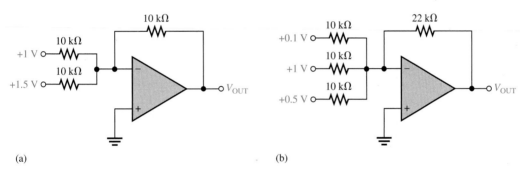

(a)   (b)

FIGURE 8–57

10. Refer to Figure 8–58. Determine the following:
    (a) V_{R1} and V_{R2}
    (b) Current through R_f
    (c) V_{OUT}

11. Find the value of R_f necessary to produce an output that is five times the sum of the inputs in Figure 8–58.

12. Design a summing amplifier that will average eight input voltages. Use input resistances of 10 kΩ each.

FIGURE 8–58

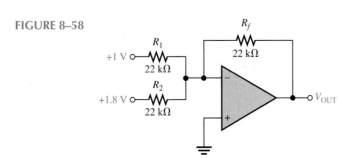

13. Find the output voltage when the input voltages shown in Figure 8–59 are applied to the scaling adder. What is the current through R_f?

FIGURE 8–59

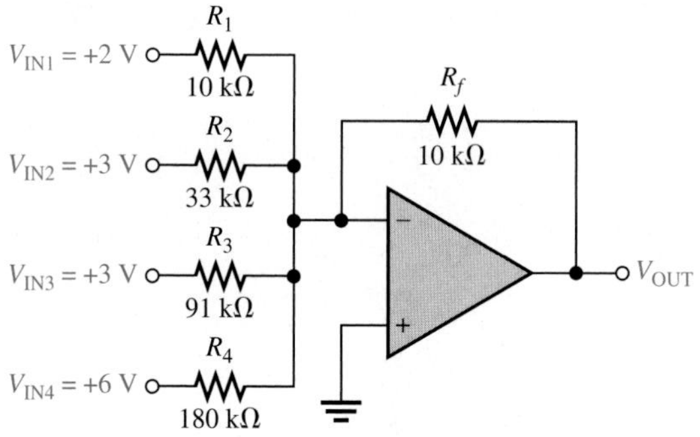

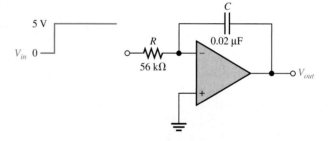

14. Determine the values of the input resistors required in a six-input scaling adder so that the lowest weighted input is 1 and each successive input has a weight twice the previous one. Use $R_f =$ 100 kΩ.

SECTION 8–3 Integrators and Differentiators

15. Determine the rate of change of the output voltage in response to the step input to the integrator in Figure 8–60.

FIGURE 8–60

16. A triangular waveform is applied to the input of the circuit in Figure 8–61 as shown. Determine what the output should be and sketch its waveform in relation to the input.

17. What is the magnitude of the capacitor current in Problem 16?

FIGURE 8–61

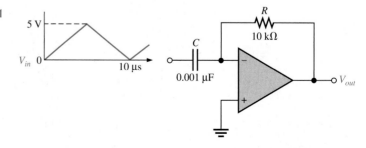

18. A triangular waveform with a peak-to-peak voltage of 2 V and a period of 1 ms is applied to the differentiator in Figure 8–62(a). What is the output voltage?

19. Beginning in position 1 in Figure 8–62(b), the switch is thrown into position 2 and held there for 10 ms, then back to position 1 for 10 ms, and so forth. Sketch the resulting output waveform. The saturated output levels of the op-amp are ±12 V.

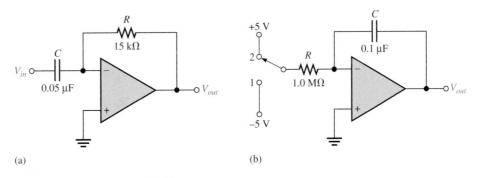

(a)   (b)

FIGURE 8–62

SECTION 8–4   Converters and Other Op-Amp Circuits

20. Determine the load current in each circuit of Figure 8–63. (Hint: Thevenize the circuit to the left of R_i.)

21. Devise a circuit for remotely sensing temperature and producing a proportional voltage that can then be converted to digital form for display. A thermistor can be used as the temperature-sensing element.

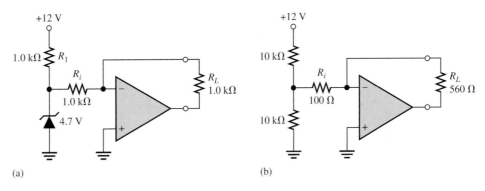

(a)   (b)

FIGURE 8–63

SECTION 8–5 Troubleshooting

22. The waveforms given in Figure 8–64(a) are observed at the indicated points in Figure 8–64(b). Is the circuit operating properly? If not, what is a likely fault?

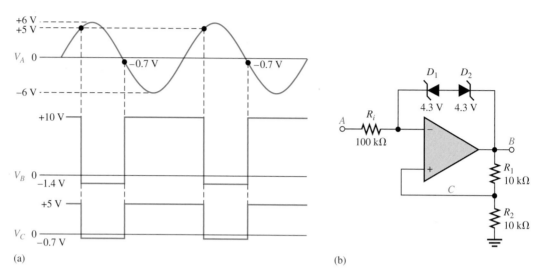

(a)

(b)

FIGURE 8–64

23. The waveforms shown for the window comparator in Figure 8–65 are measured. Determine if the output waveform is correct and, if not, specify the possible fault(s).

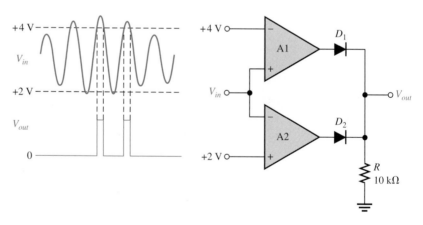

FIGURE 8–65

24. The sequences of voltage levels shown in Figure 8–66 are applied to the summing amplifier and the indicated output is observed. First, determine if this output is correct. If it is not correct, determine the fault.

25. The given ramp voltages are applied to the op-amp circuit in Figure 8–67. Is the given output correct? If it isn't, what is the problem?

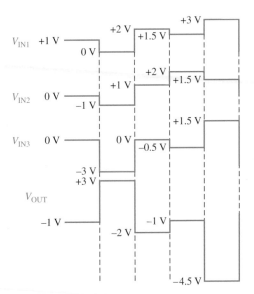
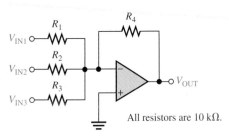

FIGURE 8–66

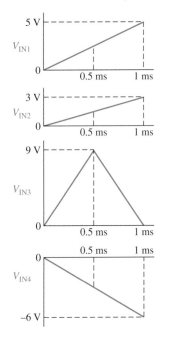
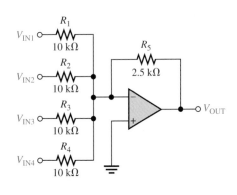
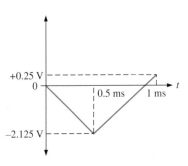

FIGURE 8–67

26. The ADC board, shown in Figure 8–68 and in the system application, has just come off the assembly line and a pass/fail test indicates that it doesn't work. The board now comes to you for troubleshooting. What is the very first thing you should do? Can you isolate the problem(s) by this first step in this case?

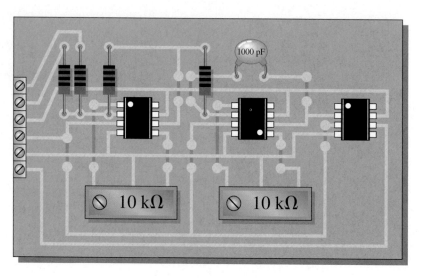

FIGURE 8–68

■ **ANSWERS TO REVIEW QUESTIONS**

Section 8–1

1. (a) $V = (10 \text{ k}\Omega/110 \text{ k}\Omega)15 \text{ V} = 1.36 \text{ V}$
   (b) $V = 22 \text{ k}\Omega/69 \text{ k}\Omega)(-12 \text{ V}) = -3.83 \text{ V}$

2. Hysteresis makes the comparator noise-free.

3. Bounding limits the output amplitude to a specified level.

Section 8–2

1. The summing point is the point where the input resistors are commonly connected.

2. $R_f/R = 1/5 = 0.2$

3. 5 kΩ

Section 8–3

1. The feedback element in an integrator is a capacitor.

2. The capacitor voltage is linear because the capacitor current is constant.

3. The feedback element in a differentiator is a resistor.

4. The output of a differentiator is proportional to the rate of change of the input.

Section 8–4

1. $I_L = 6.8 \text{ V}/10 \text{ k}\Omega = 0.68 \text{ mA}$; same value to 5 kΩ load.

2. The feedback resistor is the constant of proportionality.

Section 8–5

1. An op-amp can fail with a shorted output.

2. Replace suspected components one by one.

Section 8–6

1. Summing amplifier—IC1, integrator—IC2, comparator—IC3

2. A high slew-rate op-amp is used in the integrator to avoid slew-rate limitation of the output ramps. One is used as a comparator to achieve a fast switching time.

3. R_4 and R_6 are for eliminating output offset (nulling).

4. The board output is the comparator output. The transition of the comparator output from its positive state to its negative state notifies the control logic of the end of the variable-time interval.

5. 1/500 kHz = 2 μs

6. Because of the electronic switch, only one input voltage at a time is actually applied.

■ **ANSWERS TO PRACTICE EXERCISES FOR EXAMPLES**

8–1 1.96 V

8–2 +3.83 V, −3.83 V, V_{HYS} = 7.65 V

8–3 +1.81 V, −1.81 V

8–4 −12.5 V

8–5 −5.73 V

8–6 Changes require an additional 100 kΩ input resistor and a change of R_f to 20 kΩ.

8–7 0.45, 0.12, 0.18; V_{OUT} = −3.03 V

8–8 Change C to 5000 pF or change R to 5.0 kΩ.

8–9 Same waveform with an amplitude of 6.6 V

8–10 A pulse from −0.88 V to +7.79 V

8–11 −3.76 V

8–12 Change R_6 to 25 kΩ.

■ **ANSWERS TO SELF-TEST**

1. (c)	**2.** (a)	**3.** (c)	**4.** (e)	**5.** (b)
6. (d)	**7.** (c)	**8.** (a)	**9.** (c)	**10.** (a)
11. (b)	**12.** (c)	**13.** (b)	**14.** (d)	**15.** (d)
16. (a)	**17.** (d)	**18.** (c)		

■ **ANSWERS TO TROUBLE-SHOOTER'S QUIZ**

1. decrease	**2.** increase	**3.** increase	**4.** increase
5. increase	**6.** increase	**7.** not change	**8.** increase
9. not change	**10.** decrease	**11.** increase	**12.** not change

9

ACTIVE FILTERS

■ CHAPTER OBJECTIVES

❏ Describe the gain-versus-frequency responses of the basic filters
❏ Describe the three basic filter response characteristics and other filter parameters
❏ Understand active low-pass filters
❏ Understand active high-pass filters
❏ Understand active band-pass filters
❏ Understand active band-stop filters
❏ Discuss two methods for measuring frequency response
❏ Apply what you have learned in this chapter to a system application

■ KEY TERMS

❏ Filter
❏ Critical frequency
❏ Low-pass filter
❏ Pole
❏ Roll-off
❏ High-pass filter
❏ Band-pass filter
❏ Band-stop filter
❏ Damping factor
❏ Order

Power supply filters were introduced in Chapter 2. In this chapter, active filters used for signal processing are introduced. Filters are circuits that are capable of passing input signals with certain selected frequencies through to the output while rejecting signals with other frequencies. This property is called *selectivity*.

Active filters use devices such as transistors or op-amps combined with passive *RC*, *RL*, or *RLC* networks. The active devices provide voltage gain and the passive networks provide frequency selectivity. In terms of general response, there are four basic categories of active filters: low-pass, high-pass, band-pass, and band-stop. In this chapter, you will study active filters using op-amps and *RC* networks.

In Chapter 7, you worked with an FM stereo multiplex receiver, concentrating on the audio amplifiers. In this chapter, we again look at this same system, but this time the focus is on the filters in the left and right channel separation circuits in which several types of active filters are used. The FM stereo multiplex signal that is received is quite complex, and it is beyond the scope of our coverage to investigate the reasons it is transmitted in such a way. It is interesting, however, to see how filters, such as the ones studied in this chapter, can be used to separate out the audio signals that go to the left and right speakers.

For the system application in Section 9–8, in addition to the other topics, be sure you understand
❏ Filter responses
❏ How low-pass and band-pass filters work

www. VISIT THE COMPANION WEBSITE

Study Aids for This Chapter Are Available at

http://www.prenhall.com/floyd

9–1 ■ BASIC FILTER RESPONSES

Filters are usually categorized by the manner in which the output voltage varies with the frequency of the input voltage. The categories of active filters are low-pass, high-pass, band-pass, and band-stop. We will examine each of these general responses in this section.

After completing this section, you should be able to

❑ Describe the gain-versus-frequency responses of the basic filters
  ❑ Explain the low-pass response
  ❑ Determine the critical frequency and bandwidth of a low-pass filter
  ❑ Explain the high-pass response
  ❑ Determine the critical frequency of a high-pass filter
  ❑ Explain the band-pass response
  ❑ Explain the significance of the quality factor
  ❑ Determine the critical frequency, bandwidth, quality factor, and damping factor of a band-pass filter
  ❑ Explain the band-stop response

Low-Pass Filter Response

A **filter** is a circuit that passes certain frequencies and attenuates or rejects all other frequencies. The **passband** of a filter is the region of frequencies that are allowed to pass through the filter with minimum attenuation (usually defined as less than -3 dB of attenuation). The **critical frequency**, f_c, (also called the *cutoff frequency*) defines the end of the passband and is normally specified at the point where the response drops -3 dB (70.7%) from the passband response. Following the passband is a region called *the transition region* that leads into a region called the *stopband*. There is no precise point between the transition region and the stopband.

A **low-pass filter** is one that passes frequencies from dc to f_c and significantly attenuates all other frequencies. The passband of the ideal low-pass filter is shown in the blue shaded area of Figure 9–1(a); the response drops to zero at frequencies beyond the passband. This ideal response is sometimes referred to as a "brick-wall" because nothing gets through beyond the wall. The bandwidth of an ideal low-pass filter is equal to f_c.

$$BW = f_c \tag{9–1}$$

The ideal response shown in Figure 9–1(a) is not attainable by any practical filter. Actual filter responses depend on the number of **poles**, a term used with filters to describe the number of bypass circuits contained in the filter.[1] The most basic low-pass filter is a simple *RC* network consisting of just one resistor and one capacitor; the output is taken across the capacitor as shown in Figure 9–1(b). This basic *RC* filter has a single pole and it rolls off at -20 dB/decade beyond the critical frequency. The actual response is indicated by the colored line in Figure 9–1(a). The response is plotted on a standard log plot that is used for filters to show details of the curve as the gain drops. Notice that the gain is almost constant until the frequency is at the critical frequency; after this, the gain drops rapidly at a fixed **roll-off** rate.

---

[1] A pole is also used to describe certain complex mathematical characteristics of the transfer function for the filter.

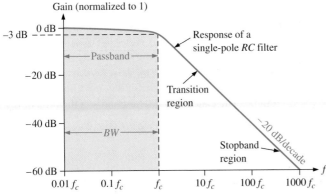

(a) Comparison of an ideal low-pass filter response with actual response

(b) Basic low-pass circuit

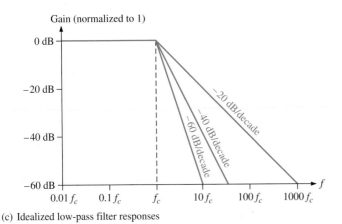

(c) Idealized low-pass filter responses

FIGURE 9–1

Low-pass filter responses.

The −20 dB/decade roll-off rate for the gain of a basic *RC* filter means that at a frequency of 10f_c, the output will be −20 dB (10%) of the input. This rather gentle roll-off is not a particularly good filter characteristic because too much of the unwanted frequencies (beyond the passband) are allowed through the filter.

The critical frequency of the simple low-pass *RC* filter occurs when $X_C = R$, where

$$f_c = \frac{1}{2\pi RC}$$

Recall from your basic dc/ac course that the output at the critical frequency is 70.7% of the input. This response is equivalent to an attenuation of −3 dB.

Figure 9–1(c) illustrates several idealized low-pass response curves including the basic one pole response (−20 dB/decade). The approximations show a *flat* response to the cutoff frequency and a roll-off at a constant rate after the cutoff frequency. Actual filters do not have a perfectly flat response to the cutoff frequency but have dropped to −3 dB at this point as described previously.

In order to produce a filter that has a steeper transition region, (and hence form a more effective filter), it is necessary to add additional circuitry to the basic filter. Responses that are steeper than −20 dB/decade in the transition region cannot be obtained by simply cascading identical *RC* stages (due to loading effects). However, by combining an op-amp with frequency-selective feedback networks, filters can be designed with roll-off rates of −40, −60, or more dB/decade. Filters that include one or more op-amps in the design are called **active filters**. These filters can optimize the roll-off rate or other attribute (such as phase response) with a particular filter design. In general, the more poles the filter uses, the steeper its transition region will be. The exact response depends on the type of filter and the number of poles.

High-Pass Filter Response

A **high-pass filter** is one that significantly attenuates or rejects all frequencies below f_c and passes all frequencies above f_c. The critical frequency is, again, the frequency at which the output is 70.7% of the input (or −3 dB) as shown in Figure 9–2(a). The ideal response, indicated by the color-shaded area, has an instantaneous drop at f_c, which, of course, is not achievable. Ideally, the passband of a high-pass filter is all frequencies above the critical frequency. The high-frequency response of practical circuits is limited by the op-amp or other components that make up the filter.

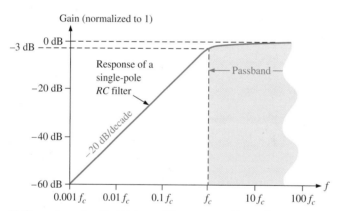

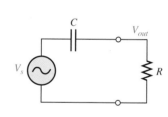

(a) Comparison of an ideal high-pass filter response with actual response

(b) Basic high-pass circuit

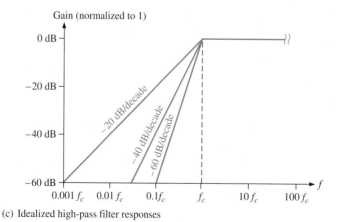

(c) Idealized high-pass filter responses

FIGURE 9–2

High-pass filter responses.

A simple *RC* network consisting of a single resistor and capacitor can be configured as a high-pass filter by taking the output across the resistor as shown in Figure 9–2(b). As in the case of the low-pass filter, the basic *RC* network has a roll-off rate of −20 dB/decade as indicated by the colored line in Figure 9–2(a). Also, the critical frequency for the basic high-pass filter occurs when $X_C = R$, where

$$f_c = \frac{1}{2\pi RC}$$

Figure 9–2(c) illustrates several idealized high-pass response curves including the basic one pole response (−20 dB/decade) for a basic *RC* network. As in the case of the low-pass filter, the approximations show a *flat* response to the cutoff frequency and a roll-off at a constant rate after the cutoff frequency. Actual high-pass filters do not have the perfectly flat response indicated or the precise roll-off rate shown. Responses that are steeper than −20 dB/decade in the transition region are also possible with active high-pass filters; the particular response depends on the type of filter and the number of poles.

Band-Pass Filter Response

A **band-pass filter** passes all signals lying within a band between a lower-frequency limit and an upper-frequency limit and essentially rejects all other frequencies that are outside this specific band. The high-frequency tuned amplifiers introduced in Section 5–2 used tuned circuits as band-pass filters. A generalized band-pass response curve is shown in Figure 9–3. The *bandwidth (BW)* is defined as the difference between the upper critical frequency (f_{c2}) and the lower critical frequency (f_{c1}).

$$BW = f_{c2} - f_{c1} \tag{9–2}$$

The critical frequencies are the points at which the response curve is 70.7% of its maximum. These critical frequencies are also called *3 dB frequencies*. The frequency about which the passband is centered is called the *center frequency, f_0*, defined as the geometric mean of the critical frequencies.

$$f_0 = \sqrt{f_{c1}f_{c2}} \tag{9–3}$$

FIGURE 9–3
General band-pass response curve.

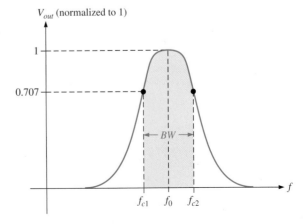

Quality Factor Recall from Section 5–2 that the quality factor *(Q)* of a band-pass filter was defined as the ratio of the center frequency to the bandwidth.

$$Q = \frac{f_0}{BW} \qquad (9\text{–}4)$$

     The value of Q is an indication of the selectivity of a band-pass filter. The higher the value of Q, the narrower the bandwidth and the better the selectivity for a given value of f_0. Band-pass filters are sometimes classified as narrow-band ($Q > 10$) or wide-band ($Q < 10$). The Q can also be expressed in terms of the damping factor *(DF)* of the filter as

$$Q = \frac{1}{DF}$$

You will study the damping factor in Section 9–2.

EXAMPLE 9–1

A certain band-pass filter has a center frequency of 15 kHz and a bandwidth of 1 kHz. Determine the Q and classify the filter as narrow-band or wide-band.

Solution
$$Q = \frac{f_0}{BW} = \frac{15 \text{ kHz}}{1 \text{ kHz}} = 15$$

Because $Q > 10$, this is a **narrow-band** filter.

*Practice Exercise** If the Q of the filter is doubled, what will the bandwidth be?

---

* Answers are at the end of the chapter.

Band-Stop Filter Response

Another category of active filter is the **band-stop filter**, also known as the *notch, band-reject,* or *band-elimination filter.* A general response curve for a band-stop filter is shown in Figure 9–4. Notice that the bandwidth is the band of frequencies between the 3 dB points, just as in the case of the band-pass filter response. You can think of the operation as oppo-

FIGURE 9–4
General band-stop filter response.

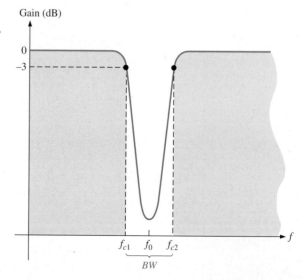

site to that of the band-pass filter because frequencies within a certain bandwidth are rejected, and frequencies outside the bandwidth are passed.

9–1 REVIEW QUESTIONS*

1. What determines the bandwidth of a low-pass filter?
2. What limits the bandwidth of an active high-pass filter?
3. How are the Q and the bandwidth of a band-pass filter related? Explain how the selectivity is affected by the Q of a filter.

* Answers are at the end of the chapter.

9–2 ■ FILTER RESPONSE CHARACTERISTICS

Each type of filter (low-pass, high-pass, band-pass, or band-stop) can be tailored by circuit component values to have either a Butterworth, Chebyshev, or Bessel characteristic. Each of these characteristics is identified by the shape of the response curve, and each has an advantage in certain applications.

After completing this section, you should be able to

❏ Describe the three basic filter response characteristics and other filter parameters
   ❏ Describe the Butterworth characteristic
   ❏ Describe the Chebyshev characteristic
   ❏ Describe the Bessel characteristic
   ❏ Define *damping factor* and discuss its significance
   ❏ Calculate the damping factor of a filter
   ❏ Discuss the order of a filter and its effect on the roll-off rate

Butterworth, Chebyshev, or Bessel response characteristics can be realized with most active filter circuit configurations by proper selection of certain component values. A general comparison of the three response characteristics for a low-pass filter response curve is shown in Figure 9–5. High-pass, band-pass, and band-stop filters can also be designed to have any one of the characteristics.

The Butterworth Characteristic The **Butterworth** characteristic provides a very flat amplitude response in the passband and a roll-off rate of −20 dB/decade/pole. The phase response is not linear, however, and the phase shift (thus, time delay) of signals passing through the filter varies nonlinearly with frequency. Therefore, a pulse applied to a filter with a Butterworth response will cause overshoots on the output because each frequency component of the pulse's rising and falling edges experiences a different time delay. Filters with the Butterworth response are normally used when all frequencies in the passband must have the same gain. The Butterworth response is often referred to as a *maximally flat response.*

The Chebyshev Characteristic Filters with the **Chebyshev** response characteristic are useful when a rapid roll-off is required because it provides a roll-off rate greater than −20 dB/decade/pole. This is a greater rate than that of the Butterworth, so filters can be

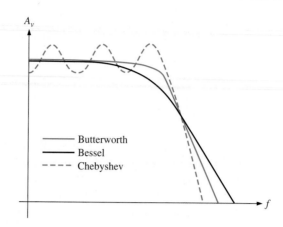

FIGURE 9–5
Comparative plots of three types of filter response characteristics.

implemented with the Chebyshev response with fewer poles and less complex circuitry for a given roll-off rate. This type of filter response is characterized by overshoot or ripples in the passband (depending on the number of poles) and an even less linear phase response than the Butterworth.

The Bessel Characteristic   The **Bessel** response exhibits a linear phase characteristic, meaning that the phase shift increases linearly with frequency. The result is almost no overshoot on the output with a pulse input. For this reason, filters with the Bessel response are used for filtering pulse waveforms without distorting the shape of the waveform.

The Damping Factor

As mentioned, an active filter can be designed to have either a Butterworth, Chebyshev, or Bessel response characteristic regardless of whether it is a low-pass, high-pass, band-pass, or band-stop type. The **damping factor *(DF)*** of an active filter circuit determines which response characteristic the filter exhibits. To explain the basic concept, a generalized active filter is shown in Figure 9–6. It includes an amplifier, a negative feedback network, and a filter section. The amplifier and feedback network are connected in a noninverting config-

FIGURE 9–6
General diagram of an active filter. Note that R_1 corresponds to R_f and R_2 corresponds to R_i as defined in Chapter 6.

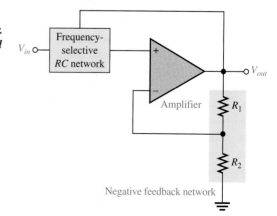

uration. The damping factor is determined by the negative feedback network and is defined by the following equation:

$$DF = 2 - \frac{R_1}{R_2} \qquad \text{(9–5)}$$

Basically, the damping factor affects the filter response by negative feedback action. Any attempted increase or decrease in the output voltage is offset by the opposing effect of the negative feedback. This tends to make the response curve flat in the passband of the filter if the value for the damping factor is precisely set. By advanced mathematics, which we will not cover, values for the damping factor have been derived for various orders of filters to achieve the maximally flat response of the Butterworth characteristic.

The value of the damping factor required to produce a desired response characteristic depends on the **order** (number of poles) of the filter. Recall that the more poles a filter has, the faster its roll-off rate is. To achieve a second-order Butterworth response, for example, the damping factor must be 1.414. To implement this damping factor, the feedback resistor ratio must be

$$\frac{R_1}{R_2} = 2 - DF = 2 - 1.414 = 0.586$$

This ratio gives the closed-loop gain of the noninverting filter amplifier, $A_{cl(NI)}$, a value of 1.586, derived as follows:

$$A_{cl(NI)} = \frac{1}{B} = \frac{1}{R_2/(R_1 + R_2)} = \frac{R_1 + R_2}{R_2} = \frac{R_1}{R_2} + 1 = 0.586 + 1 = 1.586$$

EXAMPLE 9–2

If resistor R_2 in the feedback network of an active two-pole filter of the type in Figure 9–6 is 10 kΩ, what value must R_1 be to obtain a maximally flat Butterworth response?

Solution
$$\frac{R_1}{R_2} = 0.586$$
$$R_1 = 0.586R_2 = 0.586(10 \text{ k}\Omega) = \textbf{5.86 k}\Omega$$

Using the nearest standard 5 percent value of 5600 Ω will get very close to the ideal Butterworth response.

Practice Exercise What is the damping factor for $R_2 = 10$ kΩ and $R_1 = 5.6$ kΩ?

Critical Frequency and Roll-Off Rate

The critical frequency is determined by the values of the resistor and capacitors in the *RC* network, as shown in Figure 9–6. For a single-pole (first-order) filter, as shown in Figure 9–7, the critical frequency is

$$f_c = \frac{1}{2\pi RC}$$

Although we show a low-pass configuration, the same formula is used for the f_c of a single-pole high-pass filter. The number of poles determines the roll-off rate of the filter. A Butterworth response produces −20 dB/decade/pole. So, a first-order (one-pole) filter has a roll-off of −20 dB/decade; a second-order (two-pole) filter has a roll-off rate of

FIGURE 9–7
First-order (one-pole) low-pass filter.

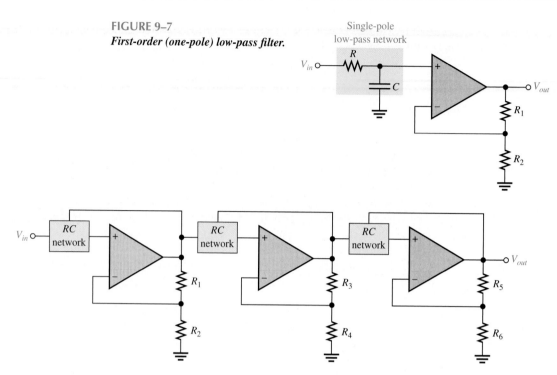

FIGURE 9–8
The number of filter poles can be increased by cascading.

-40 dB/decade; a third-order (three-pole) filter has a roll-off rate of -60 dB/decade; and so on.

Generally, to obtain a filter with three poles or more, one-pole or two-pole filters are cascaded, as shown in Figure 9–8. To obtain a third-order filter, for example, cascade a second-order and a first-order filter; to obtain a fourth-order filter, cascade two second-order filters; and so on. Each filter in a cascaded arrangement is called a *stage* or *section*.

Because of its maximally flat response, the Butterworth characteristic is the most widely used. Therefore, we will limit our coverage to the Butterworth response to illustrate basic filter concepts. Table 9–1 lists the roll-off rates, damping factors, and feedback resistor ratios for up to sixth-order Butterworth filters.

TABLE 9–1
Values for the Butterworth response.

Order	Roll-off dB/decade	1st stage			2nd stage			3rd stage		
		Poles	DF	R_1/R_2	Poles	DF	R_3/R_4	Poles	DF	R_5/R_6
1	-20	1	Optional							
2	-40	2	1.414	0.586						
3	-60	2	1.00	1	1	1.00	1			
4	-80	2	1.848	0.152	2	0.765	1.235			
5	-100	2	1.00	1	2	1.618	0.382	1	0.618	1.382
6	-120	2	1.932	0.068	2	1.414	0.586	2	0.518	1.482

9–2 REVIEW QUESTIONS

1. Explain how Butterworth, Chebyshev, and Bessel responses differ.

2. What determines the response characteristic of a filter?

3. Name the basic parts of an active filter.

9–3 ■ ACTIVE LOW-PASS FILTERS

Filters that use op-amps as the active element provide several advantages over passive filters (R, L, and C elements only). The op-amp provides gain, so that the signal is not attenuated as it passes through the filter. The high input impedance of the op-amp prevents excessive loading of the driving source, and the low output impedance of the op-amp prevents the filter from being affected by the load that it is driving. Active filters are also easy to adjust over a wide frequency range without altering the desired response.

After completing this section, you should be able to

❑ Understand active low-pass filters

    ❑ Identify a single-pole filter and determine its gain and critical frequency

    ❑ Identify a two-pole Sallen-Key filter and determine its gain and critical frequency

    ❑ Explain how a higher roll-off rate is achieved by cascading low-pass filters

A Single-Pole Filter

Figure 9–9(a) shows an active filter with a single low-pass *RC* network that provides a roll-off of −20 dB/decade above the critical frequency, as indicated by the response curve in Figure 9–9(b). The critical frequency of the single-pole filter is $f_c = 1/2\pi RC$. The op-amp

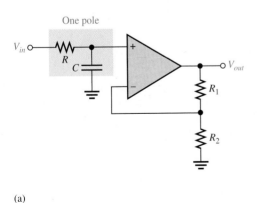

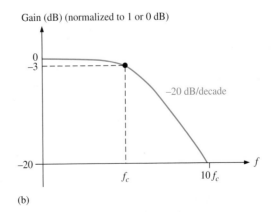

(a)   (b)

FIGURE 9–9
Single-pole active low-pass filter and response curve.

in this filter is connected as a noninverting amplifier with the closed-loop voltage gain in the passband set by the values of R_1 and R_2.

$$A_{cl(NI)} = \frac{R_1}{R_2} + 1$$

The Sallen-Key Low-Pass Filter

The Sallen-Key is one of the most common configurations for a second-order (two-pole) filter. It is also known as a VCVS (voltage-controlled voltage source) filter. A low-pass version of the Sallen-Key filter is shown in Figure 9–10. Notice that there are two low-pass RC networks that provide a roll-off of -40 dB/decade above the critical frequency (assuming a Butterworth characteristic). One RC network consists of R_A and C_A, and the second network consists of R_B and C_B. A unique feature is the capacitor C_A that provides feedback for shaping the response near the edge of the passband. The critical frequency for the second-order Sallen-Key filter is

$$f_c = \frac{1}{2\pi\sqrt{R_A R_B C_A C_B}} \tag{9-6}$$

For simplicity, the component values can be made equal so that $R_A = R_B = R$ and $C_A = C_B = C$. In this case, the expression for the critical frequency simplifies to $f_c = 1/2\pi RC$.

FIGURE 9–10

Basic Sallen-Key second-order low-pass filter.

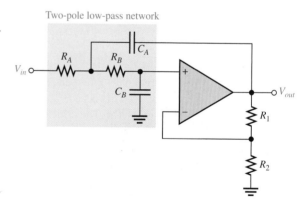

Two-pole low-pass network

As in the single-pole filter, the op-amp in the second-order Sallen-Key filter acts as a noninverting amplifier with the negative feedback provided by the R_1/R_2 network. As you have learned, the damping factor is set by the values of R_1 and R_2, thus making the filter response either Butterworth, Chebyshev, or Bessel. For example, from Table 9–1, the R_1/R_2 ratio must be 0.586 to produce the damping factor of 1.414 required for a second-order Butterworth response.

EXAMPLE 9–3

Determine the critical frequency of the low-pass filter in Figure 9–11, and set the value of R_1 for an approximate Butterworth response.

FIGURE 9–11

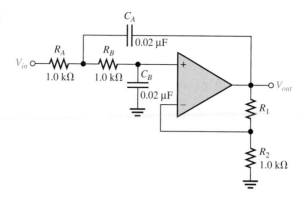

Solution Since $R_A = R_B = 1.0$ kΩ and $C_A = C_B = 0.02$ μF,

$$f_c = \frac{1}{2\pi RC} = \frac{1}{2\pi(1.0 \text{ k}\Omega)(0.02 \text{ μF})} = \textbf{7.96 kHz}$$

For a Butterworth response, $R_1/R_2 = 0.586$.

$$R_1 = 0.586R_2 = 0.586(1.0 \text{ k}\Omega) = \textbf{586 Ω}$$

Select a standard value as near as possible to this calculated value.

Practice Exercise Determine f_c for Figure 9–11 if $R_A = R_B = R_2 = 2.2$ kΩ and $C_A = C_B = 0.01$ μF. Also determine the value of R_1 for a Butterworth response.

Cascaded Low-Pass Filters Achieve a Higher Roll-Off Rate

A three-pole filter is required to get a third-order low-pass response (-60 dB/decade). This is done by cascading a two-pole low-pass filter and a single-pole low-pass filter, as shown in Figure 9–12(a). Figure 9–12(b) on the next page shows a four-pole configuration obtained by cascading two two-pole filters.

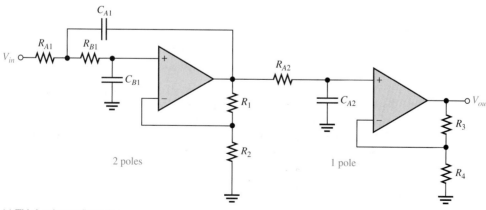

(a) Third-order configuration

FIGURE 9–12
Cascaded low-pass filters.

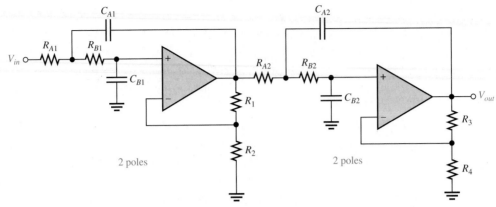

(b) Fourth-order configuration

FIGURE 9–12 (continued)

EXAMPLE 9–4

For the four-pole filter in Figure 9–12(b), determine the capacitance values required to produce a critical frequency of 2680 Hz if all the resistors in the RC low-pass networks are 1.8 kΩ. Also select values for the feedback resistors to get a Butterworth response.

Solution　Both stages must have the same f_c. Assuming equal-value capacitors,

$$f_c = \frac{1}{2\pi RC}$$

$$C = \frac{1}{2\pi R f_c} = \frac{1}{2\pi (1.8 \text{ k}\Omega)(2680 \text{ Hz})} = 0.033 \text{ } \mu\text{F}$$

$$C_{A1} = C_{B1} = C_{A2} = C_{B2} = \textbf{0.033 } \boldsymbol{\mu}\textbf{F}$$

Also select $R_2 = R_4 = 1.8$ kΩ for simplicity. Refer to Table 9–1. For a Butterworth response in the first stage, $DF = 1.848$ and $R_1/R_2 = 0.152$. Therefore,

$$R_1 = 0.152R_2 = 0.152(1800 \text{ } \Omega) = \textbf{274 } \boldsymbol{\Omega}$$

Choose $R_1 = 270$ Ω.
　In the second stage, $DF = 0.765$ and $R_3/R_4 = 1.235$. Therefore,

$$R_3 = 1.235R_4 = 1.235(1800 \text{ } \Omega) = \textbf{2.22 k}\boldsymbol{\Omega}$$

Choose $R_3 = 2.2$ kΩ.

Practice Exercise　For the filter in Figure 9–12(b), determine the capacitance values for $f_c = 1$ kHz if all the filter resistors are 680 Ω. Also specify the values for the feedback resistors to produce a Butterworth response.

1. How many poles does a second-order low-pass filter have? How many resistors and how many capacitors are used in the frequency-selective network?

2. Why is the damping factor of a filter important?

3. What is the primary purpose of cascading low-pass filters?

9–4 ■ ACTIVE HIGH-PASS FILTERS

In high-pass filters, the roles of the capacitor and resistor are reversed in the RC networks. Otherwise, the basic parameters are the same as for the low-pass filters.

After completing this section, you should be able to

❏ Understand active high-pass filters
   ❏ Identify a single-pole filter and determine its gain and critical frequency
   ❏ Identify a two-pole Sallen-Key filter and determine its gain and critical frequency
   ❏ Explain how a higher roll-off rate is achieved by cascading high-pass filters

A Single-Pole Filter

A high-pass active filter with a -20 dB/decade roll-off is shown in Figure 9–13(a). Notice that the input circuit is a single high-pass RC network. The negative feedback network is the same as for the low-pass filters previously discussed. The high-pass response curve is shown in Figure 9–13(b).

Ideally, a high-pass filter passes all frequencies above f_c without limit, as indicated in Figure 9–14(a), although in practice, this is not the case. As you have learned, all op-amps inherently have internal RC networks that limit the amplifier's response at high frequen-

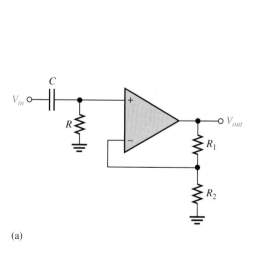

(a)

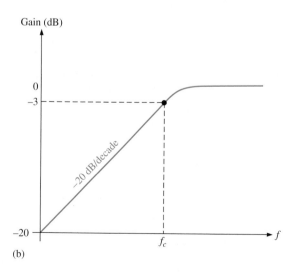

(b)

FIGURE 9–13

Single-pole active high-pass filter and response curve.

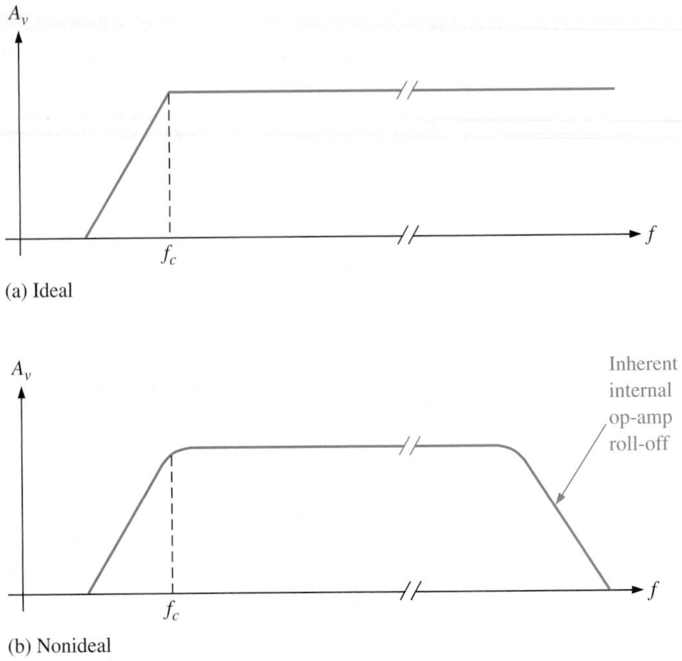

FIGURE 9–14

High-pass filter response.

A_v

(a) Ideal

f_c

f

A_v

Inherent internal op-amp roll-off

(b) Nonideal

f_c

f

cies. Therefore, there is an upper-frequency limit on the high-pass filter's response which, in effect, makes it a band-pass filter with a very wide bandwidth. In the majority of applications, the internal high-frequency limitation is so much greater than that of the filter's f_c that the limitation can be neglected. In some applications, special current-feedback op-amps or discrete transistors are used for the gain element to increase the high-frequency limitation beyond that realizable with standard op-amps.

The Sallen-Key High-Pass Filter

A high-pass second-order Sallen-Key configuration is shown in Figure 9–15. The components R_A, C_A, R_B, and C_B form the two-pole frequency-selective network. Notice that the positions of the resistors and capacitors in the frequency-selective network are opposite to those in the low-pass configuration. As with the other filters, the response characteristic can be optimized by proper selection of the feedback resistors, R_1 and R_2.

FIGURE 9–15

Basic Sallen-Key second-order high-pass filter.

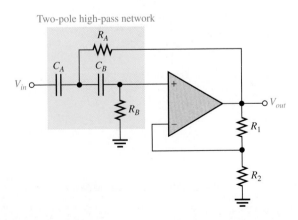

Two-pole high-pass network

EXAMPLE 9–5

Choose values for the Sallen-Key high-pass filter in Figure 9–15 to implement an equal-value second-order Butterworth response with a critical frequency of approximately 10 kHz.

Solution Start by selecting a value for R_A and R_B (R_1 or R_2 can also be the same value as R_A and R_B for simplicity).

$$R = R_A = R_B = R_2 = \textbf{3.3 k}\boldsymbol{\Omega} \qquad \text{(an arbitrary selection)}$$

Next, calculate the capacitance value from $f_c = 1/2\pi RC$.

$$C = C_A = C_B = \frac{1}{2\pi R f_c} = \frac{1}{2\pi(3.3 \text{ k}\Omega)(10 \text{ kHz})} = \textbf{0.0048 }\boldsymbol{\mu}\textbf{F}$$

For a Butterworth response, the damping factor must be 1.414 and $R_1/R_2 = 0.586$.

$$R_1 = 0.586R_2 = 0.586(3.3 \text{ k}\Omega) = \textbf{1.93 k}\boldsymbol{\Omega}$$

If you had chosen $R_1 = 3.3$ kΩ, then

$$R_2 = \frac{R_1}{0.586} = \frac{3.3 \text{ k}\Omega}{0.586} = 5.63 \text{ k}\Omega$$

Either way, an approximate Butterworth response is realized by choosing the nearest standard values.

Practice Exercise Select values for all the components in the high-pass filter of Figure 9–15 to obtain an $f_c = 300$ Hz. Use equal-value components and optimize for a Butterworth response.

Cascading High-Pass Filters

As with the low-pass configuration, first- and second-order high-pass filters can be cascaded to provide three or more poles and thereby create faster roll-off rates. Figure 9–16 shows a six-pole high-pass filter consisting of three two-pole stages. With this configuration optimized for a Butterworth response, a roll-off of -120 dB/decade is achieved.

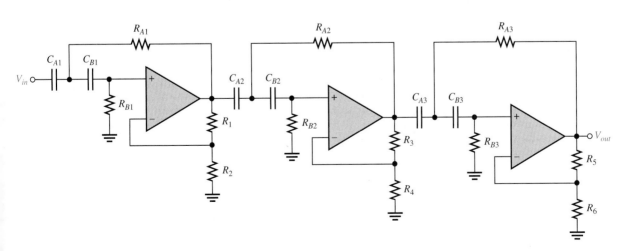

FIGURE 9–16
Sixth-order high-pass filter.

9–4 REVIEW QUESTIONS

1. How does a high-pass Sallen-Key filter differ from the low-pass configuration?

2. To increase the critical frequency of a high-pass filter, would you increase or decrease the resistor values?

3. If three two-pole high-pass filters and one single-pole high-pass filter are cascaded, what is the resulting roll-off?

9–5 ■ ACTIVE BAND-PASS FILTERS

As mentioned, band-pass filters pass all frequencies bounded by a lower-frequency limit and an upper-frequency limit and reject all others lying outside this specified band. A band-pass response can be thought of as the overlapping of a low-frequency response curve and a high-frequency response curve.

After completing this section, you should be able to

❑ Understand active band-pass filters
   ❑ Describe a band-pass filter composed of a low-pass and a high-pass filter
   ❑ Determine the critical frequencies and center frequency of a cascaded band-pass filter
   ❑ Determine center frequency, bandwidth, and gain of multiple-feedback band-pass filters
   ❑ Explain the operation of a state-variable band-pass filter

Cascaded Low-Pass and High-Pass Filters Achieve a Band-Pass Response

One way to implement a band-pass filter is a cascaded arrangement of a high-pass filter and a low-pass filter, as shown in Figure 9–17(a), as long as the critical frequencies are sufficiently separated. Each of the filters shown is a two-pole Sallen-Key Butterworth configuration so that the roll-off rates are -40 dB/decade, indicated in the composite response curve of Figure 9–17(b). The critical frequency of each filter is chosen so that the response curves overlap sufficiently, as indicated. The critical frequency of the high-pass filter must be sufficiently lower than that of the low-pass stage.

The lower frequency, f_{c1}, of the passband is the critical frequency of the high-pass filter. The upper frequency, f_{c2}, is the critical frequency of the low-pass filter. Ideally, as discussed earlier, the center frequency, f_0, of the passband is the geometric mean of f_{c1} and f_{c2}. The following formulas express the three frequencies of the band-pass filter in Figure 9–17.

$$f_{c1} = \frac{1}{2\pi\sqrt{R_{A1}R_{B1}C_{A1}C_{B1}}}$$

$$f_{c2} = \frac{1}{2\pi\sqrt{R_{A2}R_{B2}C_{A2}C_{B2}}}$$

$$f_0 = \sqrt{f_{c1}f_{c2}}$$

Of course, if equal-value components are used in implementing each filter, the critical frequency equations simplify to the form $f_c = 1/2\pi RC$.

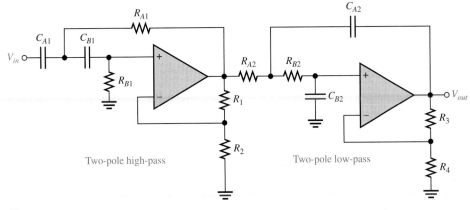

(a)

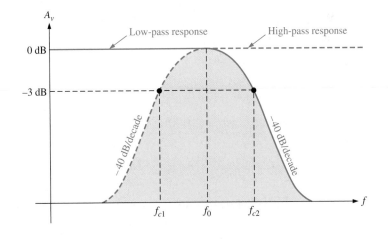

(b)

FIGURE 9–17

Band-pass filter formed by cascading a two-pole high-pass and a two-pole low-pass filter (it does not matter in which order the filters are cascaded).

Multiple-Feedback Band-Pass Filter

Another type of filter configuration, shown in Figure 9–18, is a multiple-feedback band-pass filter. The two feedback paths are through R_2 and C_1. Components R_1 and C_1 provide the low-pass response, and R_2 and C_2 provide the high-pass response. The maximum gain, A_0, occurs at the center frequency. Q values of less than 10 are typical in this type of filter. An expression for the center frequency follows, recognizing that R_1 and R_3 appear in parallel as viewed from the C_1 feedback path (with the V_{in} source replaced by a short).

$$f_0 = \frac{1}{2\pi\sqrt{(R_1\|R_3)R_2C_1C_2}}$$

Making $C_1 = C_2 = C$ gives the following formula (derived in Appendix B):

$$f_0 = \frac{1}{2\pi C}\sqrt{\frac{R_1 + R_3}{R_1R_2R_3}} \qquad\qquad (9\text{–}7)$$

FIGURE 9–18
Multiple-feedback band-pass filter.

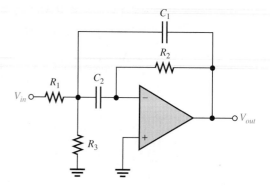

A convenient value for the capacitors is chosen; then the three resistor values are calculated based on the desired values for f_0, BW, and A_0. As you know, the Q can be determined from the relation $Q = f_0/BW$, and the resistors are found using the following formulas (stated without derivation).

$$R_1 = \frac{Q}{2\pi f_0 C A_0}$$

$$R_2 = \frac{Q}{\pi f_0 C}$$

$$R_3 = \frac{Q}{2\pi f_0 C(2Q^2 - A_0)}$$

To develop a gain expression, we solve for Q in the first two equations above.

$$Q = 2\pi f_0 A_0 C R_1$$
$$Q = \pi f_0 C R_2$$

Then,

$$2\pi f_0 A_0 C R_1 = \pi f_0 C R_2$$

An expression for the maximum gain at the center frequency is

$$A_0 = \frac{R_2}{2R_1} \qquad (9\text{–}8)$$

In order for the denominator of the equation $R_3 = Q/[2\pi f_0 C(2Q^2 - A_0)]$ to be positive, $A_0 < 2Q^2$, which imposes a limitation on the gain.

EXAMPLE 9–6

Determine the center frequency, maximum gain, and bandwidth for the filter in Figure 9–19.

FIGURE 9–19

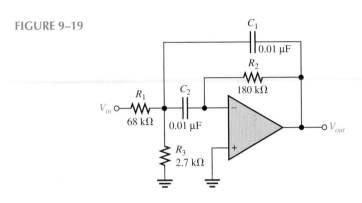

Solution

$$f_0 = \frac{1}{2\pi C} \sqrt{\frac{R_1 + R_3}{R_1 R_2 R_3}} = \frac{1}{2\pi(0.01 \ \mu\text{F})} \sqrt{\frac{68 \ \text{k}\Omega + 2.7 \ \text{k}\Omega}{(68 \ \text{k}\Omega)(180 \ \text{k}\Omega)(2.7 \ \text{k}\Omega)}} = \textbf{736 Hz}$$

$$A_0 = \frac{R_2}{2R_1} = \frac{180 \ \text{k}\Omega}{2(68 \ \text{k}\Omega)} = \textbf{1.32}$$

$$Q = \pi f_0 C R_2 = \pi(736 \ \text{Hz})(0.01 \ \mu\text{F})(180 \ \text{k}\Omega) = 4.16$$

$$BW = \frac{f_0}{Q} = \frac{736 \ \text{Hz}}{4.16} = \textbf{177 Hz}$$

Practice Exercise   If R_2 in Figure 9–19 is increased to 330 kΩ, how does this affect the gain, center frequency, and bandwidth of the filter?

State-Variable Band-Pass Filter

The state-variable or universal active filter is widely used for band-pass applications. As shown in Figure 9–20, it consists of a summing amplifier and two op-amp integrators (which act as single-pole low-pass filters) that are combined in a cascaded arrangement to form a second-order filter. Although used primarily as a band-pass (BP) filter, the state-variable configuration also provides low-pass (LP) and high-pass (HP) outputs. The center frequency is set by the *RC* networks in both integrators. When used as a band-pass filter, the critical frequencies of the integrators are usually made equal, thus setting the center frequency of the passband.

Basic Operation   At input frequencies below f_c, the input signal passes through the summing amplifier and integrators and is fed back 180° out-of-phase. Thus, the feedback signal and input signal cancel for all frequencies below approximately f_c. As the low-pass response of the integrators rolls off, the feedback signal diminishes, thus allowing the input to pass through to the band-pass output. Above f_c, the low-pass response disappears, thus preventing the input signal from passing through the integrators. As a result, the band-pass

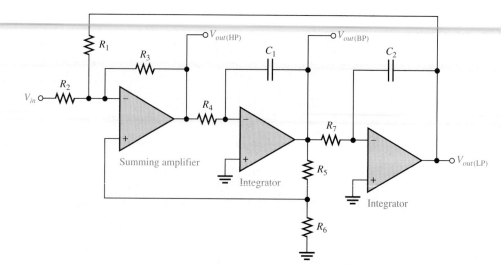

FIGURE 9–20
State-variable band-pass filter.

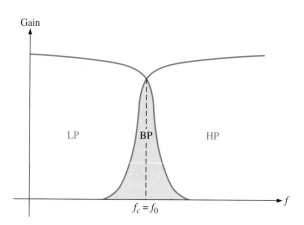

FIGURE 9–21
General state-variable response curves.

output peaks sharply at f_c, as indicated in Figure 9–21. Stable Qs up to 100 can be obtained with this type of filter. The Q is set by the feedback resistors R_5 and R_6 according to the following equation.

$$Q = \frac{1}{3}\left(\frac{R_5}{R_6} + 1\right)$$

The state-variable filter cannot be optimized for low-pass, high-pass, and band-pass performance simultaneously for this reason: To optimize for a low-pass or a high-pass Butterworth response, DF must equal 1.414. Since $Q = 1/DF$, a Q of 0.707 will result. Such a low Q provides a very poor band-pass response (large BW and poor selectivity). For optimization as a band-pass filter, the Q must be set high.

EXAMPLE 9–7   Determine the center frequency, Q, and BW for the band-pass output of the state-variable filter in Figure 9–22.

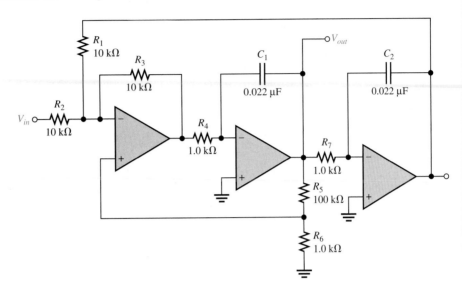

FIGURE 9–22

Solution   For each integrator,

$$f_c = \frac{1}{2\pi R_4 C_1} = \frac{1}{2\pi R_7 C_2} = \frac{1}{2\pi(1.0 \text{ k}\Omega)(0.022 \text{ }\mu\text{F})} = 7.23 \text{ kHz}$$

The center frequency is approximately equal to the critical frequencies of the integrators.

$$f_0 = f_c = \textbf{7.23 kHz}$$

$$Q = \frac{1}{3}\left(\frac{R_5}{R_6} + 1\right) = \frac{1}{3}\left(\frac{100 \text{ k}\Omega}{1.0 \text{ k}\Omega} + 1\right) = \textbf{33.7}$$

$$BW = \frac{f_0}{Q} = \frac{7.23 \text{ kHz}}{33.7} = \textbf{215 Hz}$$

Practice Exercise   Determine f_0, Q, and BW for the filter in Figure 9–22 if $R_4 = R_6 = R_7 = 330 \text{ }\Omega$ with all other component values the same as shown on the schematic.

9–5 REVIEW QUESTIONS

1. What determines selectivity in a band-pass filter?
2. One filter has a $Q = 5$ and another has a $Q = 25$. Which has the narrower bandwidth?
3. List the elements that make up a state-variable filter.

9–6 ■ ACTIVE BAND-STOP FILTERS

Band-stop filters reject a specified band of frequencies and pass all others. The response is opposite to that of a band-pass filter.

After completing this section, you should be able to

❑ Understand active band-stop filters
   ❑ Identify a multiple-feedback band-stop filter
   ❑ Explain the operation of a state-variable band-stop filter

Multiple-Feedback Band-Stop Filter

Figure 9–23 shows a multiple-feedback band-stop filter. Notice that this configuration is similar to the band-pass version except that R_3 has been moved and R_4 has been added.

FIGURE 9–23
Multiple-feedback band-stop filter.

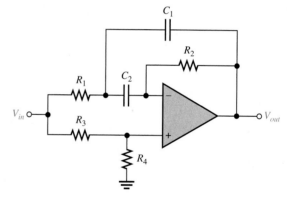

State-Variable Band-Stop Filter

Summing the low-pass and the high-pass responses of the state-variable filter covered in Section 9–5 creates a band-stop response as shown in Figure 9–24.

FIGURE 9–24
State-variable band-stop filter.

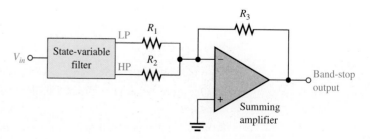

EXAMPLE 9–8   Verify that the band-stop filter in Figure 9–25 has a center frequency of 60 Hz, and optimize the filter for a Q of 10.

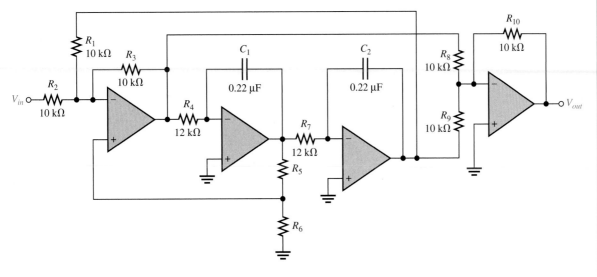

FIGURE 9–25

Solution f_0 equals the f_c of the integrator stages.

$$f_0 = \frac{1}{2\pi R_4 C_1} = \frac{1}{2\pi R_7 C_2} = \frac{1}{2\pi (12 \text{ k}\Omega)(0.22 \text{ }\mu\text{F})} = \mathbf{60.3 \text{ Hz}}$$

You can obtain a $Q = 10$ by choosing R_6 and then calculating R_5.

$$Q = \frac{1}{3}\left(\frac{R_5}{R_6} + 1\right)$$

$$R_5 = (3Q - 1)R_6$$

Choose $R_6 = \mathbf{3.3 \text{ k}\Omega}$. Then

$$R_5 = [3(10) - 1]3.3 \text{ k}\Omega = \mathbf{95.7 \text{ k}\Omega}$$

Choose 100 kΩ as the nearest standard value.

Practice Exercise   How would you change the center frequency to 120 Hz in Figure 9–25?

9–6 REVIEW QUESTIONS

1. How does a band-stop response differ from a band-pass response?
2. How is a state-variable band-pass filter converted to a band-stop filter?

9–7 ■ FILTER RESPONSE MEASUREMENTS

In this section, we discuss two methods of determining a filter's response by measurement—discrete point measurement and swept frequency measurement.

After completing this section, you should be able to

❑ Discuss two methods for measuring frequency response
  ❑ Explain the discrete point measurement method
  ❑ Explain the swept frequency measurement method

Discrete Point Measurement

Figure 9–26 shows an arrangement for taking filter output voltage measurements at discrete values of input frequency using common laboratory instruments. The general procedure is as follows:

1. Set the amplitude of the sine wave generator to a desired voltage level.

2. Set the frequency of the sine wave generator to a value well below the expected critical frequency of the filter under test. For a low-pass filter, set the frequency as near as possible to 0 Hz. For a band-pass filter, set the frequency well below the expected lower critical frequency.

3. Increase the frequency in predetermined steps sufficient to allow enough data points for an accurate response curve.

4. Maintain a constant input voltage amplitude while varying the frequency.

5. Record the output voltage at each value of frequency.

6. After recording a sufficient number of points, plot a graph of output voltage versus frequency.

If the frequencies to be measured exceed the response of the DMM, an oscilloscope may have to be used instead.

FIGURE 9–26

Test setup for discrete point measurement of the filter response. (Readings are arbitrary and for display only.)

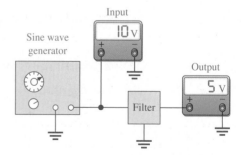

Swept Frequency Measurement

The swept frequency method requires more elaborate test equipment than does the discrete point method, but it is much more efficient and can result in a more accurate response curve. A general test setup is shown in Figure 9–27(a) using a swept frequency generator

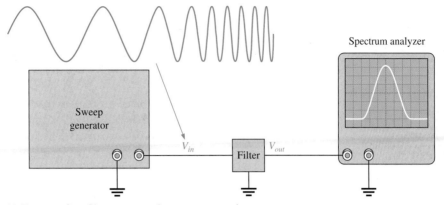

(a) Test setup for a filter response using a spectrum analyzer

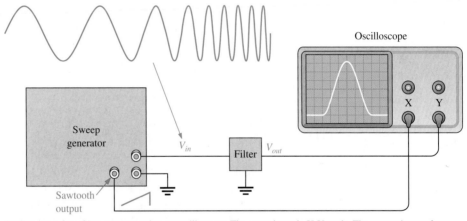

(b) Test setup for a filter response using an oscilloscope. The scope is put in X-Y mode. The sawtooth waveform from the sweep generator drives the X-channel of the oscilloscope.

FIGURE 9–27

Test setup for swept frequency measurement of the filter response.

and a spectrum analyzer. Figure 9–27(b) shows how the test can be made with an oscilloscope instead of a spectrum analyzer.

The swept frequency generator produces a constant amplitude output signal whose frequency increases linearly between two preset limits, as indicated in Figure 9–27. In part (a), the spectrum analyzer is an instrument that can be calibrated for a desired *frequency span/division* rather than for the usual *time/division* setting. Therefore, as the input frequency to the filter sweeps through a preselected range, the response curve is traced out on the screen of the spectrum analyzer. The test setup for using an oscilloscope to display the response curve is shown in part (b).

9–7 REVIEW QUESTIONS

1. What is the purpose of the two tests discussed in this section?

2. Name one disadvantage and one advantage of each test method.

9–8 ■ A SYSTEM APPLICATION

In this system application, the focus is on the filter board, which is part of the channel separation circuits in the FM stereo receiver. In addition to the active filters, the left and right channel separation circuit includes a demodulator, a frequency doubler, and a stereo matrix. Except for mentioning their purpose, we will not deal specifically with the demodulator, doubler, or matrix. However, the matrix is an interesting application of summing amplifiers, which were studied in Chapter 8 and these will be shown in detail on the schematic although we will not concentrate on them.

After completing this section, you should be able to

❑ Apply what you have learned in this chapter to a system application
    ❑ See how low-pass and band-pass active filters are used
    ❑ Use the schematic to locate and identify the components on the PC board
    ❑ Determine the operation of the filters
    ❑ Troubleshoot some common amplifier failures

A Brief Description of the System

Stereo FM (**frequency modulation**) signals are transmitted on a **carrier** frequency of 88 MHz to 108 MHz. The standard transmitted stereo signal consists of three modulating signals. These are the sum of the left and right channel audio (L + R), the difference of the left and right channel audio (L − R), and a 19 kHz pilot subcarrier.

The L + R audio extends from 30 Hz to 15 kHz and the L − R signal is contained in two sidebands extending from 23 kHz to 53 kHz as indicated in Figure 9–28. These frequencies come from the FM detector and go into the filter circuits where they are separated.

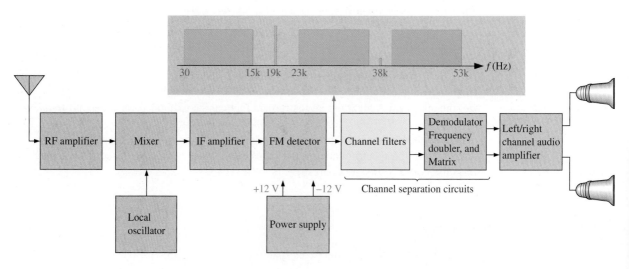

FIGURE 9–28
FM stereo receiver system.

The frequency doubler and demodulator are used to extract the audio signal from the 23 kHz to 53 kHz sidebands after which the 30 Hz to 15 kHz L − R signal is passed through a filter.

The L + R and L − R audio signals are then sent to the matrix where they are applied to the summing circuits to produce the left and right channel audio (−2L and −2R). Our focus in this application is on the filters.

Now, so that you can take a closer look at the filter board, let's take it out of the system and put it on the troubleshooter's bench.

TROUBLESHOOTER'S BENCH

■ ACTIVITY 1   Relate the PC Board to the Schematic

Locate and identify all components on the PC board in Figure 9–29, using the schematic in Figure 9–30. Label the PC board components to correspond with the schematic. Identify all inputs and outputs. Trace out the PC board to verify that it corresponds with the schematic. In this chapter, the backside of the PC board is shown instead of the x-ray view. The color-shaded areas on the schematic are the filter networks contained on the board. The other blocks and circuits are on the demodulator, frequency doubler, and matrix board located elsewhere.

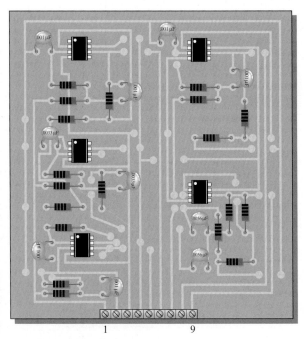

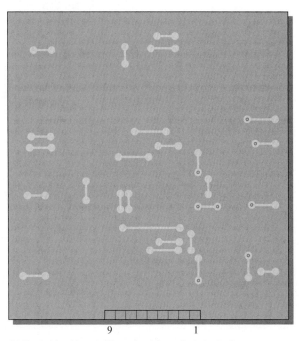

(a) Component side of board

(b) Backside of board. The pads with small circles in the centers are component feedthrough pads.

FIGURE 9–29

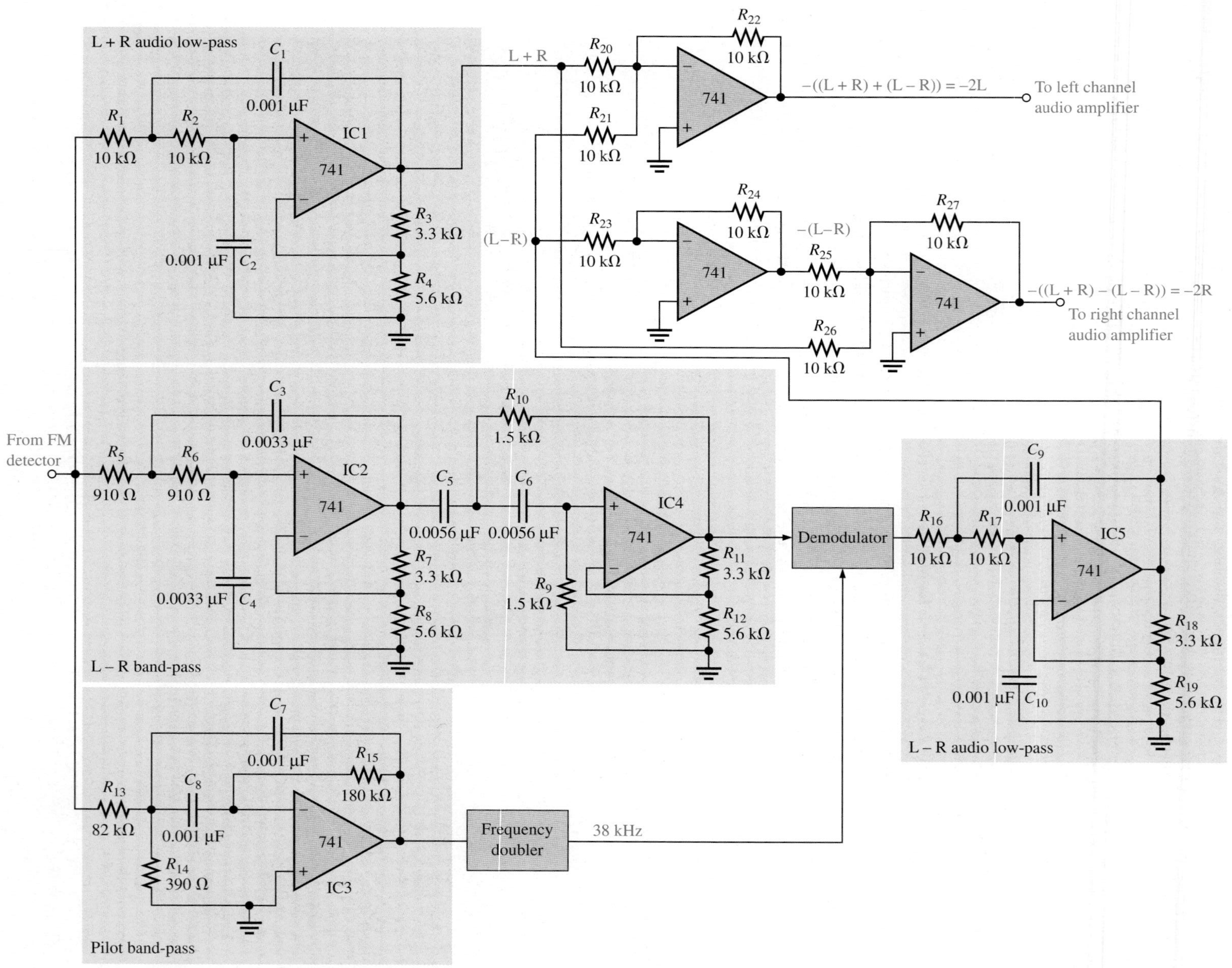
L + R audio low-pass
L – R band-pass
Pilot band-pass
L – R audio low-pass
From FM detector
IC1
IC2
IC3
IC4
IC5
741
Demodulator
Frequency doubler
38 kHz
L + R
(L–R)
–(L–R)
–((L + R) + (L – R)) = –2L
To left channel audio amplifier
–((L + R) – (L – R)) = –2R
To right channel audio amplifier
R1 10 kΩ
R2 10 kΩ
C1 0.001 μF
C2 0.001 μF
R3 3.3 kΩ
R4 5.6 kΩ
R5 910 Ω
R6 910 Ω
C3 0.0033 μF
C4 0.0033 μF
R7 3.3 kΩ
R8 5.6 kΩ
C5 0.0056 μF
C6 0.0056 μF
R9 1.5 kΩ
R10 1.5 kΩ
R11 3.3 kΩ
R12 5.6 kΩ
R13 82 kΩ
R14 390 Ω
R15 180 kΩ
C7 0.001 μF
C8 0.001 μF
R16 10 kΩ
R17 10 kΩ
C9 0.001 μF
C10 0.001 μF
R18 3.3 kΩ
R19 5.6 kΩ
R20 10 kΩ
R21 10 kΩ
R22 10 kΩ
R23 10 kΩ
R24 10 kΩ
R25 10 kΩ
R26 10 kΩ
R27 10 kΩ

FIGURE 9–30

■ **ACTIVITY 2   Analyze the Filter Networks**

Step 1: Using the component values, determine the critical frequencies of each Sallen-Key-type filter.

Step 2: Using the component values, determine the center frequency of the multiple-feedback filter.

Step 3: Determine the bandwidth of each filter.

Step 4: Determine the voltage gain of each filter.

Step 5: Verify that the Sallen-Key filters have an approximate Butterworth response characteristic.

■ **ACTIVITY 3   Write a Technical Report**

Describe each filter in detail specifying the type of filter, the frequency responses, and the function of the filter within the overall circuitry. Also, describe the overall operation of the complete channel separation circuitry.

■ **ACTIVITY 4   Troubleshoot the Filter Board**

The filter board is plugged into a test fixture that permits access to each input and output, as shown in Figure 9–31, where the socket numbers correspond to the board pin numbers. The test instruments to be used are a sweep generator, a spectrum analyzer, and a dual power supply. For the sweep generator, a minimum and a maximum frequency are selected and the instrument produces an output that repetitively sweeps through all frequencies between the minimum and maximum settings. The spectrum analyzer will plot out a frequency response curve.

Develop a basic test procedure for completely testing the board in the fixture, using general references to instrument inputs, outputs, and settings. Include a diagram of a complete test setup.

FIGURE 9–31
Filter board in a test fixture.

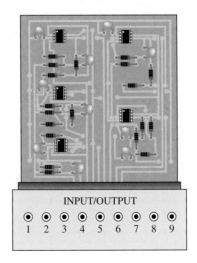

9–8 REVIEW QUESTIONS

1. What is the purpose of the filter board in this system?
2. What is the bandwidth of the L + R low-pass filter?
3. What is the bandwidth of the L − R low-pass filter?
4. Which filters on the board have approximate Butterworth responses?
5. What is the purpose of the stereo matrix circuitry?

■ SUMMARY

- The bandwidth in a low-pass filter equals the critical frequency because the response extends to 0 Hz.
- The bandwidth in a high-pass filter extends above the critical frequency and is limited only by the inherent frequency limitation of the active circuit.
- A band-pass filter passes all frequencies within a band between a lower and an upper critical frequency and rejects all others outside this band.
- The bandwidth of a band-pass filter is the difference between the upper critical frequency and the lower critical frequency.
- A band-stop filter rejects all frequencies within a specified band and passes all those outside this band.
- Filters with the Butterworth response characteristic have a very flat response in the passband, exhibit a roll-off of −20 dB/decade/pole, and are used when all the frequencies in the passband must have the same gain.
- Filters with the Chebyshev characteristic have ripples or overshoot in the passband and exhibit a faster roll-off per pole than filters with the Butterworth characteristic.
- Filters with the Bessel characteristic are used for filtering pulse waveforms. Their linear phase characteristic results in minimal waveshape distortion. The roll-off rate per pole is slower than for the Butterworth.
- In filter terminology, a single RC network is called a *pole*.
- Each pole in a Butterworth filter causes the output to roll off at a rate of −20 dB/decade.
- The quality factor Q of a band-pass filter determines the filter's selectivity. The higher the Q, the narrower the bandwidth and the better the selectivity.
- The damping factor determines the filter response characteristic (Butterworth, Chebyshev, or Bessel).

■ GLOSSARY

Key terms are in color. These terms are included in the end-of-book glossary.

Active filter   A frequency-selective circuit consisting of active devices such as transistors or op-amps coupled with reactive components.

Band-pass filter   A type of filter that passes a range of frequencies lying between a certain lower frequency and a certain higher frequency.

Band-stop filter   A type of filter that blocks or rejects a range of frequencies lying between a certain lower frequency and a certain higher frequency.

Bessel   A type of filter response having a linear phase characteristic and less than −20 dB/decade/pole roll-off.

Butterworth A type of filter response characterized by flatness in the passband and a -20 dB/decade/pole roll-off.

Carrier The high radio frequency (RF) signal that carries modulated information in AM, FM, or other communications systems.

Chebyshev A type of filter response characterized by ripples in the passband and a greater than -20 dB/decade/pole roll-off.

Critical frequency (f_c) The frequency that defines the end of the passband of a filter; also called *cutoff frequency*.

Damping factor (DF) A filter characteristic that determines the type of response.

Filter A circuit that passes certain frequencies and attenuates or rejects all other frequencies.

Frequency modulation (FM) A communication method in which a lower frequency intelligence-carrying signal modulates (varies) the frequency of a higher frequency signal.

High-pass filter A type of filter that passes frequencies above a certain frequency while rejecting lower frequencies.

Low-pass filter A type of filter that passes frequencies below a certain frequency while rejecting higher frequencies.

Order The number of poles in a filter.

Passband The region of frequencies that are allowed to pass through a filter with minimum attenuation.

Pole A network containing one resistor and one capacitor that contributes -20 dB/decade to a filter's roll-off rate.

Roll-off The rate of decrease in gain below or above the critical frequencies of a filter.

■ KEY FORMULAS

(9–1) $BW = f_c$ — Low-pass bandwidth

(9–2) $BW = f_{c2} - f_{c1}$ — Filter bandwidth of a band-pass filter

(9–3) $f_0 = \sqrt{f_{c1}f_{c2}}$ — Center frequency of a band-pass filter

(9–4) $Q = \dfrac{f_0}{BW}$ — Quality factor of a band-pass filter

(9–5) $DF = 2 - \dfrac{R_1}{R_2}$ — Damping factor

(9–6) $f_c = \dfrac{1}{2\pi\sqrt{R_A R_B C_A C_B}}$ — Critical frequency for a second-order Sallen-Key filter

(9–7) $f_0 = \dfrac{1}{2\pi C}\sqrt{\dfrac{R_1 + R_3}{R_1 R_2 R_3}}$ — Center frequency of a multiple-feedback filter

(9–8) $A_0 = \dfrac{R_2}{2R_1}$ — Gain of a multiple-feedback filter

■ SELF-TEST

Answers are at the end of the chapter.

1. The term *pole* in filter terminology refers to
   - (a) a high-gain op-amp
   - (b) one complete active filter
   - (c) a single *RC* network
   - (d) the feedback circuit

2. A single resistor and a single capacitor can be connected to form a filter with a roll-off rate of
   - (a) −20 dB/decade
   - (b) −40 dB/decade
   - (c) −6 dB/octave
   - (d) answers (a) and (c)

3. A band-pass response has
   - (a) two critical frequencies
   - (b) one critical frequency
   - (c) a flat curve in the passband
   - (d) a wide bandwidth

4. The lowest frequency passed by a low-pass filter is
   - (a) 1 Hz
   - (b) 0 Hz
   - (c) 10 Hz
   - (d) dependent on the critical frequency

5. The *Q* of a band-pass filter depends on
   - (a) the critical frequencies
   - (b) only the bandwidth
   - (c) the center frequency and the bandwidth
   - (d) only the center frequency

6. The damping factor of an active filter determines
   - (a) the voltage gain
   - (b) the critical frequency
   - (c) the response characteristic
   - (d) the roll-off rate

7. A maximally flat frequency response is known as
   - (a) Chebyshev
   - (b) Butterworth
   - (c) Bessel
   - (d) Colpitts

8. The damping factor of a filter is set by
   - (a) the negative feedback circuit
   - (b) the positive feedback circuit
   - (c) the frequency-selective circuit
   - (d) the gain of the op-amp

9. The number of poles in a filter affect the
   - (a) voltage gain
   - (b) bandwidth
   - (c) center frequency
   - (d) roll-off rate

10. Sallen-Key filters are
    - (a) single-pole filters
    - (b) second-order filters
    - (c) Butterworth filters
    - (d) band-pass filters

11. When filters are cascaded, the roll-off rate
    - (a) increases
    - (b) decreases
    - (c) does not change

12. When a low-pass and a high-pass filter are cascaded to get a band-pass filter, the critical frequency of the low-pass filter must be
    - (a) equal to the critical frequency of the high-pass filter
    - (b) less than the critical frequency of the high-pass filter
    - (c) greater than the critical frequency of the high-pass filter

13. A state-variable filter consists of
    - (a) one op-amp with multiple-feedback paths
    - (b) a summing amplifier and two integrators
    - (c) a summing amplifier and two differentiators
    - (d) three Butterworth stages

14. When the gain of a filter is minimum at its center frequency, it is a
    - (a) band-pass filter
    - (b) a band-stop filter
    - (c) a notch filter
    - (d) answers (b) and (c)

TROUBLESHOOTER'S QUIZ *Answers are at the end of the chapter.*

Refer to Figure 9–33(a).

❏ If C_1 is mistakenly replaced with a 0.15 μF capacitor instead of 0.015 μF,

    1. The bandwidth will

        (a) increase   **(b)** decrease   **(c)** not change

    2. The number of poles will

        (a) increase   **(b)** decrease   **(c)** not change

    3. The roll-off rate will

        (a) increase   **(b)** decrease   **(c)** not change

Refer to Figure 9–33(b).

❏ If C_2 is open,

    4. The ac output for a given ac input will

        (a) increase   **(b)** decrease   **(c)** not change

❏ If R_4 is 10 kΩ instead of 1.0 kΩ,

    5. The damping factor will

        (a) increase   **(b)** decrease   **(c)** not change

    6. The critical frequency will

        (a) increase   **(b)** decrease   **(c)** not change

Refer to Figure 9–33(c).

❏ If C_3 is open,

    7. The number of poles will

        (a) increase   **(b)** decrease   **(c)** not change

Refer to Figure 9–37(b).

❏ If R_3 has an incorrect value of 1.0 kΩ,

    8. The passband gain will

        (a) increase   **(b)** decrease   **(c)** not change

❏ If R_2 is less than the specified value of 150 kΩ,

    9. The center frequency will

        (a) increase   **(b)** decrease   **(c)** not change

   10. The passband gain will

        (a) increase   **(b)** decrease   **(c)** not change

   11. The bandwidth will

        (a) increase   **(b)** decrease   **(c)** not change

Refer to Figure 9–37(c).

❏ If R_5 is larger than the specified value of 560 kΩ,

   12. The bandwidth will

        (a) increase   **(b)** decrease   **(c)** not change

■ PROBLEMS

Answers to odd-numbered problems are at the end of the book.

SECTION 9–1 Basic Filter Responses

1. Identify each type of filter response (low-pass, high-pass, band-pass, or band-stop) in Figure 9–32.

2. A certain low-pass filter has a critical frequency of 800 Hz. What is its bandwidth?

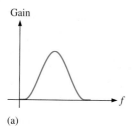
(a)

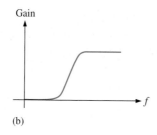
(b)

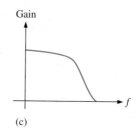
(c)

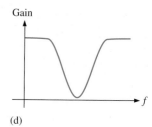
(d)

FIGURE 9–32

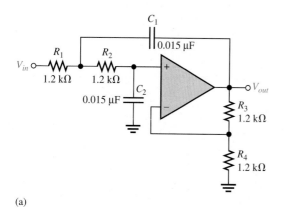
(a)

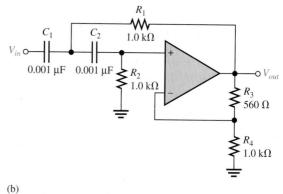
(b)

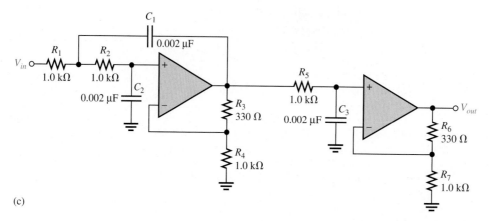
(c)

FIGURE 9–33

3. A single-pole high-pass filter has a frequency-selective network with $R = 2.2$ kΩ and $C = 0.0015$ μF. What is the critical frequency? Can you determine the bandwidth from the available information?

4. What is the roll-off rate of the filter described in Problem 3?

5. What is the bandwidth of a band-pass filter whose critical frequencies are 3.2 kHz and 3.9 kHz? What is the Q of this filter?

6. What is the center frequency of a filter with a Q of 15 and a bandwidth of 1.0 kHz?

SECTION 9–2 Filter Response Characteristics

7. What is the damping factor in each active filter shown in Figure 9–33? Which filters are approximately optimized for a Butterworth response characteristic?

8. For the filters in Figure 9–33 that do not have a Butterworth response, specify the changes necessary to convert them to Butterworth responses. (Use nearest standard values.)

9. Response curves for second-order filters are shown in Figure 9–34. Identify each as Butterworth, Chebyshev, or Bessel.

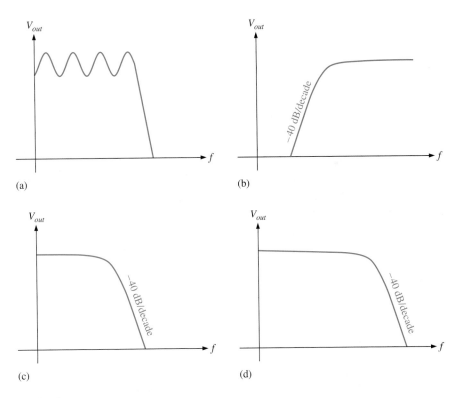

FIGURE 9–34

SECTION 9–3   Active Low-Pass Filters

10. Is the four-pole filter in Figure 9–35 approximately optimized for a Butterworth response? What is the roll-off rate?

11. Determine the critical frequency in Figure 9–35.

12. Without changing the response curve, adjust the component values in the filter of Figure 9–35 to make it an equal-value filter.

13. Modify the filter in Figure 9–35 to increase the roll-off rate to −120 dB/decade while maintaining an approximate Butterworth response.

14. Using a block diagram format, show how to implement the following roll-off rates using single-pole and two-pole low-pass filters with Butterworth responses.
   (a) −40 dB/decade   **(b)** −20 dB/decade   **(c)** −60 dB/decade
   (d) −100 dB/decade   **(e)** −120 dB/decade

FIGURE 9–35

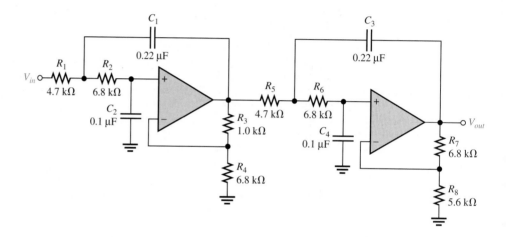

SECTION 9–4   Active High-Pass Filters

15. Convert the equal-value filter from Problem 12 to a high-pass with the same critical frequency and response characteristic.

16. Make the necessary circuit modification to reduce by half the critical frequency in Problem 15.

17. For the filter in Figure 9–36,
   (a) How would you increase the critical frequency?
   (b) How would you increase the gain?

FIGURE 9–36

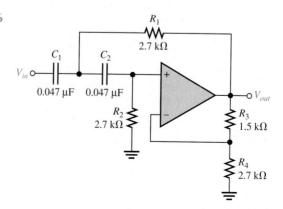

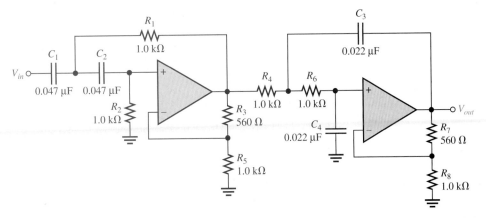

(a)

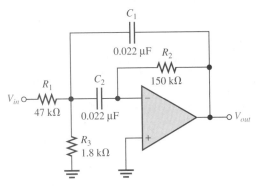

(b)

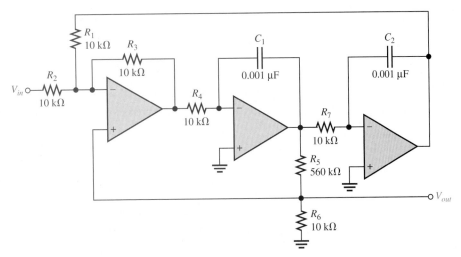

(c)

FIGURE 9–37

SECTION 9–5   Active Band-Pass Filters

18. Identify each band-pass filter configuration in Figure 9–37.

19. Determine the center frequency and bandwidth for each filter in Figure 9–37.

20. Optimize the state-variable filter in Figure 9–38 for $Q = 50$. What bandwidth is achieved?

SECTION 9–6   Active Band-Stop Filters

21. Show how to make a notch (band-stop) filter using the basic circuit in Figure 9–38.

22. Modify the band-stop filter in Problem 21 for a center frequency of 120 Hz.

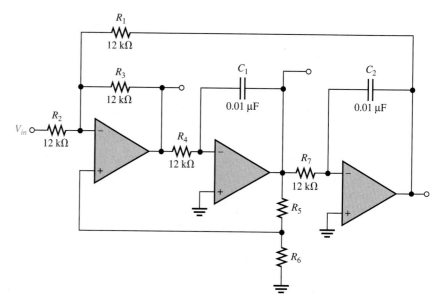

FIGURE 9–38

Section 9–1

1. The critical frequency determines the bandwidth.

2. The inherent frequency limitation of the op-amp limits the bandwidth.

3. Q and BW are inversely related. The higher the Q, the better the selectivity, and vice versa.

Section 9–2

1. Butterworth is very flat in the passband and has a -20 dB/decade/pole roll-off. Chebyshev has ripples in the passband and has greater than -20 dB/decade/pole roll-off. Bessel has a linear phase characteristic and less than -20 dB/decade/pole roll-off.

2. The damping factor determines the response characteristic.

3. Frequency-selection network, gain element, and negative feedback network are the parts of an active filter.

Section 9–3

1. A second-order filter has two poles. Two resistors and two capacitors make up the frequency-selective network.

2. The damping factor sets the response characteristic.

3. Cascading increases the roll-off rate.

Section 9–4

1. The positions of the Rs and Cs in the frequency-selection network are opposite for low-pass and high-pass configurations.
2. Decrease the R values to increase f_c.
3. -140 dB/decade

Section 9–5

1. Q determines selectivity.
2. $Q = 25$. Higher Q gives narrower BW.
3. A summing amplifier and two integrators make up a state-variable filter.

Section 9–6

1. A band-stop rejects frequencies within the stopband. A band-pass passes frequencies within the passband.
2. The low-pass and high-pass outputs are summed.

Section 9–7

1. To check the frequency response of a filter
2. Discrete point measurement—tedious and less complete; simpler equipment.
   Swept frequency measurement—uses more expensive equipment; more efficient, can be more accurate and complete.

Section 9–8

1. The filter board takes the detected FM signal and separates the L + R and L − R audio signals.
2. $BW = 15.9$ kHz
3. $BW = 15.9$ kHz
4. The L + R low-pass, the L − R low-pass, and the L − R band-pass
5. The stereo matrix combines the L + R and L − R signals and produces the separate left and right channel audio signals.

■ **ANSWERS TO PRACTICE EXERCISES FOR EXAMPLES**

9–1 500 Hz   9–2 1.44   9–3 7.23 kHz, 1.29 kΩ

9–4 $C_{A1} = C_{A2} = C_{B1} = C_{B2} = 0.234$ μF; $R_2 = R_4 = 680$ Ω, $R_1 = 103$ Ω, $R_3 = 840$ Ω

9–5 $R_A = R_B = R_2 = 10$ kΩ, $C_A = C_B = 0.053$ μF, $R_1 = 5.86$ kΩ

9–6 Gain increases to 2.43, frequency decreases to 544 Hz, and bandwidth decreases to 96.5 Hz.

9–7 $f_0 = 21.9$ kHz, $Q = 101$, $BW = 217$ Hz

9–8 Decrease the input resistors or the feedback capacitors of the two integrator stages by half.

■ **ANSWERS TO SELF-TEST**

1. (c)	**2.** (d)	**3.** (a)	**4.** (b)	**5.** (c)
6. (c)	**7.** (b)	**8.** (a)	**9.** (d)	**10.** (b)
11. (a)	**12.** (c)	**13.** (b)	**14.** (d)	

■ **ANSWERS TO TROUBLE-SHOOTER'S QUIZ**

1. decrease	**2.** not change	**3.** not change	**4.** decrease
5. increase	**6.** not change	**7.** decrease	**8.** not change
9. increase	**10.** decrease	**11.** increase	**12.** decrease

10

OSCILLATORS AND TIMERS

Courtesy Hewlett-Packard Company

■ CHAPTER OBJECTIVES

❏ Describe the basic operating principles for all oscillators
❏ Explain the operation of feedback oscillators
❏ Describe and analyze the operation of basic *RC* sinusoidal feedback oscillators
❏ Describe and analyze the operation of basic relaxation oscillators
❏ Use a 555 timer in an oscillation application
❏ Use a 555 timer as a one-shot device
❏ Apply what you have learned in this chapter to a system application

■ KEY TERMS

❏ Feedback oscillator
❏ Relaxation oscillator
❏ Positive feedback
❏ Wien-bridge oscillator
❏ Phase-shift oscillator
❏ Voltage-controlled oscillator (VCO)
❏ Astable multivibrator
❏ One-shot

■ CHAPTER INTRODUCTION

Oscillators are circuits that generate a periodic waveform to perform timing, control, or communication functions. They are found in nearly all electronic systems, including analog and digital systems, and in most test instruments such as oscilloscopes and function generators.

Oscillators require a form of positive feedback, where a portion of the output signal is fed back to the input in a way that causes it to reinforce itself and thus sustain a continuous output signal. Although an external input is not strictly necessary, many oscillators use an external signal to control the frequency or to synchronize it with another source. Oscillators are designed to produce a controlled oscillation with one of two basic methods: the unity-gain method used with feedback oscillators and the timing method used with relaxation oscillators. Both will be discussed in this chapter.

Different types of oscillators produce various types of outputs including sine waves, square waves, triangular waves, and sawtooth waves. In this chapter, several types of basic oscillator circuits using an op-amp as the gain element are introduced. Also, a very popular integrated circuit, called the 555 timer, is discussed.

■ A SYSTEM APPLICATION

The function generator shown in Figure 10–37 is a good illustration of a system application for oscillators. The oscillator is a major part of this particular system. No doubt, you are already familiar with the use of the signal or function generator in your lab. As with most types of systems, a function generator can be implemented in more than one way. The system in this chapter uses circuits with which you are already familiar without some of the refinements and features found in many commercial instruments. The system reinforces what you have studied and lets you see these circuits "at work" in a specific application.

For the system application in Section 10–7, in addition to the other topics, be sure you understand

❏ How *RC* oscillators work
❏ How a zero-level detector works
❏ How an integrator works

www.   VISIT THE COMPANION WEBSITE

Study Aids for This Chapter Are Available at

http://www.prenhall.com/floyd

10–1 ■ THE OSCILLATOR

An oscillator is a circuit that produces a periodic waveform on its output with only the dc supply voltage as a required input. A repetitive input signal is not required but is sometimes used to synchronize oscillations. The output voltage can be either sinusoidal or nonsinusoidal, depending on the type of oscillator. Two major classifications for oscillators are feedback oscillators and relaxation oscillators.

After completing this section, you should be able to

❏ Describe the basic operating principles for all oscillators
   ❏ Explain the purpose of an oscillator
   ❏ Discuss two important classifications for oscillators
   ❏ List the basic elements of a feedback oscillator

Types of Oscillators

Essentially, all **oscillators** convert electrical energy from the dc power supply to periodic waveforms that can be used for various timing, control, or signal-generating applications. A basic oscillator is illustrated in Figure 10–1. Oscillators are classified according to the technique for generating a signal.

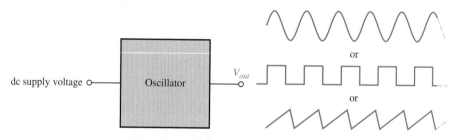

FIGURE 10–1
The basic oscillator concept showing three common types of output waveforms.

Feedback Oscillators   One type of oscillator is the **feedback oscillator** which returns a fraction of the output signal to the input with no net phase shift, resulting in a reinforcement of the output signal. After oscillations are started, the loop gain is maintained at 1.0 to maintain oscillations. A feedback oscillator consists of an amplifier for gain (either a discrete transistor or an op-amp) and a positive feedback network that produces phase shift and provides attenuation, as shown in Figure 10–2.

Relaxation Oscillators   A second type of oscillator is the **relaxation oscillator**. A relaxation oscillator uses an *RC* timing circuit to generate a waveform that is generally a square wave or other nonsinusoidal waveform. Typically, a relaxation oscillator uses a Schmitt trigger or other device that changes states to alternately charge and discharge a capacitor through a resistor. Relaxation oscillators are discussed in Section 10–4.

FIGURE 10–2
Basic elements of a feedback oscillator.

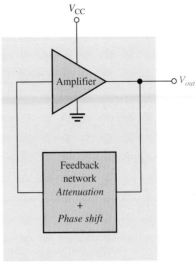

Feedback oscillator

10–2 ■ FEEDBACK OSCILLATOR PRINCIPLES

Feedback oscillator operation is based on the principle of positive feedback. In this section, we will examine this concept and look at the general conditions required for oscillation to occur. Feedback oscillators are widely used to generate sinusoidal waveforms.

After completing this section, you should be able to

❏ Explain the operation of feedback oscillators
   ❏ Explain positive feedback
   ❏ Describe the conditions for oscillation
   ❏ Discuss the start-up conditions

Positive Feedback

Positive feedback is characterized by the condition wherein an in-phase portion of the output voltage of an amplifier is fed back to the input. This basic idea is illustrated with the sinusoidal oscillator shown in Figure 10–3. As you can see, the in-phase feedback voltage is amplified to produce the output voltage, which in turn produces the feedback voltage. That is, a loop is created in which the signal sustains itself and a continuous sinusoidal output is produced. This phenomenon is called *oscillation*.

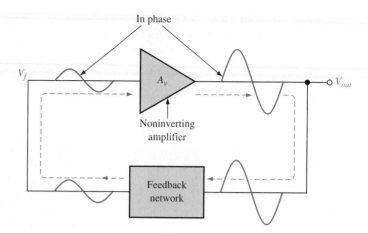

FIGURE 10–3
Positive feedback produces oscillation.

Conditions for Oscillation

Two conditions, illustrated in Figure 10–4, are required to sustain oscillations:

1. The phase shift around the feedback loop must be effectively 0°.

2. The voltage gain, A_{cl}, around the closed feedback loop (loop gain) must equal 1 (unity).

The voltage gain around the closed feedback loop, A_{cl}, is the product of the amplifier gain, A_v, and the attenuation, B, of the feedback circuit.

$$A_{cl} = A_v B$$

If a sinusoidal wave is the desired output, a loop gain greater than 1 will rapidly cause the output to saturate at both peaks of the waveform, producing unacceptable distortion. To avoid this, some form of gain control must be used to keep the loop gain at exactly 1, once oscillations have started. For example, if the attenuation of the feedback network is 0.01, the amplifier must have a gain of exactly 100 to overcome this attenuation and not

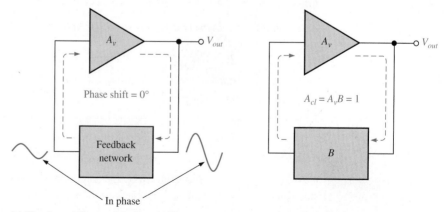

(a) The phase shift around the loop is 0°.

(b) The closed loop gain is 1.

FIGURE 10–4
Conditions for oscillation.

create unacceptable distortion ($0.01 \times 100 = 1.0$). An amplifier gain of greater than 100 will cause the oscillator to limit both peaks of the waveform.

Start-Up Conditions

So far, you have seen what it takes for an oscillator to produce a continuous sinusoidal output. Now let's examine the requirements for the oscillation to start when the dc supply voltage is turned on. As you know, the unity-gain condition must be met for oscillation to be sustained. For oscillation to *begin,* the voltage gain around the positive feedback loop must be greater than 1 so that the amplitude of the output can build up to a desired level. The gain must then decrease to 1 so that the output stays at the desired level and oscillation is sustained. (Several ways to achieve this reduction in gain after start-up are discussed in the next section.) The voltage-gain conditions for both starting and sustaining oscillation are illustrated in Figure 10–5.

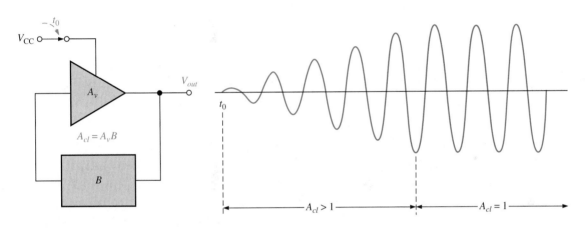

FIGURE 10–5

When oscillation starts at t_0, the condition $A_{cl} > 1$ causes the sinusoidal output voltage amplitude to build up to a desired level. Then A_{cl} decreases to 1 and maintains the desired amplitude.

A question that normally arises is this: If the oscillator is initially off and there is no output voltage, how does a feedback signal originate to start the positive feedback build-up process? Initially, a small positive feedback voltage develops from thermally produced broad-band noise in the resistors or other components or from power supply turn-on transients. The feedback circuit permits only a voltage with a frequency equal to the selected oscillation frequency to appear in phase on the amplifier's input. This initial feedback voltage is amplified and continually reinforced, resulting in a buildup of the output voltage as previously discussed.

10–2 REVIEW QUESTIONS

1. What are the conditions required for a circuit to oscillate?
2. Define positive feedback.
3. What is the voltage gain condition for oscillator start-up?

10–3 ■ SINUSOIDAL OSCILLATORS

In this section, you will learn about three types of feedback oscillators that use RC circuits to produce sinusoidal outputs: the Wien-bridge oscillator, the phase-shift oscillator, and the twin-T oscillator. Generally, RC feedback oscillators are used for frequencies up to about 1 MHz. The Wien-bridge is by far the most widely used type of RC oscillator for this range of frequencies.

After completing this section, you should be able to

❑ Describe and analyze the operation of basic *RC* sinusoidal feedback oscillators
  ❑ Identify a Wien-bridge oscillator
  ❑ Determine the resonant frequency of a Wien-bridge oscillator
  ❑ Analyze oscillator feedback conditions
  ❑ Analyze oscillator start-up conditions
  ❑ Describe a self-starting Wien-bridge oscillator
  ❑ Identify a phase-shift oscillator
  ❑ Calculate the resonant frequency and analyze the feedback conditions for a phase-shift oscillator
  ❑ Identify a twin-T oscillator and describe its operation

The Wien-Bridge Oscillator

One type of sinusoidal feedback oscillator is the **Wien-bridge oscillator**. A fundamental part of the Wien-bridge oscillator is a lead-lag network like that shown in Figure 10–6(a). R_1 and C_1 together form the lag portion of the network; R_2 and C_2 form the lead portion. The operation of this lead-lag network is as follows. At lower frequencies, the lead network dominates due to the high reactance of C_2. As the frequency increases, X_{C2} decreases, thus allowing the output voltage to increase. At some specified frequency, the response of the lag network takes over, and the decreasing value of X_{C1} causes the output voltage to decrease.

The response curve for the lead-lag network shown in Figure 10–6(b) indicates that the output voltage peaks at a frequency called the resonant frequency, f_r. At this point, the attenuation (V_{out}/V_{in}) of the network is ⅓ if $R_1 = R_2$ and $X_{C1} = X_{C2}$ as stated by the following equation, which is derived in Appendix B:

$$\frac{V_{out}}{V_{in}} = \frac{1}{3} \qquad\qquad (10\text{–}1)$$

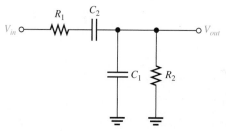

(a) Network

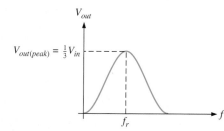

(b) Response curve

FIGURE 10–6

A lead-lag network and its response curve.

The formula for the resonant frequency is also derived in Appendix B and is

$$f_r = \frac{1}{2\pi RC} \qquad \textbf{(10–2)}$$

To summarize, the lead-lag network in the Wien-bridge oscillator has a resonant frequency, f_r, at which the phase shift through the network is $0°$ and the attenuation is $\frac{1}{3}$. Below f_r, the lead network dominates and the output leads the input. Above f_r, the lag network dominates and the output lags the input.

The Basic Circuit   The lead-lag network is used in the positive feedback loop of an op-amp, as shown in Figure 10–7(a). A voltage divider is used in the negative feedback loop. The Wien-bridge oscillator circuit can be viewed as a noninverting amplifier configuration with the input signal fed back from the output through the lead-lag network. Recall that the closed-loop gain of the amplifier is determined by the voltage divider.

$$A_{cl} = \frac{1}{B} = \frac{1}{R_2/(R_1 + R_2)} = \frac{R_1 + R_2}{R_2}$$

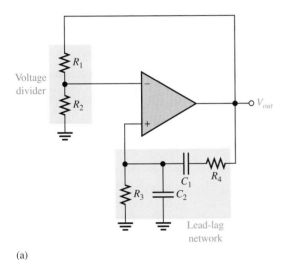

(a)

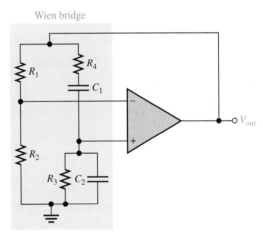

(b) Wien bridge circuit combines a voltage divider and a lead-lag network.

FIGURE 10–7
Two ways to draw the schematic of a Wien-bridge oscillator.

The circuit is redrawn in Figure 10–7(b) to show that the op-amp is connected across the bridge circuit. One leg of the bridge is the lead-lag network, and the other is the voltage divider.

Positive Feedback Conditions for Oscillation   As you know, for the circuit to produce a sustained sinusoidal output (oscillate), the phase shift around the positive feedback loop must be $0°$ and the gain around the loop must equal unity (1). The $0°$ phase-shift condition is met when the frequency is f_r because the phase shift through the lead-lag network is $0°$ and there is no inversion from the noninverting (+) input of the op-amp to the output. This is shown in Figure 10–8(a).

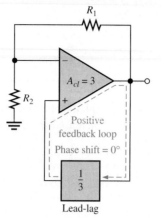

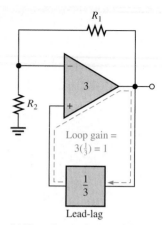

(a) The phase shift around the loop is 0°.   (b) The voltage gain around the loop is 1.

FIGURE 10–8
Conditions for oscillation.

The unity-gain condition in the feedback loop is met when

$$A_{cl} = 3$$

This offsets the ⅓ attenuation of the lead-lag network, thus making the total gain around the positive feedback loop equal to 1, as depicted in Figure 10–8(b). To achieve a closed-loop gain of 3,

$$R_1 = 2R_2$$

Then

$$A_{cl} = \frac{R_1 + R_2}{R_2} = \frac{2R_2 + R_2}{R_2} = \frac{3R_2}{R_2} = 3$$

Start-Up Conditions   Initially, the closed-loop gain of the amplifier itself must be more than 3 ($A_{cl} > 3$) until the output signal builds up to a desired level. Ideally, the gain of the amplifier must then decrease to 3 so that the total gain around the loop is 1 and the output signal stays at the desired level, thus sustaining oscillation. This is illustrated in Figure 10–9.

The circuit in Figure 10–10 illustrates a method for achieving sustained oscillations. Notice that the voltage-divider network has been modified to include an additional resistor R_3 in parallel with a back-to-back zener diode arrangement. When dc power is first applied, both zener diodes appear as opens. This places R_3 in series with R_1, thus increasing the closed-loop gain of the amplifier as follows ($R_1 = 2R_2$):

$$A_{cl} = \frac{R_1 + R_2 + R_3}{R_2} = \frac{3R_2 + R_3}{R_2} = 3 + \frac{R_3}{R_2}$$

Initially, a small positive feedback signal develops from noise or turn-on transients. The lead-lag network permits only a signal with a frequency equal to f_r to appear in phase on the noninverting input. This feedback signal is amplified and continually reinforced, resulting in a buildup of the output voltage. When the output signal reaches the zener breakdown voltage, the zeners conduct and effectively short out R_3. This lowers the amplifier's closed-loop gain to 3. At this point the total loop gain is 1 and the output signal levels off

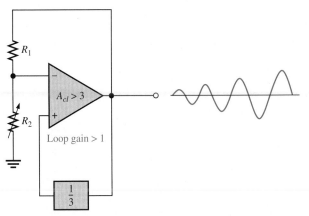

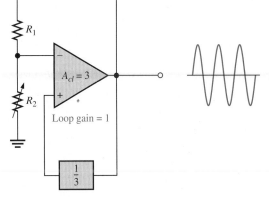

(a) Loop gain greater than 1 causes output to build up.

(b) Loop gain of 1 causes a sustained constant output.

FIGURE 10–9

Oscillator start-up conditions.

FIGURE 10–10

Self-starting Wien-bridge oscillator using back-to-back zener diodes.

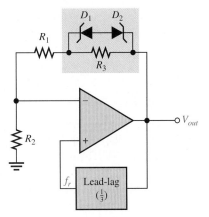

and the oscillation is sustained. All practical methods to achieve stability for feedback oscillators require the gain to be self-adjusting. This requirement is a form of automatic gain control (AGC). The zener diodes in this example limit the gain at the onset of a nonlinearity, in this case, zener conduction.

Although the zener feedback network is simple, it suffers from the fact that nonlinearity must occur to control gain; hence, it is difficult to achieve a clean sinusoidal waveform. Another method to control the gain uses a JFET as a voltage-controlled resistor in a negative feedback path. This method can produce an excellent sinusoidal waveform that is stable. Recall from Section 4–2 that a JFET operating with a small or zero V_{DS} is operating in the ohmic region. As the gate voltage increases, the drain-source resistance increases. If the JFET is placed in the negative feedback path, automatic gain control can be achieved because of this voltage-controlled resistance.

A JFET stabilized Wien-bridge oscillator is shown in Figure 10–11. The gain of the op-amp is controlled by the components shown in the shaded box, which include the JFET. The JFET's drain-source resistance depends on the gate voltage. With no output signal, the gate is at zero volts, causing the drain-source resistance to be at the minimum. With this condition, the loop gain is greater than 1. Oscillations begin and rapidly build

FIGURE 10–11
Self-starting Wien-bridge oscillator using a JFET in the negative feedback loop.

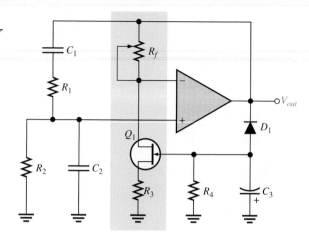

to a large output signal. Negative excursions of the output signal forward-bias D_1, causing capacitor C_3 to charge to a negative voltage. This voltage increases the drain-source resistance of the JFET and reduces the gain (and hence the output). This is classic negative feedback at work. With the proper selection of components, the gain can be stabilized at the required level. You can explore this circuit further in Experiment 27 of the laboratory manual that accompanies this text. The following example illustrates a JFET stabilized oscillator.

EXAMPLE 10–1

Determine the frequency for the Wien-bridge oscillator in Figure 10–12. Also, calculate the setting for R_f assuming the internal drain-source resistance, r'_{ds}, of the JFET is 500 Ω when oscillations are stable.

FIGURE 10–12

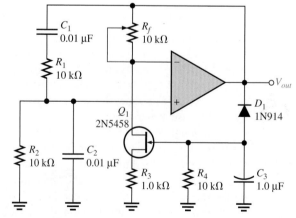

Solution For the lead-lag network, $R_1 = R_2 = R = 10 \text{ k}\Omega$ and $C_1 = C_2 = C = 0.01 \text{ μF}$. The frequency is

$$f_r = \frac{1}{2\pi RC} = \frac{1}{2\pi(10 \text{ k}\Omega)(0.01 \text{ μF})} = \mathbf{1.59 \text{ kHz}}$$

The closed-loop gain must be 3.0 for oscillations to be sustained. For an inverting amplifier, the gain is that of a noninverting amplifier.

$$A_v = \frac{R_f}{R_i} + 1$$

R_i is composed of R_3 (the source resistor) and r'_{ds}. Substituting,

$$A_v = \frac{R_f}{R_3 + r'_{ds}} + 1$$

Rearranging and solving for R_f,

$$R_f = (A_v - 1)(R_3 + r'_{ds}) = (3 - 1)(1.0 \text{ k}\Omega + 500 \text{ }\Omega) = \textbf{3.0 k}\Omega$$

Practice Exercise* What happens to the oscillations if the setting of R_f is too high? What happens if the setting is too low?

_____

* Answers are at the end of the chapter.

The Phase-Shift Oscillator

A type of sinusoidal feedback oscillator called the **phase-shift oscillator** is shown in Figure 10–13. Each of the three RC networks in the feedback loop can provide a maximum phase shift approaching 90°. Oscillation occurs at the frequency where the total phase shift through the three RC networks is 180°. The inversion of the op-amp itself provides the additional 180° to meet the requirement for oscillation of a 360° (or 0°) phase shift around the feedback loop.

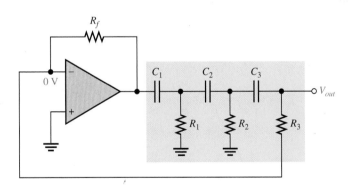

FIGURE 10–13
Op-amp phase-shift oscillator.

The attenuation B of the three-section RC feedback network is

$$B = \frac{1}{29} \tag{10–3}$$

where $B = R_3/R_f$. The derivation of this unusual result is given in Appendix B. To meet the greater-than-unity loop gain requirement, the closed-loop voltage gain of the op-amp must be greater than 29 (set by R_f and R_3). The frequency of oscillation is also derived in Appendix B and stated in the following equation, where $R_1 = R_2 = R_3 = R$ and $C_1 = C_2 = C_3 = C$.

$$f_r = \frac{1}{2\pi\sqrt{6}RC} \qquad (10\text{–}4)$$

EXAMPLE 10–2
(a) Determine the value of R_f necessary for the circuit in Figure 10–14 to operate as an oscillator.
(b) Determine the frequency of oscillation.

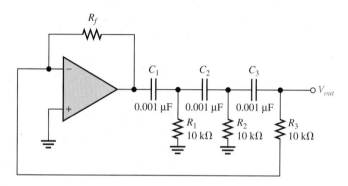

FIGURE 10–14

Solution

(a) $A_{cl} = 29$, and $B = \dfrac{1}{29} = \dfrac{R_3}{R_f}$. Therefore,

$$\frac{R_f}{R_3} = 29$$

$$R_f = 29R_3 = 29(10\ \text{k}\Omega) = \mathbf{290\ k\Omega}$$

(b) $R_1 = R_2 = R_3 = R$ and $C_1 = C_2 = C_3 = C$. Therefore,

$$f_r = \frac{1}{2\pi\sqrt{6}RC} = \frac{1}{2\pi\sqrt{6}(10\ \text{k}\Omega)(0.001\ \mu\text{F})} \cong \mathbf{6.5\ kHz}$$

Practice Exercise
(a) If R_1, R_2, and R_3 in Figure 10–14 are changed to 8.2 kΩ, what value must R_f be for oscillation?
(b) What is the value of f_r?

Twin-T Oscillator

Another type of *RC* feedback oscillator is called the *twin-T* because of the two T-type *RC* filters used in the feedback loop, as shown in Figure 10–15(a). One of the twin-T filters has a low-pass response, and the other has a high-pass response. The combined parallel filters produce a band-stop or notch response with a center frequency equal to the desired frequency of oscillation, f_r, as shown in Figure 10–15(b).

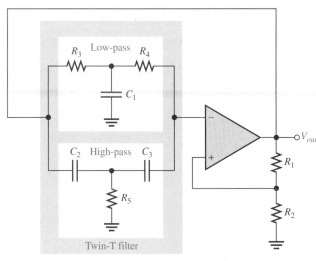

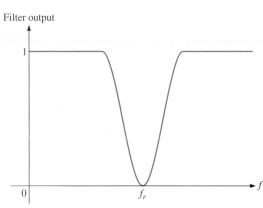

(a) Oscillator circuit

(b) Twin-T filter's frequency response curve

FIGURE 10–15
Twin-T oscillator and twin-T filter response.

Oscillation cannot occur at frequencies above or below f_r because of the negative feedback through the filters. At f_r, however, there is negligible negative feedback; thus, the positive feedback through the voltage divider (R_1 and R_2) allows the circuit to oscillate.

10–3 REVIEW QUESTIONS

1. There are two feedback loops in the Wien-bridge oscillator. What is the purpose of each?

2. A certain lead-lag network has $R_1 = R_2$ and $C_1 = C_2$. An input voltage of 5 V rms is applied. The input frequency equals the resonant frequency of the network. What is the rms output voltage?

3. Why is the phase shift through the RC feedback network in a phase-shift oscillator equal to 180°?

10–4 ■ RELAXATION OSCILLATOR PRINCIPLES

The second major category of oscillators is the relaxation oscillator. Relaxation oscillators use an RC timing circuit and a device that changes states to generate a periodic waveform. In this section, you will learn about several circuits that are used to produce nonsinusoidal waveforms.

After completing this section, you should be able to

❏ Describe and analyze the operation of basic relaxation oscillators
  ❏ Discuss the operation of basic triangular-wave oscillators
  ❏ Discuss the operation of a voltage-controlled oscillator (VCO)
  ❏ Discuss the operation of a square-wave relaxation oscillator

A Triangular-Wave Oscillator

The op-amp integrator covered in Chapter 8 can be used as the basis for a triangular-wave generator. The basic idea is illustrated in Figure 10–16(a) where a dual-polarity, switched input is used. We use the switch only to introduce the concept; it is not a practical way to implement this circuit. When the switch is in position 1, the negative voltage is applied, and the output is a positive-going ramp. When the switch is thrown into position 2, a negative-going ramp is produced. If the switch is thrown back and forth at fixed intervals, the output is a triangular wave consisting of alternating positive-going and negative-going ramps, as shown in Figure 10–16(b).

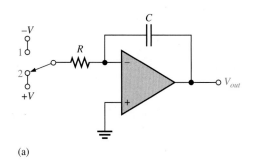

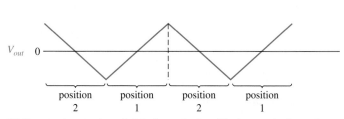

(a)

(b) Output voltage as the switch is thrown back and forth at regular intervals

FIGURE 10–16
Basic triangular-wave generator.

A Practical Triangular-Wave Oscillator One practical implementation of a triangular-wave generator utilizes an op-amp comparator to perform the switching function, as shown in Figure 10–17. The operation is as follows. To begin, assume that the output voltage of the comparator is at its maximum negative level. This output is connected to the inverting input of the integrator through R_1, producing a positive-going ramp on the output of the integrator. When the ramp voltage reaches the upper trigger point (UTP), the comparator switches to its maximum positive level. This positive level causes the integrator ramp to change to a negative-going direction. The ramp continues in this direction until the lower trigger point (LTP) of the comparator is reached. At this point, the comparator output switches back to the maximum negative level and the cycle repeats. This action is illustrated in Figure 10–18.

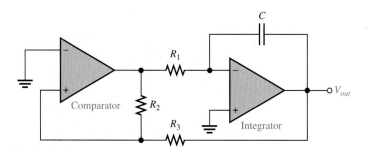

FIGURE 10–17
A triangular-wave generator using two op-amps.

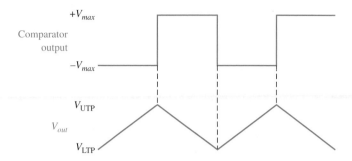

FIGURE 10–18
Waveforms for the circuit in Figure 10–17.

Since the comparator produces a square-wave output, the circuit in Figure 10–17 can be used as both a triangular-wave generator and a square-wave generator. Devices of this type are commonly known as *function generators* because they produce more than one output function. The output amplitude of the square wave is set by the output swing of the comparator, and resistors R_2 and R_3 set the amplitude of the triangular output by establishing the UTP and LTP voltages according to the following formulas:

$$V_{\text{UTP}} = +V_{max}\left(\frac{R_3}{R_2}\right)$$

$$V_{\text{LTP}} = -V_{max}\left(\frac{R_3}{R_2}\right)$$

where the comparator output levels, $+V_{max}$ and $-V_{max}$, are equal. The frequency of both waveforms depends on the R_1C time constant as well as the amplitude-setting resistors, R_2 and R_3. By varying R_1, the frequency of oscillation can be adjusted without changing the output amplitude.

$$f = \frac{1}{4R_1C}\left(\frac{R_2}{R_3}\right) \qquad\qquad (10\text{–}5)$$

EXAMPLE 10–3

Determine the frequency of the circuit in Figure 10–19. To what value must R_1 be changed to make the frequency 20 kHz?

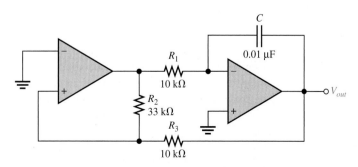

FIGURE 10–19

Solution

$$f = \frac{1}{4R_1C}\left(\frac{R_2}{R_3}\right) = \left(\frac{1}{4(10\ k\Omega)(0.01\ \mu F)}\right)\left(\frac{33\ k\Omega}{10\ k\Omega}\right) = \mathbf{8.25\ kHz}$$

To make $f = 20$ kHz,

$$R_1 = \frac{1}{4fC}\left(\frac{R_2}{R_3}\right) = \left(\frac{1}{4(20\ kHz)(0.01\ \mu F)}\right)\left(\frac{33\ k\Omega}{10\ k\Omega}\right) = \mathbf{4.13\ k\Omega}$$

Practice Exercise   What is the amplitude of the triangular wave in Figure 10–19 if the comparator output is ± 10 V?

A Voltage-Controlled Sawtooth Oscillator (VCO)

The **voltage-controlled oscillator** (VCO) is a relaxation oscillator whose frequency can be changed by a variable dc control voltage. VCOs can be either sinusoidal or nonsinusoidal. One way to build a voltage-controlled sawtooth oscillator is with an op-amp integrator that uses a switching device (PUT) in parallel with the feedback capacitor to terminate each ramp at a prescribed level and effectively "reset" the circuit. Figure 10–20(a) shows the implementation.

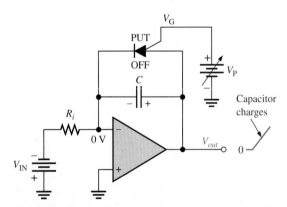

(a) Initially, the capacitor charges, the output ramp begins, and the PUT is off.

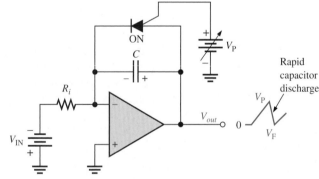

(b) The capacitor rapidly discharges when the PUT momentarily turns on.

FIGURE 10–20
Voltage-controlled sawtooth oscillator operation.

The PUT is a programmable unijunction transistor with an anode, a cathode, and a gate terminal. The gate is always biased positively with respect to the cathode. When the anode voltage exceeds the gate voltage by approximately 0.7 V, the PUT turns on and acts as a forward-biased diode. When the anode voltage falls below this level, the PUT turns off. Also, the current must be above the holding value to maintain conduction.

The operation of the sawtooth generator beings when the negative dc input voltage, $-V_{IN}$, produces a positive-going ramp on the output. During the time that the ramp is increasing, the circuit acts as a regular integrator. The PUT triggers on when the output ramp (at the anode) exceeds the gate voltage by 0.7 V. The gate is set to the approximate desired sawtooth peak voltage. When the PUT turns on, the capacitor rapidly discharges, as shown in Figure 10–20(b). The capacitor does not discharge completely to zero because of the PUT's forward voltage, V_F. Discharge continues until the PUT current falls below the holding value. At this point, the PUT turns off and the capacitor begins to charge again, thus generating a new output ramp. The cycle continually repeats, and the resulting output is a repetitive sawtooth waveform, as shown. The sawtooth amplitude and period can be adjusted by varying the PUT gate voltage.

The frequency is determined by the R_iC time constant of the integrator and the peak voltage set by the PUT. Recall that the charging rate of the capacitor is V_{IN}/R_iC. The time it takes the capacitor to charge from V_F to V_P is the period, T, of the sawtooth (neglecting the rapid discharge time).

$$T = \frac{V_P - V_F}{|V_{IN}|/R_iC}$$

From $f = 1/T$,

$$f = \frac{|V_{IN}|}{R_iC}\left(\frac{1}{V_P - V_F}\right) \tag{10–6}$$

EXAMPLE 10–4

(a) Find the peak-to-peak amplitude and frequency of the sawtooth output in Figure 10–21. Assume that the forward PUT voltage, V_F, is approximately 1 V.

(b) Sketch the output waveform.

FIGURE 10–21

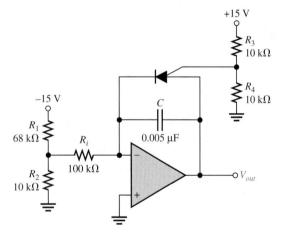

Solution

(a) First, find the gate voltage in order to establish the approximate voltage at which the PUT turns on.

$$V_G = \frac{R_4}{R_3 + R_4}(+V) = \frac{10\ k\Omega}{20\ k\Omega}(+15\ V) = 7.5\ V$$

This voltage sets the approximate maximum peak value of the sawtooth output (neglecting the 0.7 V).

$$V_P \cong 7.5\ V$$

The minimum peak value (low point) is

$$V_F \cong 1 \text{ V}$$

So the peak-to-peak amplitude is

$$V_{pp} = V_P - V_F = 7.5 \text{ V} - 1 \text{ V} = \textbf{6.5 V}$$

The frequency is determined as follows:

$$V_{IN} = \frac{R_2}{R_1 + R_2}(-V) = \frac{10 \text{ k}\Omega}{78 \text{ k}\Omega}(-15 \text{ V}) = -1.92 \text{ V}$$

$$f = \frac{|V_{IN}|}{R_i C}\left(\frac{1}{V_P - V_F}\right) = \left(\frac{1.92 \text{ V}}{(100 \text{ k}\Omega)(0.005 \text{ }\mu\text{F})}\right)\left(\frac{1}{7.5 \text{ V} - 1 \text{ V}}\right) \cong \textbf{591 Hz}$$

(b) The output waveform is shown in Figure 10–22. The period is

$$T = \frac{1}{f} = \frac{1}{591 \text{ Hz}} = 1.69 \text{ ms}$$

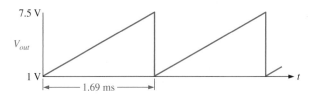

FIGURE 10–22
Output of the circuit in Figure 10–21.

Practice Exercise   If R_i is changed to 56 kΩ in Figure 10–21, what is the frequency?

A Square-Wave Oscillator

The basic square-wave oscillator shown in Figure 10–23 is a type of relaxation oscillator because its operation is based on the charging and discharging of a capacitor. Notice that the op-amp's inverting ($-$) input is the capacitor voltage and the noninverting ($+$) input is a portion of the output fed back through resistors R_2 and R_3. When the circuit is first turned on, the capacitor is uncharged, and thus the inverting input is at 0 V. This makes the output

FIGURE 10–23
A square-wave relaxation oscillator.

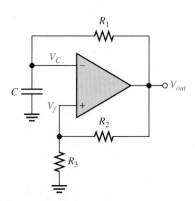

a positive maximum, and the capacitor begins to charge toward V_{out} through R_1. When the capacitor voltage (V_C) reaches a value equal to the feedback voltage (V_f) on the noninverting input, the op-amp switches to the maximum negative state. At this point, the capacitor begins to discharge from $+V_f$ toward $-V_f$. When the capacitor voltage reaches $-V_f$, the op-amp switches back to the maximum positive state. This action continues to repeat, as shown in Figure 10–24, and a square-wave output voltage is obtained.

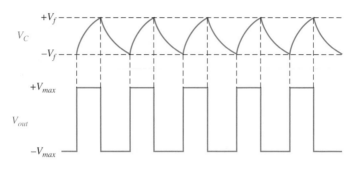

FIGURE 10–24
Waveforms for the square-wave relaxation oscillator.

10–4 REVIEW QUESTIONS

1. What is a VCO, and basically, what does it do?

2. Upon what principle does a relaxation oscillator operate?

10–5 ■ THE 555 TIMER AS AN OSCILLATOR

The 555 timer[1] is a versatile integrated circuit with many applications. In this section, you will see how the 555 is configured as an astable or free-running multivibrator, which is essentially a square-wave oscillator. The use of the 555 timer as a voltage-controlled oscillator (VCO) is also discussed.

After completing this section, you should be able to

❑ Use a 555 timer in an oscillator application
   ❑ Discuss astable operation of the 555 timer
   ❑ Explain how to use the 555 timer as a VCO

Astable Operation

A 555 timer connected to operate as an **astable multivibrator**, which is a free-running nonsinusoidal oscillator that produces a pulse waveform on its output, is shown in Figure 10–25. Notice that the threshold input (THRESH) is now connected to the trigger input

---

[1] Data sheet for LM555 available at http://www.national.com

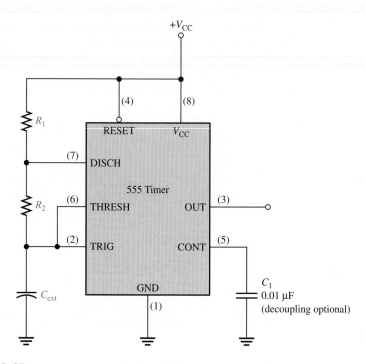

FIGURE 10–25
The 555 timer connected as an astable multivibrator.

(TRIG). The external components R_1, R_2, and C_{ext} form the timing network that sets the frequency of oscillation. The 0.01 μF capacitor connected to the control input (CONT) is strictly for decoupling and has no effect on the operation.

The frequency of oscillation is given by Equation (10–7), or it can be found using the graph in Figure 10–26.

$$f = \frac{1.44}{(R_1 + 2R_2)C_{ext}} \qquad \textbf{(10–7)}$$

FIGURE 10–26
Frequency of oscillation (free-running frequency) of a 555 timer in the astable mode as a function of C_{ext} and $R_1 + 2R_2$. The sloped lines are values of $R_1 + 2R_2$.

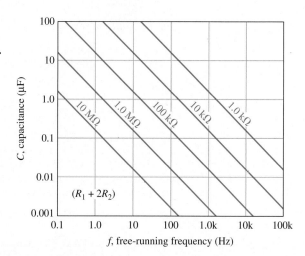

By selecting R_1 and R_2, the duty cycle of the output can be adjusted. Since C_{ext} charges through $R_1 + R_2$ and discharges only through R_2, duty cycles approaching a minimum of 50 percent can be achieved if $R_2 >> R_1$ so that the charging and discharging times are approximately equal.

A formula to calculate the duty cycle is developed as follows. The time that the output is high (t_H) is expressed as

$$t_H = 0.693(R_1 + R_2)C_{ext}$$

The time that the output is low (t_L) is expressed as

$$t_L = 0.693R_2C_{ext}$$

The period, T, of the output waveform is the sum of t_H and t_L.

$$T = t_H + t_L = 0.693(R_1 + 2R_2)C_{ext}$$

This is the reciprocal of f in Equation (10–7). Finally, the percent duty cycle is

$$\text{Duty cycle} = \left(\frac{t_H}{T}\right)100\% = \left(\frac{t_H}{t_H + t_L}\right)100\%$$

$$\text{Duty cycle} = \left(\frac{R_1 + R_2}{R_1 + 2R_2}\right)100\% \qquad \textbf{(10–8)}$$

To achieve duty cycles of less than 50 percent, the circuit in Figure 10–25 can be modified so that C_{ext} charges through only R_1 and discharges through R_2. This is achieved with a diode, D_1, placed as shown in Figure 10–27. The duty cycle can be made less than 50 percent by making R_1 less than R_2. Under this condition, the formula for the percent duty cycle is

$$\text{Duty cycle} = \left(\frac{R_1}{R_1 + R_2}\right)100\% \qquad \textbf{(10–9)}$$

FIGURE 10–27

The addition of diode D_1 allows the duty cycle of the output to be adjusted to less than 50 percent by making $R_1 < R_2$.

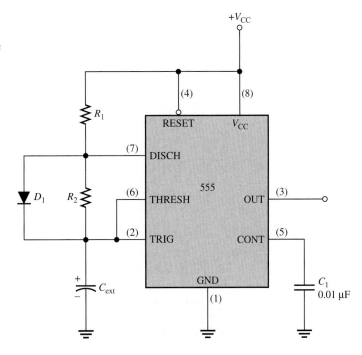

EXAMPLE 10–5   A 555 timer configured to run in the astable mode (oscillator) is shown in Figure 10–28. Determine the frequency of the output and the duty cycle.

FIGURE 10–28

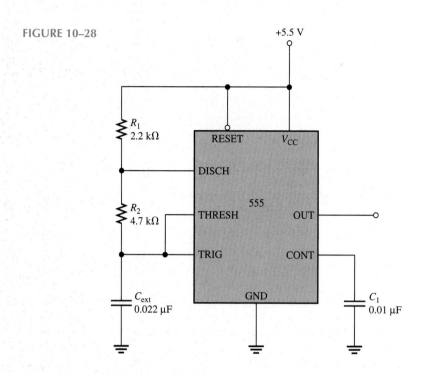

Solution

$$f = \frac{1.44}{(R_1 + 2R_2)C_{ext}} = \frac{1.44}{(2.2 \text{ k}\Omega + 9.4 \text{ k}\Omega)0.022 \text{ μF}} = \textbf{5.64 kHz}$$

$$\text{Duty cycle} = \left(\frac{R_1 + R_2}{R_1 + 2R_2}\right)100\% = \left(\frac{2.2 \text{ k}\Omega + 4.7 \text{ k}\Omega}{2.2 \text{ k}\Omega + 9.4 \text{ k}\Omega}\right)100\% = \textbf{59.5\%}$$

Practice Exercise   Determine the duty cycle in Figure 10–28 if a diode is connected across R_2 as indicated in Figure 10–27.

Operation as a Voltage-Controlled Oscillator (VCO)

A 555 timer can be set up to operate as a VCO by using the same external connections as for astable operation, with the exception that a variable control voltage is applied to the CONT input (pin 5), as indicated in Figure 10–29.

For the capacitor voltage, as shown in Figure 10–30, the upper value is V_{CONT} and the lower value is $\frac{1}{2}V_{CONT}$. When the control voltage is varied, the output frequency also varies. An increase in V_{CONT} increases the charging and discharging time of the external capacitor and causes the frequency to decrease. A decrease in V_{CONT} decreases the charging and discharging time of the capacitor and causes the frequency to increase.

An interesting application of the VCO is in phase-locked loops, which are used in various types of communications receivers to track variations in the frequency of incoming signals. You will learn about the basic operation of a phase-locked loop in Chapter 13.

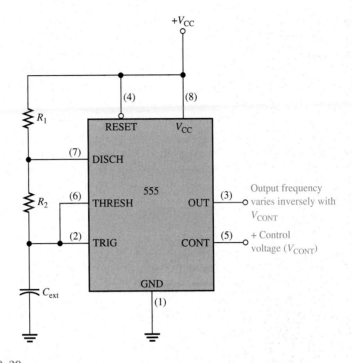

FIGURE 10–29

The 555 timer connected as a voltage-controlled oscillator (VCO). Note the variable control voltage input on pin 5.

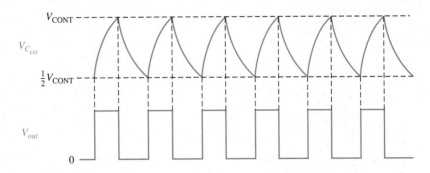

FIGURE 10–30

The VCO output frequency varies inversely with V_{CONT} because the charging and discharging time of C_{ext} is directly dependent on the control voltage.

10–5 REVIEW QUESTIONS

1. When the 555 timer is configured as an astable multivibrator, how is the duty cycle determined?

2. When the 555 timer is used as a VCO, how is the frequency varied?

10–6 ■ THE 555 TIMER AS A ONE-SHOT

A one-shot is a monostable multivibrator that produces a single output pulse for each input trigger pulse. The term monostable means that the device has only one stable state. When a one-shot is triggered, it temporarily goes to its unstable state but it always returns to its stable state. The time that it remains in its unstable state establishes the width of the output pulse and is set by the values of an external resistor and capacitor.

After completing this section, you should be able to

❑ Use a 555 timer as a one-shot device
    ❑ Discuss monostable operation
    ❑ Explain how to set the output pulse width

A 555 timer connected for **monostable** operation is shown in Figure 10–31. Compare this configuration to the one used for **astable** operation in Figure 10–25 and note the difference in the external circuit.

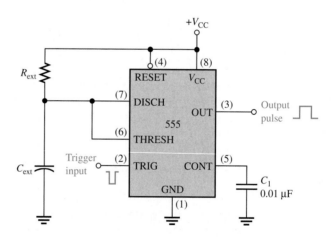

FIGURE 10–31
The 555 timer connected as a monostable multivibrator (one-shot).

Monostable Operation

A negative-going input trigger pulse produces a single output pulse with a predetermined width. Once triggered, the one-shot cannot be retriggered until it completely times out; that is, it completes a full output pulse. Once it times out, the one-shot can then be triggered again to produce another output pulse. A low level on the reset input (RESET) can be used to prematurely terminate the output pulse. The width of the output pulse is determined by the following formula:

$$t_W = 1.1R_{ext}C_{ext} \tag{10–10}$$

The graph in Figure 10–32 shows various combinations of R_{ext} and C_{ext} and the associated output pulse widths. This graph can be used to select component values for a desired pulse width.

FIGURE 10–32
555 one-shot timing.

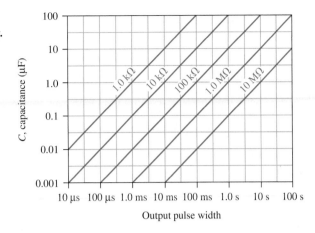

EXAMPLE 10–6

A 555 timer is connected as a one-shot with $R_{ext} = 10\ k\Omega$ and $C_{ext} = 0.1\ \mu F$. What is the pulse width of the output?

Solution   You can determine the pulse width in two ways. You can use either Equation (10–10) or the graph in Figure 10–32. Using the formula,

$$t_W = 1.1 R_{ext} C_{ext} = 1.1(10\ k\Omega)(0.1\ \mu F) = \textbf{1.1 ms}$$

To use the graph, move along the $C = 0.1\ \mu F$ line until it intersects with the sloped line corresponding to $R = 10\ k\Omega$. At that point, project down to the horizontal axis and you get a pulse width of 1.1 ms as illustrated in Figure 10–33.

FIGURE 10–33

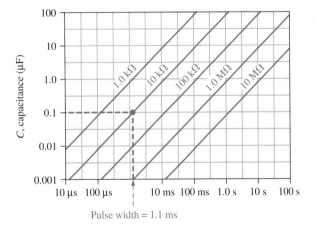

Practice Exercise   To what value must R_{ext} be changed to increase the one-shot's output pulse width to 5 ms?

Using One-Shots for Time Delay

In many applications, it is necessary to have a fixed time delay between certain events. Figure 10–34(a) shows two 555 timers connected as one-shots. The output of the first goes to the input of the second. When the first one-shot is triggered, it produces an output pulse

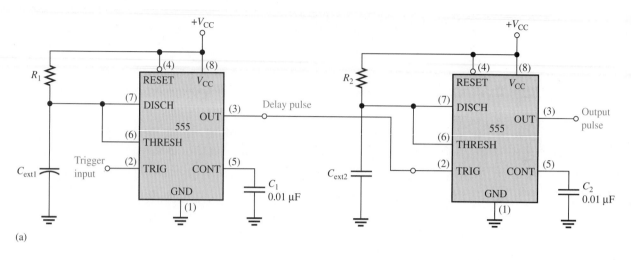

FIGURE 10–34

Two one-shots produce a delayed output pulse.

whose width establishes a time delay. At the end of this pulse, the second one-shot is triggered. Therefore, we have an output pulse from the second one-shot that is delayed from the input trigger to the first one-shot by a time equal to the pulse width of the first one-shot, as indicated in the timing diagram in Figure 10–34(b).

EXAMPLE 10–7

Determine the pulse widths and show the timing diagram (relationships of the input and output pulses) for the circuit in Figure 10–35.

Solution   The time relationship of the inputs and outputs are shown in Figure 10–36. The pulse widths for the two one-shots are

$$t_{W1} = 1.1 R_1 C_{ext1} = 1.1(100 \text{ k}\Omega)(1.0 \text{ }\mu\text{F}) = \textbf{110 ms}$$
$$t_{W2} = 1.1 R_2 C_{ext2} = 1.1(2.2 \text{ k}\Omega)(0.47 \text{ }\mu\text{F}) = \textbf{1.14 ms}$$

Practice Exercise   Suggest a way that the circuit in Figure 10–35 can be modified so that the delay can be made adjustable from 10 ms to 200 ms.

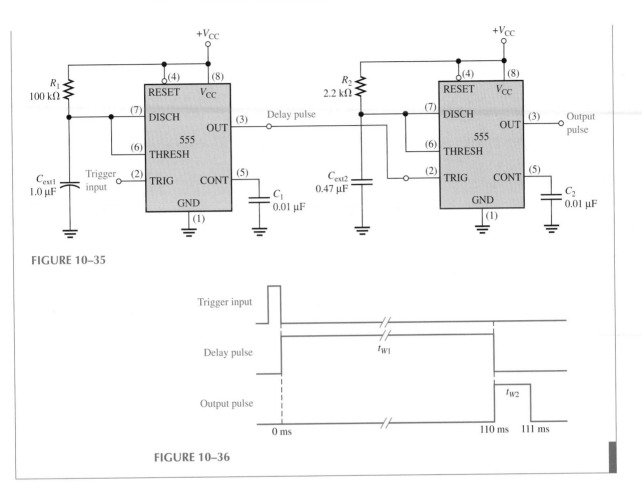

FIGURE 10–35

FIGURE 10–36

10–6 REVIEW QUESTIONS

1. How many stable states does a one-shot have?
2. A certain 555 one-shot circuit has a time constant of 5 ms. What is the output pulse width?
3. How can you decrease the pulse width of a one-shot?

10–7 ■ A SYSTEM APPLICATION

The function generator presented at the beginning of the chapter is a laboratory instrument used as a source for sine waves, square waves, and triangular waves.

After completing this section, you should be able to

❑ Apply what you have learned in this chapter to a system application
   ❑ Describe how an oscillator is used as a signal source
   ❑ State how the frequency and amplitude of the generated signal are varied
   ❑ Translate between printed circuit boards and a schematic

❑ Interconnect the front panel controls and two PC boards
❑ Troubleshoot some common system problems

A Brief Description of the System

The function generator in this system application produces either a sinusoidal wave, a square wave, or a triangular wave depending on the function selected by the front panel switches. The frequency of the selected waveform can be varied from less than 1 Hz to greater than 80 kHz using the range switches and the frequency dial. The amplitude of the output waveform can be adjusted up to approximately +10 V. Also, any dc offset voltage can be nulled out with the front panel dc offset control.

    The system block diagram is shown in Figure 10–37. The concept of this particular function generator is very simple. The oscillator produces a sinusoidal output voltage that drives a zero-level detector (comparator) to produce a square wave of the same frequency as the oscillator output. The level detector output goes to an integrator, which generates a triangular output voltage also with the same frequency as the oscillator output.

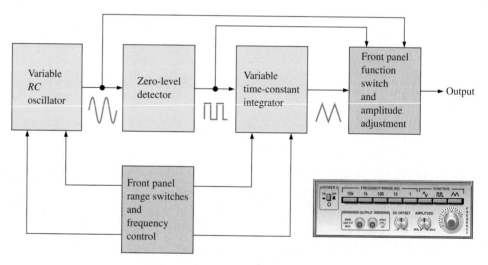

FIGURE 10–37

Function generator block diagram with front panel inset.

    The schematic of the function generator is shown in Figure 10–38, where the portions in color are the front panel components. The frequency of the sinusoidal oscillator is controlled by the selection of any two of ten capacitors (C_1 through C_{10}) in the oscillator feedback circuit. These capacitors produce the five frequency ranges indicated on the front panel switches, which are multiplication factors for the setting of the frequency dial. The adjustment of the frequency within each range is accomplished by varying resistors R_8 and R_9 in the feedback circuit of the oscillator.

    The integrator time constant is adjusted in step with the frequency by selection of the appropriate capacitor (C_{11} through C_{15}) and adjustment of resistor R_{10}. Resistors R_8, R_9,

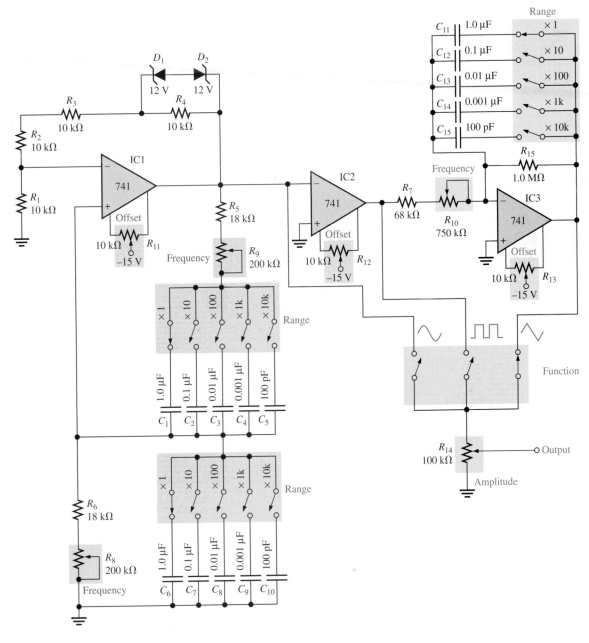

FIGURE 10–38

and R_{10} are potentiometers that are ganged together so that they change resistance together as the frequency dial is turned. For example, if the ×1k switch is selected and the frequency dial is set at 5, then the resulting output frequency for any of the three types of waveforms is 1 kHz × 5 = 5 kHz. In a practical integrator, R_{15} provides a long time constant discharge path for the capacitor; this prevents any small imbalance in the charging and discharging rate from causing the output to go into saturation.

Now, so that you can take a closer look at the oscillator boards, let's take them out of the system and put them on the troubleshooter's bench.

TROUBLESHOOTER'S BENCH

■ ACTIVITY 1   Relate the PC Boards to the Schematic

Locate and identify each component on the PC boards, shown in Figure 10–39, using the system schematic in Figure 10–38. The color-shaded portions are front panel components and are not on the boards. The range switches are mechanically linked and the frequency rheostats are mechanically linked.

Develop a board-to-board wiring list specifying which pins on the two PC boards connect to each other and also indicate which pins go to the front panel.

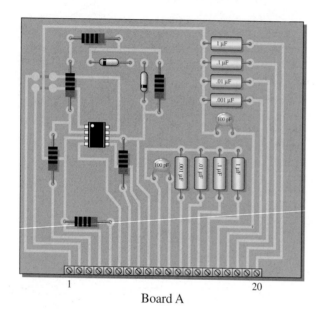

Board A

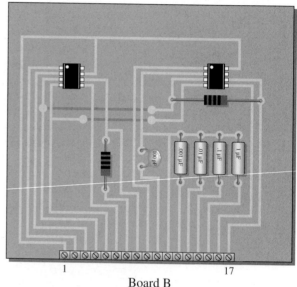

Board B

FIGURE 10–39

■ ACTIVITY 2   Analyze the System

Step 1: Determine the maximum frequency of the oscillator for each range switch position ($\times 1$, $\times 10$, and so on). Only one set of three switches corresponding to a given range setting can be closed at a time. There is a set of three switches for $\times 1$, a set of three switches for $\times 10$, and so on.

Step 2: Determine the minimum frequency of the oscillator for each range switch.

Step 3: Determine the approximate maximum peak-to-peak output voltages for each function. The dc supply voltages are $+15$ V and -15 V.

■ ACTIVITY 3   Write a Technical Report

Describe the overall operation of the function generator. Specify how each circuit works and what its purpose is. Identify the type of oscillator circuit used. Explain how the function, frequency, and amplitude are selected. Use the results of Activity 2 where appropriate.

■ **ACTIVITY 4 Troubleshoot the System for Each of the Following Problems by Stating the Probable Cause or Causes**

1. There is a square wave output when a triangular wave output is selected and only when the ×1k range is selected.
2. There is no output on any function setting.
3. There is no output when the square or triangular function is selected, but the sinusoidal output is OK.
4. Both the sinusoidal and the square wave outputs are OK, but there is no triangular wave output.

10–7 REVIEW QUESTIONS

1. What type of oscillator is used in this function generator?
2. How many frequency ranges are available?
3. List the components that determine the output frequency.
4. What is the purpose of the zener diodes in the oscillator circuit?

■ SUMMARY

- Feedback oscillators operate with positive feedback.
- The two conditions for positive feedback are the phase shift around the feedback loop must be 0° and the voltage gain around the feedback loop must equal 1.
- For initial start-up, the voltage gain around the feedback loop must be greater than 1.
- Sinusoidal *RC* oscillators include the Wien-bridge, phase-shift, and twin-T.
- A relaxation oscillator uses an *RC* timing circuit and a device that changes states to generate a periodic waveform.
- The frequency in a voltage-controlled oscillator (VCO) can be varied with a dc control voltage.
- The 555 timer is an integrated circuit that can be used as an oscillator or as a one-shot by proper connection of external components.

■ GLOSSARY

Key terms are in color. All terms are included in the end-of-book glossary.

Astable   Characterized by having no stable states; a type of oscillator.

Astable multivibrator   A type of circuit that can operate as an oscillator and produces a pulse waveform output.

Feedback oscillator   A type of oscillator that returns a fraction of output signal to the input with no net phase shift resulting in a reinforcement of the output signal.

Monostable   Characterized by having one stable state.

One-shot   A monostable multivibrator that produces a single output pulse for each input trigger pulse.

Oscillator   An electronic circuit that generates a periodic waveform to perform timing, control, or communications functions.

Phase-shift oscillator A type of sinusoidal feedback oscillator that uses three *RC* networks in the feedback loop.

Positive feedback A condition where an in-phase portion of the output voltage is fed back to the input.

Relaxation oscillator A type of oscillator that uses an *RC* timing circuit to generate a nonsinusoidal waveform.

Voltage-controlled oscillator A type of relaxation oscillator whose frequency can be changed by a variable dc voltage; also known as a VCO.

Wien-bridge oscillator A type of sinusoidal feedback oscillator that uses an *RC* lead-lag network in the feedback loop.

■ KEY FORMULAS

(10–1) $\dfrac{V_{out}}{V_{in}} = \dfrac{1}{3}$ Wien-bridge positive feedback attenuation

(10–2) $f_r = \dfrac{1}{2\pi RC}$ Wien-bridge frequency

(10–3) $B = \dfrac{1}{29}$ Phase-shift feedback attenuation

(10–4) $f_r = \dfrac{1}{2\pi\sqrt{6}RC}$ Phase-shift oscillator frequency

(10–5) $f = \dfrac{1}{4R_1C}\left(\dfrac{R_2}{R_3}\right)$ Triangular wave generator frequency

(10–6) $f = \dfrac{|V_{IN}|}{R_iC}\left(\dfrac{1}{V_P - V_F}\right)$ Sawtooth VCO frequency

(10–7) $f = \dfrac{1.44}{(R_1 + 2R_2)C_{ext}}$ 555 astable frequency

(10–8) Duty cycle $= \left(\dfrac{R_1 + R_2}{R_1 + 2R_2}\right)100\%$ 555 astable (duty cycle $\geq 50\%$)

(10–9) Duty cycle $= \left(\dfrac{R_1}{R_1 + R_2}\right)100\%$ 555 astable (duty cycle $< 50\%$)

(10–10) $t_W = 1.1R_{ext}C_{ext}$ 555 one-shot pulse width

■ SELF-TEST

Answers are at the end of the chapter.

1. An oscillator differs from an amplifier because
  (a) it has more gain **(b)** it requires no input signal
  (c) it requires no dc supply **(d)** it always has the same output

2. Wien-bridge oscillators are based on
  (a) positive feedback **(b)** negative feedback
  (c) the piezoelectric effect **(d)** high gain

3. One condition for oscillation is
  (a) a phase shift around the feedback loop of 180°
  (b) a gain around the feedback loop of one-third
  (c) a phase shift around the feedback loop of 0°
  (d) a gain around the feedback loop of less than one

4. A second condition for oscillation is
(a) no gain around the feedback loop
(b) a gain of one around the feedback loop
(c) the attenuation of the feedback network must be one-third
(d) the feedback network must be capacitive

5. In a certain oscillator, $A_v = 50$. The attenuation of the feedback network must be
(a) 1   (b) 0.01   (c) 10   (d) 0.02

6. For an oscillator to properly start, the gain around the feedback loop must initially be
(a) 1   (b) less than 1   (c) greater than 1   (d) equal to B

7. In a Wien-bridge oscillator, if the resistances in the feedback circuit are decreased, the frequency
(a) decreases   (b) increases   (c) remains the same

8. The Wien-bridge oscillator's positive feedback circuit is
(a) an RL network   (b) an LC network
(c) a voltage divider   (d) a lead-lag network

9. A phase-shift oscillator has
(a) three RC networks   (b) three LC networks
(c) a T-type network   (d) a π-type network

10. An oscillator whose frequency is changed by a variable dc voltage is known as
(a) a Wien-bridge oscillator   (b) a VCO
(c) a phase-shift oscillator   (d) an astable multivibrator

11. Which one of the following is not an input or output of the 555 timer?
(a) Threshold   (b) Control voltage   (c) Clock
(d) Trigger   (e) Discharge   (f) Reset

12. An astable multivibrator is
(a) an oscillator   (b) a one-shot
(c) a time-delay circuit   (d) characterized by having no stable states
(e) answers (a) and (d)

13. The output frequency of a 555 timer connected as an oscillator is determined by
(a) the supply voltage   (b) the frequency of the trigger pulses
(c) the external RC time constant   (d) the internal RC time constant
(e) answers (a) and (d)

14. The term *monostable* means
(a) one output   (b) one frequency
(c) one time constant   (d) one stable state

15. A 555 timer connected as a one-shot has $R_{ext} = 2.0$ kΩ and $C_{ext} = 2.0$ μF. The output pulse has a width of
(a) 1.1 ms   (b) 4 ms   (c) 4 μs   (d) 4.4 ms

TROUBLESHOOTER'S QUIZ   *Answers are at the end of the chapter.*

Refer to Figure 10–40.

❏ If D_1 suddenly opens,

1. The closed-loop gain will
(a) increase   (b) decrease   (c) not change

2. The output amplitude will
(a) increase   (b) decrease   (c) not change

❑ If D_2 has a 5.1 V breakdown voltage instead of the specified 4.7 V,

    3. The output voltage will

        (a) increase   **(b)** decrease   **(c)** not change

    4. The frequency of oscillation will

        (a) increase   **(b)** decrease   **(c)** not change

Refer to Figure 10–41.

❑ If the capacitors are 0.01 μF instead of 0.02 μF,

    5. The frequency of oscillation will

        (a) increase   **(b)** decrease   **(c)** not change

Refer to Figure 10–42.

❑ If the op-amp dc supply voltage decreases,

    6. The frequency of oscillation will

        (a) increase   **(b)** decrease   **(c)** not change

    7. The amplitude of the triangular output will

        (a) increase   **(b)** decrease   **(c)** not change

Refer to Figure 10–44.

❑ If R_2 is less than the specified value,

    8. The frequency of the output will

        (a) increase   **(b)** decrease   **(c)** not change

    9. The duty cycle of the output will

        (a) increase   **(b)** decrease   **(c)** not change

❑ If C_1 is open,

    10. The frequency of the output will

        (a) increase   **(b)** decrease   **(c)** not change

    11. The duty cycle will

        (a) increase   **(b)** decrease   **(c)** not change

    12. The amplitude of the output will

        (a) increase   **(b)** decrease   **(c)** not change

■ **PROBLEMS**   *Answers to odd-numbered problems are at the end of the book.*

SECTION 10–1 **The Oscillator**

  1. What type of input is required for an oscillator?

  2. What are the basic components of an oscillator circuit?

SECTION 10–2 **Feedback Oscillator Principles**

  3. If the voltage gain of the amplifier portion of an oscillator is 75, what must be the attenuation of the feedback circuit to sustain the oscillation?

4. Generally describe the change required in the oscillator of Problem 3 in order for oscillation to begin when the power is initially turned on.

SECTION 10–3 Sinusoidal Oscillators

5. A certain lead-lag network has a resonant frequency of 3.5 kHz. What is the rms output voltage if an input signal with a frequency equal to f_r and with an rms value of 2.2 V is applied to the input?

6. Calculate the resonant frequency of a lead-lag network with the following values: $R_1 = R_2 = 6.2$ kΩ, and $C_1 = C_2 = 0.02$ μF.

7. Determine the necessary value of R_2 in Figure 10–40 so that the circuit will oscillate. Neglect the forward resistance of the zener diodes. (*Hint:* The total gain of the circuit must be 3 when the zener diodes are conducting.)

8. Explain the purpose of R_3 in Figure 10–40.

FIGURE 10–40

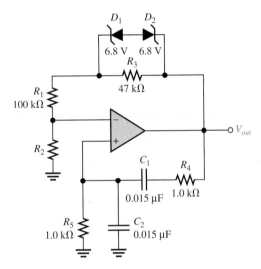

9. For the Wien-bridge in Figure 10–41, calculate the setting for R_f, assuming the internal drain-source resistance, r'_{ds}, of the JFET is 350 Ω when oscillations are stable.

10. Find the frequency of oscillation for the Wien-bridge oscillator in Figure 10–41.

FIGURE 10–41

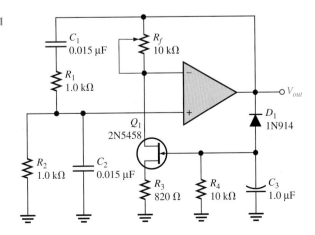

11. What value of R_f is required in Figure 10–42? What is f_r?

FIGURE 10–42

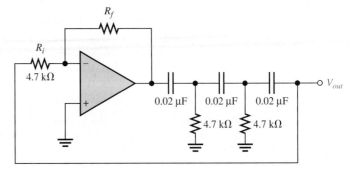

SECTION 10–4 Relaxation Oscillator Principles

12. What type of signal does the circuit in Figure 10–43 produce? Determine the frequency of the output.

13. Show how to change the frequency of oscillation in Figure 10–43 to 10 kHz.

FIGURE 10–43

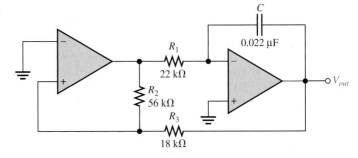

14. Determine the amplitude and frequency of the output voltage in Figure 10–44. Use 1 V as the forward PUT voltage.

15. Modify the sawtooth generator in Figure 10–44 so that its peak-to-peak output is 4 V.

FIGURE 10–44

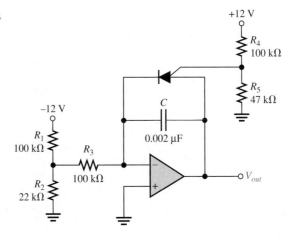

16. A certain sawtooth generator has the following parameter values: $V_{IN} = 3$ V, $R = 4.7$ kΩ, $C = 0.001$ μF, and V_F for the PUT is 1.2 V. Determine its peak-to-peak output voltage if the period is 10 μs.

SECTION 10–5 The 555 Timer as an Oscillator

17. What are the two comparator reference voltages in a 555 timer when $V_{CC} = 10$ V?

18. Determine the frequency of oscillation for the 555 astable oscillator in Figure 10–45.

19. To what value must C_{ext} be changed in Figure 10–45 to achieve a frequency of 25 kHz?

FIGURE 10–45

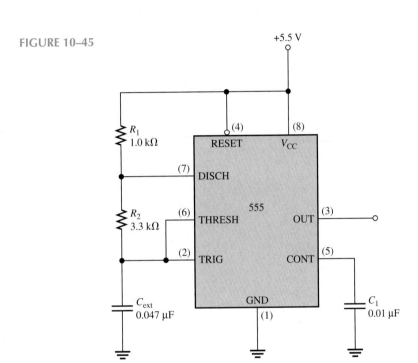

20. In an astable 555 configuration, the external resistor $R_1 = 3.3$ kΩ. What must R_2 equal to produce a duty cycle of 75 percent?

SECTION 10–6 The 555 Timer as a One-Shot

21. A 555 timer connected in the monostable configuration has a 56 kΩ external resistor and a 0.22 μF external capacitor. What is the pulse width of the output?

22. The output pulse width of a certain 555 one-shot is 12 ms. If $C_{ext} = 2.2$ μF, what is R_{ext}?

23. Suppose that you need to hook up a 555 timer as a one-shot in the lab to produce an output pulse with a width of 100 μs. Select the appropriate values for the external components.

24. Devise a circuit to produce two sequential 50 μs pulses. The first pulse must occur 100 ms after an initial trigger and the second pulse must occur 300 ms after the first pulse.

■ **ANSWERS TO REVIEW QUESTIONS**

Section 10–1

1. An oscillator is a circuit that produces a repetitive output waveform with only the dc supply voltage as an input.
2. Positive feedback
3. The feedback network provides attenuation and phase shift.

Section 10–2

1. Zero phase shift and unity voltage gain around the closed feedback
2. Positive feedback is when a portion of the output signal is fed back to the input of the amplifier such that it reinforces itself.
3. Loop gain greater than 1

Section 10–3

1. The negative feedback loop sets the closed-loop gain; the positive feedback loop sets the frequency of oscillation.
2. 1.67 V
3. The three RC networks contribute a total of 180° and the inverting amplifier contributes 180° for a total of 360° around the loop.

Section 10–4

1. A voltage-controlled oscillator exhibits a frequency that can be varied with a dc control voltage.
2. The basis of a relaxation oscillator is the charging and discharging of a capacitor.

Section 10–5

1. The duty cycle is set by the external resistors and the external capacitor.
2. The frequency of a VCO is varied by changing V_{CONT}.

Section 10–6

1. A one-shot has one stable state.
2. $t_W = 5.5$ ms
3. The pulse width can be decreased by decreasing the external resistance or capacitance.

Section 10–7

1. A Wien-bridge oscillator
2. There are five frequency ranges.
3. $R_5, R_6, R_8, R_9, C_1–C_5, C_6–C_{10}$
4. To limit the oscillator input amplitude and to help ensure start-up

■ **ANSWERS TO PRACTICE EXERCISES FOR EXAMPLES**

10–1 If R_f is too large, the output is distorted. If R_f is too small, oscillations cease.

10–2 **(a)** 238 kΩ   **(b)** 7.92 kHz

10–3 6.06 V peak-to-peak

10–4 1055 Hz

10–5 31.9%

10–6 45.5 kΩ

10–7 Replace R_1 with a potentiometer with a maximum resistance of at least 182 kΩ.

■ ANSWERS TO SELF-TEST

1. (b)	**2.** (a)	**3.** (c)	**4.** (b)	**5.** (d)
6. (c)	**7.** (b)	**8.** (d)	**9.** (a)	**10.** (b)
11. (c)	**12.** (e)	**13.** (c)	**14.** (d)	**15.** (d)

■ ANSWERS TO TROUBLE-SHOOTER'S QUIZ

1. increase	**2.** increase	**3.** increase	**4.** not change
5. increase	**6.** not change	**7.** decrease	**8.** increase
9. increase	**10.** not change	**11.** not change	**12.** not change

11

VOLTAGE REGULATORS

Courtesy Sunsweet Growers

■ CHAPTER OBJECTIVES

❏ Describe line and load regulation
❏ Discuss the principles of series voltage regulators
❏ Discuss the principles of shunt voltage regulators
❏ Discuss the principles of switching regulators
❏ Discuss integrated circuit voltage regulators
❏ Discuss applications of IC voltage regulators
❏ Apply what you have learned in this chapter to a system application

■ KEY TERMS

❏ Line regulation
❏ Load regulation
❏ Linear regulator
❏ Switching regulator

■ CHAPTER INTRODUCTION

A voltage **regulator** provides a constant dc output voltage that is practically independent of the input voltage, output load current, and temperature. The voltage regulator is one part of a power supply. Its input voltage comes from the filtered output of a rectifier derived from an ac voltage or from a battery in the case of portable systems.

Most voltage regulators fall into two broad categories—linear regulators and switching regulators. In the linear regulator category, two general types are the linear series regulator and the linear shunt regulator. These are normally available for either positive or negative output voltages. A dual regulator provides both positive and negative outputs. In the switching regulator category, three general configurations are step-down, step-up, and inverting.

Many types of integrated circuit (IC) regulators are available. The most popular types of linear regulator are the three-terminal fixed voltage regulator and the three-terminal adjustable voltage regulator. Switching regulators are also widely used. In this chapter, specific IC devices are introduced as representative of the wide range of available devices.

■ A SYSTEM APPLICATION

A dual-polarity regulated power supply is used for the FM stereo system that you worked with in Chapter 9. Two regulators, one positive and the other negative, provide the positive voltage required for the receiver circuits and the dual polarity voltages for the op-amp circuits. The regulator input voltages come from a full-wave rectifier with filtered outputs.

For the system application in Section 11–7, in addition to the other topics, be sure you understand
- ❏ How three-terminal fixed-voltage regulators are used
- ❏ The basic operation of a power supply rectifier and filter
- ❏ How to set the current limit of a regulator
- ❏ How to determine power dissipation in a pass transistor

11–1 ■ VOLTAGE REGULATION

The requirement for a reliable source of constant voltage in virtually all electronic systems has led to many advances in power supply design. Designers have used feedback and operational amplifiers, as well as pulse circuit techniques to develop reliable constant-voltage (and constant-current) power supplies. The heart of any regulated supply is the ability to establish a constant-voltage reference. In this section, you will learn more about line and load regulation (introduced in Section 2–6).

After completing this section, you should be able to

❑ Describe line and load voltage regulation
   ❑ Express line regulation as either a percentage or as a percentage per volt
   ❑ Calculate line regulation
   ❑ Express load regulation as either a percentage or as a percentage per milliamp
   ❑ Calculate load regulation from either voltage data or resistance data

Line Regulation

Line regulation was introduced in Section 2–6 and is reviewed here. **Line regulation** is a measure of the ability of a power supply to maintain a constant output for changes in the input voltage. It is typically defined as a ratio of a change in output for a corresponding change in the input and expressed as a percentage.

$$\text{Line regulation} = \left(\frac{\Delta V_{\text{OUT}}}{\Delta V_{\text{IN}}}\right)100\% \qquad (11\text{–}1)$$

This equation was given earlier as Equation (2–3). Some specification sheets show line regulation differently. It can be specified as a percentage change in the output voltage per volt divided by change in the input voltage. In this case, line regulation is defined and expressed as a percentage as

$$\text{Line regulation} = \left(\frac{\Delta V_{\text{OUT}}/V_{\text{OUT}}}{\Delta V_{\text{IN}}}\right)100\% \qquad (11\text{–}2)$$

Because this definition is different, you need to be sure which definition is used when reading specifications. The key in a specification sheet is to look at the units. If the specification is a ratio of mV/V or other pure number, then Equation (11–1) is the defining equation. If the units are shown as %/mV or %/V, then Equation (11–2) is the defining equation.

EXAMPLE 11–1 When the input to a particular voltage regulator decreases by 5 V, the output decreases by 0.25 V. The nominal output is 15 V. Determine the line regulation expressed as a percentage and in units of %/V.

Solution   From Equation (11–1), the percent line regulation is

$$\text{Line regulation} = \left(\frac{\Delta V_{\text{OUT}}}{\Delta V_{\text{IN}}}\right)100\% = \left(\frac{0.25\text{ V}}{5\text{ V}}\right)100\% = \mathbf{5\%}$$

From Equation (11–2), the percent line regulation is

$$\text{Line regulation} = \left(\frac{\Delta V_{\text{OUT}}/V_{\text{OUT}}}{\Delta V_{\text{IN}}}\right)100\% = \left(\frac{0.25\text{ V}/15\text{ V}}{5\text{ V}}\right)100\% = \mathbf{0.33\%/V}$$

*Practice Exercise**   The input of a certain regulator increases by 3.5 V. As a result, the output voltage increases by 0.42 V. The nominal output is 20 V. Determine the regulation expressed as a percentage and in units of %/V.

Load Regulation

Load regulation was introduced in Section 2–6 and is reviewed here. When the amount of current through a load changes due to a varying load resistance, the voltage regulator must maintain a nearly constant output voltage across the load. The percent load regulation specifies how much change occurs in the output voltage over a certain range of load current values, usually from minimum current (no load, NL) to maximum current (full load, FL). Ideally, the percent load regulation is 0%. It can be calculated and expressed as a percentage with the following formula:

$$\text{Load regulation} = \left(\frac{V_{\text{NL}} - V_{\text{FL}}}{V_{\text{FL}}}\right)100\% \qquad \textbf{(11–3)}$$

where V_{NL} is the output voltage with no load, and V_{FL} is the output voltage with full (maximum) load. This equation was given earlier as Equation (2–4). Equation (11–3) is expressed as a change due only to changes in load conditions; all other factors (such as input voltage and operating temperature) must remain constant. Normally, the operating temperature is specified as 25°C.

Sometimes power supply manufacturers specify the equivalent output resistance of a power supply (R_{OUT}) instead of its load regulation. Recall (Section 1–3) that an equivalent Thevenin circuit can be drawn for any two-terminal linear circuit. Figure 11–1 shows the equivalent Thevenin circuit for a power supply with a load resistor. The Thevenin voltage is the voltage from the supply with no load (V_{NL}), and the Thevenin resistance is the specified output resistance, R_{OUT}. Ideally, R_{OUT} is zero, corresponding to 0% load regulation, but in practical power supplies R_{OUT} is a small value. With the load resistor in place, the output voltage is found by applying the voltage-divider rule:

$$V_{\text{OUT}} = V_{\text{NL}}\left(\frac{R_L}{R_{\text{OUT}} + R_L}\right)$$

FIGURE 11–1
Thevenin equivalent circuit for a power supply with a load resistor.

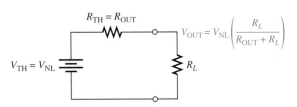

If we let R_{FL} equal the smallest-rated load resistance (largest-rated current), then the full-load output voltage (V_{FL}) is

$$V_{FL} = V_{NL}\left(\frac{R_{FL}}{R_{OUT} + R_{FL}}\right)$$

By rearranging and substituting into Equation (11–3),

$$V_{NL} = V_{FL}\left(\frac{R_{OUT} + R_{FL}}{R_{FL}}\right)$$

$$\text{Load regulation} = \frac{V_{FL}\left(\dfrac{R_{OUT} + R_{FL}}{R_{FL}}\right) - V_{FL}}{V_{FL}} \times 100\% = \left(\frac{R_{OUT} + R_{FL}}{R_{FL}} - 1\right)100\%$$

$$\text{Load regulation} = \left(\frac{R_{OUT}}{R_{FL}}\right)100\% \tag{11–4}$$

Equation (11–4) is a useful way of finding the percent load regulation when the output resistance and minimum load resistance are specified.

Alternately, the load regulation can be expressed as a percentage change in output voltage for each mA change in load current. For example, a load regulation of 0.01%/mA means that the output voltage changes 0.01 percent when the load current increases or decreases by 1 mA.

EXAMPLE 11–2

A certain voltage regulator has a +12.1 V output when there is no load ($I_L = 0$) and has a rated output current of 200 mA. With maximum current, the output voltage drops to +12.0 V. Determine the percentage load regulation and find the percent load regulation per mA change in load current.

Solution   The no-load output voltage is

$$V_{NL} = 12.1 \text{ V}$$

The full-load output voltage is

$$V_{FL} = 12.0 \text{ V}$$

The percent load regulation is

$$\text{Load regulation} = \left(\frac{V_{NL} - V_{FL}}{V_{FL}}\right)100\% = \left(\frac{12.1 \text{ V} - 12.0 \text{ V}}{12.0 \text{ V}}\right)100\% = \mathbf{0.83\%}$$

The load regulation can also be expressed as

$$\text{Load regulation} = \frac{0.83\%}{200 \text{ mA}} = \mathbf{0.0042\%/mA}$$

Practice Exercise   Prove that the results of this example are consistent with a specified output resistance of 0.5 Ω.

11–1 REVIEW QUESTIONS*

1. Define *line regulation.*

2. Define *load regulation.*

3. The input of a certain regulator increases by 3.5 V. As a result, the output voltage increases by 0.042 V. The nominal output is 20 V. Determine the line regulation in both % and in %/V.

4. If a 5.0 V power supply has an output resistance of 80 mΩ and a specified maximum output current of 1.0 A, what is the load regulation? Give the result as a % and as a %/mA.

———————————

* Answers are at the end of the chapter.

11–2 ■ BASIC SERIES REGULATORS

The fundamental classes of voltage regulators are linear regulators and switching regulators. Both of these are available in integrated circuit form. There are two basic types of linear regulator. One is the series regulator and the other is the shunt regulator. In this section, we will look at the series regulator. The shunt and switching regulators are covered in the next two sections.

After completing this section, you should be able to

❑ Discuss the principles of series voltage regulators
   ❑ Explain regulating action
   ❑ Calculate output voltage of an op-amp series regulator
   ❑ Discuss overload protection and explain how to use current limiting
   ❑ Describe a regulator with fold-back current limiting

A simple representation of a series type of **linear regulator** is shown in Figure 11–2(a), and the basic components are shown in the block diagram in Figure 11–2(b). Notice that

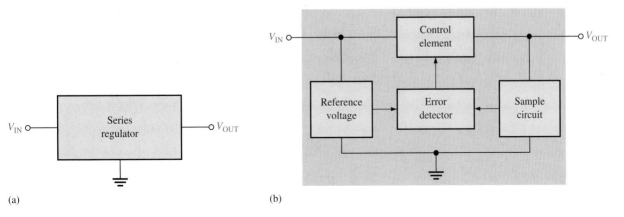

(a)   (b)

FIGURE 11–2
Simple series voltage regulator block diagram.

the control element is in series with the load between input and output. The output sample circuit senses a change in the output voltage. The error detector compares the sample voltage with a reference voltage and causes the control element to compensate in order to maintain a constant output voltage.

Voltage References

The ability of a voltage regulator to provide a constant output is dependent on the stability of a voltage reference to maintain a constant voltage for any change in temperature or other condition. Traditionally zener diodes (discussed in Section 2–8) were used as references and are shown in many of the circuits in this chapter. Zeners are designed to break down at a specific voltage and maintain a fairly constant voltage if the current in the zener is constant and the temperature does not change. The drawback to zener diodes is they tend to be noisy and the zener voltage may change slightly as the zener ages (this is called *drift*). An even more serious effect is that the zener voltage is sensitive to temperature changes; the zener voltage can change hundreds of parts per million (ppm) for a change of just 1°C in temperature. This temperature effect varies widely among different types of zeners.

Special zener diode ICs have been designed to serve as references with very low temperature drift (less than 10 ppm/°C). For low-voltage applications, zener-diode references are available that look and behave as diodes (but actually contain circuits to enhance their specifications). In the 8 V to 12 V range, two-terminal devices such as the LM329 and LM399[1] provide high stability and low noise, and have excellent temperature stability. The circuit symbol, which is the same as for a standard zener diode, and internal construction of a representative IC reference is shown in Figure 11–3. The reference shown is called a *bandgap reference.* It is designed so that positive and negative temperature coefficients cancel, producing a reference with almost no temperature coefficient. It uses a current mirror (Q_1) to set a particular current in Q_2. The output of the reference is the sum of V_{BE} (from Q_3) and the voltage drop across R_2 (V_{R2}).

---

[1] Data sheets for LM329 and LM399 available at http://www.national.com

FIGURE 11–3
An IC voltage reference. The reference shown is a bandgap type that has a very small temperature coefficient.

(a) Symbol

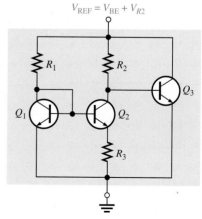

(b) Internal construction

A more complicated voltage reference is the REF102[2] precision-voltage reference. The drift is laser trimmed to 2.5 ppm/°C maximum. It is a 10.00 V reference that is within 2.5 mV of this value. It uses a zener diode and op-amp in an 8-pin package.

Regulating Action

A basic op-amp series regulator circuit is shown in Figure 11–4. The operation of the series regulator is illustrated in Figure 11–5. The resistive voltage divider formed by R_2 and R_3 senses any change in the output voltage.

FIGURE 11–4

Basic op-amp series regulator.

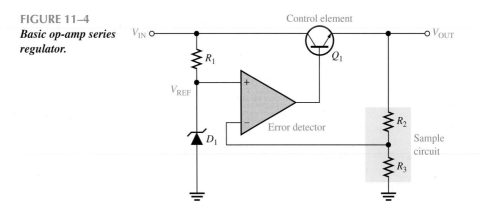

Figure 11–5(a) illustrates what happens when the output tries to decrease because of a decrease in V_{IN} or because of a change in load current. A proportional voltage decrease is applied to the op-amp's inverting input by the voltage divider. Since the zener diode (D_1) holds the other op-amp input at a nearly fixed reference voltage, V_{REF}, a small difference voltage (error voltage) is developed across the op-amp's inputs. This difference voltage is amplified, and the op-amp's output voltage increases. For highest accuracy, D_1 is replaced with an IC reference. This increase is applied to the base of Q_1, causing the emitter voltage V_{OUT} to increase until the voltage to the inverting input again equals the reference (zener) voltage. This action offsets the attempted decrease in output voltage, thus keeping it nearly constant, as shown in part (b). The power transistor, Q_1, is used with a heat sink because it must handle all of the load current.

The opposite action occurs when the output tries to increase, as indicated in Figure 11–5(c) and (d). The op-amp in the series regulator is actually connected as a noninverting amplifier where the reference voltage V_{REF} is the input at the noninverting terminal, and the R_2/R_3 voltage divider forms the negative feedback network. The closed-loop voltage gain is

$$A_{cl} = 1 + \frac{R_2}{R_3}$$

Therefore, the regulated output voltage of the series regulator is

$$V_{OUT} \cong \left(1 + \frac{R_2}{R_3}\right)V_{REF} \tag{11–5}$$

From this analysis, you can see that the output voltage is determined by the zener voltage (V_{REF}) and the feedback ratio of R_2/R_3. It is relatively independent of the input

---

[2] Data sheet for REF102 available at http://burr-brown.com

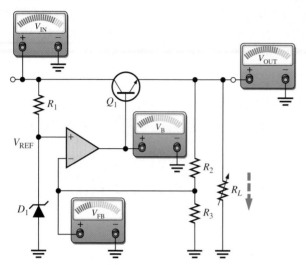

(a) When V_{IN} or R_L decreases, V_{OUT} attempts to decrease. The feedback voltage, V_{FB}, also attempts to decrease, and as a result, the op-amp's output voltage V_B attempts to increase, thus compensating for the attempted decrease in V_{OUT} by increasing the Q_1 emitter voltage. Changes in V_{OUT} are exaggerated for illustration.

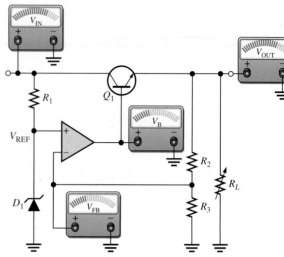

(b) When V_{IN} (or R_L) stabilizes at its new lower value, the voltages return to their original values, thus keeping V_{OUT} constant as a result of the negative feedback.

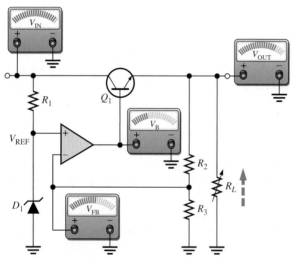

(c) When V_{IN} or R_L increases, V_{OUT} attempts to increase. The feedback voltage, V_{FB}, also attempts to increase, and as a result, V_B, applied to the base of the control transistor, attempts to decrease, thus compensating for the attempted increase in V_{OUT} by decreasing the Q_1 emitter voltage.

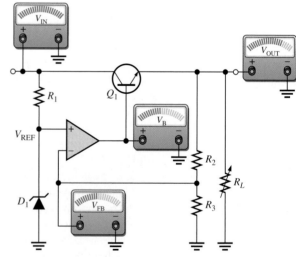

(d) When V_{IN} (or R_L) stabilizes at its new higher value, the voltages return to their original values, thus keeping V_{OUT} constant as a result of the negative feedback.

FIGURE 11–5

Illustration of series regulator action that keeps V_{OUT} constant when V_{IN} or R_L changes.

voltage, and therefore, regulation is achieved (as long as the input voltage and load current are within specified limits).

EXAMPLE 11–3   Determine the output voltage for the regulator in Figure 11–6 and the base voltage of Q_1.

FIGURE 11–6

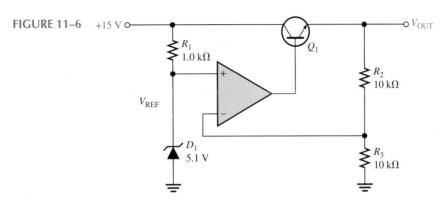

Solution $V_{REF} = 5.1$ V, the zener voltage. The regulated output voltage is therefore

$$V_{OUT} = \left(1 + \frac{R_2}{R_3}\right)V_{REF} = \left(1 + \frac{10\ k\Omega}{10\ k\Omega}\right)5.1\ V = (2)5.1\ V = \mathbf{10.2\ V}$$

The base voltage of Q_1 is

$$V_B = 10.2\ V + V_{BE} = 10.2\ V + 0.7\ V = \mathbf{10.9\ V}$$

Practice Exercise   The following changes are made in the circuit in Figure 11–6: A 3.3 V zener replaces the 5.1 V zener, $R_1 = 1.8$ kΩ, $R_2 = 22$ kΩ, and $R_3 = 18$ kΩ. What is the output voltage?

Short-Circuit or Overload Protection

If an excessive amount of load current is drawn, the series-pass transistor can be quickly damaged or destroyed. Most regulators use some type of current-limiting mechanism. Figure 11–7 shows one method of current limiting to prevent overloads called *constant-current limiting*. The current-limiting circuit consists of transistor Q_2 and resistor R_4.

FIGURE 11–7
Series regulator with constant-current limiting.

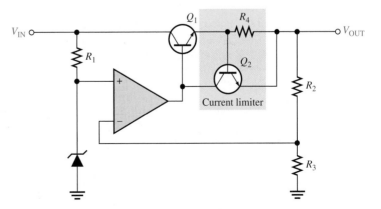

The load current through R_4 produces a voltage from base to emitter of Q_2. When I_L reaches a predetermined maximum value, the voltage drop across R_4 is sufficient to forward-bias the base-emitter junction of Q_2, thus causing it to conduct. Enough Q_1 base current is diverted into the collector of Q_2 so that I_L is limited to its maximum value $I_{L(max)}$. Since the base-to-emitter voltage of Q_2 cannot exceed about 0.7 V, the voltage across R_4 is held to this value, and the load current is limited to

$$I_{L(max)} = \frac{0.7 \text{ V}}{R_4} \tag{11-6}$$

EXAMPLE 11–4

Determine the maximum current that the regulator in Figure 11–8 can provide to a load.

FIGURE 11–8

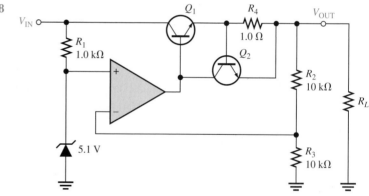

Solution

$$I_{L(max)} = \frac{0.7 \text{ V}}{R_4} = \frac{0.7 \text{ V}}{1.0 \text{ }\Omega} = \textbf{0.7 A}$$

Practice Exercise If the output of the regulator in Figure 11–8 is shorted to ground, what is the current?

Regulator with Fold-Back Current Limiting

In the previous current-limiting technique, the current is restricted to a maximum constant value. **Fold-back current limiting** is a method used particularly in high-current regulators whereby the output current under overload conditions drops to a value well below the peak load current capability to prevent excessive power dissipation.

Basic Idea The basic concept of fold-back current limiting is as follows, with reference to Figure 11–9. The circuit is similar to the constant current-limiting arrangement in Figure 11–7, with the exception of resistors R_5 and R_6. The voltage drop developed across R_4 by the load current must not only overcome the base-emitter voltage required to turn on Q_2, but it must also overcome the voltage across R_5. That is, the voltage across R_4 must be

$$V_{R4} = V_{R5} + V_{BE}$$

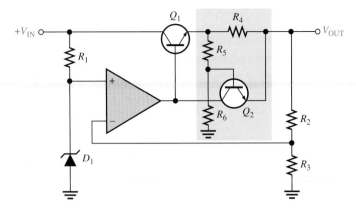

FIGURE 11–9
Series regulator with fold-back current limiting.

In an overload or short-circuit condition, the load current increases to a value, $I_{L(max)}$, that is sufficient to cause Q_2 to conduct. At this point the current can increase no further. The decrease in output voltage results in a proportional decrease in the voltage across R_5; thus less current through R_4 is required to maintain the forward-biased condition of Q_1. So, as V_{OUT} decreases, I_L decreases, as shown in the graph of Figure 11–10.

The advantage of this technique is that the regulator is allowed to operate with peak load current up to $I_{L(max)}$; but when the output becomes shorted, the current drops to a lower value to prevent overheating of the device.

FIGURE 11–10
Fold-back current limiting (output voltage versus load current).

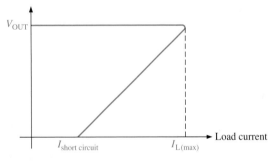

11–2 REVIEW QUESTIONS

1. What are the basic components in a series regulator?
2. A certain series regulator has an output voltage of 8 V. If the op-amp's closed loop gain is 4, what is the value of the reference voltage?

11–3 ■ BASIC SHUNT REGULATORS

The second basic type of linear voltage regulator is the shunt regulator. As you have learned, the control element in the series regulator is the series-pass transistor. In the shunt regulator, the control element is a transistor in parallel (shunt) with the load.

After completing this section, you should be able to

❑ Discuss the principles of shunt voltage regulators
   ❑ Describe the operation of a basic op-amp shunt regulator
   ❑ Compare series and shunt regulators

A simple representation of a shunt type of linear regulator is shown in Figure 11–11(a), and the basic components are shown in the block diagram in part (b) of the figure.

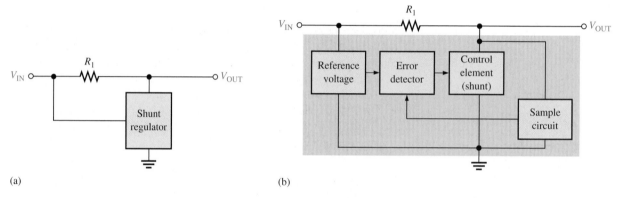

(a)   (b)

FIGURE 11–11
Simple shunt regulator and block diagram.

In the basic shunt regulator, the control element is a transistor, Q_1, in parallel with the load, as shown in Figure 11–12. A resistor, R_1, is in series with the load. The operation of the circuit is similar to that of the series regulator, except that regulation is achieved by controlling the current through the parallel transistor Q_1.

When the output voltage tries to decrease due to a change in input voltage or load current caused by a change in load resistance, as shown in Figure 11–13(a), the attempted

FIGURE 11–12
Basic op-amp shunt regulator.

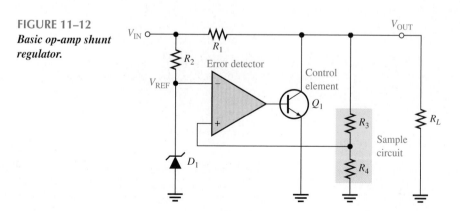

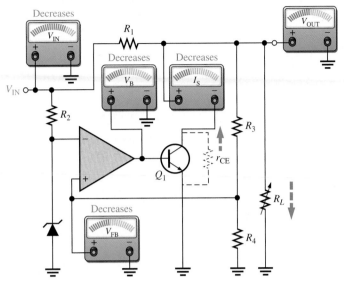

(a) Response to a decrease in V_{IN} or R_L

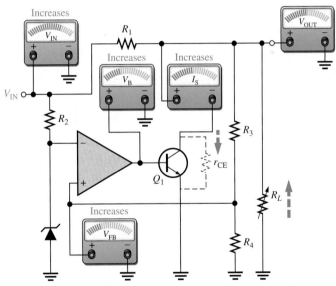

(b) Response to an increase in V_{IN} or R_L

FIGURE 11–13

Sequence of responses when V_{OUT} tries to decrease as a result of a decrease in R_L or V_{IN} (opposite responses for an attempted increase).

decrease is sensed by R_3 and R_4 and applied to the op-amp's noninverting input. The resulting difference voltage reduces the op-amp's output (V_B), driving Q_1 less, thus reducing its collector current (shunt current) and increasing its internal collector-to-emitter resistance r_{CE}. Since r_{CE} acts as a voltage divider with R_1, this action offsets the attempted decrease in V_{OUT} and maintains it at an almost constant level.

The opposite action occurs when the output tries to increase, as indicated in Figure 11–13(b). With I_L and V_{OUT} constant, a change in the input voltage produces a change in shunt current (I_S) as follows:

$$\Delta I_S = \frac{\Delta V_{IN}}{R_1}$$

With a constant V_{IN} and V_{OUT}, a change in load current causes an opposite change in shunt current.

$$\Delta I_S = -\Delta I_L$$

This formula says that if I_L increases, I_S decreases, and vice versa. The shunt regulator is less efficient than the series type but offers inherent short-circuit protection. If the output is shorted ($V_{OUT} = 0$), the load current is limited by the series resistor R_1 to a maximum value as follows ($I_S = 0$).

$$I_{L(max)} = \frac{V_{IN}}{R_1} \tag{11–7}$$

EXAMPLE 11–5

In Figure 11–14, what power rating must R_1 have if the maximum input voltage is 12.5 V?

FIGURE 11–14

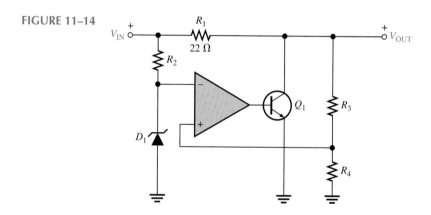

Solution The worst-case power dissipation in R_1 occurs when the output is short-circuited. $V_{OUT} = 0$, and when $V_{IN} = 12.5$ V, the voltage dropped across R_1 is $V_{IN} - V_{OUT} = 12.5$ V. The power dissipation in R_1 is

$$P_{R1} = \frac{V_{R1}^2}{R_1} = \frac{(12.5 \text{ V})^2}{22 \text{ } \Omega} = 7.1 \text{ W}$$

Therefore, a resistor with at least a **10 W** rating should be used.

Practice Exercise In Figure 11–14, R_1 is changed to 33 Ω. What must be the power rating of R_1 if the maximum input voltage is 24 V?

11–3 REVIEW QUESTIONS

1. How does the control element in a shunt regulator differ from that in a series regulator?

2. What is one advantage of a shunt regulator over a series type? What is a disadvantage?

11–4 ■ BASIC SWITCHING REGULATORS

The two types of linear regulators—series and shunt—have control elements (transistors) that are conducting all the time, with the amount of conduction varied as demanded by changes in the output voltage or current. The switching regulator is different; the control element operates as a switch. A greater efficiency can be realized with this type of voltage regulator than with the linear types because the transistor is not always conducting. Therefore, switching regulators can provide greater load currents at low voltage than linear regulators because the control transistor doesn't dissipate as much power. Three basic configurations of switching regulators are step-down, step-up, and inverting.

After completing this section, you should be able to

❏ Discuss the principles of switching regulators
   ❏ Describe the step-down configuration of a switching regulator
   ❏ Determine the output voltage of the step-down configuration
   ❏ Describe the step-up configuration of a switching regulator
   ❏ Determine the output voltage of the step-up configuration
   ❏ Describe the voltage-inverter configuration

Step-Down Configuration

In the step-down configuration, the output voltage is always less than the input voltage. A basic step-down switching regulator is shown in Figure 11–15(a), and its simplified equivalent is shown in Figure 11–15(b). Transistor Q_1 is used to switch the input voltage at a duty cycle that is based on the regulator's load requirement. The LC filter is then used to average the switched voltage. Since Q_1 is either *on* (saturated) or *off,* the power lost in the control element is relatively small. Therefore, the **switching regulator** is useful primarily in higher power applications or in applications such as computers where efficiency is of utmost concern.

The on and off intervals of Q_1 are shown in the waveform of Figure 11–16(a). The capacitor charges during the on-time (t_{on}) and discharges during the off-time (t_{off}). When the on-time is increased relative to the off-time, the capacitor charges more, thus increasing the output voltage, as indicated in Figure 11–16(b). When the on-time is decreased relative to the off-time, the capacitor discharges more, thus decreasing the output voltage, as in Figure 11–16(c). Therefore, by adjusting the duty cycle, $t_{on}/(t_{on} + t_{off})$, of Q_1, the output voltage can be varied. The inductor further smooths the fluctuations of the output voltage caused by the charging and discharging action.

FIGURE 11–15

Basic step-down switching regulator.

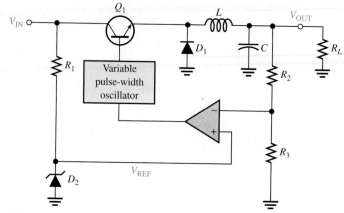

(a) Typical circuit

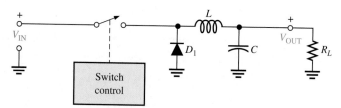

(b) Simplified equivalent circuit

FIGURE 11–16

Switching regulator waveforms. The V_C waveform is shown for no inductive filtering to illustrate the charge and discharge action (ripple). L and C smooth V_C to a nearly constant level, as indicated by the dashed line for V_{OUT}.

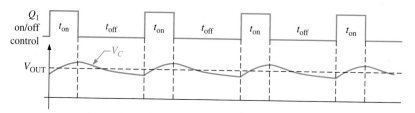

(a) V_{OUT} depends on the duty cycle.

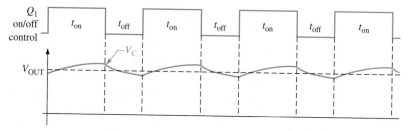

(b) Increase the duty cycle and V_{OUT} increases.

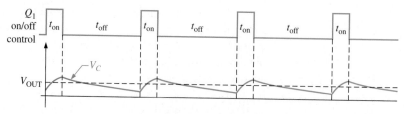

(c) Decrease the duty cycle and V_{OUT} decreases.

Ideally, the output voltage is expressed as

$$V_{\text{OUT}} = \left(\frac{t_{\text{on}}}{T}\right)V_{\text{IN}}$$

(11–8)

T is the period of the on-off cycle of Q_1 and is related to the frequency by $T = 1/f$. The period is the sum of the on-time and the off-time.

$$T = t_{\text{on}} + t_{\text{off}}$$

The ratio t_{on}/T is called the *duty cycle*.

The regulating action is as follows and is illustrated in Figure 11–17. When V_{OUT} tries to decrease, the on-time of Q_1 is increased, causing an additional charge on the

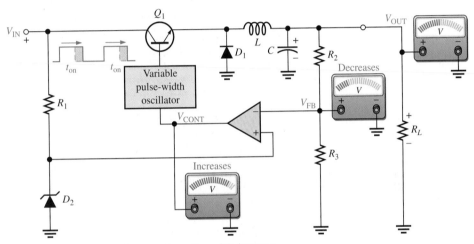

(a) When V_{OUT} attempts to decrease, the on-time of Q_1 increases.

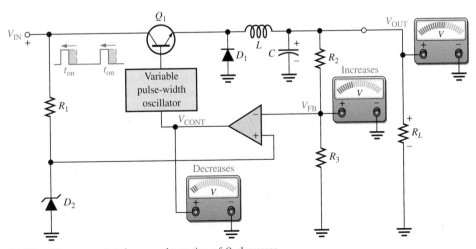

(b) When V_{OUT} attempts to increase, the on-time of Q_1 decreases.

FIGURE 11–17

Basic regulating action of a step-down switching regulator.

capacitor, C, to offset the attempted decrease. When V_{OUT} tries to increase, the on-time of Q_1 is decreased, causing C to discharge enough to offset the attempted increase.

Step-Up Configuration

A basic step-up type of switching regulator is shown in Figure 11–18, where transistor Q_1 operates as a switch to ground.

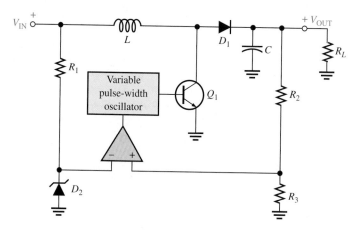

FIGURE 11–18
Basic step-up switching regulator.

The switching action is illustrated in Figures 11–19 and 11–20. When Q_1 turns on, a voltage equal to approximately V_{IN} is induced across the inductor with a polarity as indicated in Figure 11–19. During the on-time (t_{on}) of Q_1, the inductor voltage, V_L, decreases from its initial maximum and diode D_1 is reverse-biased. The longer Q_1 is on, the smaller V_L becomes. During the on-time, the capacitor only discharges an extremely small amount through the load.

When Q_1 turns off, as indicated in Figure 11–20, the inductor voltage suddenly reverses polarity and adds to V_{IN}, forward-biasing diode D_1 and allowing the capacitor to charge. The output voltage is equal to the capacitor voltage and can be larger than V_{IN} because the capacitor is charged to V_{IN} plus the voltage induced across the inductor during the off-time of Q_1.

The longer the on-time of Q_1, the more the inductor voltage will decrease and the greater the magnitude of the voltage when the inductor reverses polarity at the instant Q_1 turns off. As you have seen, this reverse polarity voltage is what charges the capacitor above V_{IN}. The output voltage is dependent on both the inductor's magnetic field action (determined by t_{on}) and the charging of the capacitor (determined by t_{off}).

Voltage regulation is achieved by the variation of the on-time of Q_1 (within certain limits) as related to changes in V_{OUT} due to changing load or input voltage. If V_{OUT} tries to increase, the on-time of Q_1 will decrease, which results in a decrease in the amount that C will charge. If V_{OUT} tries to decrease, the on-time of Q_1 will increase, which results in an increase in the amount that C will charge. This regulating action maintains V_{OUT} at an essentially constant level.

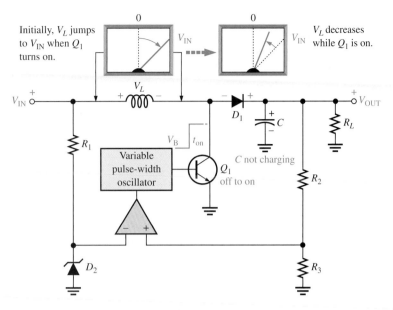

FIGURE 11–19
Basic action of a step-up regulator when Q_1 is on.

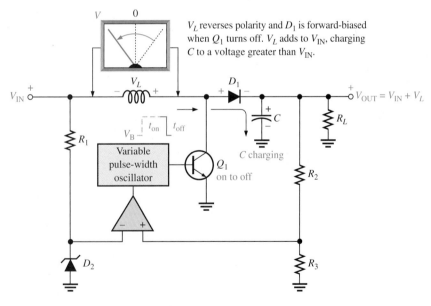

FIGURE 11–20
Basic switching action of a step-up regulator when Q_1 turns off.

Voltage-Inverter Configuration

A third type of switching regulator produces an output voltage that is opposite in polarity to the input. A basic diagram is shown in Figure 11–21.

When Q_1 turns on, the inductor voltage jumps to approximately V_{IN} and the magnetic field rapidly expands, as shown in Figure 11–22(a). While Q_1 is on, the diode is reverse-biased and the inductor voltage decreases from its initial maximum. When Q_1 turns

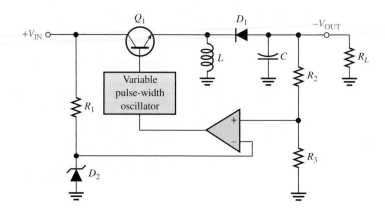

FIGURE 11–21
Basic inverting switching regulator.

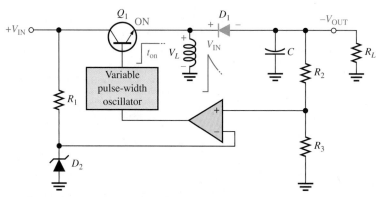

(a) When Q_1 is on, D_1 is reverse-biased.

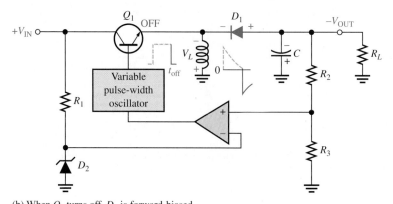

(b) When Q_1 turns off, D_1 is forward-biased.

FIGURE 11–22
Basic inverting action of an inverting switching regulator.

off, the magnetic field collapses and the inductor's polarity reverses, as shown in Figure 11–22(b). This forward-biases the diode, charges C, and produces a negative output voltage, as indicated. The repetitive on-off action of Q_1 produces a repetitive charging and discharging that is smoothed by the LC filter action.

As with the step-up regulator, the less time Q_1 is on, the greater the output voltage is, and vice versa. This regulating action is illustrated in Figure 11–23. Switching regulator efficiencies can be greater than 90 percent.

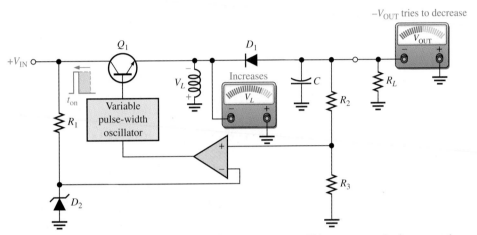

(a) When $-V_{OUT}$ tries to decrease, t_{on} decreases, causing V_L to increase. This compensates for the attempted decrease in $-V_{OUT}$.

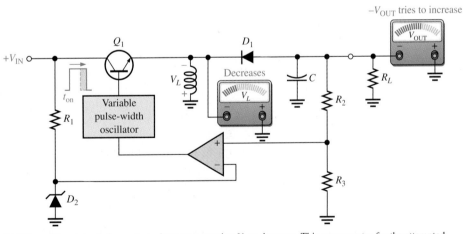

(b) When $-V_{OUT}$ tries to increase, t_{on} increases, causing V_L to decrease. This compensates for the attempted increase in $-V_{OUT}$.

FIGURE 11–23
Basic regulating action of an inverting switching regulator.

11–4 REVIEW QUESTIONS

1. What are three types of switching regulators?
2. What is the primary advantage of switching regulators over linear regulators?
3. How are changes in output voltage compensated for in the switching regulator?

11–5 ■ INTEGRATED CIRCUIT VOLTAGE REGULATORS

In the previous sections, the basic voltage regulator configurations were presented. Several types of both linear and switching regulators are available in integrated circuit (IC) form. Generally, the linear regulators are three-terminal devices that provide either positive or negative output voltages that can be either fixed or adjustable. Three-terminal regulators were introduced in Section 2–6. In this section, typical linear and switching IC regulators are covered in more detail.

After completing this section, you should be able to

❑ Discuss integrated circuit voltage regulators
  ❑ Describe the 7800 series of positive regulators
  ❑ Describe the 7900 series of negative regulators
  ❑ Describe the LM317 adjustable positive regulator
  ❑ Describe the LM337 adjustable negative regulator
  ❑ Describe IC switching regulators

Fixed Positive Linear Voltage Regulators

Although many types of IC regulators are available, the 7800[3] series of IC regulators is representative of three-terminal devices that provide a fixed positive output voltage. The three terminals are input, output, and ground as indicated in the standard fixed voltage configuration in Figure 11–24(a). The last two digits in the part number designate the output voltage. For example, the 7805 is a +5.0 V regulator. Other available output voltages are given in Figure 11–24(b). (Common IC regulator packages were shown in Figure 2–26(c).)

Capacitors are used on the input and output as indicated. The output capacitor acts basically as a line filter to improve transient response. The input capacitor is used to prevent unwanted oscillations when the regulator is some distance from the power supply filter such that the line has a significant inductance.

The 7800 series can produce output currents up to in excess of 1 A when used with an adequate heat sink. The input voltage must be at least 2 V above the output voltage in order to maintain regulation. The circuits have internal thermal overload protection and short-circuit current-limiting features.

Thermal overload occurs when the internal power dissipation becomes excessive and the temperature of the device exceeds a certain value. Thermal overload is a problem if inadequate heat sinking is provided or if the regulator is not properly secured to the heat sink. Almost all applications of regulators require a heat sink. Heat generated by the regulator must move to the heat sink and then to the surrounding air. A good heat sink is mas-

---

[3] Data sheet for the 7800 series available at www.onsemi.com

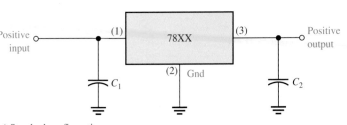

Type number	Output voltage
7805	+5.0 V
7806	+6.0 V
7808	+8.0 V
7809	+9.0 V
7812	+12.0 V
7815	+15.0 V
7818	+18.0 V
7824	+24.0 V

(a) Standard configuration   (b) The 7800 series

FIGURE 11–24
The 7800 series three-terminal fixed positive voltage regulators.

sive and has fins (to increase area). A regulator that is too hot may show symptoms of drift, excess ripple, or the output may fall out of regulation.

Fixed Negative Linear Voltage Regulators

The 7900[4] series is typical of three-terminal IC regulators that provide a fixed negative output voltage. This series is the negative-voltage counterpart of the 7800 series and shares most of the same features and characteristics. Figure 11–25 indicates the standard configuration and part numbers with corresponding output voltages that are available. Be aware that the pins on the 7900 series regulators do not have the same function as the 7800 series pins.

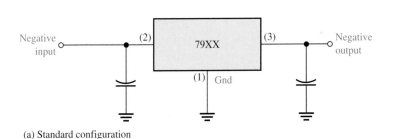

Type number	Output voltage
7905	−5.0 V
7905.2	−5.2 V
7906	−6.0 V
7908	−8.0 V
7912	−12.0 V
7915	−15.0 V
7918	−18.0 V
7924	−24.0 V

(a) Standard configuration   (b) The 7900 series

FIGURE 11–25
The 7900 series three-terminal fixed negative voltage regulators.

Adjustable Positive Linear Voltage Regulators

The LM317[5] is an example of a three-terminal positive regulator with an adjustable output voltage. A data sheet for this device is given in Appendix A. The standard configuration is shown in Figure 11–26. Input and output capacitors, C_1 and C_3 respectively, are used for the reasons discussed previously. The capacitor, C_2, at the adjustment terminal

---

[4] Data sheet for the 7900 series available at http://www.onsemi.com

[5] Data sheet for LM317 available at http://www.national.com

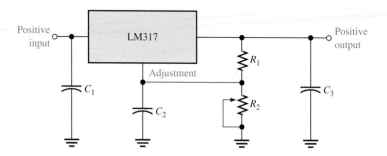

FIGURE 11–26
The LM317 three-terminal adjustable positive voltage regulator.

also acts as a filter to improve transient response. Notice that there is an input, an output, and an adjustment terminal. The external fixed resistor, R_1, and the external variable resistor, R_2, provide the output voltage adjustment. V_{OUT} can be varied from 1.2 V to 37 V depending on the resistor values. The LM317 can provide over 1.5 A of output current to a load.

The LM317 is operated as a "floating" regulator because the adjustment terminal is not connected to dc ground, but floats to whatever voltage is across R_2. This allows the output voltage to be much higher than that of a fixed-voltage regulator.

Basic Operation   As indicated in Figure 11–27, a constant 1.25 V reference voltage (V_{REF}) is maintained by the regulator between the output terminal and the adjustment terminal. This constant reference voltage produces a constant current (I_{REF}) through R_1, regardless of the value of R_2. I_{REF} is also through R_2.

$$I_{REF} = \frac{V_{REF}}{R_1} = \frac{1.25 \text{ V}}{R_1}$$

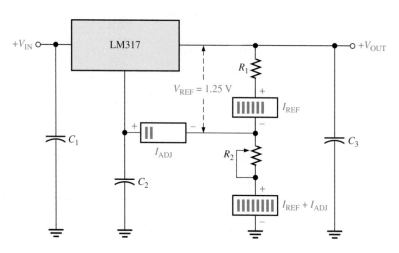

FIGURE 11–27
Operation of the LM317 adjustable voltage regulator.

In addition, there is a very small constant current into the adjustment terminal of approximately 50 μA called I_{ADJ}, which is through R_2. A formula for the output voltage is as follows:

$$V_{OUT} = V_{REF}\left(1 + \frac{R_2}{R_1}\right) + I_{ADJ}R_2 \tag{11-9}$$

As you can see, the output voltage is a function of both R_1 and R_2. Once the value of R_1 is set, the output voltage is adjusted by varying R_2.

EXAMPLE 11–6

Determine the minimum and maximum output voltages for the voltage regulator in Figure 11–28. Assume $I_{ADJ} = 50$ μA.

FIGURE 11–28

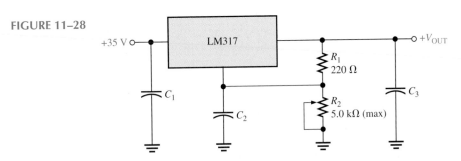

Solution

$$V_{R1} = V_{REF} = 1.25 \text{ V}$$

When R_2 is set at its minimum of 0 Ω,

$$V_{OUT(min)} = V_{REF}\left(1 + \frac{R_2}{R_1}\right) + I_{ADJ}R_2 = 1.25 \text{ V}(1) = \mathbf{1.25 \text{ V}}$$

When R_2 is set at its maximum of 5.0 kΩ,

$$V_{OUT(max)} = V_{REF}\left(1 + \frac{R_2}{R_1}\right) + I_{ADJ}R_2 = 1.25 \text{ V}\left(1 + \frac{5.0 \text{ k}\Omega}{220 \text{ }\Omega}\right) + (50 \text{ }\mu\text{A})5.0 \text{ k}\Omega$$

$$= 29.66 \text{ V} + 0.25 \text{ V} = \mathbf{29.9 \text{ V}}$$

Practice Exercise   What is the output voltage of the regulator if R_2 is set at 2.0 kΩ?

Adjustable Negative Linear Voltage Regulators

The LM337[6] is the negative output counterpart of the LM317 and is a good example of this type of IC regulator. Like the LM317, the LM337 requires two external resistors for output voltage adjustment as shown in Figure 11–29. The output voltage can be adjusted from −1.2 V to −37 V, depending on the external resistor values.

---

[6] Data sheet for LM337 available at http://www.national.com

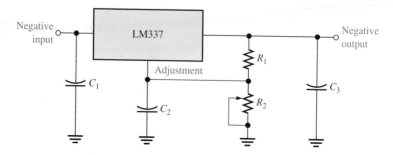

FIGURE 11–29
The LM337 three-terminal adjustable negative voltage regulator.

Troubleshooting Three-Terminal Regulators

Three-terminal regulators are very reliable devices. When problems occur, the indication is usually an incorrect voltage, high ripple, noisy or oscillating output, or drift. Troubleshooting a regulator circuit is best done with an oscilloscope as problems such as excessive ripple or noise won't show up using a DMM. Before starting, it is useful to review the possible causes of a failure (analysis) and plan measurements that will point to the failure.

If the output voltage is too low, the input voltage should be checked; the problem may be in the circuit preceding the regulator. Also check the load resistor: Does the problem go away when the load is removed? If so, it may be that the load draws too much current. A high output can occur with adjustable regulators if the feedback resistors are the wrong value or open. If there is ripple or noise on the output, check the capacitors for an open, a wrong value, or that they are installed with the proper polarity. A useful quick check of a capacitor is to place another capacitor of the same or larger size in parallel with the capacitor to be tested. If the output is oscillating, has high ripple, or drifting, check that the regulator is not too hot or supplying more than its rated current. If heat is a problem, make sure the regulator is firmly secured to the heat sink.

Switching Voltage Regulators

As an example of an IC switching voltage regulator, let's look at the 78S40[7]. This is a universal device that can be used with external components to provide step-up, step-down, and inverting operation.

The internal circuitry of the 78S40 is shown in Figure 11–30. This circuit can be compared to the basic switching regulators that were covered in Section 11–4. For example, look back at Figure 11–15(a). The oscillator and comparator functions are directly comparable. The gate and flip-flop which are digital devices were not included in the basic circuit of Figure 11–15(a), but they provide additional regulating action. Transistors Q_1 and Q_2 effectively perform the same function as Q_1 in the basic circuit. The 1.25 V reference block in the 78S40 has the same purpose as the zener diode in the basic circuit, and diode D_1 in the 78S40 corresponds to D_1 in the basic circuit.

The 78S40 also has an "uncommitted" op-amp thrown in for good measure. It is not used in any of the regulator configurations. External circuitry is required to make this device operate as a regulator, as you will see in Section 11–6.

---

[7] Data sheet for 78S40 available at http://www.national.com

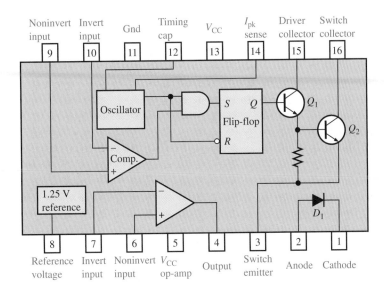

FIGURE 11–30
The 78S40 switching regulator.

11–5 REVIEW QUESTIONS

1. What are the three terminals of a fixed-voltage regulator?
2. What is the output voltage of a 7809? Of a 7915?
3. What are the three terminals of an adjustable-voltage regulator?
4. What external components are required for a basic LM317 configuration?

11–6 ■ APPLICATIONS OF IC VOLTAGE REGULATORS

In the last section, you saw several devices that are representative of the general types of IC voltage regulators. Now, several different ways these devices can be modified with external circuitry to improve or alter their performance are examined.

After completing this section, you should be able to

❑ Discuss applications of IC voltage regulators
  ❑ Explain the use of an external pass transistor
  ❑ Explain the use of current limiting
  ❑ Explain how to use a voltage regulator as a constant-current source
  ❑ Discuss some application considerations for switching regulators

The External Pass Transistor

As you know, an IC voltage regulator is capable of delivering only a certain amount of output current to a load. For example, the 7800 series regulators can handle a peak output current of 1.3 A (more under certain conditions). If the load current exceeds the maximum allowable value, there will be thermal overload and the regulator will shut down. A thermal overload condition means that there is excessive power dissipation inside the device.

If an application requires more than the maximum current that the regulator can deliver, an external pass transistor can be used. Figure 11–31 illustrates a three-terminal regulator with an external pass transistor for handling currents in excess of the output current capability of the basic regulator.

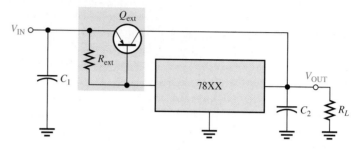

FIGURE 11–31

A 7800-series three-terminal regulator with an external pass transistor.

The value of the external current-sensing resistor R_{ext} determines the value of current at which Q_{ext} begins to conduct because it sets the base-to-emitter voltage of the transistor. As long as the current is less than the value set by R_{ext}, the transistor Q_{ext} is off, and the regulator operates normally as shown in Figure 11–32(a). This is because the voltage drop

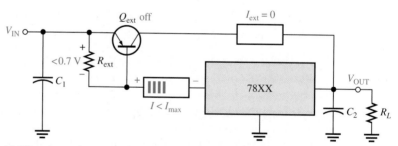

(a) When the regulator current is less than I_{max}, the external pass transistor is off and the regulator is handling all of the current.

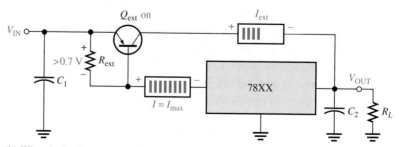

(b) When the load current exceeds I_{max}, the drop across R_{ext} turns Q_{ext} on and the transistor conducts the excess current.

FIGURE 11–32

Operation of the regulator with an external pass transistor.

across R_{ext} is less than the 0.7 V base-to-emitter voltage required to turn Q_{ext} on. R_{ext} is determined by the following formula, where I_{max} is the highest current that the voltage regulator is to handle internally.

$$R_{ext} = \frac{0.7 \text{ V}}{I_{max}}$$

When the current is sufficient to produce at least a 0.7 V drop across R_{ext}, the external pass transistor Q_{ext} turns on and conducts any current in excess of I_{max}, as indicated in Figure 11–32(b). Q_{ext} will conduct more or less, depending on the load requirements. For example, if the total load current is 3 A and I_{max} was selected to be 1 A, the external pass transistor will conduct 2 A, which is the excess over the internal regulator current I_{max}.

EXAMPLE 11–7

What value is R_{ext} if the maximum current to be handled internally by the voltage regulator in Figure 11–31 is set at 700 mA?

Solution

$$R_{ext} = \frac{0.7 \text{ V}}{I_{max}} = \frac{0.7 \text{ V}}{0.7 \text{ A}} = \mathbf{1\ \Omega}$$

Practice Exercise If R_{ext} is changed to 1.5 Ω, at what current value will Q_{ext} turn on?

The external pass transistor is typically a power transistor with heat sink that must be capable of handling a maximum power of

$$P_{ext} = I_{ext}(V_{IN} - V_{OUT})$$

EXAMPLE 11–8

What must be the minimum power rating for the external pass transistor used with a 7824 regulator in a circuit such as that shown in Figure 11–31? The input voltage is 30 V and the load resistance is 10 Ω. The maximum internal current is to be 700 mA. Assume that there is no heat sink for this calculation. Keep in mind that the use of a heat sink increases the effective power rating of the transistor and you can use a lower-rated transistor.

Solution The load current is

$$I_L = \frac{V_{OUT}}{R_L} = \frac{24 \text{ V}}{10 \text{ }\Omega} = 2.4 \text{ A}$$

The current through Q_{ext} is

$$I_{ext} = I_L - I_{max} = 2.4 \text{ A} - 0.7 \text{ A} = 1.7 \text{ A}$$

The power dissipated by Q_{ext} is

$$P_{ext(min)} = I_{ext}(V_{IN} - V_{OUT}) = 1.7 \text{ A}(30 \text{ V} - 24 \text{ V}) = 1.7 \text{ A}(6 \text{ V}) = \mathbf{10.2 \text{ W}}$$

For a safety margin, choose a power transistor with a rating greater than 10.2 W, say at least 15 W.

Practice Exercise Rework this example using a 7815 regulator.

Current Limiting

A drawback of the circuit in Figure 11–31 is that the external transistor is not protected from excessive current, such as would result from a shorted output. An additional current-limiting network (Q_{lim} and R_{lim}) can be added as shown in Figure 11–33 to protect Q_{ext} from excessive current and possible burn out.

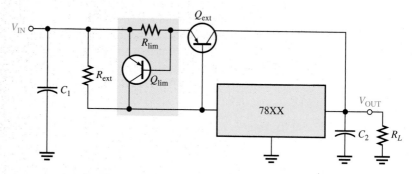

FIGURE 11–33
Regulator with current limiting.

The following describes the way the current-limiting network works. The current-sensing resistor R_{lim} sets the V_{BE} of transistor Q_{lim}. The base-to-emitter voltage of Q_{ext} is now determined by $V_{R_{ext}} - V_{R_{lim}}$ because they have opposite polarities. So, for normal operation, the drop across R_{ext} must be sufficient to overcome the opposing drop across R_{lim}. If the current through Q_{ext} exceeds a certain maximum ($I_{ext(max)}$) because of a shorted output or a faulty load, the voltage across R_{lim} reaches 0.7 V and turns Q_{lim} on. Q_{lim} now conducts current away from Q_{ext} and through the regulator, forcing a thermal overload to occur and shut down the regulator. Remember, the IC regulator is internally protected from thermal overload as part of its design.

This action is shown in Figure 11–34. In part (a), the circuit is operating normally with Q_{ext} conducting less than the maximum current that it can handle with Q_{lim} off. Part (b) shows what happens when there is a short across the load. The current through Q_{ext} suddenly increases and causes the voltage drop across R_{lim} to increase, which turns Q_{lim} on. The current is now diverted through the regulator, which causes it to shut down due to thermal overload.

A Current Regulator

The three-terminal regulator can be used as a current source when an application requires that a constant current be supplied to a variable load. The basic circuit is shown in Figure 11–35 where R_1 is the current-setting resistor. The regulator provides a fixed constant voltage, V_{OUT}, between the ground terminal (not connected to ground in this case) and the output terminal. This determines the constant current supplied to the load.

$$I_L = \frac{V_{OUT}}{R_1} + I_G$$

The current, I_G, from the ground terminal is very small compared to the output current and can often be neglected.

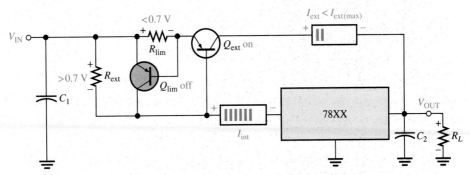

(a) During normal operation, when the load current is not excessive, Q_{lim} is off.

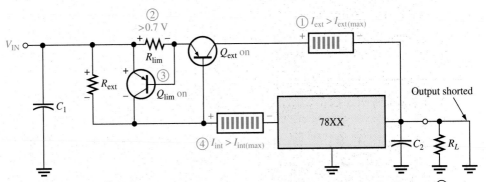

(b) When short occurs ①, the external current becomes excessive and the voltage across R_{lim} increases ② and turns on Q_{lim} ③, which then conducts current away from Q_{ext} and routes it through the regulator, causing the internal regulator current to become excessive ④ and to force the regulator into thermal shut down.

FIGURE 11–34

The current-limiting action of the regulator circuit.

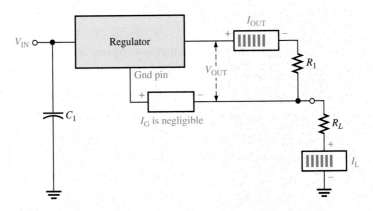

FIGURE 11–35

The three-terminal regulator as a current source.

EXAMPLE 11–9

What value of R_1 is necessary in a 7805 regulator to provide a constant current of 1 A to a variable load that can be adjusted from 0–10 Ω?

Solution   First, 1 A is within the limits of the 7805's capability (remember, it can handle at least 1.3 A without an external pass transistor).

The 7805 produces 5 V between its ground terminal and its output terminal. Therefore, if you want 1 A of current, the current-setting resistor must be (neglecting I_G)

$$R_1 = \frac{V_{OUT}}{I_L} = \frac{5\ V}{1\ A} = \mathbf{5.0\ \Omega}$$

The circuit is shown in Figure 11–36.

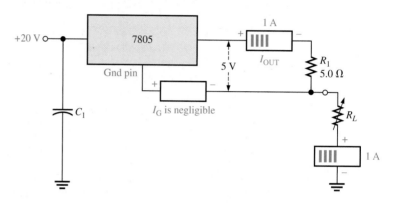

FIGURE 11–36
A 1 A constant-current source.

Practice Exercise   If a 7808 regulator is used instead of the 7805, to what value would you change R_1 to maintain a constant current of 1 A?

Switching Regulator Configurations

In Section 11–5, the 78S40 was introduced as an example of an IC switching voltage regulator. Figure 11–37 shows the external connections for a step-down configuration where the output voltage is less than the input voltage, and Figure 11–38 shows a step-up configuration in which the output voltage is greater than the input voltage. An inverting configuration is also possible, but it is not shown here.

The timing capacitor, C_T, controls the pulse width and frequency of the oscillator and thus establishes the on-time of transistor Q_2. The voltage across the current-sensing resistor R_{CS} is used internally by the oscillator to vary the duty cycle based on the desired peak load current. The voltage divider, made up of R_1 and R_2, reduces the output voltage to a nominal value equal to the reference voltage. If V_{OUT} exceeds its set value, the output of the comparator switches to its low state, disabling the gate to turn Q_2 off until the output decreases. This regulating action is in addition to that produced by the duty cycle variation of the oscillator as described in Section 11–4 in relation to the basic switching regulator.

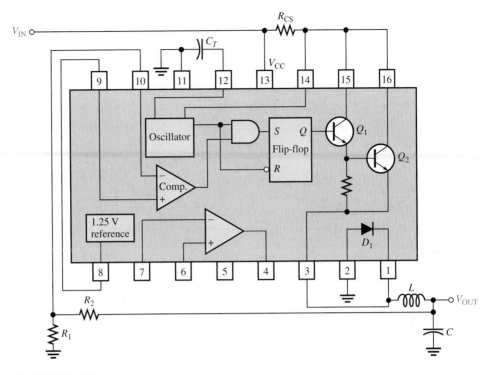

FIGURE 11–37
The step-down configuration of the 78S40 switching regulator.

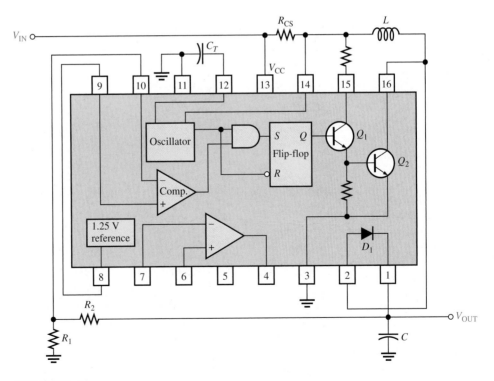

FIGURE 11–38
The step-up configuration of the 78S40 switching regulator.

11–6 REVIEW QUESTIONS

1. What is the purpose of using an external pass transistor with an IC voltage regulator?
2. What is the advantage of current limiting in a voltage regulator?
3. How can you configure a three-terminal regulator as a current source?

11–7 ■ A SYSTEM APPLICATION

In this system application, the focus is on the regulated power supply which provides the FM stereo receiver with dual polarity dc voltages. Recall from previous system applications that the op-amps in the channel separation circuits and the audio amplifiers operate from ±12 V. Both positive and negative voltage regulators are used to regulate the rectified and filtered voltages from a bridge rectifier.

After completing this section, you should be able to

❑ Apply what you have learned in this chapter to a system application
  ❑ Discuss how dual supply voltages are produced by a rectifier
  ❑ Explain how positive and negative three-terminal IC regulators are used in a power supply
  ❑ Relate a schematic to a PC board
  ❑ Analyze the operation of the power supply circuit
  ❑ Troubleshoot some common power supply failures

About the Power Supply

This power supply utilizes a full-wave bridge **rectifier** with both the positive and negative rectified voltages taken off the bridge at the appropriate points and filtered by electrolytic capacitors. A 7812 and a 7912 provide regulation.

Now, so that you can take a closer look at the dual power supply, let's take it out of the system and put it on the troubleshooter's bench.

TROUBLESHOOTER'S BENCH

■ ACTIVITY 1 Relate the PC Board to the Schematic

Develop a schematic for the power supply in Figure 11–39. Add any missing labels and include the IC pin numbers by referring to the voltage regulator data sheets in Appendix A. The rectifier diodes are type 1N4001, the filter capacitors C1 and C2 are 1000 μF, and the transformer has a turns ratio of 5:1.

■ ACTIVITY 2 Analyze the Power Supply Circuits

Step 1: Determine the approximate voltage at each of the four "corners" of the bridge with respect to ground.

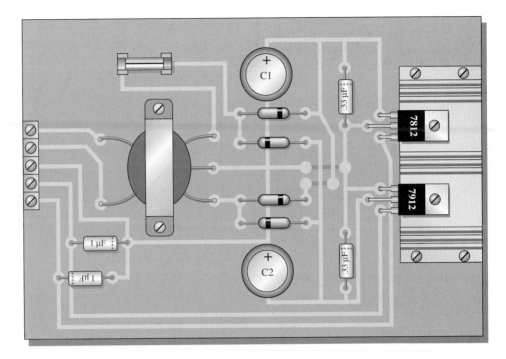

FIGURE 11–39

Step 2: Calculate the peak inverse voltage of the rectifier diodes.

Step 3: Determine the voltage at the inputs of the voltage regulators.

Step 4: In this stereo system, assume that op-amps are used only in the channel separation circuits and the channel audio amplifiers. If all of the other circuits in the receiver use +12 V and draw an average dc current of 500 mA, determine how much total current each regulator must supply. Refer to the system applications in Chapters 7 and 9. Use the appropriate data sheets.

Step 5: Based on the results in Step 4, do the IC regulators have to be attached to the heat sink or is this just for a safety margin?

■ ACTIVITY 3   Write a Technical Report

Describe the operation of the power supply with an emphasis on how both positive and negative voltages are obtained. State the purpose of each component. Use the results of Activity 2 where appropriate.

■ ACTIVITY 4   Troubleshoot the Power Supply by Stating the Probable Cause or Causes in Each Case

1. Both positive and negative output voltages are zero.
2. Positive output voltage is zero and the negative output voltage is −12 V.
3. Negative output voltage is zero and the positive output voltage is +12 V.
4. Radical voltage fluctuations on output of positive regulator.

■ **ACTIVITY 5 Troubleshooter's Bench Special Assignment**

Go to Troubleshooter's Bench 3 in the color insert section and carry out the assignment that is stated there.

11–7 REVIEW QUESTIONS

1. What should be the rating of the power supply fuse?
2. What purpose do the 0.33 μF capacitors serve?
3. Which regulator provides the negative voltage?
4. Would you recommend that an external pass transistor be used with the regulators in this power supply? Why?

■ SUMMARY

- Voltage regulators keep a constant dc output voltage when the input or load varies within limits.
- A basic voltage regulator consists of a reference voltage source, an error detector, a sampling element, and a control device. Protection circuitry is also found in most regulators.
- Two basic categories of voltage regulators are linear and switching.
- Two basic types of linear regulators are series and shunt.
- In a series linear regulator, the control element is a transistor in series with the load.
- In a shunt linear regulator, the control element is a transistor in parallel with the load.
- Three configurations for switching regulators are step-down, step-up, and inverting.
- Switching regulators are more efficient than linear regulators and are particularly useful in low-voltage, high-current applications.
- Three-terminal linear IC regulators are available for either fixed output or variable output voltages of positive or negative polarities.
- An external pass transistor increases the current capability of a regulator.
- The 7800 series are three-terminal IC regulators with fixed positive output voltage.
- The 7900 series are three-terminal IC regulators with fixed negative output voltage.
- The LM317 is a three-terminal IC regulator with a positive variable output voltage.
- The LM337 is a three-terminal IC regulator with a negative variable output voltage.
- The 78S40 is a switching voltage regulator.

■ GLOSSARY

Key terms are in color. All terms are included in the end-of-book glossary.

Fold-back current limiting A method of current limiting in voltage regulators.

Linear regulator A voltage regulator in which the control element operates in the linear region.

Line regulation The percentage change in output voltage for a given change in line (input) voltage.

Load regulation The percentage change in output voltage for a given change in load current.

Rectifier An electronic circuit that converts ac into pulsating dc.

Regulator An electronic circuit that maintains an essentially constant output voltage with a changing input voltage or load current.

Switching regulator   A voltage regulator in which the control element is a switching device.

Thermal overload   A condition in a rectifier where the internal power dissipation of the circuit exceeds a certain maximum due to excessive current.

■ **KEY FORMULAS**

(11–1) $\text{Line regulation} = \left(\dfrac{\Delta V_{\text{OUT}}}{\Delta V_{\text{IN}}}\right)100\%$ Percent line regulation

(11–2) $\text{Line regulation} = \left(\dfrac{\Delta V_{\text{OUT}}/V_{\text{OUT}}}{\Delta V_{\text{IN}}}\right)100\%$ Percent line regulation per volt

(11–3) $\text{Load regulation} = \left(\dfrac{V_{\text{NL}} - V_{\text{FL}}}{V_{\text{FL}}}\right)100\%$ Percent load regulation

(11–4) $\text{Load regulation} = \left(\dfrac{R_{\text{OUT}}}{R_{\text{FL}}}\right)100\%$ Percent load regulation given output resistance and minimum load resistance

(11–5) $V_{\text{OUT}} \cong \left(1 + \dfrac{R_2}{R_3}\right)V_{\text{REF}}$ Series regulator output

(11–6) $I_{\text{L(max)}} = \dfrac{0.7\ \text{V}}{R_4}$ Constant current limiting

(11–7) $I_{\text{L(max)}} = \dfrac{V_{\text{IN}}}{R_1}$ Maximum load current for a shunt regulator

(11–8) $V_{\text{OUT}} = \left(\dfrac{t_{\text{on}}}{T}\right)V_{\text{IN}}$ Output voltage for step-down switching regulator

(11–9) $V_{\text{OUT}} = V_{\text{REF}}\left(1 + \dfrac{R_2}{R_1}\right) + I_{\text{ADJ}}R_2$ Output voltage for IC voltage regulator

---

■ **SELF-TEST**

Answers are at the end of the chapter.

1. In the case of line regulation,
   (a) when the temperature varies, the output voltage stays constant
   (b) when the output voltage changes, the load current stays constant
   (c) when the input voltage changes, the output voltage stays constant
   (d) when the load changes, the output voltage stays constant

2. In the case of load regulation,
   (a) when the temperature varies, the output voltage stays constant
   (b) when the input voltage changes, the load current stays constant
   (c) when the load changes, the load current stays constant
   (d) when the load changes, the output voltage stays constant

3. All of the following are parts of a basic voltage regulator *except*
   (a) control element   (b) sampling circuit   (c) voltage follower
   (d) error detector   (e) reference voltage

4. The basic difference between a series regulator and a shunt regulator is
   (a) the amount of current that can be handled
   (b) the position of the control element
   (c) the type of sample circuit
   (d) the type of error detector

5. In a basic series regulator, V_{OUT} is determined by
   (a) the control element   (b) the sample circuit
   (c) the reference voltage   (d) answers (b) and (c)

6. The main purpose of current limiting in a regulator is
   (a) protection of the regulator from excessive current
   (b) protection of the load from excessive current
   (c) to keep the power supply transformer from burning up
   (d) to maintain a constant output voltage

7. In a linear regulator, the control transistor is conducting
   (a) a small part of the time   (b) half the time
   (c) all of the time   (d) only when the load current is excessive

8. In a switching regulator, the control transistor is conducting
   (a) part of the time
   (b) all of the time
   (c) only when the input voltage exceeds a set limit
   (d) only when there is an overload

9. The LM317 is an example of an IC
   (a) three-terminal negative voltage regulator   (b) fixed positive voltage regulator
   (c) switching regulator   (d) linear regulator
   (e) variable positive voltage regulator   (f) answers (b) and (d) only
   (g) answers (d) and (e) only

10. An external pass transistor is used for
    (a) increasing the output voltage
    (b) improving the regulation
    (c) increasing the current that the regulator can handle
    (d) short-circuit protection

TROUBLESHOOTER'S QUIZ   *Answers are at the end of the chapter.*

Refer to Figure 11–42.

❑ If D_1 is mistakenly replaced with a 4.7 V zener,

1. The output voltage will

   (a) increase   (b) decrease   (c) not change

2. The voltage across Q_1 from collector to emitter will

   (a) increase   (b) decrease   (c) not change

Refer to Figure 11–43.

❑ If the output current is much less than maximum and the emitter of Q_2 is open,

3. The output voltage will

   (a) increase   (b) decrease   (c) not change

4. The maximum current that can be supplied to a load will

   (a) increase   (b) decrease   (c) not change

❑ If R_2 is shorted,

5. The output voltage will

   (a) increase   (b) decrease   (c) not change

Refer to Figure 11–44.

❑ If R_1 is open,

    6. The output voltage will

        (a) increase   **(b)** decrease   **(c)** not change

Refer to Figure 11–45.

❑ If C is open,

    7. The ripple voltage on the output will

        (a) increase   **(b)** decrease   **(c)** not change

❑ If the duty cycle of the oscillator increases,

    8. The output voltage will

        (a) increase   **(b)** decrease   **(c)** not change

Refer to Figure 11–47.

❑ If R_1 is smaller than the specified value,

    9. The output voltage will

        (a) increase   **(b)** decrease   **(c)** not change

    10. The line regulation will

        (a) increase   **(b)** decrease   **(c)** not change

■ PROBLEMS

Answers to odd-numbered problems are at the end of the book.

SECTION 11–1   Voltage Regulation

1. The nominal output voltage of a certain regulator is 8 V. The output changes 2 mV when the input voltage goes from 12 V to 18 V. Determine the line regulation and express it as a percentage change.

2. Express the line regulation found in Problem 1 in units of %/V.

3. A certain regulator has a no-load output voltage of 10 V and a full-load output voltage of 9.90 V. What is the percent load regulation?

4. In Problem 3, if the full-load current is 250 mA, express the load regulation in %/mA.

SECTION 11–2   Basic Series Regulators

5. Label the functional blocks for the voltage regulator in Figure 11–40.

FIGURE 11–40

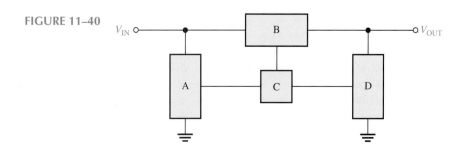

6. Determine the output voltage for the regulator in Figure 11–41.

7. Determine the output voltage for the series regulator in Figure 11–42.

8. If R_3 in Figure 11–42 is increased to 4.7 kΩ, what happens to the output voltage?

FIGURE 11–41

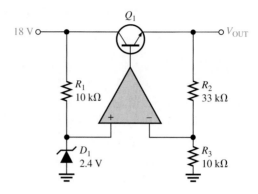

FIGURE 11–42

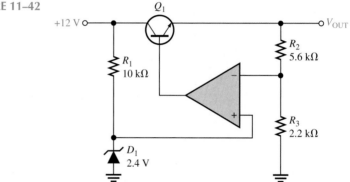

9. If the zener voltage is 2.7 V instead of 2.4 V in Figure 11–42, what is the output voltage?

10. A series voltage regulator with constant current limiting is shown in Figure 11–43. Determine the value of R_4 if the load current is to be limited to a maximum value of 250 mA. What power rating must R_4 have?

11. If the R_4 determined in Problem 10 is halved, what is the maximum load current?

FIGURE 11–43

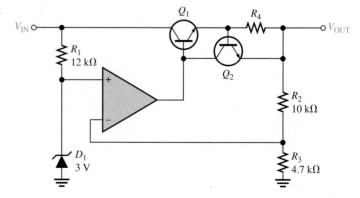

SECTION 11–3   Basic Shunt Regulators

12. In the shunt regulator of Figure 11–44, when the current through R_L increases, does Q_1 conduct more or less? Why?

13. Assume the current through R_L remains constant and V_{IN} changes by 1 V in Figure 11–44. What is the change in the collector current of Q_1?

FIGURE 11–44

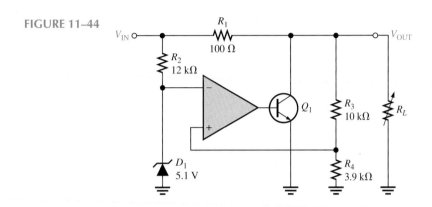

14. With a constant input voltage of 17 V, the load resistance in Figure 11–44 is varied from 1.0 kΩ to 1.2 kΩ. Neglecting any change in output voltage, how much does the shunt current through Q_1 change?

15. If the maximum allowable input voltage in Figure 11–44 is 25 V, what is the maximum possible output current when the output is short-circuited? What power rating should R_1 have?

SECTION 11–4   Basic Switching Regulators

16. A basic switching regulator is shown in Figure 11–45. If the switching frequency of the transistor is 100 Hz with an off-time of 6 ms, what is the output voltage?

17. What is the duty cycle of the transistor in Problem 16?

FIGURE 11–45

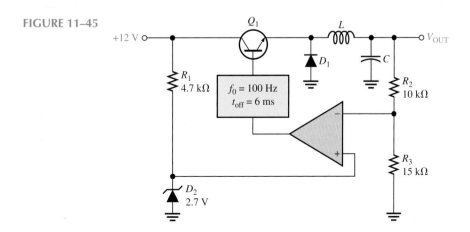

18. When does the diode D_1 in Figure 11–46 become forward-biased?

19. If the on-time of Q_1 in Figure 11–46 is decreased, does the output voltage increase or decrease?

FIGURE 11–46

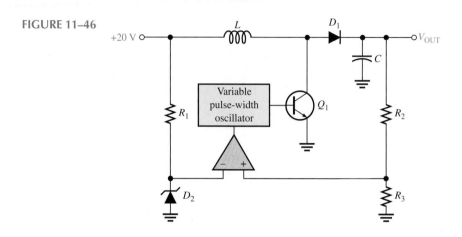

SECTION 11–5 Integrated Circuit Voltage Regulators

20. What is the output voltage of each of the following IC regulators?
(a) 7806   (b) 7905.2   (c) 7818   (d) 7924

21. Determine the output voltage of the regulator in Figure 11–47. $I_{ADJ} = 50 \ \mu A$.

FIGURE 11–47

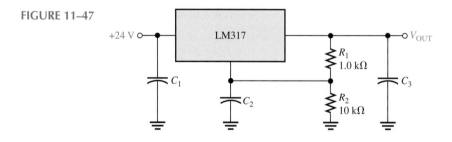

22. Determine the minimum and maximum output voltages for the circuit in Figure 11–48. $I_{ADJ} = 50 \ \mu A$.

FIGURE 11–48

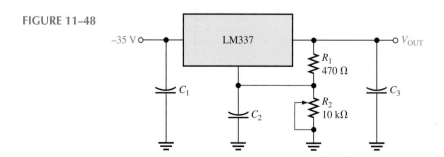

23. With no load connected, how much current is there through the regulator in Figure 11–47? Neglect the adjustment terminal current.

24. Select the values for the external resistors to be used in an LM317 circuit that is required to produce an output voltage of 12 V with an input of 18 V. The maximum regulator current with no load is to be 2 mA. There is no external pass transistor.

SECTION 11–6 Applications of IC Voltage Regulators

25. In the regulator circuit of Figure 11–49, determine R_{ext} if the maximum internal regulator current is to be 250 mA.

26. Using a 7812 voltage regulator and a 10 Ω load in Figure 11–49, how much power will the external pass transistor have to dissipate? The maximum internal regulator current is set at 500 mA by R_{ext}.

FIGURE 11–49

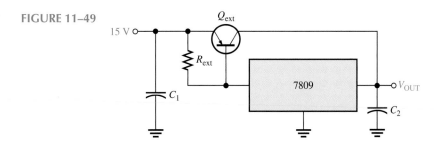

27. Show how to include current limiting in the circuit of Figure 11–49. What should the value of the limiting resistor be if the external current is to be limited to 2 A?

28. Using an LM317, design a circuit that will provide a constant current of 500 mA to a load.

29. Repeat Problem 28 using a 7908.

30. If a 78S40 switching regulator is to be used to regulate a 12V input down to a 6 V output, calculate the values of the external voltage-divider resistors.

■ **ANSWERS**
TO REVIEW
QUESTIONS

Section 11–1

1. The percentage change in the output voltage for a given change in input voltage.

2. The percentage change in output voltage for a given change in load current.

3. 1.2%; 0.06%/V

4. 1.6%; 0.0016%/mA

Section 11–2

1. Control element, error detector, sampling element, reference source

2. 2 V

Section 11–3

1. In a shunt regulator, the control element is in parallel with the load rather than in series.

2. A shunt regulator has inherent current limiting. A disadvantage is that a shunt regulator is less efficient than a series regulator.

Section 11–4

1. Step-down, step-up, inverting
2. Switching regulators operate at a higher efficiency.
3. The duty cycle varies to regulate the output.

Section 11–5

1. Input, output, and ground
2. A 7809 has a +9 V output; A 7915 has a −15 V output.
3. Input, output, adjustment
4. A two-resistor voltage divider

Section 11–6

1. A pass transistor increases the current that can be handled.
2. Current limiting prevents excessive current and prevents damage to the regulator.
3. See Figure 11–35.

Section 11–7

1. 1 A
2. Those optional capacitors on the regulator inputs prevent oscillations.
3. The 7909 is a negative-voltage regulator.
4. No. The current that either regulator must supply is less than 1 A.

■ **ANSWERS TO PRACTICE EXERCISES FOR EXAMPLES**

11–1	12%, 0.6%/V
11–2	The voltage drop of 12.1 V − 12.0 V = 0.1 V is across the specified output resistance. Since $I = 0.2$ A at 12.0 V, 0.1/0.2 = 0.5 Ω.
11–3	7.33 V
11–4	0.7 A
11–5	17.5 W
11–6	12.7 V
11–7	467 mA
11–8	12 W dissipated; choose a larger practical value (e.g., 20 W).
11–9	8 Ω

■ **ANSWERS TO SELF-TEST**

1. (c)	**2.** (d)	**3.** (c)	**4.** (b)	**5.** (d)
6. (a)	**7.** (c)	**8.** (a)	**9.** (g)	**10.** (c)

■ **ANSWERS TO TROUBLE-SHOOTER'S QUIZ**

1. increase	**2.** decrease	**3.** not change	**4.** increase
5. decrease	**6.** decrease	**7.** increase	**8.** increase
9. increase	**10.** not change		

12

SPECIAL-PURPOSE AMPLIFIERS

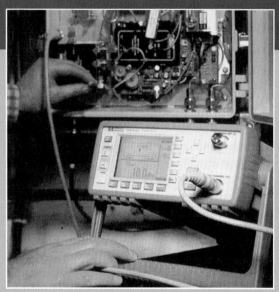

Courtesy Hewlett-Packard Company

■ CHAPTER OBJECTIVES

❏ Understand and explain the operation of an
  instrumentation amplifier (IA)
❏ Understand and explain the operation of an
  isolation amplifier
❏ Understand and explain the operation of an
  operational transconductance amplifier (OTA)
❏ Understand and explain the operation of log
  and antilog amplifiers
❏ Apply what you have learned in this chapter
  to a system application

■ KEY TERMS

❏ Instrumentation amplifier
❏ Isolation amplifier
❏ Operational transconductance amplifier
❏ Logarithm
❏ Antilog

■ CHAPTER INTRODUCTION

A general-purpose op-amp, such as the 741, is an extremely versatile and widely used device. However, many specialized IC amplifiers have been designed with certain types of applications in mind or with certain special features or characteristics. Most of these devices are actually derived from the basic op-amp. These special amplifiers include the instrumentation amplifier (IA) that is used in high-noise environments, the isolation amplifier that is used in high-voltage and medical applications, the operational transconductance amplifier (OTA) that is used as a voltage-to-current amplifier, and the logarithmic amplifiers that are used for linearizing certain types of inputs and for mathematical operations. In this chapter, you will learn about each of these devices and some of their basic applications.

■ A SYSTEM APPLICATION

Medical electronics is a very important application area for electronic devices and, without doubt, one of the most beneficial. The electrocardiograph (ECG), one of the most common and important instruments in use for medical purposes, is used to monitor the heart function of patients in order to detect any irregularities or abnormalities in the heartbeat. Sensors called electrodes are placed at points on the body to pick up the small electrical signal produced by the heart. This signal goes through an amplification process and is fed to a video monitor or chart recorder for viewing. Because of the safety hazards related to electrical equipment, it is important that the patient be protected from the possibility of unpleasant or even fatal electrical shock. For this reason, the isolation amplifier is used in medical equipment that comes in contact with the human body. The diagram in Figure 12–34 shows a basic block diagram for a simplified ECG system. Our focus in this system application is on the amplifier section.

For the system application in Section 12–5, in addition to the other topics, be sure you understand
❏ Basic op-amp operation
❏ Isolation amplifiers

12–1 ■ INSTRUMENTATION AMPLIFIERS

An instrumentation amplifier is a differential voltage-gain device that amplifies the difference between the voltages existing at its two input terminals. The main purpose of an instrumentation amplifier is to amplify small signals that are riding on large common-mode voltages. The key characteristics are high input impedance, high common-mode rejection, low output offset, and low output impedance. A basic instrumentation amplifier is made up of three op-amps and several resistors. The voltage gain is set with an external resistor. Instrumentation amplifiers (IAs) are commonly used in environments with high common-mode noise such as in data acquisition systems where remote sensing of input variables is required.

After completing this section, you should be able to

❏ Understand and explain the operation of an instrumentation amplifier (IA)
  ❏ Explain how op-amps are connected to form an IA
  ❏ Describe how the voltage gain is set
  ❏ Discuss an application
  ❏ Describe the features of the AD622 instrumentation amplifier

The Basic Instrumentation Amplifier

One of the most common problems in measuring systems is the contamination of the signal from a transducer with unwanted noise (such as 60 Hz power line interference). The transducer signal is typically a small differential signal carrying the desired information. Noise that is added to both signal conductors in the same amount is called a common-mode noise (discussed in Section 6–2). Ideally, the differential signal should be amplified and the common-mode noise should be rejected.

A second problem for measuring systems is that many transducers have a high output impedance and can easily be loaded down when connected to an amplifier. An amplifier for small transducer signals needs to have a very high input impedance to avoid this loading effect.

The solution to these measurement problems is the **instrumentation amplifier (IA)**, a specially designed differential amplifier with ultra-high input impedance and extremely good common-mode rejection (up to 130 dB) as well as being able to achieve high, stable gains. Instrumentation amplifiers can faithfully amplify low-level signals in the presence of high common-mode noise. They are used in a variety of signal-processing applications where accuracy is important and where low drift, low bias currents, precise gain, and very high CMRR are required.

Figure 12–1 shows a basic instrumentation amplifier (IA) constructed from three op-amps. Op-amps A1 and A2 are modified voltage-followers, each containing a feedback resistor (R_1 and R_2). The feedback resistors have no effect in this circuit (and could be left out) but will be used when the circuit is modified in the next step. The voltage-followers provide high input impedance with a gain of 1. Op-amp A3 is a differential amplifier that amplifies the difference between V_{out1} and V_{out2}. Although this circuit has the advantage of high input impedance, it requires extremely high precision matching of the gain resistors to achieve a high CMRR (R_3 must match R_4 and R_5 must match R_6). Further, it still has two resistors that must be changed if variable gain is desired (typically R_3 and R_4), and they must track each other with high precision over the operating temperature range.

A clever alternate configuration which solves the difficulties of the circuit in Figure 12–1 and provides high gain is the three op-amp IA shown in Figure 12–2. The inputs are

FIGURE 12–1

The basic instrumentation amplifier using three op-amps.

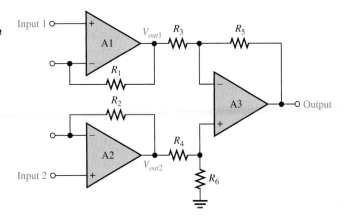

buffered by op-amp A1 and op-amp A2, providing a very high input impedance. Op-amps A1 and A2 can now provide gain. The entire assembly (except for R_G) is contained in a single IC. In this design, the common-mode gain still depends on very precisely matched resistors. However, these resistors can be critically matched during manufacture (by laser trimming) within the IC. Resistors R_3, R_4, R_5, and R_6 are generally set by the manufacturer for a gain of 1.0 for the differential amplifier. Resistors R_1 and R_2 are precision-matched resistors set equal to each other. This means the overall differential gain can be controlled by the size of just resistor R_G, supplied by the user. Appendix B derives the following equation for the output voltage:

$$V_{out} = \left(1 + \frac{2R}{R_G}\right)(V_{in2} - V_{in1}) \qquad \textbf{(12–1)}$$

where the closed-loop gain is

$$A_{cl} = 1 + \frac{2R}{R_G}$$

and where $R_1 = R_2 = R$. The last equation shows that the gain of the instrumentation amplifier can be set by the value of the external resistor R_G when R_1 and R_2 have known fixed values.

FIGURE 12–2

The instrumentation amplifier with the external gain-setting resistor R_G. Differential and common-mode signals are indicated.

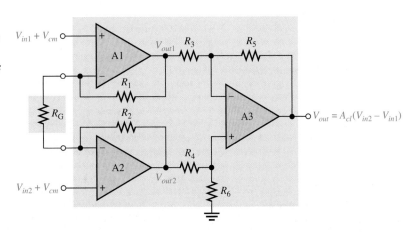

The external gain-setting resistor R_G can be calculated for a desired voltage gain by using the following formula:

$$R_G = \frac{2R}{A_{cl} - 1} \tag{12-2}$$

Instrumentation amplifiers in which the gain is set to specific values using a binary input instead of a resistor are also available.

EXAMPLE 12–1

Determine the value of the external gain-setting resistor R_G for a certain IC instrumentation amplifier with $R_1 = R_2 = 25 \text{ k}\Omega$. The voltage gain is to be 500.

Solution

$$R_G = \frac{2R}{A_{cl} - 1} = \frac{50 \text{ k}\Omega}{500 - 1} \cong \mathbf{100\ \Omega}$$

*Practice Exercise** What value of external gain-setting resistor is required for an instrumentation amplifier with $R_1 = R_2 = 39 \text{ k}\Omega$ to produce a gain of 325?

---

* Answers are at the end of the chapter.

Applications

As mentioned in the introduction to this section, the instrumentation amplifier is normally used to measure small differential signal voltages that are superimposed on a common-mode noise voltage often much larger than the signal voltage. Applications include situations where a quantity is sensed by a remote device, such as a temperature- or pressure-sensitive transducer, and the resulting small electrical signal is sent over a long line subject to electrical noise that produces common-mode voltages in the line. The instrumentation amplifier at the end of the line must amplify the small signal from the remote sensor and reject the large common-mode voltage. Figure 12–3 illustrates this.

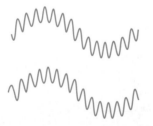
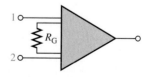
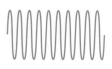

Small differential high-frequency signal riding on a larger low-frequency common-mode signal

Instrumentation amplifier

Amplified differential signal. No common-mode signal.

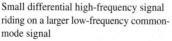

FIGURE 12–3

Illustration of the rejection of large common-mode voltages and the amplification of smaller signal voltages by an instrumentation amplifier.

A Specific Instrumentation Amplifier

Now that you have the basic idea of how an instrumentation amplifier works, let's look at a specific device. A representative device, the AD622[1], is shown in Figure 12–4. An IC package pin diagram is shown for reference. This instrumentation amplifier is based on the classic design using three op-amps as previously discussed.

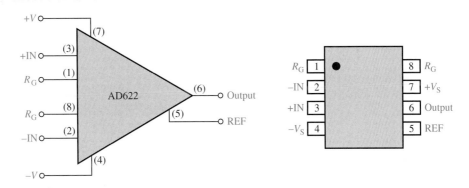

FIGURE 12–4
The AD622 instrumentation amplifier.

Some of the features of the AD622 are as follows. The voltage gain can be adjusted from 2 to 1000 with an external resistor R_G. There is unity gain with no external resistor. The input impedance is 10 GΩ. The common-mode rejection ratio (CMRR′) has a minimum value of 66 dB. Recall that a higher CMRR′ means better rejection of common-mode voltages. The AD622 has a bandwidth of 800 kHz at a gain of 10 and a slew rate of 1.2 V/μs.

Setting the Voltage Gain   For the AD622, an external resistor must be used to achieve a voltage gain greater than unity, as indicated in Figure 12–5. Resistor R_G is connected

---

[1]   Data sheet for AD622 available at http://www.analogdevices.com

FIGURE 12–5
The AD622 with a gain-setting resistor.

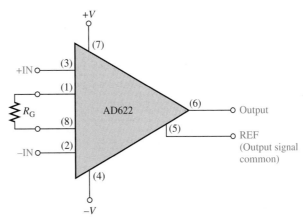

between the R_G terminals (pins 1 and 8). No resistor is required for unity gain. R_G is selected for the desired gain based on the following formula:

$$R_G = \frac{50.5 \text{ k}\Omega}{A_v - 1} \qquad \qquad (12\text{--}3)$$

Notice that this formula is the same as Equation (12–2) for the classic three-op-amp configuration where the internal resistors R_1 and R_2 are each 25.25 kΩ.

Gain versus Frequency The graph in Figure 12–6 shows how the gain varies with frequency for gains of 1, 10, 100, and 1000. As you can see, the bandwidth decreases as the gain increases.

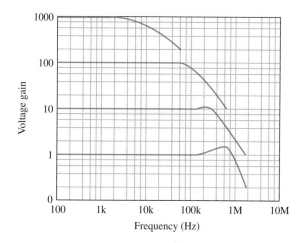

FIGURE 12–6
Gain versus frequency for the AD622 instrumentation amplifier.

EXAMPLE 12–2   Calculate the gain and determine the approximate bandwidth using the graph in Figure 12–6 for the instrumentation amplifier in Figure 12–7.

FIGURE 12–7

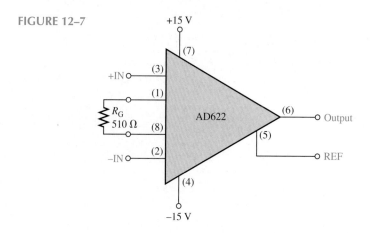

Solution   The voltage gain is determined as follows:

$$R_G = \frac{50.5 \text{ k}\Omega}{A_v - 1}$$

$$A_v - 1 = \frac{50.5 \text{ k}\Omega}{R_G}$$

$$A_v = \frac{50.5 \text{ k}\Omega}{510 \text{ }\Omega} + 1 = \mathbf{100}$$

The approximate bandwidth is determined from the graph.

$$BW \cong \mathbf{60 \text{ kHz}}$$

Practice Exercise   Modify the circuit in Figure 12–7 for a gain of approximately 45.

12–1 REVIEW QUESTIONS*

1. What is the main purpose of an instrumentation amplifier and what are three of its key characteristics?
2. What components do you need to construct a basic instrumentation amplifier?
3. How is the gain determined in a basic instrumentation amplifier?
4. In a certain AD622 configuration, $R_G = 10 \text{ k}\Omega$. What is the voltage gain?

* Answers are at the end of the chapter.

12–2 ■ ISOLATION AMPLIFIERS

An isolation amplifier provides dc isolation between input and output for the protection of human life or sensitive equipment in those applications where hazardous power-line leakage or high-voltage transients are possible. The principal areas of application for isolation amplifiers are in medical instrumentation, power plant instrumentation, industrial processing, and automated testing.

After completing this section, you should be able to

❏ Understand and explain the operation of an isolation amplifier
  ❏ Explain the basic configuration of an isolation amplifier
  ❏ Discuss an application in medical electronics
  ❏ Describe the features of the 3656KG isolation amplifier

The Basic Isolation Amplifier

In some ways, the **isolation amplifier** can be viewed as an elaborate op-amp or instrumentation amplifier. The difference is that the isolation amplifier has an input stage, an output stage, and a power supply section that are all electrically isolated from each other.

Although there are isolation amplifiers that use optical coupling or capacitive coupling to achieve isolation, the transformer-coupled device is the most commonly used. The circuit is in IC form, but the miniature, multiple-winding, toroid transformer is not fully integrated. This results in a package configuration that deviates somewhat from standard IC packages but is still designed for printed circuit board assembly.

A typical isolation amplifier is capable of operating with three independent grounds. As shown in Figure 12–8, the input and output stages and the power supplies are transformer coupled to achieve isolation of the input and output signals and the power.

The input stage contains an op-amp, a demodulator, a modulator, and a dual-polarity power supply. The output stage contains an op-amp, an oscillator, a demodulator, and a dual-polarity power supply.

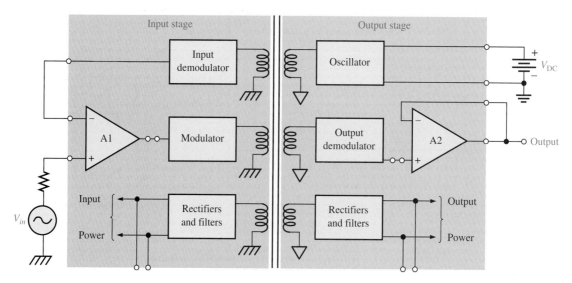

FIGURE 12–8

Diagram of a transformer-coupled isolation amplifier.

General Operation   As mentioned, there are three isolated independent grounds (indicated by three different symbols) for the input signal, the output signal, and the power supply, as shown in Figure 12–8.

The power to the input and output stages is produced as follows. An external dc supply voltage, V_{DC}, is applied to the oscillator. The oscillator converts the dc power to ac at a relatively high frequency. The ac output of the oscillator is coupled by the transformer to the input power supply (rectifiers and filters) and to the output power supply (rectifiers and filters) where it is rectified and filtered to produce dual-polarity (positive and negative) dc voltages for the input and output stages.

The oscillator output is also coupled to the modulator where it is combined with the input signal from the input op-amp A1 (modulators are covered in Chapter 13). The purpose of the modulator is to vary the amplitude of the relatively high oscillator frequency with the lower-frequency input signal. The higher modulated frequency allows a very small transformer. To couple the lower-frequency input signal without modulation would require a prohibitively large transformer.

The modulated signal is coupled to the demodulator in the output stage. The demodulator recovers the original input signal from the higher oscillator frequency. The demodulated input signal is then applied to op-amp A2. The demodulator in the input stage is part of a feedback loop that forces the signal at the inverting input of A1 to equal the original input signal at the noninverting input.

Although the isolation amplifier is a fairly complex circuit, in terms of its overall function, it is still simply an amplifier. You apply a dc voltage, put a signal in, and you get an amplified signal out. The isolation function is an unseen process.

Applications

As previously mentioned, the isolation amplifier is used in applications requiring no common grounds between a transducer and the processing circuits where interfacing to sensitive equipment is required. In chemical, nuclear, and metal-processing industries, for example, millivolt signals typically exist in the presence of large common-mode voltages that can be in the kilovolt range. In this type of environment, the isolation amplifier can amplify small signals from very noisy equipment and provide a safe output to sensitive equipment such as computers.

Another important application is in various types of medical equipment. In medical applications where body functions such as heart rate and blood pressure are monitored, the very small monitored signals are combined with large common-mode signals, such as 60 Hz power-line pickup from the skin. In these situations, without isolation, dc leakage or equipment failure could be fatal. Figure 12–9 shows a simplified diagram of an isolation amplifier in a cardiac-monitoring application. In this situation, heart signals, which are very small, are combined with much larger common-mode signals caused by muscle noise, electrochemical noise, residual electrode voltage, and 60 Hz line pickup from the skin.

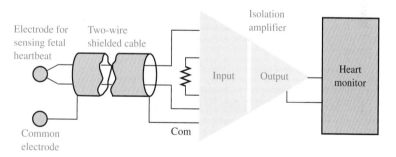

FIGURE 12–9

Fetal heartbeat monitoring using an isolation amplifier. The triangle with a "split" down the middle is one way of representing an isolation amplifier by indicating that the input and output stages are separated by transformer coupling.

The monitoring of fetal heartbeat, as illustrated, is the most demanding type of cardiac monitoring because in addition to the fetal heartbeat that typically generates 50 μV, there is also the mother's heartbeat that typically generates 1 mV. The common-mode voltages can run from about 1 mV to about 100 mV. The CMR (common-mode rejection) of the isolation amplifier separates the signal of the fetal heartbeat from that of the mother's heartbeat and from those common-mode signals. So, the signal from the fetal heartbeat is essentially all that the amplifier sends to the monitoring equipment.

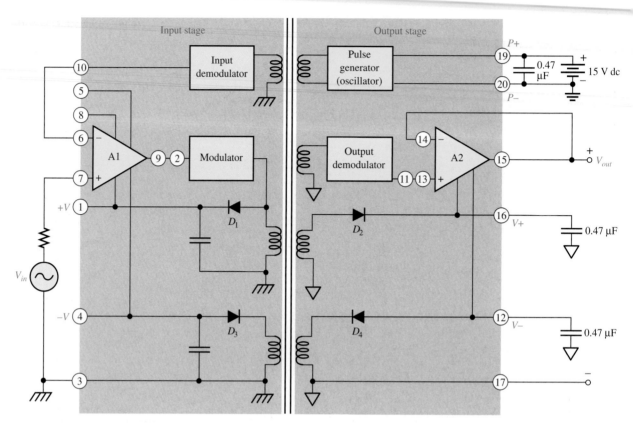

FIGURE 12–10
The 3656KG isolation amplifier.

A Specific Isolation Amplifier

Now that you know basically what an isolation amplifier is and what it does, let's look at a representative device, the Burr-Brown 3656KG.[2] As you can see in Figure 12–10, the 3656KG is similar to the basic isolation amplifier in Figure 12–8 except that it has a few more inputs and outputs and the oscillator is called a pulse generator. These additional pins provide for gain adjustments, offset adjustments, isolated dc voltage outputs, and other functions.

Isolated Power The 3656KG provides ±9 V from the filtered rectifiers (D_1 through D_4) at the +V and −V terminals for both the input and output stages. These voltages are used to power the internal circuits and are also available for powering external associated circuits such as preamplifiers and transducers. The power supply connections shown in Figure 12–10 show the three-port isolation configuration with the three separate grounds indicated by the different symbols. Filtering in the output stage is provided by two external capacitors. In the input stage, the filter capacitors are internal.

---

[2] Data sheet for 3656KG available at http://www.burr-brown.com

Instead of using the internally generated dc voltages, external supply voltages can be connected as long as they are greater than the internal voltages. This is necessary in order to reverse-bias the rectifier diodes and, essentially take them out of the picture.

Voltage Gain The voltage gains of both the input stage and the output stage can be set with external resistors, as shown in Figure 12–11. The gain of the input stage is

$$A_{v1} = \frac{R_{f1}}{R_{i1}} + 1 \qquad \text{(12–4)}$$

The gain of the output stage is

$$A_{v2} = \frac{R_{f2}}{R_{i2}} + 1 \qquad \text{(12–5)}$$

The total amplifier gain is the product of the gains of the input and output stages.

$$A_{v(tot)} = A_{v1}A_{v2}$$

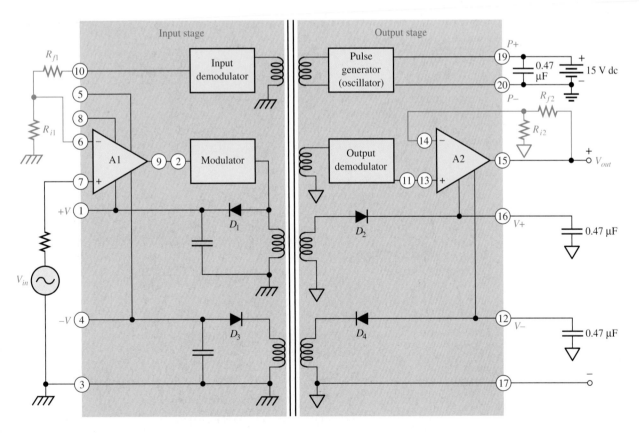

FIGURE 12–11
The 3656KG connected as a noninverting amplifier with external gain resistors (shown in color).

EXAMPLE 12–3   Determine the overall voltage gain of the 3656KG isolation amplifier in Figure 12–12.

FIGURE 12–12

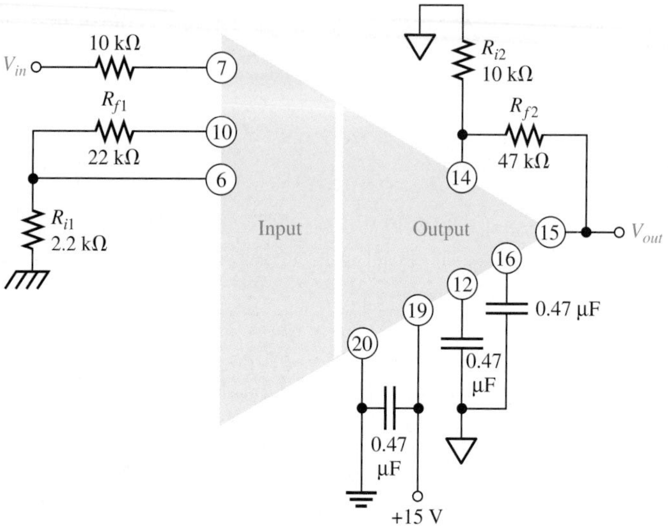

Solution   The gain of the input stage is

$$A_{v1} = \frac{R_{f1}}{R_{i1}} + 1 = \frac{22 \text{ k}\Omega}{2.2 \text{ k}\Omega} + 1 = 10 + 1 = 11$$

The gain of the output stage is

$$A_{v2} = \frac{R_{f2}}{R_{i2}} + 1 = \frac{47 \text{ k}\Omega}{10 \text{ k}\Omega} + 1 = 4.7 + 1 = 5.7$$

The total gain of the isolation amplifier is

$$A_{v(tot)} = A_{v1} A_{v2} = (11)(5.7) = \textbf{62.7}$$

Practice Exercise   Select resistor values in Figure 12–12 that will produce a total gain of approximately 100.

12–2 REVIEW QUESTIONS

1. In what types of applications are isolation amplifiers used?
2. What are the two stages in a typical isolation amplifier?
3. How are the stages in an isolation amplifier connected?
4. What is the purpose of the oscillator in an isolation amplifier?

12–3 ■ OPERATIONAL TRANSCONDUCTANCE AMPLIFIERS (OTAs)

Conventional op-amps are, as you know, primarily voltage amplifiers in which the output voltage equals the gain times the input voltage. The OTA is primarily a voltage-to-current amplifier in which the output current equals the gain times the input voltage.

After completing this section, you should be able to

❑ Understand and explain the operation of an operational transconductance amplifier (OTA)
  ❑ Identify the OTA symbol
  ❑ Discuss the relationship between transconductance and bias current
  ❑ Describe the features of the LM13700 OTA
  ❑ Discuss OTA applications

Figure 12–13 shows the symbol for an **operational transconductance amplifier (OTA)**. The double circle symbol at the output represents an output current source that is dependent on a bias current. Like the conventional op-amp, the OTA has two differential input terminals, a high input impedance, and a high CMRR. Unlike the conventional op-amp, the OTA has a bias-current input terminal, a high output impedance, and no fixed open-loop voltage gain.

FIGURE 12–13

Symbol for an operational transconductance amplifier (OTA).

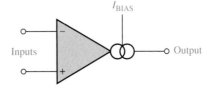

The Transconductance Is the Gain of an OTA

In general, the **transconductance** of an electronic device is the ratio of the output current to the input voltage. For an OTA, voltage is the input variable and current is the output variable; therefore, the ratio of output current to input voltage is its gain. Consequently, the voltage-to-current gain of an OTA is the transconductance, g_m.

$$g_m = \frac{I_{out}}{V_{in}}$$

In an OTA, the transconductance is dependent on a constant (K) times the bias current (I_{BIAS}) as indicated in Equation (12–6). The value of the constant is dependent on the internal circuit design.

$$g_m = KI_{BIAS} \qquad \textbf{(12–6)}$$

The output current is controlled by the input voltage and the bias current as shown by the following formula:

$$I_{out} = g_m V_{in} = KI_{BIAS}V_{in}$$

The Transconductance Is a Function of Bias Current

The relationship of the transconductance and the bias current in an OTA is an important characteristic. The graph in Figure 12–14 illustrates a typical relationship. Notice that the transconductance increases linearly with the bias current. The constant of proportionality, K, is the slope of the line and has a value of approximately 16 μS/μA.

FIGURE 12–14

Example of a transconductance versus bias current graph for a typical OTA.

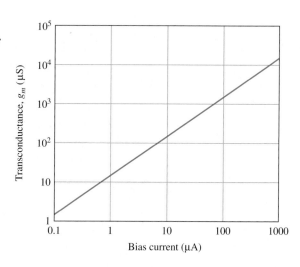

EXAMPLE 12–4

If an OTA has a g_m = 1000 μS, what is the output current when the input voltage is 50 mV?

Solution $I_{out} = g_m V_{in} = (1000 \text{ μS})(50 \text{ mV}) = \textbf{50 μA}$

Practice Exercise   Based on $K \cong 16$ μS/μA, calculate the bias current required to produce $g_m = 1000$ μS.

Basic OTA Circuits

Figure 12–15 shows the OTA used as an inverting amplifier with fixed-voltage gain. The voltage gain is set by the transconductance and the load resistance as follows.

$$V_{out} = I_{out} R_L$$

Dividing both sides by V_{in},

$$\frac{V_{out}}{V_{in}} = \left(\frac{I_{out}}{V_{in}} \right) R_L$$

Since V_{out}/V_{in} is the voltage gain and $I_{out}/V_{in} = g_m$,

$$A_v = g_m R_L$$

The transconductance of the amplifier in Figure 12–15 is determined by the amount of bias current, which is set by the dc supply voltages and the bias resistor R_{BIAS}.

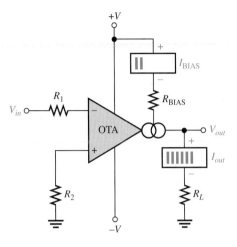

FIGURE 12–15

An OTA as an inverting amplifier with a fixed-voltage gain.

One of the most useful features of an OTA is that the voltage gain can be controlled by the amount of bias current. This can be done manually, as shown in Figure 12–16(a), by using a variable resistor in series with R_{BIAS} in the circuit of Figure 12–15. By changing the resistance, you can produce a change in I_{BIAS}, which changes the transconductance. A change in the transconductance changes the voltage gain. The voltage gain can also be controlled with an externally applied variable voltage as shown in Figure 12–16(b). A variation in the applied bias voltage causes a change in the bias current.

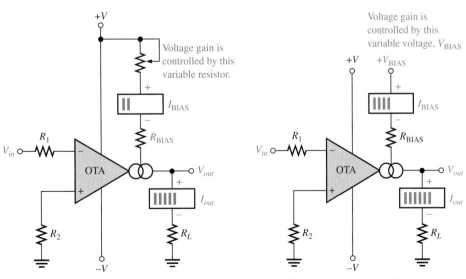

(a) Amplifier with resistance-controlled gain   (b) Amplifier with voltage-controlled gain

FIGURE 12–16

An OTA as an inverting amplifier with a variable-voltage gain.

A Specific OTA

The LM13700[3] is a typical OTA and serves as a representative device. The LM13700 is a dual-device package containing two OTAs and buffer circuits. Figure 12–17 shows the pin configuration using a single OTA in the package. The maximum dc supply voltages are ± 18 V, and its transconductance characteristic happens to be the same as indicated by the graph in Figure 12–14. For an LM13700, the bias current is determined by the following formula:

$$I_{BIAS} = \frac{+V_{BIAS} - (-V) - 1.4 \text{ V}}{R_{BIAS}} \qquad (12\text{–}7)$$

The 1.4 V is due to the internal circuit where a base-emitter junction and a diode connect the external R_{BIAS} with the negative supply voltage ($-V$). The positive bias voltage may be obtained from the positive supply voltage.

FIGURE 12–17

An LM13700 OTA. There are two in an IC package. The buffer transistors are not shown. Pin numbers for both OTAs are given in parentheses.

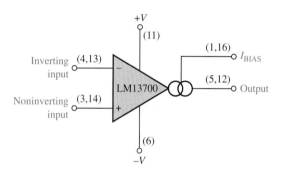

Not only does the transconductance of an OTA vary with bias current, but so does the input and output resistances. Both the input and output resistances decrease as the bias current increases, as shown in Figure 12–18.

---

[3] Data sheet for LM13700 available at http://www.national.com

FIGURE 12–18

Example of input and output resistances versus bias current.

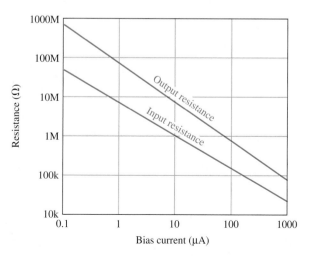

EXAMPLE 12–5   The OTA in Figure 12–19 is connected as an inverting fixed-gain amplifier. Determine the voltage gain.

FIGURE 12–19

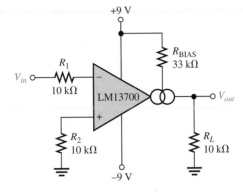

Solution   Calculate the bias current as follows:

$$I_{\text{BIAS}} = \frac{+V_{\text{BIAS}} - (-V) - 1.4 \text{ V}}{R_{\text{BIAS}}} = \frac{9 \text{ V} - (-9 \text{ V}) - 1.4 \text{ V}}{33 \text{ k}\Omega} = 503 \ \mu\text{A}$$

Using $K \cong 16 \ \mu\text{S}/\mu\text{A}$ from the graph in Figure 12–14, the value of transconductance corresponding to $I_{\text{BIAS}} = 503 \ \mu\text{A}$ is approximately

$$g_m = KI_{\text{BIAS}} \cong (16 \ \mu\text{S}/\mu\text{A})(503 \ \mu\text{A}) = 8.05 \times 10^3 \ \mu\text{S}$$

Using this value of g_m, calculate the voltage gain.

$$A_v = g_m R_L \cong (8.05 \times 10^3 \ \mu\text{S})(10 \text{ k}\Omega) = \mathbf{80.5}$$

Practice Exercise   If the OTA in Figure 12–19 is operated with dc supply voltages of ± 12 V, will this change the voltage gain and, if so, to what value?

Two OTA Applications

Amplitude Modulator   Figure 12–20 illustrates an OTA connected as an amplitude modulator. The voltage gain is varied by applying a modulation voltage to the bias input. When a constant-amplitude input signal is applied, the amplitude of the output signal will vary according to the modulation voltage on the bias input. The gain is dependent on bias current, and bias current is related to the modulation voltage by the following relationship:

$$I_{\text{BIAS}} = \frac{V_{mod} - (-V) - 1.4 \text{ V}}{R_{\text{BIAS}}}$$

This modulating action is shown in Figure 12–20 for a higher frequency sinusoidal input voltage and a lower frequency sinusoidal modulating voltage.

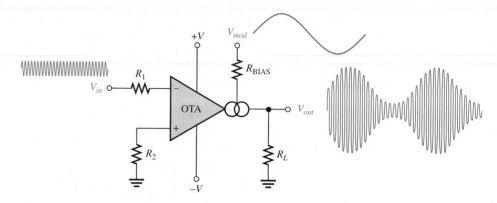

FIGURE 12–20
The OTA as an amplitude modulator.

EXAMPLE 12–6

The input to the OTA amplitude modulator in Figure 12–21 is a 50 mV peak-to-peak, 1 MHz sine wave. Determine the output signal, given the modulation voltage shown is applied to the bias input.

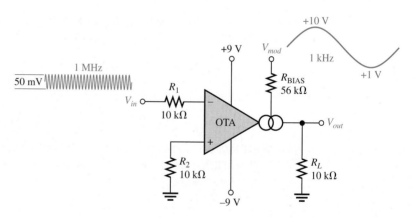

FIGURE 12–21

Solution   The maximum voltage gain is when I_{BIAS}, and thus g_m, is maximum. This occurs at the maximum peak of the modulating voltage, V_{mod}.

$$I_{BIAS(max)} = \frac{V_{mod(max)} - (-V) - 1.4 \text{ V}}{R_{BIAS}} = \frac{10 \text{ V} - (-9 \text{ V}) - 1.4 \text{ V}}{56 \text{ k}\Omega} = 314 \text{ μA}$$

From the graph in Figure 12–14, the constant K is approximately 16 μS/μA.

$$g_m = KI_{BIAS(max)} = (16 \text{ μS/μA})(314 \text{ μA}) = 5.02 \text{ mS}$$
$$A_{v(max)} = g_m R_L = (5.02 \text{ mS})(10 \text{ k}\Omega) = 50.2$$
$$V_{out(max)} = A_{v(min)} V_{in} = (50.2)(50 \text{ mV}) = 2.51 \text{ V}$$

The minimum bias current is

$$I_{\text{BIAS(min)}} = \frac{V_{mod(min)} - (-V) - 1.4 \text{ V}}{R_{\text{BIAS}}} = \frac{1 \text{ V} - (-9 \text{ V}) - 1.4 \text{ V}}{56 \text{ k}\Omega} = 154 \text{ μA}$$

$$g_m = KI_{\text{BIAS(min)}} = (16 \text{ μS/μA})(154 \text{ μA}) = 2.46 \text{ mS}$$

$$A_{v(min)} = g_m R_L = (2.46 \text{ mS})(10 \text{ k}\Omega) = 24.6$$

$$V_{out(min)} = A_{v(min)} V_{in} = (24.6)(50 \text{ mV}) = 1.23 \text{ V}$$

The resulting output voltage is shown in Figure 12–22.

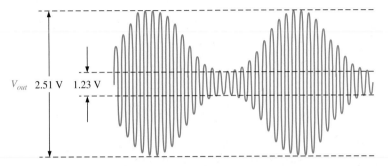

FIGURE 12–22

Practice Exercise Repeat this example with the sinusoidal modulating signal replaced by a square wave with the same maximum and minimum levels and a bias resistor of 39 kΩ.

Schmitt Trigger Figure 12–23 shows an OTA used in a Schmitt-trigger configuration. (Refer to Section 8–1.) Basically, a Schmitt trigger is a comparator with hysteresis where the input voltage drives the device into either positive or negative saturation. When the input voltage exceeds a certain threshold value or trigger point, the device switches to one of its saturated output states. When the input falls back below another threshold value, the device switches back to its other saturated output state.

FIGURE 12–23
The OTA as a Schmitt trigger.

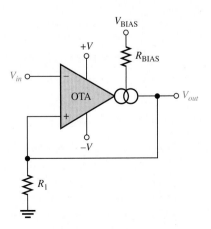

In the case of the OTA Schmitt trigger, the threshold levels are set by the current through resistor R_1. The maximum output current in an OTA equals the bias current. Therefore, in the saturated output states, $I_{out} = I_{BIAS}$. The maximum positive output voltage is $I_{out}R_1$, and this voltage is the positive threshold value or upper trigger point. When the input voltage exceeds this value, the output switches to its maximum negative voltage, which is $-I_{out}R_1$. Since $I_{out} = I_{BIAS}$, the trigger points can be controlled by the bias current. Figure 12–24 illustrates this operation.

FIGURE 12–24

Basic operation of the OTA Schmitt trigger.

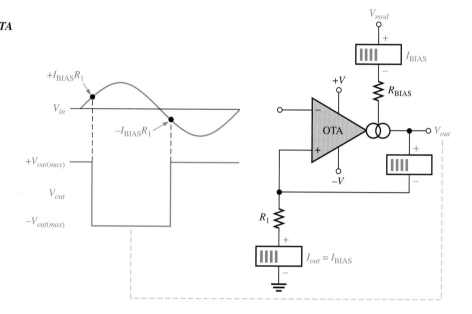

12–3 REVIEW QUESTIONS

1. What does OTA stand for?

2. If the bias current in an OTA is increased, does the transconductance increase or decrease?

3. What happens to the voltage gain if the OTA is connected as a fixed-voltage amplifier and the supply voltages are increased?

4. What happens to the voltage gain if the OTA is connected as a variable-gain voltage amplifier and the voltage at the bias terminal is decreased?

12–4 ■ LOG AND ANTILOG AMPLIFIERS

A logarithmic (log) amplifier produces an output that is proportional to the logarithm of the input. Log amplifiers are used in applications that require compression of analog input data, linearization of transducers that have exponential outputs, optical density measurements and more. An antilogarithmic (antilog) amplifier takes the antilog or inverse log of the input. In this section, the principles of these amplifiers are discussed.

After completing this section, you should be able to

❑ Understand and explain the operation of log and antilog amplifiers
    ❑ Define *logarithm* and *natural logarithm*
    ❑ Describe the feedback configurations
    ❑ Discuss signal compression with logarithmic amplifiers

Logarithms

A **logarithm** (log) is basically a power. It is defined as the power to which a base, *b,* must be raised to yield a particular number, *N*. The defining formula for a logarithm is

$$b^x = N$$

In this formula, *x* represents the log of *N*. For example, you know that $10^2 = 100$. In this example, 2 is the power of ten that yields the number 100. In other words, 2 is the log of 100 (with base ten implied).

There are two practical bases used for logarithms. Base ten is used for what are called common logs because our counting system is base ten. The abbreviation *log* in a mathematical expression or on your calculator implies base ten. Sometimes the subscript 10 is included with the abbreviation as $\log_{10}$. The second base is derived from an important mathematical series which gives the number 2.71828.[4] This number is represented by the letter *e* (mathematicians use ϵ). Base *e* is used because it is part of mathematical equations that describe natural phenomena such as the charging and discharging of a capacitor and the relationship between voltage and current in certain semiconductor devices. Logarithms that use base *e* are said to be **natural logarithms** and are shown with the abbreviation *ln* in mathematical expressions and on your calculator.

A useful conversion between the two bases is given by the equation

$$\ln x = 2.303 \log_{10} x$$

The Basic Logarithmic Amplifier

A log amp produces an output that is proportional to the logarithm of the input voltage. The key element in a basic log amplifier is a semiconductor *pn* junction in the form of either a diode or the base-emitter junction of a bipolar transistor. A *pn* junction exhibits a natural logarithmic current for many decades of input voltage. Figure 12–25(a) shows this characteristic for a typical small-signal diode, plotted as a linear plot; and Figure 12–25(b) shows the same characteristic plotted as a log plot (the *y*-axis is logarithmic). I_D is the forward diode current and V_D is the forward diode voltage. The logarithmic relationship between diode current and voltage is clearly seen in the plot in part (b). Although the plot only shows four decades of data, the actual logarithmic relationship for a diode extends over seven decades! The relationship between the current and voltage is expressed by the following general equation for a diode:

$$V_D = K \ln\left(\frac{I_D}{I_R}\right)$$

---

[4] The series is $e = \lim\limits_{n \to \inf}\left(1 + \dfrac{1}{n}\right)^n$.

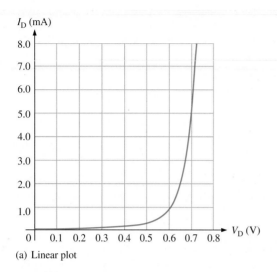

(a) Linear plot

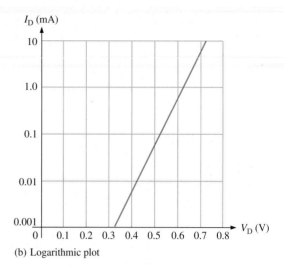

(b) Logarithmic plot

FIGURE 12–25
The characteristic curve for a typical diode.

In this equation, K is a constant that is determined by several factors including the temperature and is approximately 0.025 V at 25°C. I_R, the reverse leakage current, is a constant for a given diode.

Log Amplifier with a Diode When the feedback resistor in an inverting amplifier is replaced with a diode, the result is a basic log amp, as shown in Figure 12–26. The output voltage, V_{out}, is equal to $-V_D$. Because of the virtual ground, the input current can be expressed as V_{in}/R_1. By substituting these quantities into the diode equation, the output voltage is

$$V_{out} \cong -(0.025 \text{ V})\ln\left(\frac{V_{in}}{I_R R_1}\right) \tag{12–8}$$

From Equation (12–8), you can see that the output voltage is the negative of a logarithmic function of the input voltage. The value of the output is controlled by the value of the input voltage and the value of the resistor R_1.

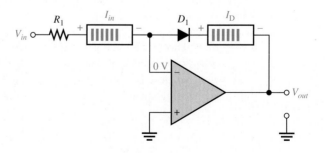

FIGURE 12–26
A basic log amplifier using a diode as the feedback element.

EXAMPLE 12–7   Determine the output voltage for the log amplifier in Figure 12–27. Assume $I_R = 50$ nA.

FIGURE 12–27

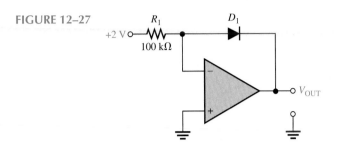

Solution   The input voltage and the resistor value are given in Figure 12–27.

$$V_{OUT} = -(0.025 \text{ V})\ln\left(\frac{V_{in}}{I_R R_1}\right) = -(0.025 \text{ V})\ln\left(\frac{2 \text{ V}}{(50 \text{ nA})(100 \text{ k}\Omega)}\right)$$

$$= -(0.025 \text{ V})\ln(400) = -(0.025 \text{ V})(5.99) = \mathbf{-0.150 \text{ V}}$$

Practice Exercise   Calculate the output voltage of the log amplifier with a +4 V input.

Log Amplifier with a BJT   The base-emitter junction of a bipolar transistor exhibits the same type of natural logarithmic characteristic as a diode because it is also a *pn* junction. A log amplifier with a BJT connected in a common-base form in the feedback loop is shown in Figure 12–28. Notice that V_{out} with respect to ground is equal to $-V_{BE}$.

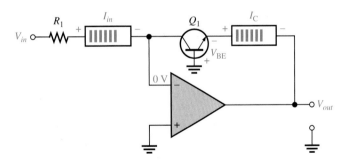

FIGURE 12–28
A basic log amplifier using a transistor as the feedback element.

The analysis for this circuit is the same as for the diode log amplifier except that $-V_{BE}$ replaces V_D, I_C replaces I_D, and I_{EBO} replaces I_R. The emitter-to-base leakage current is I_{EBO}. The expression for the output voltage is

$$V_{out} = -(0.025 \text{ V})\ln\left(\frac{V_{in}}{I_{EBO} R_1}\right) \qquad \textbf{(12–9)}$$

EXAMPLE 12–8   What is V_{out} for a transistor log amplifier with $V_{in} = 3$ V and $R_1 = 68$ kΩ? Assume $I_{EBO} = 40$ nA.

Solution
$$V_{out} = -(0.025 \text{ V})\ln\left(\frac{V_{in}}{I_{EBO}R_1}\right) = -(0.025 \text{ V})\ln\left(\frac{3 \text{ V}}{(40 \text{ nA})(68 \text{ kΩ})}\right)$$
$$= -(0.025 \text{ V})\ln(1103) = \mathbf{-0.175 \text{ V}}$$

Practice Exercise   Calculate V_{out} if R_1 is changed to 33 kΩ.

The Basic Antilog Amplifier

The antilog amplifier is the complement of a log amplifier. If you know the logarithm of a number, you know the power the base is raised to. To obtain the **antilog**, you must take the *exponential* of the logarithm.

$$x = e^{\ln x}$$

This is equivalent to saying the antilog$_e$ of ln x is just x. (Notice that the antilog is base e in this statement.) On many calculators the antilog of a base 10 logarithm is labelled $\boxed{10^x}$ and in some cases $\boxed{\text{INV}}\boxed{\text{LOG}}$. The antilog of a base e logarithm is labelled $\boxed{e^x}$ or $\boxed{\text{INV}}\boxed{\text{LN}}$.

The basic antilog amplifier is formed by reversing the position of the transistor (or diode) with the resistor in the log amp circuit. The antilog circuit is shown in Figure 12–29 using a transistor base-emitter junction as the input element and a resistor as the feedback element. The relationship between the current and voltage for a diode still applies.

$$V_D = K \ln\left(\frac{I_D}{I_R}\right)$$

FIGURE 12–29
A basic antilog amplifier.

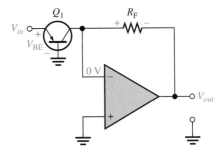

For the antilog amplifier, V_D is the negative input voltage and I_D represents the current in the feedback resistor, which by Ohm's law is V_{out}/R_F. Since a transistor is used, $I_R = I_{EBO}$. Making these substitutions in the diode equation,

$$V_{in} = -K \ln\left(\frac{V_{out}}{I_{EBO}R_F}\right)$$

Rearranging,

$$V_{out} = -I_{EBO}R_F e^{V_{in}/K}$$

Substituting $K \cong 0.025$ V and clearing the exponent,

$$V_{out} \cong -I_{\text{EBO}}R_{\text{F}}\text{antilog}_e\left(\frac{V_{in}}{25 \text{ mV}}\right) \qquad (12\text{--}10)$$

EXAMPLE 12–9

For the antilog amplifier in Figure 12–30, find the output voltage. Assume $I_{\text{EBO}} = 40$ nA.

FIGURE 12–30

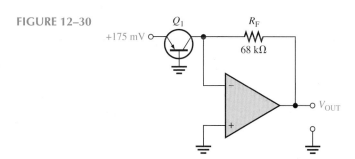

Solution First of all, notice that the input voltage in Figure 12–30 is the inverted output voltage of the log amplifier in Example 12–8. In this case, the antilog amplifier reverses the process and produces an output that is proportional to the antilog of the input. So, the output voltage of the antilog amplifier in Figure 12–30 should have the same magnitude as the input voltage of the log amplifier in Example 12–8 because all the constants are the same. Let's see if it does.

$$V_{\text{OUT}} \cong -I_{\text{EBO}}R_{\text{F}}\text{antilog}_e\left(\frac{V_{in}}{25 \text{ mV}}\right) = -(40 \text{ nA})(68 \text{ k}\Omega)\text{antilog}_e\left(\frac{0.175 \text{ V}}{25 \text{ mV}}\right)$$

$$= -(40 \text{ nA})(68 \text{ k}\Omega)(1100) = -3 \text{ V}$$

Practice Exercise Determine V_{OUT} for the amplifier in Figure 12–30 if the feedback resistor is changed to 100 kΩ.

IC Log, Log-Ratio, and Antilog Amplifiers

Several factors make the basic log amp and the basic antilog amp circuit with a diode and op-amp unsatisfactory for many applications. The basic circuit is temperature sensitive and tends to have error at very low diode currents; also, components need to be precisely matched, and the output level is not a convenient value. These problems are difficult to address with off-the-shelf components; however, manufacturers have designed precision integrated circuit logarithmic and log-ratio amplifiers with temperature compensation, low bias currents, and high accuracy that require no user adjustments. Log-ratio measurements produce an output that is proportional to the log ratio of *two* inputs.

The Burr-Brown LOG100[5] is an example of a log, log-ratio, and antilog amplifier in one 14 pin IC. It has a maximum accuracy specification of 0.37% at full-scale output

---

[5] Data sheet for LOG100 available at http://www.burr-brown.com

(FSO) and a six-decade range of input current (1 nA to 1 mA). With a few external resistors, the user can connect the amplifier as either a log, a log-ratio, or an antilog amplifier. Scaling of the output voltage can be done by simply selecting an appropriate output pin. As with most log amps, the input will function with only one polarity of input current.

Another interesting device from Burr-Brown is the 24 pin 4127[6] log amp that features the ability to accept input voltages or currents of *either* polarity, the first log amp to do so. It maintains high accuracy over a six-decade range of input current (1 nA to 1 mA) or a four-decade range of input voltage. Only a few external resistors are required to complete a log, log-ratio, or an antilog amplifier. Within this IC is an uncommitted op-amp that can be used for any purpose such as a gain stage, buffer, filter, or inverter.

Signal Compression with Logarithmic Amplifiers

In certain applications, a signal may be too large in magnitude for a particular system to handle. The term *dynamic range* is often used to describe the range of voltages contained in a signal. In these cases, the signal voltage must be scaled down by a process called **signal compression** so that it can be properly handled by the system. If a linear circuit is used to scale a signal down in amplitude, the lower voltages are reduced by the same percentage as the higher voltages. Linear signal compression often results in the lower voltages becoming obscured by noise and difficult to accurately distinguish, as illustrated in Figure 12–31(a). To overcome this problem, a signal with a large dynamic range can be com-

---

[6] Data sheet for 4127 available at http://www.burr-brown.com

FIGURE 12–31
The basic concept of signal compression with a logarithmic amplifier.

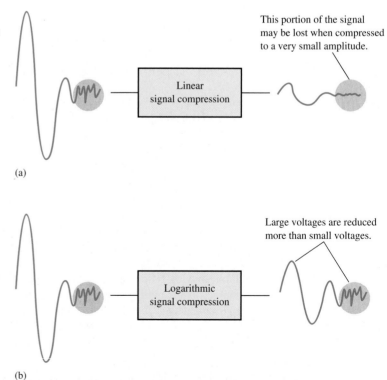

pressed using a logarithmic response, as shown in Figure 12–31(b). In logarithmic signal compression, the higher voltages are reduced more than the lower voltages, thus keeping the lower voltage signals from being lost in noise. A log amplifier preceding an 8-bit ADC can replace a more expensive 20-bit ADC because of signal compression.

A Basic Multiplier with Log and Antilog Amps

Multipliers are based on the fundamental logarithmic relationship that states that the product of two terms equals the sum of the logarithms of each term. This relationship is shown in the following formula:

$$\ln(a \times b) = \ln a + \ln b$$

This formula shows that two signal voltages are effectively multiplied if the logarithms of the signal voltages are added.

You know how to get the logarithm of a signal voltage by using a log amplifier. By summing the outputs of two log amplifiers, you get the logarithm of the product of the two original input voltages. Then, by taking the antilogarithm, you get the product of the two input voltages as indicated in the following equations:

$$\ln V_1 + \ln V_2 = \ln(V_1 V_2)$$
$$\text{antilog}_e[\ln(V_1 V_2)] = V_1 V_2$$

The block diagram in Figure 12–32 shows how the functions are connected to multiply two input voltages. Constant terms are omitted for simplicity.

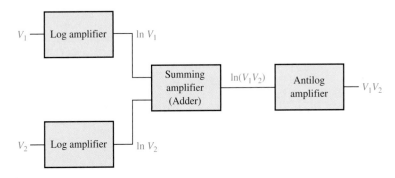

FIGURE 12–32
Basic block diagram of an analog multiplier.

Figure 12–33 shows the basic multiplier circuitry. The outputs of the log amplifiers are stated as follows:

$$V_{out(log1)} = -K_1 \ln\left(\frac{V_{in1}}{K_2}\right)$$

$$V_{out(log2)} = -K_1 \ln\left(\frac{V_{in2}}{K_2}\right)$$

where $K_1 = 0.025$ V, $K_2 = RI_{EBO}$, and $R = R_1 = R_2 = R_6$. The two output voltages from

FIGURE 12–33
A basic multiplier.

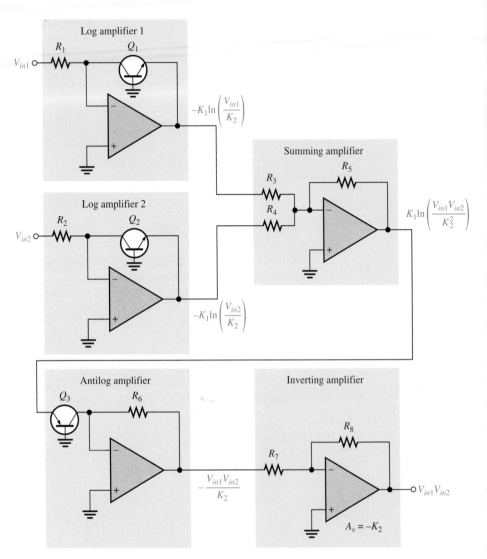

the log amplifiers are added and inverted by the unity-gain summing amplifier to produce the following result:

$$V_{out(sum)} = K_1\ln\left[\left(\frac{V_{in1}}{K_2}\right) + \ln\left(\frac{V_{in2}}{K_2}\right)\right] = K_1\ln\left(\frac{V_{in1}V_{in2}}{K_2^2}\right)$$

This expression is then applied to the antilog amplifier; the expression for the multiplier output voltage is as follows:

$$V_{out(antilog)} = -K_2\text{antilog}_e\left(\frac{V_{out(sum)}}{K_1}\right) = -K_2\text{antilog}_e\left[\frac{K_1\ln\left(\frac{V_{in1}V_{in2}}{K_2^2}\right)}{K_1}\right]$$

$$= -K_2\left(\frac{V_{in1}V_{in2}}{K_2^2}\right) = -\frac{V_{in1}V_{in2}}{K_2}$$

As you can see, the output of the antilog amplifier is a constant $(1/K_2)$ times the *product* of the input voltages. The final output is developed by an inverting amplifier with a voltage gain of $-K_2$.

$$V_{out} = -K_2\left(-\frac{V_{in1}V_{in2}}{K_2}\right)$$

$$V_{out} = V_{in1}V_{in2} \tag{12–11}$$

As in the case of log amps, analog multipliers are available in IC form. These are covered in Chapter 13.

12–4 REVIEW QUESTIONS

1. What purpose does the diode or transistor perform in the feedback loop of a log amplifier?
2. Why is the output of a basic log amplifier limited to about 0.7 V?
3. What are the factors that determine the output voltage of a basic log amplifier?
4. In terms of implementation, how does a basic antilog amplifier differ from a basic log amplifier?
5. What circuits make up a basic analog multiplier?

12–5 ■ A SYSTEM APPLICATION

The electrocardiograph, described at the beginning of the chapter, is a medical instrument used for monitoring heart signals of patients. From the output waveform of an ECG, the doctor can detect abnormalities in the heartbeat.

After completing this section, you should be able to

❏ Apply what you have learned in this chapter to a system application
  ❏ Discuss how an isolation amplifier is used in medical instrumentation
  ❏ Describe how other op-amp circuits are also used as part of the ECG system
  ❏ Translate between a printed circuit board and a schematic
  ❏ Analyze the amplifier board
  ❏ Troubleshoot some common problems

A Brief Description of the System

The human heart produces an electrical signal that can be picked up by electrodes in contact with the skin. When the heart signal is displayed on a chart recorder or on a video monitor, it is called an electrocardiograph or ECG. Typically, the heart signal picked up by the electrode is about 1 mV and has significant frequency components from less than 1 Hz to about 100 Hz.

As indicated in the block diagram in Figure 12–34, an ECG system has at least three electrodes. There is a right-arm (RA) electrode, a left-arm (LA) electrode, and a right-leg

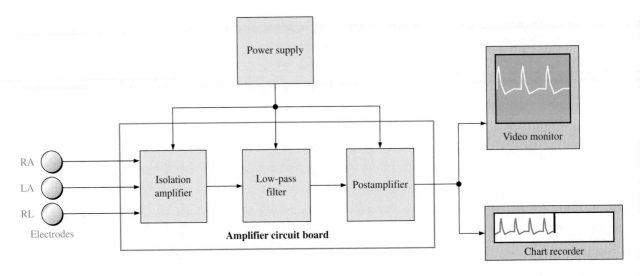

FIGURE 12–34
ECG block diagram.

(RL) electrode that is used as the common. The isolation amplifier provides for differential inputs from the electrodes, provides a high CMR to eliminate the relatively high common-mode noise voltages associated with heart signals, and provides electrical isolation for protection of the patient. The low-pass active filter rejects frequencies above those contained in the heart signal. The postamplifier provides most of the amplification in the system and drives a video monitor and/or a chart recorder. The three op-amp circuits—the isolation amplifier, the low-pass filter, and the postamplifier—are on a single PC board called the amplifier board.

Now, so that you can take a closer look at the amplifier board, let's take it out of the system and put it on the troubleshooter's bench.

TROUBLESHOOTER'S BENCH

A Brief Description of the Circuits

The inputs from the electrode sensors come into the amplifier board shown in Figure 12–35 with a shielded cable to prevent noise pickup. The schematic for the amplifier board is shown in Figure 12–36. The shielded cable is basically a twisted pair of wires surrounded by a braided metal sheathing that is covered by an insulated sheathing. The braided metal shield serves as the conduit for the common connection. The incoming differential signal is amplified by the fixed gain of the 3656KG isolation amplifier. The 3656KG package in this system is a 20-pin package beginning with pin 1 at the "square" corner. Pin 20 is directly across from pin 1.

The low-pass filter is a Sallen-Key two-pole filter, and the postamplifier is an inverting amplifier with adjustable gain. The inverting input of the postamplifier also serves as a summing point for the signal voltage and a dc voltage used for adding a dc level to the output for purposes of adjusting the vertical position of the display.

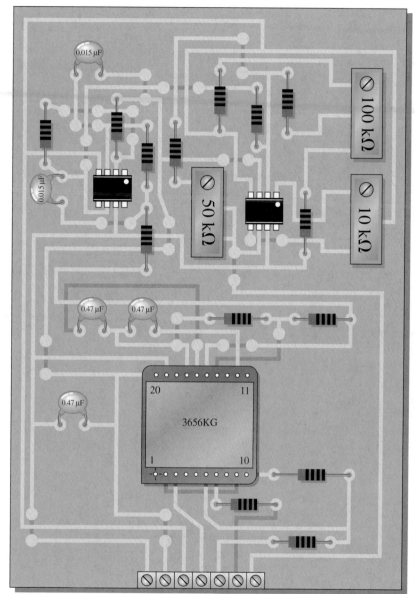

+15 GND –15 RL LA RA Output

FIGURE 12–35

■ ACTIVITY 1 Relate the PC Board to the Schematic

Locate and identify each component and each input/output pin on the PC board in Figure 12–35 using the schematic in Figure 12–36. Verify that the board and the schematic agree.

■ ACTIVITY 2 Analyze the System

Step 1: Determine the voltage gain of the isolation amplifier.

Step 2: Determine the bandwidth of the active filter.

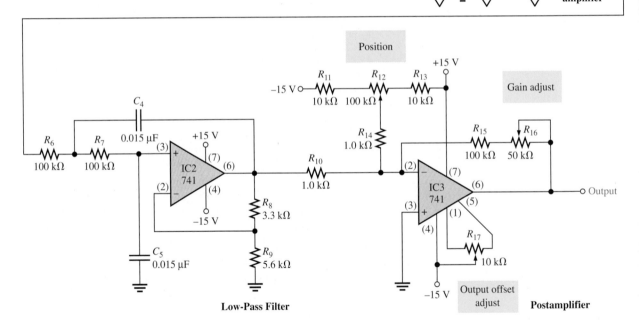

FIGURE 12–36

Step 3: Determine the minimum and maximum voltage gain of the postamplifier.

Step 4: Determine the overall gain range of the amplifier board.

Step 5: Determine the voltage range at the wiper of the position adjustment potentiometer.

■ **ACTIVITY 3** **Write a Technical Report**

Describe the overall operation of the amplifier board. Specify how each circuit works and what its purpose is. Explain how the gain is adjusted and how the dc level of the output can be changed. Use the results of Activity 2 as appropriate.

■ **ACTIVITY 4 Troubleshoot the System for Each of the Following Problems by Stating the Probable Cause or Causes**

1. There is no final output voltage when there is a verified 1 mV input signal.
2. There is a 10 mV signal at the output of IC1, but no signal at pin 3 of IC2.
3. There is a 15 mV signal at the output of IC2, but no signal at pin 2 of IC3.
4. With a valid input signal, IC3 is being driven into its saturated states and is basically acting as a comparator.

12–5 REVIEW QUESTIONS

1. Which resistors in Figure 12–36 determine the voltage gain of the isolation amplifier?
2. What is the voltage gain of the output stage of IC1?
3. What are the lower and upper critical frequencies of the active filter IC2?
4. What is the gain of the postamplifier IC3 if R_{16} is set at its midpoint resistance?

■ SUMMARY

- A basic instrumentation amplifier is formed by three op-amps and seven resistors, including the gain-setting resistor, R_G.
- An instrumentation amplifier has high input impedance, high CMRR, low output offset, and low output impedance.
- The voltage gain of a basic instrumentation amplifier is set by a single external resistor.
- An instrumentation amplifier is useful in applications where small signals are embedded in large common-mode noise.
- A basic isolation amplifier has three electrically isolated parts: input, output, and power.
- Most isolation amplifiers use transformer coupling for isolation.
- Isolation amplifiers are used to interface sensitive equipment with high-voltage environments and to provide protection from electrical shock in certain medical applications.
- The operational transconductance amplifier (OTA) is a voltage-to-current amplifier.
- The output current of an OTA is the input voltage times the transconductance.
- In an OTA, transconductance varies with bias current; therefore, the gain of an OTA can be varied with a bias voltage or a variable resistor.
- The operation of log and antilog amplifiers is based on the nonlinear (logarithmic) characteristic of a *pn* junction.
- A log amplifier has a BJT in the feedback loop.
- An antilog amplifier has a BJT in series with the input.
- Logarithmic amplifiers are used for signal compression, analog multiplication, and log-ratio measurements.
- An analog multiplier is based on the mathematical principle that states the logarithm of the product of two variables equals the sum of the logarithms of the variables.

■ GLOSSARY

Key terms are in color. All terms are included in the end-of-book glossary.

Antilog The number corresponding to a given logarithm.

Instrumentation amplifier A differential voltage-gain device that amplifies the differences between the voltages existing at its two input terminals.

Isolation amplifier   An amplifier in which the input and output stages are not electrically connected.

Logarithm   An exponent; the logarithm of a quantity is the exponent or power to which a given number called the base must be raised in order to equal the quantity.

Natural logarithm   The exponent to which the base e ($e = 2.71828$) must be raised in order to equal a given quantity.

Operational transconductance amplifier   An amplifier in which the output current is the gain times the input voltage.

Signal compression   The process of scaling down the amplitude of a signal voltage.

Transconductance   The ratio of output current to input voltage.

■ KEY FORMULAS

Instrumentation Amplifier

(12–1) $$V_{out} = \left(\frac{1 + 2R}{R_G}\right)(V_{in2} - V_{in1})$$

(12–2) $$R_G = \frac{2R}{A_{cl} - 1}$$

(12–3) $$R_G = \frac{50.5 \text{ k}\Omega}{A_v - 1} \qquad \text{(for the AD622)}$$

Isolation Amplifier

(12–4) $$A_{v(input)} = \frac{R_{f1}}{R_{i1}} + 1$$

(12–5) $$A_{v(output)} = \frac{R_{f2}}{R_{i2}} + 1$$

Operational Transconductance Amplifier (OTA)

(12–6) $$g_m = KI_{BIAS}$$

(12–7) $$I_{BIAS} = \frac{+V_{BIAS} - (-V) - 1.4 \text{ V}}{R_{BIAS}} \qquad \text{(for the LM13700)}$$

Logarithmic Amplifier

(12–8) $$V_{out} \cong -(0.025 \text{ V})\ln\left(\frac{V_{in}}{I_R R_1}\right)$$

(12–9) $$V_{out} = -(0.025 \text{ V})\ln\left(\frac{V_{in}}{I_{EBO} R_1}\right)$$

(12–10) $$V_{out} = -I_{EBO} R_F \text{antilog}_e\left(\frac{V_{in}}{25 \text{ mV}}\right)$$

(12–11) $$V_{out} = V_{in1} V_{in2}$$

■ SELF-TEST

Answers are at the end of the chapter.

 1. To make a basic instrumentation amplifier, it takes
   (a) one op-amp with a certain feedback arrangement
   (b) two op-amps and seven resistors

   (c) three op-amps and seven capacitors
   (d) three op-amps and seven resistors

2. Typically, an instrumentation amplifier has an external resistor used for
   (a) establishing the input impedance   (b) setting the voltage gain
   (c) setting the current gain   (d) for interfacing with an instrument

3. Instrumentation amplifiers are used primarily in
   (a) high-noise environments   (b) medical equipment
   (c) test instruments   (d) filter circuits

4. Isolation amplifiers are used primarily in
   (a) remote, isolated locations
   (b) systems that isolate a single signal from many different signals
   (c) applications where there are high voltages and sensitive equipment
   (d) applications where human safety is a concern
   (e) answers (c) and (d)

5. The three parts of a basic isolation amplifier are
   (a) amplifier, filter, and power   (b) input, output, and coupling
   (c) input, output, and power   (d) gain, attenuation, and offset

6. The stages of most isolation amplifiers are connected by
   (a) copper strips   (b) transformers
   (c) microwave links   (d) current loops

7. The characteristic that allows an isolation amplifier to amplify small signal voltages in the presence of much greater noise voltages is its
   (a) CMRR   (b) high gain
   (c) high input impedance   (d) magnetic coupling between input and output

8. The term *OTA* means
   (a) operational transistor amplifier
   (b) operational transformer amplifier
   (c) operational transconductance amplifier
   (d) output transducer amplifier

9. In an OTA, the transconductance is controlled by
   (a) the dc supply voltage   (b) the input signal voltage
   (c) the manufacturing process   (d) a bias current

10. The voltage gain of an OTA circuit is set by
   (a) a feedback resistor
   (b) the transconductance only
   (c) the transconductance and the load resistor
   (d) the bias current and supply voltage

11. An OTA is basically a
   (a) voltage-to-current amplifier   (b) current-to-voltage amplifier
   (c) current-to-current amplifier   (d) voltage-to-voltage amplifier

12. The operation of a logarithmic amplifier is based on
   (a) the nonlinear operation of an op-amp
   (b) the logarithmic characteristic of a *pn* junction
   (c) the reverse breakdown characteristic of a *pn* junction
   (d) the logarithmic charge and discharge of an *RC* circuit

13. If the input to a log amplifier is *x,* the output is proportional to
   (a) e^x (b) $\ln x$ (c) $\log_{10}x$ (d) $2.3 \log_{10}x$
   (e) answers (a) and (c)   (f) answers (b) and (d)

14. If the input to an antilog amplifier is x, the output is proportional to

    (a) $e^{\ln x}$ **(b)** e^x **(c)** $\ln x$ **(d)** e^{-x}

15. The logarithm of the product of two numbers is equal to the

    (a) sum of the two numbers

    (b) sum of the logarithms of each of the numbers

    (c) difference of the logarithms of each of the numbers

    (d) ratio of the logarithms of the numbers

TROUBLESHOOTER'S QUIZ   *Answers are at the end of the chapter.*

Refer to Figure 12–37.

❑ If R_5 opens,

    1. The output signal voltage will

        (a) increase   **(b)** decrease   **(c)** not change

❑ If the output of op-amp 3 opens,

    2. The output signal voltage will

        (a) increase   **(b)** decrease   **(c)** not change

Refer to Figure 12–38.

❑ If R_G opens,

    3. The voltage gain will

        (a) increase   **(b)** decrease   **(c)** not change

❑ If the value of R_G is larger than the specified value,

    4. The bandwidth will

        (a) increase   **(b)** decrease   **(c)** not change

Refer to Figure 12–39(a).

❑ If there is a short across the 18 kΩ resistor,

    5. The voltage gain will

        (a) increase   **(b)** decrease   **(c)** not change

❑ If the 150 kΩ resistor shorts,

    6. The output signal voltage will

        (a) increase   **(b)** decrease   **(c)** not change

❑ If the capacitor between pins 19 and 20 is open,

    7. The gain will

        (a) increase   **(b)** decrease   **(c)** not change

Refer to Figure 12–40.

❑ If the value of R_{BIAS} is less than the specified value,

    8. The output voltage will

        (a) increase   **(b)** decrease   **(c)** not change

☐ If the negative dc supply voltage decreases (less negative),

  9. The output voltage will

     (a) increase   **(b)** decrease   **(c)** not change

☐ If the positive supply voltage increases,

  10. The transconductance will

     (a) increase   **(b)** decrease   **(c)** not change

■ PROBLEMS

Answers to odd-numbered problems are at the end of the book.

SECTION 12–1 Instrumentation Amplifiers

1. Determine the voltage gains of op-amps A1 and A2 for the instrumentation amplifier configuration in Figure 12–37.

2. Find the overall voltage gain of the instrumentation amplifier in Figure 12–37.

3. The following voltages are applied to the instrumentation amplifier in Figure 12–37. $V_{in1} = 5$ mV, $V_{in2} = 10$ mV, and $V_{cm} = 225$ mV. Determine the final output voltage.

4. What value of R_G must be used to change the gain of the instrumentation amplifier in Figure 12–37 to 1000?

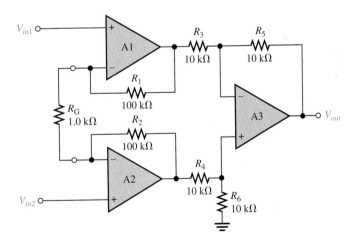

FIGURE 12–37

5. What is the voltage gain of the AD622 instrumentation amplifier in Figure 12–38?

6. Determine the approximate bandwidth of the amplifier in Figure 12–38 if the voltage gain is set to 10. Use the graph in Figure 12–6.

7. Specify what you must do to change the gain of the amplifier in Figure 12–38 to approximately 24.

8. Determine the value of R_G in Figure 12–38 for a voltage gain of 20.

FIGURE 12–38

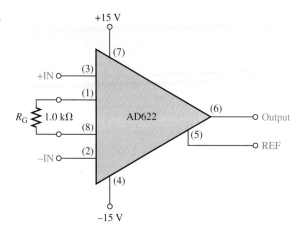

FIGURE 12–38

SECTION 12–2   Isolation Amplifiers

9. The op-amp in the input stage of a certain isolation amplifier has a voltage gain of 30. The output stage is set for a gain of 10. What is the overall voltage gain of this device?

10. Determine the overall voltage gain of each 3656KG in Figure 12–39.

11. Specify how you would change the overall gain of the amplifier in Figure 12–39(a) to approximately 100 by changing only the gain of the input stage.

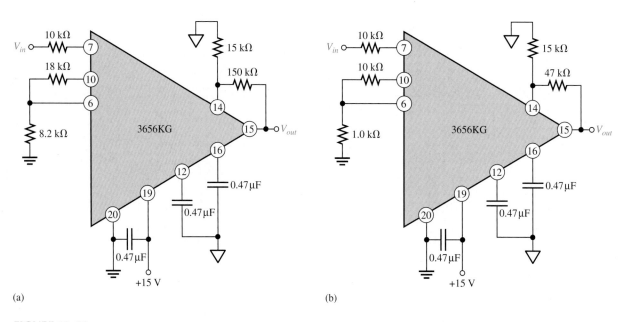

(a)   (b)

FIGURE 12–39

12. Specify how you would change the overall gain in Figure 12–39(b) to approximately 440 by changing only the gain of the output stage.

13. Specify how you would connect each amplifier in Figure 12–39 for unity gain.

SECTION 12–3 Operational Transconductance Amplifiers (OTAs)

14. A certain OTA has an input voltage of 10 mV and an output current of 10 μA. What is the transconductance?

15. A certain OTA with a transconductance of 5000 μS has a load resistance of 10 kΩ. If the input voltage is 100 mV, what is the output current? What is the output voltage?

16. The output voltage of a certain OTA with a load resistance is determined to be 3.5 V. If its transconductance is 4000 μS and the input voltage is 100 mV, what is the value of the load resistance?

FIGURE 12–40

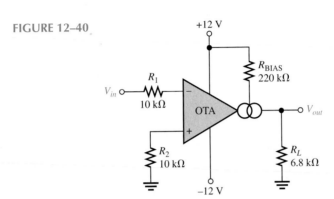

17. Determine the voltage gain of the OTA in Figure 12–40. Assume $K = 16$ μS/μA for the graph in Figure 12–41.

18. If a 10 kΩ rheostat is added in series with the bias resistor in Figure 12–40, what are the minimum and maximum voltage gains?

FIGURE 12–41

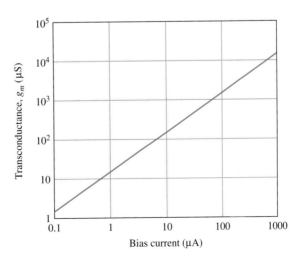

FIGURE 12–42

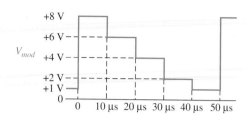

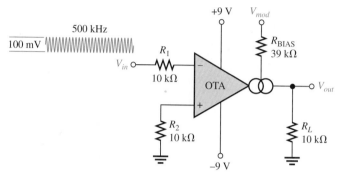

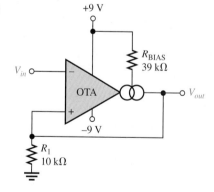

19. The OTA in Figure 12–42 functions as an amplitude modulation circuit. Determine the output voltage waveform for the given input waveforms assuming $K = 16 \ \mu S/\mu A$.

20. Determine the trigger points for the Schmitt-trigger circuit in Figure 12–43.

21. Determine the output voltage waveform for the Schmitt trigger in Figure 12–43 in relation to a 1 kHz sine wave with peak values of ± 10 V.

FIGURE 12–43

SECTION 12–4   Log and Antilog Amplifiers

22. Using your calculator, find the natural logarithm (ln) of each of the following numbers:
    (a) 0.5   (b) 2   (c) 50   (d) 130

23. Repeat Problem 22 for $\log_{10}$.

24. What is the antilog of 1.6?

25. Explain why the output of a log amplifier is limited to approximately 0.7 V.

26. What is the output voltage of a certain log amplifier with a diode in the feedback path when the input voltage is 3 V? The input resistor is 82 kΩ and the reverse leakage current is 100 nA.

27. Determine the output voltage for the amplifier in Figure 12–44. Assume $I_{EBO} = 60$ nA.

28. Determine the output voltage for the amplifier in Figure 12–45. Assume $I_{EBO} = 60$ nA.

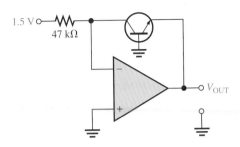

FIGURE 12–44

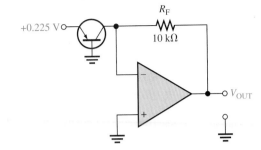

FIGURE 12–45

29. Signal compression is one application of logarithmic amplifiers. Suppose an audio signal with a maximum voltage of 1 V and a minimum voltage of 100 mV is applied to the log amplifier in Figure 12–44. What will be the maximum and minimum output voltages? What conclusion can you draw from this result?

SECTION 12–5   A System Application

30. With a 1 mV, 50 Hz signal applied to the ECG amplifier board in Figure 12–46, what voltage would you expect to see at each of the probed points? All voltages are measured with respect to ground (not shown). Assume that the offset voltage is nulled out and the position control is adjusted for zero deflection. The gain adjustment potentiometer is set at midrange. Refer to the schematic in Figure 12–36.

31. Repeat Problem 30 for a 2 mV, 1 kHz input signal.

FIGURE 12–46

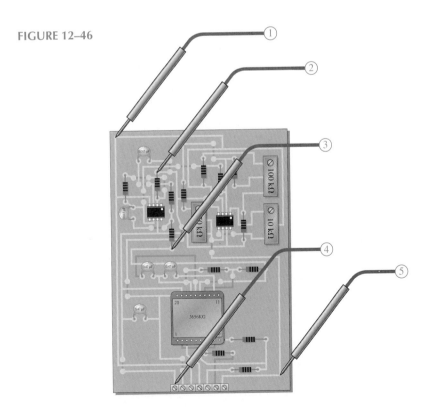

■ ANSWERS TO REVIEW QUESTIONS

Section 12–1

1. The main purpose of an instrumentation amplifier is to amplify small signals that occur on large common-mode voltages. The key characteristics are high input impedance, high CMRR, low output impedance, and low output offset.

2. Three op-amps and seven resistors are required to construct a basic instrumentation amplifier (see Figure 12–2).

3. The gain is set by the external resistor R_G.

4. $A_v \cong 6$

Section 12–2

1. Isolation amplifiers are used in medical equipment, power plant instrumentation, industrial processing, and automated testing.

2. The two stages of an isolation amplifier are input and output.

3. The stages are connected by transformer coupling.

4. The oscillator acts as a dc to ac converter so that the dc power can be ac coupled to the input and output stages.

Section 12–3

1. OTA stands for Operational Transconductance Amplifier.

2. Transconductance increases with bias current.

3. Assuming that the bias input is connected to the supply voltage, the voltage gain increases when the supply voltage is increased because this increases the bias current.

4. The gain decreases as the bias voltage decreases.

Section 12–4

1. A diode or transistor in the feedback loop provides the exponential (nonlinear) characteristic.

2. The output of a basic log amplifier is limited to the barrier potential of the *pn* junction (about 0.7 V).

3. The output voltage is determined by the input voltage, the input resistor, and the emitter-to-base leakage current.

4. The transistor in an antilog amplifier is in series with the input rather than in the feedback loop.

5. A multiplier is made of two log amplifiers, a summing amplifier, an antilog amplifier, and an inverting amplifier.

Section 12–5

1. The gain of the input stage is set by R_2 and R_3. The gain of the output stage is set by R_4 and R_5.

2. $R_5/R_4 + 1 = 120 \text{ k}\Omega/100 \text{ k}\Omega + 1 = 2.2$

3. Lower: 0 Hz; Upper: $1/2\pi RC = 1/2\pi(0.015 \text{ μF})(100 \text{ k}\Omega) = 106$ Hz

4. $A_{v(I)} = -R_f/R_i = -(100 \text{ k}\Omega + 25 \text{ k}\Omega)/1.0 \text{ k}\Omega = -(125 \text{ k}\Omega/1.0 \text{ k}\Omega) = -125$

■ ANSWERS TO PRACTICE EXERCISES FOR EXAMPLES

12–1 240 Ω

12–2 Make $R_G = 1.1$ kΩ.

12–3 Many combinations are possible. Here is one: $R_{f1} = 10$ kΩ, $R_{i1} = 1.0$ kΩ, $R_{f2} = 10$ kΩ, $R_{i2} = 1.0$ kΩ

12–4   62.5 μA. Note the scale is logarithmic.

12–5   Yes. Approximately 110.

12–6 $V_{out(max)} = 3.61$ V; $V_{out(min)} = 1.76$ V

12–7   −0.167 V

12–8   −0.193 V

12–9   −4.4 V

■ **ANSWERS TO SELF-TEST**				
1. (d)	**2.** (b)	**3.** (a)	**4.** (e)	**5.** (c)
6. (b)	**7.** (a)	**8.** (c)	**9.** (d)	**10.** (c)
11. (a)	**12.** (b)	**13.** (f)	**14.** (b)	**15.** (b)

■ **ANSWERS TO TROUBLE-SHOOTER'S QUIZ**			
1. increase	**2.** decrease	**3.** decrease	**4.** increase
5. decrease	**6.** decrease	**7.** not change	**8.** increase
9. decrease	**10.** increase		

13

COMMUNICATIONS CIRCUITS

Courtesy Hewlett-Packard Company

■ **CHAPTER OBJECTIVES**

❏ Describe basic superheterodyne receivers
❏ Discuss the function of a linear multiplier
❏ Discuss the fundamentals of amplitude modulation
❏ Discuss the basic function of a mixer
❏ Describe AM demodulation
❏ Describe IF and audio amplifiers
❏ Describe frequency modulation
❏ Describe the phase-locked loop (PLL)
❏ Apply what you have learned in this chapter to a system application

■ **KEY TERMS**

❏ Amplitude modulation (AM)
❏ Frequency modulation (FM)
❏ Four-quadrant multiplier
❏ Balanced modulation
❏ Mixer
❏ Voltage-controlled oscillator (VCO)
❏ Phase-locked loop (PLL)
❏ Modem

■ CHAPTER INTRODUCTION

Communications electronics encompasses a wide range of systems, including both analog (linear) and digital. Any system that sends information from one point to another over relatively long distances can be classified as a communications system. Some of the categories of communications systems are radio (broadcast, ham, CB, marine), television, telephony, radar, navigation, satellite, data (digital), and telemetry.

Many communications systems use either amplitude modulation (AM) or frequency modulation (FM) to send information. Other modulation methods include pulse modulation, phase modulation, and frequency shift keying (FSK) as well as more specialized techniques. By necessity, the scope of this chapter is limited and is intended to introduce you to basic AM and FM communications systems and circuits. You will cover communications electronics more thoroughly in another course.

■ A SYSTEM APPLICATION

Digital data consisting of a series of binary digits (1s and 0s) are commonly sent from one computer to another over the telephone lines. Two voltage levels are used to represent the two types of bits, a high-voltage level and a low-voltage level. The data stream is made up of time intervals when the voltage has a constant high value or a constant low value with very fast transitions from one level to the other. In other words, the data stream contains very low frequencies (constant-voltage intervals) and very high frequencies (transitions). Since the standard telephone system has a bandwidth of approximately 300 Hz to 3000 Hz, it cannot handle the very low and the very high frequencies that make up a typical data stream without losing most of the information. Because of the bandwidth limitation of the telephone system, it is necessary to modify digital data before they are sent out; and one method of doing this is with frequency shift keying (FSK), which is a form of frequency modulation.

A simplified block diagram of a digital communications equipment (DCE) system for interfacing digital terminal equipment (DTE), such as a computer, to the telephone network is shown in Figure 13–56. The system FSK-modulates digital data before they are transmitted over the phone line and demodulates FSK signals received from another computer. Because the DTE's basic function is to *mod*ulate and *demod*ulate, it is called a *modem*. Although the modem performs many associated functions, as indicated by the different blocks, in this system application our focus will be on the modulation and demodulation circuits.

For the system application in Section 13–9, in addition to the other topics, be sure you understand
❑ The basic operation of a VCO
❑ The basic operation of a PLL
❑ How to use an LM565 PLL

13–1 ■ BASIC RECEIVERS

Receivers based on the superheterodyne principle are standard in one form or another in most types of communications systems and are found in familiar systems such as standard broadcast radio, stereo, and television. In several of the system applications in previous chapters, we presented the superheterodyne receiver in order to focus on a given circuit; now we cover it from a system viewpoint. This section provides a basic introduction to amplitude modulation and frequency modulation and an overview of the complete AM and FM receiver.

After completing this section, you should be able to

❑ Describe basic superheterodyne receivers
   ❑ Define *AM* and *FM*
   ❑ Discuss the major functional blocks of an AM receiver
   ❑ Discuss the major functional blocks of an FM receiver

Amplitude Modulation (AM)

Amplitude modulation (AM) is a method for sending audible information, such as voice and music, by electromagnetic waves that are broadcast through the atmosphere. In AM, the amplitude of a signal with a specific frequency (f_c), called the *carrier,* is varied according to a modulating signal which can be an audio signal (such as voice or music), as shown in Figure 13–1. The carrier frequency permits the receiver to be tuned to a specific known frequency. The resulting AM waveform contains the carrier frequency, an upper-side frequency equal to the carrier frequency plus the modulation frequency ($f_c + f_m$), and a lower-side frequency equal to the carrier frequency minus the modulation frequency ($f_c - f_m$). Harmonics of these frequencies are also present. For example, if a 1 MHz carrier is amplitude modulated with a 5 kHz audio signal, the frequency components in the AM waveform are 1 MHz (carrier), 1 MHz + 5 kHz = 1,005,000 Hz (upper side), and 1 MHz − 5 kHz = 995,000 Hz (lower side).

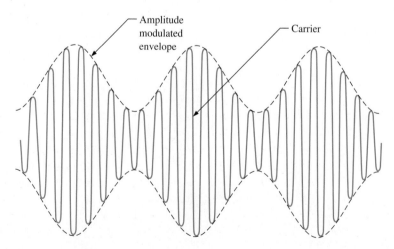

FIGURE 13–1

An example of an amplitude modulated signal. In this case, the higher-frequency carrier is modulated by a lower-frequency sinusoidal signal.

The frequency band for AM broadcast receivers is 540 kHz to 1640 kHz. This means that an AM receiver can be tuned to pick up a specific carrier frequency that lies in the broadcast band. Each AM radio station transmits at a specific carrier frequency that is different from any other station in the area, so you can tune the receiver to pick up any desired station.

The Superheterodyne AM Receiver

A block diagram of a superheterodyne AM receiver is shown in Figure 13–2. The receiver shown consists of an antenna, an RF (radio frequency) amplifier, a mixer, a local oscillator (LO), an IF (intermediate frequency) amplifier, a detector, an audio amplifier and a power amplifier, and a speaker.

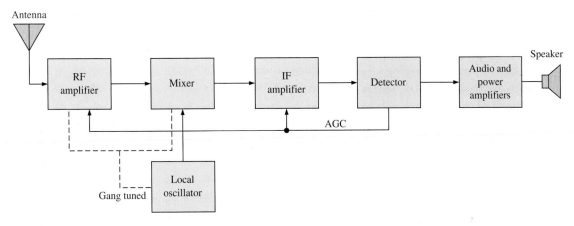

FIGURE 13–2
Superheterodyne AM receiver block diagram.

Antenna   The antenna picks up all radiated signals and feeds them into the RF amplifier. These signals are very small (usually only a few microvolts).

RF Amplifier   This circuit can be adjusted (tuned) to select and amplify any frequency within the AM broadcast band. Only the selected frequency and its two side bands pass through the amplifier. (Some AM receivers do not have a separate RF amplifier stage.)

Local Oscillator   This circuit generates a steady sine wave at a frequency 455 kHz above the selected RF frequency.

Mixer   This circuit accepts two inputs, the amplitude modulated RF signal from the output of the RF amplifier (or the antenna when there is no RF amplifier) and the sinusoidal output of the local oscillator. These two signals are then "mixed" by a nonlinear process called *heterodyning* to produce sum and difference frequencies. For example, if the RF carrier has a frequency of 1000 kHz, the LO frequency is 1455 kHz and the sum and difference frequencies out of the mixer are 2455 kHz and 455 kHz, respectively. The difference frequency is always 455 kHz no matter what the RF carrier frequency.

IF Amplifier   The input to the IF amplifier is the 455 kHz AM signal, a replica of the original AM carrier signal except that the frequency has been lowered to 455 kHz. The IF amplifier significantly increases the level of this signal.

Detector   This circuit recovers the modulating signal (audio signal) from the 455 kHz IF. At this point the IF is no longer needed, so the output of the detector consists of only the audio signal.

Audio and Power Amplifiers   This circuit amplifies the detected audio signal and drives the speaker to produce sound.

AGC   The automatic gain control (AGC) provides a dc level out of the detector that is proportional to the strength of the received signal. This level is fed back to the IF amplifier, and sometimes to the mixer and RF amplifier, to adjust the gains so as to maintain constant signal levels throughout the system over a wide range of incoming carrier signal strengths.

   Figure 13–3 shows the signal flow through an AM superheterodyne receiver. The receiver can be tuned to accept any frequency in the AM band. The RF amplifier, mixer, and local oscillator are tuned simultaneously so that the LO frequency is always 455 kHz above the incoming RF signal frequency. This is called *gang tuning*.

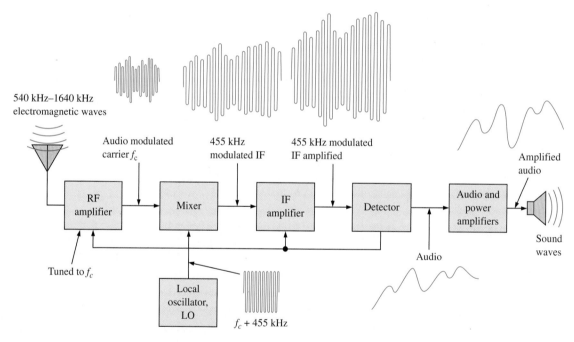

FIGURE 13–3
Illustration of signal flow through an AM receiver.

Frequency Modulation (FM)

In this method of **modulation**, the modulating signal (audio) varies the frequency of a carrier as opposed to the amplitude, as in the case of AM. Figure 13–4 illustrates basic **frequency modulation (FM)**. The standard FM broadcast band consists of carrier frequencies from 88 MHz to 108 MHz, which is significantly higher than AM. The FM receiver is similar to the AM receiver in many ways, but there are several differences.

The Superheterodyne FM Receiver

A block diagram of a superheterodyne FM receiver is shown in Figure 13–5. Notice that it includes an RF amplifier, mixer, local oscillator, and IF amplifier just as in the AM receiver. These circuits must, however, operate at higher frequencies than in the AM system.

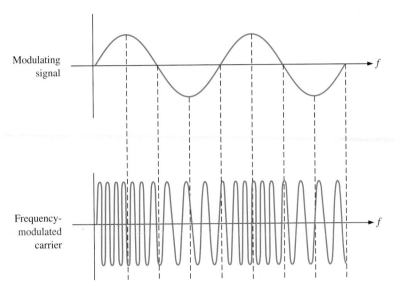

FIGURE 13–4
An example of frequency modulation.

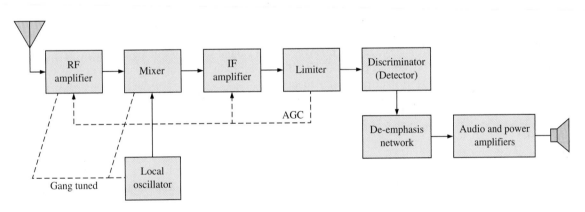

FIGURE 13–5
Superheterodyne FM receiver block diagram.

A significant difference in FM is the way the audio signal must be recovered from the modulated IF. This is accomplished by the limiter, discriminator, and de-emphasis network. Figure 13–6 depicts the signal flow through an FM receiver.

RF Amplifier   This circuit must be capable of amplifying any frequency between 88 MHz and 108 MHz. It is highly selective so that it passes only the selected carrier frequency and significant side-band frequencies that contain the **audio**.

Local Oscillator   This circuit produces a sine wave at a frequency 10.7 MHz above the selected RF frequency.

Mixer   This circuit performs the same function as in the AM receiver, except that its output is a 10.7 MHz FM signal regardless of the RF carrier frequency.

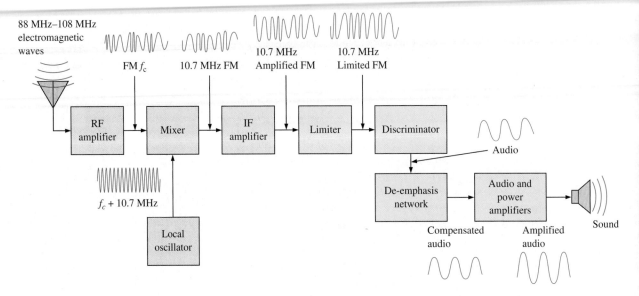

FIGURE 13–6
Example of signal flow through an FM receiver.

IF Amplifier   This circuit amplifies the 10.7 MHz FM signal.

Limiter   The limiter removes any unwanted variations in the amplitude of the FM signal as it comes out of the IF amplifier and produces a constant amplitude FM output at the 10.7 MHz intermediate frequency.

Discriminator   This circuit performs the equivalent function of the detector in an AM system and is often called a detector rather than a discriminator. The **discriminator** recovers the audio from the FM signal.

De-emphasis Network   For certain reasons, the higher modulating frequencies are amplified more than the lower frequencies at the transmitting end of an FM system by a process called *preemphasis*. The de-emphasis circuit in the FM receiver brings the high-frequency audio signals back to the proper amplitude relationship with the lower frequencies.

Audio and Power Amplifiers   This circuit is the same as in the AM system and can be shared when there is a dual AM/FM configuration.

13–1 REVIEW QUESTIONS*

1. What do *AM* and *FM* mean?
2. How do AM and FM differ?
3. What are the standard broadcast frequency bands for AM and FM?

_____

* Answers are at the end of the chapter.

13–2 ■ THE LINEAR MULTIPLIER

The linear multiplier is a key circuit in many types of communications systems. In this section, you will examine the basic principles of IC linear multipliers and look at a few applications that are found in communications as well as in other areas. In the following sections, we will concentrate on multiplier applications in AM and FM systems.

After completing this section, you should be able to

❑ Discuss the function of a linear multiplier
  ❑ Describe multiplier quadrants and transfer characteristic
  ❑ Discuss scale factor
  ❑ Show how to use a multiplier circuit as a multiplier, squaring circuit, divide circuit, square root circuit, and mean square circuit

Multiplier Quadrants

There are one-quadrant, two-quadrant, and **four-quadrant multipliers**. The quadrant classification indicates the number of input polarity combinations that the multiplier can handle. A graphical representation of the quadrants is shown in Figure 13–7. A four-quadrant multiplier can accept any of the four possible input polarity combinations and produce an output with the corresponding polarity.

FIGURE 13–7
Four-quadrant polarities and their products.

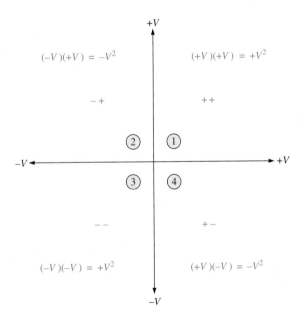

The Multiplier Transfer Characteristic

Figure 13–8 shows the transfer characteristic for a typical IC linear multiplier. To find the output voltage from the transfer characteristic graph, you find the intersection of the two input voltages V_X and V_Y. Values of V_X run along the horizontal axis and values of V_Y are the sloped lines. The output voltage is found by projecting the point of intersection over to the vertical axis. An example will illustrate this.

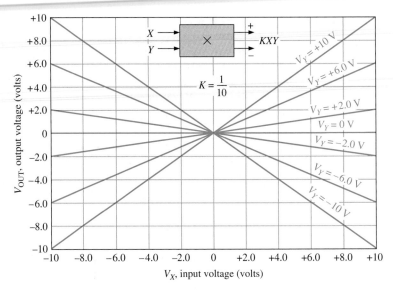

FIGURE 13–8
A four-quadrant multiplier transfer characteristic.

EXAMPLE 13–1

Determine the output voltage for a four-quadrant linear multiplier whose transfer characteristic is given in Figure 13–8. The input voltages are $V_X = -4$ V and $V_Y = +10$ V.

Solution   The output voltage is **−4 V** as illustrated in Figure 13–9. For this transfer characteristic, the output voltage is a factor of ten smaller than the actual product of the two input voltages. This is due to the *scale factor* of the multiplier, which is discussed next.

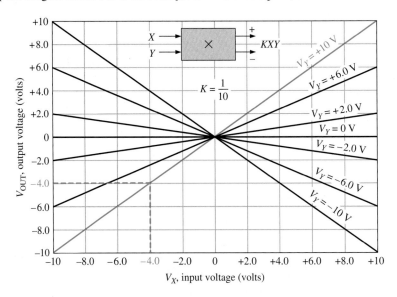

FIGURE 13–9

Practice Exercise*   Find V_{OUT} if $V_X = -6$ V and $V_Y = +6$ V.

---

* Answers are at the end of the chapter.

The Scale Factor, K

The scale factor, K, is basically an internal attenuation that reduces the output by a fixed amount. The scale factor on most IC multipliers is adjustable and has a typical value of 0.1. Figure 13–10 shows an MC1495[1] configured as a basic multiplier. The scale factor is determined by external resistors, which include two equal load resistors, according to the following formula:

$$K = \frac{2R_L}{R_X R_Y I_{R2}}$$

The current I_{R2} is set by internal and external parameters according to this formula:

$$I_{R2} = \frac{|-V| - 0.7\text{ V}}{R_2 + 500\ \Omega}$$

where R_2 is the combination of the fixed resistor and the potentiometer. The potentiometer provides for fine adjustment by controlling I_{R2}.

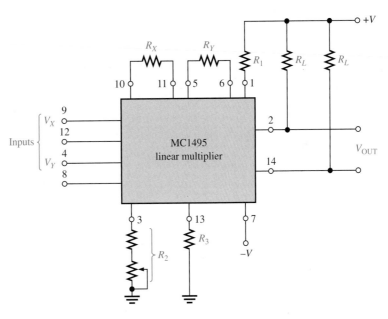

FIGURE 13–10
Basic MC1495 linear multiplier with external circuitry for setting the scale factor.

The expression for the output voltage of the IC linear multiplier includes the scale factor, K, as indicated in Equation (13–1).

$$V_{\text{OUT}} = K V_X V_Y \qquad \qquad \textbf{(13–1)}$$

---

[1] Data sheet for MC1495 available at http://www.onsemi.com

EXAMPLE 13–2

Determine the scale factor for the basic MC1495 multiplier in Figure 13–11. Assume the 5 kΩ potentiometer portion of R_2 is set to 2.5 kΩ. Also, determine the output voltage for the given inputs.

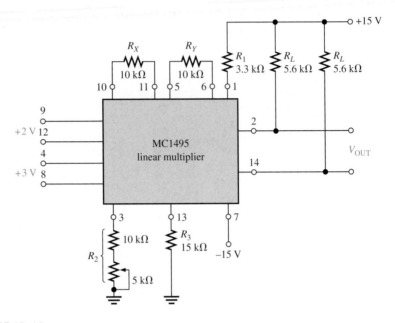

FIGURE 13–11

Solution Calculate I_{R2} as follows:

$$I_{R2} = \frac{|-V| - 0.7 \text{ V}}{R_2 + 500 \text{ }\Omega} = \frac{15 \text{ V} - 0.7 \text{ V}}{12.5 \text{ k}\Omega + 500 \text{ }\Omega} = \frac{14.3 \text{ V}}{13 \text{ k}\Omega} = 1.1 \text{ mA}$$

The scale factor is

$$K = \frac{2R_L}{R_X R_Y I_{R2}} = \frac{2(5.6 \text{ k}\Omega)}{(10 \text{ k}\Omega)(10 \text{ k}\Omega)(1.1 \text{ mA})} = \mathbf{0.102}$$

The output voltage is

$$V_{OUT} = KV_X V_Y = 0.102(+2 \text{ V})(+3 \text{ V}) = \mathbf{0.611 \text{ V}}$$

Practice Exercise What is the output voltage in Figure 13–11 if the 5 kΩ potentiometer is set to its maximum resistance?

Offset Adjustment

Due to internal mismatches, generally small offset voltages are at the inputs and the output of an IC linear multiplier. External circuits to null out the offset voltages are shown in Figure 13–12. The resistive voltage dividers on the inputs allow the actual input voltages to be greater than the recommended maximum for the device. For example, the MC1495 has a maximum input voltage of 5 V. The voltage dividers allow a maximum of 10 V to be applied if the resistors are of equal value. The zener diodes in the input offset adjust circuit keep the inputs on pins 8 and 12 from exceeding the maximum of 5 V.

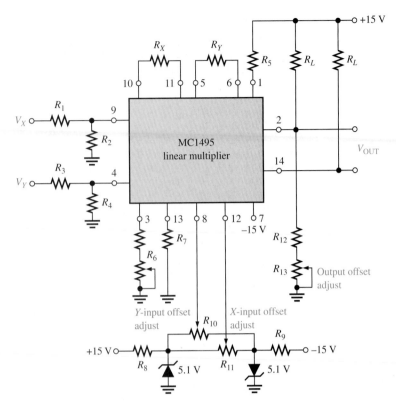

FIGURE 13–12
Basic MC1495 multiplier with both scale factor and offset circuitry.

Basic Applications of the Multiplier

Applications of linear multipliers are numerous. Some basic applications are now presented.

Multiplier   The most obvious application of a linear multiplier is, of course, to multiply two voltages as indicated in Figure 13–13.

FIGURE 13–13
Multiplier.

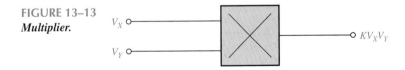

Squaring Circuit   A special case of the multiplier is a squaring circuit that is realized by simply applying the same variable to both inputs by connecting the inputs together as shown in Figure 13–14.

FIGURE 13–14
Squaring circuit.

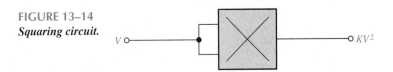

Divide Circuit   The circuit in Figure 13–15 shows the multiplier placed in the feedback loop of an op-amp. The basic operation is as follows. There is a virtual ground at the inverting ($-$) input of the op-amp and therefore the current at the inverting input is negligible. Therefore, I_1 and I_2 are equal. Since the inverting input voltage is 0 V, the voltage across R_1 is KV_YV_{OUT} and the current through R_1 is

$$I_1 = \frac{KV_YV_{OUT}}{R_1}$$

The voltage across R_2 is V_X, so the current through R_2 is

$$I_2 = \frac{V_X}{R_2}$$

Since $I_1 = -I_2$,

$$\frac{KV_YV_{OUT}}{R_1} = -\frac{V_X}{R_2}$$

Solving for V_{OUT},

$$V_{OUT} = -\frac{V_XR_1}{KV_YR_2}$$

If $R_1 = KR_2$,

$$V_{OUT} = -\frac{V_X}{V_Y}$$

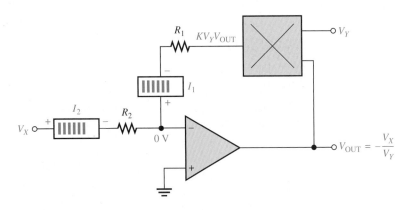

FIGURE 13–15
Divide circuit.

Square Root Circuit   The square root circuit is a special case of the divide circuit where V_{OUT} is applied to both inputs of the multiplier as shown in Figure 13–16.

Mean Square Circuit   In this application, the multiplier is used as a squaring circuit with its output connected to an op-amp integrator as shown in Figure 13–17. The integrator produces the average or mean value of the squared input over time, as indicated by the integration sign ($\int$).

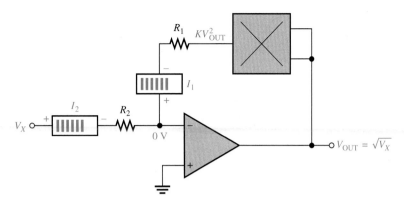

FIGURE 13–16
Square root circuit.

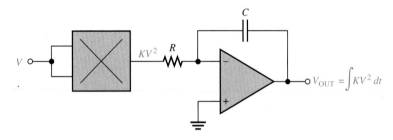

FIGURE 13–17
Mean square circuit.

13–2 ■ REVIEW QUESTIONS

1. Compare a four-quadrant multiplier to a one-quadrant multiplier in terms of the inputs that can be handled.

2. If 5 V and 1 V are applied to the inputs of a multiplier and its output is 0.5 V, what is the scale factor? What must the scale factor be for an output of 5 V?

3. How do you convert a basic multiplier to a squaring circuit?

13–3 ■ AMPLITUDE MODULATION

Amplitude modulation (AM) is an important method for transmitting information. Of course, the AM superheterodyne receiver is designed to receive transmitted AM signals. In this section, we further define amplitude modulation and show how the linear multiplier can be used as an amplitude-modulated device.

After completing this section, you should be able to

❑ Discuss the fundamentals of amplitude modulation
   ❑ Explain how AM is basically a multiplication process
   ❑ Describe sum and difference frequencies

❑ Discuss balanced modulation
❑ Describe the frequency spectra
❑ Explain standard AM

As you learned in Section 13–1, amplitude modulation is the process of varying the amplitude of a signal of a given frequency (carrier) with another signal of much lower frequency (modulating signal). One reason that the higher-frequency carrier signal is necessary is because audio or other signals with relatively low frequencies cannot be transmitted with antennas of a practical size. The basic concept of standard amplitude modulation is illustrated in Figure 13–18.

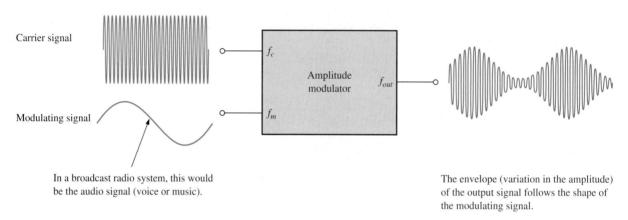

FIGURE 13–18
Basic concept of amplitude modulation.

A Multiplication Process

If a signal is applied to the input of a variable-gain device, the resulting output is an amplitude-modulated signal because $V_{out} = A_v V_{in}$. The output voltage is the input voltage multiplied by the voltage gain. For example, if the gain of an amplifier is made to vary sinusoidally at a certain frequency and an input signal is applied at a higher frequency, the output signal will have the higher frequency. However, its amplitude will vary according to the variation in gain as illustrated in Figure 13–19. Amplitude modulation is basically a multiplication process (input voltage multiplied by a variable gain).

Sum and Difference Frequencies

If the expressions for two sinusoidal signals of different frequencies are multiplied mathematically, a term containing both the difference and the sum of the two frequencies is produced. Recall from ac circuit theory that a sinusoidal voltage can be expressed as

$$v = V_p \sin 2\pi f t$$

where V_p is the peak voltage and f is the frequency. Two different sinusoidal signals can be expressed as follows:

$$v_1 = V_{1(p)} \sin 2\pi f_1 t$$
$$v_2 = V_{2(p)} \sin 2\pi f_2 t$$

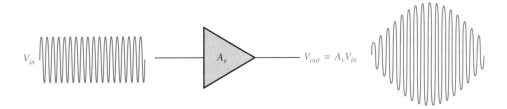

Gain variation

FIGURE 13–19
The amplitude of the output voltage varies according to the gain and is the product of voltage gain and input voltage.

Multiplying these two sinusoidal wave terms,

$$v_1 v_2 = (V_{1(p)}\sin 2\pi f_1 t)(V_{2(p)}\sin 2\pi f_2 t)$$
$$= V_{1(p)} V_{2(p)}(\sin 2\pi f_1 t)(\sin 2\pi f_2 t)$$

The basic trigonometric identity for the product of two sinusoidal functions is

$$(\sin A)(\sin B) = \frac{1}{2}[\cos(A - B) - \cos(A + B)]$$

Applying this identity to the previous formula for $v_1 v_2$,

$$v_1 v_2 = \frac{V_{1(p)} V_{2(p)}}{2}[(\cos 2\pi f_1 t - 2\pi f_2 t) - (\cos 2\pi f_1 t + 2\pi f_2 t)]$$

$$= \frac{V_{1(p)} V_{2(p)}}{2}[(\cos 2\pi(f_1 - f_2)t) - (\cos 2\pi(f_1 + f_2)t)]$$

$$v_1 v_2 = \frac{V_{1(p)} V_{2(p)}}{2}\cos 2\pi(f_1 - f_2)t - \frac{V_{1(p)} V_{2(p)}}{2}\cos 2\pi(f_1 + f_2)t \qquad \textbf{(13–2)}$$

You can see in Equation (13–2) that the product of the two sinusoidal voltages V_1 and V_2 contains a difference frequency $(f_1 - f_2)$ and a sum frequency $(f_1 + f_2)$. The fact that the product terms are cosine simply indicates a 90° phase shift in the multiplication process.

Analysis of Balanced Modulation

Since amplitude modulation is simply a multiplication process, the preceding analysis is now applied to carrier and modulating signals. The expression for the sinusoidal carrier signal can be written as

$$v_c = V_{c(p)}\sin 2\pi f_c t$$

Assuming a sinusoidal modulating signal, it can be expressed as

$$v_m = V_{m(p)}\sin 2\pi f_m t$$

Substituting these two signals in Equation (13–2),

$$v_c v_m = \frac{V_{c(p)}V_{m(p)}}{2}\cos 2\pi(f_c - f_m)t - \frac{V_{c(p)}V_{m(p)}}{2}\cos 2\pi(f_c + f_m)t$$

An output signal described by this expression for the product of two sinusoidal signals is produced by a linear multiplier. Notice that there is a difference frequency term $(f_c - f_m)$ and a sum frequency term $(f_c + f_m)$, but the original frequencies, f_c and f_m, do not appear alone in the expression. Thus, the product of two sinusoidal signals contains no signal with the carrier frequency, f_c, or with the modulating frequency, f_m. This form of amplitude modulation is called **balanced modulation** because there is no carrier frequency in the output. The carrier frequency is "balanced out."

The Frequency Spectra of a Balanced Modulator

A graphical picture of the frequency content of a signal is called its frequency spectrum (see Sec. 1–2). A frequency spectrum shows voltage on a frequency base rather than on a time base as a waveform diagram does. The frequency spectra of the product of two sinusoidal signals are shown in Figure 13–20. Part (a) shows the two input frequencies and part (b) shows the output frequencies. In communications terminology, the sum frequency is called the *upper-side frequency* and the difference frequency is called the *lower-side frequency* because the frequencies appear on each side of the missing carrier frequency.

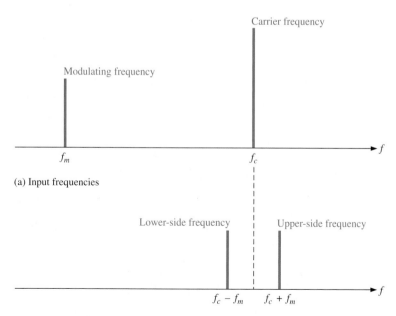

FIGURE 13–20
Illustration of the input and output frequency spectra for a linear multiplier.

The Linear Multiplier as a Balanced Modulator

As mentioned, the linear multiplier acts as a balanced modulator when a carrier signal and a modulating signal are applied to its inputs, as illustrated in Figure 13–21. A balanced modulator produces an upper-side frequency and a lower-side frequency, but it does not produce a carrier frequency. Since there is no carrier signal, balanced modulation is sometimes known as *suppressed-carrier modulation*. Balanced modulation is used in certain types of communications such as single side-band systems, but it is not used in standard AM broadcast systems.

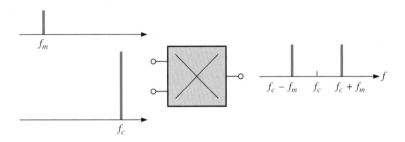

FIGURE 13–21
The linear multiplier as a balanced modulator.

EXAMPLE 13–3

Determine the frequencies contained in the output signal of the balanced modulator in Figure 13–22.

FIGURE 13–22

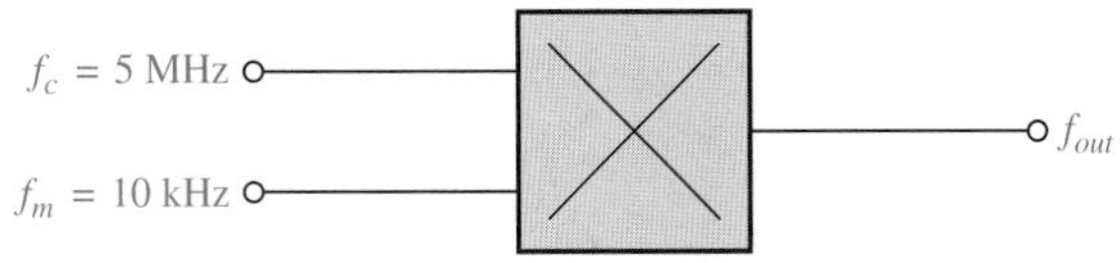

$f_c = 5$ MHz

$f_m = 10$ kHz

f_{out}

Solution   The upper-side frequency is

$$f_c + f_m = 5 \text{ MHz} + 10 \text{ kHz} = \textbf{5.01 MHz}$$

The lower-side frequency is

$$f_c - f_m = 5 \text{ MHz} - 10 \text{ kHz} = \textbf{4.99 MHz}$$

Practice Exercise   Explain how the separation between the side frequencies can be increased using the same carrier frequency.

Standard Amplitude Modulation (AM)

In standard AM systems, the output signal contains the carrier frequency as well as the sum and difference side frequencies. Standard amplitude modulation is illustrated by the frequency spectrum in Figure 13–23.

FIGURE 13–23
The output frequency spectrum of a standard amplitude modulator.

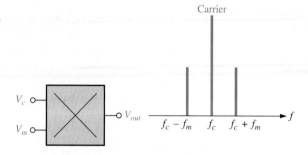

The expression for a standard amplitude-modulated signal is

$$V_{out} = V_{c(p)}^2 \sin 2\pi f_c t + \frac{V_{c(p)}V_{m(p)}}{2}\cos 2\pi(f_c - f_m)t - \frac{V_{c(p)}V_{m(p)}}{2}\cos 2\pi(f_c + f_m)t \qquad \textbf{(13–3)}$$

Notice in Equation (13–3) that the first term is for the carrier frequency and the other two terms are for the side frequencies. Let's see how the carrier-frequency term gets into the equation.

If a dc voltage equal to the peak of the carrier voltage is added to the modulating signal before the modulating signal is multiplied by the carrier signal, a carrier-signal term appears in the final result as shown in the following steps. Add the peak carrier voltage to the modulating signal, and you get the following expression:

$$V_{c(p)} + V_{m(p)}\sin 2\pi f_m t$$

Multiply by the carrier signal.

$$V_{out} = (V_{c(p)}\sin 2\pi f_c t)(V_{c(p)} + V_{m(p)}\sin 2\pi f_m t)$$
$$= \underbrace{V_{c(p)}^2 \sin 2\pi f_c t}_{\text{carrier term}} + \underbrace{V_{c(p)}V_{m(p)}(\sin 2\pi f_c t)(\sin 2\pi f_m t)}_{\text{product term}}$$

Apply the basic trigonometric identity to the product term.

$$V_{out} = V_{c(p)}^2 \sin 2\pi f_c t + \frac{V_{c(p)}V_{m(p)}}{2}\cos 2\pi(f_c - f_m)t - \frac{V_{c(p)}V_{m(p)}}{2}\cos 2\pi(f_c + f_m)t$$

This result shows that the output of the multiplier contains a carrier term and two side-frequency terms. Figure 13–24 illustrates how a standard amplitude modulator can be implemented by a summing circuit followed by a linear multiplier. Figure 13–25 shows a possible implementation of the summing circuit.

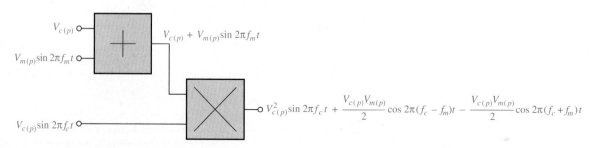

FIGURE 13–24
Basic block diagram of an amplitude modulator.

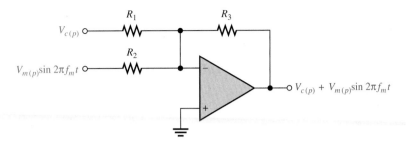

FIGURE 13–25
Implementation of the summing circuit in the amplitude modulator.

EXAMPLE 13–4

A carrier frequency of 1200 kHz is modulated by a sinusoidal wave with a frequency of 25 kHz by a standard amplitude modulator. Determine the output frequency spectrum.

Solution   The lower-side frequency is

$$f_c - f_m = 1200 \text{ kHz} - 25 \text{ kHz} = \textbf{1175 kHz}$$

The upper-side frequency is

$$f_c + f_m = 1200 \text{ kHz} + 25 \text{ kHz} = \textbf{1225 kHz}$$

The output contains the carrier frequency and the two side frequencies as shown in Figure 13–26.

FIGURE 13–26

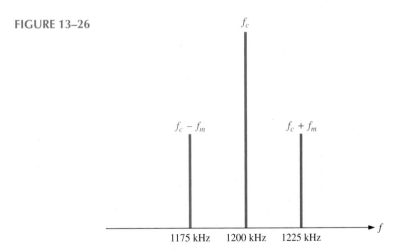

Practice Exercise   Compare the output frequency spectrum in this example to that of a balanced modulator having the same inputs.

Amplitude Modulation with Voice or Music

To this point in our discussion, we have considered the modulating signal to be a pure sinusoidal signal just to keep things fairly simple. If you receive an AM signal modulated by a pure sinusoidal signal in the audio frequency range, you will hear a single tone from the receiver's speaker.

A voice or music signal consists of many sinusoidal components within a range of frequencies from about 20 Hz to 20 kHz. For example, if a carrier frequency is amplitude modulated with voice or music with frequencies from 100 Hz to 10 kHz, the frequency spectrum is as shown in Figure 13–27. Instead of one lower-side and one upper-side frequency as in the case of a single-frequency modulating signal, a band of lower-side frequencies and a band of upper-side frequencies correspond to the sum and difference frequencies of each sinusoidal component of the voice or music signal.

FIGURE 13–27

Example of a frequency spectrum for a voice or music signal.

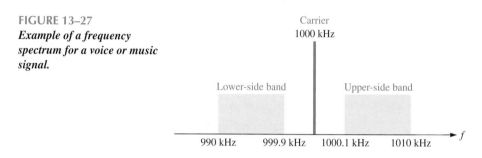

13–3 REVIEW QUESTIONS

1. What is amplitude modulation?
2. What is the difference between balanced modulation and standard AM?
3. What two input signals are used in amplitude modulation? Explain the purpose of each signal.
4. What are the upper-side frequency and the lower-side frequency?
5. How can a balanced modulator be changed to a standard amplitude modulator?

13–4 ■ THE MIXER

The mixer in the receiver system discussed in Section 13–1 can be implemented with a linear multiplier as you will see in this section. The basic principles of linear multiplication of sinusoidal signals are covered, and you will see how sum and difference frequencies are produced. The difference frequency is a critical part of the operation of many types of receiver systems.

After completing this section, you should be able to

❏ Discuss the basic function of a mixer
    ❏ Explain why a mixer is a linear multiplier
    ❏ Describe the frequencies in the mixer and IF portion of a receiver

The **mixer** is basically a frequency converter because it changes the frequency of a signal to another value. The mixer in a receiver system takes the incoming modulated RF signal (which is sometimes amplified by an RF amplifier and sometimes not) along with the signal from the local oscillator and produces a modulated signal with a frequency equal to the difference of its two input frequencies (RF and LO). The mixer also produces a frequency equal to the sum of the input frequencies. The mixer function is illustrated in Figure 13–28.

FIGURE 13–28
The mixer function.

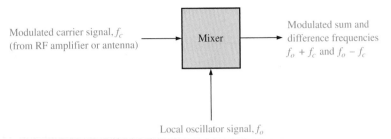

The Mixer Is a Linear Multiplier

In the case of receiver applications, the mixer must produce an output that has a frequency component equal to the difference of its input frequencies. From the mathematical analysis in Section 13–3, you can see that if two sinusoidal signals are multiplied, the product contains the difference frequency and the sum frequency. Thus, the mixer is actually a linear multiplier as indicated in Figure 13–29.

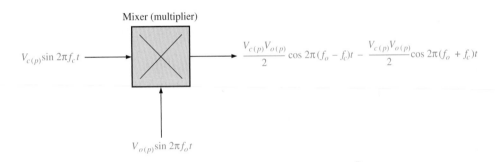

FIGURE 13–29
The mixer as a linear multiplier.

EXAMPLE 13–5

Determine the output expression for a multiplier with one sinusoidal input having a peak voltage of 5 mV and a frequency of 1200 kHz and the other input having a peak voltage of 10 mV and a frequency of 1655 kHz.

Solution The two input expressions are

$$v_1 = (5 \text{ mV})\sin 2\pi(1200 \text{ kHz})t$$
$$v_2 = (10 \text{ mV})\sin 2\pi(1655 \text{ kHz})t$$

Multiplying,

$$v_1 v_2 = (5 \text{ mV})(10 \text{ mV})[\sin 2\pi(1200 \text{ kHz})t][\sin 2\pi(1655 \text{ kHz})t]$$

Applying the trigonometric identity, $(\sin A)(\sin B) = \frac{1}{2}[\cos(A - B) - \cos(A + B)]$,

$$V_{out} = \frac{(5 \text{ mV})(10 \text{ mV})}{2}\cos 2\pi(1655 \text{ kHz} - 1200 \text{ kHz})t$$

$$- \frac{(5 \text{ mV})(10 \text{ mV})}{2}\cos 2\pi(1655 \text{ kHz} + 1200 \text{ kHz})t$$

$$V_{out} = (25 \text{ } \mu\text{V})\cos 2\pi(455 \text{ kHz})t - (25 \text{ } \mu\text{V})\cos 2\pi(2855 \text{ kHz})t$$

Practice Exercise What is the value of the peak amplitude and frequency of the difference frequency component in this example?

In the receiver system, both the sum and difference frequencies from the mixer are applied to the IF (intermediate frequency) amplifier. The IF amplifier is actually a tuned amplifier that is designed to respond to the difference frequency while rejecting the sum frequency. You can think of the IF amplifier section of a receiver as a band-pass filter plus an amplifier because it uses resonant circuits to provide the frequency selectivity. This is illustrated in Figure 13–30.

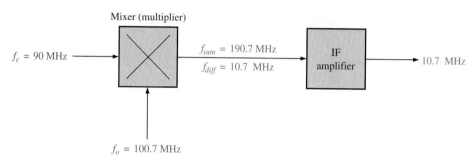

FIGURE 13–30
Example of frequencies in the mixer and IF portion of a receiver.

EXAMPLE 13–6 Determine the output frequency of the IF amplifier for the conditions shown in Figure 13–31.

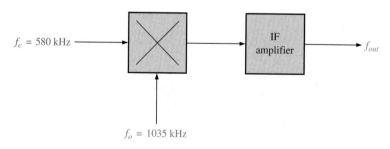

FIGURE 13–31

Solution The IF amplifier produces only the difference frequency signal on its output.

$$f_{out} = f_{diff} = f_o - f_c = 1035 \text{ kHz} - 580 \text{ kHz} = \textbf{455 kHz}$$

Practice Exercise Based on your basic knowledge of the superheterodyne receiver from Section 13–1, determine the IF output frequency when the incoming RF signal changes to 1550 kHz.

13–4 REVIEW QUESTIONS

1. What is the purpose of the mixer in a superheterodyne receiver?
2. How does the mixer produce its output?
3. If a mixer has 1000 kHz on one input and 350 kHz on the other, what frequencies appear on the output?

13–5 ■ AM DEMODULATION

The linear multiplier can be used to demodulate or detect an AM signal as well as to perform the modulation process that was discussed in Section 13–3. **Demodulation** *can be thought of as reverse modulation. The purpose is to get back the original modulating signal (voice or music in the case of standard AM receivers). The detector in the AM receiver can be implemented using a multiplier, although another method using peak envelope detection is common.*

After completing this section, you should be able to

❑ Describe AM demodulation
   ❑ Discuss a basic AM demodulator
   ❑ Discuss the frequency spectra

The Basic AM Demodulator

An AM demodulator can be implemented with a linear multiplier followed by a low-pass filter, as shown in Figure 13–32. The critical frequency of the filter is the highest audio frequency that is required for a given application (15 kHz, for example).

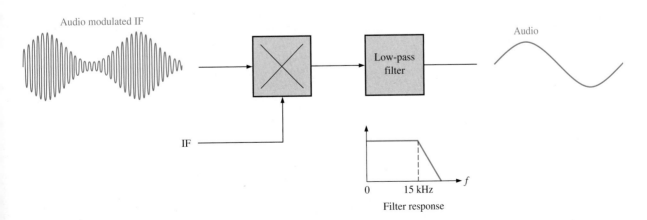

FIGURE 13–32
Basic AM demodulator.

Operation in Terms of the Frequency Spectra

Let's assume a carrier modulated by a single tone with a frequency of 10 kHz is received and converted to a modulated intermediate frequency of 455 kHz, as indicated by the frequency spectra in Figure 13–33. Notice that the upper-side and lower-side frequencies are separated from both the carrier and the IF by 10 kHz.

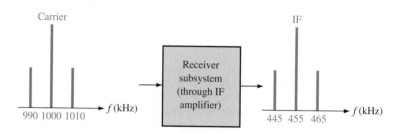

FIGURE 13–33
An AM signal converted to IF.

When the modulated output of the IF amplifier is applied to the demodulator along with the IF, sum and difference frequencies for each input frequency are produced as shown in Figure 13–34. Only the 10 kHz audio frequency is passed by the filter. A drawback to this type of AM detection is that a pure IF must be produced to mix with the modulated IF.

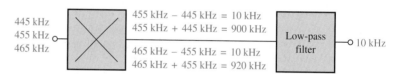

FIGURE 13–34
Example of demodulation.

13–5 REVIEW QUESTIONS

1. What is the purpose of the filter in the linear multiplier demodulator?
2. If a 455 kHz IF modulated by a 1 kHz audio frequency is demodulated, what frequency or frequencies appear on the output of the demodulator?

13–6 ■ IF AND AUDIO AMPLIFIERS

In this section, IC amplifiers for intermediate and audio frequencies are introduced. A typical IF amplifier is discussed and audio preamplifiers and power amplifiers are covered. As you have learned, the IF amplifier in a communications receiver provides amplification of the modulated IF signal out of the mixer before it is applied to the detector. After the audio signal is recovered by the detector, it goes to the audio preamp where it is amplified and applied to the power amplifier that drives the speaker.

After completing this section, you should be able to

❑ Describe IF and audio amplifiers
  ❑ Discuss the function of an IF amplifier
  ❑ Explain how the local oscillator and mixer operate with the IF amplifier
  ❑ Discuss the MC1350 IF amplifier
  ❑ State the purpose of the audio amplifier
  ❑ Discuss the LM386 audio power amplifier

The Basic Function of the IF Amplifier

The IF amplifier in a receiver is a tuned amplifier with a specified bandwidth operating at a center frequency of 455 kHz for AM and 10.7 MHz for FM. The IF amplifier is one of the key features of a superheterodyne receiver because it is set to operate at a single resonant frequency that remains the same over the entire band of carrier frequencies that can be received. Figure 13–35 illustrates the basic function of an IF amplifier in terms of the frequency spectra.

Assume, for example, that the received carrier frequency of $f_c = 1$ MHz is modulated by an audio signal with a maximum frequency of $f_m = 5$ kHz, indicated in Figure 13–35 by

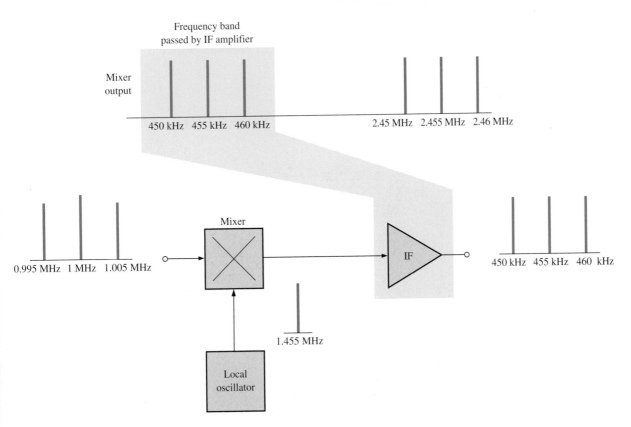

FIGURE 13–35

An illustration of the basic function of the IF amplifier in an AM receiver.

the frequency spectrum on the input to the mixer. For this frequency, the local oscillator is at a frequency of

$$f_o = 1 \text{ MHz} + 455 \text{ kHz} = 1.455 \text{ MHz}$$

The mixer produces the following sum and difference frequencies as indicated in Figure 13–35.

$$f_o + f_c = 1.455 \text{ MHz} + 1 \text{ MHz} = 2.455 \text{ MHz}$$
$$f_o - f_c = 1.455 \text{ MHz} - 1 \text{ MHz} = 455 \text{ kHz}$$
$$f_o + (f_c + f_m) = 1.455 \text{ MHz} + 1.005 \text{ MHz} = 2.46 \text{ MHz}$$
$$f_o + (f_c - f_m) = 1.455 \text{ MHz} + 0.995 \text{ MHz} = 2.45 \text{ MHz}$$
$$f_o - (f_c + f_m) = 1.455 \text{ MHz} - 1.005 \text{ MHz} = 450 \text{ kHz}$$
$$f_o - (f_c - f_m) = 1.455 \text{ MHz} - 0.995 \text{ MHz} = 460 \text{ kHz}$$

Since the IF amplifier is a frequency-selective circuit, it responds only to 455 kHz and any side frequencies lying in the 10 kHz band centered at 455 kHz. So, all of the frequencies out of the mixer are rejected except the 455 kHz IF, all lower-side frequencies down to 450 kHz, and all upper-side frequencies up to 460 kHz. This frequency spectrum is the audio modulated IF.

A Basic IF Amplifier

Although the detailed circuitry of the IF amplifier may differ from one system to another, it always has a tuned (resonant) circuit on the input or on the output or on both. Figure 13–36(a) shows a basic IF amplifier with tuned transformer coupling at the input and output. The general frequency response curve is shown in Figure 13–36(b).

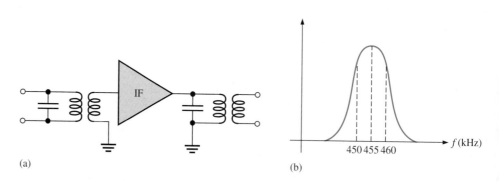

(a)   (b)

FIGURE 13–36
A basic IF amplifier with a tuned circuit on the input and output.

The MC1350   This device is representative of integrated circuit IF amplifiers. It can be used in either AM or FM systems and has a typical power gain of 62 dB at 455 kHz. Figure 13–37 shows packaging and a typical circuit diagram for application in an AM receiver. This configuration has a single-tuned transformer-coupled output. The AGC input is normally fed back from the detector in an AM receiver and is used to keep the IF gain at a constant level so that variations in the strength of the incoming RF signal does not cause the audio output to vary significantly. When the AGC voltage increases, the IF gain decreases and when the AGC voltage decreases, the IF gain increases.

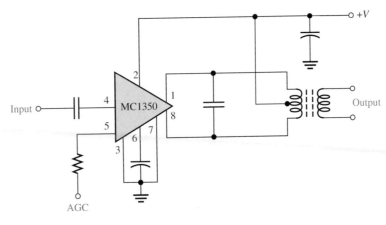

FIGURE 13–37
A typical circuit configuration using the MC1350 IF amplifier.

Audio Amplifiers

Audio amplifiers are used in a receiver system following the detector to provide amplification of the recovered audio signal and audio power to drive the speaker(s), as indicated in Figure 13–38. Audio amplifiers typically have bandwidths of 3 kHz to 15 kHz depending on the requirements of the system. IC audio amplifiers are available with a range of capabilities. Previously, in Section 5–7, the fixed-gain, LM384 IC power amplifier was introduced. To complete the radio, the more versatile LM386 is selected here.

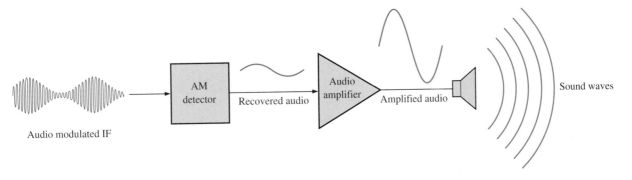

FIGURE 13–38
The audio amplifier in a receiver system.

The LM386 Audio-Power Amplifier[2]   This device is an example of a low-power audio amplifier that is capable of providing several hundred milliwatts to an 8 Ω speaker. It operates from any dc supply voltage in the 4 V to 12 V range, making it a good choice for battery operation. The pin configuration of the LM386 is shown in Figure 13–39(a). The voltage gain of the LM386 is 20 without external connections to the gain terminals, as shown in Figure 13–39(b). A voltage gain of 200 is achieved by connecting a 10 μF capacitor from pin 1 to pin 8, as shown in Figure 13–39(c). Voltage gains between 20 and 200 can be realized by a resistor (R_G) and capacitor (C_G) connected in series from pin 1 to pin

---

[2] Data sheet for LM386 available at http://www.national.com

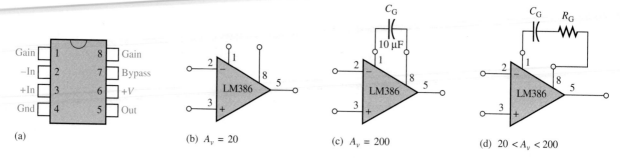

FIGURE 13–39

Pin configuration and gain connections for the LM386 audio amplifier.

8 as shown in Figure 13–39(d). These external components are effectively placed in parallel with an internal gain-setting resistor.

A typical application of the LM386 as a power amplifier in a radio receiver is shown in Figure 13–40. Here the detected AM signal is fed to the inverting input through the volume control potentiometer, R_1, and resistor R_2. C_1 is the input coupling capacitor and C_2 is the power supply decoupling capacitor. R_2 and C_3 filter out any residual RF or IF signal that may be on the output of the detector. R_3 and C_6 provide additional filtering before the audio signal is applied to the speaker through the coupling capacitor C_7.

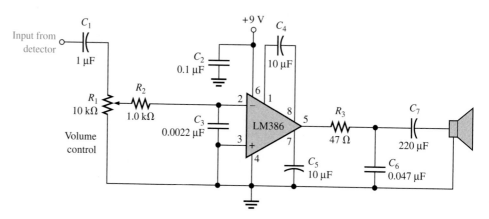

FIGURE 13–40

The LM386 as an AM audio power amplifier.

13–6 REVIEW QUESTIONS

1. What is the purpose of the IF amplifier in an AM receiver?
2. What is the center frequency of an AM IF amplifier?
3. Why is the bandwidth of an AM receiver IF amplifier 10 kHz?
4. Why must the audio amplifier follow the detector in a receiver system?
5. Compare the frequency response of the IF amplifier to that of the audio amplifier.

13–7 ■ FREQUENCY MODULATION

As you have seen, modulation is the process of varying a parameter of a carrier signal with an information signal. Recall that in amplitude modulation the parameter of amplitude is varied. In frequency modulation (FM), the frequency of a carrier is varied above and below its normal or at-rest value by a modulating signal. This section provides a basic introduction to FM and discusses the differences between an AM and an FM receiver.

After completing this section, you should be able to

❑ Describe frequency modulation
   ❑ Discuss the voltage-controlled oscillator
   ❑ Describe the MC2833D FM modulator
   ❑ Explain frequency demodulation

In a frequency-modulated (FM) signal, the carrier frequency is increased or decreased according to the modulating signal. The amount of deviation above or below the carrier frequency depends on the amplitude of the modulating signal. The rate at which the frequency deviation occurs depends on the frequency of the modulating signal.

Figure 13–41 illustrates both a square wave and a sine wave modulating the frequency of a carrier. The carrier frequency is highest when the modulating signal is at its maximum positive amplitude and is lowest when the modulating signal is at its maximum negative amplitude.

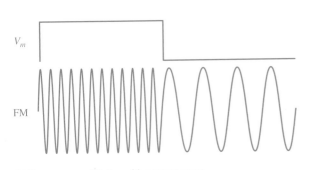

(a) Frequency modulation with a square wave

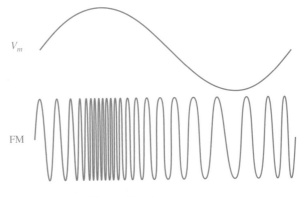

(b) Frequency modulation with a sine wave

FIGURE 13–41
Examples of frequency modulation.

A Basic Frequency Modulator

Frequency modulation is achieved by varying the frequency of an oscillator with the modulating signal. A **voltage-controlled oscillator (VCO)** is typically used for this purpose, as illustrated in Figure 13–42.

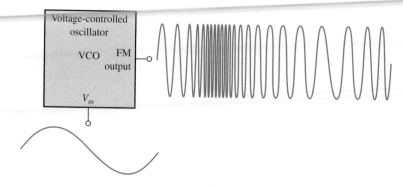

FIGURE 13–42
Frequency modulation with a voltage-controlled oscillator.

Generally, a variable-reactance type of voltage-controlled oscillator is used in FM applications. The variable-reactance VCO uses the varactor diode as a voltage-variable capacitance, as illustrated in Figure 13–43, where the capacitance is varied with the modulating voltage, V_m.

FIGURE 13–43
Basic variable-reactance VCO.

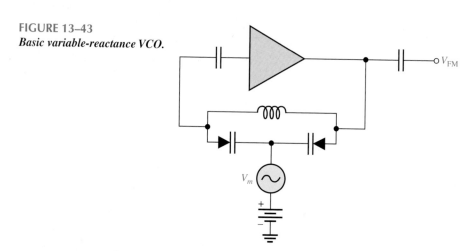

An Integrated Circuit FM Transmitter

An example of a single-chip FM transmitter is the MC2833D[3], which is designed for cordless telephone and other FM communications equipment. The 16-pin package configuration showing the basic functional blocks is illustrated in Figure 13–44. The microphone amplifier (mic amp) amplifies the low-level input from a microphone and feeds its output to the variable-reactance circuit which uses a varactor diode tuning circuit to control the RF oscillator. The reference voltage circuit (V_{REF}) provides stable bias to the reactance circuit. The two individual transistors can be connected as tuned amplifiers to boost the power output.

A typical VHF narrow-band FM transmitter using the MC2833D appropriate external circuitry is shown in Figure 13–45. This particular implementation has an output frequency of 49.7 MHz. The frequency of the oscillator is set by the external 16.5667 MHz crystal.

---

[3] Information about the MC2833D is shown on the MC13110A data sheet available at http://www.motorola.com

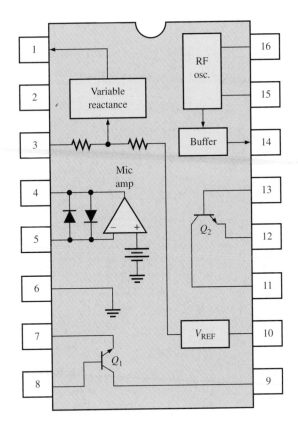

FIGURE 13–44
The MC2833D FM modulator.

The reactance circuit deviates this frequency with the amplified audio input to produce an FM signal. The 16.5667 MHz output of the oscillator goes to the buffer and is then applied to the input of Q_2, which is operated as a frequency tripler (16.5667 MHz $\times$ 3 = 49.7 MHz). A frequency tripler is basically a class C amplifier with the output tuned to a resonant frequency equal to three times the input. The signal from the Q_2 frequency tripler goes to the input of Q_1, which functions as a linear amplifier that drives the transmitting antenna from its resonant output circuit.

FM Demodulation

Except for the higher frequencies, the standard broadcast FM receiver is basically the same as the AM receiver up through the IF amplifier. The main difference between an FM receiver and an AM receiver is the method used to recover the audio signal from the modulated IF.

There are several methods for demodulating an FM signal. These include slope detection, phase-shift discrimination, ratio detection, quadrature detection, and phase-locked loop demodulation. Most of these methods are covered in detail in communications courses. However, because of its importance in many types of applications, we will cover the phase-locked loop (PLL) demodulation in the next section.

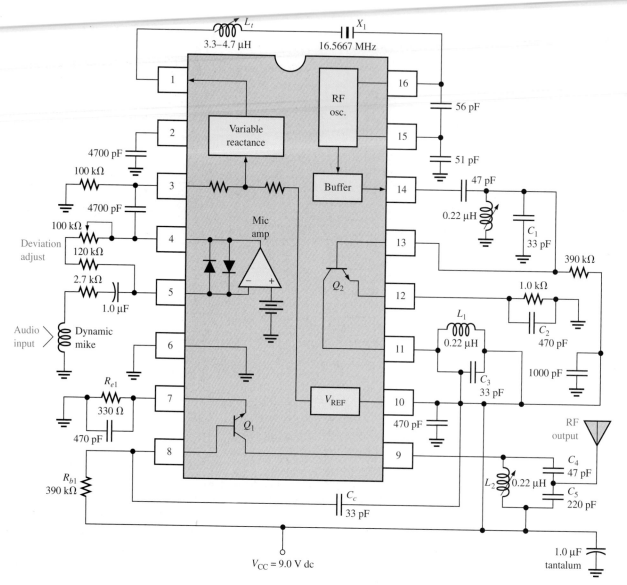

FIGURE 13-45
The MC2833D connected as a 49.7 MHz FM transmitter.

13-7 REVIEW QUESTIONS

1. How does an FM signal carry information?
2. What does VCO stand for?
3. On what principle are most VCOs used in FM based?

13–8 ■ THE PHASE-LOCKED LOOP (PLL)

In the last section, the PLL was mentioned as a way to demodulate an FM signal. In addition to FM demodulation, PLLs are used in a wide variety of communications applications, which include TV receivers, tone decoders, telemetry receivers, modems, and data synchronizers, to name a few. Many of these applications are covered in an electronic communications course. In fact, entire books have been written on the finer points of PLL operation, analysis, and applications. The approach in this section is intended only to present the basic concept and give you an intuitive idea of how PLLs work and how they are used in FM demodulation. A specific PLL integrated circuit is also introduced.

After completing this section, you should be able to

❑ Describe the phase-locked loop (PLL)
   ❑ Draw a basic block diagram for the PLL
   ❑ Discuss the phase detector and state its purpose
   ❑ State the purpose of the VCO
   ❑ State the purpose of the low-pass filter
   ❑ Explain lock range and capture range
   ❑ Discuss the LM565 PLL and explain how it can be used as an FM demodulator

The Basic PLL Concept

The **phase-locked loop (PLL)** is a feedback circuit consisting of a phase detector, a low-pass filter, and a voltage-controlled oscillator (VCO). Some PLLs also include an amplifier in the loop, and in some applications the filter is not used.

The PLL is capable of locking onto or synchronizing with an incoming signal. When the phase of the incoming signal changes, indicating a change in frequency, the phase detector's output increases or decreases just enough to keep the VCO frequency the same as the frequency of the incoming signal. A basic PLL block diagram is shown in Figure 13–46.

The general operation of the PLL is as follows. The phase detector compares the phase difference between the incoming signal, V_i, and the VCO signal, V_o. When the

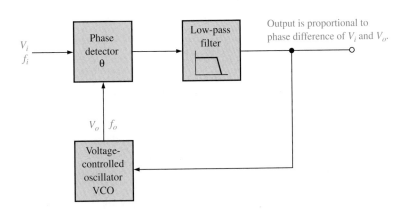

FIGURE 13–46
Basic PLL block diagram.

frequency of the incoming signal, f_i, is different from that of the VCO frequency, f_o, the phase angle between the two signals is also different. The output of the phase detector and the filter is proportional to the phase difference of the two signals. This proportional voltage is fed to the VCO, forcing its frequency to move toward the frequency of the incoming signal until the two frequencies are equal. At this point, the PLL is locked onto the incoming frequency. If f_i changes, the phase difference also changes, forcing the VCO to track the incoming frequency.

The Phase Detector

The phase-detector circuit in a PLL is basically a linear multiplier. The following analysis illustrates how it works in a PLL application. The incoming signal, V_i, and the VCO signal, V_o, applied to the phase detector can be expressed as

$$v_i = V_i \sin(2\pi f_i t + \theta_i)$$
$$v_o = V_o \sin(2\pi f_o t + \theta_o)$$

where θ_i and θ_o are the relative phase angles of the two signals. The phase detector multiplies these two signals and produces a sum and difference frequency output, V_d, as follows:

$$V_d = V_i \sin(2\pi f_i t + \theta_i) \times V_o \sin(2\pi f_o t + \theta_o)$$
$$= \frac{V_i V_o}{2} \cos[(2\pi f_i t + \theta_i) - (2\pi f_o t + \theta_o)] - \frac{V_i V_o}{2} \cos[(2\pi f_i t + \theta_i) + (2\pi f_o t + \theta_o)]$$

When the PLL is locked,

$$f_i = f_o$$

and

$$2\pi f_i t = 2\pi f_o t$$

Therefore, the detector output voltage is

$$V_d = \frac{V_i V_o}{2}[\cos(\theta_i - \theta_o) - \cos(4\pi f_i t + \theta_i + \theta_o)]$$

The second cosine term in the above equation is a second harmonic term ($2 \times 2\pi f_i t$) and is filtered out by the low-pass filter. The control voltage on the output of the filter is expressed as

$$V_c = \frac{V_i V_o}{2}\cos \theta_e \qquad \text{(13–4)}$$

where $\theta_e = \theta_i - \theta_o$, where θ_e is the *phase error*. The filter output voltage is proportional to the phase difference between the incoming signal and the VCO signal and is used as the control voltage for the VCO. This operation is illustrated in Figure 13–47.

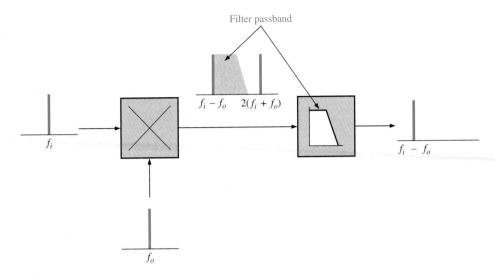

FIGURE 13–47
Basic phase detector/filter operation.

EXAMPLE 13–7

A PLL is locked onto an incoming signal with a frequency of 1 MHz at a phase angle of 50°. The VCO signal is at a phase angle of 20°. The peak amplitude of the incoming signal is 0.5 V and that of the VCO output signal is 0.7 V.
(a) What is the VCO frequency?
(b) What is the value of the control voltage being fed back to the VCO at this point?

Solution
(a) Since the PLL is in lock, $f_i = f_o = $ **1 MHz.**
(b) $\theta_e = \theta_i - \theta_o = 50° - 20° = 30°$

$$V_c = \frac{V_i V_o}{2}\cos\theta_e = \frac{(0.5\text{ V})(0.7\text{ V})}{2}\cos 30° = (0.175\text{ V})\cos 30° = \mathbf{0.152\text{ V}}$$

Practice Exercise   If the phase angle of the incoming signal changes instantaneously to 30°, indicating a change in frequency, what is the instantaneous VCO control voltage?

The Voltage-Controlled Oscillator (VCO)

Voltage-controlled oscillators can take many forms. A VCO can be some type of *LC* or crystal oscillator as was shown in Section 13–7 or it can be some type of *RC* oscillator or multivibrator. No matter the exact type, most VCOs employed in PLLs operate on the principle of *variable reactance* using the varactor diode as a voltage-variable capacitor.

The capacitance of a varactor diode varies inversely with reverse-bias voltage. The capacitance decreases as reverse voltage increases and vice versa.

In a PLL, the control voltage fed back to the VCO is applied as a reverse-bias voltage to the varactor diode within the VCO. The frequency of oscillation is inversely related to capacitance for an *RC* type oscillator by the formula

$$f_o = \frac{1}{2\pi RC}$$

and for an *LC* type oscillator by the formula

$$f_o = \frac{1}{2\pi\sqrt{LC}}$$

These formulas show that frequency increases as capacitance decreases and vice versa.

Capacitance decreases as reverse voltage (control voltage) increases. Therefore, an increase in control voltage to the VCO causes an increase in frequency and vice versa. Basic VCO operation is illustrated in Figure 13–48. The graph in part (b) shows that at the nominal control voltage, $V_{c(nom)}$, the oscillator is running at its nominal or free-running frequency, $f_{o(nom)}$. An increase in V_c above the nominal value forces the oscillator frequency to increase, and a decrease in V_c below the nominal value forces the oscillator frequency to decrease. There are, of course, limits on the operation as indicated by the minimum and maximum points. The transfer function or conversion gain, K, of the VCO is normally expressed as a certain frequency deviation per unit change in control voltage.

$$K = \frac{\Delta f_o}{\Delta V_c}$$

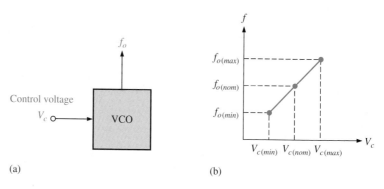

(a) (b)

FIGURE 13–48
Basic VCO operation.

EXAMPLE 13–8

The output frequency of a certain VCO changes from 50 kHz to 65 kHz when the control voltage increases from 0.5 V to 1 V. What is the conversion gain, K?

Solution

$$K = \frac{\Delta f_o}{\Delta V_c} = \frac{65 \text{ kHz} - 50 \text{ kHz}}{1 \text{ V} - 0.5 \text{ V}} = \frac{15 \text{ kHz}}{0.5 \text{ V}} = \textbf{30 kHz/V}$$

Practice Exercise If the conversion gain of a certain VCO is 20 kHz/V, how much frequency deviation does a change in control voltage from 0.8 V to 0.5 V produce? If the VCO frequency is 250 kHz at 0.8 V, what is the frequency at 0.5 V?

Basic PLL Operation

When the PLL is locked, the incoming frequency, f_i, and the VCO frequency, f_o, are equal. However, there is always a phase difference between them called the *static phase error*. The phase error, θ_e, is the parameter that keeps the PLL locked in. As you have seen, the filtered voltage from the phase detector is proportional to θ_e (Equation (13–4)). This voltage controls the VCO frequency and is always just enough to keep $f_o = f_i$.

Figure 13–49 shows the PLL and two sinusoidal signals of the same frequency but with a phase difference, θ_e. For this condition the PLL is in lock and the VCO control voltage is constant. If f_i decreases, θ_e increases to θ_{e1} as illustrated in Figure 13–50. This increase in θ_e is sensed by the phase detector causing the VCO control voltage to decrease, thus decreasing f_o until $f_o = f_i$ and keeping the PLL in lock. If f_i increases, θ_e decreases to θ_{e1} as illustrated in Figure 13–51. This decrease in θ_e causes the VCO control voltage to increase, thus increasing f_o until $f_o = f_i$ and keeping the PLL in lock.

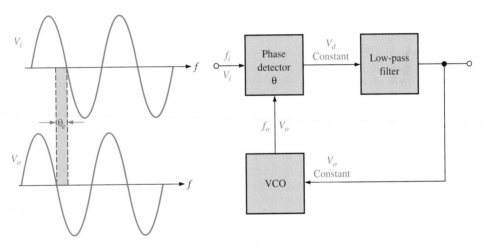

FIGURE 13–49
PLL in lock under static condition ($f_o = f_i$ and constant θ_e).

Lock Range Once the PLL is locked, it will track frequency changes in the incoming signal. The range of frequencies over which the PLL can maintain lock is called the *lock* or *tracking range*. Limitations on the hold-in range are the maximum frequency deviations of the VCO and the output limits of the phase detector. The hold-in range is independent of the bandwidth of the low-pass filter because, when the PLL is in lock, the difference frequency $(f_i - f_o)$ is zero or a very low instantaneous value that falls well within the bandwidth. The hold-in range is usually expressed as a percentage of the VCO frequency.

Capture Range Assuming the PLL is not in lock, the range of frequencies over which it can acquire lock with an incoming signal is called the *capture range*. Two basic conditions are required for a PLL to acquire lock. First, the difference frequency $(f_o - f_i)$ must be low enough to fall within the filter's bandwidth. This means that the incoming frequency must not be separated from the nominal or free-running frequency of the VCO by more than the bandwidth of the low-pass filter. Second, the maximum deviation, Δf_{max}, of the VCO frequency must be sufficient to allow f_o to increase or decrease to a value equal to

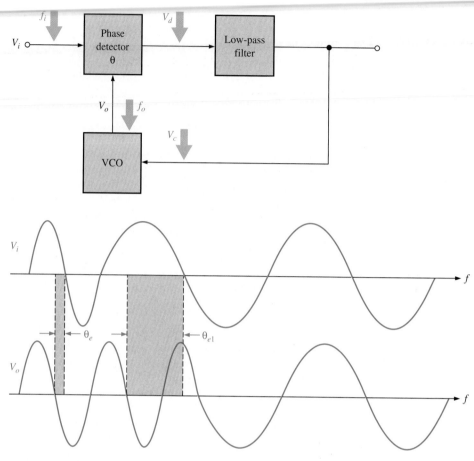

FIGURE 13–50
PLL action when f_i decreases.

f_i. These conditions are illustrated in Figure 13–52; and when they exist, the PLL will "pull" the VCO frequency toward the incoming frequency until $f_o = f_i$.

The LM565 Phase-Locked Loop[4]

The LM565 is a good example of an integrated circuit PLL. The circuit consists of a VCO, phase detector, a low-pass filter formed by an internal resistor and an external capacitor, and an amplifier. The free-running VCO frequency can be set with external components. A block diagram is shown in Figure 13–53. The LM565 can be used for the frequency range from 0.001 Hz to 500 kHz.

The free-running frequency of the VCO is set by the values of R_1 and C_1 in Figure 13–53 according to the following formula. The frequency is in hertz when the resistance is in ohms and the capacitance is in farads.

$$f_o \cong \frac{1.2}{4R_1C_1} \qquad (13\text{–}5)$$

---

[4] Data sheet for LM565 available at http://www.national.com

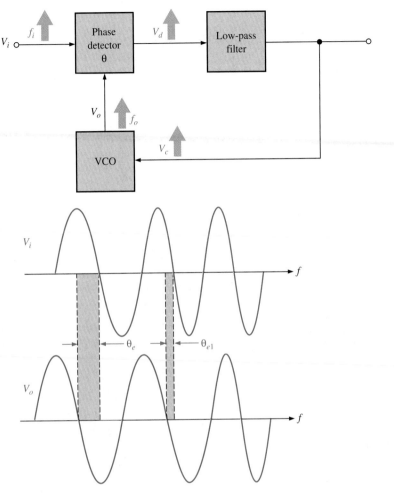

FIGURE 13–51
PLL action when f_i increases.

The lock range is given by

$$f_{lock} = \pm\frac{8f_o}{V_{CC}} \tag{13-6}$$

where V_{CC} is the total voltage between the positive and negative dc supply voltage terminals.

The capture range is given by

$$f_{cap} \cong \pm\frac{1}{2\pi}\sqrt{\frac{2\pi f_{lock}}{(3600)C_2}} \tag{13-7}$$

The 3600 is the value of the internal filter resistor in ohms. You can see that the capture range is dependent on the filter bandwidth as determined by the internal resistor and the external capacitor C_2.

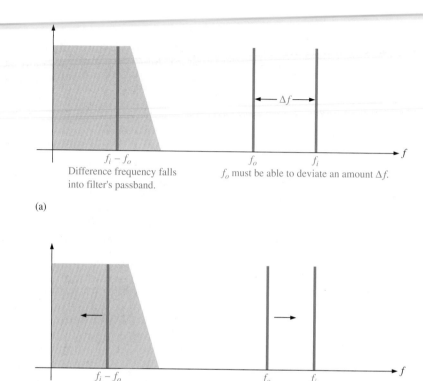

(a)

$f_i - f_o$
Difference frequency falls
into filter's passband.

f_o f_i
f_o must be able to deviate an amount Δf.

$f_i - f_o$

f_o f_i

(b) $f_i - f_o$ decreases as f_o deviates towards f_i.

FIGURE 13–52
Illustration of the conditions for a PLL to acquire lock.

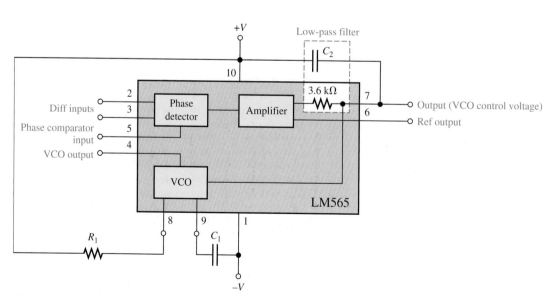

FIGURE 13–53
Block diagram of the LM565 PLL.

The PLL as an FM Demodulator

As you have seen, the VCO control voltage in a PLL depends on the deviation of the incoming frequency. The PLL will produce a voltage proportional to the frequency of the incoming signal which, in the case of FM, is the original modulating signal.

Figure 13–54 shows a typical connection for the LM565 as an FM demodulator. If the IF input is frequency modulated by a sinusoidal signal, you get a sinusoidal signal on the output as indicated. Since the maximum operating frequency is 500 kHz, this device must be used in double-conversion FM receivers. A double-conversion FM receiver is one in which essentially two mixers are used to first convert the RF to a 10.7 MHz IF and then convert this to a 455 kHz IF.

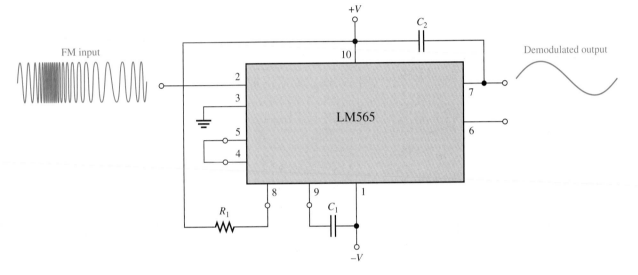

FIGURE 13–54
The LM565 as an FM demodulator.

The free-running frequency of the VCO is adjusted to approximately 455 kHz, which is the center of the modulated IF range. C_1 can be any value, but R_1 should be in the range from 2 kΩ to 20 kΩ. The input can be directly coupled as long as there is no dc voltage difference between pins 2 and 3. The VCO is connected to the phase detector by an external wire between pins 4 and 5.

EXAMPLE 13–9

Determine the values for R_1, C_1, and C_2 for the LM565 in Figure 13–54 for a free-running frequency of 455 kHz and a capture range of ±10 kHz. The dc supply voltages are ±6 V.

Solution   Use Equation (13–5) to calculate C_1. Choose $R_1 = $ **4.7 kΩ**.

$$f_o \cong \frac{1.2}{4R_1C_1}$$

$$C_1 \cong \frac{1.2}{4R_1f_o} = \frac{1.2}{4(4700\ \Omega)(455 \times 10^3\ \text{Hz})} = 140 \times 10^{-12}\ \text{F} = \textbf{140 pF}$$

The lock range and capture range must be determined before C_2 can be calculated. The lock range is

$$f_{lock} = \pm\frac{8f_o}{V_{CC}} = \pm\frac{8(455 \text{ kHz})}{12 \text{ V}} = \pm303 \text{ kHz}$$

Use Equation (13–7) to calculate C_2.

$$f_{cap} \cong \pm\frac{1}{2\pi}\sqrt{\frac{2\pi f_{lock}}{(3600)C_2}}$$

$$f_{cap}^2 \cong \left(\frac{1}{2\pi}\right)^2\frac{2\pi f_{lock}}{(3600)C_2}$$

Therefore,

$$C_2 \cong \left(\frac{1}{2\pi}\right)^2\frac{2\pi f_{lock}}{(3600)f_{cap}^2} = \left(\frac{1}{2\pi}\right)^2\frac{2\pi(303 \times 10^3 \text{ Hz})}{(3600)(10 \times 10^3 \text{ Hz})^2} = 0.134 \times 10^{-6} \text{ F} = \textbf{0.134 } \boldsymbol{\mu}\textbf{F}$$

Practice Exercise What can you do to increase the capture range from ±10 kHz to ±15 kHz?

13–8 REVIEW QUESTIONS

1. List the three basic components in a phase-locked loop.
2. What is another circuit used in some PLLs other than the three listed in Question 1?
3. What is the basic function of a PLL?
4. What is the difference between the lock range and the capture range of a PLL?
5. Basically, how does a PLL track the incoming frequency?

13–9 ■ A SYSTEM APPLICATION

The DCE (data communications equipment) system introduced at the opening of this chapter includes an FSK (frequency shift keying) modem (modulator/demodulator). FSK is one method for modulating digital data for transmission over voice phone lines and is basically a form of frequency modulation. In this system application, the focus is on the low-speed modulator/demodulator (modem) board, which is implemented with a VCO for transmitting FSK signals and a PLL for receiving FSK signals.

After completing this section, you should be able to

❑ Apply what you have learned in this chapter to a system application
   ❑ Describe how a VCO and a PLL can be used in a communications system
   ❑ Discuss how FSK is used to send digital information over phone lines
   ❑ Translate between a printed circuit board and a schematic
   ❑ Analyze the modem circuitry
   ❑ Troubleshoot some common problems

A Brief Description of the System

The FSK modem interfaces a computer with the telephone network so that digital data, which are incompatible with the standard phone system because of bandwidth limitations, can be transmitted and received over regular phone lines, thus allowing computers to communicate with each other. Figure 13–55 shows a diagram of a simple data communications system in which a modem at each end of the phone line provides interfacing for a computer.

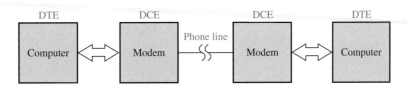

FIGURE 13–55
A data communications system.

The modem (DCE) consists of three basic functional blocks as shown in Figure 13–56: the FSK modem circuits, the phone line interface circuits, and the timing and control circuits. The dual polarity power supply is not shown. Although the focus of this system application is the FSK modem board, we will briefly look at each of the other parts to give you a basic idea of the overall system function.

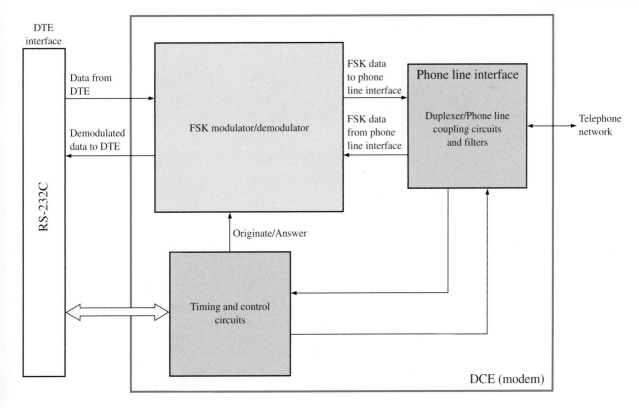

FIGURE 13–56
Basic block diagram of a modem.

The Phone Line Interface   The main purposes of this circuitry are to couple the phone line to the modem by proper impedance matching, to provide necessary filtering, and to accommodate full-duplex transmission of data. *Full-duplex* means essentially that information can be going both ways on a single phone line at the same time. This allows a computer, connected to a modem, to be sending data and receiving data simultaneously without the transmitted data interfering with the received data. Full-duplexing is implemented by assigning the transmitted data one bandwidth and the received data another separate bandwidth within the 300 Hz to 3 kHz overall bandwidth of the phone network.

Timing and Control   One basic function of the timing and control circuits is to determine the proper mode of operation for the modem. The two modes are the originate mode and the answer mode. Another function is to provide a standard interface (such as RS-232C) with the DTE (computer). The RS-232C standard requires certain defined command and control signals, data signals, and voltage levels for each signal.

Digital Data   Before we get into FSK, let's briefly review digital data. A detailed knowledge of binary numbers is not necessary for this system application. Information is represented in digital form by 1s and 0s, which are the binary digits or bits. In terms of voltage waveforms, a 1 is generally represented by a high level and a 0 by a low level. A stream of serial data consists of a sequence of bits as illustrated by an example in Figure 13–57(a).

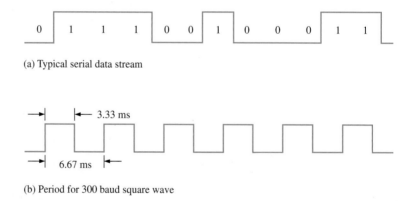

(a) Typical serial data stream

(b) Period for 300 baud square wave

FIGURE 13–57
A serial stream of digital data.

Baud Rate   A low-speed modem, such as the one we are focusing on, sends and receives digital data at a rate of 300 bits/s or 300 baud.[5] For example, if we have an alternating sequence of 1s and 0s (highs and lows), as indicated in Figure 13–57(b), each bit takes 3.33 ms. Since it takes two bits, a 1 and a 0, to make up the period of this particular waveform, the fundamental frequency of this format is 1/6.67 ms = 150 Hz. This is the maximum frequency of a 300 baud data stream because normally there may be several consecutive 1s and/or several consecutive zeros in a sequence, thus reducing the frequency. As mentioned earlier, the telephone network has a 300 Hz minimum frequency response, so the fundamental frequency of the 300 baud data stream will fall outside of the telephone bandwidth. This prevents sending digital data in its pure form over the phone lines.

---

[5] Technically, bit rate and baud rate are not the same. Baud rate indicates how many frequency shifts are sent per second. Each frequency shift can represent more than one bit; thus, a 14,400 bits/s modem actually transmits at 2400 baud.

Frequency-Shift Keying (FSK)   FSK is one method used to overcome the bandwidth limitation of the telephone system so that digital data can be sent over the phone lines. The basic idea of FSK is to represent 1s and 0s by two different frequencies within the telephone bandwidth. By the way, any frequency within the telephone bandwidth is an audible tone. The standard frequencies for a full-duplex 300 baud modem in the originate mode are 1070 Hz for a 0 (called a space) and 1270 Hz for a 1 (called a mark). In the answer mode, 2025 Hz is a 0 and 2225 Hz is a 1. The relationship of these FSK frequencies and the telephone bandwidth is illustrated in Figure 13–58. Signals in both the originate and answer bands can exist at the same time on the phone line and not interfere with each other because of the frequency separation.

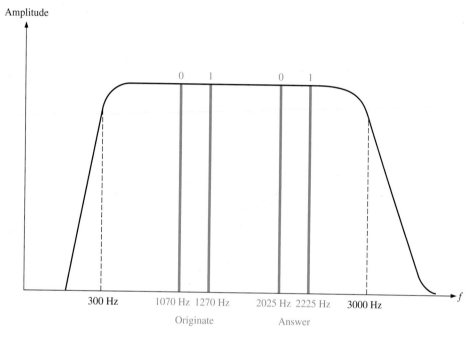

FIGURE 13–58
Frequencies for 300 baud, full-duplex data transmission.

An example of a digital data stream converted to FSK by a modem is shown in Figure 13–59.

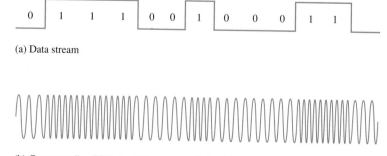

(a) Data stream

(b) Corresponding FSK signal (frequency relationships are not exact)

FIGURE 13–59
Example of FSK data.

Modem Circuit Operation

The FSK modem circuits, shown in Figure 13–60, contain an LM565 PLL and a VCO integrated circuit. The VCO can be a device such as the 4046 (not covered specifically in this chapter), which is a PLL device in which the VCO portion can be used by itself because all of the necessary inputs and outputs are available. The VCO in the LM565 cannot be used independently of the PLL because there is no input pin for the control voltage.

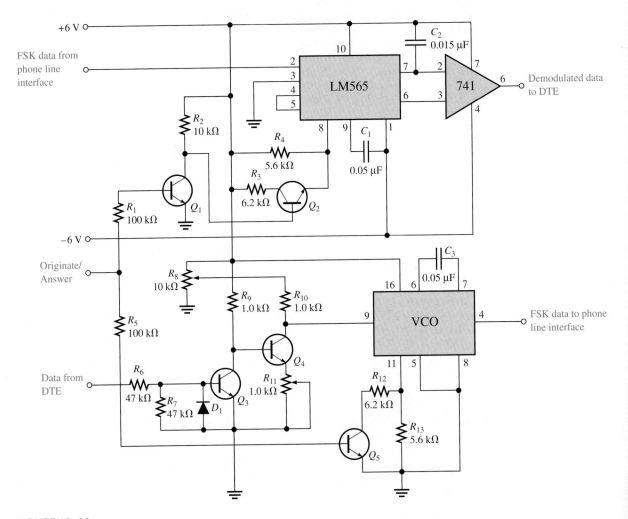

FIGURE 13–60
FSK modulator/demodulator circuit.

The function of the VCO is to accept digital data from a DTE and provide FSK modulation. The VCO is always the transmitting device. The digital data come in on the control voltage input (pin 9) of the VCO via a level-shifting circuit formed by Q_3 and Q_4. This circuit is used because the data from the RS-232C interface are dual polarity with a positive voltage representing a 0 and a negative voltage representing a 1. Potentiometer R_8 is for adjusting the high level of the control voltage and R_{11} is for adjusting the low level for the purpose of fine-tuning the frequency. Transistor Q_5 provides for originate/answer mode frequency selection by changing the value of the frequency-selection resistance from pin 11 to ground. Transistors Q_1 and Q_2 perform a similar function for the PLL.

When the digital data are at high levels, corresponding to logic 0s, the VCO oscillates at 1070 Hz in the originate mode and 2025 Hz in the answer mode. When the digital data are at low levels, corresponding to logic 1s, the VCO oscillates at 1270 Hz in the originate mode and 2225 Hz in the answer mode. An example of the originate mode is when a DTE issues a request for data and transmits that request to another DTE. An example of the answer mode is when the receiving DTE responds to a request and sends data back to the originating DTE.

The function of the PLL is to accept incoming FSK-modulated data and convert it to a digital data format for use by the DTE. The PLL is always a receiving device. When the modem is in the originate mode, the PLL is receiving answer-mode data from the other modem. When the modem is in the answer mode, the PLL is receiving originate-mode data from the other modem. The 741 op-amp is connected as a comparator that changes the data levels from the PLL to a dual-polarity format for compatibility with the RS-232C interface.

Now, so that you can take a closer look at the FSK modem board, let's take it out of the system and put it on the troubleshooter's bench.

TROUBLESHOOTER'S BENCH

■ ACTIVITY 1 Relate the PC Board to the Schematic

Locate and identify each component and each input/output pin on the PC board in Figure 13–61 using the schematic in Figure 13–60. Verify that the board and the schematic agree. If the PC board and the schematic do not agree, indicate the problem.

FIGURE 13–61

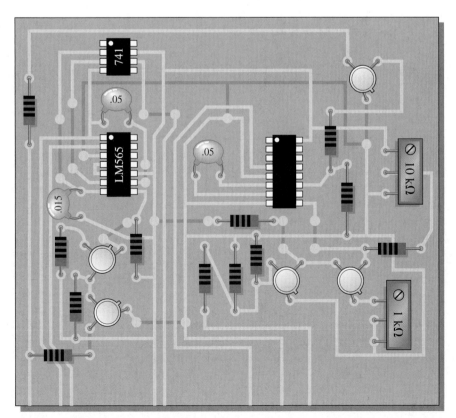

■ ACTIVITY 2   Analyze the Circuits

For this application, the free-running frequencies of both the PLL and the VCO circuits are determined by the formula in Equation (13–5).

Step 1:   Verify that the free-running frequency for the PLL IC is approximately 1070 Hz in the originate mode and approximately 1270 Hz in the answer mode.

Step 2:   Repeat Step 1 for the VCO.

Step 3:   Determine the approximate minimum and maximum output voltages for the 741 comparator.

Step 4:   Determine the maximum high-level voltage on pin 9 of the VCO.

Step 5:   If a 300 Hz square wave that varies from +5 V to −5 V is applied to the data from the DTE input, what should you observe on pin 4 of the VCO?

Step 6:   When the data from the DTE are low, pin 9 of the VCO is at approximately 0 V. At this level, the VCO oscillates at 1070 Hz or 2025 Hz. When the data from the DTE go high, to what value should the voltage at pin 9 be adjusted to produce a 1270 Hz or 2225 Hz frequency if the transfer function of the VCO is 50 Hz/V?

■ ACTIVITY 3   Write a Technical Report

Describe the overall operation of the FSK modem board. Specify how each circuit works and what its purpose is. Identify the function of each component. Use the results of Activity 2 as appropriate.

■ ACTIVITY 4   Troubleshoot the System for Each of the Following Problems By Stating the Probable Cause or Causes

1. There is no demodulated data output voltage when there are verified FSK data from the phone line interface.

2. The LM565 properly demodulates 1070 Hz and 1270 Hz FSK data but does not properly demodulate 2025 Hz and 2225 Hz data.

3. The VCO produces no FSK output.

4. The VCO produces a continuous 1070 Hz tone in the originate mode and a continuous 2025 Hz tone in the answer mode when there are proper data from the DTE.

13–9 REVIEW QUESTIONS

1. The originate/answer input to the modem is *low*. In what mode is the system?

2. What is the purpose of diode D_1 in the FSK modem circuit?

3. The VCO is transmitting 1070 Hz and 1270 Hz FSK signals. To what frequencies does the PLL respond from another modem?

4. If the VCO is transmitting a constant 2225 Hz tone, what does this correspond to in terms of digital data? In what mode is the modem?

■ **SUMMARY**

- In amplitude modulation (AM), the amplitude of a higher-frequency carrier signal is varied by a lower-frequency modulating signal (usually an audio signal).

- A basic superheterodyne AM receiver consists of an RF amplifier (not always), a mixer, a local oscillator, an IF (intermediate frequency) amplifier, an AM detector, and audio and power amplifiers.

- The IF in a standard AM receiver is 455 kHz.

- The AGC (automatic gain control) in a receiver tends to keep the signal strength constant within the receiver to compensate for variations in the received signal.

- In frequency modulation (FM), the frequency of a carrier signal is varied by a modulating signal.

- A superheterodyne FM receiver is basically the same as an AM receiver except that it requires a limiter to keep the IF amplitude constant, a different kind of detector or discriminator, and a de-emphasis network. The IF is 10.7 MHz.

- A four-quadrant linear multiplier can handle any combination of voltage polarities on its inputs.

- Amplitude modulation is basically a multiplication process.

- The multiplication of sinusoidal signals produces sum and difference frequencies.

- The output spectrum of a balanced modulator includes upper-side and lower-side frequencies, but no carrier frequency.

- The output spectrum of a standard amplitude modulator includes upper-side and lower-side frequencies and the carrier frequency.

- A linear multiplier is used as the mixer in receiver systems.

- A mixer converts the RF signal down to the IF signal. The radio frequency varies over the AM or FM band. The intermediate frequency is constant.

- One type of AM demodulator consists of a multiplier followed by a low-pass filter.

- The audio and power amplifiers boost the output of the detector or discriminator and drive the speaker.

- A voltage-controlled oscillator (VCO) produces an output frequency that can be varied by a control voltage. Its operation is based on a variable reactance.

- A VCO is a basic frequency modulator when the modulating signal is applied to the control voltage input.

- A phase-locked loop (PLL) is a feedback circuit consisting of a phase detector, a low-pass filter, a VCO, and sometimes an amplifier.

- The purpose of a PLL is to lock onto and track incoming frequencies.

- A linear multiplier can be used as a phase detector.

- A modem is a modulator/demodulator.

- DTE stands for digital terminal equipment.

- DCE stands for digital communications equipment.

■ **GLOSSARY**

Key terms are in color. All terms are included in the end-of-book glossary.

Amplitude modulation (AM)   A communication method in which a lower-frequency signal modulates (varies) the amplitude of a higher-frequency signal (carrier).

Audio   Related to the range of frequencies that can be heard by the human ear and generally considered to be in the 20 Hz to 20 kHz range.

Balanced modulation   A form of amplitude modulation in which the carrier is suppressed; sometimes known as *suppressed-carrier modulation*.

Demodulation   The process in which the information signal is recovered from the IF carrier signal; the reverse of modulation.

Discriminator   A type of FM demodulator.

Four-quadrant multiplier   A linear device that produces an output voltage proportional to the product of two input voltages.

Freqency modulation (FM)   A communication method in which a lower-frequency intelligence-carrying signal modulates (varies) the frequency of a higher-frequency signal.

Mixer   A device for down-converting frequencies in a receiver system.

Modem   A device that converts signals produced by one type of device to a form compatible with another; *mo*dulator/*dem*odulator.

Modulation   The process in which a signal containing information is used to modify the amplitude, frequency, or phase of a much higher-frequency signal called the carrier.

Phase-locked loop (PLL)   A device for locking onto and tracking the frequency of an incoming signal.

Voltage-controlled oscillator (VCO)   An oscillator for which the output frequency is dependent on a controlling input voltage.

■ **KEY FORMULAS**

(13–1) $V_{OUT} = KV_X V_Y$ Multiplier output voltage

(13–2) $v_1 v_2 = \dfrac{V_{1(p)} V_{2(p)}}{2} \cos 2\pi (f_1 - f_2)t$ Sum and difference frequencies

$$- \dfrac{V_{1(p)} V_{2(p)}}{2} \cos 2\pi (f_1 + f_2)t$$

(13–3) $V_{out} = V^2_{c(p)} \sin 2\pi f_c t$ Standard AM

$$+ \dfrac{V_{c(p)} V_{m(p)}}{2} \cos 2\pi (f_c - f_m)t$$

$$- \dfrac{V_{c(p)} V_{m(p)}}{2} \cos 2\pi (f_c + f_m)t$$

(13–4) $V_c = \dfrac{V_i V_o}{2} \cos \theta_e$ PLL control voltage

(13–5) $f_o \cong \dfrac{1.2}{4R_1 C_1}$ Output frequency LM565

(13–6) $f_{lock} = \pm \dfrac{8f_o}{V_{CC}}$ Lock range LM565

(13–7) $f_{cap} \cong \pm \dfrac{1}{2\pi} \sqrt{\dfrac{2\pi f_{lock}}{(3600)C_2}}$ Capture range LM565

■ **SELF-TEST**   *Answers are at the end of the chapter.*

1. In amplitude modulation, the pattern produced by the peaks of the carrier signal is called the
   (a) index   **(b)** envelope   **(c)** audio signal   **(d)** upper-side frequency

2. Which of the following is not a part of an AM superheterodyne receiver?
   (a) Mixer   **(b)** IF amplifier   **(c)** DC restorer
   (d) Detector   **(e)** Audio amplifier   **(f)** Local oscillator

3. In an AM receiver, the local oscillator always produces a frequency that is above the incoming RF by
   (a) 10.7 kHz   (b) 455 MHz   (c) 10.7 MHz   (d) 455 kHz

4. An FM receiver has an IF frequency that is
   (a) in the 88 MHz to 108 MHz range
   (b) in the 540 kHz to 1640 kHz range
   (c) 455 kHz
   (d) greater than the IF in an AM receiver

5. The detector or discriminator in an AM or an FM receiver
   (a) detects the difference frequency from the mixer
   (b) changes the RF to IF
   (c) recovers the audio signal
   (d) maintains a constant IF amplitude

6. In order to handle all combinations of input voltage polarities, a multiplier must have
   (a) four-quadrant capability   (b) three-quadrant capability
   (c) four inputs   (d) dual-supply voltages

7. The internal attenuation of a multiplier is called the
   (a) transconductance   (b) scale factor   (c) reduction factor

8. When the two inputs of a multiplier are connected together, the device operates as a
   (a) voltage doubler   (b) square root circuit
   (c) squaring circuit   (d) averaging circuit

9. Amplitude modulation is basically a
   (a) summing of two signals   (b) multiplication of two signals
   (c) subtraction of two signals   (d) nonlinear process

10. The frequency spectrum of a balanced modulator contains
    (a) a sum frequency   (b) a difference frequency   (c) a carrier frequency
    (d) answers (a), (b), and (c)   (e) answers (a) and (b)   (f) answers (b) and (c)

11. The IF in a receiver is the
    (a) sum of the local oscillator frequency and the RF carrier frequency
    (b) local oscillator frequency
    (c) difference of the local oscillator frequency and the carrier RF frequency
    (d) difference of the carrier frequency and the audio frequency

12. When a receiver is tuned from one RF frequency to another,
    (a) the IF changes by an amount equal to the LO (local oscillator) frequency
    (b) the IF stays the same
    (c) the LO frequency changes by an amount equal to the audio frequency
    (d) both the LO and the IF frequencies change

13. The output of the AM detector goes directly to the
    (a) IF amplifier   (b) mixer   (c) audio amplifier   (d) speaker

14. If the control voltage to a VCO increases, the output frequency
    (a) decreases   (b) does not change   (c) increases

15. A PLL maintains lock by comparing
    (a) the phase of two signals
    (b) the frequency of two signals
    (c) the amplitude of two signals

TROUBLESHOOTER'S QUIZ   *Answers are at the end of the chapter.*

Refer to Figure 13–65.

❏ If R_X opens,

1. The output voltage willl

(a) increase   (b) decrease   (c) not change

❏ If one of the R_Ls is larger than specified,

2. The output voltage will

(a) increase   (b) decrease   (c) not change

❏ If there is a short across the 12 kΩ resistor connected to pin 3,

3. The output signal voltage will

(a) increase   (b) decrease   (c) not change

Refer to Figure 13–72.

❏ If the R_1 potentiometer is 20 kΩ instead of 10 kΩ,

4. The output signal range will

(a) increase   (b) decrease   (c) not change

❏ If C_3 opens,

5. A low-frequency output signal voltage will

(a) increase   (b) decrease   (c) not change

6. A high-frequency cutoff will

(a) increase   (b) decrease   (c) not change

❏ If C_2 opens,

7. The amplifier gain will

(a) increase   (b) decrease   (c) not change

Refer to Figure 13–74.

❏ If the value of R_1 is less than the specified value,

8. The free-running frequency will

(a) increase   (b) decrease   (c) not change

❏ If the positive dc supply voltage decreases,

9. The lock range of the PLL will

(a) increase   (b) decrease   (c) not change

❏ If the positive dc supply voltage increases,

10. The capture range will

(a) increase   (b) decrease   (c) not change

■ PROBLEMS

Answers to odd-numbered problems are at the end of the book.

SECTION 13–1 Basic Receivers

1. Label each block in the AM receiver in Figure 13–62.
2. Label each block in the FM receiver in Figure 13–63.

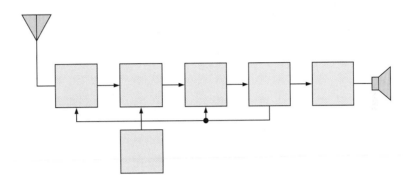

FIGURE 13–62

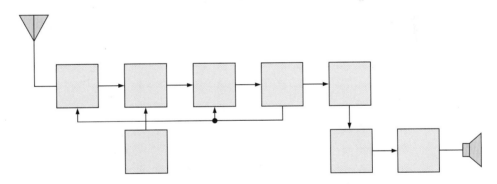

FIGURE 13–63

3. An AM receiver is tuned to a transmitted frequency of 680 kHz. What is the local oscillator (LO) frequency?
4. An FM receiver is tuned to a transmitted frequency of 97.2 MHz. What is the LO frequency?
5. The LO in an FM receiver is running at 101.9 MHz. What is the incoming RF? What is the IF?

SECTION 13–2 The Linear Multiplier

6. From the graph in Figure 13–64, determine the multiplier output voltage for each of the following pairs of input voltages.
   (a) $V_X = -4$ V, $V_Y = +6$ V   (b) $V_X = +8$ V, $V_Y = -2$ V
   (c) $V_X = -5$ V, $V_Y = -2$ V   (d) $V_X = +10$ V, $V_Y = +10$ V

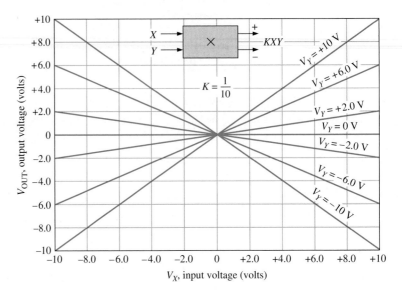

FIGURE 13–64

7. How much pin 3 current is there for the multiplier in Figure 13–65? The potentiometer is set at 2.8 kΩ.

8. Determine the scale factor for the multiplier in Figure 13–65.

9. If a certain multiplier has a scale factor of 0.8 and the inputs are +3.5 V and −2.9 V, what is the output voltage?

10. Show the connections for the multiplier in Figure 13–65 in order to implement a squaring circuit.

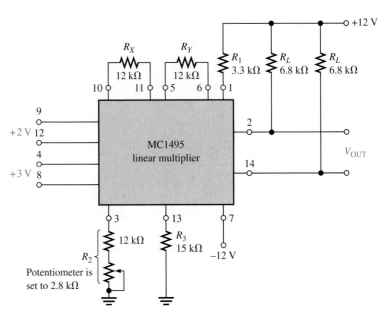

FIGURE 13–65

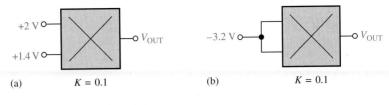

(a) $K = 0.1$ (b) $K = 0.1$

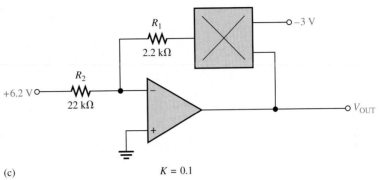

(c) $K = 0.1$

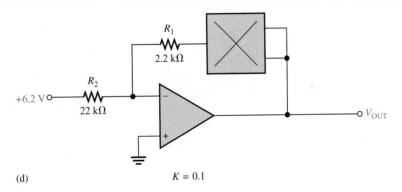

(d) $K = 0.1$

FIGURE 13–66

11. Determine the output voltage for each circuit in Figure 13–66.

SECTION 13–3 Amplitude Modulation

12. If a 100 kHz signal and a 30 kHz signal are applied to a balanced modulator, what frequencies will appear on the output?

13. What are the frequencies on the output of the balanced modulator in Figure 13–67?

FIGURE 13–67

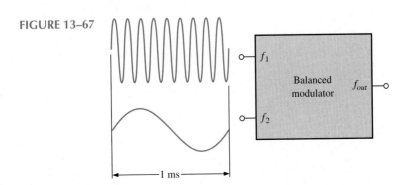

14. If a 1000 kHz signal and a 3 kHz signal are applied to a standard amplitude modulator, what frequencies will appear on the output?

15. What are the frequencies on the output of the standard amplitude modulator in Figure 13–68?

FIGURE 13–68

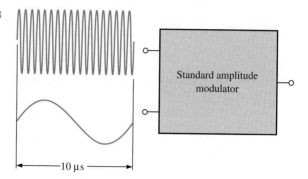

16. The frequency spectrum in Figure 13–69 is for the output of a standard amplitude modulator. Determine the carrier frequency and the modulating frequency.

17. The frequency spectrum in Figure 13–70 is for the output of a balanced modulator. Determine the carrier frequency and the modulating frequency.

18. A voice signal ranging from 300 Hz to 3 kHz amplitude modulates a 600 kHz carrier. Develop the frequency spectrum.

FIGURE 13–69

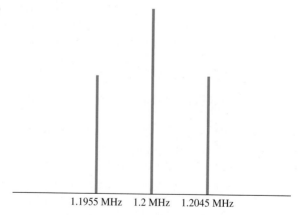

FIGURE 13–70

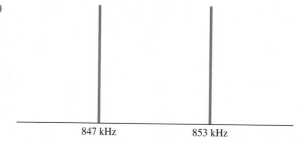

SECTION 13–4   The Mixer

19. Determine the output expression for a multiplier with one sinusoidal input having a peak voltage of 0.2 V and a frequency of 2200 kHz and the other input having a peak voltage of 0.15 V and a frequency of 3300 kHz.

20. Determine the output frequency of the IF amplifier for the frequencies shown in Figure 13–71.

FIGURE 13–71

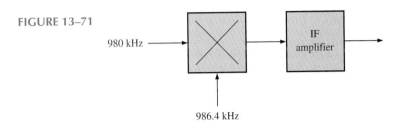

980 kHz

IF amplifier

986.4 kHz

SECTION 13–5   AM Demodulation

21. The input to a certain AM receiver consists of a 1500 kHz carrier and two side frequencies separated from the carrier by 20 kHz. Determine the frequency spectrum at the output of the mixer amplifier.

22. For the same conditions stated in Problem 21, determine the frequency spectrum at the output of the IF amplifier.

23. For the same conditions stated in Problem 21, determine the frequency spectrum at the output of the AM detector (demodulator).

SECTION 13–6   IF and Audio Amplifiers

24. For a carrier frequency of 1.2 MHz and a modulating frequency of 8.5 kHz, list all of the frequencies on the output of the mixer in an AM receiver.

25. In a certain AM receiver, one amplifier has a passband from 450 kHz to 460 kHz and another has a passband from 10 Hz to 5 kHz. Identify these amplifiers.

26. Determine the maximum and minimum output voltages for the audio power amplifier in Figure 13–72.

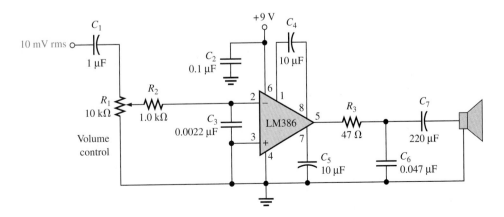

FIGURE 13–72

SECTION 13–7 **Frequency Modulation**

27. Explain how a VCO is used as a frequency modulator.

28. How does an FM signal differ from an AM signal?

29. What is the variable reactance element shown in the MC2833D diagram in Figure 13–44?

SECTION 13–8 **The Phase-Locked Loop (PLL)**

30. Label each block in the PLL diagram of Figure 13–73.

31. A PLL is locked onto an incoming signal with a peak amplitude of 250 mV and a frequency of 10 MHz at a phase angle of 30°. The 400 mV peak VCO signal is at a phase angle of 15°.
   (a) What is the VCO frequency?
   (b) What is the value of the control voltage being fed back to the VCO at this point?

FIGURE 13–73

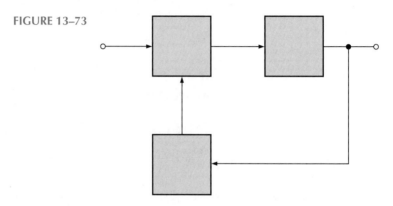

32. What is the conversion gain of a VCO if a 0.5 V increase in the control voltage causes the output frequency to increase by 3.6 kHz?

33. If the conversion gain of a certain VCO is 1.5 kHz per volt, how much does the frequency change if the control voltage increases 0.67 V?

34. Name two conditions for a PLL to acquire lock.

35. Determine the free-running frequency, the lock range, and the capture range for the PLL in Figure 13–74.

FIGURE 13–74

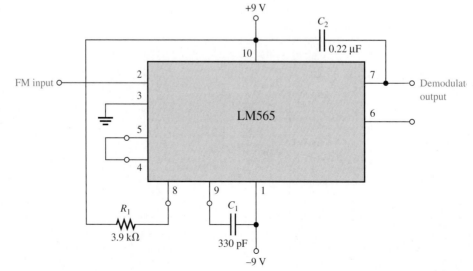

■ **ANSWERS**
TO REVIEW
QUESTIONS

Section 13–1

1. AM is amplitude modulation. FM is frequency modulation.

2. In AM, the modulating signal varies the amplitude of a carrier. In FM, the modulating signal varies the frequency of a carrier.

3. AM: 540 kHz to 1640 kHz; FM: 88 MHz to 108 MHz

Section 13–2

1. A four-quadrant multiplier can handle any combination (4) of positive and negative inputs. A one-quadrant multiplier can only handle two positive inputs, for example.

2. $K = 0.1$. K must be 1 for an output of 5 V.

3. Connect the two inputs together and apply a single input variable.

Section 13–3

1. Amplitude modulation is the process of varying the amplitude of a carrier signal with a modulating signal.

2. Balanced modulation produces no carrier frequency on the output, whereas standard AM does.

3. The carrier signal is the modulated signal and has a sufficiently high frequency for transmission. The modulating signal is a lower-frequency signal that contains information and varies the carrier amplitude according to its waveshape.

4. The upper-side frequency is the sum of the carrier frequency and the modulating frequency. The lower-side frequency is the difference of the carrier frequency and the modulating frequency.

5. By summing the peak carrier voltage and the modulating signal before mixing with the carrier signal

Section 13–4

1. The mixer produces (among other frequencies) a signal representing the difference between the incoming carrier frequency and the local oscillator frequency. This is called the intermediate frequency.

2. The mixer multiplies the carrier and the local oscillator signals.

3. 1000 kHz + 350 kHz = 1350 kHz, 1000 kHz − 350 kHz = 650 kHz

Section 13–5

1. The filter removes all frequencies except the audio.

2. Only the 1 kHz

Section 13–6

1. To amplify the 455 kHz amplitude modulated IF coming from the mixer

2. The IF center frequency is 455 kHz.

3. The 10 kHz bandwidth allows the upper-side and lower-side frequencies that contain the information to pass.

4. The audio amplifier follows the detector because the detector is the circuit that recovers the audio from the modulated IF.

5. The IF has a response of approximately 455 kHz ± 5 kHz. The typical audio amplifier has a maximum bandwidth from tens of hertz up to about 15 kHz although for many amplifiers, the bandwidth can be much less than this typical maximum.

Section 13–7

1. The frequency variation of an FM signal bears the information.

2. VCO is voltage-controlled oscillator.

3. VCOs are based on the principle of voltage-variable reactance.

Section 13–8

1. Phase detector, low-pass filter, and VCO

2. Sometimes a PLL uses an amplifier in the loop.

3. A PLL locks onto and tracks a variable incoming frequency.

4. The lock range specifies how much a lock-on frequency can deviate without the PLL losing lock. The capture range specifies how close the incoming frequency must be from the free-running VCO frequency in order for the PLL to lock.

5. The PLL detects a change in the phase of the incoming signal compared to the VCO signal that indicates a change in frequency. The positive feedback then causes the VCO frequency to change along with the incoming frequency.

Section 13–9

1. A *low* on the originate/answer input puts the modem in the originate mode.

2. The diode clips excess negative voltage to protect the base-emitter junction of the transistor.

3. The PLL responds to 2025 Hz and 2225 Hz because the other modem is transmitting in the answer mode.

4. A constant 2225 Hz represents a continuous string of 1s; answer mode

---

■ ANSWERS
TO PRACTICE
EXERCISES
FOR
EXAMPLES

13–1 -3.6 V from the graph in Figure 13–9

13–2 0.728 V

13–3 Modulate the carrier with a higher-frequency signal.

13–4 The balanced modulator output has the same side frequencies but does not have a carrier frequency.

13–5 $V_p = 0.025$ mV, $f = 455$ kHz

13–6 455 kHz

13–7 0.172 V

13–8 A decrease of 6 kHz; 244 kHz

13–9 Decrease C_2 to 0.0595 μF.

■ **ANSWERS TO SELF-TEST**

1. (b)	**2.** (c)	**3.** (d)	**4.** (d)	**5.** (c)
6. (a)	**7.** (b)	**8.** (c)	**9.** (b)	**10.** (e)
11. (c)	**12.** (b)	**13.** (c)	**14.** (c)	**15.** (a)

■ **ANSWERS TO TROUBLE-SHOOTER'S QUIZ**

1. decrease	**2.** increase	**3.** decrease	**4.** not change
5. not change	**6.** increase	**7.** not change	**8.** increase
9. increase	**10.** decrease		

14

DATA CONVERSION CIRCUITS

Courtesy Hewlett-Packard Company

■ CHAPTER OBJECTIVES

❑ Explain analog switches and identify each type
❑ Discuss the operation of sample-and-hold amplifiers
❑ Discuss analog and digital quantities and general interfacing considerations
❑ Describe the operation of digital-to-analog converters (DACs)
❑ Describe A/D conversion
❑ Discuss the operation of analog-to-digital converters (ADCs)
❑ Discuss the basic operation of V/F converters and F/V converters
❑ Troubleshoot DACs and ADCs
❑ Apply what you have learned in this chapter to a system application

■ KEY TERMS

❑ Analog switch
❑ Sample-and-hold
❑ Acquisition time
❑ Analog-to-digital converter (ADC)
❑ Digital-to-analog converter (DAC)
❑ Resolution
❑ Quantization
❑ Flash
❑ Successive approximation

■ CHAPTER INTRODUCTION

Data conversion circuits make interfacing between analog and digital systems possible. Most things in nature occur in analog form. For example, your voice is analog, time is analog, temperature and pressure are analog, and the speed of your vehicle is analog. These quantities and others are first sensed or measured with analog (linear) circuits and are then frequently converted to digital form to facilitate storage, processing, or display.

Also, in many applications, information in digital form must be converted back to analog form. An example of this is digitized music that is stored on a CD. Before you can hear the sounds, the digital information must be converted to its original analog form.

In this chapter, you will study several basic types of circuits found in applications that require data conversion.

■ A SYSTEM APPLICATION

This particular system consists of four large solar cell arrays (see Figure 14–61) that can be individually positioned for proper orientation to the sun's rays. Both the azimuth and the elevation of each array are controlled by stepping motors. The angular position in both azimuth and elevation is sensed by position potentiometers that produce voltages proportional to the angular positions. Many systems use synchros or resolvers as angular transducers, but our system uses potentiometers for simplicity.

The electronic control circuits utilize an analog multiplexer to obtain the analog position from each solar array. The output of the analog multiplexer is then converted to digital form by the analog-to-digital converter for processing by the digital controller. The computer in the digital controller determines how much each array must be rotated in both azimuth and elevation based on stored information about the sun's position. The digital controller then sends the appropriate number of pulses to step the motors to the proper position. Basically, the system controls the solar arrays so that they track the sun each day to maintain an approximate 90° angle to the sun's rays.

This chapter's system application focuses on the analog multiplexer and ADC circuits. For the system application in Section 14–9, in addition to the other topics, be sure you understand
❑ The basic operation of an analog switch
❑ The basic operation of an analog multiplexer
❑ The basic operation of analog-to-digital converters (ADCs)

14–1 ■ ANALOG SWITCHES

Analog switches are important in many types of electronic systems where it is necessary to switch signals on and off electronically. Major applications of analog switches are in signal selection, routing, and processing. Analog switches usually incorporate a FET as the basic switching element.

After completing this section, you should be able to

❏ Explain analog switches and identify each type
  ❏ Identify a single-pole–single-throw analog switch
  ❏ Identify a single-pole–double-throw analog switch
  ❏ Identify a double-pole–single-throw analog switch
  ❏ Describe the ADG202A analog switch IC
  ❏ Discuss multiple-channel analog switches

Types of Analog Switches

Three basic types of **analog switches** in terms of their functional operation are

❏ Single pole–single throw (SPST)

❏ Single pole–double throw (SPDT)

❏ Double pole–single throw (DPST)

Figure 14–1 illustrates these three basic types of analog switches. As you can see, the analog switch consists of a control element and one or more input-to-output paths called *switch channels*.

FIGURE 14–1

Basic types of analog switches.

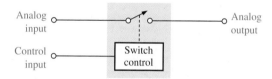

(a) Single pole–single throw (SPST)

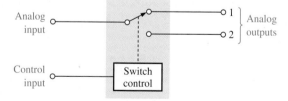

(b) Single pole–double throw (SPDT)

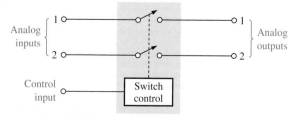

(c) Double pole–single throw (DPST)

An example of an analog switch is the ADG202A.[1] This IC device has four independently operated normally closed SPST switches as shown in Figure 14–2(a). Typical packages are shown in Figure 14–2(b).

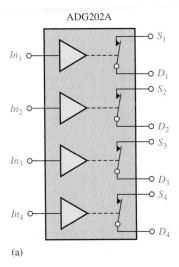

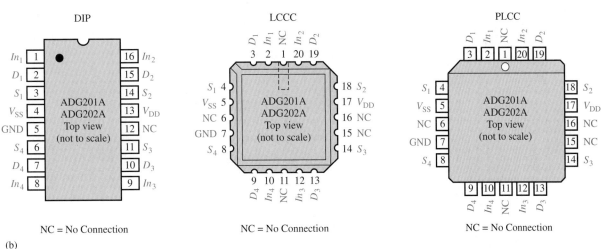

FIGURE 14–2
The ADG202A Quad SPST switches.

When the control input is at a high-level voltage (at least 2.4 V for the ADG202A), the switch is closed (on). When the control input is at a low-level voltage (no greater than 0.8 V for this device), the switch is open (off). The switches themselves are typically implemented with MOSFETs.

---

[1] Data sheet for ADG202A available at http://www.analogdevices.com

EXAMPLE 14–1 Determine the output waveform of the analog switch in Figure 14–3(a) for the control voltage and analog input voltage shown.

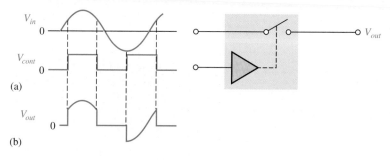

(a)

(b)

FIGURE 14–3

Solution   When the control voltage is high, the switch is closed and the analog input passes through to the output. When the control voltage is low, the switch is open and there is no output voltage. The output waveform is shown in Figure 14–3(b) in relation to the other voltages.

*Practice Exercise**   What will be the output waveform in Figure 14–3 if the frequency of the control voltage is doubled but keeping the same duty cycle?

_____

* Answers are at the end of the chapter.

Multiple-Channel Analog Switches

In data acquisition systems where inputs from several different sources must be independently converted to digital form for processing, a technique called *multiplexing* is used. A separate analog switch is used for each analog source as illustrated in Figure 14–4 for a four-channel system. In this type of application, all of the outputs of the analog switches

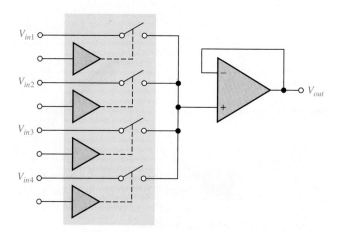

FIGURE 14–4
A four-channel analog multiplexer.

are connected together to form a common output and only one switch can be closed at a given time. The common switch outputs are connected to the input of a voltage-follower as indicated.

A good example of an IC analog multiplexer is the AD9300[2] shown in Figure 14–5. This device contains four analog switches that are controlled by a channel decoder. The inputs A_0 and A_1 determine which one of the four switches is on. If A_0 and A_1 are both low, input In_1 is selected. If A_0 is high and A_1 is low, input In_2 is selected. If A_0 is low and A_1 is high, input In_3 is selected. If A_0 and A_1 are both high, input In_4 is selected. The Enable input controls the switch that connects or disconnects the output. The AD9300 is capable of switching 4 channels of video for applications including video routing, medical imaging, radar systems, and data acquisition systems.

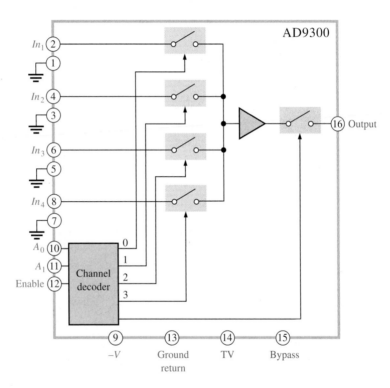

FIGURE 14–5
The AD9300 analog multiplexer.

---

[2] Data sheet for AD9300 is available at http://www.analogdevices.com and Appendix A.

EXAMPLE 14–2   Determine the output waveform of the analog multiplexer in Figure 14–6 for the control inputs and the analog inputs shown.

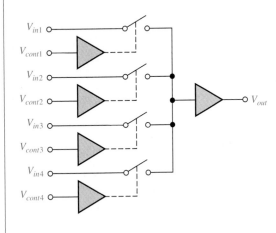

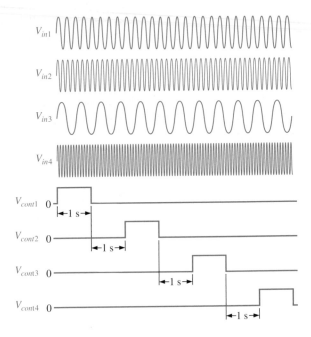

FIGURE 14–6

Solution   When a control input is a high level, the corresponding switch is closed and the analog voltage on its input is switched through to the output. Notice that only one control voltage is high at a time. The inputs to the switches are sinusoidal waves, each having a different frequency. The resulting output is a sequence of different sinusoidal waves that last for one second and that are separated by a one-second interval, as indicated in Figure 14–7.

FIGURE 14–7

V_{out}

Practice Exercise   How is the output waveform in Figure 14–7 affected if the time interval between the control voltage pulses is decreased?

14–1 REVIEW QUESTIONS*

1. What is the purpose of an analog switch?
2. What is the basic function of an analog multiplexer?

* Answers are at the end of the chapter.

14–2 ■ SAMPLE-AND-HOLD AMPLIFIERS

A sample-and-hold amplifier samples an analog input voltage at a certain point in time and retains or holds the sampled voltage for an extended time after the sample is taken. The sample-and-hold process keeps the sampled analog voltage constant for the length of time necessary to allow an analog-to-digital converter (ADC) to convert the voltage to digital form.

After completing this section, you should be able to

❑ Discuss the operation of sample-and-hold amplifiers
   ❑ Describe tracking in a sample-and-hold amplifier
   ❑ Define *aperture time, aperture jitter, acquisition time, droop,* and *feedthrough*
   ❑ Describe the AD585 sample-and-hold amplifier

A Basic Sample-and-Hold Circuit

A basic **sample-and-hold** circuit consists of an analog switch, a capacitor, and input and output buffer amplifiers as shown in Figure 14–8. The analog switch samples the analog input voltage through the input buffer amplifier, the capacitor (C_H) stores or holds the sampled voltage for a period of time, and the output buffer amplifier provides a high input impedance to prevent the capacitor from discharging quickly.

FIGURE 14–8
A basic sample-and-hold circuit.

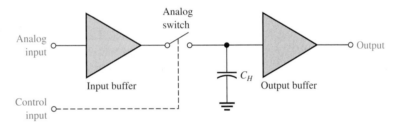

As illustrated in Figure 14–9, a relatively narrow control voltage pulse closes the analog switch and allows the capacitor to charge to the value of the input voltage. The switch then opens, and the capacitor holds the voltage for a long period of time because of the very high impedance discharge path through the op-amp input. So basically, the

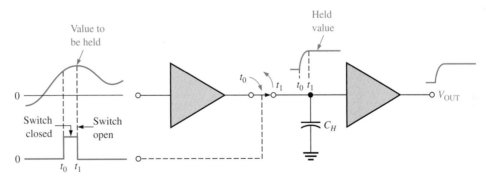

FIGURE 14–9
Basic action of a sample-and-hold.

sample-and-hold circuit converts an instantaneous value of the analog input voltage to a dc voltage.

Tracking During Sample Time

Perhaps a more appropriate designation for a sample-and-hold amplifier is sample/track-and-hold because the circuit actually tracks the input voltage during the sample interval. As indicated in Figure 14–10, the output follows the input during the time that the control voltage is high; and when the control voltage goes low, the last voltage is held until the next sample interval.

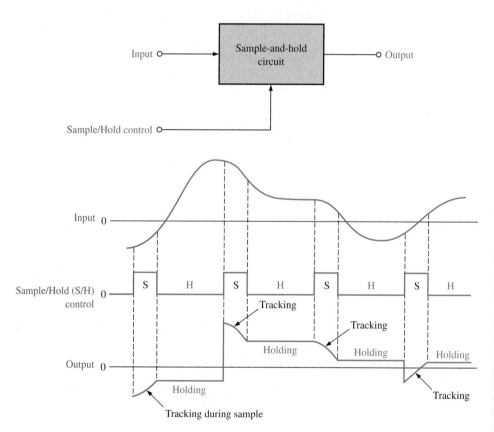

FIGURE 14–10
Example of tracking during a sample-and-hold sequence.

EXAMPLE 14–3 Determine the output voltage waveform for the sample/track-and-hold amplifier in Figure 14–11, given the input and control voltage waveforms.

Solution During the time that the control voltage is high, the analog switch is closed and the circuit is tracking the input. When the control voltage goes low, the analog switch opens; and the last voltage value is held at a constant level until the next time the control voltage goes high. This is shown in Figure 14–12.

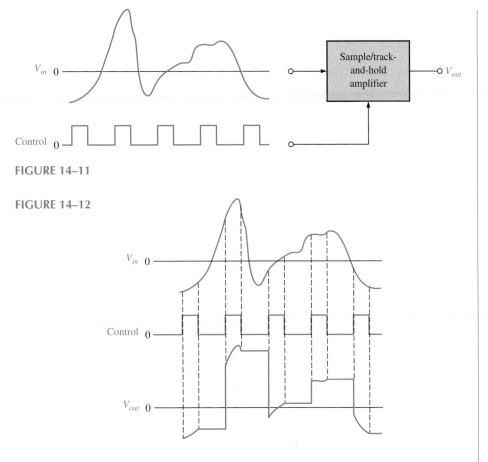

FIGURE 14–11

FIGURE 14–12

Practice Exercise   Sketch the output voltage waveform for Figure 14–11 if the control voltage frequency is reduced by half.

Performance Specifications

In addition to specifications similar to those of a closed-loop op-amp that were discussed in Chapter 6, several specifications are peculiar to sample-and-hold amplifiers. These include the aperture time, aperture jitter, acquisition time, droop, and feedthrough.

❏ **Aperture time**—the time for the analog switch to fully open after the control voltage switches from its sample level to its hold level. Aperture time produces a delay in the effective sample point.

❏ **Aperture jitter**—the uncertainty in the aperture time.

❏ **Acquisition time**—the time required for the device to reach its final value when the control voltage switches from its hold level to its sample level.

❏ **Droop**—the change in voltage from the sampled value during the hold interval because of charge leaking off of the hold capacitor.

❏ **Feedthrough**—the component of the output voltage that follows the input signal after the analog switch is opened. The inherent capacitance from the input to the output of the switch causes feedthrough.

Each of these parameters is illustrated in Figure 14–13 for an example input voltage waveform.

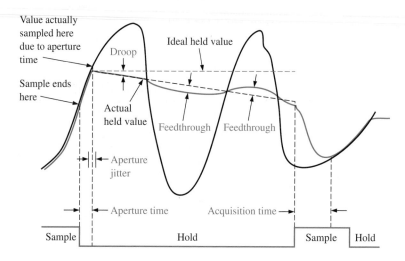

FIGURE 14–13

Sample-and-hold amplifier specifications. The effects are exaggerated for clarity. The black curve is input voltage waveform; the colored curve is output voltage.

A Specific Device

An example of a basic sample-and-hold amplifier is the AD585[3]. The circuit and pin configuration are shown in Figure 14–14. As shown in the figure, this particular device consists of two buffer amplifiers and an analog switch that is controlled by a logic gate. The internal hold capacitor has a value of 100 pF. An additional capacitor can be connected externally in parallel, if necessary, between pins 7 and 8.

FIGURE 14–14

The AD585 sample-and-hold amplifier.

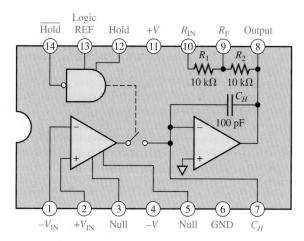

The control voltage for establishing the sample/hold intervals is applied between pins 14 and 13 or pins 12 and 13. The input signal to be sampled is applied to pin 2. A potentiometer for nulling out the offset voltage can be connected between pins 3 and 5, and

---

[3] Data sheet for AD585 available at http://www.analogdevices.com

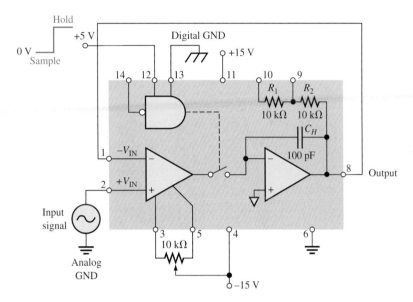

(a) Sample and hold with $A_v = +1$ and an optional offset null adjustment

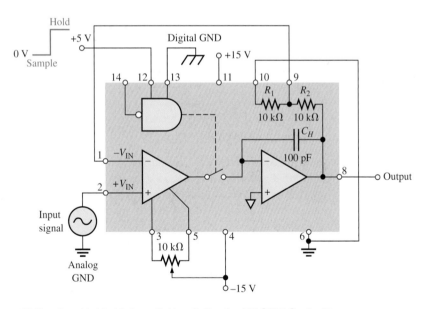

(b) Sample and hold with $A_v = +2$. $(A_v = R_2/R_1 + 1 = 10\ k\Omega/10\ k\Omega + 1 = 2)$

FIGURE 14–15
Two possible configurations of the AD585 sample-and-hold amplifier.

the overall gain of the device can be set to pin 1 or pin 2 without external feedback connections. Two typical configurations are shown in Figure 14–15. Other values of gain can be achieved using external resistors.

14–2 REVIEW QUESTIONS

1. What is the basic function of a sample-and-hold amplifier?
2. In reference to the output of a sample-and-hold amplifier, what does droop mean?
3. Define *aperture time*.
4. What is acquisition time?

14–3 ■ INTERFACING THE ANALOG AND DIGITAL WORLDS

Analog quantities are sometimes called real-world quantities because most physical quantities are analog in nature. Many applications of computers and other digital systems require the input of real-world quantities, such as temperature, speed, position, pressure, and force. Real-world quantities can even include graphic images. Also, digital systems are often used to control real-world quantities. A basic familiarity with the binary number system is assumed for this and the next sections.

After completing this section, you should be able to

❑ Discuss digital and analog quantities and general interfacing considerations
  ❑ Describe an analog quantity
  ❑ Describe a digital quantity
  ❑ Discuss examples of real-world analog/digital interfacing

Digital and Analog Signals

An analog quantity is one that has a continuous set of values over a given range, as contrasted with discrete values for the digital case. Almost any measurable quantity is analog in nature, such as temperature, pressure, speed, and time. To further illustrate the difference between an analog and a digital representation of a quantity, let's take the case of a voltage that varies over a range from 0 V to +15 V. The analog representation of this quantity takes in all values between 0 and +15 V of which there is an infinite number.

In the case of *digital* representation using a 4-bit binary code, only sixteen values can be defined. More values between 0 and +15 can be represented by using more bits in the digital code. So an analog quantity can be represented to a high degree of accuracy with a digital code that specifies discrete values within the range. This concept is illustrated in Figure 14–16, where the analog function shown is a smoothly changing curve that takes on values between 0 V and +15 V. If a 4-bit code is used to represent this curve, each binary number represents a discrete point on the curve.

In Figure 14–16 the voltage on the analog curve is measured, or sampled, at each of thirty-five equal intervals. The voltage at each of these intervals is represented by a 4-bit code as indicated. At this point, we have a series of binary numbers representing various voltage values along the analog curve. This is the basic idea of analog-to-digital (A/D) conversion.

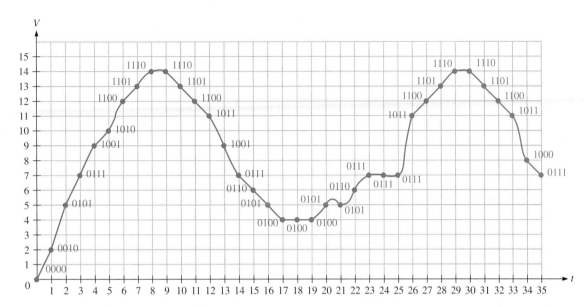

FIGURE 14–16
Discrete (digital) points on an analog curve.

An approximation of the analog function in Figure 14–16 can be reconstructed from the sequence of digital numbers that has been generated. Obviously, there will be some error in the reconstruction because only certain values are represented (thirty-six in this example) and not the continuous set of values. If the digital values at all of the thirty-six points are graphed as shown in Figure 14–17, we have a reconstructed function. As you can see, the graph only approximates the original curve because values between the points are not known.

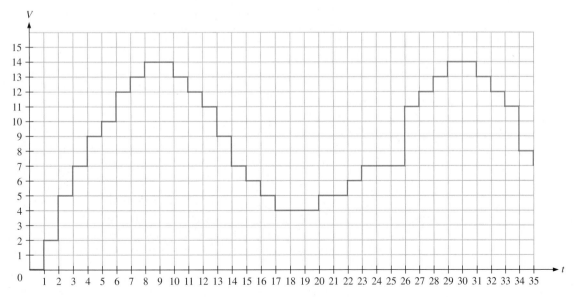

FIGURE 14–17
A rough digital reproduction of an analog curve.

Interfacing Applications

To interface between the digital and analog worlds, two basic processes are required. These are analog-to-digital (A/D) conversion and digital-to-analog (D/A) conversion. The following two system examples illustrate the application of these conversion processes.

An Electronic Thermostat   A simplified block diagram of a microprocessor-based electronic thermostat is shown in Figure 14–18. The room temperature sensor produces an analog voltage that is proportional to the temperature. This voltage is increased by the linear amplifier and applied to the **analog-to-digital converter (ADC)**, where it is converted to a digital code and periodically sampled by the microprocessor. For example, suppose the room temperature is 67°F. A specific voltage value corresponding to this temperature appears on the ADC input and is converted to an 8-bit binary number, 01000011.

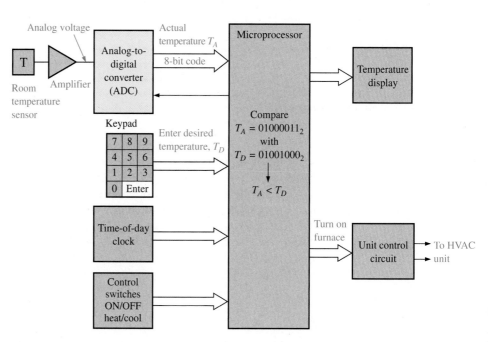

FIGURE 14–18
An electronic thermostat that uses an ADC.

Internally, the microprocessor compares this binary number with a binary number representing the desired temperature (say 01001000 for 72°F). This desired value has been previously entered from the keypad and stored in a register. The comparison shows that the actual room temperature is less than the desired temperature. As a result, the microprocessor instructs the unit control circuit to turn the furnace on. As the furnace runs, the microprocessor continues to monitor the actual temperature via the ADC. When the actual temperature equals or exceeds the desired temperature, the microprocessor turns the furnace off.

A Digital Audio Tape (DAT) Player/Recorder   Another system example that includes both A/D and D/A conversion is the DAT player/recorder. A basic block diagram is shown in Figure 14–19.

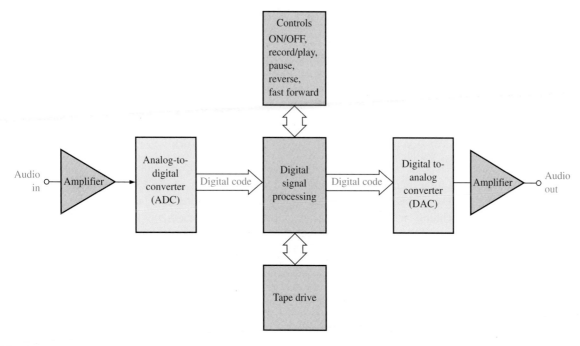

FIGURE 14–19
Basic block diagram of a DAT system.

An audio signal, of course, is an analog quantity. In the record mode, sound is picked up, amplified, and converted to digital form by the ADC. The digital codes representing the audio signal are processed and recorded on the tape.

 In the play mode, the digitized audio signal is read from the tape, processed, and converted back to analog form by the **digital-to-analog converter (DAC)**. It is then amplified and sent to the speaker system.

14–3 REVIEW QUESTIONS

1. In what form do quantities appear naturally?
2. Explain the basic purpose of A/D conversion.
3. Explain the basic purpose of D/A conversion.

14–4 ■ DIGITAL-TO-ANALOG (D/A) CONVERSION

D/A conversion is an important part of many systems. In this section, we will examine two basic types of digital-to-analog converters (DACs) and learn about their performance characteristics. The binary-weighted-input DAC was introduced in Chapter 8 as an example of a scaling adder application and is covered more thoroughly in this section. Also, a more commonly used configuration called the R/2R ladder DAC is introduced.

After completing this section, you should be able to

❑ Describe the operation of digital-to-analog converters (DACs)
  ❑ Describe the binary-weighted-input DAC
  ❑ Describe the *R/2R* ladder DAC
  ❑ Discuss resolution, accuracy, linearity, monotonicity, and settling time

Binary-Weighted-Input Digital-to-Analog Converter

The binary-weighted-input DAC uses a resistor network with resistance values that represent the binary weights of the input bits of the digital code. Figure 14–20 shows a 4-bit DAC of this type. Each of the input resistors will either have current or have no current, depending on the input voltage level. If the input voltage is zero (binary 0), the current is also zero. If the input voltage is high (binary 1), the amount of current depends on the input resistor value and is different for each input resistor, as indicated by the meters.

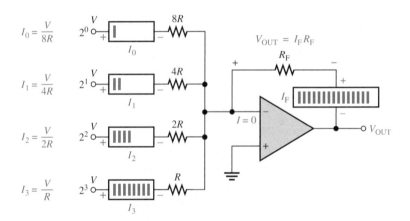

FIGURE 14–20
A 4-bit DAC with binary-weighted inputs.

Since there is practically no current at the op-amp inverting $(-)$ input, all of the input currents sum together and go through R_F. Since the inverting input is at 0 V (virtual ground), the drop across R_F is equal to the output voltage, so $V_{OUT} = I_F R_F$. Notice that V_{OUT} is negative in this example.

The values of the input resistors are chosen to be inversely proportional to the binary weights of the corresponding input bits. The lowest-value resistor (R) corresponds to the highest binary-weighted input (2^3). The other resistors are multiples of R (that is, $2R$, $4R$, and $8R$) and correspond to the binary weights 2^2, 2^1, and 2^0, respectively. The input currents are also proportional to the binary weights. Thus, the output voltage is proportional to the sum of the binary weights because the sum of the currents is through R_F.

One of the disadvantages of this type of DAC is the number of different resistor values. For example, an 8-bit converter requires eight resistors, ranging from some value of R to $128R$ in binary-weighted steps. This range of resistors requires tolerances of one part in 255 (less than 0.5%) to accurately convert the input, making this type of DAC very difficult to mass-produce.

EXAMPLE 14–4   Determine the output of the DAC in Figure 14–21(a) if the waveforms representing a sequence of 4-bit binary numbers in Figure 14–21(b) are applied to the inputs. Input D_0 is the least significant bit (LSB).

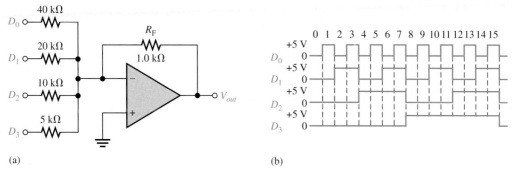

(a)   (b)

FIGURE 14–21

Solution   First, determine the current for each of the weighted inputs. Since the inverting ($-$) input of the op-amp is at 0 V (virtual ground) and a binary 1 corresponds to $+5$ V, the current through any of the input resistors is 5 V divided by the resistance value.

$$I_0 = \frac{5 \text{ V}}{40 \text{ k}\Omega} = 0.125 \text{ mA}$$

$$I_1 = \frac{5 \text{ V}}{20 \text{ k}\Omega} = 0.25 \text{ mA}$$

$$I_2 = \frac{5 \text{ V}}{10 \text{ k}\Omega} = 0.5 \text{ mA}$$

$$I_3 = \frac{5 \text{ V}}{5 \text{ k}\Omega} = 1.0 \text{ mA}$$

There is essentially no current at the inverting op-amp input because of its extremely high impedance. Therefore, assume that all of the current goes through the feedback resistor R_F. Since one end of R_F is at 0 V (virtual ground), the drop across R_F equals the output voltage, which is negative with respect to virtual ground.

$$V_{\text{OUT}(D0)} = (1.0 \text{ k}\Omega)(-0.125 \text{ mA}) = -0.125 \text{ V}$$
$$V_{\text{OUT}(D1)} = (1.0 \text{ k}\Omega)(-0.25 \text{ mA}) = -0.25 \text{ V}$$
$$V_{\text{OUT}(D2)} = (1.0 \text{ k}\Omega)(-0.5 \text{ mA}) = -0.5 \text{ V}$$
$$V_{\text{OUT}(D3)} = (1.0 \text{ k}\Omega)(-1.0 \text{ mA}) = -1.0 \text{ V}$$

From Figure 14–21(b), the first binary input code is 0000, which produces an output voltage of 0 V. The next input code is 0001, which produces an output voltage of -0.125 V. The next code is 0010, which produces an output voltage of -0.25 V. The next code is 0011, which produces an output voltage of -0.125 V $+$ -0.25 V $=$ -0.375 V. Each successive binary code increases the output voltage by -0.125 V, so for this particular straight binary sequence on the inputs, the output is a stairstep waveform going from 0 V to -1.875 V in -0.125 V steps. This is shown in Figure 14–22.

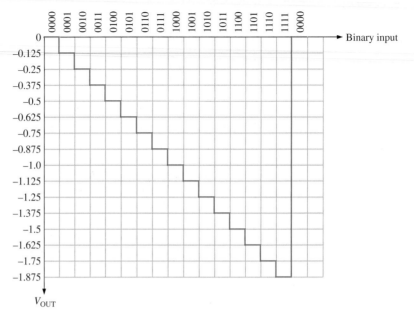

FIGURE 14–22
Output of the DAC in Figure 14–21.

Practice Exercise   What size are the output steps of the DAC if the feedback resistance is changed to 2.0 kΩ?

The *R/2R* Ladder Digital-to-Analog Converter

Another method of D/A conversion is the *R/2R* ladder, as shown in Figure 14–23 for four bits. It overcomes one of the problems in the binary-weighted-input DAC in that it requires only two resistor values.

Start by assuming that the D_3 input is at a high level (+5 V) and the others are at a low level (ground, 0 V). This condition represents the binary number 1000. A circuit analysis will show that this reduces to the equivalent form shown in Figure 14–24(a). Essentially

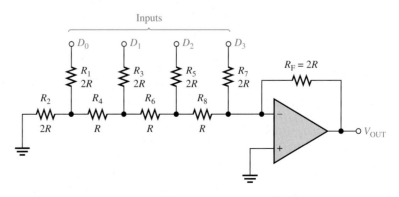

FIGURE 14–23
An R/2R ladder DAC.

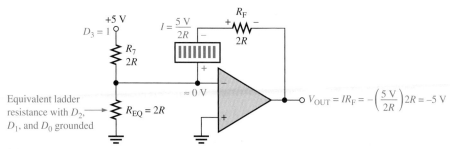

(a) Equivalent circuit for $D_3 = 1, D_2 = 0, D_1 = 0, D_0 = 0$

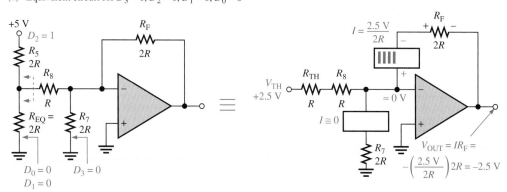

(b) Equivalent circuit for $D_3 = 0, D_2 = 1, D_1 = 0, D_0 = 0$

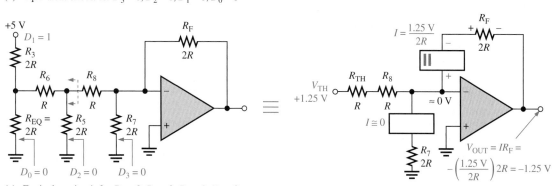

(c) Equivalent circuit for $D_3 = 0, D_2 = 0, D_1 = 1, D_0 = 0$

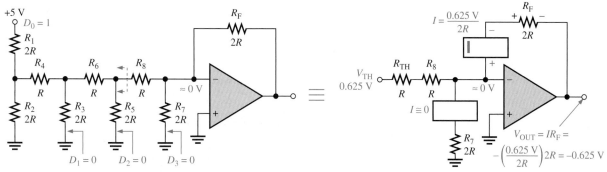

(d) Equivalent circuit for $D_3 = 0, D_2 = 0, D_1 = 0, D_0 = 1$

FIGURE 14–24
Analysis of the R/2R ladder DAC.

no current goes through the $2R$ equivalent resistance because the inverting input is at virtual ground. Thus, all of the current ($I = 5\ \text{V}/2R$) through R_7 also goes through R_F, and the output voltage is -5 V.

Figure 14–24(b) shows the equivalent circuit when the D_2 input is at $+5$ V and the others are at ground. This condition represents 0100. If we thevenize looking from R_8, we get 2.5 V in series with R, as shown in part (b).[4] This results in a current through R_F of $I = 2.5\ \text{V}/2R$, which gives an output voltage of -2.5 V. Keep in mind that there is no current at the op-amp inverting input and that there is no current through the equivalent resistance to ground because it has 0 V across it, due to the virtual ground.

Figure 14–24(c) shows the equivalent circuit when the D_1 input is at $+5$ V and the others are at ground. Again thevenizing looking from R_8, we get 1.25 V in series with R as shown. This results in a current through R_F of $I = 1.25\ \text{V}/2R$, which gives an output voltage of -1.25 V.

In part (d) of Figure 14–24, the equivalent circuit representing the case where D_0 is at $+5$ V and the other inputs are at ground is shown. Thevenizing from R_8 gives an equivalent of 0.625 V in series with R as shown. The resulting current through R_F is $I = 0.625\ \text{V}/2R$, which gives an output voltage of -0.625 V.

Notice that each successively lower-weighted input produces an output voltage that is halved, so that the output voltage is proportional to the binary weight of the input bits.

Performance Characteristics of Digital-to-Analog Converters

The performance characteristics of a DAC include resolution, accuracy, linearity, monotonicity, and settling time, each of which is discussed in the following list:

❑ **Resolution.** The resolution of a DAC is the reciprocal of the maximum number of discrete steps in the output. Resolution, of course, is dependent on the number of input bits. For example, a 4-bit DAC has a resolution of one part in $2^4 - 1$ (one part in fifteen). Expressed as a percentage, this is $(1/15)100 = 6.67\%$. The total number of discrete steps equals $2^n - 1$, where n is the number of bits. Resolution can also be expressed as the number of bits that are converted.

❑ **Accuracy.** Accuracy is a comparison of the actual output of a DAC with the expected output. It is expressed as a percentage of a full-scale, or maximum, output voltage. For example, if a converter has a full-scale output of 10 V and the accuracy is $\pm0.1\%$, then the maximum error for any output voltage is $(10\ \text{V})(0.001) = 10$ mV. Ideally, the accuracy should be no worse than $\pm\frac{1}{2}$ of a least significant bit (LSB). For an 8-bit converter, the least significant bit is 0.39% of full scale. The accuracy should be approximately $\pm0.2\%$.

❑ **Linearity.** A linear error is a deviation from the ideal straight-line output of a DAC. A special case is an offset error, which is the amount of output voltage when the input bits are all zeros.

❑ **Monotonicity.** A DAC is monotonic if it does not miss any steps when it is sequenced over its entire range of input bits.

❑ **Settling time.** Settling time is normally defined as the time it takes a DAC to settle within $\pm\frac{1}{2}$ LSB of its final value when a change occurs in the input code.

---

[4] Section 1–3 describes the Thevenin equivalent circuit and how to thevenize.

EXAMPLE 14–5   Determine the resolution, expressed as a percentage, of **(a)** an 8-bit DAC and **(b)** a 12-bit DAC.

Solution
(a) For the 8-bit converter,

$$\frac{1}{2^8 - 1} \times 100 = \frac{1}{255} \times 100 = \mathbf{0.392\%}$$

(b) For the 12-bit converter,

$$\frac{1}{2^{12} - 1} \times 100 = \frac{1}{4095} \times 100 = \mathbf{0.0244\%}$$

Practice Exercise   Determine the percent resolution for an 18-bit converter.

14–4 REVIEW QUESTIONS

1. What is the disadvantage of the DAC with binary-weighted inputs?
2. What is the resolution of a 4-bit DAC?

14–5 ■ BASIC CONCEPTS OF ANALOG-TO-DIGITAL (A/D) CONVERSION

As you have seen, analog-to-digital conversion is the process by which an analog quantity is converted to digital form. A/D conversion is necessary when measured quantities must be in digital form for processing in a computer or for display or storage. Basic concepts of A/D conversion including resolution, conversion time, sampling theory, and quantization error are introduced in this section.

After completing this section, you should be able to

❏ Describe A/D conversion
  ❏ Define *resolution*
  ❏ Explain conversion time
  ❏ Discuss sampling theory
  ❏ Define *quantization error*

Resolution

An analog-to-digital converter (ADC) translates a continuous analog signal into a series of binary numbers. Each binary number represents the value of the analog signal at the time of conversion. The **resolution** of an ADC can be expressed as the number of bits (binary digits) used to represent each value of the analog signal. A 4-bit ADC can represent sixteen different values of an analog signal because $2^4 = 16$. An 8-bit ADC can represent 256 different values of an analog signal because $2^8 = 256$. A 12-bit ADC can represent 4096 different values of the analog signal because $2^{12} = 4096$. The more bits, the more accurate is the conversion and the greater is the resolution because more values of a given analog signal can be represented.

Resolution is basically illustrated in Figure 14–25 using the analog voltage ramp in part (a). For the case of 3-bit resolution as shown in part (b), only eight values of the voltage ramp can be represented by binary numbers. D/A reconstruction of the ramp using the eight binary values results in the stair-step approximation shown. For the case of 4-bit resolution as shown in part (c), sixteen values can be represented, and D/A reconstruction results in a more accurate 16-step approximation as shown. For the case of 5-bit resolution as shown in part (d), D/A reconstruction produces an even more accurate 32-step approximation of the ramp.

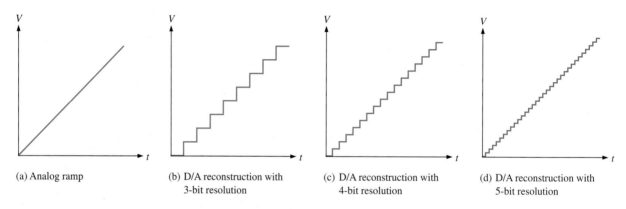

(a) Analog ramp   (b) D/A reconstruction with 3-bit resolution   (c) D/A reconstruction with 4-bit resolution   (d) D/A reconstruction with 5-bit resolution

FIGURE 14–25
Illustration of the effect of resolution on the representation of an analog signal (a ramp in this case).

Conversion Time

In addition to resolution, another important characteristic of ADCs is conversion time. The conversion of a value on an analog waveform into a digital quantity is not an instantaneous event, but it is a process that takes a certain amount of time. The conversion time can range from microseconds for fast converters to milliseconds for slower devices. Conversion time is illustrated in a basic way in Figure 14–26. As you can see, the value of the analog voltage to be converted occurs at time t_0 but the conversion is not complete until time t_1.

Sampling Theory

In A/D conversion, an analog waveform is sampled at a given point and the sampled value is then converted to a binary number. Since it takes a certain interval of time to accomplish the conversion, the number of samples of an analog waveform during a given period of time is limited. For example, if a certain ADC can make one conversion in 1 ms, it can make 1000 conversions in one second. That is, it can convert 1000 different analog values to digital form in a one-second interval.

In order to represent an analog waveform, the minimum sample rate must be greater than twice the maximum frequency component of the analog signal. This minimum sampling rate is known as the **Nyquist rate**. At the Nyquist rate, an analog signal is sampled and converted more than two times per cycle, which establishes the fundamental frequency

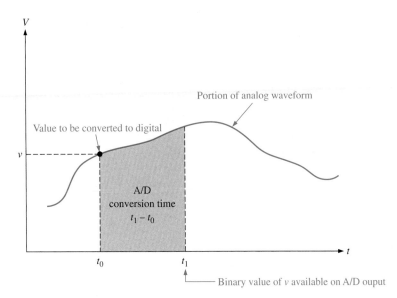

FIGURE 14–26
An illustration of A/D conversion time.

of the analog signal. Filtering can be used to obtain a facsimile of the original signal after D/A conversion. Obviously, a greater number of conversions per cycle of the analog signal results in a more accurate representation of the analog signal. This is illustrated in Figure 14–27 for two different sample rates. The lower waveforms are the D/A reconstructions for various sample rates.

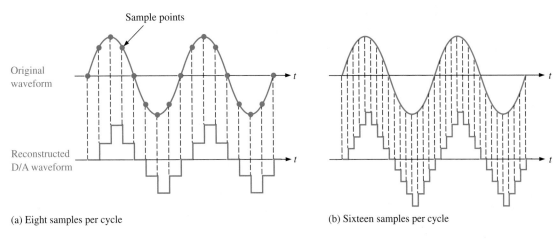

(a) Eight samples per cycle   (b) Sixteen samples per cycle

FIGURE 14–27
Illustration of two sampling rates.

Quantization Error

The term **quantization** in this context refers to determining a value for an analog quantity. Ideally, we would like to determine a value at a given instant and convert it immediately to digital form. This is, of course, impossible because of the conversion time of ADCs. Since an analog signal may change during a conversion time, its value at the end of the conversion time may not be the same as it was at the beginning (unless the input is a constant dc). This change in the value of the analog signal during the conversion time produces what is called the **quantization error**, as illustrated in Figure 14–28.

One way to avoid or at least minimize quantization error is to use a sample-and-hold circuit at the input to the ADC. As you learned in Section 14–2, a sample-and-hold amplifier quickly samples the analog input and then holds the sampled value for a certain time.

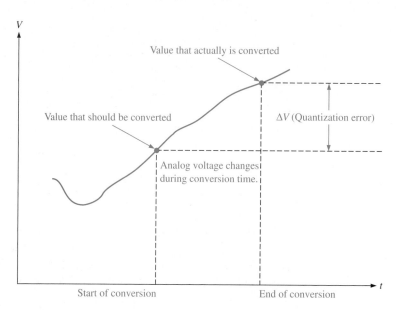

FIGURE 14–28
Illustration of quantization error in A/D conversion.

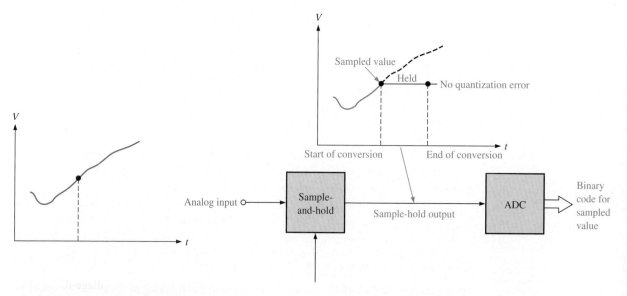

FIGURE 14–29
Using a sample-and-hold amplifier to avoid quantization error.

When used in conjunction with an ADC, the sample-and-hold is held constant for the duration of the conversion time. This allows the ADC to convert a constant value to digital form and avoids the quantization error. A basic illustration of this process is shown in Figure 14–29. When compared to the conversion in Figure 14–28, you can see that a more accurate representation of the analog input at the desired sample point is achieved.

14–5 REVIEW QUESTIONS

1. What is conversion time?
2. According to sampling theory, what is the minimum sampling rate for a 100 Hz sine wave?
3. Basically, how does a sample-and-hold circuit avoid quantization error in A/D conversion?

14–6 ■ ANALOG-TO-DIGITAL (A/D) CONVERSION METHODS

Now that you are familiar with some basic A/D conversion concepts, we will look at several methods for A/D conversion. These methods are flash (simultaneous), stairstep ramp, tracking, single-slope, dual-slope, and successive approximation. The flash and dual-slope methods were introduced in Chapter 8 as examples of op-amp applications. Some of that material is reviewed and expanded upon in this section.

After completing this section, you should be able to

❏ Discuss the operation of analog-to-digital converters (ADCs)
   ❏ Describe the flash ADC
   ❏ Describe the stairstep-ramp ADC
   ❏ Describe the tracking ADC
   ❏ Describe the single-slope ADC
   ❏ Describe the dual-slope ADC
   ❏ Describe the successive-approximation ADC

Flash (Simultaneous) Analog-to-Digital Converter

The flash (simultaneous) method utilizes comparators that compare reference voltages with the analog input voltage. When the analog voltage exceeds the reference voltage for a given comparator, a high-level output is generated. Figure 14–30 shows a 3-bit converter that uses seven comparator circuits; a comparator is not needed for the all-0s condition. A 4-bit converter of this type requires fifteen comparators. In general, $2^n - 1$ comparators are required for conversion to an n-bit binary code. The large number of comparators necessary for a reasonable-sized binary number is one of the disadvantages of the flash ADC. Its chief advantage is that it provides a fast conversion time.

The reference voltage for each comparator is set by the resistive voltage-divider network. The output of each comparator is connected to an input of the priority encoder. The encoder is sampled by a pulse on the Enable input, and a 3-bit binary code representing the

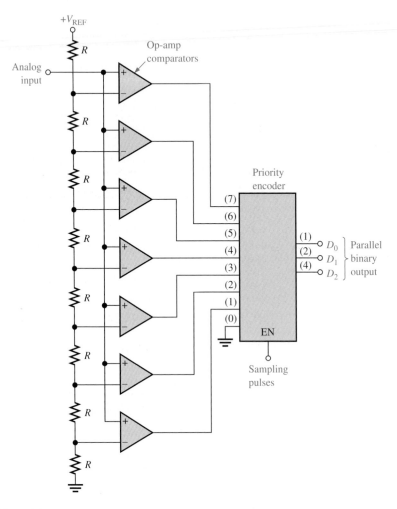

FIGURE 14–30
A 3-bit flash ADC.

value of the analog input appears on the encoder's outputs. The binary code is determined by the highest-order input having a high level.

The sampling rate determines the accuracy with which the sequence of digital codes represents the analog input of the ADC. The more samples taken in a given unit of time, the more accurately the analog signal is represented in digital form.

The following example illustrates the basic operation of the flash ADC in Figure 14–30.

EXAMPLE 14–6   Determine the binary code output of the 3-bit flash ADC for the analog input signal in Figure 14–31 and the sampling pulses (encoder Enable) shown. For this example, $V_{REF} = +8$ V.

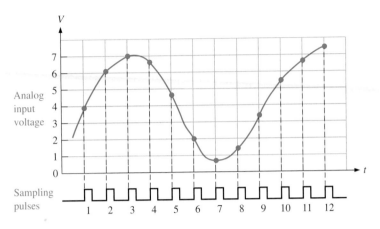

FIGURE 14–31
Sampling of values on an analog waveform for conversion to digital form.

Solution The resulting A/D output sequence is listed as follows and shown in the waveform diagram of Figure 14–32 in relation to the sampling pulses.

100, 110, 111, 110, 100, 010, 000, 001, 011, 101, 110, 111

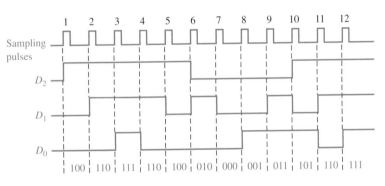

FIGURE 14–32
Resulting digital outputs for sampled values. Output D_0 is the least significant bit (LSB).

Practice Exercise If the amplitude of the analog voltage in Figure 14–31 is reduced by half, what will the A/D output sequence be?

Stairstep-Ramp Analog-to-Digital Converter

The stairstep-ramp method of A/D conversion is also known as the *digital-ramp* or the *counter* method. It employs a DAC and a binary counter to generate the digital value of an analog input. Figure 14–33 shows a diagram of this type of converter.

Assume that the counter begins in the reset state (all 0s) and the output of the DAC is zero. Now assume that an analog voltage is applied to the input. When it exceeds the reference voltage (output of DAC), the comparator switches to a high-level output state and enables the AND gate. The clock pulses begin advancing the counter through its binary states, producing a stairstep reference voltage from the DAC. The counter continues to

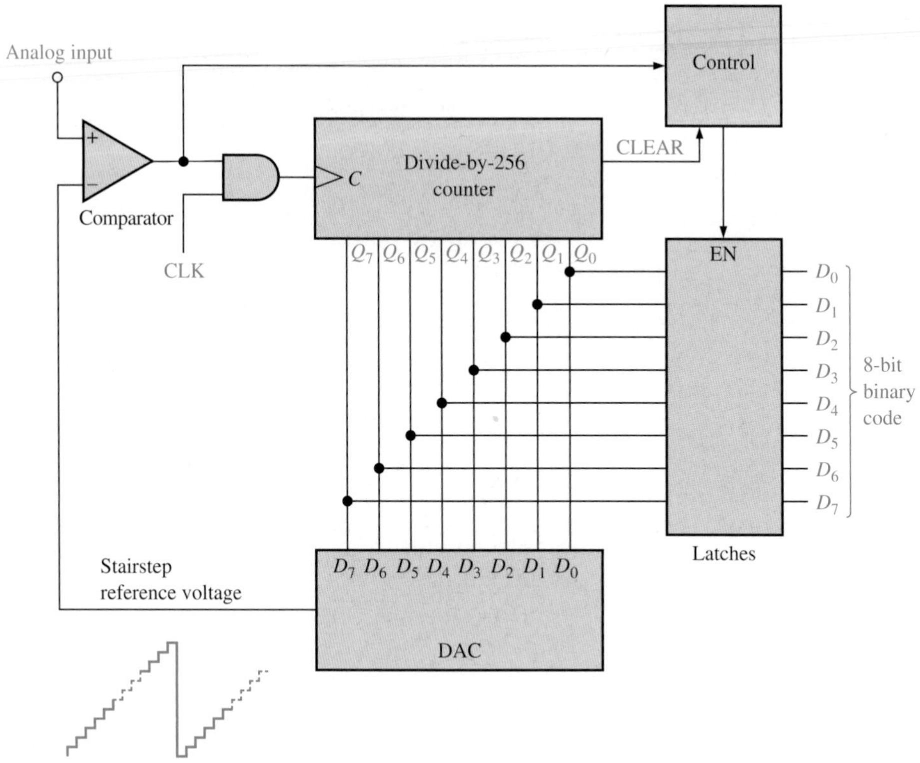

FIGURE 14–33
Stairstep-ramp ADC (8 bits).

advance from one binary state to the next, producing successively higher steps in the reference voltage. When the stairstep reference voltage reaches the analog input voltage, the comparator output will go to its low level and disable the AND gate, thus cutting off the clock pulses to stop the counter. The binary state of the counter at this point equals the number of steps in the reference voltage required to make the reference equal to or greater than the analog input. This binary number, of course, represents the value of the analog input. The control logic loads the binary count into the latches and resets the counter, thus beginning another count sequence to sample the input value.

The stairstep-ramp method is slower than the flash method because, in the worst case of maximum input, the counter must sequence through its maximum number of states before a conversion occurs. For an 8-bit conversion, this means a maximum of 256 counter states. Figure 14–34 illustrates a conversion sequence for a 4-bit conversion. Notice that for each sample, the counter must count from zero up to the point at which the stairstep reference voltage reaches the analog input voltage. The conversion time varies, depending on the analog voltage.

Tracking Analog-to-Digital Converter

The tracking method uses an up/down counter (a counter that can go either way in a binary sequence) and is faster than the stairstep-ramp method because the counter is not reset after each sample, but rather tends to track the analog input. Figure 14–35 shows a typical 8-bit tracking ADC.

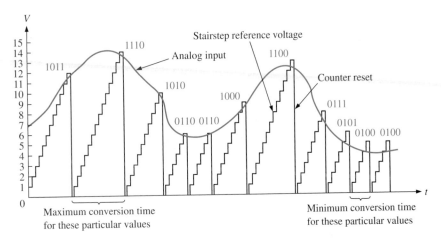

FIGURE 14–34

Example of a 4-bit conversion, showing an analog input and the stairstep reference voltage.

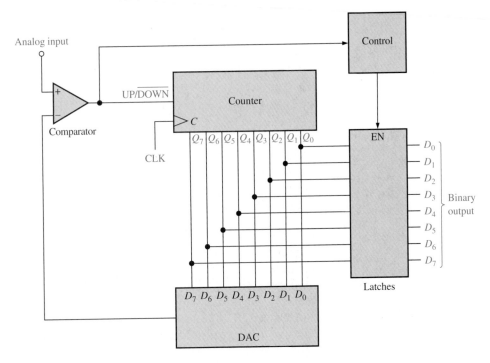

FIGURE 14–35
An 8-bit tracking ADC.

As long as the DAC reference voltage is less than the analog input, the comparator output level is high, putting the counter in the UP mode, which causes it to produce an up sequence of binary counts. This causes an increasing stairstep reference voltage out of the DAC, which continues until the stairstep reaches the value of the input voltage.

When the reference voltage equals the analog input, the comparator's output switches to its low level and puts the counter in the DOWN mode, causing it to back up one count. If the analog input is decreasing, the counter will continue to back down in its sequence and effectively track the input. If the input is increasing, the counter will back down one count after the comparison occurs and then will begin counting up again. When the input is constant, the counter backs down one count when a comparison occurs. The reference output is now less than the analog input, and the comparator output goes to its high level, causing the counter to count up. As soon as the counter increases one state, the reference voltage becomes greater than the input, switching the comparator to its low-output state. This causes the counter to back down one count. This back-and-forth action continues as long as the analog input is a constant value, thus causing an oscillation between two binary states in the ADC output. This is a disadvantage of this type of converter.

Figure 14–36 illustrates the tracking action of this type of ADC for a 4-bit conversion.

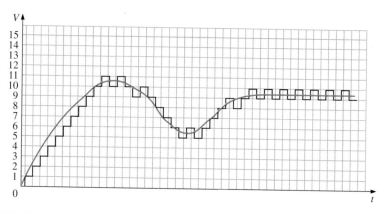

FIGURE 14–36
Tracking action of an ADC.

Single-Slope Analog-to-Digital Converter

Unlike the previous two methods, the single-slope converter does not require a DAC. It uses a linear ramp generator to produce a constant-slope reference voltage. A diagram is shown in Figure 14–37.

At the beginning of a conversion cycle, the counter is in the reset state and the ramp generator output is 0 V. The analog input is greater than the reference voltage at this point and therefore produces a high-level output from the comparator. This high level enables the clock to the counter and starts the ramp generator.

Assume that the slope of the ramp is 1 V/ms. The ramp will increase until it equals the analog input; at this point the ramp is reset, and the binary count is stored in the latches by the control logic. Let's assume that the analog input is 2 V at the point of comparison. This means that the ramp is also 2 V and has been running for 2 ms. Since the comparator output has been at its high level for 2 ms, 200 clock pulses have been allowed to pass through the gate to the counter (assuming a clock frequency of 100 kHz). At the point of comparison, the counter is in the binary state that represents decimal 200. With proper scaling and decoding, this number can be displayed as 2.00 V. This basic concept is used in some digital voltmeters.

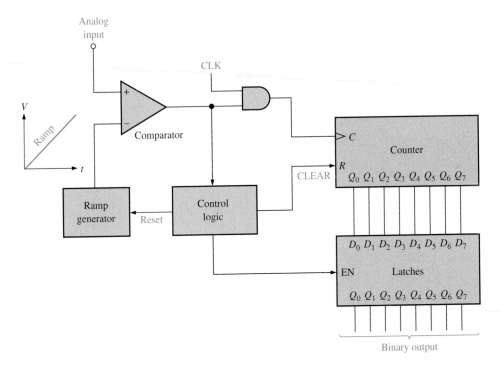

FIGURE 14–37
Single-slope ADC.

Dual-Slope Analog-to-Digital Converter

The operation of the dual-slope ADC is similar to that of the single-slope type except that a variable-slope ramp and a fixed-slope ramp are both used. This type of converter is common in digital voltmeters.

A ramp generator (integrator), A1, is used to produce the dual-slope characteristic. A block diagram of a dual-slope ADC is shown in Figure 14–38 for reference.

Figure 14–39 illustrates dual-slope conversion. Let's assume that the counter is reset and the output of the integrator is zero. Now assume that a positive input voltage is applied to the input through the switch (SW) as selected by the control logic. Since the inverting ($-$) input of A1 is at virtual ground, and assuming that V_{in} is constant for a period of time, there will be constant current through the input resistor R and therefore through the capacitor C. Capacitor C will charge linearly because the current is constant, and as a result, there will be a negative-going linear voltage ramp on the output of A1, as illustrated in Figure 14–39(a).

When the counter reaches a specified count, it will be reset, and the control logic will switch the negative reference voltage ($-V_{REF}$) to the input of A1, as shown in Figure 14–39(b). At this point the capacitor is charged to a negative voltage ($-V$) proportional to the input analog voltage.

Now the capacitor discharges linearly because of the constant current from the $-V_{REF}$, as shown in Figure 14–39(c). This linear discharge produces a positive-going ramp on the A1 output, starting at $-V$ and having a constant slope that is independent of the charge voltage. As the capacitor discharges, the counter advances from its reset state.

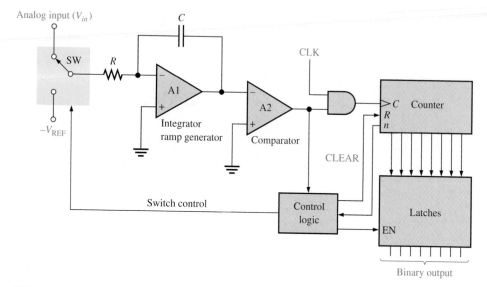

FIGURE 14–38
Dual-slope ADC.

The time it takes the capacitor to discharge to zero depends on the initial voltage $-V$ (proportional to V_{in}) because the discharge rate (slope) is constant. When the integrator (A1) output voltage reaches zero, the comparator (A2) switches to its low state and disables the clock to the counter. The binary count is latched, thus completing one conversion cycle. The binary count is proportional to V_{in} because the time it takes the capacitor to discharge depends only on $-V$, and the counter records this interval of time.

Successive-Approximation Analog-to-Digital Converter

Perhaps the most widely used method of A/D conversion is **successive approximation**. It has a much faster conversion time than the other methods with the exception of the flash method. It also has a fixed conversion time that is the same for any value of the analog input.

Figure 14–40 shows a basic block diagram of a 4-bit successive-approximation ADC. It consists of a DAC, a successive-approximation register (SAR), and a comparator. The basic operation is as follows. The bits of the DAC are enabled one at a time, starting with the most significant bit (MSB). As each bit is enabled, the comparator produces an output that indicates whether the analog input voltage is greater or less than the output of the DAC. If the DAC output is greater than the analog input, the comparator's output is low, causing the bit in the register to reset. If the DAC output is less than the analog input, the bit is retained in the register. The system does this with the MSB first, then the next most significant bit, then the next, and so on. After all the bits of the DAC have been tried, the conversion cycle is complete.

In order to better understand the operation of the successive-approximation ADC, let's take a specific example of a 4-bit conversion. Figure 14–41 illustrates the step-by-step conversion of a given analog input voltage (5 V in this case). Let's assume that the DAC has the following output characteristic: $V_{OUT} = 8$ V for the 2^3 bit (MSB), $V_{OUT} = 4$ V for the 2^2 bit, $V_{OUT} = 2$ V for the 2^1 bit, and $V_{OUT} = 1$ V for the 2^0 bit (LSB).

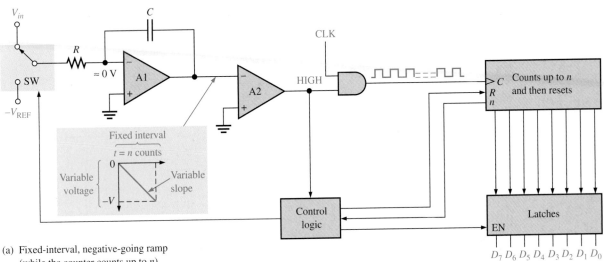

(a) Fixed-interval, negative-going ramp
(while the counter counts up to n)

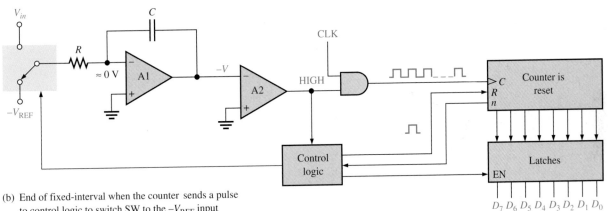

(b) End of fixed-interval when the counter sends a pulse
to control logic to switch SW to the $-V_{REF}$ input

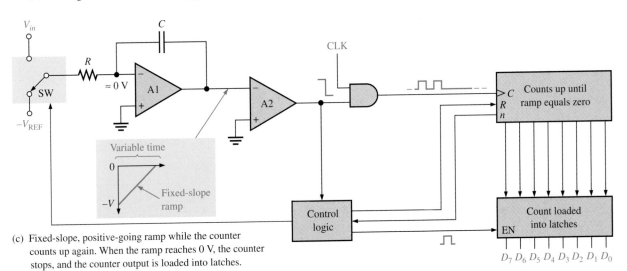

(c) Fixed-slope, positive-going ramp while the counter
counts up again. When the ramp reaches 0 V, the counter
stops, and the counter output is loaded into latches.

FIGURE 14–39
Dual-slope conversion.

FIGURE 14–40

Successive-approximation ADC.

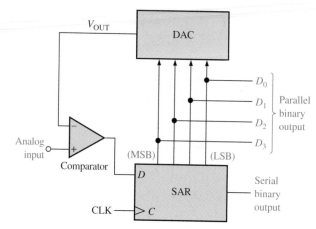

Figure 14–41(a) shows the first step in the conversion cycle with the MSB = 1. The output of the DAC is 8 V. Since this is greater than the analog input of 5 V, the output of the comparator is low, causing the MSB in the SAR to be reset to a 0.

Figure 14–41(b) shows the second step in the conversion cycle with the 2^2 bit equal to a 1. The output of the DAC is 4 V. Since this is less than the analog input of 5 V, the output of the comparator switches to its high level, causing this bit to be retained in the SAR.

Figure 14–41(c) shows the third step in the conversion cycle with the 2^1 bit equal to a 1. The output of the DAC is 6 V because there is a 1 on the 2^2 bit input and on the 2^1 bit input; 4 V + 2 V = 6 V. Since this is greater than the analog input of 5 V, the output of the comparator switches to its low level, causing this bit to be reset to a 0.

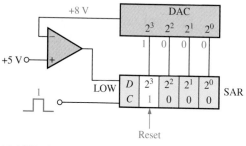

(a) MSB trial

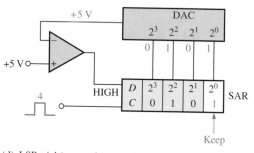

(b) 2^2-bit trial

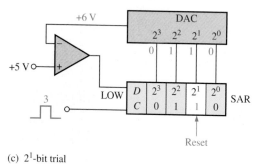

(c) 2^1-bit trial

(d) LSB trial (conversion complete)

FIGURE 14–41

Successive-approximation conversion process.

Figure 14–41(d) shows the fourth and final step in the conversion cycle with the 2^0 bit equal to a 1. The output of the DAC is 5 V because there is a 1 on the 2^2 bit input and on the 2^0 bit input; 4 V + 1 V = 5 V.

The four bits have all been tried, thus completing the conversion cycle. At this point the binary code in the register is 0101, which is the binary value of the analog input of 5 V. Another conversion cycle now begins, and the basic process is repeated. The SAR is cleared at the beginning of each cycle.

A Specific Analog-to-Digital Converter

The ADC0804[5] is an example of a successive-approximation ADC. A block diagram is shown in Figure 14–42. This device operates from a +5 V supply and has a resolution of eight bits with a conversion time of 100 μs. Also, it has guaranteed monotonicity and an on-chip clock generator. The data outputs are tristate so that they can be interfaced with a microprocessor bus system.

FIGURE 14–42
The ADC0804 analog-to-digital converter.

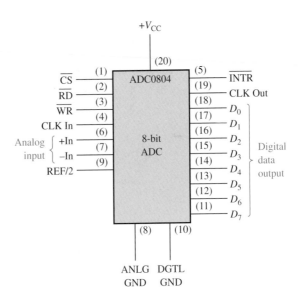

A detailed logic diagram of the ADC0804 is shown in Figure 14–43, and the basic operation of the device is as follows. The ADC0804 contains the equivalent of a 256-resistor DAC network. The successive-approximation logic sequences the network to match the analog differential input voltage $(+V_{IN} - (-V_{IN}))$ with an output from the resistive network. The MSB is tested first. After eight comparisons (sixty-four clock periods), an 8-bit binary code is transferred to an output latch, and the interrupt ($\overline{INTR}$) output goes low. The device can be operated in a free-running mode by connecting the $\overline{INTR}$ output to the write ($\overline{WR}$) input and holding the conversion start ($\overline{CS}$) low. To ensure start-up under all conditions, a low $\overline{WR}$ input is required during the power-up cycle. Taking $\overline{CS}$ low anytime after that will interrupt the conversion process.

When the $\overline{WR}$ input goes low, the internal successive-approximation register (SAR) and the 8-bit shift register are reset. As long as both $\overline{CS}$ and $\overline{WR}$ remain low, the analog-to-digital converter remains in a reset state. Conversion starts one to eight clock periods after $\overline{CS}$ or $\overline{WR}$ makes a low-to-high transition.

[5] Data sheet for ADC0804 available at http://www.intersil.com

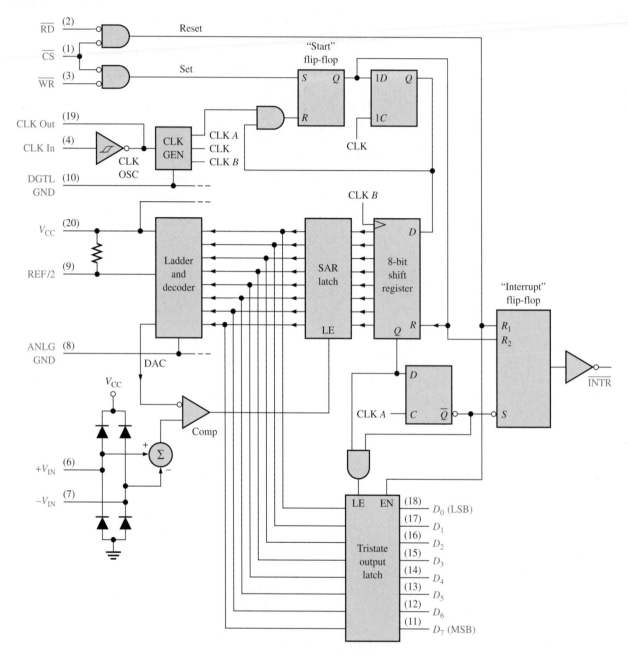

FIGURE 14–43
Logic diagram of the ADC0804 ADC.

When the $\overline{CS}$ and $\overline{WR}$ inputs are low, the start flip-flop is set, and the interrupt flip-flop and 8-bit register are reset. The high is ANDed with the next clock pulse, which puts a high on the reset input of the start flip-flop. If either $\overline{CS}$ or $\overline{WR}$ has gone high, the set signal to the start flip-flop is removed, causing it to be reset. A high is placed on the D input of the 8-bit shift register, and the conversion process is started. If the $\overline{CS}$ and $\overline{WR}$ inputs are still low, the start flip-flop, the 8-bit shift register, and the SAR remain reset. This action allows for wide $\overline{CS}$ and $\overline{WR}$ inputs, with conversion starting from one to eight clock periods after one of the inputs has gone high.

When the high input has been clocked through the 8-bit shift register, completing the SAR search, it is applied to an AND gate controlling the output latches and to the D input of a flip-flop. On the next clock pulse, the digital word is transferred to the tristate output latches, and the interrupt flip-flop is set. The output of the interrupt flip-flop is inverted to provide an $\overline{INTR}$ output that is high during conversion and low when conversion is complete.

When a low is at both the $\overline{CS}$ and $\overline{RD}$ inputs, the tristate output latch is enabled, the output code is applied to the D_0 through D_7 lines, and the interrupt flip-flop is reset. When either the $\overline{CS}$ or the $\overline{RD}$ input returns to a high, the D_0 through D_7 outputs are disabled. The interrupt flip-flop remains reset.

A few additional IC analog-to-digital converters are listed in Table 14–1.

TABLE 14–1
Several popular ADCs.

Device	Description	Resolution	Conversion Time	Supply Voltages
AD673[1]	Successive Approximation	8 bits	20 μs	+5 V, −12 V
AD578[1]	Successive Approximation	12 bits	3 μs	+5 V to ±15 V
ADC0802[2]	Successive Approximation	8 bits	100 μs	+5 V
ADC0803[2]	Successive Approximation	8 bits	100 μs	+5 V
H3286[3]	Flash Conversion	8 bits	—	+5 V
TLC5510[4]	Flash Conversion	8 bits	25 ns	+5 V
TLC320AD57[4]	Sigma Delta	16/18 bits	20 μs	+5 V

[1] Data sheet available at http://www.analogdevices.com

[2] Data sheet available at http://www.national.com

[3] Data sheet available at http://www.intersil.com

[4] Data sheet available at http://www.ti.com

14–6 REVIEW QUESTIONS

1. What is the fastest method of analog-to-digital conversion?
2. Which A/D conversion method uses an up/down counter?
3. The successive-approximation converter has a fixed conversion time. (True or false)

14–7 ■ VOLTAGE-TO-FREQUENCY (*V/F*) AND FREQUENCY-TO-VOLTAGE (*F/V*) CONVERTERS

Voltage-to-frequency converters convert an analog input voltage to a pulse stream or square wave in such a way that there is a linear relationship between the analog voltage and the frequency of the pulse stream. Frequency-to-voltage converters perform the inverse operation by converting a pulse stream to a voltage that is proportional to the pulse stream frequency. Actually, V/F and F/V converters can be used as ADCs and DACs in certain applications. In other applications, V/F and F/V converters are used, for example, in high-noise immunity digital transmission and in digital voltmeters.

After completing this section, you should be able to

❑ Discuss the basic operation of *V/F* and *F/V* converters
   ❑ Describe the AD650 *V/F* converter
   ❑ Discuss *V/F* and *F/V* applications

A Basic Voltage-to-Frequency Converter

The concept of voltage-to-frequency converters is shown in Figure 14–44. An analog voltage on the input is converted to a pulse signal with a frequency that is directly proportional to the amplitude of the input voltage. There are several ways to implement a *V/F* converter. For example, the VCO (voltage-controlled oscillator) with which you are already familiar can be used as one type of *V/F* converter. In this section, we will look at a relatively common implementation called the *charge-balance V/F converter.*

    Figure 14–45 shows the diagram of a basic charge-balance *V/F* converter. It consists of an integrator, a comparator, a one-shot, a current source, and an electronic switch. The input resistor R_{in}, the integration capacitor C_{int}, and the one-shot timing capacitor C_{os} are components whose values are selected based on desired performance.

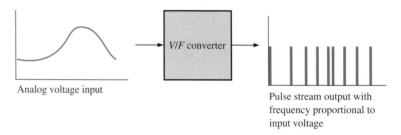

        Analog voltage input

                                      Pulse stream output with
                                      frequency proportional to
                                      input voltage

FIGURE 14–44
The basic V/F concept.

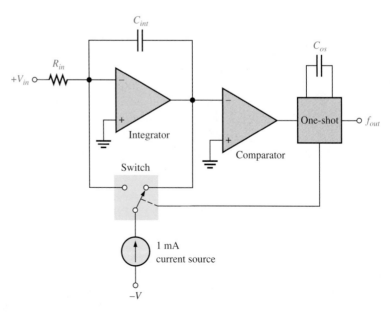

FIGURE 14–45
A basic voltage-to-frequency converter.

The basic operation of the *V/F* converter in Figure 14–46 is as follows. A positive input voltage produces an input current ($I_{in} = V_{in}/R_{in}$) which charges the capacitor C_{int}, as indicated in Figure 14–46(a). During this integrate mode, the integrator output voltage is a

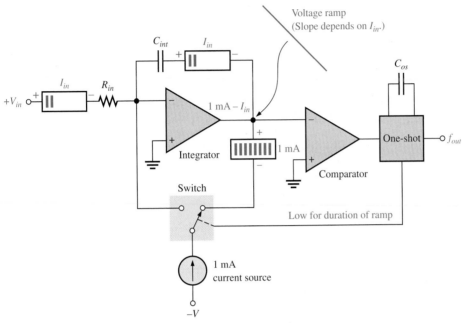

(a) *V/F* converter in the integrate mode

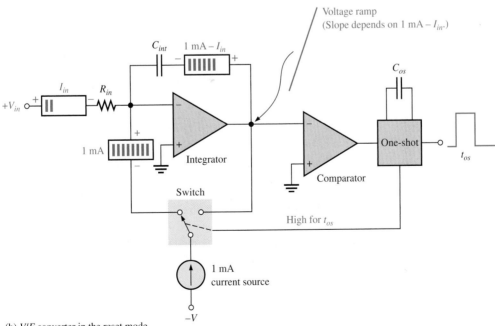

(b) *V/F* converter in the reset mode

FIGURE 14–46
Basic operation of a V/F converter for a constant input voltage.

downward ramp, as shown. When the integrator output voltage reaches zero, the comparator triggers the one-shot. The one-shot produces a pulse with a fixed width, t_{os}, that switches the 1 mA current source to the input of the integrator and initiates the reset mode.

During the reset mode, current through the capacitor is in the opposite direction from the integrate mode, as indicated in Figure 14–46(b). This produces an upward ramp on the integrator output as indicated. After the one-shot times out, the current source is switched back to the integrator output, initiating another integrate mode and the cycle repeats.

If the input voltage is held constant, the output waveform of the integrator is as shown in Figure 14–47(a), where the amplitude and the integrate time remain constant. The final output of the *V/F* converter is taken off the one-shot, as indicated in Figure 14–46. As long as the input voltage is constant, the output pulse stream has a constant frequency as indicated in Figure 14–47(b).

FIGURE 14–47

V/F converter waveforms for a constant input voltage.

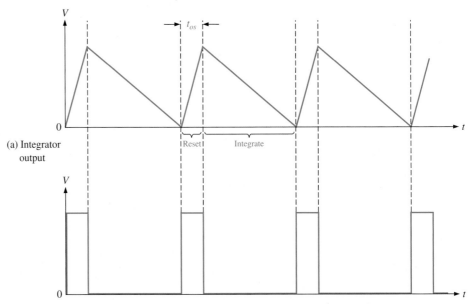

(a) Integrator output

(b) Final output (one-shot)

When the Input Voltage Increases   An increase in the input voltage, V_{in}, causes the input current, I_{in}, to increase. In the basic relationship $I_C = (V_C/t)C$, the term V_C/t is the slope of the capacitor voltage. If the current increases, V_C/t also increases since C is constant. As applied to the *V/F* converter, this means that if the input current (I_{in}) increases, then the slope of the integrator output during the integrate mode will also increase and reduce the period of the final output voltage. Also, during the reset mode, the opposite current through the capacitor, 1 mA $- I_{in}$, is smaller, thus decreasing the slope of the upward ramp and reducing the amplitude of the integrator output voltage. This is illustrated in the waveform diagram of Figure 14–48 where the input voltage, and thus the input current, takes a step increase from one value to another. Notice that during reset, the positive-going slope of the integrator voltage is less, so it reaches a smaller amplitude during the time t_{os}. Remember, t_{os} does not change. Notice also that during integration, the negative-going slope of the integrator voltage is greater, so it reaches zero quicker. The net result of this

FIGURE 14–48
The output frequency increases when the input voltage increases.

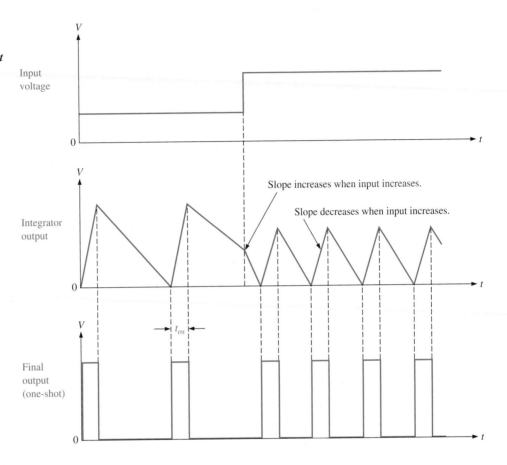

increase in input voltage is that the output frequency increases an amount proportional to the increase in the input voltage. So, as the input voltage varies, the output frequency varies proportionally.

The AD650 Integrated Circuit *V/F* Converter[6]

The AD650 is a good example of a *V/F* converter very similar to the basic device we just discussed. The main differences in the AD650 are the output transistor and the comparator threshold voltage of -0.6 V instead of ground, as shown in Figure 14–49. The input resistor, integrating capacitor, one-shot capacitor, and the output pull-up resistor are external components, as indicated.

The values of the external components determine the operating characteristics of the device. The pulse width of the one-shot output is set by the following formula:

$$t_{os} = C_{os}(6.8 \times 10^3 \text{ s/F}) + 3 \times 10^{-7} \text{ s} \tag{14–1}$$

---

[6] Data sheet for AD650 available at http://www.analogdevices.com

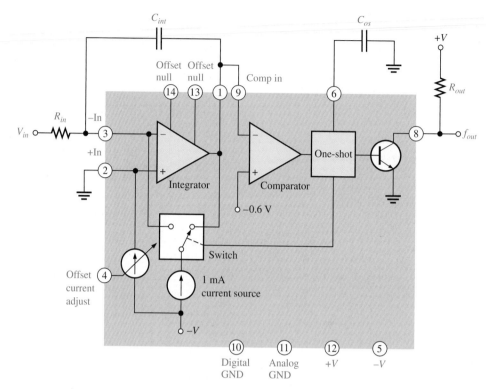

FIGURE 14–49
The AD650 V/F converter.

During the reset interval, the integrator output voltage increases by an amount expressed as

$$\Delta V = \frac{(1 \text{ mA} - I_{in})t_{os}}{C_{int}} \tag{14–2}$$

The duration of the integrate interval when the integrator output is sloping downward is

$$t_{int} = \frac{\Delta V}{I_{in}/C_{int}} = \frac{t_{os}(1 \text{ mA} - I_{in})/C_{int}}{I_{in}/C_{int}}$$

$$t_{int} = \left(\frac{1 \text{ mA}}{I_{in}} - 1\right)t_{os} \tag{14–3}$$

The period of a full cycle consists of the reset interval plus the integrate interval.

$$T = t_{os} + t_{int} = t_{os} + \left(\frac{1 \text{ mA}}{I_{in}} - 1\right)t_{os} = \left(1 + \frac{1 \text{ mA}}{I_{in}} - 1\right)t_{os} = \left(\frac{1 \text{ mA}}{I_{in}}\right)t_{os}$$

Therefore, the output frequency can be expressed as

$$f_{out} = \frac{I_{in}}{t_{os}(1 \text{ mA})}$$ **(14–4)**

As you can see in Equation (14–4), the output frequency is directly proportional to the input current; and since $I_{in} = V_{in}/R_{in}$, it is also directly proportional to the input voltage and inversely proportional to the input resistance. The output frequency is also inversely proportional to t_{os}, which depends on the value of C_{os}.

EXAMPLE 14–7 Determine the output frequency for the AD650 *V/F* converter in Figure 14–50 when a constant input voltage of 5 V is applied.

FIGURE 14–50

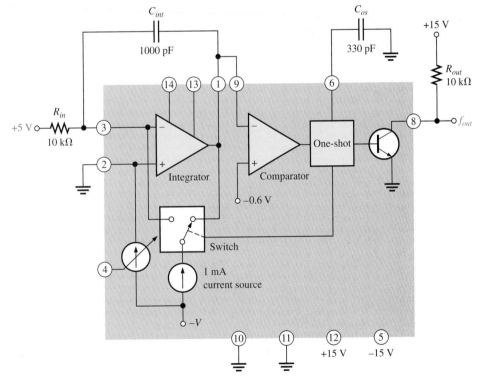

Solution

$$t_{os} = C_{os}(6.8 \times 10^3 \text{ s/F}) + 3 \times 10^{-7} \text{ s}$$
$$= 330 \text{ pF}(6.8 \times 10^3 \text{ s/F}) + 3 \times 10^{-7} \text{ s} = 2.5 \text{ μs}$$

$$I_{in} = \frac{V_{in}}{R_{in}} = \frac{5 \text{ V}}{10 \text{ kΩ}} = 500 \text{ μA}$$

$$f_{out} = \frac{I_{in}}{t_{os}(1 \text{ mA})} = \frac{500 \text{ μA}}{(2.5 \text{ μs})(1 \text{ mA})} = \textbf{200 kHz}$$

Practice Exercise What are the minimum and maximum output frequencies for the *V/F* converter in Figure 14–50 when a triangular wave with a minimum peak value of 1 V and a maximum peak value of 6 V is applied to the input?

A Basic *F/V* Converter

Figure 14–51 shows a basic frequency-to-voltage converter. The elements are the same as those in the voltage-to-frequency converter of Figure 14–45, but they are connected differently.

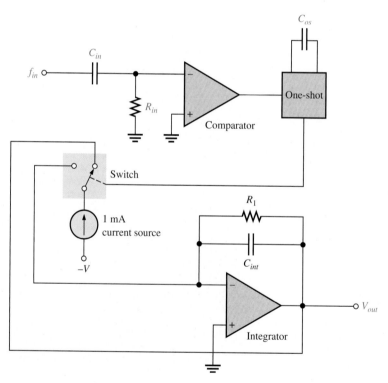

FIGURE 14–51
A basic frequency-to-voltage (F/V) converter.

When an input frequency is applied to the comparator input, it triggers the one-shot which produces a fixed pulse width (t_{os}) determined by C_{os}. This switches the 1 mA current source to the integrator input and C_{int} charges. Between one-shot pulses, C_{int} discharges through R_1. The higher the input frequency, the closer the one-shot pulses are together and

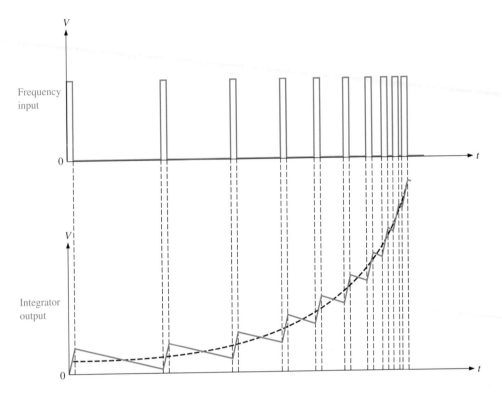

FIGURE 14–52
An example of frequency-to-voltage conversion.

the less C_{int} discharges. This causes the integrator output to increase as input frequency increases and to decrease as the input frequency decreases. The integrator output is the final voltage output of the *F/V* converter. *F/V* conversion action is illustrated by the waveforms in Figure 14–52. C_{int} and R_1 act as a filter and tend to smooth out the ripples on the integrator output as indicated by the dashed curve.

Figure 14–53 shows the AD650 connected to function as a frequency-to-voltage converter. Compare this configuration with the voltage-to-frequency connection in Figure 14–49.

An Application

One application of *V/F* and *F/V* converters is in the remote sensing of a quantity (temperature, pressure, level) that is converted to an analog voltage by a transducer. The analog voltage is then converted to a pulse frequency by a *V/F* converter which is then transmitted by some method (radio link, fiber-optical link, telemetry) to a base unit receiver that includes an *F/V* converter. This basic application of *V/F* and *F/V* conversion is illustrated in Figure 14–54.

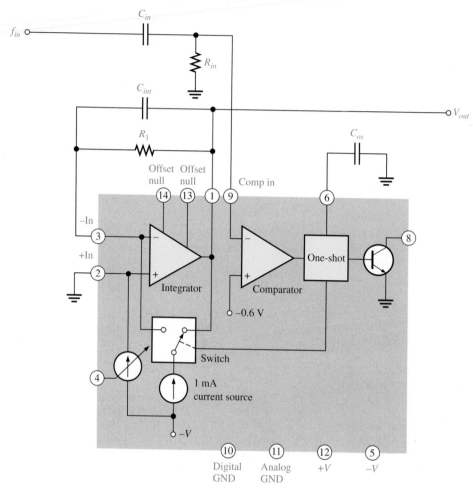

FIGURE 14–53

The AD650 connected as a frequency-to-voltage (F/V) converter.

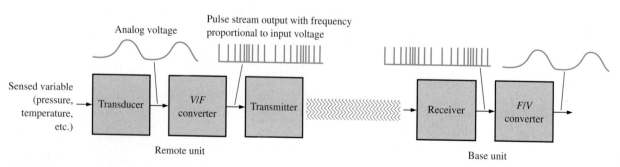

FIGURE 14–54

Basic application of V/F and F/V conversion.

14–7 REVIEW QUESTIONS

1. List the basic components in a typical *V/F* converter.

2. In a *V/F* converter, if the input voltage changes from 1 V to 6.5 V, what happens to the output?

3. Describe the basic differences between a *V/F* and a *F/V* converter in terms of inputs and outputs.

14–8 ■ TROUBLESHOOTING

Basic testing of DACs and ADCs includes checking their performance characteristics, such as monotonicity, offset, linearity, and gain, and checking for missing or incorrect codes. In this section, the fundamentals of testing these analog interfaces are introduced.

After completing this section, you should be able to

❏ Troubleshoot DACs and ADCs
  ❏ Identify D/A conversion errors
  ❏ Identify A/D conversion errors

Testing Digital-to-Analog Converters

The concept of DAC testing is illustrated in Figure 14–55. In this basic method, a sequence of binary codes is applied to the inputs, and the resulting output is observed. The binary code sequence extends over the full range of values from 0 to $2^n - 1$ in ascending order, where n is the number of bits.

FIGURE 14–55
Basic test setup for a DAC.

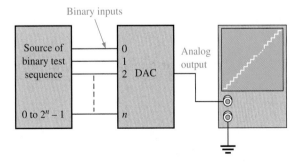

The ideal output is a straight-line stairstep as indicated. As the number of bits in the binary code is increased, the resolution is improved. That is, the number of discrete steps increases, and the output approaches a straight-line linear ramp.

D/A Conversion Errors

Several D/A conversion errors to be checked for are shown in Figure 14–56, which uses a 4-bit conversion for illustration purposes. A 4-bit conversion produces fifteen discrete steps. Each graph in the figure includes an ideal stairstep ramp for comparison with the faulty outputs.

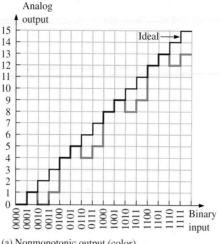

(a) Nonmonotonic output (color)

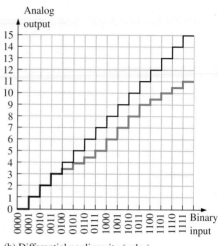

(b) Differential nonlinearity (color)

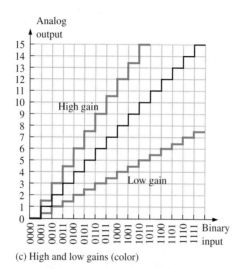

(c) High and low gains (color)

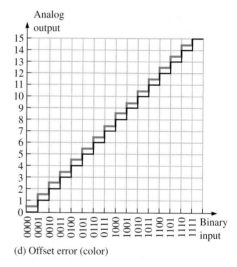

(d) Offset error (color)

FIGURE 14–56
Illustrations of several D/A conversion errors.

Nonmonotonicity The step reversals in Figure 14–56(a) indicate **nonmonotonic** performance, which is a form of nonlinearity. In this particular case, the error occurs because the 2^1 bit in the binary code is interpreted as a constant 0. That is, a short is causing the bit input line to be stuck in the low state.

Differential Nonlinearity Figure 14–56(b) illustrates differential nonlinearity in which the step amplitude is less than it should be for certain input codes. This particular output could be caused by the 2^2 bit having an insufficient weight, perhaps because of a faulty input resistor. We could also see steps with amplitudes greater than normal if a particular binary weight were greater than it should be.

Low or High Gain Output errors caused by low or high gain are illustrated in Figure 14–56(c). In the case of low gain, all of the step amplitudes are less than ideal. In the case of high gain, all of the step amplitudes are greater than ideal. This situation may be caused by a faulty feedback resistor in the op-amp circuit.

Offset Error   An offset error is illustrated in Figure 14–56(d). Notice that when the binary input is 0000, the output voltage is nonzero and that this amount of offset is the same for all steps in the conversion. A faulty op-amp may be the culprit in this situation.

EXAMPLE 14–8

The DAC output in Figure 14–57 is observed when a straight 4-bit binary sequence is applied to the inputs. Identify the type of error, and suggest an approach to isolate the fault.

FIGURE 14–57

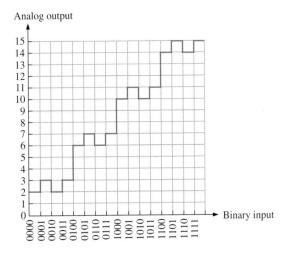

Solution   The DAC in this case is nonmonotonic. Analysis of the output reveals that the device is converting the following sequence, rather than the actual binary sequence applied to the inputs.

0010, 0011, 0010, 0011, 0110, 0111, 0110, 0111, 1010, 1011, 1010,
1011, 1110, 1111, 1110, 1111

Apparently, the 2^1 bit (second from right) is stuck in the high (1) state. To find the problem, first monitor the bit input pin of the device. If it is changing states, the fault is internal, most likely an open. If the external pin is not changing states and is always high, check for an external short to $+V$ that may be caused by a solder bridge somewhere on the circuit board. If no problem is found here, disconnect the source output from the DAC input pin, and see if the output signal is correct. If these checks produce no results, the fault is most likely internal to the DAC, perhaps a short to the supply voltage.

Practice Exercise   Graph the output of a DAC if a straight 4-bit binary sequence is applied to the inputs and the most significant bit input of the DAC is stuck high.

Testing Analog-to-Digital Converters

One method for testing ADCs is shown in Figure 14–58. A DAC is used as part of the test setup to convert the ADC output back to analog form for comparison with the test input.

A test input in the form of a linear ramp is applied to the input of the ADC. The resulting binary output sequence is then applied to the DAC test unit and converted to a stairstep ramp. The input and output ramps are compared for any deviation.

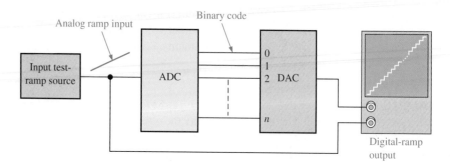

FIGURE 14–58
A method for testing ADCs.

A/D Conversion Errors

Again, a 4-bit conversion is used to illustrate the principles. Let's assume that the test input is an ideal linear ramp.

Missing Code   The stairstep output in Figure 14–59(a) indicates that the binary code 1001 does not appear on the output of the ADC. Notice that the 1000 value stays for two intervals and then the output jumps to the 1010 value.

In a flash ADC, for example, a failure of one of the comparators can cause a missing-code error.

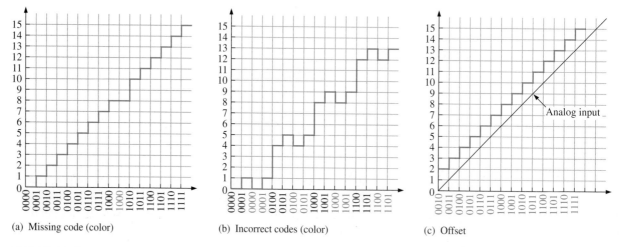

(a) Missing code (color)   (b) Incorrect codes (color)   (c) Offset

FIGURE 14–59
Illustrations of A/D conversion errors.

Incorrect Codes   The stairstep output in Figure 14–59(b) indicates that several of the binary code words coming out of the ADC are incorrect. Analysis indicates that the 2^1-bit line is stuck in the low state in this particular case.

Offset   Offset conditions are shown in 14–59(c). In this situation, the ADC interprets the analog input voltage as greater than its actual value. This error is probably due to a faulty comparator circuit.

EXAMPLE 14–9   A 4-bit flash ADC is shown in Figure 14–60(a). It is tested with a setup like the one in Figure 14–58. The resulting reconstructed analog output is shown in Figure 14–60(b). Identify the problem and the most probable fault.

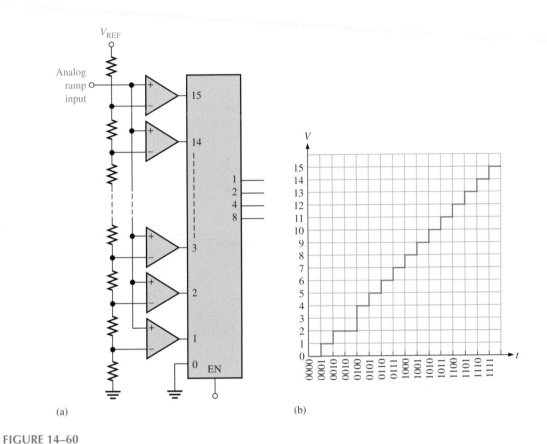

(a)   (b)

FIGURE 14–60

Solution   The binary code 0011 is missing from the ADC output, as indicated by the missing step. Most likely, the output of comparator 3 is stuck in its inactive state (low).

Practice Exercise   If the output of comparator 15 is stuck in the high state, what will be the reconstructed analog output when the ADC is tested in a setup like the one in Figure 14–58?

14–8 REVIEW QUESTIONS

1. How do you detect nonmonotonic behavior in a DAC?
2. What effect does low gain have on a DAC output?
3. Name two types of output errors in an ADC.

14–9 ■ A SYSTEM APPLICATION

The solar panel control system introduced at the opening of this chapter includes analog multiplexers and ADCs. The analog multiplexers accept position information from the potentiometers located on each solar unit and send it to the ADCs where the analog position information is converted to digital form. The digital outputs of the ADCs go to the digital controller for processing and control signals are sent back to the solar units to keep them properly positioned.

After completing this section, you should be able to

❏ Apply what you have learned in this chapter to a system application
    ❏ Describe how an analog multiplexer can be used to collect analog data
    ❏ Explain how an ADC is used in a system application
    ❏ Translate between a printed circuit board and a schematic
    ❏ Analyze the analog board
    ❏ Troubleshoot some common problems

A Brief Description of the System

The solar panel control system maintains each solar panel at approximately a 90° angle with respect to the sun's rays. Two angular positions are required to properly align the panels. The azimuth position is along a curve from east to west parallel with the horizon. The elevation position is from the horizon to directly overhead. These angular movements for a solar unit are illustrated in Figure 14–61. There are two stepping motors that drive the solar panel to the proper position, one for azimuth and one for elevation. Also, there are two position transducers (potentiometers are used as the sensors in this case) that produce voltages proportional to the positions of the panel as determined by the angular positions of the motor shafts.

    In this application, the movement of the solar panel is very slow. The azimuth angle begins at an easterly orientation at sunrise and turns through approximately 180° by sunset. The elevation angle tracks the arc of the sun each day and also adjusts for seasonal varia-

FIGURE 14–61
A solar panel with position controls and sensors for azimuth and elevation.

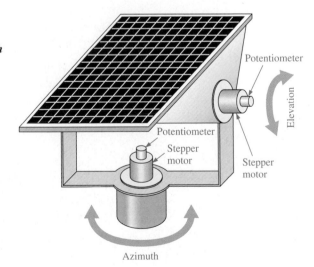

tion of the sun's relative position in the sky. On the first day of winter it tracks through the lowest arc, and on the first day of summer it tracks through the highest arc.

A basic block diagram of the control electronics is shown in Figure 14–62. There are two 4-input analog multiplexers, one for the azimuth and one for the elevation. Each input has a dc voltage coming from the associated position potentiometer. Each voltage is proportional to the current angular position of the solar panel as measured by the potentiometer. Every five minutes the digital controller causes the multiplexer to quickly sequence through the four azimuth and the four elevation position voltages so that one voltage at a time is applied to the ADC. Each resulting digital code is processed by comparing the angle it represents to angular information about solar position based on time of day and date that is permanently stored in the digital controller's memory. Based on its computations, the digital controller issues the proper control signal to the appropriate stepping motor to update its position.

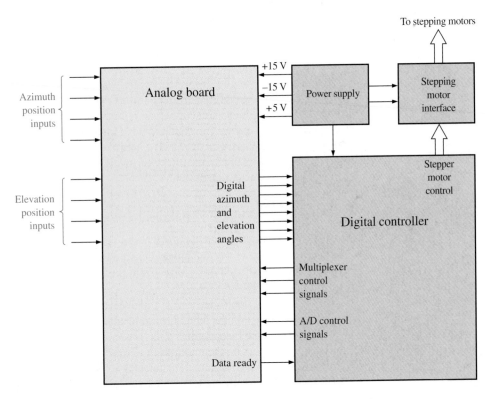

FIGURE 14–62
Basic block diagram of the solar panel control system.

The Potentiometer   In this application, the potentiometer is used as a position transducer to convert the angular shaft position of the stepping motor to a proportional dc voltage. The potentiometer is mechanically linked to the associated motor so as the motor shaft turns, the wiper of the potentiometer slides along the resistive element. It is calibrated so that the smallest angle produces the smallest voltage. In Figure 14–63, a simplified diagram shows the basic construction, and the schematic indicates minimum and maximum angles corresponding to the wiper position.

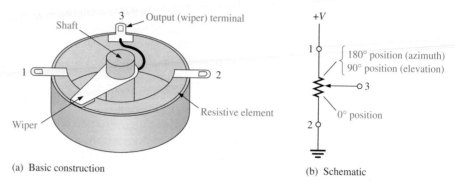

FIGURE 14–63

The potentiometer as an angular transducer.

Stepping Motors Although our focus is on the analog electronics in this system, a basic familiarization with stepping motors will help you have a better understanding of the overall system operation. A stepping motor is one in which the rotor shaft can be rotated in a series of incremental moves called steps or step angles. Stepping motors are available with step angles ranging from 0.9° to 30°. The motors used in this system are assumed to have a 2° step angle (180 steps/revolution). Rotation is achieved by applying a series of pulses to the windings. Generally, one pulse will rotate the shaft by one step angle (2° in our case). The sequence in which the pulses are applied to the windings determine the direction of rotation (CW or CCW). The speed of the motor is controlled by the rate at which the stepping pulses are applied. Obviously, in this system the motors move very slowly so speed is not a consideration.

Basic Operation of the Analog Circuits

A block diagram of the analog board is shown in Figure 14–64. DC voltages from the four azimuth potentiometers (remember there are four solar panels) are applied to the inputs of the upper AD9300, and dc voltages from the four elevation potentiometers are applied to the inputs of the lower AD9300.

The digital controller issues a high-level enable signal to the azimuth multiplexer. While the Enable input is high, the digital controller sequentially switches each of the four inputs to the output by applying a 2-bit binary code to the channel select inputs. As each input voltage appears on the multiplexer output, the digital controller issues a convert signal to the AD673[7] ADC to initiate a conversion. After the conversion is complete, the ADC then sends a data ready signal to the digital controller. The controller responds with a data enable signal, which places the 8-bit binary code on the outputs allowing the binary number to be processed by the controller.

After the first conversion, the digital controller advances to the next azimuth multiplexer input and repeats the operation. After all four azimuth inputs have been converted to digital and processed, the controller disables the azimuth multiplexer and enables the elevation multiplexer for conversion of its four inputs to digital. The 200 Ω variable resistors are used to adjust the input voltage to the ADC so that the highest output code (all 1s) corresponds to the maximum input voltage. Therefore, the full range of azimuth and elevation angles are converted into 256 discrete values.

---

[7] Data sheet for AD673 available at http://www.analogdevices.com and Appendix A.

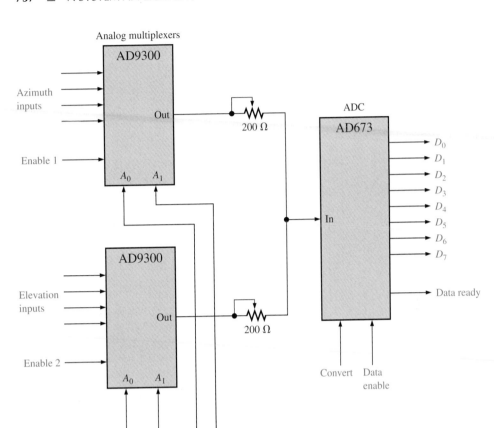

FIGURE 14–64
Simplified block diagram of the analog board.

The digital controller repeats the conversion sequence every eight minutes in this system. The basic timing diagram in Figure 14–65 graphically illustrates the operation.

Now, so that you can take a closer look at the analog board, let's take it out of the system and put it on the troubleshooter's bench.

TROUBLESHOOTER'S BENCH

■ ACTIVITY 1   Relate the PC Board to the Schematic

Using the pin configuration diagrams on the data sheets for the AD9300 and the AD673 in Appendix A, complete the diagram in Figure 14–64 by adding pin numbers, ground connections, and supply voltage connections. Use the completed schematic to locate and identify the components and the input and output functions on the analog board in Figure 14–66.

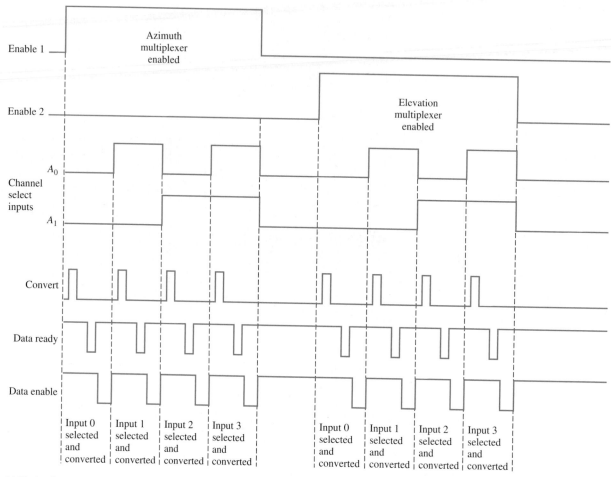

(a) Timing for one series of A/D conversions

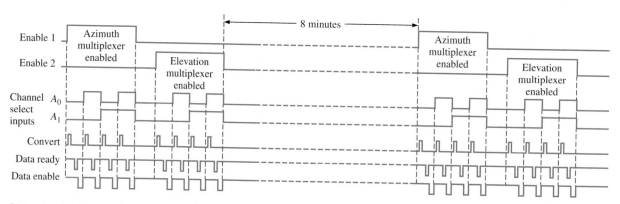

(b) A series of A/D conversions occurs every eight minutes.

FIGURE 14–65

Timing diagram for the A/D conversion sequence.

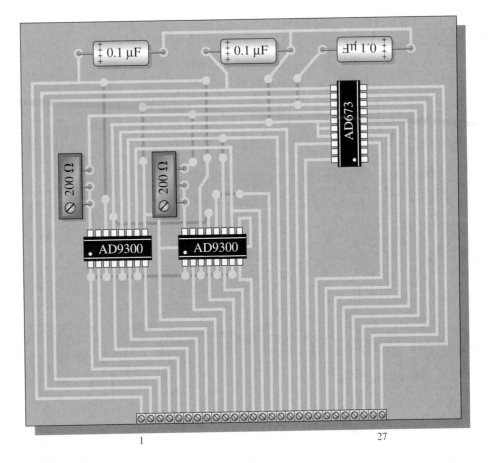

FIGURE 14–66

■ ACTIVITY 2   Analyze the System

For this application, assume that, at its limits, the azimuth potentiometer produces 0 V for a 0° orientation (due east) and 10 V for a 180° orientation (due west). At its limits, the elevation potentiometer produces 0 V for a 0° orientation (horizontal) and 10 V for a 90° orientation (vertical). The potentiometers rotate in 2° increments.

Step 1:   Calculate the voltage increment that will occur on the azimuth multiplexer input when the azimuth motor makes one step.

Step 2:   Calculate the voltage increment that will occur on the elevation multiplexer input when the elevation motor makes one step.

Step 3:   Determine the resolution in degrees to which the ADC can represent the input voltage range for both azimuth and elevation. Is this resolution adequate? Explain.

■ ACTIVITY 3   Write a Technical Report

Describe the overall operation of the solar panel control system. List the functional requirements for each block in the system. Discuss the operation of the analog board and its role in the overall system. Use the results of Activity 2 if appropriate.

■ **ACTIVITY 4   Troubleshoot the System for Each of the Following Problems By Stating the Probable Cause or Causes**

1. No voltage on one of the azimuth input lines to the analog multiplexer when there are 2.22 V on each of the other three inputs.

2. All four solar units are stuck at a given elevation position, but their azimuth position advances properly.

3. One solar unit is stuck at a given azimuth position, but its elevation position advances properly.

14–9 REVIEW QUESTIONS

1. What is the purpose of the three capacitors on the analog board?

2. How many discrete values of the analog position voltage can be converted to digital?

3. Explain why this system repeats the conversion sequence every eight minutes.

■ SUMMARY

- There are three basic types of analog switches: single pole–single throw (SPST), single pole–double throw (SPDT), and double pole–single throw (DPST).

- An analog switch is typically a MOSFET that is opened or closed with a control input.

- A sample-and-hold amplifier samples a voltage at a certain point in time and retains or holds that voltage for an interval of time.

- An analog quantity is one that has a continuous set of values over time.

- A digital quantity is one that has a set of discrete values over time.

- Two basic types of digital-to-analog converters (DACs) are the binary-weighted-input converter and the $R/2R$ ladder converter.

- The $R/2R$ ladder DAC is easier to implement because only two resistor values are required compared to a different value for each input in the binary-weighted-input DAC.

- The number of bits in an analog-to-digital converter (ADC) determines its resolution.

- The minimum sampling rate for A/D conversion is twice the maximum frequency component of the analog signal.

- The flash or simultaneous method of A/D conversion is the fastest.

- The successive-approximation method of A/D conversion is the most widely used.

- Other common methods of A/D conversion are single-slope, dual-slope, tracking, and stairstep ramp (counter method).

- In a voltage-to-frequency converter (V/F), the output frequency is directly proportional to the amplitude of the analog input voltage.

- In a frequency-to-voltage converter (F/V), the amplitude of the output voltage is directly proportional to the input frequency.

- Types of D/A conversion errors include nonmonotonicity, differential nonlinearity, low or high gain, and offset error.

- Types of A/D conversion errors include missing code, incorrect code, and offset.

■ GLOSSARY

Key terms are in color. All terms are included in the end-of-book glossary.

Accuracy   In relation to DACs or ADCs, a comparison of the actual output with the expected output, expressed as a percentage.

Acquisition time   In an analog switch, the time required for the device to reach its final value when switched from hold to sample.

Analog switch   A type of semiconductor switch that connects an analog signal from input to output with a control input.

Analog-to-digital converter (ADC)   A device used to convert an analog signal to a sequence of digital codes.

Aperture jitter   In an analog switch, the uncertainty in the aperture time.

Aperture time   In an analog switch, the time to fully open after being switched from sample to hold.

Digital-to-analog converter (DAC)   A device in which information in digital form is converted to an analog form.

Droop   In an analog switch, the change in the sampled value during the hold interval.

Feedthrough   In an analog switch, the component of the output voltage which follows the input voltage after the switch opens.

Flash   A method of A/D conversion.

Linearity   A straight-line relationship. A linear error is a deviation from the ideal straight-line output of a DAC.

Monotonicity   In relation to DACs, the presence of all steps in the output when sequenced over the entire range of input bits.

Nonmonotonicity   In relation to DACs, a step reversal or missing step in the output when sequenced over the entire range of input bits.

Nyquist rate   In sampling theory, the minimum rate at which an analog voltage can be sampled for A/D conversion. The sample rate must be equal to more than twice the maximum frequency component of the input signal.

Quantization   The determination of a value for an analog quantity.

Quantization error   The error resulting from the change in the analog voltage during the A/D conversion time.

Resolution   In relation to DACs or ADCs, the number of bits involved in the conversion. Also, for DACs, the reciprocal of the maximum number of discrete steps in the output.

Sample-and-hold   The process of taking the instantaneous value of a quantity at a specific point in time and storing it on a capacitor.

Settling time   The time it takes a DAC to settle within $\pm\frac{1}{2}$ LSB of its final value when a change occurs in the input code.

Successive approximation   A method of A/D conversion.

■ **KEY FORMULAS**

V/F Converters

(14–1) $\quad t_{os} = C_{os}(6.8 \times 10^3 \text{ s/F}) + 3 \times 10^{-7} \text{ s}$ — One-shot time

(14–2) $\quad \Delta V = \dfrac{(1 \text{ mA} - I_{in})t_{os}}{C_{int}}$ — Integrator output increase in reset interval

(14–3) $\quad t_{int} = \left(\dfrac{1 \text{ mA}}{I_{in}} - 1\right)t_{os}$ — Integrate interval

(14–4) $\quad f_{out} = \dfrac{I_{in}}{t_{os}(1 \text{ mA})}$ — Output frequency

■ SELF-TEST

Answers are at the end of the chapter.

1. An analog switch
   (a) changes an analog signal to digital
   (b) connects or disconnects an analog signal to the output
   (c) stores the value of an analog voltage at a certain point
   (d) combines two or more analog signals onto a single line

2. An analog multiplexer
   (a) produces the sum of several analog voltages on an output line
   (b) connects two or more analog signals to an output at the same time
   (c) connects two or more analog signals to an output one at a time in sequence
   (d) distributes two or more analog signals to different outputs in sequence

3. A basic sample-and-hold circuit contains
   (a) an analog switch and an amplifier
   (b) an analog switch, a capacitor, and an amplifier
   (c) an analog multiplexer and a capacitor
   (d) an analog switch, a capacitor, and input and output buffer amplifiers

4. In a sample/track-and-hold amplifier,
   (a) the voltage at the end of the sample interval is held
   (b) the voltage at the beginning of the sample interval is held
   (c) the average voltage during the sample interval is held
   (d) the output follows the input during the sample interval
   (e) answers (a) and (d)

5. In an analog switch, the aperture time is the time it takes for the switch to
   (a) fully open after the control switches from hold to sample
   (b) fully close after the control switches from sample to hold
   (c) fully open after the control switches from sample to hold
   (d) fully close after the control switches from hold to sample

6. In a binary-weighted-input digital-to-analog converter (DAC),
   (a) all of the input resistors are of equal value
   (b) there are only two input resistor values required
   (c) the number of different input resistor values equals the number of inputs

7. In a 4-bit binary-weighted-input DAC, if the lowest-valued input resistor is 1.0 kΩ, the highest-valued input resistor is
   (a) 2 kΩ   (b) 4 kΩ   (c) 8 kΩ   (d) 16 kΩ

8. The advantage of an *R/2R* ladder DAC is
   (a) it is more accurate   (b) it uses only two resistor values
   (c) it uses only one resistor value   (d) it can handle more inputs

9. In a DAC, monotonicity means that
   (a) the accuracy is within one-half of a least significant bit
   (b) there are no missing steps in the output
   (c) there is one bit missing from the input
   (d) there are no linear errors

10. An 8-bit analog-to-digital converter (ADC) can represent
    (a) 144 discrete values of an analog input
    (b) 4096 discrete values of an analog input
    (c) a continuous set of values of an analog input
    (d) 256 discrete values of an analog input

11. An analog signal must be sampled at a minimum rate greater than
    (a) twice the maximum frequency
    (b) twice the minimum frequency
    (c) the maximum frequency
    (d) the minimum frequency

12. Quantization error in an ADC is due to
   (a) poor resolution
   (b) nonlinearity of the input
   (c) a missing bit in the output
   (d) a change in the input voltage during the conversion time

13. Quantization error can be avoided by
   (a) using a higher resolution ADC
   (b) using a sample-and-hold prior to the ADC
   (c) shortening the conversion time
   (d) using a flash ADC

14. The type of ADC with the fastest conversion time is the
   (a) dual-slope   **(b)** single-slope
   (c) simultaneous   **(d)** successive-approximation

15. The output of a *V/F* converter
   (a) has an amplitude proportional to the frequency of the input
   (b) is a digital reproduction of the input voltage
   (c) has a frequency that is inversely proportional to the amplitude of the input
   (d) has a frequency that is directly proportional to the amplitude of the input

16. An element not found in the typical *V/F* converter is
   (a) a linear amplifier   **(b)** a one-shot
   (c) an integrator   **(d)** a comparator

TROUBLESHOOTER'S QUIZ   *Answers are at the end of the chapter.*

Refer to Figure 14–71(a).

❑ If the gate input from pin 12 is stuck high,

   1. The output voltage will

   (a) increase   **(b)** decrease   **(c)** not change

❑ If the external capacitor $C_{H(ext)}$ opens,

   2. The sampling rate will

   (a) increase   **(b)** decrease   **(c)** not change

   3. The device's ability to hold the sampled value will

   (a) increase   **(b)** decrease   **(c)** not change

Refer to Figure 14–71(b).

❑ If the external 10 kΩ resistor opens,

   4. The voltage gain will

   (a) increase   **(b)** decrease   **(c)** not change

   5. The offset voltage compensation will

   (a) increase   **(b)** decrease   **(c)** not change

Refer to Figure 14–73.

❑ If the 10 kΩ resistor opens,

   6. The magnitude of the output signal voltage will

   (a) increase   **(b)** decrease   **(c)** not change

7. The amplifier open-loop gain will

    (a) increase   **(b)** decrease   **(c)** not change

Refer to Figure 14–77.

❑ If R_1 has a value smaller than specified,

    8. The output frequency will

        (a) increase   **(b)** decrease   **(c)** not change

❑ If C_2 is larger than its specified value,

    9. The output frequency will

        (a) increase   **(b)** decrease   **(c)** not change

❑ If the positive dc supply voltage increases,

    10. The amplitude of the output voltage will

        (a) increase   **(b)** decrease   **(c)** not change

■ **PROBLEMS**   *Answers to odd-numbered problems are at the end of the book.*

SECTION 14–1   Analog Switches

1. Determine the output waveform for the analog switch in Figure 14–67(a) for each set of waveforms in parts (b), (c), and (d).

2. Determine the output of the 4-channel analog multiplexer in Figure 14–68 for the signal and control inputs shown.

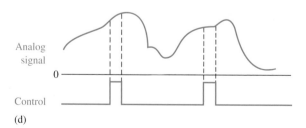

(a)   (b)

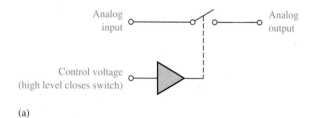

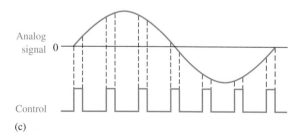

(c)   (d)

FIGURE 14–67

FIGURE 14–68

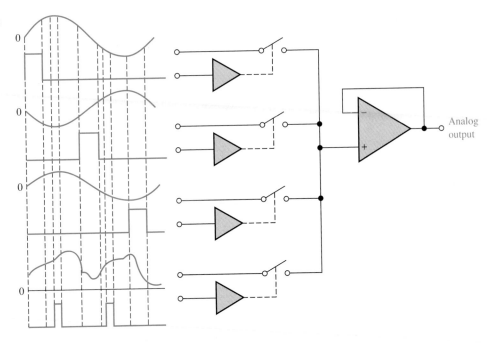

SECTION 14–2 Sample-and-Hold Amplifiers

3. Determine the output voltage waveform for the sample/track-and-hold amplifier in Figure 14–69 given the analog input and the control voltage waveforms shown. Sample is the high control level.

FIGURE 14–69

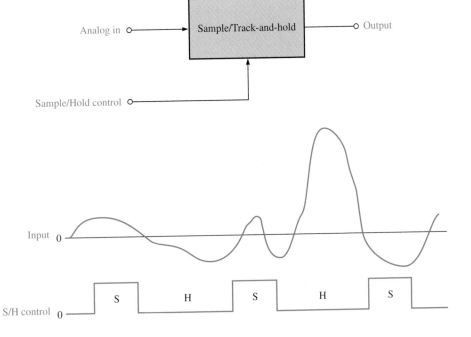

FIGURE 14–70

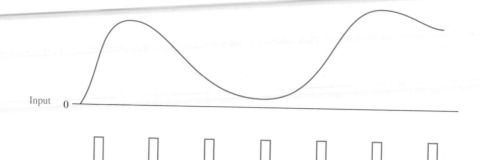

FIGURE 14–71

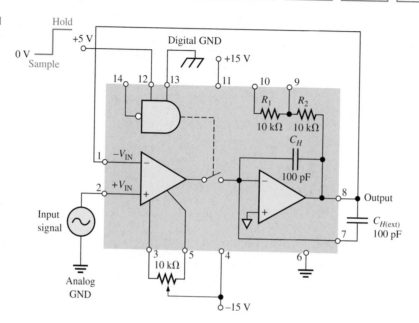

(a)

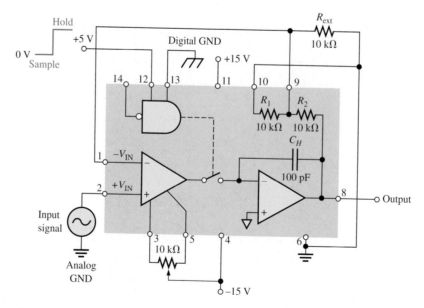

(b)

4. Repeat Problem 3 for the waveforms in Figure 14–70.

5. Determine the gain of each AD585 sample-and-hold amplifier in Figure 14–71.

SECTION 14–3 Interfacing the Analog and Digital Worlds

6. The analog signal in Figure 14–72 is sampled at 2 ms intervals. Represent the signal by a series of 4-bit binary numbers.

7. Sketch the digital reproduction of the analog curve represented by the series of binary numbers developed in Problem 6.

8. Graph the analog signal represented by the following sequence of binary numbers: 1111, 1110, 1101, 1100, 1010, 1001, 1000, 0111, 0110, 0101, 0100, 0101, 0110, 0111, 1000, 1001, 1010, 1011, 1100, 1100, 1011, 1010, 1010.

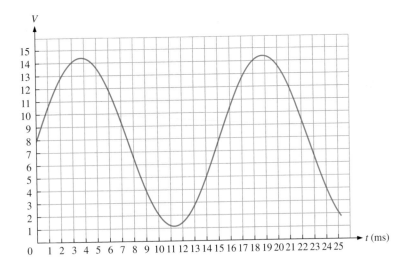

FIGURE 14–72

SECTION 14–4 Digital-to-Analog (D/A) Conversion

9. In a certain 4-bit DAC, the lowest-weighted resistor has a value of 10 kΩ. What should the values of the other input resistors be?

10. Determine the output of the DAC in Figure 14–73(a) if the sequence of 4-bit numbers in part (b) is applied to the inputs.

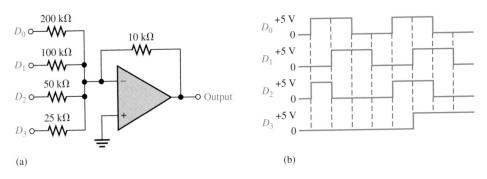

(a)

(b)

FIGURE 14–73

11. Repeat Problem 10 for the inputs in Figure 14–74.

12. Determine the resolution expressed as a percentage for each of the following DACs:
  (a) 3-bit   (b) 10-bit   (c) 18-bit

FIGURE 14–74
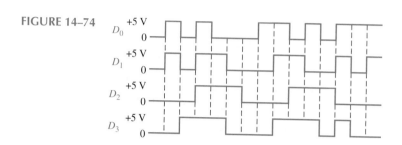

SECTION 14–5   Basic Concepts of Analog-to-Digital (A/D) Conversion

13. How many discrete values of an analog signal can each of the following ADCs represent?
  (a) 4-bit   (b) 5-bit   (c) 8-bit   (d) 16-bit

14. Determine the Nyquist rate for sinusoidal voltages with each of the following periods:
  (a) 10 s   (b) 1 ms   (c) 30 μs   (d) 1000 ns

15. What is the quantization error expressed in volts of an ADC with a sample-and-hold input for a sampled value of 3.2 V if the sample-and-hold has a droop of 100 mV/s? Assume that the conversion time of the ADC is 10 ms.

SECTION 14–6   Analog-to-Digital (A/D) Conversion Methods

16. Determine the binary output sequence of a 3-bit flash ADC for the analog input signal in Figure 14–75. The sampling rate is 100 kHz and V_{REF} = 8 V.

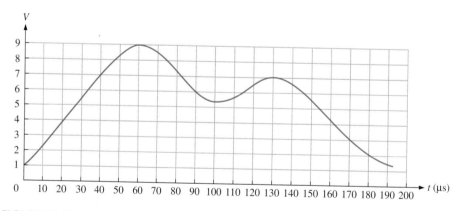

FIGURE 14–75

17. Repeat Problem 16 for the analog waveform in Figure 14–76.

18. For a certain 4-bit successive-approximation ADC, the maximum ladder output is +8 V. If a constant +6 V is applied to the analog input, determine the sequence of binary states for the SAR.

SECTION 14–7   Voltage-to-Frequency (*V/F*) and Frequency-to-Voltage (*F/V*) Converters

19. The analog input to a *V/F* converter increases from 0.5 V to 3.5 V. Does the output frequency increase, decrease, or remain unchanged?

20. Assume that when the input to a certain *V/F* converter is 0 V, there is no output signal (0 Hz). Also, when a constant +2 V is applied to the input, the corresponding output frequency is 1 kHz. Now, if the input takes a step up to +4 V, what is the output frequency?

21. Calculate the value of the timing capacitor required to produce a 5 μs pulse width in an AD650 *V/F* converter.

22. Determine the increase in the integrator output voltage during the reset interval in the AD650 shown in Figure 14–77.

FIGURE 14–76

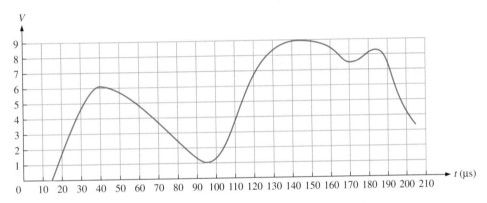

FIGURE 14–77

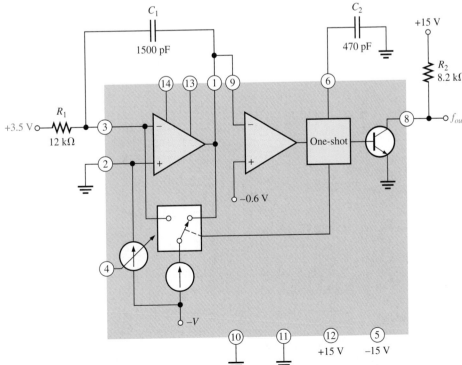

FIGURE 14–78

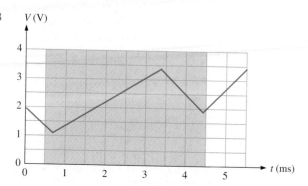

23. Determine the minimum and maximum output frequencies of the AD650 in Figure 14–77 for the portion of the input voltage shown within the shaded area in Figure 14–78.

SECTION 14–8 Troubleshooting

24. A 4-bit DAC has failed in such a way that the MSB is stuck at 0. Draw the analog output when a straight binary sequence is applied to the inputs.

25. A straight binary sequence is applied to a 4-bit DAC and the output in Figure 14–79 is observed. What is the problem?

26. An ADC produces the following sequence of binary numbers when a certain analog signal is applied to its input: 0000, 0001, 0010, 0011, 0100, 0101, 0110, 0111, 0110, 0101, 0100, 0011, 0010, 0001, 0000.

    (a) Reconstruct the input from the digital codes as a DAC would.

    (b) If the ADC failed so that the code 0111 were missing from the output, what would the reconstructed output look like?

FIGURE 14–79   Analog output

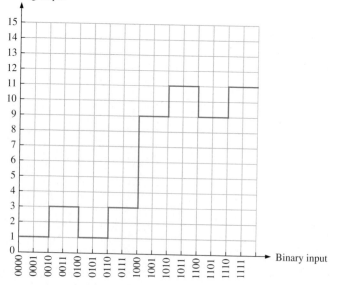

Section 14–1

1. To switch analog signals on or off electronically.

2. An analog multiplexer switches analog voltages from several lines onto a common output line in a time sequence.

Section 14–2

1. A sample-and-hold retains the value of an analog signal taken at a given point.
2. Droop is the decrease in the held voltage due to capacitor leakage.
3. Aperture time is the time required for an analog switch to fully open at the end of a sample pulse.
4. Acquisition time is the time required for the device to reach final value at the start of the sample pulse.

Section 14–3

1. Natural quantities are in analog form.
2. A/D conversion changes an analog quantity into digital form.
3. D/A conversion changes a digital quantity into analog form.

Section 14–4

1. Each input resistor must have a different value.
2. 6.67%

Section 14–5

1. The time for a sampled analog value to be converted to digital is the conversion time.
2. Greater than 200 Hz.
3. The sample-and-hold keeps the sampled value constant during conversion.

Section 14–6

1. Flash is the fastest method of A/D conversion.
2. Tracking A/D conversion uses an up-down counter.
3. True

Section 14–7

1. V/F components: integrator, comparator, one-shot, current source, and switch
2. The output frequency increases proportionally.
3. The V/F converter has a voltage input and a frequency output. The F/V converter has a frequency input and a voltage output.

Section 14–8

1. Nonmonotonicity is indicated by a step reversal.
2. Step amplitudes are less than ideal.
3. Missing code, incorrect code, offset (any two)

Section 14–9

1. The capacitors are for power supply decoupling.
2. $2^8 = 256$ values can be converted to digital.
3. Eight minutes is the minimum interval between 2-degree steps required to track the sun through a $180°$ arc for 12 hours.

FIGURE 14–80

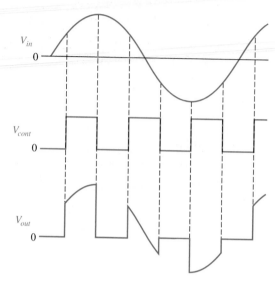

FIGURE 14–81

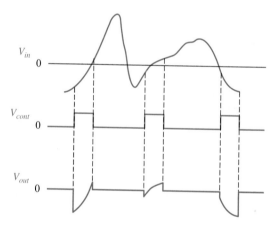

FIGURE 14–82

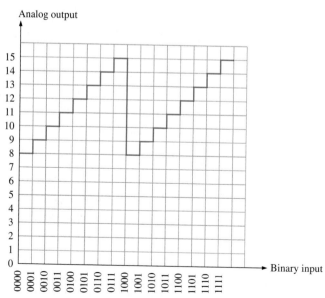

■ **ANSWERS TO PRACTICE EXERCISES FOR EXAMPLES**

14–1 See Figure 14–80.

14–2 The same tones will be closer together.

14–3 See Figure 14–81.

14–4 0.25 V

14–5 0.00038%

14–6 010, 011, 011, 011, 010, 001, 000, 000, 001, 010, 011, 011

14–7 $f_{min} = 40$ kHz, $f_{max} = 240$ kHz

14–8 See Figure 14–82.

14–9 A constant 15 V output.

■ **ANSWERS TO SELF-TEST**

1. (b)	**2.** (c)	**3.** (d)	**4.** (e)	**5.** (c)
6. (c)	**7.** (c)	**8.** (b)	**9.** (b)	**10.** (d)
11. (a)	**12.** (d)	**13.** (b)	**14.** (c)	**15.** (d)
16. (a)				

■ **ANSWERS TO TROUBLE-SHOOTER'S QUIZ**

1. decrease	**2.** not change	**3.** decrease	**4.** decrease
5. not change	**6.** increase	**7.** not change	**8.** increase
9. decrease	**10.** increase		

15

MEASUREMENT AND CONTROL CIRCUITS

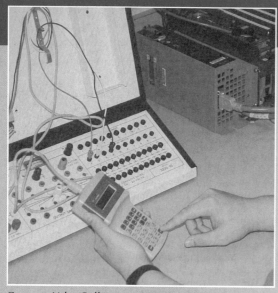

Courtesy Yuba College

■ CHAPTER OUTLINE

■ CHAPTER OBJECTIVES

❑ Describe the basic operation of rms-to-dc converters
❑ Discuss angle measurement using a synchro
❑ Discuss the operation of three types of temperature-measuring circuits
❑ Describe methods of measuring strain, pressure, and motion
❑ Describe how power to a load is controlled
❑ Apply what you have learned in this chapter to a system application

■ KEY TERMS

❑ Root mean square
❑ Transducer
❑ Synchro
❑ Resolver
❑ Synchro-to-digital converter (SDC)
❑ Resolver-to-digital converter (RDC)
❑ Thermocouple
❑ Resistance temperature detector (RTD)
❑ Thermistor
❑ Strain gage
❑ Thyristor
❑ Silicon-controlled rectifier (SCR)
❑ Triac
❑ Zero-voltage switching

This chapter introduces several types of transducers and related circuits for measuring basic physical analog parameters such as angular position, temperature, strain, pressure, and flow rate.

A transducer is a device that converts a physical parameter into another form. Transducers can be used at the input (a microphone) or output (a speaker) of a system. With electronic-measuring systems, the input transducer converts a quantity to be measured (temperature, humidity, flow rate, weight) into an electrical parameter (voltage, current, resistance, capacitance) that can be processed by an electronic instrument or system.

Transducers and their associated circuits are important in many applications. The measurement of angular position is critical in robotics, radar, and industrial machine control. Temperature-measuring and pressure-measuring circuits are used in industry for monitoring temperatures and pressures of various fluids or gases in tanks and pipes, and they are used in automotive applications for measuring temperatures and pressures in various parts of the automobile. Strain measurement is important for testing the strength of materials under stress in such areas as aircraft design. Also, the silicon-controlled rectifier, triac, and zero-voltage switch, which are important power control applications, are introduced in this chapter.

■ A SYSTEM APPLICATION

The system application in Section 15–6 focuses on the measurement of wind speed and direction. The input from the wind speed measurement part of the system (anemometer) is generated by a type of flow meter in the form of a propeller arrangement mounted on a wind vane. The wind causes the flow meter blades to rotate on a shaft at a rate proportional to the wind speed. A magnetic device senses each rotation and the circuitry produces a pulse. The frequency of the pulses indicates the speed of the wind. The input from the wind direction measurement part of the system is generated by the wind vane attached to the shaft of a resolver. The wind vane aligns itself with the direction of the wind, and the resolver produces electrical signals proportional to the angular position. Figure 15–47 shows a simplified diagram for the wind-measuring system that will be the focus of the system application. Notice that some circuits from previous chapters are utilized in this application.

For the system application in Section 15–6, in addition to the other topics, be sure you understand

❏ Basic resolver and RDC operation
❏ The 555 timer, frequency-to-voltage converters, and ADCs

15–1 ■ RMS-TO-DC CONVERTERS

One important application of rms-to-dc converters is in noise measurement for determining thermal noise, transistor, and switch contact noise. Another application is in the measurement of signals from mechanical phenomena such as strain, vibration, and expansion or contraction. RMS-to-dc converters are also useful for accurate measurements of low-frequency, low duty-cycle pulse trains.

After completing this section, you should be able to

❏ Describe the basic operation of rms-to-dc converters
  ❏ Define *rms*
  ❏ Explain the rms-to-dc conversion process
  ❏ List the basic circuits in an rms-to-dc converter
  ❏ Discuss the difference between explicit and implicit rms-to-dc converters
  ❏ Give examples of rms-to-dc converter applications

Definition of RMS

RMS stands for **root mean square** and is related to the amplitude of an ac signal. In practical terms, the rms value of an ac voltage is equal to the value of a dc voltage required to produce the same heating effect in a resistance. For this reason, it is sometimes called the *effective* value of an ac voltage. For example, an ac voltage with an rms value of 1 V produces the same amount of heat in a given resistor as a 1 V dc voltage. Mathematically, the rms value is found by taking the square root of the average (**mean**) of the signal voltage squared, as expressed in the following formula:

$$V_{rms} = \sqrt{\mathrm{avg}(V_{in}^2)}$$ (15–1)

RMS-to-DC Conversion

RMS-to-dc converters are electronic circuits that continuously compute the square of the input signal voltage, average it, and take the square root of the result. The output of an rms-to-dc converter is a dc voltage that is proportional to the rms value of the input signal. The block diagram in Figure 15–1 illustrates the basic conversion process.

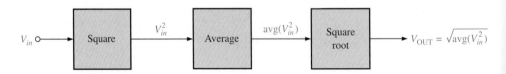

FIGURE 15–1
The rms-to-dc conversion process.

The Squaring Circuit   The squaring circuit is generally a linear multiplier with the signal applied to both inputs as shown in Figure 15–2. Linear multipliers were introduced in Chapter 13.

FIGURE 15–2
The squaring circuit is a linear multiplier.

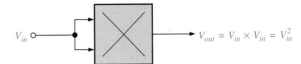

The Averaging Circuit

The simplest type of averaging circuit is a single-pole low-pass filter on the input of an op-amp voltage-follower, as shown in Figure 15–3. The RC filter passes only the dc component (average value) of the squared input voltage. The overbar designates an average value.

FIGURE 15–3
A basic averaging circuit.

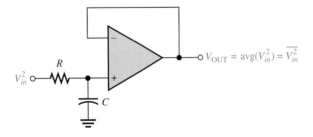

The Square Root Circuit

Recall from Chapter 13 that a square root circuit uses a linear multiplier in an op-amp feedback loop as shown in Figure 15–4.

FIGURE 15–4
A square root circuit.

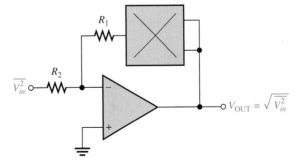

A Complete RMS-to-DC Converter

Figure 15–5 shows the three functional circuits combined to form an rms-to-dc converter. This combination is often referred to as an *explicit* rms-to-dc converter because of the straightforward method used in determining the rms value.

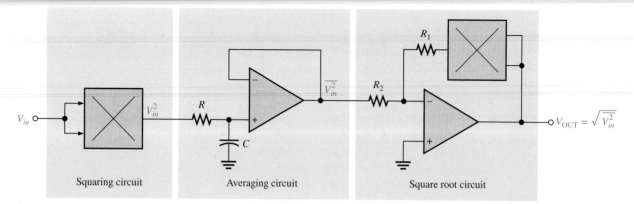

FIGURE 15–5
Explicit type of rms-to-dc converter.

Another method for achieving rms-to-dc conversion, sometimes called the *implicit* method, uses feedback to perform the square root operation. A basic circuit is shown in Figure 15–6. The first block squares the input voltage and divides by the output voltage. The averaging circuit produces the final dc output voltage, which is fed back to the squarer/divider circuit.

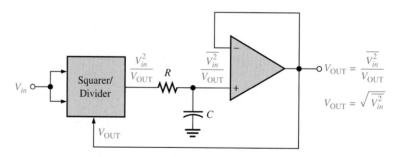

FIGURE 15–6
Implicit type of rms-to-dc converter.

The operation of the circuit in Figure 15–6 can be understood better by going through the mathematical steps performed by the circuit as follows. The expression for the output of the squarer/divider is

$$\frac{V_{in}^2}{V_{OUT}}$$

The voltage at the noninverting (+) input to the voltage-follower is

$$V_{in(NI)} = \frac{\overline{V_{in}^2}}{V_{OUT}}$$

where the overbar indicates average value. The final output voltage is

$$V_{OUT} = \frac{\overline{V_{in}^2}}{V_{OUT}}$$

$$V_{OUT}^2 = \overline{V_{in}^2}$$

$$V_{OUT} = \sqrt{\overline{V_{in}^2}} \qquad\qquad (15\text{--}2)$$

The AD637 IC RMS-to-DC Converter[1]

As an example of a specific IC device, let's look at the AD637 rms-to-dc converter. This device is essentially an implicit type of converter except that it has an absolute value circuit at the input and it uses an inverting low-pass filter for averaging, as indicated in Figure 15–7. The averaging capacitor, C_{avg}, is an external component that can be selected to provide a minimum averaging error under various input conditions.

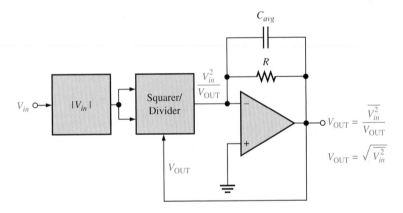

FIGURE 15–7
Basic diagram of the AD637 rms-to-dc converter.

The absolute value circuit in the first block is simply a full-wave rectifier that changes all the negative portions of an input voltage to positive. The squarer/divider circuit is actually implemented with log and antilog circuits as shown in Figure 15–8.

Notice that the second block in Figure 15–8 produces the log of the square of the input by taking the log of V_{in} and then multiplying it by two.

$$2 \log V_{in} = \log V_{in}^2$$

This equation is based on the fundamental rule of logarithms that states the log of a variable squared is equal to twice the log of the variable. The third block is a subtracter that subtracts the logarithm of the output voltage from the log of the input squared.

---

[1] Data sheet for AD637 available at http://www.analogdevices.com

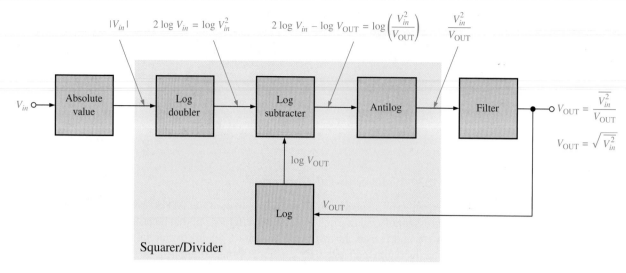

FIGURE 15–8

Internal function diagram of the AD637.

$$2 \log V_{in} - \log V_{OUT} = \log V_{in}^2 - \log V_{OUT} = \log\left(\frac{V_{in}^2}{V_{OUT}}\right)$$

This equation is based on the fundamental rule of logarithms that states that the difference of two logarithmic terms equals the logarithm of the quotient of the two terms. The antilog circuit takes the antilog of $\log(V_{in}^2/V_{OUT})$ and produces an output equal to $\overline{V_{in}^2}/V_{OUT}$, as indicated in the figure. The low-pass filter averages the output of the antilog circuit and produces the final output.

Examples of RMS-to-DC Converter Applications

In addition to the measurement applications mentioned in the section introduction, rms-to-dc converters are used in a variety of system applications. A couple of typical applications are in automatic gain control (AGC) circuits and rms voltmeters. Let's look at each of these in a general way.

AGC Circuits Figure 15–9 shows a general diagram of an AGC circuit that incorporates an rms-to-dc converter. AGC circuits are used in audio systems to keep the output amplitude constant when the input signal level varies over a certain range. They are also used in signal generators to keep the output amplitude constant with variations in waveform, duty cycle, and frequency.

RMS Voltmeters Figure 15–10 basically illustrates the rms-to-dc converter in an rms voltmeter. The rms-to-dc converter produces a dc output that is the rms value of the input signal. This rms value is then converted to digital form by an ADC and displayed.

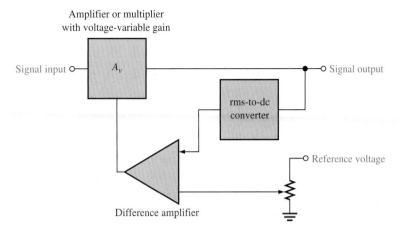

FIGURE 15–9
A simplified AGC circuit using an rms-to-dc converter.

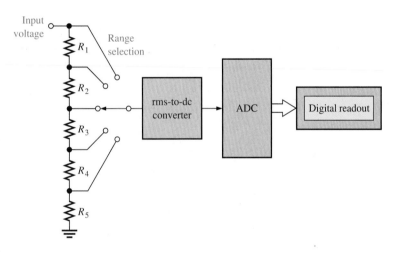

FIGURE 15–10
A simplified rms voltmeter.

15–1 REVIEW QUESTIONS*

1. What is the basic purpose of an rms-to-dc converter?

2. What are the three internal functions performed by an rms-to-dc converter?

15–2 ■ ANGLE-MEASURING CIRCUITS

In many applications, the angular position of a shaft or other mechanical mechanism must be measured and converted to an electrical signal for processing or display. Examples of this mechanical-to-electrical interfacing are found in radar and satellite antennas, wind vanes, solar systems, industrial machines including robots, and military fire control systems, to name a few. In this section, the circuits for interfacing angular position transducers, called synchros, are introduced. Transducers, *in general, are devices that convert a physical parameter from one form to another. Before we get into the circuits used in angular measurements, a brief introduction to synchros will provide some background.*

After completing this section, you should be able to

❑ Discuss angle measurement using a synchro
    ❑ Define *synchro* and explain the basic operation
    ❑ Define *resolver* and explain the basic operation
    ❑ Discuss synchro-to-digital converters and resolver-to-digital converters
    ❑ Describe the basic operation of an RDC
    ❑ Show how angles can be represented by digital codes
    ❑ Discuss an RDC application

Synchros

A **synchro** is an electromechanical transducer used for shaft angle measurement and positioning. There are several different types of synchros, but all can be thought of as basically as rotating transformers. In physical appearance, a synchro resembles a small ac motor as shown in Figure 15–11(a) with a diameter ranging from about 0.5 in. to about 4 in.

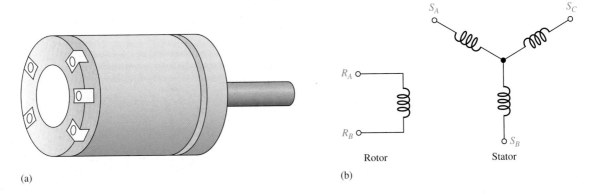

(a)   Rotator   (b)   Stator

FIGURE 15–11
A typical synchro and its basic winding structure.

The basic synchro consists of a **rotor,** which can revolve within a fixed **stator** assembly. A shaft is connected to the rotor so that when the shaft rotates, the rotor also rotates. In most synchros, there is a rotor winding and three stator windings. The stator windings are connected as shown in Figure 15–11(b) and are separated by 120° around the stator. The windings are brought out to a terminal block at one end of the housing.

Synchro Voltages When a reference sinusoidal voltage is applied across the rotor winding, the voltage induced across any one of the stator windings is proportional to the sine of the angle (θ) between the rotor winding and the stator winding. The angle θ is dependent on the shaft position.

The voltage induced across any two stator windings (between any two stator terminals) is the sum or difference of the two stator voltages. These three voltages, called *synchro format voltages,* are represented in Figure 15–12 and are derived using a basic trigonometric identity. The important thing is that each of the three synchro format voltages is a function of the shaft angle, θ, and can be used to determine the angular position at any time. As the shaft rotates, the format voltages change proportionally.

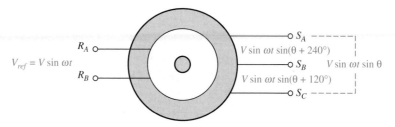

FIGURE 15–12
Synchro format voltages with a reference voltage applied to the rotor.

Resolvers

The **resolver** is a particular type of synchro that is often used in rotational systems to transduce the angular position. Resolvers differ from regular synchros in that the rotor and two stator windings are separated from each other by 90° rather than by 120°. The basic winding configuration of a simple resolver is shown in Figure 15–13.

FIGURE 15–13
Simple resolver winding configuration.

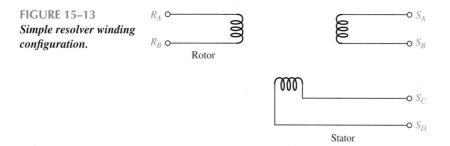

Resolver Voltages   If a reference sinusoidal voltage is applied to the rotor winding, the resulting voltages across the stator windings are as given in Figure 15–14. These voltages are a function of the shaft angle θ and are called *resolver format voltages.* One of the voltages is proportional to the sine of θ and the other voltage is proportional to the cosine of θ. Notice that the resolver has a four-terminal output compared to the three-terminal output of the standard synchro.

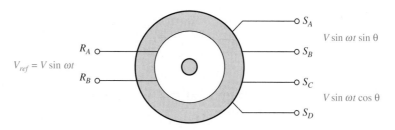

FIGURE 15–14
Resolver format voltages with a reference voltage applied to the rotor.

Basic Operation of Synchro-to-Digital and Resolver-to-Digital Converters

Synchro-to-digital converters (SDCs) and **resolver-to-digital converters (RDCs)** are electronic circuits used to convert the format voltages from a synchro or resolver to a digital format. These devices may be considered a very specialized form of analog-to-digital converter.

All converters, both SDCs and RDCs, operate internally with resolver format voltages. Therefore, the output format voltages of a synchro must first be transformed into resolver format by a special type of transformer called the *Scott-T transformer,* as illustrated in Figure 15–15.

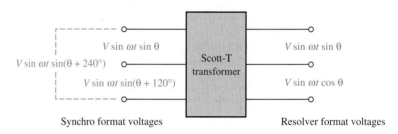

FIGURE 15–15
Inputs and outputs of a Scott-T transformer.

Some SDCs have internal Scott-T transformers, but others require a separate transformer. Other than the transformer, the basic operation and internal circuitry of SDCs and RDCs are the same, so let's focus on RDCs. A simplified block diagram of a tracking RDC is shown in Figure 15–16.

The two resolver format voltages, $V_1 = V \sin \omega t \sin \theta$ and $V_2 = V \sin \omega t \cos \theta$, are applied to the RDC inputs as indicated in Figure 15–16 (θ is the current shaft angle of the resolver). These resolver voltages go through buffers to special multiplier circuits. Let's assume that the current state of the up/down counter represents some angle, ϕ. The digital code representing ϕ is applied to the multiplier circuits along with the resolver voltages. The cosine multiplier takes the cosine of ϕ and multiplies it times the resolver voltage V_1. The sine multiplier takes the sine of ϕ and multiplies it times the resolver voltage V_2. The resulting output of the cosine multiplier is

$$V_1\cos \phi = V \sin \omega t \sin \theta \cos \phi$$

The resulting output of the sine multiplier is

$$V_2\sin \phi = V \sin \omega t \cos \theta \sin \phi$$

These two voltages are subtracted by the error amplifier to produce the following error voltage:

$$V \sin \omega t \sin \theta \cos \phi - V \sin \omega t \cos \theta \sin \phi = V \sin \omega t(\sin \theta \cos \phi - \cos \theta \sin \phi)$$

A basic trigonometric identity reduces the error voltage expression to

$$V \sin \omega t \sin(\theta - \phi)$$

The phase-sensitive detector produces a dc error voltage proportional to $\sin(\theta - \phi)$, which is applied to the integrator. The output of the integrator drives a voltage-controlled

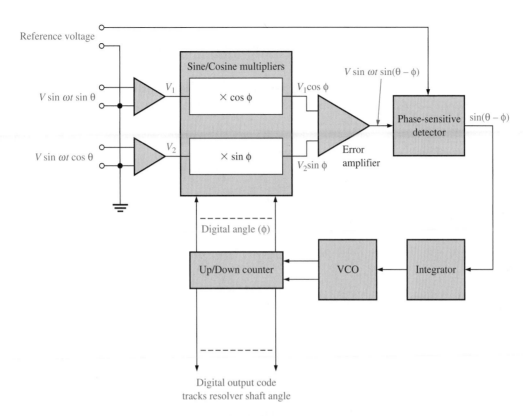

FIGURE 15–16
Simplified diagram of a resolver-to-digital converter (RDC).

oscillator (VCO), which provides clock pulses to the up/down counter. When the counter reaches the value of the current shaft angle θ, then $\phi = \theta$ and

$$\sin(\theta - \phi) = 0$$

If the sine is zero, then the difference of the angles is 0°.

$$\theta - \phi = 0°$$

At this point, the angle stored in the counter equals the resolver shaft angle.

$$\phi = \theta$$

When the shaft angle changes, the counter will count up or down until its count equals the new shaft angle. Therefore, the RDC will continuously track the resolver shaft angle and produce an output digital code that equals the angle at all times.

Representation of Angles with a Digital Code

The most common method of representing an angular measurement with a digital code is given in Table 15–1 for word lengths up to 16 bits. A binary 1 in any bit position means that the corresponding angle is included, and a 0 means that the corresponding angle is not included.

TABLE 15–1
Bit weights for resolver-to-digital conversion.

Bit Position	Angle (Degrees)
1 (MSB)	180.00000
2	90.00000
3	45.00000
4	22.50000
5	11.25000
6	5.62500
7	2.81250
8	1.40625
9	0.70313
10	0.35156
11	0.17578
12	0.08790
13	0.04395
14	0.02197
15	0.01099
16	0.00549

EXAMPLE 15–1

A certain RDC has an 8-bit digital output. What is the angle being measured if the output code is 01001101? The left-most bit is the MSB.

Solution

Bit Position	Bit	Angle (Degrees)
1	0	0
2	1	90.00000
3	0	0
4	0	0
5	1	11.25000
6	1	5.62500
7	0	0
8	1	1.40625

To get the angle represented, add all the included angles (as indicated by the presence of a 1 in the output code).

$$90.00000° + 11.25000° + 5.62500° + 1.40625° = \mathbf{108°}$$

Although more digits are carried in the calculation to show the process, the answer is rounded to the nearest degree.

*Practice Exercise** What is the angular shaft position measured by a 12-bit RDC when it has a binary code of 100000100001 on its outputs?

_____
* Answers are at the end of the chapter.

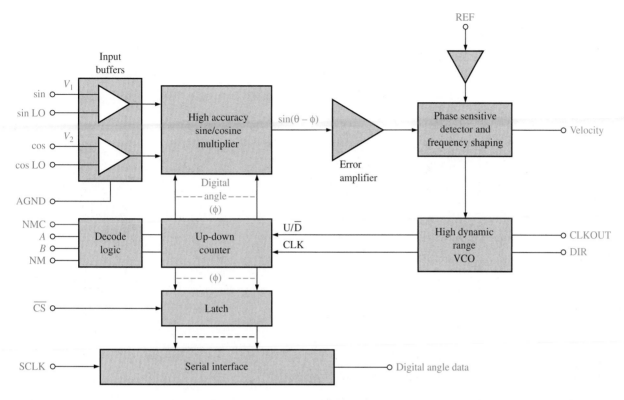

FIGURE 15–17
Diagram of the AD2S90 resolver-to-digital converter.

A Specific Resolver-to-Digital Converter

To illustrate a typical IC device, let's look at the AD2S90[2], which is a 12-bit converter. The diagram for this device is shown in Figure 15–17, and as you can see, it is basically the same as the general RDC in Figure 15–16 with some additions. Additional circuits include the latch and serial interface for controlling the data transfer to and interfacing with other digital systems. These additional circuits do not affect the conversion process.

Additional outputs include the Direction (DIR) output which indicates the direction of rotation of the resolver shaft. The Velocity output is proportional to the rate of change of the input angle.

A Simple Resolver-to-Digital Converter Application

The measurement of wind direction is one example of how an RDC can be used. As shown in Figure 15–18, a wind vane is fixed to the shaft of a resolver. As the wind vane moves to align with the direction of the wind, the resolver shaft rotates and its angle indicates the wind direction. The resolver output is applied to an RDC and the resulting digital output code, which represents the wind direction, drives a digital readout.

---

[2] Data sheet for AD2S90 available at http://www.analogdevices.com

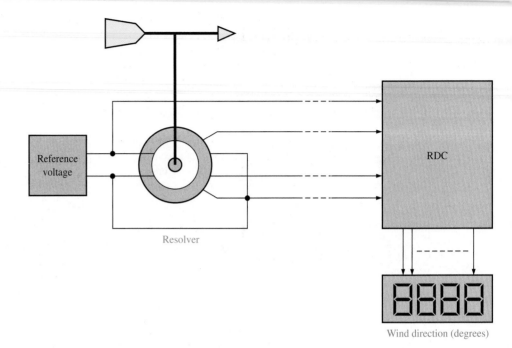

FIGURE 15–18
Measurement and display of wind direction with a resolver and an RDC.

15–2 REVIEW QUESTIONS

1. What is a transducer that converts a mechanical shaft position into electrical signals called?

2. What type of input does an RDC accept?

3. What type of output does an RDC produce?

4. What is the function of an RDC?

15–3 ■ TEMPERATURE-MEASURING CIRCUITS

Temperature is perhaps the most common physical parameter that is measured and converted to electrical form. Several types of temperature sensors respond to temperature and produce a corresponding indication by a change or alteration in a physical characteristic that can be detected by an electronic circuit. Common types of temperature sensors are thermocouples, resistance temperature detectors (RTDs), and thermistors. In this section, we will look at each of these sensors and at signal conditioning circuits that are required to interface the transducers to electronic equipment.

After completing this section, you should be able to

❑ Discuss the operation of three types of temperature-measuring circuits
  ❑ Describe the thermocouple and how to interface it with an electronic circuit
  ❑ Describe the resistance temperature detector (RTD) and circuit interfacing
  ❑ Describe the thermistor and circuit interfacing

The Thermocouple

The **thermocouple** is formed by joining two dissimilar metals. A small voltage, called the *Seebeck voltage,* is produced across the junction of the two metals when heated, as illustrated in Figure 15–19. The amount of voltage produced is dependent on the types of metals and is directly proportional to the temperature of the junction (positive temperature coefficient); however, this voltage is generally much less than 100 mV. The voltage versus temperature characteristic of thermocouples is somewhat nonlinear, but the amount of nonlinearity is predictable. Thermocouples are widely used in certain industries because they have a wide temperature range and can be used to measure very high temperatures.

FIGURE 15–19

A voltage proportional to temperature is generated when a thermocouple is heated.

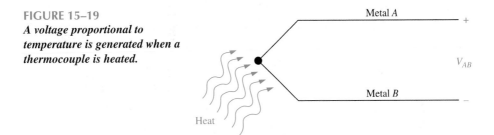

Some common metal combinations used in commercial thermocouples are chromel-alumel (chromel is a nickel-chromium alloy and alumel is a nickel-aluminum alloy), iron-constantan (constantan is a copper-nickel alloy), chromel-aluminum, tungsten-rhenium alloys, and platinum-10% Rh/Pt. Each of these types of thermocouple has a different temperature range, coefficient, and voltage characteristic and is designated by the letter *E, J, K, W,* and *S,* respectively. The overall temperature range covered by thermocouples is from −250°C to 2000°C. Each type covers a different portion of this range, as shown in Figure 15–20.

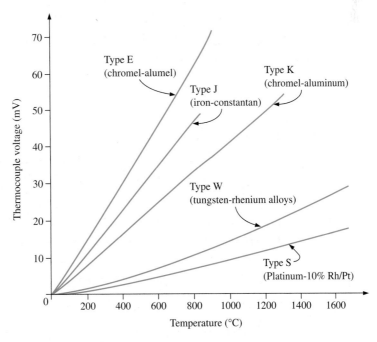

FIGURE 15–20

Output of some common thermocouples with 0°C as the reference temperature.

Thermocouple-to-Electronics Interface   When a thermocouple is connected to a signal-conditioning circuit, as illustrated in Figure 15–21, an *unwanted* thermocouple is effectively created at the point(s) where one or both of the thermocouple wires connect to the circuit terminals made of a dissimilar metal. The unwanted thermocouple junction is sometimes referred to as a **cold junction** in some references because it is normally at a significantly lower temperature than that being measured by the measuring thermocouple. These unwanted thermocouples can have an unpredictable effect on the overall voltage that is sensed by the circuit because the voltage produced by the unwanted thermocouple opposes the measured thermocouple voltage and its value depends on ambient temperature.

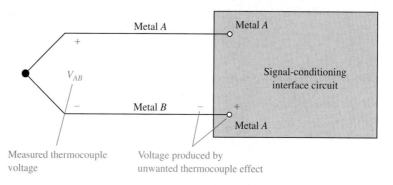

FIGURE 15–21
Creation of an unwanted thermocouple in a thermocouple-to-electronics interface.

Example of a Thermocouple-to-Electronics Interface   As shown in Figure 15–22, a copper/constantan thermocouple (known as type *T*) is used, in this case, to measure the temperature in an industrial temperature chamber. The copper thermocouple wire is connected to a copper terminal on the circuit board and the constantan wire is also connected to a copper terminal on the circuit board. The copper-to-copper connection is no problem because the metals are the same. The constantan-to-copper connection acts as an

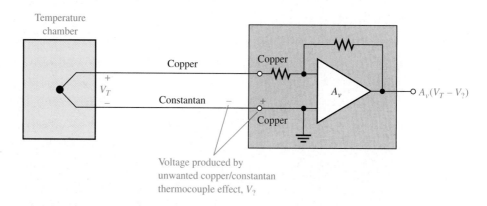

FIGURE 15–22
A simplified temperature-measuring circuit with an unwanted thermocouple at the junction of the constantan wire and the copper terminal.

unwanted thermocouple that will produce a voltage in opposition to the thermocouple voltage because the metals are dissimilar.

Since the unwanted thermocouple connection is not at a fixed temperature, its effects are unpredictable and it will introduce inaccuracy into the measured temperature. One method for eliminating an unwanted thermocouple effect is to add a reference thermocouple at a constant known temperature (usually 0°C). Figure 15–23 shows that by using a reference thermocouple that is held at a constant known temperature, the unwanted thermocouple at the circuit terminal is eliminated because both contacts to the circuit terminals are now copper-to-copper. The voltage produced by the reference thermocouple is a known constant value and can be compensated for in the circuitry.

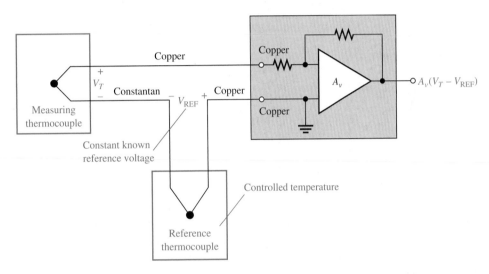

FIGURE 15–23
Using a reference thermocouple in a temperature-measuring circuit.

EXAMPLE 15–2

Suppose the thermocouple in Figure 15–22 is measuring 200°C in an industrial oven. The circuit board is in an area where the ambient temperature can vary from 15°C to 35°C. Using Table 15–2 for a type-T (copper/constantan) thermocouple, determine the voltage across the circuit input terminals at the ambient temperature extremes. What is the maximum percent error in the voltage at the circuit input terminals?

TABLE 15–2
Type-T thermocouple voltage.

Temperature (°C)	Output (mV)
−200	−5.603
−100	−3.378
0	0.000
+100	4.277
+200	9.286
+300	14.860
+400	20.869

Solution From Table 15–2, you know that the measuring thermocouple is producing 9.286 mV. To determine the voltage that the unwanted thermocouple is creating at 15°C, you must interpolate from the table. Since 15°C is 15% of 100°C, a linear interpolation gives the following voltage:

$$0.15(4.277 \text{ mV}) = 0.642 \text{ mV}$$

Since 35°C is 35% of 100°C, the voltage is

$$0.35(4.277 \text{ mV}) = 1.497 \text{ mV}$$

The voltage across the circuit input terminals at 15°C is

$$9.286 \text{ mV} - 0.642 \text{ mV} = \textbf{8.644 mV}$$

The voltage across the circuit input terminals at 35°C is

$$9.286 \text{ mV} - 1.497 \text{ mV} = \textbf{7.789 mV}$$

The maximum percent error in the voltage at the circuit input terminals is

$$\left(\frac{9.286 \text{ mV} - 7.789 \text{ mV}}{9.286 \text{ mV}} \right) 100\% = \textbf{16.1\%}$$

You can never be sure how much it is off because you have no control over the ambient temperature. Also, the linear interpolation may or may not be accurate depending on the linearity of the temperature characteristic of the unwanted thermocouple.

Practice Exercise In the case of the circuit in Figure 15–22, if the temperature being measured goes up to 300°C, what is the maximum percent error in the voltage across the circuit input terminals?

EXAMPLE 15–3

Refer to the thermocouple circuit in Figure 15–23. Suppose the thermocouple is measuring 200°C. Again, the circuit board is in an area where the ambient temperature can vary from 15°C to 35°C. The reference thermocouple is held at exactly 0°C. Determine the voltage across the circuit input terminals at the ambient temperature extremes.

Solution From Table 15–2 in Example 15–2, the thermocouple voltage is 0 V at 0°C. Since the reference thermocouple produces no voltage at 0°C and is completely independent of ambient temperature, there is no error in the measured voltage over the ambient temperature range. Therefore, the voltage across the circuit input terminals at both temperature extremes equals the measuring thermocouple voltage, which is **9.286 mV**.

Practice Exercise If the reference thermocouple were held at −100°C instead of 0°C, what would be the voltage across the circuit input terminals if the measuring thermocouple were at 400°C?

Compensation It is bulky and expensive to maintain a reference thermocouple at a fixed temperature (usually an ice bath is required). Another approach is to compensate for the unwanted thermocouple effect by adding a compensation circuit as shown in Figure 15–24. This is sometimes referred to as *cold-junction compensation*. The compensation circuit consists of a resistor and an integrated circuit temperature sensor with a temperature coefficient that matches that of the unwanted thermocouple.

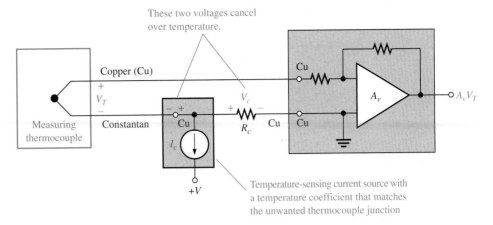

FIGURE 15–24
Simplified circuit for compensation of unwanted thermocouple effect.

The current source in the temperature sensor produces a current that creates a voltage drop, V_c, across the compensation resistor, R_c. The resistance is adjusted so that this voltage drop is equal and opposite the voltage produced by the unwanted thermocouple at a given temperature. When the ambient temperature changes, the current changes proportionally, so that the voltage across the compensation resistor is always approximately equal to the unwanted thermocouple voltage. Since the compensation voltage, V_c, is opposite in polarity to the unwanted thermocouple voltage, the unwanted voltage is effectively cancelled.

The functions shown in the circuit of Figure 15–24 plus others are available in IC packages and hybrid modules known as *thermocouple signal conditioners*. The AD596, 1B51, and 3B47[3] are examples of this type of circuit. They are designed for interfacing a thermocouple with various types of electronic systems and provide gain, compensation, isolation, common-mode rejection, and other features in one package.

Resistance Temperature Detectors (RTDs)

A second major type of temperature transducer is the **resistance temperature detector (RTD)**. The RTD is a resistive device in which the resistance changes directly with temperature (positive temperature coefficient). The RTD is more nearly linear than the thermocouple. RTDs are constructed in either a wire-wound configuration or by a metal-film technique. The most common RTDs are made of either platinum, nickel, or nickel alloys.

Generally, RTDs are used to sense temperature in two basic ways. First, as shown in Figure 15–25(a), the RTD is driven by a current source and, since the current is constant, the change in voltage across it is proportional (by Ohm's law) to the change in its resistance with temperature. Second, as shown in Figure 15–25(b), the RTD is connected in a 3-wire bridge circuit; and the bridge output voltage is used to sense the change in the RTD resistance and, thus, the temperature. An example of the RTD in IC form is the 1B41.[4]

---

[3] Data sheets for AD596, 1B51, and 3B47 are available at http://www.analogdevices.com

[4] Data sheet for 1B41 available at http://www.analogdevices.com

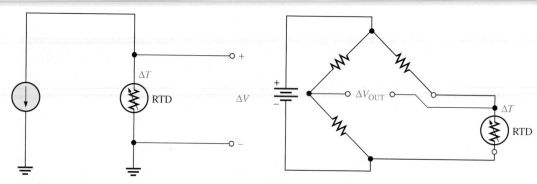

(a) A change in temperature, ΔT, produces a change in voltage, ΔV, across the RTD proportional to the change in RTD resistance when the current is constant.

(b) A change in temperature, ΔT, produces a change in bridge output voltage, ΔV, proportional to the change in RTD resistance.

FIGURE 15–25

Basic methods of employing an RTD in a temperature-sensing circuit.

Theory of the 3-wire Bridge   To avoid subjecting the three bridge resistors to the same temperature that the RTD is sensing, the RTD is usually remotely located to the point where temperature variations are to be measured and connected to the rest of the bridge by long wires. The resistance of the three bridge resistors must remain constant. The long extension wires to the RTD have resistance that can affect the accurate operation of the bridge.

Figure 15–26(a) shows the RTD connected in the bridge with a 2-wire configuration. Notice that the resistance of both of the long connecting wires appear in the same leg of the bridge as the RTD. Recall from your study of basic circuits that $V_{OUT} = 0$ V and the bridge is balanced when $R_{RTD} = R_3$ if $R_1 = R_2$. The wire resistances will throw the bridge off balance when $R_{RTD} = R_3$ and will cause an error in the output voltage for any value of the RTD resistance because they are in series with the RTD in the same leg of the bridge.

The 3-wire configuration in Figure 15–26(b) overcomes the wire resistance problem. By connecting a third wire to one end of the RTD as shown, the resistance of wire A is now placed in the same leg of the bridge as R_3 and the resistance of wire B is placed in the same leg of the bridge as the RTD. Because the wire resistances are now in opposite legs of the bridge, their effects will cancel if both wire resistances are the same (equal lengths of same type of wire). The resistance of the third wire has no effect; essentially no current goes through it because the output terminals of the bridge are open or are connected across a very high impedance. The balance condition is expressed as

$$R_{RTD} + R_B = R_3 + R_A$$

If $R_A = R_B$, then they cancel in the equation and the balance condition is completely independent of the wire resistances.

$$R_{RTD} = R_3$$

The method described here is important in many measurements that use a sensitive transducer and a bridge. It is often used in strain-gage measurements (described in Section 15–4).

Basic RTD Temperature-Sensing Circuits   Two simplified RTD measurement circuits are shown in Figure 15–27. The circuit in part (a) is one implementation of an RTD driven by a constant current. The operation is as follows. From your study of basic op-amp circuits, recall that the input current and the current through the feedback path are essentially equal because the input impedance of the op-amp is ideally infinite. Therefore,

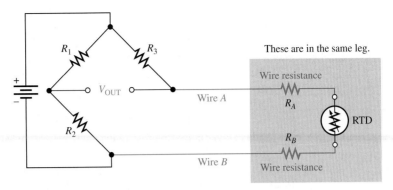

(a) Two-wire bridge connection

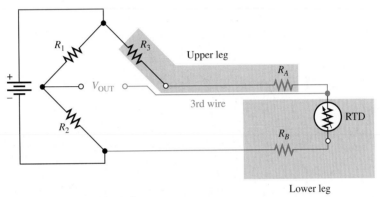

(b) Three-wire bridge connection

FIGURE 15–26
Comparison of 2-wire and 3-wire bridge connections in an RTD circuit.

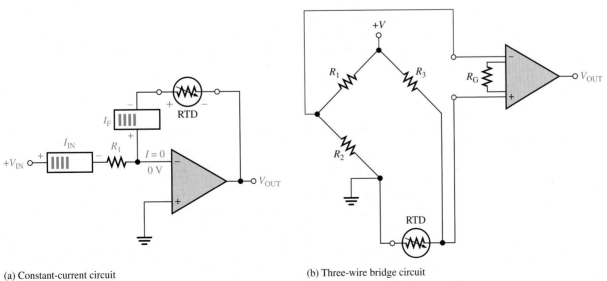

(a) Constant-current circuit

(b) Three-wire bridge circuit

FIGURE 15–27
Basic RTD temperature-measuring circuits.

the constant current through the RTD is set by the constant input voltage, V_{IN}, and the input resistance, R_1, because the inverting input is at virtual ground. The RTD is in the feedback path and, therefore, the output voltage of the op-amp is equal to the voltage across the RTD. As the resistance of the RTD changes with temperature, the voltage across the RTD also changes because the current is constant.

The circuit in Figure 15–27(b) shows a basic circuit in which an instrumentation amplifier is used to amplify the voltage across the 3-wire bridge circuit. The RTD forms one leg of the bridge; and as its resistance changes with temperature, the bridge output voltage also changes proportionally. The bridge is adjusted for balance ($V_{OUT} = 0$ V) at some reference temperature, say 0°C. This means that R_3 is selected to equal the resistance of the RTD at this reference temperature.

EXAMPLE 15–4

Determine the output voltage of the instrumentation amplifier in the RTD circuit in Figure 15–28 if the resistance of the RTD is 1320 Ω at the temperature being measured.

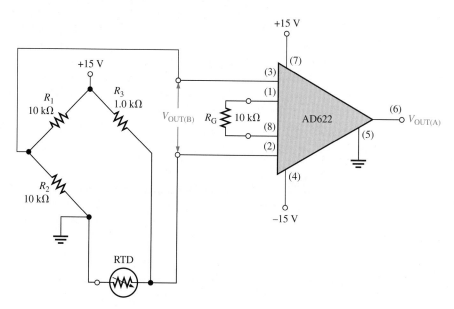

FIGURE 15–28

Solution The bridge output voltage is

$$V_{OUT(B)} = \left(\frac{R_{RTD}}{R_3 + R_{RTD}}\right)15 \text{ V} - \left(\frac{R_2}{R_1 + R_2}\right)15 \text{ V} = \left(\frac{1320 \ \Omega}{2320 \ \Omega}\right)15 \text{ V} - \left(\frac{10 \text{ k}\Omega}{20 \text{ k}\Omega}\right)15 \text{ V}$$

$$= 8.53 \text{ V} - 7.5 \text{ V} = 1.03 \text{ V}$$

From Equation (12–3), the voltage gain of the AD622 instrumentation amplifier is

$$R_G = \frac{50.5 \text{ k}\Omega}{A_v - 1}$$

$$A_v = \frac{50.5 \text{ k}\Omega}{R_G} + 1 = 5.05 + 1 = 6.05$$

The output voltage from the amplifier is

$$V_{OUT(A)} = (6.05)(1.03 \text{ V}) = \textbf{6.23 V}$$

Practice Exercise   What must be the nominal resistance of the RTD in Figure 15–28 to balance the bridge at 25°C? What is the amplifier output voltage when the bridge is balanced?

Thermistors

A third major type of temperature transducer is the **thermistor**, which is a resistive device made from a semiconductive material such as nickel oxide or cobalt oxide. The resistance of a thermistor changes inversely with temperature (negative temperature coefficient). The temperature characteristic for thermistors is more nonlinear than that for thermocouples or RTDs; in fact, a thermistor's temperature characteristic is essentially logarithmic. Also, like the RTD, the temperature range of a thermistor is more limited than that of a thermocouple. Thermistors have the advantage of a greater sensitivity than either thermocouples or RTDs and are generally less expensive. This means that their change in resistance per degree change in temperature is greater. Since they are both variable-resistance devices, the thermistor and the RTD can be used in similar circuits.

Like the RTD, thermistors can be used in constant-current-driven configurations or in bridges. In Figure 15–29, the general response of a thermistor in a constant-current op-amp circuit is compared to that of an RTD in a similar circuit. Both the RTD and the thermistor are exposed to the same temperature environment as indicated. It is assumed that at some reference temperature, the RTD and the thermistor have the same resistance and produce the same output voltage. In the RTD circuit, as the temperature increases from the reference value, the op-amp's output voltage decreases from the reference value because the resistance of the RTD increases. In the thermistor circuit, as the temperature increases, the op-amp's output voltage increases from the reference value because the thermistor's resistance decreases due to its negative temperature coefficient. Also, for the same temperature change, the change in the output voltage of the thermistor circuit is greater

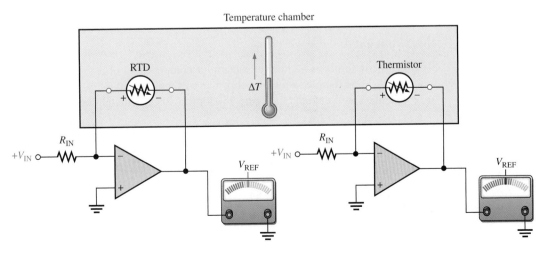

FIGURE 15–29
General comparison of the responses of a thermistor circuit to a similar RTD circuit.

than the corresponding change in the output voltage of the RTD circuit because of the greater sensitivity of the thermistor.

15–3 REVIEW QUESTIONS

1. What is a thermocouple?
2. How can temperature be measured with a thermocouple?
3. What is an RTD and how does its operation differ from a thermocouple?
4. What is the primary operational difference between an RTD and a thermistor?
5. Of the three devices introduced in this section, which one would most likely be used to measure extremely high temperatures?

15–4 ■ STRAIN-MEASURING, PRESSURE-MEASURING, AND MOTION-MEASURING CIRCUITS

In this section, methods of measuring three types of force-related parameters (strain, pressure, and motion) are examined. A variety of applications require the measurement of these three parameters. Also, other parameters, such as the flow rate of a fluid, can be measured indirectly by measuring strain, pressure, or motion.

After completing this section, you should be able to

❏ Describe methods of measuring strain, pressure, and motion
   ❏ Explain how a strain gage operates
   ❏ Discuss strain gage circuits
   ❏ Explain how pressure transducers work
   ❏ Discuss pressure-measuring circuits
   ❏ List several pressure transducer applications
   ❏ Explain displacement transducers, velocity transducers, and acceleration transducers

The Strain Gage

Strain is the deformation, either expansion or compression, of a material due to a force acting on it. For example, a metal rod or bar will lengthen slightly when an appropriate force is applied as illustrated in Figure 15–30(a). Also, if a metal plate is bent, there is an

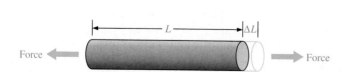

(a) Strain occurs as length changes from L to $L + \Delta L$ when force is applied.

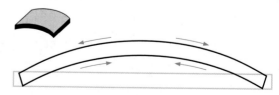

(b) Strain occurs when the flat plate is bent, causing the upper surface to expand and the lower surface to contract.

FIGURE 15–30
Examples of strain.

expansion of the upper surface, called *tensile strain,* and a compression of the lower surface, called *compressive strain,* as shown in Figure 15–30(b).

Strain gages are based on the principle that the resistance of a material increases if its length increases and decreases if its length decreases. This is expressed by the following formula (which you should recall from your dc/ac circuits course).

$$R = \frac{\rho L}{A} \tag{15–3}$$

This formula states that the resistance of a material, such as a length of wire, depends directly on the resistivity (ρ) and the length (L) and inversely on the cross-sectional area (A).

A **strain gage** is basically a long very thin strip of resistive material that is bonded to the surface of an object on which strain is to be measured, such as a wing or tail section of an airplane under test. When a force acts on the object to cause a slight elongation, the strain gage also lengthens proportionally and its resistance increases. Most strain gages are formed in a pattern similar to that in Figure 15–31(a) to achieve enough length for a sufficient resistance value in a smaller area. It is then placed along the line of strain as indicated in Figure 15–31(b).

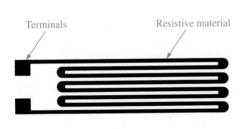

(a) Typical strain gage configuration.

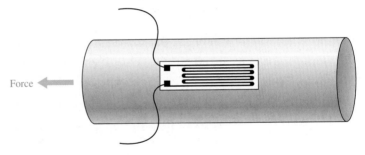

(b) The strain gage is bonded to the surface to be measured along the line of force. When the surface lengthens, the strain gage stretches.

FIGURE 15–31
A simple strain gage and its placement.

The Gage Factor of a Strain Gage An important characteristic of strain gages is the **gage factor (GF),** which is defined as the ratio of the fractional change in resistance to the fractional change in length along the axis of the gage. For metallic strain gages, the *GF*s are typically around 2. The concept of gage factor is illustrated in Figure 15–32 and expressed in Equation (15–4) where R is the nominal resistance and ΔR is the change in resistance due to strain. The fractional change in length ($\Delta L/L$) is designated strain (ϵ) and is usually expressed in parts per million, called *microstrain* (designated $\mu\epsilon$).

$$GF = \frac{\Delta R/R}{\Delta L/L} \tag{15–4}$$

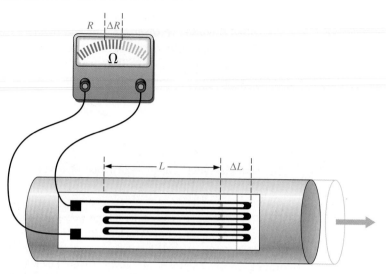

FIGURE 15–32
Illustration of gage factor. The ohmmeter symbol is not intended to represent a practical method for measuring ΔR.

EXAMPLE 15–5

A certain material being measured under stress undergoes a strain of 5 parts per million (5 $\mu\epsilon$). The strain gage has a nominal (unstrained) resistance of 320 Ω and a gage factor of 2.0. Determine the resistance change in the strain gage.

Solution

$$GF = \frac{\Delta R/R}{\Delta L/L} = \frac{\Delta R/R}{\epsilon}$$

$$\Delta R = (GF)(R)(\epsilon) = 2.0(320 \ \Omega)(5 \times 10^{-6}) = \textbf{3.2 m}\Omega$$

Practice Exercise If the strain in this example is 8 $\mu\epsilon$, how much does the resistance change?

Basic Strain Gage Circuits

Because a strain gage exhibits a resistance change when the quantity it is sensing changes, it is typically used in circuits similar to those used for RTDs. The basic difference is that strain instead of temperature is being measured. Therefore, strain gages are usually applied in bridge circuits or in constant-current-driven circuits, as shown in Figure 15–33. They can be used in applications in the same way as RTDs and thermistors. The 1B31[5] is an example of a strain gage signal conditioner.

Pressure Transducers

Pressure transducers are devices that exhibit a change in resistance proportional to a change in pressure. Basically, pressure sensing is accomplished using a strain gage bonded to a flexible diaphragm as shown in Figure 15–34(a). Figure 15–34(b) shows the

---

[5] Data sheet for 1B31 available at http://www.analogdevices.com

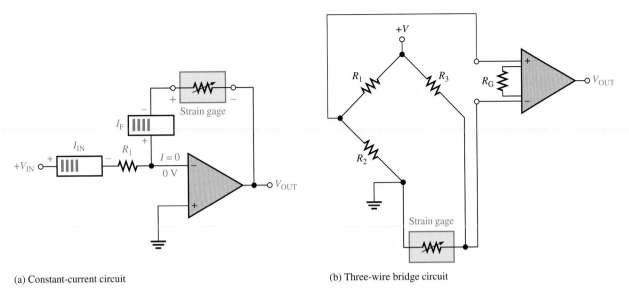

(a) Constant-current circuit

(b) Three-wire bridge circuit

FIGURE 15–33

Basic strain-measuring circuits.

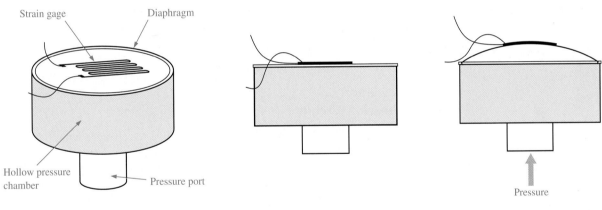

(a) Basic pressure gage construction

(b) With no net pressure on diaphragm, strain gage resistance is at its nominal value (side view).

(c) Net pressure forces diaphragm to expand, causing elongation of the strain gage and thus an increase in its resistance.

FIGURE 15–34

A simplified pressure sensor constructed with a strain gage bonded to a flexible diaphragm.

diaphragm with no net pressure exerted on it. When a net positive pressure exists on one side of the diaphragm, as shown in Figure 15–34(c), the diaphragm is pushed upward and its surface expands. This expansion causes the strain gage to lengthen and its resistance to increase.

Pressure transducers typically are manufactured using a foil strain gage bonded to a stainless steel diaphragm or by integrating semiconductor strain gages (resistors) in a silicon diaphragm. Either way, the basic principle remains the same.

Pressure transducers come in three basic configurations in terms of relative pressure measurement. The absolute pressure transducer measures applied pressure relative to a vacuum as illustrated in Figure 15–35(a). The gage pressure transducer measures applied pressure relative to the pressure of the surroundings (ambient pressure) as illustrated in Figure 15–35(b). The differential pressure transducer measures one applied pressure relative to another applied pressure as shown in Figure 15–35(c). Some transducer configurations include circuitry such as bridge completion circuits and op-amps within the same package as the sensor itself, as indicated.

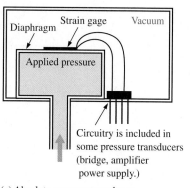

(a) Absolute pressure transducer

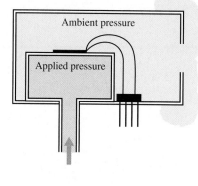

(b) Gage pressure transducer

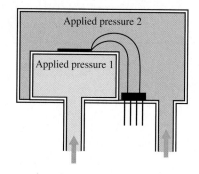

(c) Differential pressure transducer

FIGURE 15–35
Three basic types of pressure transducers.

Pressure-Measuring Circuits

Because pressure transducers are devices in which the resistance changes with the quantity being measured, they are usually in a bridge configuration as shown by the basic op-amp bridge circuit in Figure 15–36(a). In some cases, the complete circuitry is built into the transducer package, and in other cases the circuitry is external to the sensor. The symbols in parts (b) through (d) of Figure 15–36 are sometimes used to represent the complete pressure transducer with an amplified output. The symbol in part (b) represents the absolute pressure transducer, the symbol in part (c) represents the gage pressure transducer, and the symbol in part (d) represents the differential pressure transducer.

Flow Rate Measurement   One common method of measuring the flow rate of a fluid through a pipe is the differential-pressure method. A flow restriction device such as a Venturi section (or other type of restriction such as an orifice) is placed in the flow stream. The Venturi section is formed by a narrowing of the pipe, as indicated in Figure 15–37. Although the velocity of the fluid increases as it flows through the narrow channel, the volume of fluid per minute (volumetric flow rate) is constant throughout the pipe.

Because the velocity of the fluid increases as it goes through the restricted area, the pressure also increases. If pressure is measured at a wide point and at a narrow point, the flow rate can be determined because flow rate is proportional to the square root of the differential pressure, as shown in Figure 15–37.

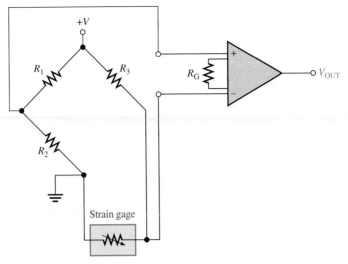

(a) Basic bridge circuit

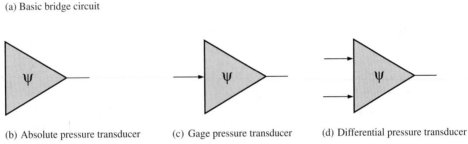

(b) Absolute pressure transducer   (c) Gage pressure transducer   (d) Differential pressure transducer

FIGURE 15–36
A basic pressure transducer circuit and symbols.

FIGURE 15–37
A basic method of flow rate measurement.

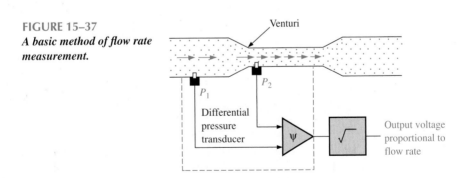

Pressure Transducer Applications Pressure transducers are used anywhere there is a need to determine the pressure of a substance. In medical applications, pressure transducers are used for blood pressure measurement; in aircraft, pressure transducers are used for altitude pressure, cabin pressure, and hydraulic pressure; in automobiles, pressure transducers are used for fuel flow, oil pressure, brake line pressure, manifold pressure, and steering system pressure, to name a few applications.

Motion-Measuring Circuits

Displacement Transducers   *Displacement* is a quantity that indicates the change in position of a body or point. Angular displacement refers to a rotation that can be measured in degrees or radians. Displacement transducers can be either contacting or noncontacting.

Contacting transducers typically use a sensing shaft with a coupling device to follow the position of the measured quantity. A contacting type of displacement sensor that relates a change in inductance to displacement is the linear variable differential transformer (LVDT). The sensing shaft is connected to a moving magnetic core inside a specially wound transformer. A typical LVDT is shown in Figure 15–38. The primary of the transformer is in line and located between two identical secondaries. The primary winding is excited with ac (usually in the range of 1 to 5 kHz). When the core is centered, the voltage induced in each secondary is equal. As the core moves off center, the voltage in one secondary will be greater than the other. With the demodulator circuit shown, the polarity of the output changes as the core passes the center position. The transducer has excellent sensitivity, linearity, and repeatability.

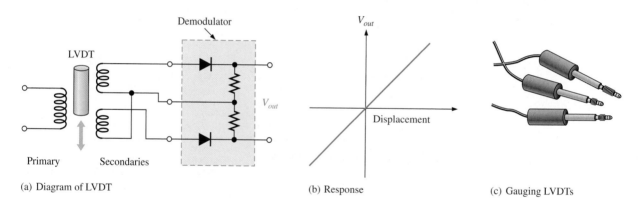

(a) Diagram of LVDT

(b) Response

(c) Gauging LVDTs

FIGURE 15–38
LVDT displacement transducers.

Noncontacting displacement transducers include optical and capacitive transducers. Photocells can be arranged to observe light through holes in an encoding disk or to count fringes painted on the surface to be measured. Optical systems are fast; but noise, including background light sources, can produce spurious signals in optical sensors. It is useful to build hysteresis into the system if noise is a problem (see Section 8–1).

Fiber-optic sensors make excellent proximity detectors for close ranges. Reflective sensors use two fiber bundles, one for transmitting light and the other for receiving light from a reflective surface, as illustrated in Figure 15–39. Light is transmitted in the fiber bundle without any significant attenuation. When it leaves the transmitting fiber bundle, it forms a spot on the target that is inversely proportional to the square of the distance. The receiving bundle is aimed at the spot and collects the reflected light to an optical sensor. The light intensity detected by the receiving bundle depends on the physical size and arrangement of the fibers as well as the distance to the spot and the reflecting surface, but the technique can respond to distances approaching 1 microinch. The major disadvantage is limited dynamic range.

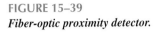

FIGURE 15–39
Fiber-optic proximity detector.

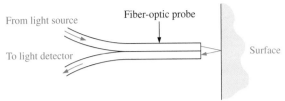

Capacitive sensors can be made into very sensitive displacement and proximity transducers. The capacitance is varied by moving one of the plates of a capacitor with respect to the second plate. The moving plate can be any metallic surface such as the diaphragm of a capacitive microphone or a surface that is being measured. The capacitor can be used to control the frequency of a resonant circuit to convert the capacitive change into a usable electrical output. (Experiment 37 of the lab book shows an example.)

Velocity Transducers   *Velocity* is defined as the rate of change of displacement. It follows that velocity can be determined indirectly with a displacement sensor and measuring the time between two positions. A direct measurement of velocity is possible with certain transducers that have an output proportional to the velocity to be measured. These transducers can respond to either linear or angular velocity. Linear velocity transducers can be constructed using a permanent magnet inside a concentric coil, forming a simple motor by generating an emf proportional to the velocity. Either the coil or the magnet can be fixed and the other moved with respect to the fixed component. The output is taken from the coil.

There are a variety of transducers that are designed to measure angular velocity. Tachometers, a class of angular velocity transducers, provide a dc or ac voltage output. A dc tachometer is basically a small generator with a coil that rotates in a constant magnetic field. A voltage is induced in the coil as it rotates in the magnetic field. The average value of the induced voltage is proportional to the speed of rotation, and the polarity is indicative of the direction of rotation, an advantage with dc tachometers. AC tachometers can be designed as generators that provide an output frequency that is proportional to the rotational speed.

Another technique for measuring angular velocity is to rotate a shutter over a photosensitive element. The shutter interrupts a light source from reaching the photocells, causing the output of the photocells to vary at a rate proportional to the rotational speed.

Acceleration Transducers   Acceleration is usually measured by use of a spring-supported seismic mass, mounted in a suitable enclosure as shown in Figure 15–40. Damping is provided by a dashpot, which is a mechanical device to reduce the vibration. The relative motion between the case and the mass is proportional to the acceleration. A secondary transducer such as a resistive displacement transducer or an LVDT is used to convert the relative motion to an electrical output. Ideally, the mass does not move when the case accelerates because of its inertia; in practice, it does because of forces applied to it through the spring. The accelerometer has a natural frequency, the period of which should be shorter than the time required for the measured acceleration to change. Accelerometers used to measure vibration should also be used at frequencies less than the natural frequency.

An accelerometer that uses the basic principle of the LVDT can be constructed to measure vibration. The mass is made from a magnet that is surrounded with a coil. Voltage induced in the coil is a function of the acceleration.

Another type of accelerometer uses a piezoelectric crystal in contact with the seismic mass. The crystal generates an output voltage in response to forces induced by the acceleration of the mass. Piezoelectric crystals are small in size and have a natural frequency that

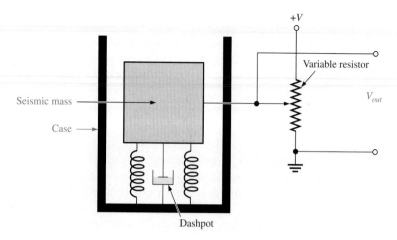

FIGURE 15–40
A basic accelerometer. Motion is converted to a voltage by the variable resistor.

is very high; they can be used to measure high-frequency vibration. The drawback to piezoelectric crystals is that the output is very low and the impedance of the crystal is high, making it subject to problems from noise.

15–4 REVIEW QUESTIONS

1. Describe a basic strain gage.
2. Describe a basic pressure gage.
3. List three types of pressure gages.
4. **(a)** What is an LVDT?   **(b)**   What does it measure?

15–5 ■ POWER-CONTROL CIRCUITS

A useful application of electronic circuits is to control power to a load. In this section, you will learn about two devices that are widely used in power control applications—the SCR and the triac. These devices are members of a class of devices known as thyristors, which are widely used in industrial controls for motors, heaters, phase controls, and many other applications. A thyristor can be thought of as an electronic switch that can rapidly turn on or off a large current to a load. Integrated circuits are frequently used to determine the time to turn on or off the SCR or triac.

After completing this section, you should be able to

❏ Describe how power to a load is controlled
   ❏ Describe the SCR and triac
   ❏ Explain how to turn an SCR on or off
   ❏ Explain the term *zero-voltage switching*
   ❏ Define *microcontroller*

The Silicon-Controlled Rectifier

A **thyristor** is a semiconductor switch composed of four or more layers of alternating *pnpn* material. There are various types of thyristors; the type principally depends on the number of layers and the particular connections to the layers. When a connection is made to the first, second, and fourth layer of a four-layer thyristor, a form of gated diode known as a **silicon-controlled rectifier (SCR)** is formed. This is one of the most important devices in the thyristor family because it acts like a diode that can be turned on when required. The basic structure and schematic symbol for an SCR is shown in Figure 15–41. For an SCR, the three connections are labeled the anode (A), cathode (K), and gate (G) as shown.

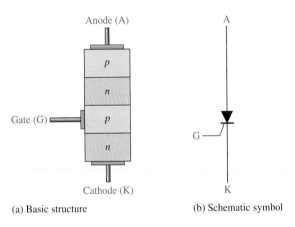

(a) Basic structure   (b) Schematic symbol

FIGURE 15–41
The silicon-controlled rectifier (SCR).

The characteristic curve for an SCR is shown in Figure 15–42(a) for a gate current of zero. There are a total of four regions of the characteristic curve of interest. The reverse characteristic (plotted in quadrant 3) is the same as a normal diode with regions called the reverse-blocking region and a reverse-avalanche region. The reverse-blocking region is equivalent to an open switch. The reverse voltage that must be applied to an SCR to drive it into the avalanche region is typically several hundred volts or more. SCRs are normally not operated in the reverse-avalanche region.

The forward characteristic (plotted in quadrant 1) is divided into two regions. There is a forward-blocking region, where the SCR is basically off and the very high resistance between the anode and cathode can be approximated by an open switch. The second region is the forward-conduction region, where anode current occurs as in a normal diode. To move an SCR into this region, the forward-breakover voltage, $V_{BR(F)}$, must be exceeded. When an SCR is operated in the forward conduction region, it approximates a closed switch between anode and cathode. Notice the similarity to a normal diode characteristic (see Figure 1–31) except for the forward-blocking region.

Turning the SCR On   There are two ways to move an SCR into the forward-conduction region. In both cases, the anode to cathode must be forward-biased; that is, the anode must be positive with respect to the cathode. The first method has already been mentioned and requires the application of forward voltage that exceeds the forward breakover voltage, $V_{BR(F)}$. Breakover voltage triggering is not normally used as a triggering method. The second method requires a positive pulse of current (trigger) on the gate. This pulse reduces the forward-breakover voltage, as shown in Figure 15–42(b) and the SCR conducts. The

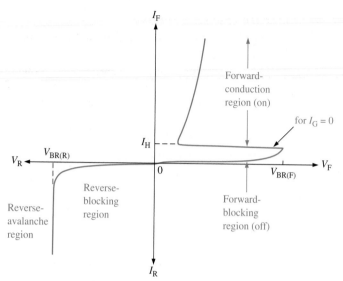

(a) When $I_G = 0$, $V_{BR(F)}$ must be exceeded to move into the conduction region.

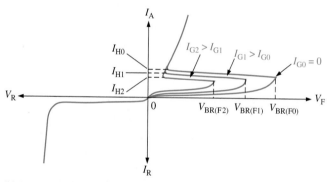

(b) I_G controls the value of $V_{BR(F)}$ required for turn on.

FIGURE 15–42
SCR characteristic curves.

greater the gate current, the lower the value of $V_{BR(F)}$. This is the normal method for turning on an SCR.

Once the SCR is turned on, the gate loses control. In effect, the SCR is latched and will continue to approximate a closed switch as long as anode current is maintained. When the anode current drops below a value of current called the holding current, the SCR will drop out of conduction. The holding current is indicated in Figure 15–42.

Turning the SCR Off There are two basic methods for turning off an SCR: anode current interruption and forced commutation. The anode current can be interrupted by opening the path in the anode circuit, causing the anode current to drop to zero, turning off the SCR. One common "automatic" method to interrupt the anode current is to connect the SCR in an ac circuit. The negative cycle of the ac waveform will turn off the SCR.

The forced commutation method requires momentarily forcing current through the SCR in the direction opposite to the forward conduction so that forward current is reduced below the holding value. This can be implemented by various circuits. Probably the simplest is to electronically switch a charged capacitor across the SCR in the reverse direction.

The Triac

The **triac** is a thyristor with the ability to pass current bidirectionally and is therefore an ac power control device. Although it is one device, its performance is equivalent to two SCRs connected in parallel in opposite directions but with a common gate terminal. The basic characteristic curves for a triac are illustrated in Figure 15–43. Because a triac is like two back-to-back SCRs, there is no reverse characteristic.

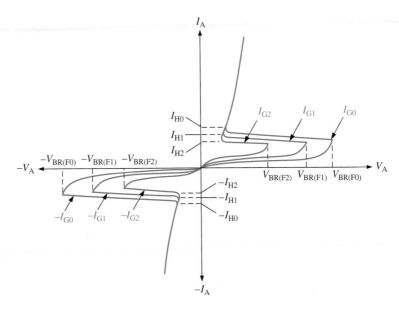

FIGURE 15–43
Triac characteristic curves.

    As in the case of the SCR, gate triggering is the usual method for turning on a triac. Application of current to the triac gate initiates the latching mechnaism discussed in the previous section. Once conduction has been initiated, the triac will conduct on with either polarity, hence it is useful as an ac controller. A triac can be triggerted such that ac power is supplied to the load for a portion of the ac cycle. This enables the triac to provide more or less power to the load depending on the trigger point. This basic operation is illustrated with the circuit in Figure 15–44.

FIGURE 15–44
Basic triac phase control. The timing of the gate trigger determines the portion of the ac cycle passed to the load.

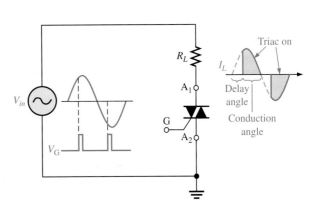

The Zero-Voltage Switch

One problem that arises with triggering an SCR or triac when it is switched on during the ac cycle is generation of RFI (radio frequency interference) due to switching transients. If the SCR or triac is suddenly switched on near the peak of the ac cycle, for example, there would be a sudden inrush of current to the load. When there is a sudden transition of voltage or current, many high-frequency components are generated. These high-frequency components can radiate into sensitive electronic circuits, creating serious disturbances, even catastrophic failures. By switching the SCR or triac on when the voltage across it is zero, the sudden increase in current is prevented because the current will increase sinusoidally with the ac voltage. **Zero-voltage switching** also prevents thermal shock to the load which, depending on the type of load, may shorten its life.

Not all applications can use zero-voltage switching, but when it is possible, noise problems are greatly reduced. For example, the load might be a resistive heating element, and the power is typically turned on for several cycles of the ac and then turned off for several cycles to maintain a certain temperature. The zero-voltage switch uses a sensing circuit to determine when to turn power on. The idea of zero-voltage switching is illustrated in Figure 15–45.

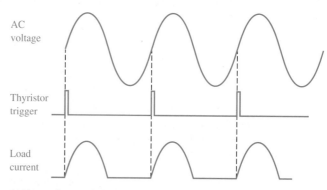

(a) Zero-voltage switching of load current

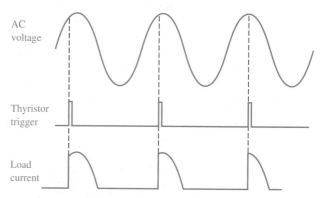

(b) Nonzero switching of load current produces current transients that cause RFI.

FIGURE 15–45

Comparison of zero-voltage switching to nonzero switching of power to a load.

A basic circuit that can provide a trigger as the ac waveform crosses the zero axis in the positive direction is shown in Figure 15–46. Resistor R_1 and diodes D_1 and D_2 protect the input of the comparator from excessive voltage swings. The output voltage level of the comparator is a square wave. C_1 and R_2 form a differentiating circuit to convert the square wave output to trigger pulses. Diode D_3 limits the output to positive triggers only.

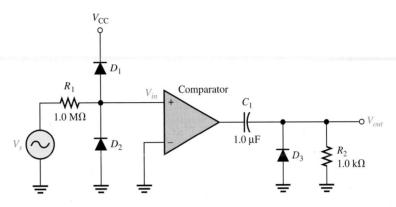

FIGURE 15–46

A circuit that can provide triggers when the ac waveform crosses the zero axis in the positive direction.

Microcontrollers

SCRs and triacs are often used in systems that have many additional requirements. For instance, a system as basic as a washing machine requires timing functions, speed or torque regulation, motor protection, sequence generation, display control and so on. Systems like this can be controlled by a special class of computers called **microcontrollers**. A microcontroller is constructed as a single integrated circuit with all of the basic features found in a microprocessor with special input/output (I/O) circuits, ADCs (analog-to-digital converters), counters, timers, oscillators, memory, and other features. Microcontrollers can be configured for a specific system and offer an inexpensive alternative to older methods for providing a trigger to an SCR or triac.

A specific microcontroller is the Texas Instruments MSP430[6]. The MSP430 is a low-cost 16-bit controller with a reduced instruction set (RISC). It can operate at high speed with extremely low power consumption. It can perform all of the control fucntions required in a small system and can be used to directly drive the gate of a small triac or SCR. It is possible to build a zero-crossing input for the MSP430. Essentially, the same input protection circuit shown in Figure 15–46 is connected to one of the input ports of the MSP430.

15–5 REVIEW QUESTIONS

1. How does an SCR differ from a triac in terms of delivering power to a load?
2. Explain the basic purpose in zero-voltage switching.

---

[6] Data sheet for MSP430 available at http://www.ti.com

15–6 ■ A SYSTEM APPLICATION

The wind speed and direction measurement system is a type of instrument that is typically found at a meteorological data-gathering facility. This system is actually two systems in one because it measures two parameters, wind speed and wind direction, independently. In this system, you will find circuits that you learned about in this chapter and some that were studied in previous chapters.

After completing this section, you should be able to

❏ Apply what you have learned in this chapter to a system application
   ❏ Explain how resolvers and RDCs are used to measure wind direction
   ❏ Descibe how wind speed can be measured
   ❏ Translate between the printed circuit board and a schematic
   ❏ Analyze the measurement circuit board
   ❏ Troubleshoot some common problems

A Brief Description of the System

As you can see in the system block diagram in Figure 15–47, there are two transducers—the anemometer flow meter and the resolver. The flow meter used in this system is basically a propeller-type instrument in which the blades revolve as the wind blows across

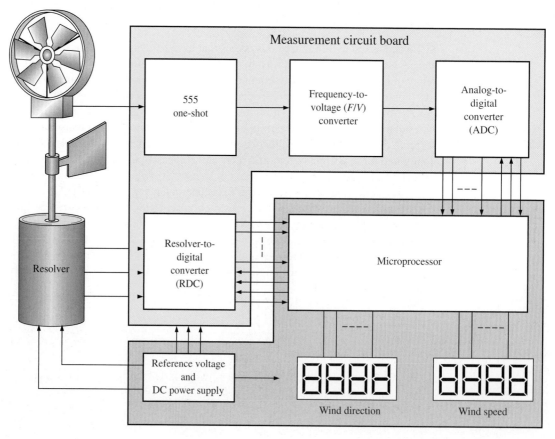

FIGURE 15–47
Block diagram of the wind-measuring system.

them. The faster the wind blows, the faster the blades revolve. A magnetic sensor detects each time one revolution is completed and produces a short duration pulse that triggers the 555 one-shot. The frequency of the pulse train produced by the one-shot increases as the wind speed increases. A frequency-to-voltage converter produces an output voltage that is proportional to the frequency and thus the wind speed. This voltage is converted to digital form and the resulting digital code goes to the microprocessor, which translates it to a binary number corresponding to the wind speed and produces a digital readout. A resolver and a resolver-to-digital converter (RDC) are used to measure the wind direction. The digital output of the RDC goes to the microprocessor where it is translated to an appropriate binary number and displayed.

As indicated in Figure 15–47, one circuit board in this system contains the measurement circuitry and another board contains the microprocessor, display circuitry, and power supply. Our focus in this section is on the measurement circuit board.

Now, so that you can take a closer look at the measurement circuit board, let's take it out of the system and put it on the troubleshooter's bench.

TROUBLESHOOTER'S BENCH

■ ACTIVITY 1   Relate the PC Board to the Schematic

Locate and identify each component and each input/output pin on the PC board in Figure 15–48 after all of the inputs and outputs on the schematic in Figure 15–49 have been identified. Verify that the board and the schematic agree.

FIGURE 15–48
Pin 1 of the AD2S90 is at the dot and the pin numbers increase counterclockwise to pin 20, which is above and adjacent to pin 1.

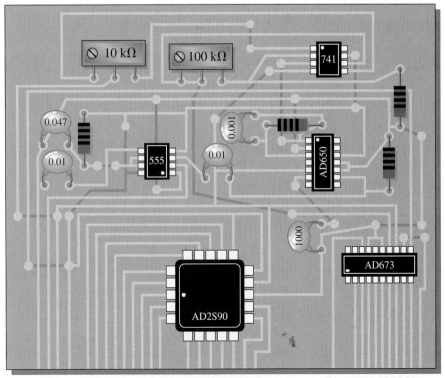

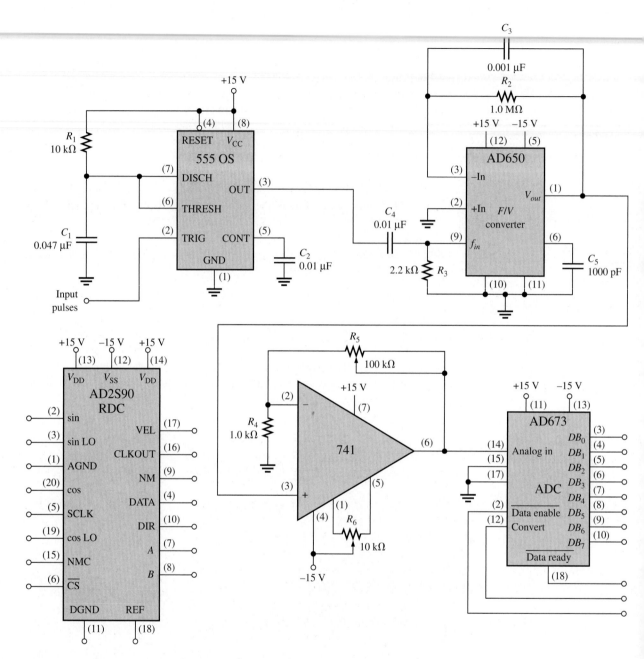

FIGURE 15–49

■ ACTIVITY 2   Write a Technical Report

Describe the overall operation of the measurement circuit board. Specify how each circuit works and its purpose.

■ ACTIVITY 3   Troubleshoot the Circuit Board for Each of the Following Problems by Stating the Probable Cause or Causes

1. No pulses out of the one-shot.

2. There is a 100 mV level out of the *F/V* converter, but the output of the ADC indicates zero.

3. There are pulses out of the one-shot but no voltage out of the *F/V* converter.

4. For one complete revolution of the resolver shaft, the maximum angle represented by the RDC output code is 180.

15–6 REVIEW QUESTIONS

1. Which components determine the pulse width of the one-shot?
2. What is the purpose of the 741 op-amp circuit?

■ SUMMARY

- An rms-to-dc converter performs three basic functions: squaring, averaging, and taking the square root.
- Squaring is usually implemented with a linear multiplier.
- A simple averaging circuit is a low-pass filter that passes only the dc component of the input.
- A square root circuit utilizes a linear multiplier in the feedback loop of an op-amp.
- A synchro is a shaft angle transducer having three stator windings.
- A resolver is a type of synchro which, in its simplest form, has two stator windings.
- The output voltages of a synchro or resolver are called *format voltages* and are proportional to the shaft angle.
- A resolver-to-digital converter (RDC) converts resolver format voltages to a digital code that represents the angular position of the shaft.
- A thermocouple is a type of temperature transducer formed by the junction of two dissimilar metals.
- When the thermocouple junction is heated, a voltage is generated across the junction that is proportional to the temperature.
- Thermocouples can be used to measure very high temperatures.
- The resistance temperature detector (RTD) is a temperature transducer in which the resistance changes directly with temperature. It has a positive temperature coefficient.
- RTDs are typically used in bridge circuits or in constant-current circuits to measure temperature. They have a more limited temperature range than thermocouples.
- The thermistor is a temperature transducer in which the resistance changes inversely with temperature. It has a negative temperature coefficient.
- Thermistors are more sensitive than RTDs or thermocouples, but their temperature range is limited.
- The strain gage is based on the fact that the resistance of a material increases when its length increases.
- The gage factor of a strain gage is the fractional change in resistance to the fractional change in length.
- Pressure transducers are constructed with strain gages bonded to a flexible diaphragm.
- An absolute pressure transducer measures pressure relative to a vacuum.
- A gage pressure transducer measures pressure relative to ambient pressure.
- A differential pressure transducer measures one pressure relative to another pressure.

■ The flow rate of a liquid can be measured using a differential pressure gage.

■ A zero-voltage switch generates pulses at the zero crossings of an ac voltage for triggering a thyristor used in power control.

■ Motion-measuring circuits include LVDT displacement transducers, velocity transducers, and accelerometers.

■ The SCR and triac are two types of thyristors used in power control circuits.

■ GLOSSARY

Key Terms are in color. All terms are included in the end-of-book glossary.

Cold junction   A reference thermocouple held at a fixed temperature and used for compensation in thermocouple circuits.

Gage factor (*GF*)   The ratio of the fractional change in resistance to the fractional change in length along the axis of the gage.

Mean   Average value.

Microcontroller   A specialized microprocessor designed for control functions.

Radio frequency interference (RFI)   High frequencies produced when high values of current and voltage are rapidly switched on and off.

Resistance temperature detector (RTD)   A type of temperature transducer in which resistance is directly proportional to temperature.

Resolver   A type of synchro.

Resolver-to-digital converter (RDC)   An electronic circuit that converts resolver voltages to a digital format which represents the angular position of the rotor shaft.

Root mean square (rms)   The value of an ac voltage that corresponds to a dc voltage that produces the same heating effect in a resistance.

Rotor   The part of a synchro that is attached to the shaft and rotates. The rotor winding is located on the rotor.

Silicon-controlled rectifier   A type of three-terminal thyristor that conducts current when triggered on and remains on until the anode current falls below a specific value.

Stator   The part of a synchro that is fixed. The stator windings are located on the stator.

Strain   The expansion or compression of a material caused by stress forces acting on it.

Strain gage   A transducer formed by a resistive material in which a lengthening or shortening due to stress produces a proportional change in resistance.

Synchro   An electromechanical transducer used for shaft angle measurement and control.

Synchro-to-digital converter (SDC)   An electronic circuit that converts synchro voltages to a digital format which represents the angular position of the rotor shaft.

Thermistor   A type of temperature transducer in which resistance is inversely proportional to temperature.

Thermocouple   A type of temperature transducer formed by the junction of two dissimilar metals which produces a voltage proportional to temperature.

Thyristor   A class of four-layer (*pnpn*) semiconductor devices.

Transducer   A device that converts a physical parameter into an electrical quantity.

Triac   A three-terminal thyristor that can conduct current in either direction when properly activated.

Zero-voltage switching   The process of switching power to a load at the zero crossings of an ac voltage to minimize RF noise generation.

■ KEY FORMULAS

(15–1) $V_{rms} = \sqrt{\text{avg}(V_{in}^2)}$ Root-mean-square value

(15–2) $V_{OUT} = \sqrt{\overline{V_{in}^2}}$ RMS-to-dc converter output

(15–3) $R = \dfrac{\rho L}{A}$ Resistance of a material

(15–4) $GF = \dfrac{\Delta R/R}{\Delta L/L}$ Gage factor of a strain gage

■ SELF-TEST

Answers are at the end of the chapter.

1. The rms value of an ac signal is equal to
 (a) the peak value
 (b) the dc value that produces the same heating effect
 (c) the square root of the average value
 (d) answers (b) and (c)

2. An explicit type of rms-to-dc converter contains
 (a) a squaring circuit **(b)** an averaging circuit
 (c) a square root circuit **(d)** a squarer/divider circuit
 (e) all of the above **(f)** answers (a), (b), and (c) only

3. A synchro produces
 (a) three format voltages **(b)** two format voltages
 (c) one format voltage **(d)** one reference voltage

4. A resolver produces
 (a) three format voltages **(b)** two format voltages
 (c) one format voltage **(d)** none of these

5. A Scott-T transformer is used for
 (a) coupling the reference voltage to a synchro or resolver
 (b) changing resolver format voltages to synchro format voltages
 (c) changing synchro format voltages to resolver format voltages
 (d) isolating the rotor winding from the stator windings

6. The output of an RDC is a
 (a) sine wave with an amplitude proportional to the angular position of the resolver shaft
 (b) digital code representing the angular position of the stator housing
 (c) digital code representing the angular position of the resolver shaft
 (d) sine wave with a frequency proportional to the angular position of the resolver shaft

7. A thermocouple
 (a) produces a change in resistance for a change in temperature
 (b) produces a change in voltage for a change in temperature
 (c) is made of two dissimilar metals
 (d) answers (b) and (c)

8. In a thermocouple circuit, where each of the thermocouple wires is connected to a copper circuit board terminal,
 (a) an unwanted thermocouple is produced **(b)** compensation is required
 (c) a reference thermocouple must be used **(d)** answers (a), (b), and (c)
 (e) answers (a) and (c)

9. A thermocouple signal conditioner is designed to provide
 (a) gain **(b)** compensation
 (c) isolation **(d)** common-mode rejection
 (e) all of the answers

10. An RTD
    (a) produces a change in resistance for a change in temperature
    (b) has a negative temperature coefficient
    (c) has a wider temperature range than a thermocouple
    (d) all of these

11. The purpose of a 3-wire bridge is to eliminate
    (a) nonlinearity of an RTD
    (b) the effects of wire resistance in an RTD circuit
    (c) noise from the RTD resistance
    (d) none of these

12. A thermistor has
    (a) less sensitivity than an RTD
    (b) a greater temperature range than a thermocouple
    (c) a negative temperature coefficient
    (d) a positive temperature coefficient

13. Both RTDs and thermistors are used in
    (a) circuits that measure resistance   (b) circuits that measure temperature
    (c) bridge circuits   (d) constant-current-driven circuits
    (e) answers (b), (c), and (d)   (f) answers (b) and (c) only

14. When the length of a strain gage increases,
    (a) it produces more voltage   (b) its resistance increases
    (c) its resistance decreases   (d) it produces an open circuit

15. A higher gage factor indicates that the strain gage is
    (a) less sensitive to a change in length
    (b) more sensitive to a change in length
    (c) has more total resistance
    (d) made of a physically larger conductor

16. Many types of pressure transducers are made with
    (a) thermistors   (b) RTDs
    (c) strain gages   (d) none of these

17. Gage pressure is measured relative to
    (a) ambient pressure   (b) a vacuum
    (c) a reference pressure

18. The flow rate of a liquid can be measured
    (a) with a string
    (b) with a temperature sensor
    (c) with an absolute pressure transducer
    (d) with a differential pressure transducer

19. Zero-voltage switching is commonly used in
    (a) determining thermocouple voltage   (b) SCR and triac power control circuits
    (c) in balanced bridge circuits   (d) RFI generation

20. A major disadvantage of nonzero switching of power to a load is
    (a) lack of efficiency   (b) possible damage to the thyristor
    (c) RF noise generation

TROUBLESHOOTER'S QUIZ *Answers are at the end of the chapter.*

Refer to Figure 15–50.

❑ If the 220 kΩ resistor opens,

   1. The closed-loop gain will

      (a) increase   **(b)** decrease   **(c)** not change

Refer to Figure 15–51.

❑ If the bridge is balanced and the dc supply voltage is disconnected,

   2. The output voltage will

      (a) increase   **(b)** decrease   **(c)** not change

❑ If the RTD opens,

   3. The magnitude of the voltage across the output terminals will

      (a) increase   **(b)** decrease   **(c)** not change

Refer to Figure 15–53.

❑ If R_G is larger than specified,

   4. The amplifier output voltage will

      (a) increase   **(b)** decrease   **(c)** not change

   5. The bridge output voltage will

      (a) increase   **(b)** decrease   **(c)** not change

Refer to Figure 15–55.

❑ If the gate of the SCR opens and the input does not exceed the breakover voltage,

   6. The output voltage will

      (a) increase   **(b)** decrease   **(c)** not change

❑ If R opens,

   7. The output voltage will

      (a) increase   **(b)** decrease   **(c)** not change

❑ If the gate trigger voltage V_G increases in amplitude,

   8. The output voltage will

      (a) increase   **(b)** decrease   **(c)** not change

Refer to Figure 15–56.

❑ If the input voltage increases in amplitude,

   9. The output voltage will

      (a) increase   **(b)** decrease   **(c)** not change

❑ If D_3 is open,

   10. The amplitude of the positive triggers will

      (a) increase   **(b)** decrease   **(c)** not change

❑ If D_3 is reversed,

   11. The amplitude of the positive triggers will

      (a) increase   **(b)** decrease   **(c)** not change

■ PROBLEMS

Answers to odd-numbered problems are at the end of the book.

SECTION 15–1   RMS-to-DC Converters

1. A 5 V dc voltage is applied across a 1.0 kΩ resistor. To achieve the same power in the 1.0 kΩ resistor as produced by the dc voltage, what must be the rms value of a sinusoidal voltage?

2. Based on the fundamental definition of rms, determine the rms value for a symmetrical square wave with an amplitude of ±1 V.

SECTION 15–2   Angle-Measuring Circuits

3. A certain RDC has an 8-bit digital output. What is the angle that is being measured if the output code is 10000111?

4. Repeat Problem 3 for an RDC output code of 00010101.

5. How many bits does the latch hold in an AD2S90 RDC?

6. Explain the Direction and Velocity outputs on an AD2S90 RDC.

SECTION 15–3   Temperature-Measuring Circuits

7. Three identical thermocouples are each exposed to a different temperature as follows: Thermocouple *A* is exposed to 450°C, thermocouple *B* is exposed to 420°C, and thermocouple *C* is exposed to 1200°C. Which thermocouple produces the most voltage?

8. You have two thermocouples. One is a *K* type and the other is a *T* type. In general, what do these letter designations tell you?

9. Determine the output voltage of the op-amp in Figure 15–50 if the thermocouple is measuring a temperature of 400°C and the circuit itself is at 25°C. Refer to Table 15–2.

10. What should be the output voltage in Problem 9 if the circuit is properly compensated?

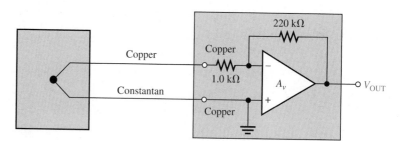

FIGURE 15–50

11. At what resistance value of the RTD will the bridge circuit in Figure 15–51 be balanced if the wires running to the RTD each have a resistance of 10 Ω?

12. At what resistance value of the RTD will the bridge circuit in Figure 15–52 be balanced if the wires running to the RTD each have a resistance of 10 Ω?

13. Explain the difference in the results of Problems 11 and 12.

14. Determine the output voltage of the instrumentation amplifier in Figure 15–53 if the resistance of the RTD is 697 Ω at the temperature being measured.

FIGURE 15–51

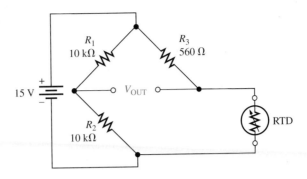

FIGURE 15–52

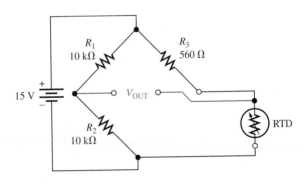

FIGURE 15–53

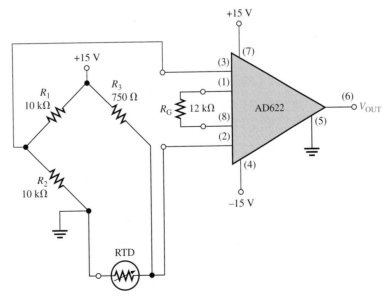

SECTION 15–4 Strain-Measuring, Pressure-Measuring, and Motion-Measuring Circuits

15. A certain material being measured undergoes a strain of 3 parts per million. The strain gage has a nominal resistance of 600 Ω and a gage factor of 2.5. Determine the resistance change in the strain gage.

16. Explain how a strain gage can be used to measure pressure.

17. Identify and compare the three symbols in Figure 15–54.

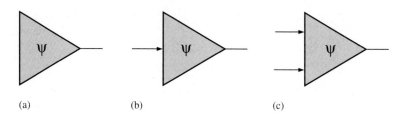

(a) (b) (c)

FIGURE 15–54

SECTION 15–5 Power-Control Circuits

18. Name two ways an SCR can be placed in the forward-conduction region.

19. Sketch the V_R waveform for the circuit in Figure 15–55, given the indicated relationship of the input waveforms.

FIGURE 15–55

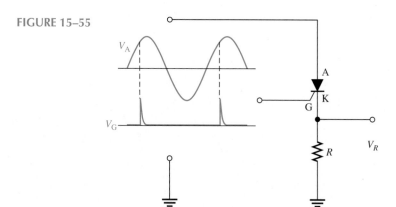

20. For the circuit in Figure 15–56, sketch the waveform at the output of the comparator and at the output of the circuit in relation to the input. Assume the input is a 115 V rms sine wave and the comparator and the power supply voltages for the comparator are ± 10 V.

21. What change to the circuit in Figure 15–56 would you make if you wanted to have positive triggers on the negative slope of the input waveform?

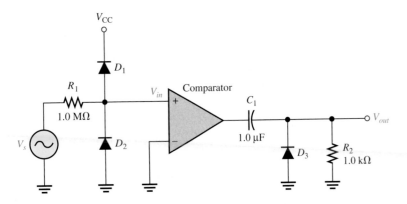

FIGURE 15–56

■ ANSWERS TO REVIEW QUESTIONS

Section 15–1

1. An rms-to-dc converter produces a dc output voltage that is equal to the rms value of the ac input voltage.

2. Internally, an rms-to-dc converter squares, averages, and takes the square root.

Section 15–2

1. Synchro

2. An RDC accepts resolver format voltages on its inputs.

3. An RDC produces a digital code representing the angular shaft position of the resolver.

4. An RDC converts the angular shaft position of a resolver into a digital code.

Section 15–3

1. A thermocouple is a temperature transducer formed by the junction of two dissimilar metals.

2. A voltage proportional to the temperature is produced across the junction of two dissimilar metals.

3. An RTD is a resistance temperature detector in which the resistance is proportional to the temperature, whereas the thermocouple produces a voltage.

4. An RTD has a positive temperature coefficient and a thermistor has a negative temperature coefficient.

5. The thermocouple has a greater temperature range than the RTD or thermistor.

Section 15–4

1. Basically, a strain gage is a resistive element whose dimensions can be altered by an applied force to produce a change in resistance.

2. Basically, a pressure gage is a strain gage bonded to a flexible diaphragm.

3. Absolute, gage, and differential

4. **(a)** A linear variable differential transformer
   (b) Displacement

Section 15–5

1. An SCR is unidirectional and therefore allows current through the load only during half of the ac cycle. A triac is bidirectional and allows current during the complete cycle.

2. Zero-voltage switching eliminates fast transitions in the current to a load, thus reducing RFI emissions and thermal shock to the load element.

Section 15–6

1. R_1 and C_1 set the pulse width of the one-shot.

2. The noninverting op-amp provides gain for the output of the F/V converter and permits adjustment of the input to the ADC for calibration purposes.

■ **ANSWERS TO PRACTICE EXERCISES FOR EXAMPLES**

15–1	183°
15–2	10.1%
15–3	24.247 mV
15–4	1.0 kΩ; 0 V
15–5	5.12 mΩ

■ **ANSWERS TO SELF-TEST**	**1.** (b)	**2.** (f)	**3.** (a)	**4.** (b)	**5.** (c)
	6. (c)	**7.** (d)	**8.** (d)	**9.** (e)	**10.** (a)
	11. (b)	**12.** (c)	**13.** (e)	**14.** (b)	**15.** (b)
	16. (c)	**17.** (a)	**18.** (d)	**19.** (b)	**20.** (c)

■ **ANSWERS TO TROUBLE-SHOOTER'S QUIZ**	**1.** increase	**2.** not change	**3.** increase	**4.** decrease
	5. not change	**6.** decrease	**7.** decrease	**8.** not change
	9. not change	**10.** not change	**11.** decrease	

A DATA SHEETS

**1.5KE6.8, A thru 1.5KE250, A
See Page 4-59**

Designers Data Sheet

500-MILLIWATT HERMETICALLY SEALED GLASS SILICON ZENER DIODES

- Complete Voltage Range — 2.4 to 110 Volts
- DO-35 Package — Smaller than Conventional DO-7 Package
- Double Slug Type Construction
- Metallurgically Bonded Construction
- Oxide Passivated Die

Designer's Data for "Worst Case" Conditions

The Designer's Data sheets permit the design of most circuits entirely from the information presented. Limit curves — representing boundaries on device characteristics — are given to facilitate "worst case" design.

**1N746 thru 1N759
1N957A thru 1N986A
1N4370 thru 1N4372**

**GLASS ZENER DIODES
500 MILLIWATTS
2.4-110 VOLTS**

MAXIMUM RATINGS

Rating	Symbol	Value	Unit
DC Power Dissipation @ $T_L \leq 50^{\circ}C$, Lead Length = 3/8''	P_D		
*JEDEC Registration		400	mW
*Derate above $T_L = 50^{\circ}C$		3.2	mW/$^{\circ}C$
Motorola Device Ratings		500	mW
Derate above $T_L = 50^{\circ}C$		3.33	mW/$^{\circ}C$
Operating and Storage Junction Temperature Range	T_J, T_{stg}		$^{\circ}C$
*JEDEC Registration		–65 to +175	
Motorola Device Ratings		–65 to +200	

*Indicates JEDEC Registered Data.

MECHANICAL CHARACTERISTICS

MAXIMUM LEAD TEMPERATURE FOR SOLDERING PURPOSES: 230°C, 1/16'' from case for 10 seconds

FINISH: All external surfaces are corrosion resistant with readily solderable leads.

POLARITY: Cathode indicated by color band. When operated in zener mode, cathode will be positive with respect to anode.

MOUNTING POSITION: Any

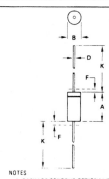

NOTES
1. PACKAGE CONTOUR OPTIONAL WITHIN A AND B. HEAT SLUGS, IF ANY, SHALL BE INCLUDED WITHIN THIS CYLINDER, BUT NOT SUBJECT TO THE MINIMUM LIMIT OF B.
2. LEAD DIAMETER NOT CONTROLLED IN ZONE F TO ALLOW FOR FLASH, LEAD FINISH BUILDUP AND MINOR IRREGULARITIES OTHER THAN HEAT SLUGS.
3. POLARITY DENOTED BY CATHODE BAND.
4. DIMENSIONING AND TOLERANCING PER ANSI Y14.5, 1973.

DIM	MILLIMETERS		INCHES	
	MIN	MAX	MIN	MAX
A	3.05	5 08	0.120	0.200
B	1.52	2 29	0.060	0.090
D	0.46	0.56	0.018	0.022
F		1.27	–	0.050
K	25.40	38.10	1.000	1.500

All JEDEC dimensions and notes apply.

**CASE 299-02
DO-204AH
GLASS**

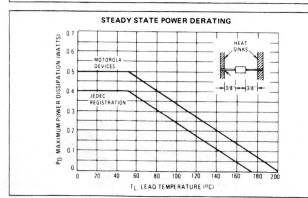

STEADY STATE POWER DERATING

HEAT SINKS

MOTOROLA DEVICES

JEDEC REGISTRATION

P_D, MAXIMUM POWER DISSIPATION (WATTS)

T_L, LEAD TEMPERATURE ($^{\circ}C$)

1N746 thru 1N759, 1N957A thru 1N986A, 1N4370 thru 1N4372

ELECTRICAL CHARACTERISTICS (T_A = 25°C, V_F = 1.5 V max at 200 mA for all types)

Type Number (Note 1)	Nominal Zener Voltage V_Z @ I_{ZT} (Note 2) Volts	Test Current I_{ZT} mA	Maximum Zener Impedance Z_{ZT} @ I_{ZT} (Note 3) Ohms	*Maximum DC Zener Current I_{ZM} (Note 4) mA		Maximum Reverse Leakage Current	
						T_A = 25°C I_R @ V_R = 1 V μA	T_A = 150°C I_R @ V_R = 1 V μA
1N4370	2.4	20	30	150	190	100	200
1N4371	2.7	20	30	135	165	75	150
1N4372	3.0	20	29	120	150	50	100
1N746	3.3	20	28	110	135	10	30
1N747	3.6	20	24	100	125	10	30
1N748	3.9	20	23	95	115	10	30
1N749	4.3	20	22	85	105	2	30
1N750	4.7	20	19	75	95	2	30
1N751	5.1	20	17	70	85	1	20
1N752	5.6	20	11	65	80	1	20
1N753	6.2	20	7	60	70	0.1	20
1N754	6.8	20	5	55	65	0.1	20
1N755	7.5	20	6	50	60	0.1	20
1N756	8.2	20	8	45	55	0.1	20
1N757	9.1	20	10	40	50	0.1	20
1N758	10	20	17	35	45	0.1	20
1N759	12	20	30	30	35	0.1	20

Type Number (Note 1)	Nominal Zener Voltage V_Z (Note 2) Volts	Test Current I_{ZT} mA	Maximum Zener Impedance (Note 3) Z_{ZT} @ I_{ZT} Ohms	Z_{ZK} @ I_{ZK} Ohms	I_{ZK} mA	*Maximum DC Zener Current I_{ZM} (Note 4) mA		Maximum Reverse Current I_R Maximum μA	Test Voltage Vdc 5% V_R	10%
1N957A	6.8	18.5	4.5	700	1.0	47	61	150	5.2	4.9
1N958A	7.5	16.5	5.5	700	0.5	42	55	75	5.7	5.4
1N959A	8.2	15	6.5	700	0.5	38	50	50	6.2	5.9
1N960A	9.1	14	7.5	700	0.5	35	45	25	6.9	6.6
1N961A	10	12.5	8.5	700	0.25	32	41	10	7.6	7.2
1N962A	11	11.5	9.5	700	0.25	28	37	5	8.4	8.0
1N963A	12	10.5	11.5	700	0.25	26	34	5	9.1	8.6
1N964A	13	9.5	13	700	0.25	24	32	5	9.9	9.4
1N965A	15	8.5	16	700	0.25	21	27	5	11.4	10.8
1N966A	16	7.8	17	700	0.25	19	37	5	12.2	11.5
1N967A	18	7.0	21	750	0.25	17	23	5	13.7	13.0
1N968A	20	6.2	25	750	0.25	15	20	5	15.2	14.4
1N969A	22	5.6	29	750	0.25	14	18	5	16.7	15.8
1N970A	24	5.2	33	750	0.25	13	17	5	18.2	17.3
1N971A	27	4.6	41	750	0.25	11	15	5	20.6	19.4
1N972A	30	4.2	49	1000	0.25	10	13	5	22.8	21.6
1N973A	33	3.8	58	1000	0.25	9.2	12	5	25.1	23.8
1N974A	36	3.4	70	1000	0.25	8.5	11	5	27.4	25.9
1N975A	39	3.2	80	1000	0.25	7.8	10	5	29.7	28.1
1N976A	43	3.0	93	1500	0.25	7.0	9.6	5	32.7	31.0
1N977A	47	2.7	105	1500	0.25	6.4	8.8	5	35.8	33.8
1N978A	51	2.5	125	1500	0.25	5.9	8.1	5	38.8	36.7
1N979A	56	2.2	150	2000	0.25	5.4	7.4	5	42.6	40.3
1N980A	62	2.0	185	2000	0.25	4.9	6.7	5	47.1	44.6
1N981A	68	1.8	230	2000	0.25	4.5	6.1	5	51.7	49.0
1N982A	75	1.7	270	2000	0.25	1.0	5.5	5	56.0	54.0
1N983A	82	1.5	330	3000	0.25	3.7	5.0	5	62.2	59.0
1N984A	91	1.4	400	3000	0.25	3.3	4.5	5	69.2	65.5
1N985A	100	1.3	500	3000	0.25	3.0	4.5	5	76	72
1N986A	110	1.1	750	4000	0.25	2.7	4.1	5	83.6	79.2

NOTE 1. TOLERANCE AND VOLTAGE DESIGNATION

Tolerance Designation

The type numbers shown have tolerance designations as follows:

1N4370 series: ± 10%, suffix A for ± 5% units,
C for ± 2%, D for ± 1%.

1N746 series: ± 10%, suffix A for ± 5% units,
C for ± 2%, D for ± 1%.

1N957 series: ± 10%, suffix A for ± 10% units,
C for ± 2%, D for ± 1%,
suffix B for ± 5% units,
C for ± 2%, D for ± 1%.

MOTOROLA
■ SEMICONDUCTOR ■
TECHNICAL DATA

**1N1183A
thru
1N1190A**

MEDIUM-CURRENT RECTIFIERS

. . . for applications requiring low forward voltage drop and rugged construction.

- High Surge Handling Ability
- Rugged Construction
- Reverse Polarity Available; Eliminates Need for Insulating Hardware in Many Cases
- Hermetically Sealed

**20-AMP
RECTIFIERS**

SILICON
DIFFUSED-JUNCTION

---

*MAXIMUM RATINGS

Rating	Symbol	1N1183A	1N1184A	1N1186A	1N1188A	1N1190A	Unit
Peak Repetitive Reverse Voltage	V_RRM V_RWM V_R	50	100	200	400	600	Volts
Average Half-Wave Rectified Forward Current With Resistive Load @ T_A = 150°C	I_O	40	40	40	40	40	Amp
Peak One Cycle Surge Current (60 Hz and 150°C Case Temperature)	I_FSM	800	800	800	800	800	Amp
Operating Junction Temperature	T_J	−65 to +200					°C
Storage Temperature	T_stg	−65 to +200					°C

*ELECTRICAL CHARACTERISTICS (All Types) at 25°C Case Temperature

Characteristic	Symbol	Value	Unit
Maximum Forward Voltage at 100 Amp DC Forward Current	V_F	1.1	Volts
Maximum Reverse Current at Rated DC Reverse Voltage	I_R	5.0	mAdc

THERMAL CHARACTERISTICS

Characteristic	Symbol	Typical	Unit
Thermal Resistance, Junction to Case	R_θJC	1.0	°C/W

*Indicates JEDEC registered data.

	MILLIMETERS		INCHES	
DIM	MIN	MAX	MIN	MAX
A	—	20.07	—	0.790
B	16.94	17.45	0.669	0.687
C	—	11.43	—	0.450
D	—	9.53	—	0.375
E	2.92	5.08	0.115	0.200
F	—	2.03	—	0.080
J	10.72	11.51	0.422	0.453
K	19.05	25.40	0.750	1.00
L	3.96	—	0.156	—
P	5.59	6.32	0.220	0.249
Q	3.56	4.45	0.140	0.175
R	—	16.94	—	0.667
S	—	2.26	—	0.089

**CASE 42A-01
DO-203AB
METAL**

MECHANICAL CHARACTERISTICS

CASE: Welded, hermetically sealed construction
FINISH: All external surfaces corrosion-resistant and the terminal lead is readily solderable
WEIGHT: 25 grams (approx.)
POLARITY: Cathode connected to case (reverse polarity available denoted by Suffix R, i.e.: 1N3212R)
MOUNTING POSITION: Any
MOUNTING TORQUE: 25 in-lb max

Axial Lead
Standard Recovery Rectifiers

This data sheet provides information on subminiature size, axial lead mounted rectifiers for general–purpose low–power applications.

Mechanical Characteristics

- Case: Epoxy, Molded
- Weight: 0.4 gram (approximately)
- Finish: All External Surfaces Corrosion Resistant and Terminal Leads are Readily Solderable
- Lead and Mounting Surface Temperature for Soldering Purposes: 220°C Max. for 10 Seconds, 1/16″ from case
- Shipped in plastic bags, 1000 per bag.
- Available Tape and Reeled, 5000 per reel, by adding a "RL" suffix to the part number
- Polarity: Cathode Indicated by Polarity Band
- Marking: 1N4001, 1N4002, 1N4003, 1N4004, 1N4005, 1N4006, 1N4007

**1N4001
thru
1N4007**

1N4004 and 1N4007 are
Motorola Preferred Devices

**LEAD MOUNTED
RECTIFIERS
50–1000 VOLTS
DIFFUSED JUNCTION**

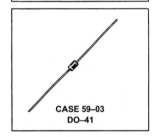

**CASE 59–03
DO–41**

MAXIMUM RATINGS

Rating	Symbol	1N4001	1N4002	1N4003	1N4004	1N4005	1N4006	1N4007	Unit
*Peak Repetitive Reverse Voltage Working Peak Reverse Voltage DC Blocking Voltage	V_{RRM} V_{RWM} V_R	50	100	200	400	600	800	1000	Volts
*Non–Repetitive Peak Reverse Voltage (halfwave, single phase, 60 Hz)	V_{RSM}	60	120	240	480	720	1000	1200	Volts
*RMS Reverse Voltage	$V_{R(RMS)}$	35	70	140	280	420	560	700	Volts
*Average Rectified Forward Current (single phase, resistive load, 60 Hz, see Figure 8, T_A = 75°C)	I_O	1.0							Amp
*Non–Repetitive Peak Surge Current (surge applied at rated load conditions, see Figure 2)	I_{FSM}	30 (for 1 cycle)							Amp
Operating and Storage Junction Temperature Range	T_J T_{stg}	– 65 to +175							°C

ELECTRICAL CHARACTERISTICS*

Rating	Symbol	Typ	Max	Unit
Maximum Instantaneous Forward Voltage Drop (i_F = 1.0 Amp, T_J = 25°C) Figure 1	v_F	0.93	1.1	Volts
Maximum Full–Cycle Average Forward Voltage Drop (I_O = 1.0 Amp, T_L = 75°C, 1 inch leads)	$V_{F(AV)}$	—	0.8	Volts
Maximum Reverse Current (rated dc voltage) (T_J = 25°C) (T_J = 100°C)	I_R	0.05 1.0	10 50	μA
Maximum Full–Cycle Average Reverse Current (I_O = 1.0 Amp, T_L = 75°C, 1 inch leads)	$I_{R(AV)}$	—	30	μA

*Indicates JEDEC Registered Data

Preferred devices are Motorola recommended choices for future use and best overall value.

PACKAGE DIMENSIONS

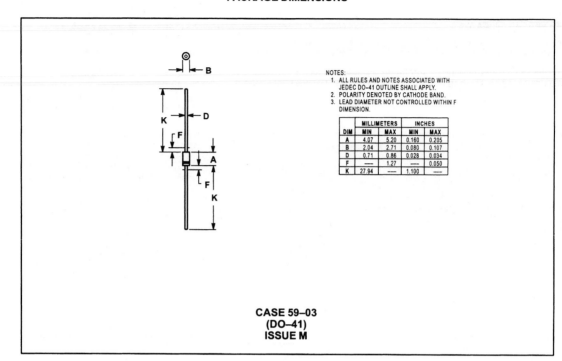

NOTES:
1. ALL RULES AND NOTES ASSOCIATED WITH JEDEC DO–41 OUTLINE SHALL APPLY.
2. POLARITY DENOTED BY CATHODE BAND.
3. LEAD DIAMETER NOT CONTROLLED WITHIN F DIMENSION.

DIM	MILLIMETERS		INCHES	
	MIN	MAX	MIN	MAX
A	4.07	5.20	0.160	0.205
B	2.04	2.71	0.080	0.107
D	0.71	0.86	0.028	0.034
F	——	1.27	——	0.050
K	27.94	——	1.100	——

CASE 59–03
(DO–41)
ISSUE M

How to reach us:
USA/EUROPE/Locations Not Listed : Motorola Literature Distribution;
P.O. Box 5405, Denver, Colorado 80217. 303–675–2140 or 1–800–441–2447

Mfax™: RMFAX0@email.sps.mot.com – TOUCHTONE 6 02–244–6609
                                                – US & Canada ONLY 1–800–774–1848

INTERNET: http://motorola.com/sps

JAPAN: Nippon Motorola Ltd.: SPD, Strategic Planning Office, 4–32–1, Nishi–Gotanda, Shinagawa–ku, Tokyo 141, Japan. 81–3–5487–8488

ASIA/PACIFIC: Motorola Semiconductors H.K. Ltd.; 8B Tai Ping Industrial Park, 51 Ting Kok Road, Tai Po, N.T., Hong Kong. 852–26629298

 **MOTOROLA**

◊

1N4001/D

MAXIMUM RATINGS

Rating	Symbol	2N2219 2N2222	2N2218A 2N2219A 2N2221A 2N2222A	Unit
Collector-Emitter Voltage	V_{CEO}	30	40	Vdc
Collector-Base Voltage	V_{CBO}	60	75	Vdc
Emitter-Base Voltage	V_{EBO}	5.0	6.0	Vdc
Collector Current — Continuous	I_C	800	800	mAdc

		2N2218A 2N2219,A	2N2221A 2N2222,A	
Total Device Dissipation @ T_A = 25°C Derate above 25°C	P_D	0.8 4.57	0.4 2.28	Watt mW/°C
Total Device Dissipation @ T_C = 25°C Derate above 25°C	P_D	3.0 17.1	1.2 6.85	Watts mW/°C
Operating and Storage Junction Temperature Range	T_J, T_{stg}	−65 to +200		°C

THERMAL CHARACTERISTICS

Characteristic	Symbol	2N2218A 2N2219,A	2N2221A 2N2222,A	Unit
Thermal Resistance, Junction to Ambient	$R_{\theta JA}$	219	145.8	°C/W
Thermal Resistance, Junction to Case	$R_{\theta JC}$	58	437.5	°C/W

2N2218A,2N2219,A
2N2221A,2N2222,A

JAN, JTX, JTXV AVAILABLE

2N2218, A/2N2219, A
CASE 79-04
TO-39 (TO-205AD)
STYLE 1

2N2221, A/2N2222, A
CASE 22-03
TO-18 (TO-206AA)
STYLE 1

3 Collector
2 Base
1 Emitter

GENERAL PURPOSE TRANSISTORS
NPN SILICON

ELECTRICAL CHARACTERISTICS (T_A = 25°C unless otherwise noted.)

Characteristic		Symbol	Min	Max	Unit
OFF CHARACTERISTICS					
Collector-Emitter Breakdown Voltage (I_C = 10 mAdc, I_B = 0)	Non-A Suffix A-Suffix	$V_{(BR)CEO}$	30 40	— —	Vdc
Collector-Base Breakdown Voltage (I_C = 10 µAdc, I_E = 0)	Non-A Suffix A-Suffix	$V_{(BR)CBO}$	60 75	— —	Vdc
Emitter-Base Breakdown Voltage (I_E = 10 µAdc, I_C = 0)	Non-A Suffix A-Suffix	$V_{(BR)EBO}$	5.0 6.0	— —	Vdc
Collector Cutoff Current (V_{CE} = 60 Vdc, $V_{EB(off)}$ = 3.0 Vdc)	A-Suffix	I_{CEX}	—	10	nAdc
Collector Cutoff Current (V_{CB} = 50 Vdc, I_E = 0) (V_{CB} = 50 Vdc, I_E = 0) (V_{CB} = 50 Vdc, I_E = 0, T_A = 150°C) (V_{CB} = 60 Vdc, I_E = 0, T_A = 150°C)	Non-A Suffix A-Suffix Non-A Suffix A-Suffix	I_{CBO}	— — — —	0.01 0.01 10 10	µAdc
Emitter Cutoff Current (V_{EB} = 3.0 Vdc, I_C = 0)	A-Suffix	I_{EBO}	—	10	nAdc
Base Cutoff Current (V_{CE} = 60 Vdc, $V_{EB(off)}$ = 3.0 Vdc)	A-Suffix	I_{BL}	—	20	nAdc
ON CHARACTERISTICS					
DC Current Gain (I_C = 0.1 mAdc, V_{CE} = 10 Vdc)	2N2218A, 2N2221A(1) 2N2219,A, 2N2222,A(1)	h_{FE}	20 35	— —	—
(I_C = 1.0 mAdc, V_{CE} = 10 Vdc)	2N2218A, 2N2221A 2N2219,A, 2N2222,A		25 50	— —	
(I_C = 10 mAdc, V_{CE} = 10 Vdc)	2N2218A, 2N2221A(1) 2N2219,A, 2N2222,A(1)		35 75	— —	
(I_C = 10 mAdc, V_{CE} = 10 Vdc, T_A = −55°C)	2N2218A, 2N2221A 2N2219,A, 2N2222,A		15 35	— —	
(I_C = 150 mAdc, V_{CE} = 10 Vdc)(1)	2N2218A, 2N2221A 2N2219,A, 2N2222,A		40 100	120 300	

ELECTRICAL CHARACTERISTICS (continued) (T_A = 25°C unless otherwise noted.)

Characteristic		Symbol	Min	Max	Unit
(I_C = 150 mAdc, V_{CE} = 1.0 Vdc)(1)	2N2218A, 2N2221A 2N2219,A, 2N2222,A		20 50	— —	
(I_C = 500 mAdc, V_{CE} = 10 Vdc)(1)	2N2219, 2N2222 2N2218A, 2N2221A, 2N2219A, 2N2222A		30 25 40	— — —	
Collector-Emitter Saturation Voltage(1) (I_C = 150 mAdc, I_B = 15 mAdc)	Non-A Suffix A-Suffix	$V_{CE(sat)}$	— —	0.4 0.3	Vdc
(I_C = 500 mAdc, I_B = 50 mAdc)	Non-A Suffix A-Suffix		— —	1.6 1.0	
Base-Emitter Saturation Voltage(1) (I_C = 150 mAdc, I_B = 15 mAdc)	Non-A Suffix A-Suffix	$V_{BE(sat)}$	0.6 0.6	1.3 1.2	Vdc
(I_C = 500 mAdc, I_B = 50 mAdc)	Non-A Suffix A-Suffix		— —	2.6 2.0	

SMALL-SIGNAL CHARACTERISTICS

Characteristic		Symbol	Min	Max	Unit
Current Gain — Bandwidth Product(2) (I_C = 20 mAdc, V_{CE} = 20 Vdc, f = 100 MHz)	All Types, Except 2N2219A, 2N2222A	f_T	250 300	— —	MHz
Output Capacitance(3) (V_{CB} = 10 Vdc, I_E = 0, f = 1.0 MHz)		C_{obo}	—	8.0	pF
Input Capacitance(3) (V_{EB} = 0.5 Vdc, I_C = 0, f = 1.0 MHz)	Non-A Suffix A-Suffix	C_{ibo}	— —	30 25	pF
Input Impedance (I_C = 1.0 mAdc, V_{CE} = 10 Vdc, f = 1.0 kHz)	2N2218A, 2N2221A 2N2219A, 2N2222A	h_{ie}	1.0 2.0	3.5 8.0	kohms
(I_C = 10 mAdc, V_{CE} = 10 Vdc, f = 1.0 kHz)	2N2218A, 2N2221A 2N2219A, 2N2222A		0.2 0.25	1.0 1.25	
Voltage Feedback Ratio (I_C = 1.0 mAdc, V_{CE} = 10 Vdc, f = 1.0 kHz)	2N2218A, 2N2221A 2N2219A, 2N2222A	h_{re}	— —	5.0 8.0	X 10^{-4}
(I_C = 10 mAdc, V_{CE} = 10 Vdc, f = 1.0 kHz)	2N2218A, 2N2221A 2N2219A, 2N2222A		— —	2.5 4.0	
Small-Signal Current Gain (I_C = 1.0 mAdc, V_{CE} = 10 Vdc, f = 1.0 kHz)	2N2218A, 2N2221A 2N2219A, 2N2222A	h_{fe}	30 50	150 300	—
(I_C = 10 mAdc, V_{CE} = 10 Vdc, f = 1.0 kHz)	2N2218A, 2N2221A 2N2219A, 2N2222A		50 75	300 375	
Output Admittance (I_C = 1.0 mAdc, V_{CE} = 10 Vdc, f = 1.0 kHz)	2N2218A, 2N2221A 2N2219A, 2N2222A	h_{oe}	3.0 5.0	15 35	μmhos
(I_C = 10 mAdc, V_{CE} = 10 Vdc, f = 1.0 kHz)	2N2218A, 2N2221A 2N2219A, 2N2222A		10 15	100 200	
Collector Base Time Constant (I_E = 20 mAdc, V_{CB} = 20Vdc, f = 31.8 MHz)	A-Suffix	$rb'C_c$	—	150	ps
Noise Figure (I_C = 100 μAdc, V_{CE} = 10 Vdc, R_S = 1.0 kohm, f = 1.0 kHz)	2N2222A	NF	—	4.0	dB
Real Part of Common-Emitter High Frequency Input Impedance (I_C = 20 mAdc, V_{CE} = 20 Vdc, f = 300 MHz)	2N2218A, 2N2219A 2N2221A, 2N2222A	Re(h_{ie})	—	60	Ohms

(1) Pulse Test: Pulse Width ≤ 300 μs, Duty Cycle ≤ 2.0%.
(2) f_T is defined as the frequency at which $|h_{fe}|$ extrapolates to unity.
(3) 2N5581 and 2N5582 are Listed C_{cb} and C_{eb} for these conditions and values.

General Purpose Transistors
NPN Silicon

COLLECTOR
3

2
BASE

1
EMITTER

2N3903
2N3904*

*Motorola Preferred Device

CASE 29–04, STYLE 1
TO–92 (TO–226AA)

MAXIMUM RATINGS

Rating	Symbol	Value	Unit
Collector–Emitter Voltage	V_{CEO}	40	Vdc
Collector–Base Voltage	V_{CBO}	60	Vdc
Emitter–Base Voltage	V_{EBO}	6.0	Vdc
Collector Current — Continuous	I_C	200	mAdc
Total Device Dissipation @ T_A = 25°C Derate above 25°C	P_D	625 5.0	mW mW/°C
Total Device Dissipation @ T_C = 25°C Derate above 25°C	P_D	1.5 12	Watts mW/°C
Operating and Storage Junction Temperature Range	T_J, T_{stg}	−55 to +150	°C

THERMAL CHARACTERISTICS[1]

Characteristic	Symbol	Max	Unit
Thermal Resistance, Junction to Ambient	$R_{\theta JA}$	200	°C/W
Thermal Resistance, Junction to Case	$R_{\theta JC}$	83.3	°C/W

ELECTRICAL CHARACTERISTICS (T_A = 25°C unless otherwise noted)

Characteristic	Symbol	Min	Max	Unit
OFF CHARACTERISTICS				
Collector–Emitter Breakdown Voltage [2] (I_C = 1.0 mAdc, I_B = 0)	$V_{(BR)CEO}$	40	—	Vdc
Collector–Base Breakdown Voltage (I_C = 10 μAdc, I_E = 0)	$V_{(BR)CBO}$	60	—	Vdc
Emitter–Base Breakdown Voltage (I_E = 10 μAdc, I_C = 0)	$V_{(BR)EBO}$	6.0	—	Vdc
Base Cutoff Current (V_{CE} = 30 Vdc, V_{EB} = 3.0 Vdc)	I_{BL}	—	50	nAdc
Collector Cutoff Current (V_{CE} = 30 Vdc, V_{EB} = 3.0 Vdc)	I_{CEX}	—	50	nAdc

1. Indicates Data in addition to JEDEC Requirements.
2. Pulse Test: Pulse Width ≤ 300 μs; Duty Cycle ≤ 2.0%.

Preferred devices are Motorola recommended choices for future use and best overall value.

ELECTRICAL CHARACTERISTICS (T_A = 25°C unless otherwise noted) (Continued)

Characteristic		Symbol	Min	Max	Unit
ON CHARACTERISTICS					
DC Current Gain[1]		h_{FE}			—
(I_C = 0.1 mAdc, V_{CE} = 1.0 Vdc)	2N3903		20	—	
	2N3904		40	—	
(I_C = 1.0 mAdc, V_{CE} = 1.0 Vdc)	2N3903		35	—	
	2N3904		70	—	
(I_C = 10 mAdc, V_{CE} = 1.0 Vdc)	2N3903		50	150	
	2N3904		100	300	
(I_C = 50 mAdc, V_{CE} = 1.0 Vdc)	2N3903		30	—	
	2N3904		60	—	
(I_C = 100 mAdc, V_{CE} = 1.0 Vdc)	2N3903		15	—	
	2N3904		30	—	
Collector–Emitter Saturation Voltage[1]		$V_{CE(sat)}$			Vdc
(I_C = 10 mAdc, I_B = 1.0 mAdc)			—	0.2	
(I_C = 50 mAdc, I_B = 5.0 mAdc)			—	0.3	
Base–Emitter Saturation Voltage[1]		$V_{BE(sat)}$			Vdc
(I_C = 10 mAdc, I_B = 1.0 mAdc)			0.65	0.85	
(I_C = 50 mAdc, I_B = 5.0 mAdc)			—	0.95	
SMALL–SIGNAL CHARACTERISTICS					
Current–Gain — Bandwidth Product		f_T			MHz
(I_C = 10 mAdc, V_{CE} = 20 Vdc, f = 100 MHz)	2N3903		250	—	
	2N3904		300	—	
Output Capacitance		C_{obo}	—	4.0	pF
(V_{CB} = 5.0 Vdc, I_E = 0, f = 1.0 MHz)					
Input Capacitance		C_{ibo}	—	8.0	pF
(V_{EB} = 0.5 Vdc, I_C = 0, f = 1.0 MHz)					
Input Impedance		h_{ie}			k Ω
(I_C = 1.0 mAdc, V_{CE} = 10 Vdc, f = 1.0 kHz)	2N3903		1.0	8.0	
	2N3904		1.0	10	
Voltage Feedback Ratio		h_{re}			X 10^{-4}
(I_C = 1.0 mAdc, V_{CE} = 10 Vdc, f = 1.0 kHz)	2N3903		0.1	5.0	
	2N3904		0.5	8.0	
Small–Signal Current Gain		h_{fe}			—
(I_C = 1.0 mAdc, V_{CE} = 10 Vdc, f = 1.0 kHz)	2N3903		50	200	
	2N3904		100	400	
Output Admittance		h_{oe}	1.0	40	μmhos
(I_C = 1.0 mAdc, V_{CE} = 10 Vdc, f = 1.0 kHz)					
Noise Figure		NF			dB
(I_C = 100 μAdc, V_{CE} = 5.0 Vdc, R_S = 1.0 k Ω, f = 1.0 kHz)	2N3903		—	6.0	
	2N3904		—	5.0	

SWITCHING CHARACTERISTICS

			Symbol	Min	Max	Unit
Delay Time	(V_{CC} = 3.0 Vdc, V_{BE} = 0.5 Vdc,		t_d	—	35	ns
Rise Time	I_C = 10 mAdc, I_{B1} = 1.0 mAdc)		t_r	—	35	ns
Storage Time	(V_{CC} = 3.0 Vdc, I_C = 10 mAdc,	2N3903	t_s	—	175	ns
	I_{B1} = I_{B2} = 1.0 mAdc)	2N3904		—	200	
Fall Time			t_f	—	50	ns

1. Pulse Test: Pulse Width ≤ 300 μs; Duty Cycle ≤ 2.0%.

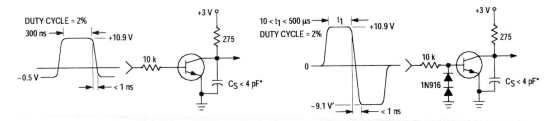

* Total shunt capacitance of test jig and connectors

**Figure 1. Delay and Rise Time
Equivalent Test Circuit**

**Figure 2. Storage and Fall Time
Equivalent Test Circuit**

TYPICAL TRANSIENT CHARACTERISTICS

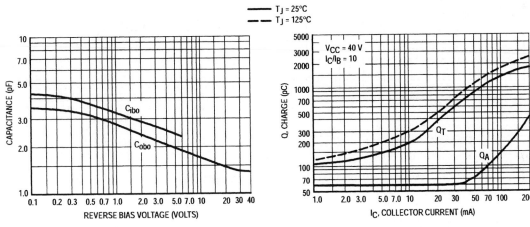

Figure 3. Capacitance

Figure 4. Charge Data

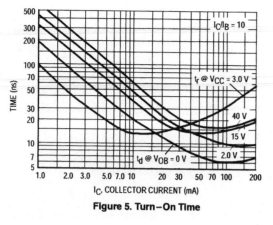

Figure 5. Turn–On Time

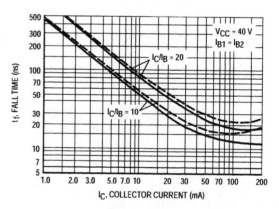

Figure 6. Rise Time

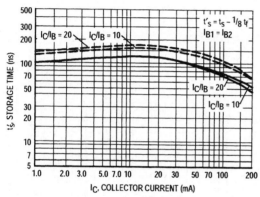

Figure 7. Storage Time

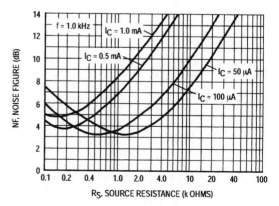

Figure 8. Fall Time

TYPICAL AUDIO SMALL–SIGNAL CHARACTERISTICS
NOISE FIGURE VARIATIONS
(V_{CE} = 5.0 Vdc, T_A = 25°C, Bandwidth = 1.0 Hz)

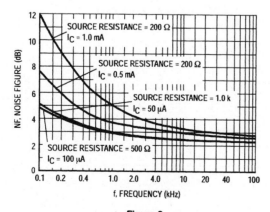

Figure 9.

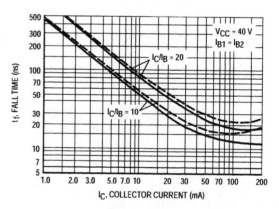

Figure 10.

h PARAMETERS
(V_{CE} = 10 Vdc, f = 1.0 kHz, T_A = 25°C)

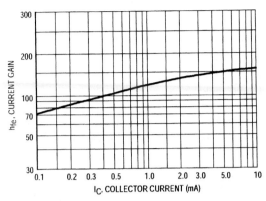

Figure 11. Current Gain

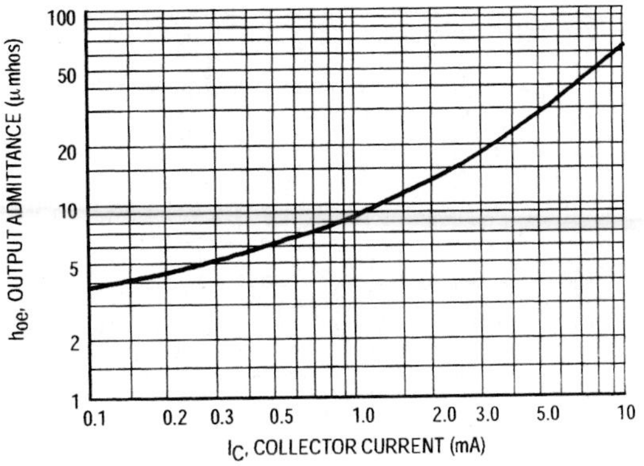

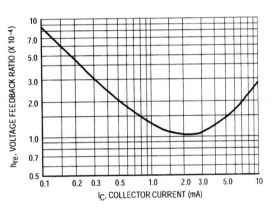

Figure 12. Output Admittance

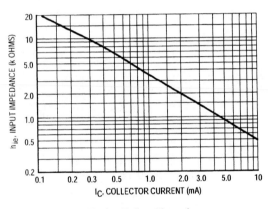

Figure 13. Input Impedance

Figure 14. Voltage Feedback Ratio

TYPICAL STATIC CHARACTERISTICS

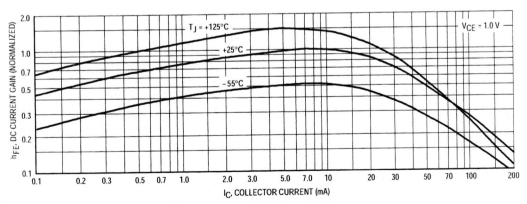

Figure 15. DC Current Gain

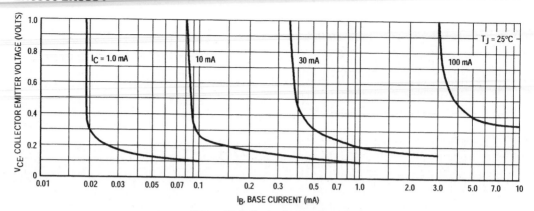

Figure 16. Collector Saturation Region

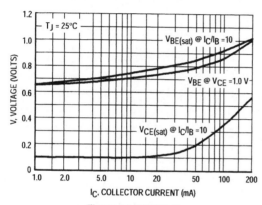

Figure 17. "ON" Voltages

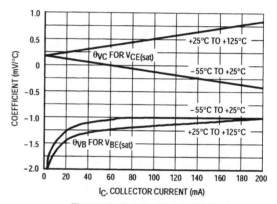

Figure 18. Temperature Coefficients

General Purpose Transistors

PNP Silicon

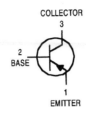

COLLECTOR
3

2
BASE

1
EMITTER

2N3905
2N3906*

*Motorola Preferred Device

CASE 29–04, STYLE 1
TO–92 (TO–226AA)

MAXIMUM RATINGS

Rating	Symbol	Value	Unit
Collector–Emitter Voltage	V_{CEO}	40	Vdc
Collector–Base Voltage	V_{CBO}	40	Vdc
Emitter–Base Voltage	V_{EBO}	5.0	Vdc
Collector Current — Continuous	I_C	200	mAdc
Total Device Dissipation @ $T_A = 25°C$ Derate above 25°C	P_D	625 5.0	mW mW/°C
Total Power Dissipation @ $T_A = 60°C$	P_D	250	mW
Total Device Dissipation @ $T_C = 25°C$ Derate above 25°C	P_D	1.5 12	Watts mW/°C
Operating and Storage Junction Temperature Range	T_J, T_{stg}	−55 to +150	°C

THERMAL CHARACTERISTICS[1]

Characteristic	Symbol	Max	Unit
Thermal Resistance, Junction to Ambient	$R_{\theta JA}$	200	°C/W
Thermal Resistance, Junction to Case	$R_{\theta JC}$	83.3	°C/W

ELECTRICAL CHARACTERISTICS ($T_A = 25°C$ unless otherwise noted)

Characteristic	Symbol	Min	Max	Unit
OFF CHARACTERISTICS				
Collector–Emitter Breakdown Voltage [2] ($I_C = 1.0$ mAdc, $I_B = 0$)	$V_{(BR)CEO}$	40	—	Vdc
Collector–Base Breakdown Voltage ($I_C = 10$ μAdc, $I_E = 0$)	$V_{(BR)CBO}$	40	—	Vdc
Emitter–Base Breakdown Voltage ($I_E = 10$ μAdc, $I_C = 0$)	$V_{(BR)EBO}$	5.0	—	Vdc
Base Cutoff Current ($V_{CE} = 30$ Vdc, $V_{EB} = 3.0$ Vdc)	I_{BL}	—	50	nAdc
Collector Cutoff Current ($V_{CE} = 30$ Vdc, $V_{EB} = 3.0$ Vdc)	I_{CEX}	—	50	nAdc

1. Indicates Data in addition to JEDEC Requirements.
2. Pulse Test: Pulse Width ≤ 300 μs; Duty Cycle ≤ 2.0%.

Preferred devices are Motorola recommended choices for future use and best overall value.

REV 2

MOTOROLA

ELECTRICAL CHARACTERISTICS (T_A = 25°C unless otherwise noted) (Continued)

Characteristic		Symbol	Min	Max	Unit
ON CHARACTERISTICS(1)					
DC Current Gain		h_{FE}			—
(I_C = 0.1 mAdc, V_{CE} = 1.0 Vdc) 2N3905			30	—	
2N3906			60	—	
(I_C = 1.0 mAdc, V_{CE} = 1.0 Vdc) 2N3905			40	—	
2N3906			80	—	
(I_C = 10 mAdc, V_{CE} = 1.0 Vdc) 2N3905			50	150	
2N3906			100	300	
(I_C = 50 mAdc, V_{CE} = 1.0 Vdc) 2N3905			30	—	
2N3906			60	—	
(I_C = 100 mAdc, V_{CE} = 1.0 Vdc) 2N3905			15	—	
2N3906			30	—	
Collector–Emitter Saturation Voltage		$V_{CE(sat)}$			Vdc
(I_C = 10 mAdc, I_B = 1.0 mAdc)			—	0.25	
(I_C = 50 mAdc, I_B = 5.0 mAdc)			—	0.4	
Base–Emitter Saturation Voltage		$V_{BE(sat)}$			Vdc
(I_C = 10 mAdc, I_B = 1.0 mAdc)			0.65	0.85	
(I_C = 50 mAdc, I_B = 5.0 mAdc)			—	0.95	
SMALL–SIGNAL CHARACTERISTICS					
Current–Gain — Bandwidth Product		f_T			MHz
(I_C = 10 mAdc, V_{CE} = 20 Vdc, f = 100 MHz) 2N3905			200	—	
2N3906			250	—	
Output Capacitance		C_{obo}	—	4.5	pF
(V_{CB} = 5.0 Vdc, I_E = 0, f = 1.0 MHz)					
Input Capacitance		C_{ibo}	—	10.0	pF
(V_{EB} = 0.5 Vdc, I_C = 0, f = 1.0 MHz)					
Input Impedance		h_{ie}			kΩ
(I_C = 1.0 mAdc, V_{CE} = 10 Vdc, f = 1.0 kHz) 2N3905			0.5	8.0	
2N3906			2.0	12	
Voltage Feedback Ratio		h_{re}			X 10^{-4}
(I_C = 1.0 mAdc, V_{CE} = 10 Vdc, f = 1.0 kHz) 2N3905			0.1	5.0	
2N3906			0.1	10	
Small–Signal Current Gain		h_{fe}			—
(I_C = 1.0 mAdc, V_{CE} = 10 Vdc, f = 1.0 kHz) 2N3905			50	200	
2N3906			100	400	
Output Admittance		h_{oe}			μmhos
(I_C = 1.0 mAdc, V_{CE} = 10 Vdc, f = 1.0 kHz) 2N3905			1.0	40	
2N3906			3.0	60	
Noise Figure		NF			dB
(I_C = 100 μAdc, V_{CE} = 5.0 Vdc, R_S = 1.0 kΩ, f = 1.0 kHz) 2N3905			—	5.0	
2N3906			—	4.0	

SWITCHING CHARACTERISTICS

		Symbol	Min	Max	Unit
Delay Time	(V_{CC} = 3.0 Vdc, V_{BE} = 0.5 Vdc,	t_d	—	35	ns
Rise Time	I_C = 10 mAdc, I_{B1} = 1.0 mAdc)	t_r	—	35	ns
Storage Time 2N3905		t_s	—	200	ns
2N3906	(V_{CC} = 3.0 Vdc, I_C = 10 mAdc,		—	225	
Fall Time 2N3905	I_{B1} = I_{B2} = 1.0 mAd	t_f	—	60	ns
2N3906			—	75	

1. Pulse Test: Pulse Width ≤ 300 μs; Duty Cycle ≤ 2.0%.

842

JFETs — General Purpose
N–Channel — Depletion

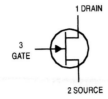

1 DRAIN

3 GATE

2 SOURCE

2N5457

*Motorola Preferred Device

CASE 29–04, STYLE 5
TO–92 (TO–226AA)

MAXIMUM RATINGS

Rating	Symbol	Value	Unit
Drain–Source Voltage	V_{DS}	25	Vdc
Drain–Gate Voltage	V_{DG}	25	Vdc
Reverse Gate–Source Voltage	V_{GSR}	−25	Vdc
Gate Current	I_G	10	mAdc
Total Device Dissipation @ T_A = 25°C Derate above 25°C	P_D	310 2.82	mW mW/°C
Junction Temperature Range	T_J	125	°C
Storage Channel Temperature Range	T_{stg}	−65 to +150	°C

ELECTRICAL CHARACTERISTICS (T_A = 25°C unless otherwise noted)

Characteristic	Symbol	Min	Typ	Max	Unit
OFF CHARACTERISTICS					
Gate–Source Breakdown Voltage (I_G = −10 µAdc, V_{DS} = 0)	$V_{(BR)GSS}$	−25	—	—	Vdc
Gate Reverse Current (V_{GS} = −15 Vdc, V_{DS} = 0) (V_{GS} = −15 Vdc, V_{DS} = 0, T_A = 100°C)	I_{GSS}	— —	— —	−1.0 −200	nAdc
Gate–Source Cutoff Voltage (V_{DS} = 15 Vdc, I_D = 10 nAdc)	$V_{GS(off)}$	−0.5	—	−6.0	Vdc
Gate–Source Voltage (V_{DS} = 15 Vdc, I_D = 100 µAdc)	V_{GS}	—	−2.5	—	Vdc
ON CHARACTERISTICS					
Zero–Gate–Voltage Drain Current (1) (V_{DS} = 15 Vdc, V_{GS} = 0)	I_{DSS}	1.0	3.0	5.0	mAdc
SMALL–SIGNAL CHARACTERISTICS					
Forward Transfer Admittance Common Source (1) (V_{DS} = 15 Vdc, V_{GS} = 0, f = 1.0 kHz)	$|y_{fs}|$	1000	—	5000	µmhos
Output Admittance Common Source (1) (V_{DS} = 15 Vdc, V_{GS} = 0, f = 1.0 kHz)	$|y_{os}|$	—	10	50	µmhos
Input Capacitance (V_{DS} = 15 Vdc, V_{GS} = 0, f = 1.0 MHz)	C_{iss}	—	4.5	7.0	pF
Reverse Transfer Capacitance (V_{DS} = 15 Vdc, V_{GS} = 0, f = 1.0 MHz)	C_{rss}	—	1.5	3.0	pF

1. Pulse Test; Pulse Width ≤ 630 ms, Duty Cycle ≤ 10%.

Preferred devices are Motorola recommended choices for future use and best overall value.

 **MOTOROLA**

TYPICAL CHARACTERISTICS

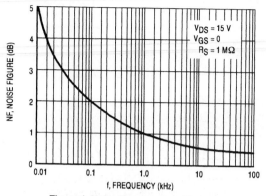

Figure 1. Noise Figure versus Frequency

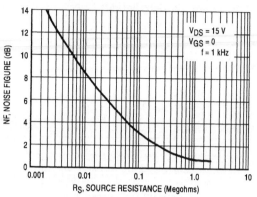

Figure 2. Noise Figure versus Source
Resistance

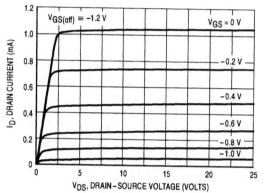

Figure 3. Typical Drain Characteristics

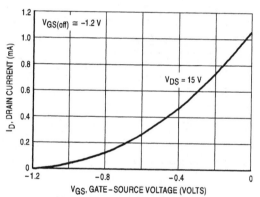

Figure 4. Common Source Transfer
Characteristics

TYPICAL CHARACTERISTICS

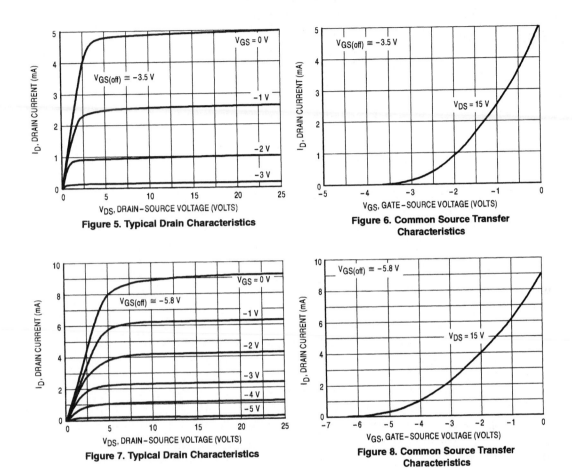

Figure 5. Typical Drain Characteristics

Figure 6. Common Source Transfer Characteristics

Figure 7. Typical Drain Characteristics

Figure 8. Common Source Transfer Characteristics

Note: Graphical data is presented for dc conditions. Tabular data is given for pulsed conditions (Pulse Width = 630 ms, Duty Cycle = 10%). Under dc conditions, self heating in higher I_{DSS} units reduces I_{DSS}.

MOTOROLA

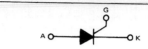

SILICON CONTROLLED RECTIFIERS

. . . designed primarily for half-wave ac control applications, such as motor controls, heating controls and power supplies; or wherever half-wave silicon gate-controlled, solid-state devices are needed.

● Glass Passivated Junctions and Center Gate Fire for Greater Parameter Uniformity and Stability

● Small, Rugged, Thermowatt▲ Construction for Low Thermal Resistance, High Heat Dissipation and Durability

● Blocking Voltage to 800 Volts

THYRISTORS

12 AMPERES RMS
50-800 VOLTS

*MAXIMUM RATINGS

Rating	Symbol	Value	Unit
Peak Reverse Voltage (1)	V_{RRM}		Volts
2N6394		50	
2N6395		100	
2N6396		200	
MCR220-5		300	
2N6397		400	
MCR220-7		500	
2N6398		600	
MCR220-9		700	
2N6399		800	
Forward Current RMS T_J = 125°C (All Conduction Angles)	$I_{T(RMS)}$	12	Amps
Peak Forward Surge Current (1/2 cycle, Sine Wave, 60 Hz, T_J = 125°C)	I_{TSM}	100	Amps
Circuit Fusing Considerations (T_J = –40 to +125°C, t = 1.0 to 8.3 ms)	I^2t	40	A^2s
Forward Peak Gate Power	P_{GM}	20	Watts
Forward Average Gate Power	$P_{G(AV)}$	0.5	Watt
Forward Peak Gate Current	I_{GM}	2.0	Amps
Operating Junction Temperature Range	T_J	–40 to +125	°C
Storage Temperature Range	T_{stg}	–40 to +150	°C

THERMAL CHARACTERISTICS

Characteristic	Symbol	Max	Unit
Thermal Resistance, Junction to Case	$R_{\theta JC}$	2.0	°C/W

(1) V_{RRM} for all types can be applied on a continuous dc basis without incurring damage. Ratings apply for zero or negative gate voltage. Devices should not be tested for blocking capability in a manner such that the voltage supplied exceeds the rated blocking voltage.

* Indicates JEDEC Registered Data.
▲ Trademark of Motorola Inc.

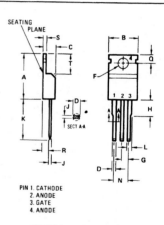

PIN 1. CATHODE
2. ANODE
3. GATE
4. ANODE

All JEDEC dimensions and notes apply

DIM	MILLIMETERS		INCHES	
	MIN	MAX	MIN	MAX
A	14.23	15.87	0.560	0.625
B	9.66	10.66	0.380	0.420
C	3.56	4.82	0.140	0.190
D	0.51	1.14	0.020	0.045
F	3.531	3.733	0.139	0.147
G	2.29	2.79	0.090	0.110
H	–	6.35	–	0.250
J	0.31	1.14	0.012	0.045
K	12.70	14.27	0.500	0.562
L	1.14	1.77	0.045	0.070
N	4.83	5.33	0.190	0.210
Q	2.54	3.04	0.100	0.120
R	2.04	2.92	0.080	0.115
S	0.51	1.39	0.020	0.055
T	5.85	6.85	0.230	0.270

CASE 221-02
TO 220 AB

ELECTRICAL CHARACTERISTICS ($T_C = 25^oC$ unless otherwise noted.)

Characteristic	Symbol	Min	Typ	Max	Unit
*Peak Forward Blocking Voltage	V_{DRM}				Volts
($T_J = 125^oC$) 2N6394		50	—	—	
2N6395		100	—	—	
2N6396		200	—	—	
MCR220-5		300	—	—	
2N6397		400	—	—	
MCR220-7		500	—	—	
2N6398		600	—	—	
MCR220-9		700	—	—	
2N6399		800	—	—	
* Peak Forward Blocking Current (Rated V_{DRM} @ $T_J = 125^oC$)	I_{DRM}	—	—	2.0	mA
* Peak Reverse Blocking Current (Rated V_{RRM} @ $T_J = 125^oC$)	I_{RRM}	—	—	2.0	mA
* Forward "On" Voltage (I_{TM} = 24 A Peak)	V_{TM}	—	1.7	2.2	Volts
* Gate Trigger Current (Continuous dc) (Anode Voltage = 12 Vdc, R_L = 100 Ohms)	I_{GT}	—	5.0	30	mA
* Gate Trigger Voltage (Continuous dc) (Anode Voltage = 12 Vdc, R_L = 100 Ohms)	V_{GT}	—	0.7	1.5	Volts
* Gate Non-Trigger Voltage (Anode Voltage = Rated V_{DRM}, R_L = 100 Ohms, $T_J = 125^oC$)	V_{GD}	0.2	—	—	Volts
* Holding Current (Anode Voltage = 12 Vdc)	I_H	—	6.0	40	mA
Turn-On Time (I_{TM} = 12 A, I_{GT} = 40 mAdc)	t_{gt}	—	1.0	2.0	µs
Turn-Off Time (V_{DRM} = rated voltage) (I_{TM} = 12 A, I_R = 12 A) (I_{TM} = 12 A, I_R = 12 A, $T_J = 125^oC$)	t_q	— —	15 35	— —	µs
Forward Voltage Application Rate ($T_J = 125^oC$)	dv/dt	—	50	—	V/µs

*Indicates JEDEC Registered Data.

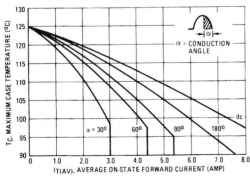

FIGURE 1 – AVERAGE CURRENT DERATING

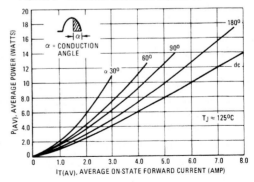

FIGURE 2 – MAXIMUM ON-STATE POWER DISSIPATION

 MOTOROLA *Semiconductor Products Inc.*

LM117
LM217
LM317

THREE-TERMINAL ADJUSTABLE POSITIVE VOLTAGE REGULATORS

SILICON MONOLITHIC INTEGRATED CIRCUIT

THREE-TERMINAL ADJUSTABLE OUTPUT POSITIVE VOLTAGE REGULATORS

The LM117/217/317 are adjustable 3-terminal positive voltage regulators capable of supplying in excess of 1.5 A over an output voltage range of 1.2 V to 37 V. These voltage regulators are exceptionally easy to use and require only two external resistors to set the output voltage. Further, they employ internal current limiting, thermal shutdown and safe area compensation, making them essentially blow-out proof.

The LM117 series serve a wide variety of applications including local, on card regulation. This device can also be used to make a programmable output regulator, or by connecting a fixed resistor between the adjustment and output, the LM117 series can be used as a precision current regulator.

- Output Current in Excess of 1.5 Ampere in K and T Suffix Packages
- Output Current in Excess of 0.5 Ampere in H Suffix Package
- Output Adjustable between 1.2 V and 37 V
- Internal Thermal Overload Protection
- Internal Short-Circuit Current Limiting Constant with Temperature
- Output Transistor Safe-Area Compensation
- Floating Operation for High Voltage Applications
- Standard 3-lead Transistor Packages
- Eliminates Stocking Many Fixed Voltages

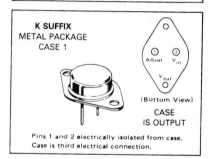

K SUFFIX
METAL PACKAGE
CASE 1

(Bottom View)
CASE IS OUTPUT

Pins 1 and 2 electrically isolated from case.
Case is third electrical connection.

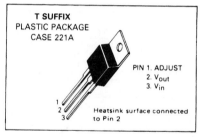

T SUFFIX
PLASTIC PACKAGE
CASE 221A

PIN 1. ADJUST
2. V_{out}
3. V_{in}

Heatsink surface connected to Pin 2

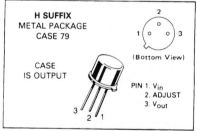

H SUFFIX
METAL PACKAGE
CASE 79

CASE IS OUTPUT

(Bottom View)

PIN 1. V_{in}
2. ADJUST
3. V_{out}

STANDARD APPLICATION

LM117

V_{in} V_{out}

I_{Adj} Adjust

R_1 240

* C_{in} 0.1 μF

** C_O 1 μF

R_2

* = C_{in} is required if regulator is located an appreciable distance from power supply filter.

** = C_O is not needed for stability, however it does improve transient response.

$$V_{out} = 1.25 \text{ V} \left(1 + \frac{R_2}{R_1}\right) + I_{Adj} R_2$$

Since I_{Adj} is controlled to less than 100 μA, the error associated with this term is negligible in most applications.

ORDERING INFORMATION

Device	Tested Operating Temperature Range	Package
LM117H LM117K	$T_J = -55°C$ to $+150°C$	Metal Can Metal Power
LM217H LM217K	$T_J = -25°C$ to $+150°C$	Metal Can Metal Power
LM317H LM317K LM317T	$T_J = 0°C$ to $+125°C$	Metal Can Metal Power Plastic Power
LM317BT#	$T_J = -40°C$ to $+125°C$	Plastic Power

\#Automotive temperature range selections are available with special test conditions and additional tests.
Contact your local Motorola sales office for information.

LM117, LM217, LM317

MAXIMUM RATINGS

Rating	Symbol	Value	Unit
Input-Output Voltage Differential	V_I-V_O	40	Vdc
Power Dissipation	P_D	Internally Limited	
Operating Junction Temperature Range LM117 LM217 LM317	T_J	-55 to $+150$ -25 to $+150$ 0 to $+150$	°C
Storage Temperature Range	T_{stg}	-65 to $+150$	°C

ELECTRICAL CHARACTERISTICS (V_I-V_O = 5.0 V; I_O = 0.5 A for K and T packages; I_O = 0.1 A for H package; T_J = T_{low} to T_{high} [see Note 1]; I_{max} and P_{max} per Note 2; unless otherwise specified.)

Characteristic	Figure	Symbol	LM117/217			LM317			Unit
			Min	Typ	Max	Min	Typ	Max	
Line Regulation (Note 3) T_A = 25°C, 3.0 V ≤ V_I-V_O ≤ 40 V	1	Reg_{line}	—	0.01	0.02	—	0.01	0.04	%/V
Load Regulation (Note 3) T_A = 25°C, 10 mA ≤ I_O ≤ I_{max} V_O ≤ 5.0 V V_O ≥ 5.0 V	2	Reg_{load}	 — —	 5.0 0.1	 15 0.3	 — —	 5.0 0.1	 25 0.5	 mV %/V_O
Thermal Regulation (T_A = +25°C) 20 ms Pulse	—	—	—	0.02	0.07	—	0.03	0.07	%/W
Adjustment Pin Current	3	I_{Adj}	—	50	100	—	50	100	μA
Adjustment Pin Current Change 2.5 V ≤ V_I-V_O ≤ 40 V 10 mA ≤ I_L ≤ I_{max}, P_D ≤ P_{max}	1,2	ΔI_{Adj}	—	0.2	5.0	—	0.2	5.0	μA
Reference Voltage (Note 4) 3.0 V ≤ V_I-V_O ≤ 40 V 10 mA ≤ I_O ≤ I_{max}, P_D ≤ P_{max}	3	V_{ref}	1.2	1.25	1.3	1.2	1.25	1.3	V
Line Regulation (Note 3) 3.0 V ≤ V_I-V_O ≤ 40 V	1	Reg_{line}	—	0.02	0.05	—	0.02	0.07	%/V
Load Regulation (Note 3) 10 mA ≤ I_O ≤ I_{max} V_O ≤ 5.0 V V_O ≥ 5.0 V	2	Reg_{load}	 — —	 20 0.3	 50 1.0	 — —	 20 0.3	 70 1.5	 mV %/V_O
Temperature Stability (T_{low} ≤ T_J ≤ T_{high})	3	T_S	—	0.7	—	—	0.7	—	%/V_O
Minimum Load Current to Maintain Regulation (V_I-V_O = 40 V)	3	I_{Lmin}	—	3.5	5.0	—	3.5	10	mA
Maximum Output Current V_I-V_O ≤ 15 V, P_D ≤ P_{max} K and T Packages H Package V_I-V_O = 40 V, P_D ≤ P_{max}, T_A = 25°C K and T Packages H Package	3	I_{max}	 1.5 0.5 0.25 —	 2.2 0.8 0.4 0.07	 — — — —	 1.5 0.5 0.15 —	 2.2 0.8 0.4 0.07	 — — — —	A
RMS Noise, % of V_O T_A = 25°C, 10 Hz ≤ f ≤ 10 kHz	—	N	—	0.003	—	—	0.003	—	%/V_O
Ripple Rejection, V_O = 10 V, f = 120 Hz (Note 5) Without C_{Adj} C_{Adj} = 10 μF	4	RR	 — 66	 65 80	 — —	 — 66	 65 80	 — —	dB
Long-Term Stability, T_J = T_{high} (Note 6) T_A = 25°C for Endpoint Measurements	3	S	—	0.3	1.0	—	0.3	1.0	%/1.0 k Hrs.
Thermal Resistance Junction to Case H Package K Package T Package	—	$R_{\theta JC}$	 — — —	 12 2.3 —	 15 3.0 —	 — — —	 12 2.3 5.0	 15 3.0 —	°C/W

NOTES: (1) T_{low} = -55°C for LM117 T_{high} = $+150$°C for LM117
      = -25°C for LM217   = $+150$°C for LM217
      =   0°C for LM317   = $+125$°C for LM317
    (2) I_{max} = 1.5 A for K and T Packages
           = 0.5 A for H Package
       P_{max} = 20 W for K Package
            = 20 W for T Package
            = 2.0 W for H Package
    (3) Load and line regulation are specified at constant junction temperature. Changes in V_O due to heating effects must

be taken into account separately. Pulse testing with low duty cycle is used.
(4) Selected devices with tightened tolerance reference voltage available.
(5) C_{ADJ}, when used, is connected between the adjustment pin and ground.
(6) Since Long-Term Stability cannot be measured on each device before shipment, this specification is an engineering estimate of average stability from lot to lot.

February 2000

LM555
Timer

General Description

The LM555 is a highly stable device for generating accurate time delays or oscillation. Additional terminals are provided for triggering or resetting if desired. In the time delay mode of operation, the time is precisely controlled by one external resistor and capacitor. For astable operation as an oscillator, the free running frequency and duty cycle are accurately controlled with two external resistors and one capacitor. The circuit may be triggered and reset on falling waveforms, and the output circuit can source or sink up to 200mA or drive TTL circuits.

Features

- Direct replacement for SE555/NE555
- Timing from microseconds through hours
- Operates in both astable and monostable modes
- Adjustable duty cycle
- Output can source or sink 200 mA
- Output and supply TTL compatible
- Temperature stability better than 0.005% per °C
- Normally on and normally off output
- Available in 8-pin MSOP package

Applications

- Precision timing
- Pulse generation
- Sequential timing
- Time delay generation
- Pulse width modulation
- Pulse position modulation
- Linear ramp generator

Schematic Diagram

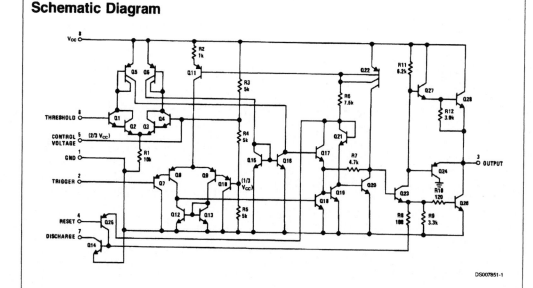

DS007851-1

www.national.com

Connection Diagram

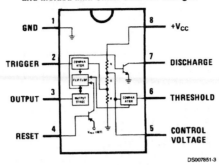

Dual-In-Line, Small Outline
and Molded Mini Small Outline Packages

GND 1   8 +V_CC

TRIGGER 2   7 DISCHARGE

OUTPUT 3   6 THRESHOLD

RESET 4   5 CONTROL
                 VOLTAGE

DS007851-3

Top View

Ordering Information

Package	Part Number	Package Marking	Media Transport	NSC Drawing
8-Pin SOIC	LM555CM	LM555CM	Rails	M08A
	LM555CMX	LM555CM	2.5k Units Tape and Reel	
8-Pin MSOP	LM555CMM	Z55	1k Units Tape and Reel	MUA08A
	LM555CMMX	Z55	3.5k Units Tape and Reel	
8-Pin MDIP	LM555CN	LM555CN	Rails	N08E

851

Absolute Maximum Ratings (Note 2)

If Military/Aerospace specified devices are required, please contact the National Semiconductor Sales Office/ Distributors for availability and specifications.

Supply Voltage	+18V
Power Dissipation (Note 3)	
LM555CM, LM555CN	1180 mW
LM555CMM	613 mW
Operating Temperature Ranges	
LM555C	0˚C to +70˚C
Storage Temperature Range	−65˚C to +150˚C

Soldering Information
  Dual-In-Line Package
    Soldering (10 Seconds)   260˚C
  Small Outline Packages
    (SOIC and MSOP)
    Vapor Phase (60 Seconds)   215˚C
    Infrared (15 Seconds)   220˚C
See AN-450 "Surface Mounting Methods and Their Effect on Product Reliability" for other methods of soldering surface mount devices.

Electrical Characteristics (Notes 1, 2)

(T_A = 25˚C, V_{CC} = +5V to +15V, unless otherwise specified)

Parameter	Conditions	Limits LM555C			Units
		Min	Typ	Max	
Supply Voltage		4.5		16	V
Supply Current	V_{CC} = 5V, R_L = ∞		3	6	
	V_{CC} = 15V, R_L = ∞		10	15	mA
	(Low State) (Note 4)				
Timing Error, Monostable					
Initial Accuracy			1		%
Drift with Temperature	R_A = 1k to 100kΩ,		50		ppm/˚C
	C = 0.1µF, (Note 5)				
Accuracy over Temperature			1.5		%
Drift with Supply			0.1		%/V
Timing Error, Astable					
Initial Accuracy			2.25		%
Drift with Temperature	R_A, R_B = 1k to 100kΩ,		150		ppm/˚C
	C = 0.1µF, (Note 5)				
Accuracy over Temperature			3.0		%
Drift with Supply			0.30		%/V
Threshold Voltage			0.667		x V_{CC}
Trigger Voltage	V_{CC} = 15V		5		V
	V_{CC} = 5V		1.67		V
Trigger Current			0.5	0.9	µA
Reset Voltage		0.4	0.5	1	V
Reset Current			0.1	0.4	mA
Threshold Current	(Note 6)		0.1	0.25	µA
Control Voltage Level	V_{CC} = 15V	9	10	11	V
	V_{CC} = 5V	2.6	3.33	4	
Pin 7 Leakage Output High			1	100	nA
Pin 7 Sat (Note 7)					
Output Low	V_{CC} = 15V, I_7 = 15mA		180		mV
Output Low	V_{CC} = 4.5V, I_7 = 4.5mA		80	200	mV

Electrical Characteristics (Notes 1, 2) (Continued)

(T_A = 25°C, V_{CC} = +5V to +15V, unless othewise specified)

Parameter	Conditions	Limits LM555C			Units
		Min	Typ	Max	
Output Voltage Drop (Low)	V_{CC} = 15V				
	I_{SINK} = 10mA		0.1	0.25	V
	I_{SINK} = 50mA		0.4	0.75	V
	I_{SINK} = 100mA		2	2.5	V
	I_{SINK} = 200mA		2.5		V
	V_{CC} = 5V				
	I_{SINK} = 8mA				V
	I_{SINK} = 5mA		0.25	0.35	V
Output Voltage Drop (High)	I_{SOURCE} = 200mA, V_{CC} = 15V		12.5		V
	I_{SOURCE} = 100mA, V_{CC} = 15V	12.75	13.3		V
	V_{CC} = 5V	2.75	3.3		V
Rise Time of Output			100		ns
Fall Time of Output			100		ns

Note 1: All voltages are measured with respect to the ground pin, unless otherwise specified.

Note 2: Absolute Maximum Ratings indicate limits beyond which damage to the device may occur. Operating Ratings indicate conditions for which the device is functional, but do not guarantee specific performance limits. Electrical Characteristics state DC and AC electrical specifications under particular test conditions which guarantee specific performance limits. This assumes that the device is within the Operating Ratings. Specifications are not guaranteed for parameters where no limit is given, however, the typical value is a good indication of device performance.

Note 3: For operating at elevated temperatures the device must be derated above 25°C based on a +150°C maximum junction temperature and a thermal resistance of 106°C/W (DIP), 170°C/W (SO-8), and 204°C/W (MSOP) junction to ambient.

Note 4: Supply current when output high typically 1 mA less at V_{CC} = 5V.

Note 5: Tested at V_{CC} = 5V and V_{CC} = 15V.

Note 6: This will determine the maximum value of $R_A + R_B$ for 15V operation. The maximum total ($R_A + R_B$) is 20MΩ.

Note 7: No protection against excessive pin 7 current is necessary providing the package dissipation rating will not be exceeded.

Note 8: Refer to RETS555X drawing of military LM555H and LM555J versions for specifications.

4

853

N *National Semiconductor*

August 2000

LM741
Operational Amplifier

General Description

The LM741 series are general purpose operational amplifiers which feature improved performance over industry standards like the LM709. They are direct, plug-in replacements for the 709C, LM201, MC1439 and 748 in most applications.

The amplifiers offer many features which make their application nearly foolproof: overload protection on the input and output, no latch-up when the common mode range is exceeded, as well as freedom from oscillations.

The LM741C is identical to the LM741/LM741A except that the LM741C has their performance guaranteed over a 0˚C to +70˚C temperature range, instead of −55˚C to +125˚C.

Connection Diagrams

Metal Can Package

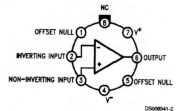

DS009341-2

Note 1: LM741H is available per JM38510/10101

Order Number LM741H, LM741H/883 (Note 1),
LM741AH/883 or LM741CH
See NS Package Number H08C

Dual-In-Line or S.O. Package

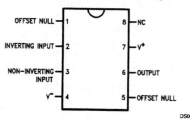

DS009341-3

Order Number LM741J, LM741J/883, LM741CN
See NS Package Number J08A, M08A or N08E

Ceramic Flatpak

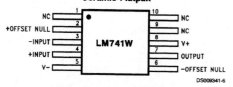

DS009341-6

Order Number LM741W/883
See NS Package Number W10A

Typical Application

Offset Nulling Circuit

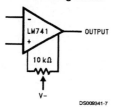

DS009341-7

---

www.national.com

Absolute Maximum Ratings (Note 2)

If Military/Aerospace specified devices are required, please contact the National Semiconductor Sales Office/ Distributors for availability and specifications.

(Note 7)

	LM741A	LM741	LM741C
Supply Voltage	±22V	±22V	±18V
Power Dissipation (Note 3)	500 mW	500 mW	500 mW
Differential Input Voltage	±30V	±30V	±30V
Input Voltage (Note 4)	±15V	±15V	±15V
Output Short Circuit Duration	Continuous	Continuous	Continuous
Operating Temperature Range	−55°C to +125°C	−55°C to +125°C	0°C to +70°C
Storage Temperature Range	−65°C to +150°C	−65°C to +150°C	−65°C to +150°C
Junction Temperature	150°C	150°C	100°C
Soldering Information			
N-Package (10 seconds)	260°C	260°C	260°C
J- or H-Package (10 seconds)	300°C	300°C	300°C
M-Package			
Vapor Phase (60 seconds)	215°C	215°C	215°C
Infrared (15 seconds)	215°C	215°C	215°C

See AN-450 "Surface Mounting Methods and Their Effect on Product Reliability" for other methods of soldering surface mount devices.

| ESD Tolerance (Note 8) | 400V | 400V | 400V |

Electrical Characteristics (Note 5)

Parameter	Conditions	LM741A			LM741			LM741C			Units
		Min	Typ	Max	Min	Typ	Max	Min	Typ	Max	
Input Offset Voltage	T_A = 25°C										
	$R_S \leq 10$ kΩ					1.0	5.0		2.0	6.0	mV
	$R_S \leq 50\Omega$		0.8	3.0							mV
	$T_{AMIN} \leq T_A \leq T_{AMAX}$										
	$R_S \leq 50\Omega$			4.0							mV
	$R_S \leq 10$ kΩ						6.0			7.5	mV
Average Input Offset Voltage Drift				15							µV/°C
Input Offset Voltage Adjustment Range	T_A = 25°C, V_S = ±20V	±10				±15			±15		mV
Input Offset Current	T_A = 25°C		3.0	30		20	200		20	200	nA
	$T_{AMIN} \leq T_A \leq T_{AMAX}$			70		85	500			300	nA
Average Input Offset Current Drift				0.5							nA/°C
Input Bias Current	T_A = 25°C		30	80		80	500		80	500	nA
	$T_{AMIN} \leq T_A \leq T_{AMAX}$			0.210			1.5			0.8	µA
Input Resistance	T_A = 25°C, V_S = ±20V	1.0	6.0		0.3	2.0		0.3	2.0		MΩ
	$T_{AMIN} \leq T_A \leq T_{AMAX}$, V_S = ±20V	0.5									MΩ
Input Voltage Range	T_A = 25°C							±12	±13		V
	$T_{AMIN} \leq T_A \leq T_{AMAX}$				±12	±13					V

Electrical Characteristics (Note 5) (Continued)

Parameter	Conditions	LM741A			LM741			LM741C			Units
		Min	Typ	Max	Min	Typ	Max	Min	Typ	Max	
Large Signal Voltage Gain	$T_A = 25°C$, $R_L \geq 2$ kΩ										
	$V_S = \pm20V$, $V_O = \pm15V$	50									V/mV
	$V_S = \pm15V$, $V_O = \pm10V$				50	200		20	200		V/mV
	$T_{AMIN} \leq T_A \leq T_{AMAX}$,										
	$R_L \geq 2$ kΩ,										
	$V_S = \pm20V$, $V_O = \pm15V$	32									V/mV
	$V_S = \pm15V$, $V_O = \pm10V$				25			15			V/mV
	$V_S = \pm5V$, $V_O = \pm2V$	10									V/mV
Output Voltage Swing	$V_S = \pm20V$										
	$R_L \geq 10$ kΩ	±16									V
	$R_L \geq 2$ kΩ	±15									V
	$V_S = \pm15V$										
	$R_L \geq 10$ kΩ				±12	±14		±12	±14		V
	$R_L \geq 2$ kΩ				±10	±13		±10	±13		V
Output Short Circuit	$T_A = 25°C$	10	25	35		25			25		mA
Current	$T_{AMIN} \leq T_A \leq T_{AMAX}$	10		40							mA
Common-Mode	$T_{AMIN} \leq T_A \leq T_{AMAX}$										
Rejection Ratio	$R_S \leq 10$ kΩ, $V_{CM} = \pm12V$				70	90		70	90		dB
	$R_S \leq 50\Omega$, $V_{CM} = \pm12V$	80	95								dB
Supply Voltage Rejection	$T_{AMIN} \leq T_A \leq T_{AMAX}$,										
Ratio	$V_S = \pm20V$ to $V_S = \pm5V$										
	$R_S \leq 50\Omega$	86	96								dB
	$R_S \leq 10$ kΩ				77	96		77	96		dB
Transient Response	$T_A = 25°C$, Unity Gain										
Rise Time			0.25	0.8		0.3			0.3		µs
Overshoot			6.0	20		5			5		%
Bandwidth (Note 6)	$T_A = 25°C$	0.437	1.5								MHz
Slew Rate	$T_A = 25°C$, Unity Gain	0.3	0.7			0.5			0.5		V/µs
Supply Current	$T_A = 25°C$					1.7	2.8		1.7	2.8	mA
Power Consumption	$T_A = 25°C$										
	$V_S = \pm20V$		80	150							mW
	$V_S = \pm15V$					50	85		50	85	mW
LM741A	$V_S = \pm20V$										
	$T_A = T_{AMIN}$			165							mW
	$T_A = T_{AMAX}$			135							mW
LM741	$V_S = \pm15V$										
	$T_A = T_{AMIN}$					60	100				mW
	$T_A = T_{AMAX}$					45	75				mW

Note 2: "Absolute Maximum Ratings" indicate limits beyond which damage to the device may occur. Operating Ratings indicate conditions for which the device is functional, but do not guarantee specific performance limits.

Electrical Characteristics (Note 5) (Continued)

Note 3: For operation at elevated temperatures, these devices must be derated based on thermal resistance, and T_j max. (listed under "Absolute Maximum Ratings"). $T_j = T_A + (\theta_{jA} P_D)$.

Thermal Resistance	Cerdip (J)	DIP (N)	HO8 (H)	SO-8 (M)
θ_{jA} (Junction to Ambient)	100°C/W	100°C/W	170°C/W	195°C/W
θ_{jC} (Junction to Case)	N/A	N/A	25°C/W	N/A

Note 4: For supply voltages less than ±15V, the absolute maximum input voltage is equal to the supply voltage.

Note 5: Unless otherwise specified, these specifications apply for $V_S = \pm 15V$, $-55°C \leq T_A \leq +125°C$ (LM741/LM741A). For the LM741C/LM741E, these specifications are limited to $0°C \leq T_A \leq +70°C$.

Note 6: Calculated value from: BW (MHz) = 0.35/Rise Time(µs).

Note 7: For military specifications see RETS741X for LM741 and RETS741AX for LM741A.

Note 8: Human body model, 1.5 kΩ in series with 100 pF.

Schematic Diagram

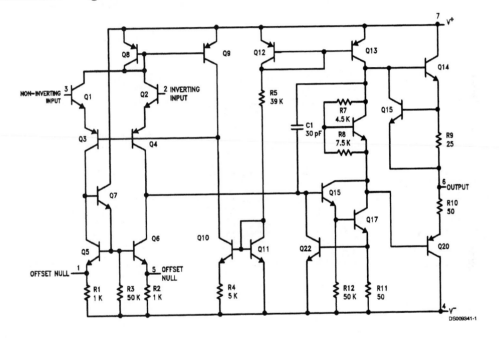

DS009341-1

MC7800, MC7800A, LM340, LM340A Series

Three-Terminal Positive Voltage Regulators

These voltage regulators are monolithic integrated circuits designed as fixed–voltage regulators for a wide variety of applications including local, on–card regulation. These regulators employ internal current limiting, thermal shutdown, and safe–area compensation. With adequate heatsinking they can deliver output currents in excess of 1.0 A. Although designed primarily as a fixed voltage regulator, these devices can be used with external components to obtain adjustable voltages and currents.

- Output Current in Excess of 1.0 A
- No External Components Required
- Internal Thermal Overload Protection
- Internal Short Circuit Current Limiting
- Output Transistor Safe–Area Compensation
- Output Voltage Offered in 2% and 4% Tolerance
- Available in Surface Mount D^{2}PAK and Standard 3–Lead Transistor Packages

ON Semiconductor

http://onsemi.com

TO–220
T SUFFIX
CASE 221A

Heatsink surface connected to Pin 2.

D^{2}PAK
D2T SUFFIX
CASE 936

Pin 1. Input
2. Ground
3. Output

Heatsink surface (shown as terminal 4 in case outline drawing) is connected to Pin 2.

MAXIMUM RATINGS (T_A = 25°C, unless otherwise noted.)

Rating	Symbol	Value	Unit
Input Voltage (5.0 – 18 V) (24 V)	V_I	35 40	Vdc
Power Dissipation Case 221A T_A = 25°C	P_D	Internally Limited	W
Thermal Resistance, Junction–to–Ambient	$R_{\theta JA}$	65	°C/W
Thermal Resistance, Junction–to–Case	$R_{\theta JC}$	5.0	°C/W
Case 936 (D^{2}PAK) T_A = 25°C	P_D	Internally Limited	W
Thermal Resistance, Junction–to–Ambient	$R_{\theta JA}$	See Figure 13	°C/W
Thermal Resistance, Junction–to–Case	$R_{\theta JA}$	5.0	°C/W
Storage Junction Temperature Range	T_{stg}	–65 to +150	°C
Operating Junction Temperature	T_J	+150	°C

NOTE: ESD data available upon request.

STANDARD APPLICATION

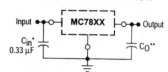

A common ground is required between the input and the output voltages. The input voltage must remain typically 2.0 V above the output voltage even during the low point on the input ripple voltage.

XX, These two digits of the type number indicate nominal voltage.

  * C_{in} is required if regulator is located an appreciable distance from power supply filter.

  ** C_O is not needed for stability; however, it does improve transient response. Values of less than 0.1 μF could cause instability.

ORDERING INFORMATION

See detailed ordering and shipping information in the package dimensions section on page 16 of this data sheet.

DEVICE MARKING INFORMATION

See general marking information in the device marking section on page 18 of this data sheet.

1

Publication Order Number:
MC7800/D

MC7800, MC7800A, LM340, LM340A Series

ELECTRICAL CHARACTERISTICS (V_{in} = 10 V, I_O = 500 mA, T_J = T_{low} to T_{high} [Note 1.], unless otherwise noted.)

Characteristic	Symbol	MC7805B			MC7805C/LM340T–5			Unit
		Min	Typ	Max	Min	Typ	Max	
Output Voltage (T_J = 25°C)	V_O	4.8	5.0	5.2	4.8	5.0	5.2	Vdc
Output Voltage (5.0 mA ≤ I_O ≤ 1.0 A, P_D ≤ 15 W) 7.0 Vdc ≤ V_{in} ≤ 20 Vdc 8.0 Vdc ≤ V_{in} ≤ 20 Vdc	V_O	–	–	–	4.75	5.0	5.25	Vdc
		4.75	5.0	5.25	–	–	–	
Line Regulation (Note 2.) 7.5 Vdc ≤ V_{in} ≤ 20 Vdc, 1.0 A 8.0 Vdc ≤ V_{in} ≤ 12 Vdc	Reg_{line}	– –	5.0 1.3	100 50	– –	0.5 0.8	20 10	mV
Load Regulation (Note 2.) 5.0 mA ≤ I_O ≤ 1.0 A 5.0 mA ≤ I_O ≤ 1.5 A (T_A = 25°C)	Reg_{load}	– –	1.3 0.15	100 50	– –	1.3 1.3	25 25	mV
Quiescent Current	I_B	–	3.2	8.0	–	3.2	6.5	mA
Quiescent Current Change 7.0 Vdc ≤ V_{in} ≤ 25 Vdc 5.0 mA ≤ I_O ≤ 1.0 A (T_A = 25°C)	ΔI_B	– –	– –	– 0.5	– –	0.3 0.08	1.0 0.8	mA
Ripple Rejection 8.0 Vdc ≤ V_{in} ≤ 18 Vdc, f = 120 Hz	RR	–	68	–	62	83	–	dB
Dropout Voltage (I_O = 1.0 A, T_J = 25°C)	$V_I - V_O$	–	2.0	–	–	2.0	–	Vdc
Output Noise Voltage (T_A = 25°C) 10 Hz ≤ f ≤ 100 kHz	V_n	–	10	–	–	10	–	μV/V_O
Output Resistance f = 1.0 kHz	r_O	–	0.9	–	–	0.9	–	mΩ
Short Circuit Current Limit (T_A = 25°C) V_{in} = 35 Vdc	I_{SC}	–	0.2	–	–	0.6	–	A
Peak Output Current (T_J = 25°C)	I_{max}	–	2.2	–	–	2.2	–	A
Average Temperature Coefficient of Output Voltage	TCV_O	–	–0.3	–	–	–0.3	–	mV/°C

ELECTRICAL CHARACTERISTICS (V_{in} = 10 V, I_O = 1.0 A, T_J = T_{low} to T_{high} [Note 1.], unless otherwise noted.)

Characteristic	Symbol	MC7805AC/LM340AT–5			Unit
		Min	Typ	Max	
Output Voltage (T_J = 25°C)	V_O	4.9	5.0	5.1	Vdc
Output Voltage (5.0 mA ≤ I_O ≤ 1.0 A, P_D ≤ 15 W) 7.5 Vdc ≤ V_{in} ≤ 20 Vdc	V_O	4.8	5.0	5.2	Vdc
Line Regulation (Note 2.) 7.5 Vdc ≤ V_{in} ≤ 25 Vdc, I_O = 500 mA 8.0 Vdc ≤ V_{in} ≤ 12 Vdc, I_O = 1.0 A 8.0 Vdc ≤ V_{in} ≤ 12 Vdc, I_O = 1.0 A, T_J = 25°C 7.3 Vdc ≤ V_{in} ≤ 20 Vdc, I_O = 1.0 A, T_J = 25°C	Reg_{line}	– – – –	0.5 0.8 1.3 4.5	10 12 4.0 10	mV
Load Regulation (Note 2.) 5.0 mA ≤ I_O ≤ 1.5 A, T_J = 25°C 5.0 mA ≤ I_O ≤ 1.0 A 250 mA ≤ I_O ≤ 750 mA	Reg_{load}	– – –	1.3 0.8 0.53	25 25 15	mV
Quiescent Current	I_B	–	3.2	6.0	mA
Quiescent Current Change 8.0 Vdc ≤ V_{in} ≤ 25 Vdc, I_O = 500 mA 7.5 Vdc ≤ V_{in} ≤ 20 Vdc, T_J = 25°C 5.0 mA ≤ I_O ≤ 1.0 A	ΔI_B	– – –	0.3 – 0.08	0.8 0.8 0.5	mA

1. T_{low} = 0°C for MC78XXAC, C, LM340AT–XX, LM340T–XX T_{high} = +125°C for MC78XXAC, C, LM340AT–XX, LM340T–XX
   = –40°C for MC78XXB
2. Load and line regulation are specified at constant junction temperature. Changes in V_O due to heating effects must be taken into account separately. Pulse testing with low duty cycle is used.

 **MOTOROLA**

Three-Terminal Negative Voltage Regulators

MC7900 Series

THREE–TERMINAL NEGATIVE FIXED VOLTAGE REGULATORS

The MC7900 series of fixed output negative voltage regulators are intended as complements to the popular MC7800 series devices. These negative regulators are available in the same seven–voltage options as the MC7800 devices. In addition, one extra voltage option commonly employed in MECL systems is also available in the negative MC7900 series.

Available in fixed output voltage options from −5.0 V to −24 V, these regulators employ current limiting, thermal shutdown, and safe–area compensation – making them remarkably rugged under most operating conditions. With adequate heatsinking they can deliver output currents in excess of 1.0 A.

- No External Components Required
- Internal Thermal Overload Protection
- Internal Short Circuit Current Limiting
- Output Transistor Safe–Area Compensation
- Available in 2% Voltage Tolerance (See Ordering Information)

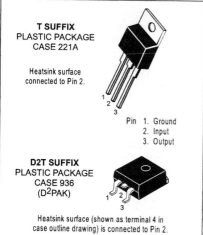

T SUFFIX
PLASTIC PACKAGE
CASE 221A

Heatsink surface connected to Pin 2.

Pin 1. Ground
    2. Input
    3. Output

D2T SUFFIX
PLASTIC PACKAGE
CASE 936
(D2PAK)

Heatsink surface (shown as terminal 4 in case outline drawing) is connected to Pin 2.

Representative Schematic Diagram

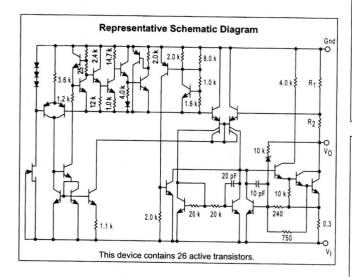

This device contains 26 active transistors.

STANDARD APPLICATION

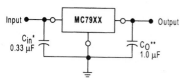

A common ground is required between the input and the output voltages. The input voltage must remain typically 2.0 V above more negative even during the high point of the input ripple voltage.

XX, These two digits of the type number indicate nominal voltage.
 * C_{in} is required if regulator is located an appreciable distance from power supply filter.
 ** C_O improve stability and transient response.

ORDERING INFORMATION

Device	Output Voltage Tolerance	Operating Temperature Range	Package
MC79XXACD2T	2%		Surface Mount
MC79XXCD2T	4%	$T_J = 0°$ to $+125°C$	Surface Mount
MC79XXACT	2%		Insertion Mount
MC79XXCT	4%		Insertion Mount
MC79XXBD2T	4%	$T_J = −40°$ to $+125°C$	Surface Mount
MC79XXBT	4%		Insertion Mount

XX indicates nominal voltage.

DEVICE TYPE/NOMINAL OUTPUT VOLTAGE

MC7905	5.0 V	MC7912	12 V
MC7905.2	5.2 V	MC7915	15 V
MC7906	6.0 V	MC7918	18 V
MC7908	8.0 V	MC7924	24 V

Rev 6

MC7900

MAXIMUM RATINGS (T$_A$ = +25°C, unless otherwise noted.)

Rating	Symbol	Value	Unit
Input Voltage (−5.0 V ≥ V$_O$ ≥ −18 V) (24 V)	V$_I$	−35 −40	Vdc

Power Dissipation
Case 221A
  T$_A$ = +25°C
  Thermal Resistance, Junction−to−Ambient
  Thermal Resistance, Junction−to−Case
Case 936 (D^{2}PAK)
  T$_A$ = +25°C
  Thermal Resistance, Junction−to−Ambient
  Thermal Resistance, Junction−to−Case
	
	P$_D$
	θ$_{JA}$
	θ$_{JC}$
	
	P$_D$
	θ$_{JA}$
	θ$_{JC}$
		
		Internally Limited
		65
		5.0
		
		Internally Limited
		70
		5.0
			
			W
			°C/W
			°C/W
			
			W
			°C/W
			°C/W

Storage Junction Temperature Range	T$_{stg}$	−65 to +150	°C
Junction Temperature	T$_J$	+150	°C

THERMAL CHARACTERISTICS

Characteristics	Symbol	Max	Unit
Thermal Resistance, Junction−to−Ambient	R$_{θJA}$	65	°C/W
Thermal Resistance, Junction−to−Case	R$_{θJC}$	5.0	°C/W

MC7905C

ELECTRICAL CHARACTERISTICS (V$_I$ = −10 V, I$_O$ = 500 mA, 0°C < T$_J$ < +125°C, unless otherwise noted.)

Characteristics	Symbol	Min	Typ	Max	Unit
Output Voltage (T$_J$ = +25°C)	V$_O$	−4.8	−5.0	−5.2	Vdc

Line Regulation (Note 1)
  (T$_J$ = +25°C, I$_O$ = 100 mA)
    −7.0 Vdc ≥ V$_I$ ≥ −25 Vdc
    −8.0 Vdc ≥ V$_I$ ≥ −12 Vdc
  (T$_J$ = +25°C, I$_O$ = 500 mA)
    −7.0 Vdc ≥ V$_I$ ≥ −25 Vdc
    −8.0 Vdc ≥ V$_I$ ≥ −12 Vdc
	Reg$_{line}$	
		
		−
		−
		
		−
		−
			
			7.0
			2.0
			
			35
			8.0
				
				50
				25
				
				100
				50
					mV

Load Regulation, T$_J$ = +25°C (Note 1)
  5.0 mA ≤ I$_O$ ≤ 1.5 A
  250 mA ≤ I$_O$ ≤ 750 mA
	Reg$_{load}$	
		−
		−
			
			11
			4.0
				
				100
				50
					mV

Output Voltage −7.0 Vdc ≥ V$_I$ ≥ −20 Vdc, 5.0 mA ≤ I$_O$ ≤ 1.0 A, P ≤ 15 W	V$_O$	−4.75	−	−5.25	Vdc
Input Bias Current (T$_J$ = +25°C)	I$_{IB}$	−	4.3	8.0	mA

Input Bias Current Change
  −7.0 Vdc ≥ V$_I$ ≥ −25 Vdc
  5.0 mA ≤ I$_O$ ≤ 1.5 A
	ΔI$_{IB}$	
		−
		−
			
			−
			−
				
				1.3
				0.5
					mA

Output Noise Voltage (T$_A$ = +25°C, 10 Hz ≤ f ≤ 100 kHz)	V$_n$	−	40	−	µV
Ripple Rejection (I$_O$ = 20 mA, f = 120 Hz)	RR	−	70	−	dB
Dropout Voltage I$_O$ = 1.0 A, T$_J$ = +25°C	V$_I$–V$_O$	−	2.0	−	Vdc
Average Temperature Coefficient of Output Voltage I$_O$ = 5.0 mA, 0°C ≤ T$_J$ ≤ +125°C	ΔV$_O$/ΔT	−	−1.0	−	mV/°C

NOTE: 1. Load and line regulation are specified at constant junction temperature. Changes in V$_O$ due to heating effects must be taken into account separately. Pulse testing with low duty cycle is used.

3656

Transformer Coupled
ISOLATION AMPLIFIER

FEATURES

● **INTERNAL ISOLATED POWER**
● **8000V ISOLATION TEST VOLTAGE**
● **0.5μA MAX LEAKAGE AT 120V, 60Hz**
● **3-PORT ISOLATION**
● **IMR: 125dB REJECTION AT 60Hz**
● **1" x 1" x 0.25" CERAMIC PACKAGE**

APPLICATIONS

● **MEDICAL**
  Patient Monitoring and Diagnostic
  Instrumentation
● **INDUSTRIAL**
  Ground Loop Elimination and
  Off-ground Signal Measurement
● **NUCLEAR**
  Input/Output/Power Isolation

DESCRIPTION

The 3656 was the first amplifier to provide a total isolation function, both signal and power isolation, in integrated circuit form. This remarkable advancement in analog signal processing capability was accomplished by use of a patented modulation technique and miniature hybrid transformer.

Versatility and performance are outstanding features of the 3656. It is capable of operating with three completely independent grounds (three-port isolation). In addition, the isolated power generated is available to power external circuitry at either the input or output. The uncommitted op amps at the input and the output allow a wide variety of closed-loop configurations to match the requirements of many different types of isolation applications.

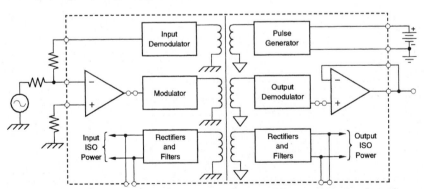

International Airport Industrial Park • Mailing Address: PO Box 11400, Tucson, AZ 85734 • Street Address: 6730 S. Tucson Blvd., Tucson, AZ 85706 • Tel: (520) 746-1111 • Twx: 910-952-1111
Internet: http://www.burr-brown.com/ • FAXLine: (800) 548-6133 (US/Canada Only) • Cable: BBRCORP • Telex: 066-6491 • FAX: (520) 889-1510 • Immediate Product Info: (800) 548-6132

©1987 Burr-Brown Corporation   PDS-403G   Printed in U.S.A. January, 1997

SPECIFICATIONS

ELECTRICAL

At +25°C, V± = 15VDC and 15VDC between P+ and P–, unless otherwise specified.

| PARAMETER | CONDITIONS | 3656AG, BG, HG, JG, KG | | | UNITS |
		MIN	TYP	MAX	
ISOLATION					
Voltage					
Rated Continuous[1], DC		3500 (1000)			VDC
Test, 10s[1]		8000 (3000)			VDC
Test, 60s[1]	G_1 = 10V/V	2000 (700)			Vrms
Rejection					
DC			160		dB
60Hz, < 100Ω in I/P Com[2]			125		dB
60Hz, 5kΩ in I/P Com[2]					
3656HG		108			dB
3656AG, BG, JG, KG		112			dB
Capacitance[1]			6 (6.3)		pF
Resistance[1]			10^{12} (10^{12})		Ω
Leakage Current	120V, 60Hz		0.28	0.5	µA
GAIN					
Equations	See Text				
Accuracy of Equations					
Initial[3] 3656HG	G < 100V/V			1.5	%
3656AG, JG, KG				1	%
3656BG				0.3	%
vs Temperature 3656HG				480	ppm/°C
3656AG, JG				120	ppm/°C
3656BG, KG				60	ppm/°C
vs Time			0.02 (1 + log khrs.)		%
Nonlinearity	$R_A + R_F = R_B \geq 2M\Omega$				
External Supplies Used at					
Pins 12 and 16, 3656HG	Unipolar or Bipolar Output			±0.15	%
3656AG, JG, KG				±0.1	%
3656BG				±0.05	%
Internal Supplies Used for	Bipolar Output Voltage				
Output Stage	Swing, Full Load[4]		±0.15		%
OFFSET VOLTAGE[5], RTI					
Initial[3], 3656HG	15Vp between P+ and P–			±[4 + (40/G_1)]	mV
3656AG, JG				±[2 + (20/G_1)]	mV
3656BG, KG				±[1 + (10/G_1)]	mV
vs Temperature, 3656HG				±[200 + (1000/G_1)]	µV/°C
3656JG				±[50 + (750/G_1)]	µV/°C
3656AG				±[25 + (500/G_1)]	µV/°C
3656KG				±[10 + (350/G_1)]	µV/°C
3656BG				±[5 + (350/G_1)]	µV/°C
vs Supply Voltage	Supply between P+ and P–				
3656HG				±[0.6 + (3.5/G_1)]	mV/V
3656AG, BG, JG, KG				±[0.3 + (2.1/G_1)]	mV/V
vs Current[6]			±[0.1 + (10/G_1)]	±[0.2 + (20/G_1)]	mV/mA
vs Time		±[10 + (100/G_1)] • (1 + log khrs.)			µV
AMPLIFIER PARAMETERS, Apply to A_1 and A_2					
Bias Current[7]					
Initial				100	nA
vs Temperature			0.5		nA/°C
vs Supply			0.2		nA/V
Offset Current[7]			5	20	nA
Impedance	Common-Mode		100 || 5		MΩ || pF
Input Noise Voltage	f_B = 0.05Hz to 100Hz		5		µVp-p
	f_B = 10Hz to 10kHz		5		µVrms
Input Voltage Range[8]					
Linear Operation	Internal Supply			±5	V
	External Supply			Supply –5	V
Output Current	V_{OUT} = ±5V				
	±15V External Supply	±5			mA
	Internal Supply	±2.5			mA
	V_{OUT} = ±10V				
	±15V External Supply	±2.5			mA
	V_{OUT} = ±2V, $V_{P+, P-}$ = 8.5V				
	Internal Supply		±1		mA
Quiescent Current			150	450	µA

ELECTRICAL

At +25°C, V± = 15VDC and 15VDC between P+ and P−, unless otherwise specified.

PARAMETER	CONDITIONS	3656AG, BG, HG, JG, KG			UNITS
		MIN	TYP	MAX	
FREQUENCY RESPONSE					
±3dB Response	Small Signal		30		kHz
Full Power			1.3		kHz
Slew Rate	Direction Measured at Output	+0.1, −0.04			V/µs
Settling Time	to 0.05%		500		µs
OUTPUT					
Noise Voltage (RTI)	f_B = 0.05Hz to 100Hz		$\sqrt{(5)^2 + (22/G_1)^2}$		µVp-p
	f_B = 10Hz to 10kHz		$\sqrt{(5)^2 + (11/G_1)^2}$		µVrms
Residual Ripple[9]			5		mVp-p
POWER SUPPLY IN, at P+, P−					
Rated Performance			15		VDC
Voltage Range[10]	Derated Performance	8.5		16	VDC
Ripple Current[9]			10	25	mAp-p
Quiescent Current[11]	Average		14	18	mA/DC
Current vs Load Current[12]	vs Current from +V, −V, V+, V−		0.7		mA/mA
ISOLATED POWER OUT, At +V, −V, V+, V− pins[13]					
Voltage, No Load	15V Between P+ and P−	8.5	9	9.5	V
Voltage, Full Load	±5mA (10mA sum) Load[12]	7	8	9	V
Voltage vs Power Supply	vs Supply Between P+ and P−		0.66		V/V
Ripple Voltage[9]					
No Load			40		mVp-p
Full Load	±5mA Load		80	200	mVp-p
TEMPERATURE RANGE					
Specification 3656AG, BG		−25		+85	°C
3656HG, JG, KG		0		+70	°C
Operation[10]		−55		+100	°C
Storage[14]		−65		+125	°C

NOTES: (1) Ratings in parenthesis are between P− (pin 20) and O/P Com (pin 17). Other isolation ratings are between I/P Com and O/P Com or I/P Com and P−. (2) See Performance Curves. (3) May be trimmed to zero. (4) If output swing is unipolar, or if the output is not loaded, specification same as if external supply were used. (5) Includes effects of A_1 and A_2 offset voltages and bias currents if recommended resistors used. (6) Versus the sum of all external currents drawn from V+, V−, +V, −V (= ISO). (7) Effects of A_1 and A_2 bias currents and offset currents are included in Offset Voltage specifications. (8) With respect to I/P Com (pin 3) for A_1 and with respect to O/P Com (pin 17) for A_2. CMR for A_1 and A_2 is 100dB, typical. (9) In configuration of Figure 3. Ripple frequency approximately 750kHz. Measurement bandwidth is 30kHz. (10) Decreases linearly from 16VDC at 85°C to 12VDC at 100°C. (11) Instantaneous peak current required from pins 19 and 20 at turn-on is 100mA for slow rising voltages (50ms) and 300mA for fast rises (50µs). (12) Load current is sum drawn form +V, −V, V+, V− (= I_{ISO}). (13) Maximum voltage rating at pins 1 and 4 is ±18VDC; maximum voltage rating at pins 12 and 16 is ±18VDC. (14) Isolation ratings may degrade if exposed to 125°C for more than 1000 hours or 90°C for more than 50,000 hours.

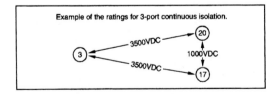

Example of the ratings for 3-port continuous isolation.

PACKAGE INFORMATION

PRODUCT	PACKAGE	PACKAGE DRAWING NUMBER[1]
3656	20-Lead ISO Omni	102A

NOTE: (1) For detailed drawing and dimension table, please see end of data sheet, or Appendix C of Burr-Brown IC Data Book.

PIN DESIGNATIONS

NO.	DESCRIPTION	NO.	DESCRIPTION
1	+V	11	Output DEMOD
2	MOD Input	12	V−
3	Input DEMOD COM	13	A_2 Noninverting Input
4	−V	14	A_2 Inverting Input
5	Balance	15	A_2 Output
6	A_1 Inverting Input	16	V+
7	A_1 Noninverting Input	17	Output DEMOD COM
8	Balance	18	No Pin
9	A_1 Output	19	P+
10	Input DEMOD	20	P−

ABSOLUTE MAXIMUM RATINGS

Supply Without Damage	16V
Input Voltage Range Using Internal Supply	±8V
Input Voltage Range Using External Supply	Supply
Continuous Isolation Voltage[1]	3500, (1000) VDC
Storage Temperature	−65°C to +125°C
Lead Temperature, (soldering, 10s)	+300°C

NOTE: (1) Ratings in parenthesis are between P− (pin 20) and O/P Com (pin 17). Other isolation ratings are between I/P Com and O/P Com or I/P Com and P−.

BURR-BROWN® **3656** **BB**

Low Cost
Instrumentation Amplifier

AD622

CONNECTION DIAGRAM

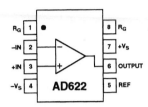

FEATURES
Easy to Use
Low Cost Solution
Higher Performance than Two or Three Op Amp Design
Unity Gain with No External Resistor
Optional Gains with One External Resistor
  (Gain Range 2 to 1000)
Wide Power Supply Range (±2.6 V to ±15 V)
Available in 8-Lead PDIP and SOIC
Low Power, 1.5 mA max Supply Current

GOOD DC PERFORMANCE
0.15% Gain Accuracy (G = 1)
125 μV max Input Offset Voltage
1.0 μV/°C max Input Offset Drift
5 nA max Input Bias Current
66 dB min Common-Mode Rejection Ratio (G = 1)

NOISE
12 nV/√Hz @ 1 kHz Input Voltage Noise
0.60 μV p-p Noise (0.1 Hz to 10 Hz, G = 10)

EXCELLENT AC CHARACTERISTICS
800 kHz Bandwidth (G = 10)
10 μs Settling Time to 0.1% @ G = 1–100
1.2 V/μs Slew Rate

APPLICATIONS
Transducer Interface
Low Cost Thermocouple Amplifier
Industrial Process Controls
Difference Amplifier
Low Cost Data Acquisition

PRODUCT DESCRIPTION
The AD622 is a low cost, moderately accurate instrumentation amplifier that requires only one external resistor to set any gain between 2 and 1,000. Or for a gain of 1, no external resistor is required. The AD622 is a complete difference or subtracter amplifier "system" while providing superior linearity and common-mode rejection by incorporating precision laser trimmed resistors.

The AD622 replaces low cost, discrete, two or three op amp instrumentation amplifier designs and offers good common-mode rejection, superior linearity, temperature stability, reliability, and board area consumption. The low cost of the AD622 eliminates the need to design discrete instrumentation amplifiers to meet stringent cost targets. While providing a lower cost solution, it also provides performance and space improvements.

REV. C

One Technology Way, P.O. Box 9106, Norwood, MA 02062-9106, U.S.A.
Tel: 781/329-4700   World Wide Web Site: http://www.analog.com
Fax: 781/326-8703   © Analog Devices, Inc., 1999

Model	Conditions	AD622 Min	AD622 Typ	AD622 Max	Units
GAIN	$G = 1 + (50.5\ k/R_G)$				
Gain Range		1		1000	
Gain Error[1]	$V_{OUT} = \pm 10$ V				
G = 1			0.05	0.15	%
G = 10			0.2	0.50	%
G = 100			0.2	0.50	%
G = 1000			0.2	0.50	%
Nonlinearity,	$V_{OUT} = \pm 10$ V				
G = 1–1000	$R_L = 10$ kΩ		10		ppm
G = 1–100	$R_L = 2$ kΩ		10		ppm
Gain vs. Temperature	Gain = 1			10	ppm/°C
	Gain >1[1]			−50	ppm/°C
VOLTAGE OFFSET	(Total RTI Error = $V_{OSI} + V_{OSO}/G$)				
Input Offset, V_{OSI}	$V_S = \pm 5$ V to ± 15 V		60	125	μV
Average TC	$V_S = \pm 5$ V to ± 15 V			1.0	μV/°C
Output Offset, V_{OSO}	$V_S = \pm 5$ V to ± 15 V		600	1500	μV
Average TC	$V_S = \pm 5$ V to ± 15 V			15	μV/°C
Offset Referred to the					
Input vs.					
Supply (PSR)	$V_S = \pm 5$ V to ± 15 V				
G = 1		80	100		dB
G = 10		95	120		dB
G = 100		110	140		dB
G = 1000		110	140		dB
INPUT CURRENT					
Input Bias Current			2.0	5.0	nA
Average TC			3.0		pA/°C
Input Offset Current			0.7	2.5	nA
Average TC			2.0		pA/°C
INPUT					
Input Impedance					
Differential			10‖2		GΩ‖pF
Common-Mode			10‖2		GΩ‖pF
Input Voltage Range[2]	$V_S = \pm 2.6$ V to ± 5 V	$-V_S + 1.9$		$+V_S - 1.2$	V
Over Temperature		$-V_S + 2.1$		$+V_S - 1.3$	V
	$V_S = \pm 5$ V to ± 18 V	$-V_S + 1.9$		$+V_S - 1.4$	V
Over Temperature		$-V_S + 2.1$		$+V_S - 1.4$	V
Common-Mode Rejection					
Ratio DC to 60 Hz with					
1 kΩ Source Imbalance	$V_{CM} = 0$ V to ± 10 V				
G = 1		66	78		dB
G = 10		86	98		dB
G = 100		103	118		dB
G = 1000		103	118		dB
OUTPUT					
Output Swing	$R_L = 10$ kΩ,				
	$V_S = \pm 2.6$ V to ± 5 V	$-V_S + 1.1$		$+V_S - 1.2$	V
Over Temperature		$-V_S + 1.4$		$+V_S - 1.3$	V
	$V_S = \pm 5$ V to ± 18 V	$-V_S + 1.2$		$+V_S - 1.4$	V
Over Temperature		$-V_S + 1.6$		$+V_S - 1.5$	V
Short Current Circuit			± 18		mA

REV. C

Model	Conditions	AD622			Units
		Min	Typ	Max	
DYNAMIC RESPONSE					
Small Signal –3 dB Bandwidth					
G = 1			1000		kHz
G = 10			800		kHz
G = 100			120		kHz
G = 1000			12		kHz
Slew Rate			1.2		V/μs
Settling Time to 0.1%	10 V Step				
G = 1–100			10		μs
NOISE					
Voltage Noise, 1 kHz	$Total\ RTI\ Noise = \sqrt{(e^2_{ni}) + (e_{no}/G)^2}$				
Input, Voltage Noise, e_{ni}			12		nV/√Hz
Output, Voltage Noise, e_{no}			72		nV/√Hz
RTI, 0.1 Hz to 10 Hz					
G = 1			4.0		μV p-p
G = 10			0.6		μV p-p
G = 100–1000			0.3		μV p-p
Current Noise	f = 1 kHz		100		fA/√Hz
0.1 Hz to 10 Hz			10		pA p-p
REFERENCE INPUT					
R_{IN}			20		kΩ
I_{IN}	V_{IN+}, V_{REF} = 0		+50	+60	μA
Voltage Range		$-V_S$ + 1.6		$+V_S$ – 1.6	V
Gain to Output			1 ± 0.0015		
POWER SUPPLY					
Operating Range[3]		±2.6		±18	V
Quiescent Current	V_S = ±2.6 V to ±18 V		0.9	1.3	mA
Over Temperature			1.1	1.5	mA
TEMPERATURE RANGE					
For Specified Performance			–40 to +85		°C

NOTES
[1]Does not include effects of external resistor R_G.
[2]One input grounded. G = 1.
[3]This is defined as the same supply range that is used to specify PSR.
Specifications subject to change without notice.

Voltage-to-Frequency and Frequency-to-Voltage Converter

AD650

FEATURES
V/F Conversion to 1MHz
Reliable Monolithic Construction
Very Low Nonlinearity
0.002% typ at 10kHz
0.005% typ at 100kHz
0.07% typ at 1MHz
Input Offset Trimmable to Zero
CMOS or TTL Compatible
Unipolar, Bipolar, or Differential V/F
V/F or F/V Conversion
Available in Surface Mount
MIL-STD-883-Compliant Versions Available

PIN CONFIGURATION

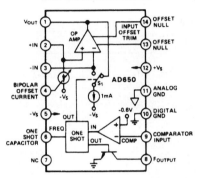

PRODUCT DESCRIPTION
The AD650 V/F/V (voltage-to-frequency or frequency-to-voltage converter) provides a combination of high frequency operation and low nonlinearity previously unavailable in monolithic form. The inherent monotonicity of the V/F transfer function makes the AD650 useful as a high-resolution analog-to-digital converter. A flexible input configuration allows a wide variety of input voltage and current formats to be used, and an open-collector output with separate digital ground allows simple interfacing to either standard logic families or opto-couplers.

The linearity error of the AD650 is typically 20ppm (0.002% of full scale) and 50ppm (0.005%) maximum at 10kHz full scale. This corresponds to approximately 14-bit linearity in an analog-to-digital converter circuit. Higher full-scale frequencies or longer count intervals can be used for higher resolution conversions. The AD650 has a useful dynamic range of six decades allowing extremely high resolution measurements. Even at 1MHz full scale, linearity is guaranteed less than 1000ppm (0.1%) on the AD650KN, KP, BD and SD grades.

In addition to analog-to-digital conversion, the AD650 can be used in isolated analog signal transmission applications, phased-locked-loop circuits, and precision stepper motor speed controllers. In the F/V mode, the AD650 can be used in precision tachometer and FM demodulator circuits.

The input signal range and full-scale output frequency are user-programmable with two external capacitors and one resistor. Input offset voltage can be trimmed to zero with an external potentiometer.

The AD650JN and AD650KN are offered in a plastic 14-pin DIP package. The AD650JP and AD650KP are available in a 20-pin plastic leaded chip carrier (PLCC). Both plastic packaged versions of the AD650 are specified for the commerical (0 to +70°C) temperature range. For industrial temperature range (−25°C to +85°C) applications, the AD650AD and AD650BD are offered in a ceramic package. The AD650SD is specified for the full −55°C to +125°C extended temperature range.

PRODUCT HIGHLIGHTS
1. In addition to very high linearity, the AD650 can operate at full scale output frequency up to 1MHz. The combination of these two features makes the AD650 an inexpensive solution for applications requiring high resolution monotonic A/D conversion.

2. The AD650 has a very versatile architecture that can be configured to accommodate bipolar, unipolar, or differential input voltages, or unipolar input currents.

3. TTL or CMOS compatibility is achieved using an open collector frequency output. The pullup resistor can be connected to voltages up to + 30V, or + 15V or + 5V for conventional CMOS or TTL logic levels.

4. The same components used for V/F conversion can also be used for F/V conversion by adding a simple logic biasing network and reconfiguring the AD650.

5. The AD650 provides separate analog and digital grounds. This feature allows prevention of ground loops in real-world applications.

6. The AD650 is available in versions compliant with MIL-STD-883. Refer to the Analog Devices Military Products Databook or current AD650/883B data sheet for detailed specifications.

AD650—SPECIFICATIONS (@ +25°C with $V_S = \pm 15V$ unless otherwise noted)

Model	AD650J/AD650A			AD650K/AD650B			AD650S			Units
	Min	Typ	Max	Min	Typ	Max	Min	Typ	Max	
DYNAMIC PERFORMANCE										
Full Scale Frequency Range			1			1			1	MHz
Nonlinearity[1] f_{max} = 10kHz		0.002	0.005		0.002	0.005		0.002	0.005	%
100kHz		0.005	**0.02**		0.005	**0.02**		0.005	**0.02**	%
500kHz		0.02	0.05		0.02	0.05		0.02	0.05	%
1MHz		0.1			0.05	**0.1**		0.05	**0.1**	%
Full Scale Calibration Error[2], 100kHz		±5			±5			±5		%
1MHz		±10			±10			±5		%
vs. Supply[3]	0.015		+0.015	−0.015		+0.015	−0.015		+0.015	% of FSR/V
vs. Temperature										
A, B, and S Grades										
at 10kHz			±75			±75			±75	ppm/°C
at 100kHz			±150			±150			±150	ppm/°C
J and K Grades										
at 10kHz		±75			±75					ppm/°C
at 100kHz		±150			±150					ppm/°C
BIPOLAR OFFSET CURRENT										
Activated by 1.24kΩ between pins 4 and 5	0.45	0.5	0.55	0.45	0.5	0.55	0.45	0.5	0.55	mA
DYNAMIC RESPONSE										
Maximum Settling Time for Full Scale										
Step Input	1 Pulse of New Frequency Plus 1µs			1 Pulse of New Frequency Plus 1µs			1 Pulse of New Frequency Plus 1µs			
Overload Recovery Time										
Step Input	1 Pulse of New Frequency Plus 1µs			1 Pulse of New Frequency Plus 1µs			1 Pulse of New Frequency Plus 1µs			
ANALOG INPUT AMPLIFIER (V/F Conversion)										
Current Input Range (Figure 1)	0		+0.6	0		+0.6	0		+0.6	mA
Voltage Input Range (Figure 5)	10		0	10		0	10		0	V
Differential Impedance	2MΩ‖10pF			2MΩ‖10pF			2MΩ‖10pF			
Common Mode Impedance	1000MΩ‖10pF			1000MΩ‖10pF			1000MΩ‖10pF			
Input Bias Current										
Noninverting Input		40	100		40	100		40	100	nA
Inverting Input		±8	±20		±8	±20		±8	±20	nA
Input Offset Voltage			±4			±4			±4	mV
(Trimmable to Zero)										
vs. Temperature (T_{min} to T_{max})		±30			±30			±30		µV/°C
Safe Input Voltage		±V_S			±V_S			±V_S		V
COMPARATOR (F/V Conversion)										
Logic "0" Level	V_S		1	V_S		1	V_S		+1	V
Logic "1" Level	0		+V_S	0		+V_S	0		+V_S	V
Pulse Width Range[4]	0.1		(0.3 · t_{OS})	0.1		(0.3 · t_{OS})	0.1		(0.3 · t_{OS})	µs
Input Impedance		250			250			250		kΩ
OPEN COLLECTOR OUTPUT (V/F Conversion)										
Output Voltage in Logic "0"										
I_{SINK} = 8mA, T_{min} to T_{max}			0.4			0.4			0.4	V
Output Leakage Current in Logic "1"			100			100			100	nA
Voltage Range[5]	0		+36	0		+36	0		+36	V
AMPLIFIER OUTPUT (F/V Conversion)										
Voltage Range (1500Ω min load resistance)	0		+10	0		+10	0		+10	V
Source Current (750Ω max load resistance)	10			10			10			mA
Capacitive Load (Without Oscillation)			100			100			100	pF
POWER SUPPLY										
Voltage, Rated Performance	±9		±18	±9		±18	±9		±18	V
Quiescent Current		8			8			8		mA
TEMPERATURE RANGE										
Rated Performance – N Package	0		+70	0		+70	55		+125	°C
D Package	25		+85	25		+85				°C
Storage – N Package	25		+85	25		+85				°C
D Package	65		+150	65		+150	65		+150	°C
PACKAGE OPTIONS[6]										
PLCC (P-20A)	AD650JP			AD650KP						
Plastic DIP (N-14)	AD650JN			AD650KN						
Ceramic DIP (D-14)	AD650AD			AD650BD			AD650SD			

NOTES
[1] Nonlinearity is defined as deviation from a straight line from zero to full scale, expressed as a fraction of full scale.
[2] Full scale calibration error adjustable to zero.
[3] Measured at full scale output frequency of 100kHz.
[4] Refer to F/V conversion section of the text.
[5] Referred to digital ground.
[6] D = Ceramic DIP; N = Plastic DIP; P = Plastic Leaded Chip Carrier. For outline information see Package Information section.

Specifications subject to change without notice.

Specifications shown in boldface are tested on all production units at final electrical test. Results from those tests are used to calculate outgoing quality levels. All min and max specifications are guaranteed, although only those shown in boldface are tested on all production units.

ABSOLUTE MAXIMUM RATINGS

Total Supply Voltage $+V_S$ to $-V_S$ 36V
Storage Temperature Ceramic . . . . $-55°C$ to $+165°C$
                  Plastic   . . . . $-25°C$ to $+125°C$
Differential Input Voltage (Pins 2 & 3) . . . . $\pm10V$
Maximum Input Voltage . . . . $\pm V_S$
Open Collector Output   Voltage Above Digital GND . . 36V
                     Current 50mA
Amplifier Short Ckt to Ground . . . . Indefinite
Comparator Input Voltage (Pin 9) . . . . $\pm V_S$

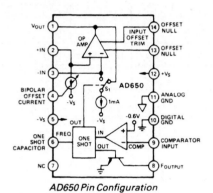

AD650 Pin Configuration

ORDERING GUIDE

Part[1] Number	Gain Tempco ppm/°C 100kHz	1MHz Linearity	Specified Temperature Range °C	Package
AD650JN	150 typ	0.1% typ	0 to +70	Plastic DIP
AD650KN	150 typ	0.1% max	0 to +70	Plastic DIP
AD650JP	150 typ	0.1% typ	0 to +70	PLCC
AD650KP	150 typ	0.1% max	0 to +70	PLCC
AD650AD	150 max	0.1% typ	−25 to +85	Ceramic
AD650BD	150 max	0.1% max	−25 to +85	Ceramic
AD650SD	150 max	0.1% max	−55 to +125	Ceramic

NOTE
[1]For details on grade and package offerings screened in accordance with MIL-STD-883, refer to the Analog Devices Military Products Databook or current AD650/883B data sheet.

8-Bit A/D Converter

AD673*

FEATURES
Complete 8-Bit A/D Converter with Reference, Clock
and Comparator
30µs Maximum Conversion Time
Full 8- or 16-Bit Microprocessor Bus Interface
Unipolar and Bipolar Inputs
No Missing Codes Over Temperature
Operates on +5V and −12V to −15V Supplies
MIL-STD-883 Compliant Version Available

FUNCTIONAL BLOCK DIAGRAM

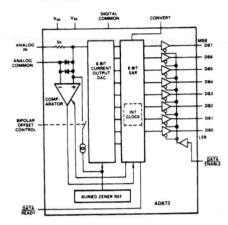

PRODUCT DESCRIPTION
The AD673 is a complete 8-bit successive approximation analog-to-digital converter consisting of a DAC, voltage reference, clock, comparator, successive approximation register (SAR) and 3 state output buffers–all fabricated on a single chip. No external components are required to perform a full accuracy 8-bit conversion in 20µs.

The AD673 incorporates advanced integrated circuit design and processing technologies. The successive approximation function is implemented with I²L (integrated injection logic). Laser trimming of the high stability SiCr thin film resistor ladder network insures high accuracy, which is maintained with a temperature compensated sub-surface Zener reference.

Operating on supplies of +5V and −12V to −15V, the AD673 will accept analog inputs of 0 to +10V or −5V to +5V. The trailing edge of a positive pulse on the CONVERT line initiates the 20µs conversion cycle. DATA READY indicates completion of the conversion.

The AD673 is available in two versions. The AD673J as specified over the 0 to +70°C temperature range and the AD673S guarantees ±½LSB relative accuracy and no missing codes from −55°C to +125°C.

Two package configurations are offered. All versions are also offered in a 20-pin hermetically sealed ceramic DIP. The AD673J is also available in a 20-pin plastic DIP.

PRODUCT HIGHLIGHTS
1. The AD673 is a complete 8-bit A/D converter. No external components are required to perform a conversion.

2. The AD673 interfaces to many popular microprocessors without external buffers or peripheral interface adapters.

3. The device offers true 8-bit accuracy and exhibits no missing codes over its entire operating temperature range.

4. The AD673 adapts to either unipolar (0 to +10V) or bipolar (−5V to +5V) analog inputs by simply grounding or opening a single pin.

5. Performance is guaranteed with +5V and −12V or −15V supplies.

6. The AD673 is available in a version compliant with MIL-STD-883. Refer to the Analog Devices Military Products Databook or current /883B data sheet for detailed specifications.

*Protected by U.S. Patent Nos. 3,940,760; 4,213,806; 4,136,349; 4,400,689; and 4,400,690

Model	AD673J			AD673S			Units
	Min	Typ	Max	Min	Typ	Max	
RESOLUTION		8			8		Bits
RELATIVE ACCURACY,[1]			±1/2			±1/2	LSB
$T_A = T_{min}$ to T_{max}			±1/2			±1/2	LSB
FULL SCALE CALIBRATION[2]		±2			±2		LSB
UNIPOLAR OFFSET			±1/2			±1/2	LSB
BIPOLAR OFFSET			±1/2			±1/2	LSB
DIFFERENTIAL NONLINEARITY,[3]	8			8			Bits
$T_A = T_{min}$ to T_{max}	8			8			Bits
TEMPERATURE RANGE	0		+70	-55		+125	°C
TEMPERATURE COEFFICIENTS							
Unipolar Offset			±1			±1	LSB
Bipolar Offset			±1			±1	LSB
Full Scale Calibration[2]			±2			±2	LSB
POWER SUPPLY REJECTION							
Positive Supply							
$+4.5 \leq V+ \leq +5.5V$			±2			±2	LSB
Negative Supply							
$-15.75V \leq V- \leq -14.25V$			±2			±2	LSB
$-12.6V \leq V- \leq -11.4V$			±2			±2	LSB
ANALOG INPUT IMPEDANCE	3.0	5.0	7.0	3.0	5.0	7.0	kΩ
ANALOG INPUT RANGES							
Unipolar	0		+10	0		+10	V
Bipolar	-5		+5	-5		+5	V
OUTPUT CODING							
Unipolar	Positive True Binary			Positive True Binary			
Bipolar	Positive True Offset Binary			Positive True Offset Binary			
LOGIC OUTPUT							
Output Sink Current							
(V_{OUT} = 0.4V max, T_{min} to T_{max})	3.2			3.2			mA
Output Source Current[4]							
(V_{OUT} = 2.4V min, T_{min} to T_{max})	0.5			0.5			mA
Output Leakage			±40			±40	μA
LOGIC INPUTS							
Input Current			±100			±100	μA
Logic "1"	2.0			2.0			V
Logic "0"			0.8			0.8	V
CONVERSION TIME, T_A and							
T_{min} to T_{max}	10	20	30	10	20	30	μs
POWER SUPPLY							
V+	+4.5	+5.0	+7.0	+4.5	+5.0	+7.0	V
V-	-11.4	-15	-16.5	-11.4	-15	-16.5	V
OPERATING CURRENT							
V+		15	20		15	20	mA
V-		9	15		9	15	mA

NOTES

[1] Relative accuracy is defined as the deviation of the code transition points from the ideal transfer point on a straight line from the zero to the full scale of the device.

[2] Full scale calibration is guaranteed trimmable to zero with an external 200Ω potentiometer in place of the 15Ω fixed resistor.

Full scale is defined as 10 volts minus 1LSB, or 9.961 volts.

[3] Defined as the resolution for which no missing codes will occur.

[4] The data output lines have active pull-ups to source 0.5mA. The DATA READY line is open collector with a nominal 6kΩ internal pull-up resistor.

Specifications subject to change without notice.

Specifications shown in boldface are tested on all production units at final electrical test. Results from those tests are used to calculate outgoing quality levels. All min and max specifications are guaranteed, although only those shown in boldface are tested on all production units.

ABSOLUTE MAXIMUM RATINGS

V+ to Digital Common . . . . 0 to +7V
V– to Digital Common . . . . 0 to –16.5V
Analog Common to Digital Common . . . . ±1V
Analog Input to Analog Common . . . . ±15V
Control Inputs . . . . 0 to V+
Digital Outputs (High Impedance State) . . . . . 0 to V+
Power Dissipation 800mW

Model	Temperature Range	Relative Accuracy	Package Options[1]
AD673JN	0 to +70°C	±1/2LSB max	Plastic DIP (N-20)
AD673JD	0 to +70°C	±1/2LSB max	Ceramic DIP (D-20)
AD673SD[2]	–55°C to +125°C	±1/2LSB max	Ceramic DIP (D-20)
AD673JP	0 to +70°C	±1/2LSB max	PLCC (P-20A)

NOTES
[1]D = Ceramic DIP; N = Plastic DIP; P = Plastic Leaded Chip Carrier. For outline information see Package Information section.
[2]For details on grade and package offering screened in accordance with MIL-STD-883, refer to Analog Devices Military Products Databook.

FUNCTIONAL DESCRIPTION

A block diagram of the AD673 is shown in Figure 1. The positive CONVERT pulse must be at least 500ns wide. $\overline{DR}$ goes high within 1.5μs after the leading edge of the convert pulse indicating that the internal logic has been reset. The negative edge of the CONVERT pulse initiates the conversion. The internal 8-bit current output DAC is sequenced by the integrated injection logic (I^2L) successive approximation register (SAR) from its most significant bit to least significant bit to provide an output current which accurately balances the input signal current through the 5kΩ resistor. The comparator determines whether the addition of each successively weighted bit current causes the DAC current sum to be greater or less than the input current; if the sum is more, the bit is turned off. After testing all bits, the SAR contains a 8-bit binary code which accurately represents the input signal to within (0.05% of full scale).

The temperature compensated buried Zener reference provides the primary voltage reference to the DAC and ensures excellent stability with both time and temperature. The bipolar offset input controls a switch which allows the positive bipolar offset current (exactly equal to the value of the MSB less ½LSB) to be injected into the summing (+) node of the comparator to offset the DAC output. Thus the nominal 0 to +10V unipolar input range becomes a –5V to +5V range. The 5kΩ thin film input resistor is trimmed so that with a full scale input signal, an input current will be generated which exactly matches the DAC output with all bits on.

UNIPOLAR CONNECTION

The AD673 contains all the active components required to perform a complete A/D conversion. Thus, for many applications, all that is necessary is connection of the power supplies (+5V and –12V to –15V), the analog input and the convert pulse. However, there are some features and special connections which should be considered for achieving optimum performance. The functional pin-out is shown in Figure 2.

The standard unipolar 0 to +10V range is obtained by shorting the bipolar offset control pin (pin 16) to digital common (pin 17).

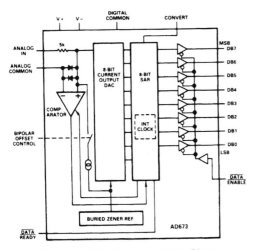

Figure 1. AD673 Functional Block Diagram

The SAR drives $\overline{DR}$ low to indicate that the conversion is complete and that the data is available to the output buffers. $\overline{DATA}$ $\overline{ENABLE}$ can then be activated to enable the 8-bits of data desired. $\overline{DATA}$ $\overline{ENABLE}$ should be brought high prior to the next conversion to place the output buffers in the high impedance state.

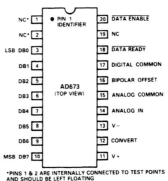

*PINS 1 & 2 ARE INTERNALLY CONNECTED TO TEST POINTS AND SHOULD BE LEFT FLOATING

Figure 2. AD673 Pin Connections

4 × 1 Wideband Video Multiplexer

AD9300

FEATURES
34MHz Full Power Bandwidth
± 0.1dB Gain Flatness to 8MHz
72dB Crosstalk Rejection @ 10MHz
0.03°/0.01% Differential Phase/Gain
Cascadable for Switch Matrices
MIL-STD-883 Compliant Versions Available

APPLICATIONS
Video Routing
Medical Imaging
Electro-Optics
ECM Systems
Radar Systems
Data Acquisition

FUNCTIONAL BLOCK DIAGRAM
(Based on Cerdip)

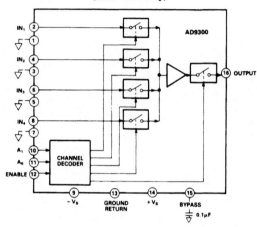

GENERAL DESCRIPTION
The AD9300 is a monolithic high-speed video signal multiplexer useable in a wide variety of applications.

Its four channels of video input signals can be randomly switched at megahertz rates to the single output. In addition, multiple devices can be configured in either parallel or cascade arrangements to form switch matrices. This flexibility in using the AD9300 is possible because the output of the device is in a high-impedance state when the chip is not enabled; when the chip is enabled, the unit acts as a buffer with a high input impedance and low output impedance.

An advanced bipolar process provides fast, wideband switching capabilities while maintaining crosstalk rejection of 72dB at 10MHz. Full power bandwidth is a minimum 27MHz. The device can be operated from ± 10V to ± 15V power supplies.

The AD9300K is available in a 16-pin ceramic DIP and a 20-pin PLCC and is designed to operate over the commercial temperature range of 0 to + 70°C. The AD9300TQ is a hermetic 16-pin ceramic DIP for military temperature range (− 55°C to + 125°C) applications. This part is also available processed to MIL-STD-883. The AD9300 is available in a 20-pin LCC as the model AD9300TE, which operates over a temperature range of − 55°C to + 125°C.

The AD9300 Video Multiplexer is available in versions compliant with MIL-STD-883. Refer to the Analog Devices *Military Products Databook* or current AD9300/883B data sheet for detailed specifications.

PIN DESIGNATIONS

DIP

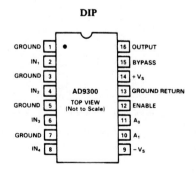

AD9300
TOP VIEW
(Not to Scale)

LCC and PLCC

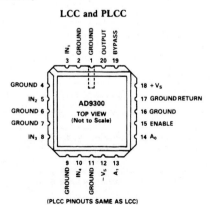

AD9300
TOP VIEW
(Not to Scale)

(PLCC PINOUTS SAME AS LCC)

AD9300–SPECIFICATIONS

ELECTRICAL CHARACTERISTICS ($\pm V_S = \pm 12V \pm 5\%$; $C_L = 10pF$; $R_L = 2k\Omega$, unless otherwise noted)

Parameter (Conditions)	Temp	Test Level	COMMERCIAL 0°C to +70°C AD9300KQ/KP Min	Typ	Max	Units
INPUT CHARACTERISTICS						
Input Offset Voltage	+25°C	I		3	10	mV
Input Offset Voltage	Full	VI			14	mV
Input Offset Voltage Drift[2]	Full	V		75		µV/°C
Input Bias Current	+25°C	I		15	37	µA
Input Bias Current	Full	VI			55	µA
Input Resistance	+25°C	V		3.0		MΩ
Input Capacitance	+25°C	V		2		pF
Input Noise Voltage (dc to 8MHz)	+25°C	V		16		µV rms
TRANSFER CHARACTERISTICS						
Voltage Gain[3]	+25°C	I	0.990	0.994		V/V
Voltage Gain[3]	Full	VI	0.985			V/V
DC Linearity[4]	+25°C	V		0.01		%
Gain Tolerance ($V_{IN} = \pm 1V$)						
dc to 5MHz	+25°C	I		0.05	0.1	dB
5MHz to 8MHz	+25°C	I		0.1	0.3	dB
Small-Signal Bandwidth	+25°C	V		350		MHz
($V_{IN} = 100mV$ p-p)						
Full Power Bandwidth[5]	+25°C	I	27	34		MHz
($V_{IN} = 2V$ p-p)						
Output Swing	Full	VI	±2			V
Output Current (Sinking @ = 25°C)	+25°C	V		5		mA
Output Resistance	+25°C	IV, V		9	15	Ω
DYNAMIC CHARACTERISTICS						
Slew Rate[6]	+25°C	I	170	215		V/µs
Settling Time						
(to 0.1% on ±2V Output)	+25°C	IV		70	100	ns
Overshoot						
To T-Step[7]	+25°C	V		<0.1		%
To Pulse[8]	+25°C	V		<10		%
Differential Phase[9]	+25°C	IV		0.03	0.1	°
Differential Gain[9]	+25°C	IV		0.01	0.1	%
Crosstalk Rejection						
Three Channels[10]	+25°C	IV	68	72		dB
One Channel[11]	+25°C	IV	70	76		dB
SWITCHING CHARACTERISTICS[12]						
A_X Input to Channel HIGH Time[13]	+25°C	I		40	50	ns
(t_{HIGH})						
A_X Input to Channel LOW Time[15]	+25°C	I		35	45	ns
(t_{LOW})						
Enable to Channel ON Time[15]	+25°C	I		35	45	ns
(t_{ON})						
Enable to Channel OFF Time[16]	+25°C	I		35	45	ns
(t_{OFF})						
Switching Transient[17]	+25°C	V		60		mV

EXPLANATION OF TEST LEVELS

Test Level I   –   100% production tested.
Test Level II   –   100% production tested at +25°C, and sample tested at specified temperatures.
Test Level III   –   Sample tested only.
Test Level IV   –   Parameter is guaranteed by design and characterization testing.
Test Level V   –   Parameter is a typical value only.
Test Level VI   –   All devices are 100% production tested at +25°C. 100% production tested at temperature extremes for military temperature devices; sample tested at temperature extremes for commercial/industrial devices.

Parameter (Conditions)	Temp	Test Level	COMMERCIAL 0°C to +70°C AD9300KQ/KP			Units
			Min	Typ	Max	
DIGITAL INPUTS						
Logic "1" Voltage	Full	VI	2			V
Logic "0" Voltage	Full	VI			0.8	V
Logic "1" Current	Full	VI			5	μA
Logic "0" Current	Full	VI			1	μA
POWER SUPPLY						
Positive Supply Current (+12V)	+25°C	I		13	16	mA
Positive Supply Current (+12V)	Full	VI		13	16	mA
Negative Supply Current (−12V)	+25°C	I		12.5	15	mA
Negative Supply Current (−12V)	Full	VI		12.5	16	mA
Power Supply Rejection Ratio ($\pm V_S = \pm 12V \pm 5\%$)	Full	VI	67	75		dB
Power Dissipation ($\pm 12V$)[19]	+25°C	V		306		mW

NOTES

[1]Permanent damage may occur if any one absolute maximum rating is exceeded. Functional operation is not implied, and device reliability may be impaired by exposure to higher-than-recommended voltages for extended periods of time.

[2]Measured at extremes of temperature range.

[3]Measured as slope of V_{OUT} versus V_{IN} with $V_{IN} = \pm 1V$.

[4]Measured as worst deviation from end-point fit with $V_{IN} = \pm 1V$.

[5]Full Power Bandwith (FPBW) based on Slew Rate (SR). FPBW = SR $2\pi V_{PEAK}$

[6]Measured between 20% and 80% transition points of $\pm 1V$ output.

[7]T-Step = $Sin^2 X$ Step, when Step between 0V and +700mV points has 10%-to-90% risetime = 125ns.

[8]Measured with a pulse input having slew rate >250V μs.

[9]Measured at output between 0.28Vdc and 1.0Vdc with $V_{IN} = 284mV$ p-p at 3.58MHz and 4.43MHz.

[10]This specification is critically dependent on circuit layout. Value shown is measured with selected channel grounded and 10MHz 2V p-p signal applied to remaining three channels. If selected channel is grounded through 75Ω, value is approximately 6dB higher.

[11]This specification is critically dependent on circuit layout. Value shown is measured with selected channel grounded and 10MHz 2V p-p signal applied to one other channel. If selected channel is grounded through 75Ω, value is approximately 6dB higher. Minimum specification in ( ) applies to DIPs.

[12]Consult system timing diagram.

[13]Measured from address change to 90% point of −2V to +2V output LOW-to-HIGH transition.

[14]Measured from address change to 90% point of +2V to −2V output HIGH-to-LOW transition.

[15]Measured from 50% transition point of ENABLE input to 90% transition of 0V to −2V and 0V to +2V output.

[16]Measured from 50% transition point of ENABLE input to 10% transition of +2V to 0V and −2V to 0V output.

[17]Measured while switching between two grounded channels.

[18]Maximum power dissipation is a package-dependent parameter related to the following typical thermal impedances:

16-Pin Ceramic	$\theta_{JA} = 87°C/W$;	$\theta_{JC} = 25°C/W$
20-Pin LCC	$\theta_{JA} = 74°C/W$;	$\theta_{JC} = 10°C/W$
20-Pin PLCC	$\theta_{JA} = 71°C/W$;	$\theta_{JC} = 26°C/W$

Specifications subject to change without notice.

ABSOLUTE MAXIMUM RATINGS[1]

Supply Voltages ($\pm V_S$) . . . . $\pm 16V$

Analog Input Voltage Each Input
(IN$_1$ thru IN$_4$) . . . . $\pm 3.5V$

Differential Voltage Between Any Two
Inputs (IN$_1$ thru IN$_4$) 5V

Digital Input Voltages (A$_0$, A$_1$, ENABLE) . −0.5V to +5.5V

Output Current
Sinking 6.0mA
Sourcing 6.0mA

Operating Temperature Range
AD9300KQ/KP . . . . 0°C to +70°C

Storage Temperature Range . . . . −65°C to +150°C

Junction Temperature . . . . +175°C

Lead Soldering (10sec) . . . . +300°C

ORDERING INFORMATION

Device	Temperature Range	Description	Package Option[1]
AD9300KQ	0 to +70°C	16-Pin Cerdip, Commercial	Q-16
AD9300TE/883B[2]	−55°C to +125°C	20-Pin LCC, Military Temperature	E-20A
AD9300TQ/883B[2]	−55°C to +125°C	16-Pin Cerdip, Military Temperature	Q-16
AD9300KP	0 to +70°C	20-Pin PLCC, Commercial	P-20A

NOTES

[1]E = Ceramic Leadless Chip Carrier; P = Plastic Leaded Chip Carrier; Q = Cerdip. For outline information see Package Information section.

[2]For specifications, refer to Analog Devices *Military Products Databook*.

BURR - BROWN®

LOG100

Precision
LOGARITHMIC AND LOG RATIO AMPLIFIER

FEATURES

- **ACCURACY**
  0.37% FSO max Total Error
  Over 5 Decades
- **LINEARITY**
  0.1% max Log Conformity
  Over 5 Decades
- **EASY TO USE**
  Pin-selectable Gains
  Internal Laser-trimmed Resistors
- **WIDE INPUT DYNAMIC RANGE**
  6 Decades, 1nA to 1mA
- **HERMETIC CERAMIC DIP**

APPLICATIONS

- **LOG, LOG RATIO AND ANTILOG COMPUTATIONS**
- **ABSORBANCE MEASUREMENTS**
- **DATA COMPRESSION**
- **OPTICAL DENSITY MEASUREMENTS**
- **DATA LINEARIZATION**
- **CURRENT AND VOLTAGE INPUTS**

DESCRIPTION

The LOG100 uses advanced integrated circuit technologies to achieve high accuracy, ease of use, low cost, and small size. It is the logical choice for your logarithmic-type computations. The amplifier has guaranteed maximum error specifications over the full six-decade input range (1nA to 1mA) and for all possible combinations of I_1 and I_2. Total error is guaranteed so that involved error computations are not necessary.

The circuit uses a specially designed compatible thin-film monolithic integrated circuit which contains amplifiers, logging transistors, and low drift thin-film

resistors. The resistors are laser-trimmed for maximum precision. FET input transistors are used for the amplifiers whose low bias currents (1pA typical) permit signal currents as low as 1nA while maintaining guaranteed total errors of 0.37% FSO maximum.

Because scaling resistors are self-contained, scale factors of 1V, 3V or 5V per decade are obtained simply by pin selections. No other resistors are required for log ratio applications. The LOG100 will meet its guaranteed accuracy with no user trimming. Provisions are made for simple adjustments of scale factor, offset voltage, and bias current if enhanced performance is desired.

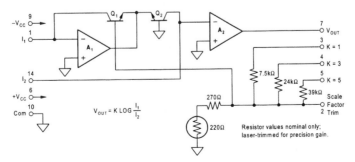

$$V_{OUT} = K \, LOG \frac{I_1}{I_2}$$

Resistor values nominal only;
laser-trimmed for precision gain.

International Airport Industrial Park • Mailing Address: PO Box 11400 • Tucson, AZ 85734 • Street Address: 6730 S. Tucson Blvd. • Tucson, AZ 85706
Tel: (520) 746-1111 • Twx: 910-952-1111 • Cable: BBRCORP • Telex: 066-6491 • FAX: (520) 889-1510 • Immediate Product Info: (800) 548-6132

PDS-437E

Printed in U.S.A. January, 1995

SPECIFICATIONS

ELECTRICAL

T_A = +25°C and $\pm V_{CC}$ = ±15V, after 15 minute warm-up, unless otherwise specified.

PARAMETER	CONDITIONS	LOG100JP MIN	LOG100JP TYP	LOG100JP MAX	UNITS
TRANSFER FUNCTION			V_{OUT} = K Log (I_1/I_2)		
Log Conformity Error[1]	Either I_1 or I_2				
Initial	1nA to 100µA (5 decades)		0.04	0.1	%
	1nA to 1mA (6 decades)		0.15	0.25	%
Over Temperature	1nA to 100µA (5 decades)		0.002		%/°C
	1nA to 1mA (6 decades)		0.001		%/°C
K Range[2]			1, 3, 5		V/decade
Accuracy			0.3		%
Temperature Coefficient			0.03		%/°C
ACCURACY					
Total Error[3]	K = 1,[4] Current Input Operation				
Initial	I_1, I_2 = 1mA			±55	mV
	I_1, I_2 = 100µA			±30	mV
	I_1, I_2 = 10µA			±25	mV
	I_1, I_2 = 1µA			±20	mV
	I_1, I_2 = 100nA			±25	mV
	I_1, I_2 = 10nA			±30	mV
	I_1, I_2 = 1nA			±37	mV
vs Temperature	I_1, I_2 = 1mA		±0.20		mV/°C
	I_1, I_2 = 100µA		±0.37		mV/°C
	I_1, I_2 = 10µA		±0.28		mV/°C
	I_1, I_2 = 1µA		±0.033		mV/°C
	I_1, I_2 = 100nA		±0.28		mV/°C
	I_1, I_2 = 10nA		±0.51		mV/°C
	I_1, I_2 = 1nA		±1.26		mV/°C
vs Supply	I_1, I_2 = 1mA		±4.3		mV/V
	I_1, I_2 = 100µA		±1.5		mV/V
	I_1, I_2 = 10µA		±0.37		mV/V
	I_1, I_2 = 1µA		±0.11		mV/V
	I_1, I_2 = 100nA		±0.61		mV/V
	I_1, I_2 = 10nA		±0.91		mV/V
	I_1, I_2 = 1nA		±2.6		mV/V
INPUT CHARACTERISTICS (of Amplifiers A_1 and A_2)					
Offset Voltage					
Initial			±0.7	±5	mV
vs Temperature			±80		µV/°C
Bias Current					
Initial			1	5[5]	pA
vs Temperature			Doubles Every 10°C		
Voltage Noise	10Hz to 10kHz, RTI		3		µVrms
Current Noise	10Hz to 10kHz, RTI		0.5		pArms
AC PERFORMANCE					
3dB Response[6], I_2 = 10µA					
1nA	C_C = 4500pF		0.11		kHz
1µA	C_C = 150pF		38		kHz
10µA	C_C = 150pF		27		kHz
1mA	C_C = 50pF		45		kHz
Step Response[6]					
Increasing	C_C = 150pF				
1µA to 1mA			11		µs
100nA to 1µA			7		µs
10nA to 100nA			110		µs
Decreasing	C_C = 150pF				
1mA to 1µA			45		µs
1µA to 100nA			20		µs
100nA to 10nA			550		µs
OUTPUT CHARACTERISTICS					
Full Scale Output (FSO)		±10			V
Rated Output					
Voltage	I_{OUT} = ±5mA	±10			V
Current	V_{OUT} = ±10V	±5			mA
Current Limit					
Positive			12.5		mA
Negative			15		mA
Impedance			0.05		Ω

LOG100

2

878

SPECIFICATIONS (CONT)

ELECTRICAL

T_A = +25°C and $\pm V_{cc}$ = ±15V, after 15 minute warm-up, unless otherwise specified.

PARAMETER	CONDITIONS	LOG100JP MIN	LOG100JP TYP	LOG100JP MAX	UNITS
POWER SUPPLY REQUIREMENTS					
Rated Voltage			±15		VDC
Operating Range	Derated Performance	±12		±18	VDC
Quiescent Current			±7	±9	mA
AMBIENT TEMPERATURE RANGE					
Specification		0		+70	°C
Operating Range	Derated Performance	−25		+85	°C
Storage		−40		+85	°C

NOTES: (1) Log Conformity Error is the peak deviation from the best-fit straight line of the V_{OUT} vs Log I_{IN} curve expressed as a percent of peak-to-peak full scale output. (2) May be trimmed to other values. See Applications section. (3) The worst-case Total Error for any ratio of I_1/I_2 is the largest of the two errors when I_1 and I_2 are considered separately. (4) Total Error at other values of K is K times Total Error for K = 1. (5) Guaranteed by design. Not directly measurable due to amplifier's committed configuration. (6) 3dB and transient response are a function of both the compensation capacitor and the level of input current. See Typical Performance Curves.

ABSOLUTE MAXIMUM RATINGS

Supply	±18V
Internal Power Dissipation	600mV
Input Current	10mA
Input Voltage Range	±18V
Storage Temperature Range	−40°C to +85°C
Lead Temperature (soldering, 10s)	+300°C
Output Short-circuit Duration	Continuous to ground
Junction Temperature	175°C

PIN CONFIGURATION

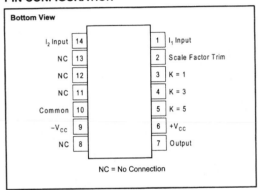

NC = No Connection

SCALE FACTOR PIN CONNECTIONS

K, V/DECADE	CONNECTIONS
5	5 to 7
3	4 to 7
1.9	4 and 5 to 7
1	3 to 7
0.85	3 and 5 to 7
0.77	3 and 4 to 7
0.68	3 and 4 and 5 to 7

ELECTROSTATIC DISCHARGE SENSITIVITY

Any integral circuit can be damaged by ESD. Burr-Brown recommends that all integrated circuits be handled with appropriate precautions. Failure to observe proper handling and installation procedures can cause damage.

ESD damage can range from subtle performance degradation to complete device failure. Precision integrated circuits may be more susceptible to damage because very small parametric changes could cause the device not to meet published specifications.

FREQUENCY COMPENSATION

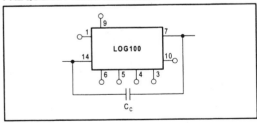

ORDERING INFORMATION

MODEL	PACKAGE	SPECIFIED TEMPERATURE RANGE
LOG100JP	14-Pin Hermetic Ceramic DIP	0°C to +70°C

PACKAGE INFORMATION

MODEL	PACKAGE	PACKAGE DRAWING NUMBER[1]
LOG100JP	14-Pin Hermetic Ceramic DIP	148[2]

NOTES: (1) For detailed drawing and dimension table, please see end of data sheet, or Appendix D of Burr-Brown IC Data Book. (2) During 1994, the package was changed from plastic to hermetic ceramic. Pinout, model number, and specifications remained unchanged. The metal lid of the new package is internally connected to common, pin 10.

BURR - BROWN ®

TYPICAL PERFORMANCE CURVES

$T_A = +25°C$, $V_{CC} = \pm15VDC$, unless otherwise noted.

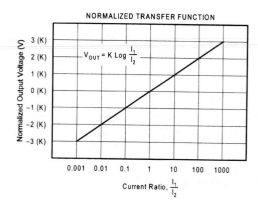

NORMALIZED TRANSFER FUNCTION

$$V_{OUT} = K \log \frac{I_1}{I_2}$$

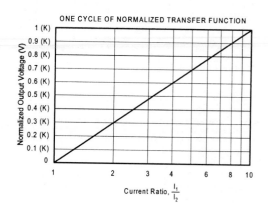

ONE CYCLE OF NORMALIZED TRANSFER FUNCTION

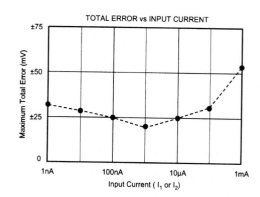

TOTAL ERROR vs INPUT CURRENT

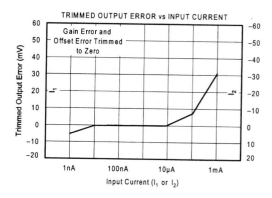

TRIMMED OUTPUT ERROR vs INPUT CURRENT

Gain Error and Offset Error Trimmed to Zero

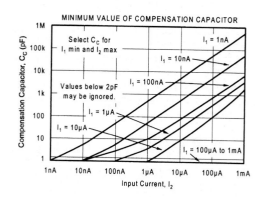

MINIMUM VALUE OF COMPENSATION CAPACITOR

Select C_C for I_1 min and I_2 max

Values below 2pF may be ignored.

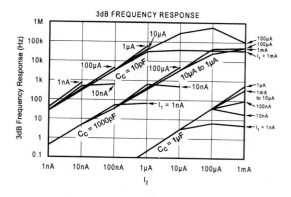

3dB FREQUENCY RESPONSE

Low Cost, Complete 12-Bit Resolver-to-Digital Converter

AD2S90

FEATURES
Complete Monolithic Resolver-to-Digital Converter
Incremental Encoder Emulation (1024-Line)
Absolute Serial Data (12-Bit)
Differential Inputs
12-Bit Resolution
Industrial Temperature Range
20-Lead PLCC
Low Power (50 mW)

APPLICATIONS
Industrial Motor Control
Servo Motor Control
Industrial Gauging
Encoder Emulation
Automotive Motion Sensing and Control
Factory Automation
Limit Switching

FUNCTIONAL BLOCK DIAGRAM

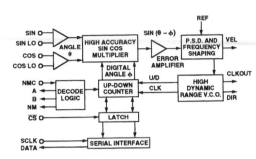

GENERAL DESCRIPTION

The AD2S90 is a complete 12-bit resolution tracking resolver-to-digital converter. No external components are required to operate the device.

The converter accepts 2 V rms ± 10% input signals in the range 3 kHz–20 kHz on the SIN, COS and REF inputs. A Type II servo loop is employed to track the inputs and convert the input SIN and COS information into a digital representation of the input angle. The bandwidth of the converter is set internally at 1 kHz within the tolerances of the device. The guaranteed maximum tracking rate is 500 rps.

Angular position output information is available in two forms, absolute serial binary and incremental A quad B.

The absolute serial binary output is 12-bit (1 in 4096). The data output pin is high impedance when Chip Select $\overline{CS}$ is logic HI. This allows the connection of multiple converters onto a common bus. Absolute angular information in serial pure binary form is accessed by $\overline{CS}$ followed by the application of an external clock (SCLK) with a maximum rate of 2 MHz.

The encoder emulation outputs A, B and NM continuously produce signals equivalent to a 1024 line encoder. When decoded this corresponds to 12 bits of resolution. Three common north marker pulsewidths are selected via a single pin (NMC).

An analog velocity output signal provides a representation of velocity from a rotating resolver shaft traveling in either a clockwise or counterclockwise direction.

The AD2S90 operates on ±5 V dc ± 5% power supplies and is fabricated on Analog Devices' Linear Compatible CMOS process (LC²MOS). LC²MOS is a mixed technology process that combines precision bipolar circuits with low power CMOS logic circuits.

PRODUCT HIGHLIGHTS

Complete Resolver-Digital Interface. The AD2S90 provides the complete solution for digitizing resolver signals (12-bit resolution) without the need for external components.

Dual Format Position Data. Incremental encoder emulation in standard A QUAD B format with selectable North Marker width. Absolute serial 12-bit angular binary position data accessed via simple 3-wire interface.

Single High Accuracy Grade in Low Cost Package. ±10.6 arc minutes of angular accuracy available in a 20-lead PLCC.

Low Power. Typically 50 mW power consumption.

REV. D

One Technology Way, P.O. Box 9106, Norwood, MA 02062-9106, U.S.A.
Tel: 781/329-4700   World Wide Web Site: http://www.analog.com
Fax: 781/326-8703   © Analog Devices, Inc., 1999

Parameter	Min	Typ	Max	Units	Test Condition
SIGNAL INPUTS					
Voltage Amplitude	1.8	2.0	2.2	V rms	Sinusoidal Waveforms, Differential SIN to SINLO, COS to COSLO
Frequency	3		20	kHz	
Input Bias Current			100	nA	V_{IN} = 2 ± 10% V rms
Input Impedance	1.0			MΩ	V_{IN} = 2 ± 10% V rms
Common-Mode Volts[1]			100	mV peak	CMV @ SINLO, COSLO w.r.t.
CMRR	60			dB	AGND @ 10 kHz
REFERENCE INPUT					
Voltage Amplitude	1.8	2.0	3.35	V rms	Sinusoidal Waveform
Frequency	3		20	kHz	
Input Bias Current			100	nA	
Input Impedance	100			kΩ	
Permissible Phase Shift	–10		+10	Degrees	Relative to SIN, COS Inputs
CONVERTER DYNAMICS					
Bandwidth	700	840	1000	Hz	
Maximum Tracking Rate	500			rps	
Maximum VCO Rate (CLKOUT)	2.048			MHz	
Settling Time					
1° Step		2	7	ms	
179° Step			20	ms	
ACCURACY					
Angular Accuracy[2]			±10.6 + 1 LSB	arc min	
Repeatability[3]			1	LSB	
VELOCITY OUTPUT					
Scaling	120	150	180	rps/V dc	
Output Voltage at 500 rps	±2.78	±3.33	±4.17	V dc	
Load Drive Capability			±250	μA	V_{OUT} = ±2.5 V dc (typ), R_L ≥ 10 kΩ
LOGIC INPUTS SCLK, $\overline{CS}$					
Input High Voltage (V_{INH})	3.5			V dc	V_{DD} = +5 V dc, V_{SS} = –5 V dc
Input Low Voltage (V_{INL})			1.5	V dc	V_{DD} = +5 V dc, V_{SS} = –5 V dc
Input Current (I_{IN})			10	μA	
Input Capacitance			10	pF	
LOGIC OUTPUTS DATA, A, B,[4]					V_{DD} = +5 V dc, V_{SS} = –5 V dc
NM, CLKOUT, DIR					
Output High Voltage	4.0			V dc	I_{OH} = 1 mA
Output Low Voltage			1.0	V dc	I_{OL} = 1 mA
			0.4	V dc	I_{OL} = 400 μA
SERIAL CLOCK (SCLK)					
SCLK Input Rate			2	MHz	
NORTH MARKER CONTROL (NMC)					
90°	+4.75	+5.0	+5.25	V dc	North Marker Width Relative to "A" Cycle
180°	–0.75	DGND	+0.75	V dc	
360°	–4.75	–5.0	–5.25	V dc	
POWER SUPPLIES					
V_{DD}	+4.75	+5.00	+5.25	V dc	
V_{SS}	–4.75	–5.00	–5.25	V dc	
I_{DD}			10	mA	
I_{SS}			10	mA	

NOTES
[1]If the tolerance on signal inputs = ±5%, then CMV = 200 mV.
[2]1 LSB = 5.3 arc minute.
[3]Specified at constant temperature.
[4]Output load drive capability.
Specifications subject to change without notice.

REV. D

Wide Bandwidth Strain Gage Signal Conditioner
1B31

FEATURES
Low Cost
Complete Signal-Conditioning Solution
Small Package: 28-Pin Double DIP
Internal Half-Bridge Completion Resistors
Remote Sensing
High Accuracy
  Low Drift: ±0.25µV/°C
  Low Noise: 0.3µV p-p
  Low Nonlinearity: ±0.005% max
  High CMR: 140dB min (60Hz, G = 1000V/V)
Programmable Bridge Excitation: +4V to +15V
Adjustable Low Pass Filter: f$_C$ = 10Hz to 20kHz

APPLICATIONS
Measurement of: Strain, Torque, Force, Pressure
Instrumentation: Indicators, Recorders, Controllers
Data Acquisition Systems
Microcomputer Analog I/O

FUNCTIONAL BLOCK DIAGRAM

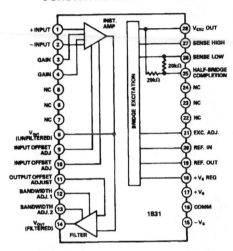

GENERAL DESCRIPTION
Model 1B31 is a high performance strain gage signal-conditioning component that offers the industry's best price/performance solution for applications involving high-accuracy interface to strain gage transducers and load cells. Packaged in a 28-pin double DIP using hybrid technology, the 1B31 is a compact and highly reliable product. Functionally, the signal conditioner consists of three sections: a precision instrumentation amplifier, a two-pole low pass filter, and an adjustable transducer excitation.

The instrumentation amplifier (IA) section features low input offset drift of ±0.25µV/°C (RTI, G = 1000V/V) and excellent nonlinearity of ±0.005% max. In addition, the IA exhibits low noise of 0.3µV p-p typ (0.1Hz-10Hz), and outstanding 140dB min common-mode rejection (G = 1000V/V, 60Hz). The gain is programmable from 2V/V up to 5000V/V by one external resistor.

The two-pole low pass filter offers a 40dB/decade roll-off from 1kHz to reduce high frequency noise and improve system signal-to-noise ratio. The corner frequency is adjustable downwards by external capacitors and upwards to 20kHz by three resistors. The output voltage can also be offset by ±10V with an external potentiometer to null out dead weight.

The 1B31's regulated transducer excitation stage features low output drift (±0.004%/°C typ) and can drive 120Ω or higher resistance load cells. The excitation is preset at +10V and is adjustable from +4V and +15V. This section also has remote sensing capability to allow for lead-wire compensation in 6-wire bridge configurations. For half-bridge strain gages, a matched

pair of thin-film 20kΩ resistors is connected across the excitation outputs. This assures temperature tracking of ±5ppm/°C max and reduces part count.

The 1B31 is available in a plastic package specified over the industrial (−40°C to +85°C) temperature range and will be available soon in a bottom-brazed ceramic package specified over the military (−55°C to +125°C) temperature range.

DESIGN FEATURES AND USER BENEFITS
Ease of Use: Direct transducer interface with minimum external parts required, convenient offset and span adjustment capability.

Half-Bridge Completion: Matched resistor pair tracking to ±5ppm/°C max for half-bridge strain gage applications.

Remote Sensing: Voltage drops across the excitation lead-wires are compensated by the regulated supply, making 6-wire load-cell interfacing straightforward.

Programmable Transducer Excitation: Excitation source preset for +10V dc operation without external components. User-programmable from a +4V to +15V dc to optimize transducer performance.

Adjustable Low Pass Filter: The two-pole active filter (f$_C$ = 1kHz) reduces noise bandwidth and aliasing errors with provisions for external adjustment of cutoff frequency (10Hz to 20kHz).

1B31 — SPECIFICATIONS (typical @ +25°C and $V_S = \pm15V$ unless otherwise noted)

Model	1B31AN	1B31SD†
GAIN[1]		
Gain Range	2 to 5000V/V	*
Gain Equation	$R_G = \dfrac{80k\Omega}{G-2}$	*
Gain Equation Accuracy, G≤1000V/V	±3%	*
Gain Temperature Coefficient[2]	±15ppm/°C (±25ppm/°C max)	*
Nonlinearity	±0.005% max	*
OFFSET VOLTAGES[1]		
Total Offset Voltage, Referred to Input		
Initial; @ +25°C (Adjustable to Zero)		
G = 2V/V	±2mV (±10mV max)	*
G = 1000V/V	±50µV (±200µV max)	*
Warm-Up Drift, 5 min., G = 1000V/V	Within ±1µV of final value	*
vs. Temperature		
G = 2V/V	±25µV/°C (±50µV/°C max)	*
G = 1000V/V	±0.25µV/°C (±2µV/°C max)	*
At Other Gains	$\left(\pm2 \pm \dfrac{100}{G}\right)\mu V/°C$	*
vs. Supply		
G = 2V/V	±50µV/V	*
G = 1000V/V	±0.5µV/V	*
Output Offset Adjust Range	±10V min	*
INPUT BIAS CURRENT		
Initial @25°C	±10nA (±50nA max)	*
vs. Temperature	±25pA/°C	*
INPUT DIFFERENCE CURRENT		
Initial @ +25°C	±5nA (±20nA max)	*
vs. Temperature	±10pA/°C	*
INPUT IMPEDANCE		
Differential	1GΩ‖4pF	*
Common Mode	1GΩ‖4pF	*
INPUT VOLTAGE RANGE		
Linear Differential Input (V_D)	±5V	*
Maximum CMV Input	$\pm\left(12 - \dfrac{G \times V_D}{4}\right)V\ max$	*
CMR, 1kΩ Source Imbalance		
G = 2V/V, dc to 60Hz	86dB	*
G = 100V/V to 5000V/V		
1kHz Bandwidth[3]		
@ dc to 60Hz	110dB min	*
10Hz Bandwidth[4]		
@ dc	110dB min	*
@ 60Hz	140dB min	*
INPUT NOISE		
Voltage, G = 1000V/V		
0.1Hz to 10Hz	0.3µV p-p	*
10Hz to 100Hz	1µV p-p	*
Current, G = 1000V/V		
0.1Hz to 10Hz	60pA p-p	*
10Hz to 100Hz	100pA p-p	*
RATED OUTPUT[1]		
Voltage, 2kΩ Load, min	±10V	*
Current	±5mA	*
Impedance, dc to 2Hz, G = 2V/V to 1000V/V	0.5Ω	*
Load Capacitance	1000pF	*
Output Short-Circuit Duration	Indefinite	*
DYNAMIC RESPONSE[1]		
Small Signal Bandwidth −3dB, G = 2V/V to 1000V/V	1kHz	*
Slew Rate	0.05V/µs	*
Full Power	350Hz	*
Settling Time, G = 2V/V to 1000V/V, ±10V Output, Step to ±0.1%	2ms	*
LOW PASS FILTER		
Number of Poles	2	*
Gain (Pass Band)	−2V/V	*
Cutoff Frequency (−3dB Point)	1kHz	*
Roll-Off	40dB/decade	*

OUTLINE DIMENSIONS
Dimensions shown in inches and (mm).

Plastic Package (N)

Ceramic Package (D)

NOTES:
1. LEAD NO. 1 IDENTIFIED BY DOT OR NOTCH.

PIN DESIGNATIONS

PIN	FUNCTION	PIN	FUNCTION
1	+INPUT	15	−V_S
2	−INPUT	16	COMMON
3	GAIN	17	+V_S
4	GAIN	18	+V_S REGULATOR
8	V_{OUT} (UNFILTERED)	19	REF OUT
9	INPUT OFFSET ADJ.	20	REF IN
10	INPUT OFFSET ADJ.	21	EXCITATION ADJ.
11	OUTPUT OFFSET ADJ.	25	HALF-BRIDGE COMP.
12	BANDWIDTH ADJ. 1	26	SENSE LOW
13	BANDWIDTH ADJ. 2	27	SENSE HIGH
14	V_{OUT} (FILTERED)	28	V_{EXC} OUT

−2−

Isolated mV/Thermocouple Signal Conditioner

1B51

FEATURES
Functionally Complete Precision Conditioner
High Accuracy
  Low Input Offset Tempco: $\pm0.1\mu$V/°C
  Low Nonlinearity: $\pm0.025\%$
  High CMR: 160dB (60Hz, G=1000V/V)
High CMV Isolation: 1500V rms Continuous
240V rms Input Protection
Small Package: 1.0"×2.1"×0.35" DIP
Isolated Power
Low Pass Filter ($f_C=3$Hz)
Pin Compatible with 1B41 Isolated RTD Conditioner

APPLICATIONS
Multichannel Thermocouple Temperature
Measurement
Low Level Data Acquisition Systems
Industrial Measurement & Control Systems

GENERAL DESCRIPTION
The 1B51 is a precision, mV/thermocouple signal conditioner that incorporates a circuit design utilizing transformer based isolation and automated surface mount manufacturing technology. It provides an unbeatable combination of versatility and performance in a compact plastic package. Designed for measurement and control applications, it is specially suited for harsh environments with extremely high common-mode interference. Unlike costlier solutions that require separate dc/dc converters, each 1B51 generates its own input side power, providing true, low cost channel-to-channel isolation.

Functionally, the signal conditioner consists of three basic sections: chopper stabilized amplifier, isolation and output filter. The chopper amplifier features a highly stable offset tempco of $\pm0.1\mu$V/°C and resistor programmable gains from 2 to 1000. Wide range zero suppression can be implemented at this stage.

The isolation section has complete input to output galvanic isolation of 1500V rms continuous using transformer coupling techniques. Isolated power of 2mA at ±6.2V is provided for ancillary circuits such as zero suppression and open-input detection. Filtering at 3Hz is implemented by a passive antialiasing filter at

FUNCTIONAL BLOCK DIAGRAM

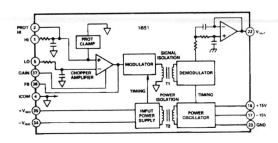

the front end and a two-pole active filter at the output. Overall NMR is 60dB and CMR is 160dB min @ 60Hz, G = 1000.

The 1B51 is specified over -25°C to $+85$°C and operates over the industrial (-40°C to $+85$°C) temperature range.

DESIGN FEATURES AND USER BENEFITS
High Noise Rejection: The combination of a chopper stabilized front end with a low pass filter provides high system accuracy in harsh industrial environments as well as excellent rejection of 50/60Hz noise.

Input Protection: The input is internally protected against continuous application of 240V rms.

Low Cost: The 1B51 offers a very low cost per channel for high performance, isolated, low level signal conditioners.

Wide Range Zero Suppression: This input referred function is a convenient way to null large input offsets.

Low Pass Filter: The three pole active filter ($f_C=3$Hz) reduces 60Hz noise and aliasing errors.

Small Size: The 1B51 package size (1.0"×2.1"×0.35") and functional completeness make it an excellent choice in systems with limited board space and clearance.

1B51—SPECIFICATIONS (typical @ +25°C and V$_S$ = ±15V unless otherwise noted)

Model	1B51AN	1B51BN
GAIN		
Gain Equation	$G=\left[1 + \dfrac{R_{FB}}{R_G}\right] \times 2$	*
Gain Error	1% max	*
Gain Temperature Coefficient[1]	50ppm/°C	*
Gain Nonlinearity	±0.035% (±0.05% max)	±0.025% (±0.04% max)
OFFSET VOLTAGES		
Input Offset Voltage		
Initial, @ +25°C (Adjustable to Zero)	25μV (100μV max)	*
vs. Temperature	±0.1μV/°C (±0.5μV/°C max)	*
vs. Time, Noncumulative	±1μV/month max	*
Output Offset Voltage		
Initial	−50mV	−25mV
vs. Temperature	−175μV/°C	−50μV/°C
INPUT OFFSET CURRENT		
Initial	0.6nA (2.5nA max)	*
vs. Temperature	±2.5pA/°C (12.5pA/°C max)	*
INPUT BIAS CURRENT		
Initial @ +25°C	10nA	*
vs. Temperature	10pA/°C	*
INPUT IMPEDANCE		
Power On	50MΩ	*
Power Off	40kΩ min	*
INPUT VOLTAGE RANGE		
Linear Differential Input	±10mV to ±5V	*
Max CMV, Input to Output		
ac, 60Hz, Continuous	1500V rms	*
Continuous, dc	±2000V	*
CMR @ 60Hz, 1kΩ Source Imbalance, G = 1000	160dB min	*
NMR @ 60Hz	60dB min	*
Transient Protection	IEEE-STD 472 (SWC)	*
INPUT NOISE		
Voltage, 0.1Hz to 10Hz, 1kΩ Source Imbalance	1μV p-p	*
RATED OUTPUT		
Voltage, 2kΩ Load, min	±10V	*
Current	±5mA	*
Output Noise, dc to 100kHz	1mV p-p	*
Impedance, dc	0.1Ω	*
FREQUENCY RESPONSE		
Bandwidth, −3dB	dc to 3Hz	*
ISOLATED POWER		
Voltage, No Load	±6.2V±5%	*
Current	2mA	*
Regulation, No Load to Full Load	7.5%	*
Ripple	250mV p-p	*
POWER SUPPLY		
Voltage, Rated Performance	±15V dc	*
Voltage, Operating	±13.5V to ±18V	*
Current, Quiescent	+12mA @ +15V, −4mA @ −15V	*
PSRR	0.1%/V	*
ENVIRONMENTAL		
Temperature Range		
Rated Performance	−25°C to +85°C	*
Operating	−40°C to +85°C	*
Storage	−40°C to +85°C	*
Relative Humidity	0 to 95% @ +60°C	*
CASE SIZE	1.00"×2.10"×0.35" (25.4×53.3×8.9)mm	*

NOTES
*Specifications same as 1B51AN.
[1]See graph in text.
Specifications subject to change without notice.

OUTLINE DIMENSIONS
Dimensions shown in inches and (mm).

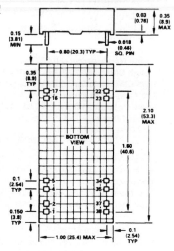

PIN DESIGNATIONS

PIN	DESIGNATION
1	HI
2	PROT HI
4	ICOM
5	LO
16	+15V
17	−15V
22	V$_O$
23	GND
34	−V$_{ISO}$
35	+V$_{ISO}$
37	GAIN
38	FB

−2−

B DERIVATIONS OF SELECTED EQUATIONS

EQUATION (3–9)

The Shockley equation for the base-emitter pn junction is

$$I_E = I_R(e^{VQ/kT} - 1)$$

where I_E = the total forward current across the base-emitter junction
$\quad I_R$ = the reverse saturation current
$\quad V$ = the voltage across the depletion region
$\quad Q$ = the charge on an electron
$\quad k$ = a number known as Boltzmann's constant
$\quad T$ = the absolute temperature

At ambient temperature, $Q/kT \cong 40$, so

$$I_E = I_R(e^{40V} - 1)$$

Differentiating,

$$\frac{dI_E}{dV} = 40I_R e^{40V}$$

Since $I_R e^{40V} = I_E + I_R$,

$$\frac{dI_E}{dV} = 40(I_E + I_R)$$

Assuming $I_R \ll I_E$,

$$\frac{dI_E}{dV} \cong 40I_E$$

The ac resistance r_e' of the base-emitter junction can be expressed as dV/dI_E.

$$r_e' = \frac{dV}{dI_E} \cong \frac{1}{40I_E} \cong \frac{25 \text{ mV}}{I_E}$$

EQUATION (4–10)

The gain of a CD amplifier is

$$A_v = \frac{R_s}{r_s' + R_s}$$

Substituting $1/g_m$ for r_s' gives

$$A_v = \frac{R_s}{\dfrac{1}{g_m} + R_s} = \frac{g_m R_s}{1 + g_m R_s}$$

EQUATION (7–4)

The formula for open-loop gain in Equation (7–2) can be expressed in complex notation as

$$A_{ol} = \frac{A_{ol(mid)}}{1 + jf/f_{c(ol)}}$$

Substituting the above expression into the equation $A_{cl} = A_{ol}/(1 + BA_{ol})$, we get a formula for the total closed-loop gain.

$$A_{cl} = \frac{A_{ol(mid)}/(1 + jf/f_{c(ol)})}{1 + BA_{ol(mid)}/(1 + jf/f_{c(ol)})}$$

Multiplying the numerator and denominator by $1 + jf/f_{c(ol)}$ yields

$$A_{cl} = \frac{A_{ol(mid)}}{1 + BA_{ol(mid)} + jf/f_{c(ol)}}$$

Dividing the numerator and denominator by $1 + BA_{ol(mid)}$ gives

$$A_{cl} = \frac{A_{ol(mid)}/(1 + BA_{ol(mid)})}{1 + j[f/(f_{c(ol)}(1 + BA_{ol(mid)}))]}$$

The above expression is of the form of the first equation

$$A_{cl} = \frac{A_{cl(mid)}}{1 + jf/f_{c(cl)}}$$

where $f_{c(cl)}$ is the closed-loop critical frequency. Thus,

$$f_{c(cl)} = f_{c(ol)}(1 + BA_{ol(mid)})$$

EQUATION (9–7)

The center frequency equation is

$$f_0 = \frac{1}{2\pi\sqrt{(R_1\|R_3)R_2 C_1 C_2}}$$

Substituting C for C_1 and C_2 and rewriting $R_1 \| R_3$ as the product-over-sum produces

$$f_0 = \frac{1}{2\pi C\sqrt{\left(\dfrac{R_1 R_3}{R_1 + R_3}\right)R_2}}$$

Rearranging,

$$f_0 = \frac{1}{2\pi C} \sqrt{\left(\frac{R_1 + R_3}{R_1 R_2 R_3}\right)}$$

EQUATION (10–1)

$$\frac{V_{out}}{V_{in}} = \frac{R(-jX)/(R - jX)}{(R - jX) + R(-jX)/(R - jX)}$$

$$= \frac{R(-jX)}{(R - jX)^2 - jRX}$$

Multiplying the numerator and denominator by j,

$$\frac{V_{out}}{V_{in}} = \frac{RX}{j(R - jX)^2 + RX}$$

$$= \frac{RX}{RX + j(R^2 - j2RX - X^2)}$$

$$= \frac{RX}{RX + jR^2 + 2RX - jX^2}$$

$$\frac{V_{out}}{V_{in}} = \frac{RX}{3RX + j(R^2 - X^2)}$$

For a 0° phase angle there can be no j term. Recall from complex numbers in ac theory that a *nonzero* angle is associated with a complex number having a j term. Therefore, at f_r the j term is 0.

$$R^2 - X^2 = 0$$

Thus,

$$\frac{V_{out}}{V_{in}} = \frac{RX}{3RX}$$

Cancelling,

$$\frac{V_{out}}{V_{in}} = \frac{1}{3}$$

EQUATION (10–2)

From the derivation of Equation (10–1),

$$R^2 - X^2 = 0$$
$$R^2 = X^2$$
$$R = X$$

Since $X = \dfrac{1}{2\pi f_r C}$,

$$R = \frac{1}{2\pi f_r C}$$

$$f_r = \frac{1}{2\pi RC}$$

EQUATIONS (10–3) AND (10–4)

The feedback network in the phase-shift oscillator consists of three RC stages, as shown in Figure B–1. An expression for the attenuation is derived using the mesh analysis method for the loop assignment shown. All Rs are equal in value, and all Cs are equal in value.

FIGURE B–1

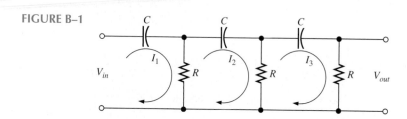

$$(R - j1/2\pi fC)I_1 - RI_2 + 0I_3 = V_{in}$$
$$-RI_1 + (2R - j1/2\pi fC)I_2 - RI_3 = 0$$
$$0I_1 - RI_2 + (2R - j1/2\pi fC)I_3 = 0$$

In order to get V_{out}, we must solve for I_3 using determinants:

$$I_3 = \frac{\begin{vmatrix} (R - j1/2\pi fC) & -R & V_{in} \\ -R & (2R - j1/2\pi fC) & 0 \\ 0 & -R & 0 \end{vmatrix}}{\begin{vmatrix} (R - j1/2\pi fC) & -R & 0 \\ -R & (2R - j1/2\pi fC) & -R \\ 0 & -R & (2R - j1/2\pi fC) \end{vmatrix}}$$

$$I_3 = \frac{R^2 V_{in}}{(R - j1/2\pi fC)(2R - j1/2\pi fC)^2 - R^2(2R - j1/2\pi fC) - R^2(R - 1/2\pi fC)}$$

$$\frac{V_{out}}{V_{in}} = \frac{RI_3}{V_{in}}$$

$$= \frac{R^3}{(R - j1/2\pi fC)(2R - j1/2\pi fC)^2 - R^3(2 - j1/2\pi fRC) - R^3(1 - 1/2\pi fRC)}$$

$$= \frac{R^3}{R^3(1 - j1/2\pi fRC)(2 - j1/2\pi fRC)^2 - R^3[(2 - j1/2\pi fRC) - (1 - j1/2\pi fRC)]}$$

$$= \frac{R^3}{R^3(1 - j1/2\pi fRC)(2 - j1/2\pi fRC)^2 - R^3(3 - j1/2\pi fRC)}$$

$$= \frac{1}{(1 - j1/2\pi fRC)(2 - j1/2\pi fRC)^2 - (3 - j1/2\pi fRC)}$$

Expanding and combining the real terms and the j terms separately,

$$\frac{V_{out}}{V_{in}} = \frac{1}{\left(1 - \dfrac{5}{4\pi^2 f^2 R^2 C^2}\right) - j\left(\dfrac{6}{2\pi fRC} - \dfrac{1}{(2\pi f)^3 R^3 C^3}\right)}$$

For oscillation in the phase-shift amplifier, the phase shift through the RC network must equal $180°$. For this condition to exist, the j term must be 0 at the frequency of oscillation f_r.

$$\frac{6}{2\pi f_r RC} - \frac{1}{(2\pi f_r)^3 R^3 C^3} = 0$$

$$\frac{6(2\pi)^2 f_r^2 R^2 C^2 - 1}{(2\pi)^3 f_r^3 R^3 C^3} = 0$$

$$6(2\pi)^2 f_r^2 R^2 C^2 - 1 = 0$$

$$f_r^2 = \frac{1}{6(2\pi)^2 R^2 C^2}$$

$$f_r = \frac{1}{2\pi \sqrt{6} RC}$$

Since the j term is 0,

$$\frac{V_{out}}{V_{in}} = \frac{1}{1 - \dfrac{5}{4\pi^2 f_r^2 R^2 C^2}} = \frac{1}{1 - \dfrac{5}{\left(\dfrac{1}{\sqrt{6}RC}\right)^2 R^2 C^2}}$$

$$= \frac{1}{1 - 30} = -\frac{1}{29}$$

The negative sign results from the $180°$ inversion. Thus, the value of attenuation for the feedback network is

$$B = \frac{1}{29}$$

EQUATION (12–1)

The output voltage of the upper op-amp is called V_{out1} and the output voltage of the lower op-amp is called V_{out2}. The difference in these two voltages sets up a current in the two feedback resistors, R, and R_G, given by Ohm's law.

$$i = \frac{V_{out1} - V_{out2}}{2R + R_G}$$

Because of negative feedback, ideally, the input voltage is across R_G (no voltage drop across the op-amp inputs). Applying Ohm's law again,

$$i = \frac{V_{in1} - V_{in2}}{R_G}$$

The current in the feedback resistors (R) and the gain resistor (R_G) are the same, since the op-amp inputs (ideally) draw no current. Equating the currents,

$$\frac{V_{out1} - V_{out2}}{R_G + 2R} = \frac{V_{in1} - V_{in2}}{R_G}$$

The third op-amp is set up as a unity-gain differential amplifier. Its output is

$$V_{out} = -(V_{out1} - V_{out2})$$

Substituting this result into the previous equation,

$$\frac{-V_{out}}{R_G + 2R} = \frac{V_{in1} - V_{in2}}{R_G}$$

Rearranging, changing signs, and simplifying,

$$V_{out} = \left(1 + \frac{2R}{R_G}\right)(V_{in2} - V_{in1})$$

The Troubleshooter's Bench assignments in this color section and the circuit board in Chapter 4 are each related to a different System Application as indicated in these graphics. These are representative of the circuit boards with which you will be working. Each Troubleshooter's Bench in this section contains a circuit board page and an instrumentation page. The assignment is stated on the circuit board page of each Troubleshooter's Bench.

Use this section only when directed by the special assignment activity in the System Application section of a chapter. You may use the special worksheets available from your instructor to facilitate performing these assignments as well as the other activities in the System Application.

Each Troubleshooter's Bench contains a two-page spread and is related to one of the System Application sections in the text. The left page contains the assignment, the circuit board with connections and/or labeled points, and a schematic of the board. The right page contains certain instruments that are connected to the board or are to be connected to the board.

Connections between the circuit board and the instruments are indicated by corresponding numbers in color-coded circles. For example, a probe attached to the board with a *red circle 2* goes to the instrument input(s) or output(s) also labeled with a *red circle 2*.

The assignments in this section are of two basic types:

1. Evaluation of indicated instrument settings and readings for the purpose of troubleshooting the board.
2. Selection of instrument setting and proper readings for specified inputs to circuit boards.

Instrument readings that involve circular dial, rotary switch, push-buttons switch, or slide switch settings should be evident. Examples are the *SEC/DIV* control switch on the oscilloscope and the *frequency* dial on the function generator.

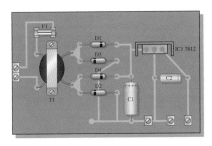

TROUBLESHOOTER'S BENCH 1, Chapter 2
DC Power Supply Board

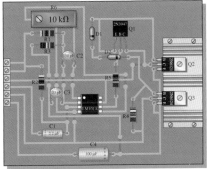

TROUBLESHOOTER'S BENCH 2, Chapter 7
Stereo Amplifier Board

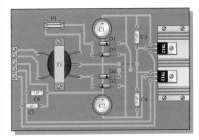

TROUBLESHOOTER'S BENCH 3, Chapter 11
Dual Power Supply Board

Assignment

Evaluate the instrument settings and readings for this test setup and determine if the circuit is operating properly. If it is not operating properly, isolate the fault(s). The colored circled numbers indicate connections between the PC board and the instruments: 1 (blue) goes to 1 (blue), 2 (red) goes to 2 (red), etc. Assume a 200 Ω test load is connected to the output.

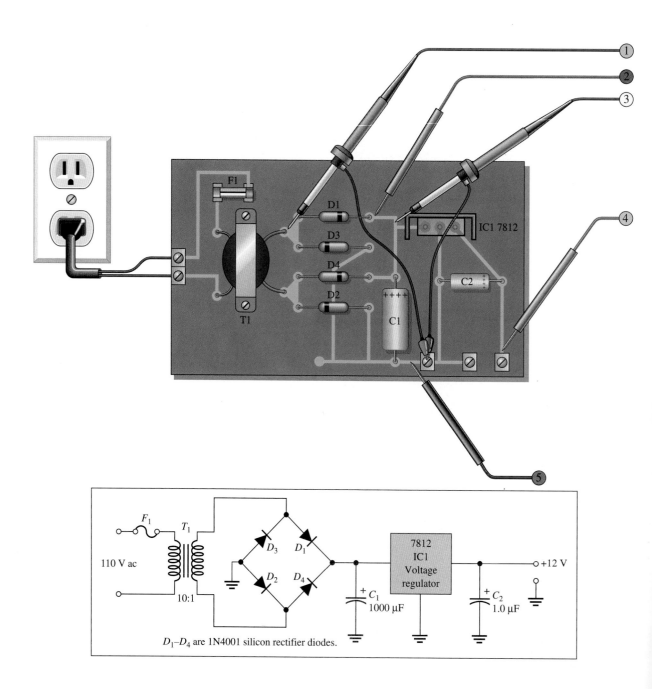

D_1–D_4 are 1N4001 silicon rectifier diodes.

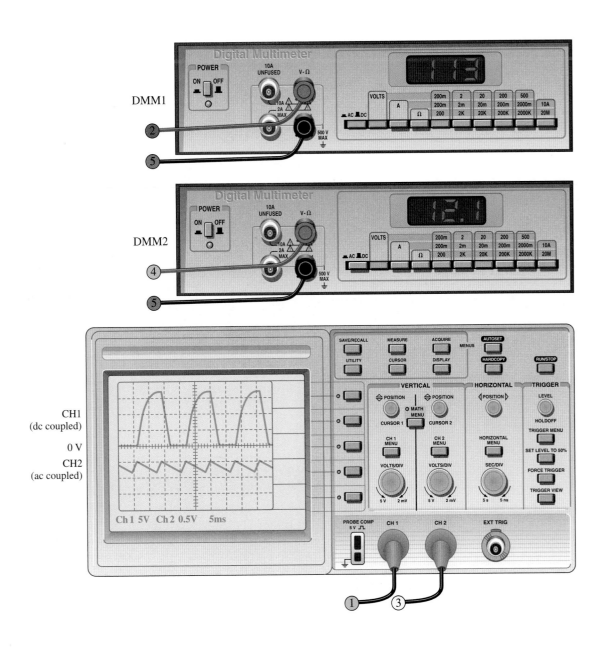

DMM1

DMM2

CH1
(dc coupled)

0 V

CH2
(ac coupled)

Ch 1 5V   Ch 2 0.5V   5ms

Assignment

Evaluate the instrument settings and readings for this test setup and determine if the circuit is operating properly. If it is not operating properly, determine the most probable fault(s). The colored circled numbers indicate connections between the PC board and the instruments: 1 (blue) goes to 1 (blue), 2 (red) goes to 2 (red), etc. The top waveform on the scope is CH 1.

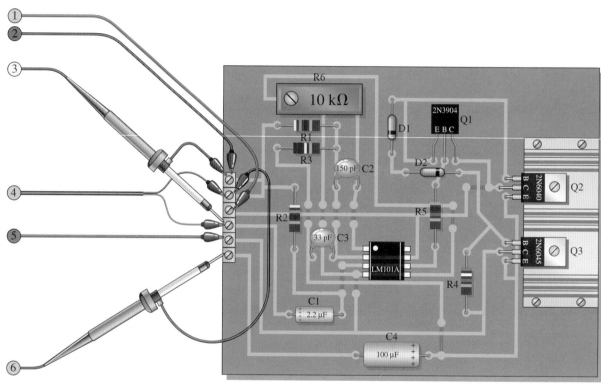

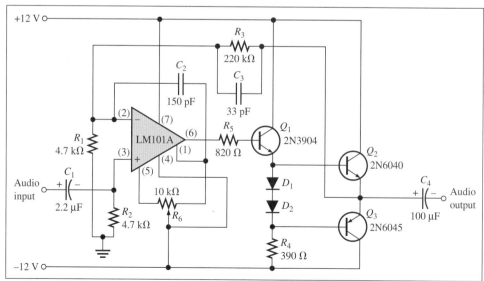

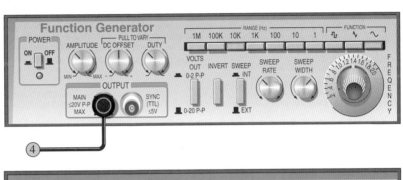

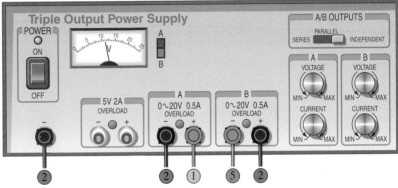

CH1
(ac coupled)

CH2
(ac coupled)

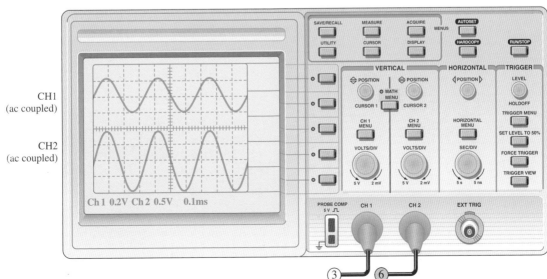

Assignment

1. Set up the range and function of controls on the instruments for observing the voltages at the probed points. Existing settings may not be correct.
2. Indicate the readings and displays you should observe if the power supply is working properly.
3. Indicate the readings and displays you should observe if diode D1 is open.
4. Indicate the readings and displays you should observe if capacitor C2 is open.

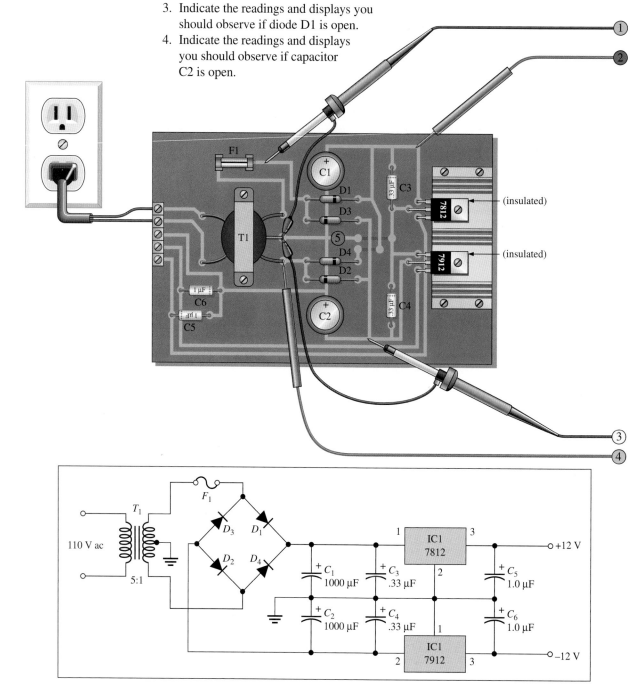

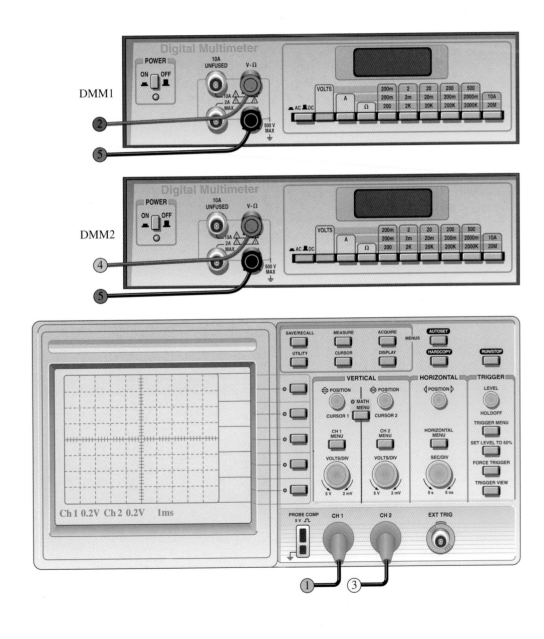

The circuit board shown here is a full-color version of the board in the Chapter 4 Application Assignment. Refer to Chapter 4 for instructions.

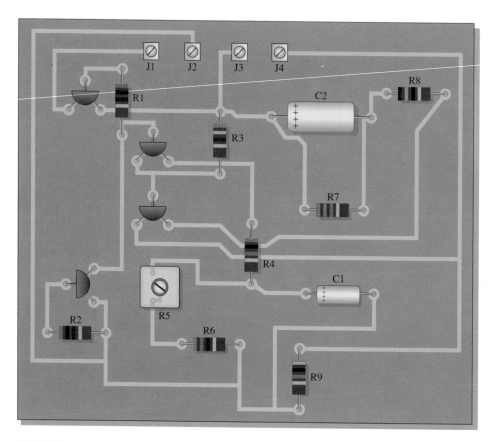

FIGURE 4-56
Preamplifier PC board.

ANSWERS TO ODD-NUMBERED PROBLEMS

Chapter 1

1. 45.4 mS

3. $\dfrac{\Delta V}{\Delta I} = \dfrac{0.75 - 0.65 \text{ V}}{8 - 3.2 \text{ mA}} = 21\ \Omega$

5. **(a)** $V_p = 100$ V, $V_{avg} = 63.7$ V, $\omega = 200$ rad/s
   (b) 79.6 V

7. 37 kHz

9. 1.11

11. Odd harmonics

13. Voltage across 1.0 kΩ load = 1.65;
   voltage across 2.7 kΩ load = 3.25 V;
   Voltage across 3.6 kΩ load = 3.79 V

15. See Figure ANS–1.

FIGURE ANS–1 $I\,(\mu\text{A})$

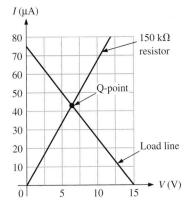

17. See Figure ANS–2.

19. 4.0 V

21. 51 dB

FIGURE ANS–2

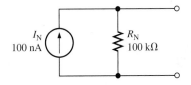

23. −60 dB

25. **(a)** −10 dB   **(b)** 10 V

27. The supply is common to both channels, so it is not the problem. Start by reversing the channels at the input of the amplifier. If the Channel 2 is still bad, the problem is most likely the amplifier or the Ch-2 speaker. Speakers can be tested by reversing them.

   If the problem changes channels when the first test is done, the problem is before the amplifier inputs and could be the A_2 microphone or a problem in wiring including the battery lead at the microphone. Test by changing SW to the B microphones. If this corrects the problem, check the A_2 microphone; otherwise look for continuity to the switch and check the switch itself.

29. Use a static-safe wrist strap (and static free work station, if possible).

Chapter 2

1. See Figure ANS–3.

3. **(a)** Full-wave rectifier
   (b) 28.3 V (total)
   (c) 14.1 V (reference is center tap)
   (d) See Figure ANS–4 (offset approximation).
   (e) 13.4 mA (offset approximation)
   (f) 28.3 V (ideal approximation)

5. $V_p = 50$ V/0.637 = 78.5 V; PIV = 78.5 V

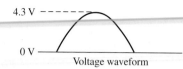

4.3 V

0 V

Voltage waveform

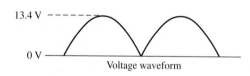

43 mA

0 mA

Current waveform

FIGURE ANS–3

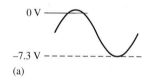

13.4 V

0 V

Voltage waveform

FIGURE ANS–4

7. 60 μV

9. 11.94 V

11. 9.06 V

13. See Figure ANS–5.

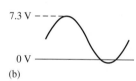

0 V

–7.3 V

(a)

7.3 V

0 V

(b)

FIGURE ANS–5

15. See Figure ANS–6.

FIGURE ANS–6

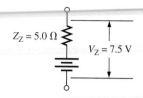

$Z_Z = 5.0\ \Omega$

$V_Z = 7.5$ V

17. 2.6%

19. 2.0 V. Note: since the plot is logarithmic, 25 pF is 70% of the linear distance between 20 pF and 30 pF.

21. 2.0 V

23. Dark current

25. $V_{RRM} = 400$ V

27. DMM1 is correct but DMM2 is reading the rectified average voltage rather than the peak voltage that it would show if the capacitor was in the circuit. DMM3, indicating no voltage, implies an open circuit between the bridge and the output. The most likely cause is an open path along the output line between the bridge and the filter capacitor.

29. (a) Readings are correct
   (b) Open zener diode
   (c) Open switch or fuse blown
   (d) Open capacitor
   (e) Open transformer winding (less likely: more than one diode open)

31. See Figure ANS–7. The output voltages are $V_{OUT1} = 6.8$ V and $V_{OUT2} = 24$ V.

33. A turns ratio of $N_{pri}{:}N_{sec} = 5{:}1$ is a reasonable choice based on the 24 V output.

FIGURE ANS–7

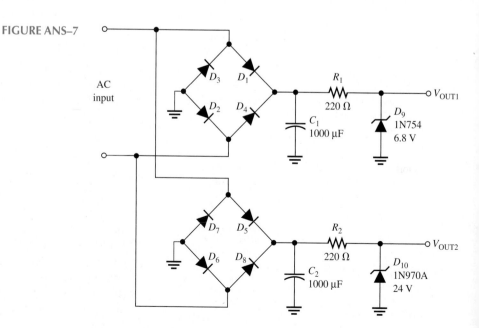

Chapter 3

1. 5.29 mA

3. 29.4 mA

5. $I_B = 0.276$ mA; $I_C = 20.7$ mA; $V_C = 15.1$ V

7. $I_B = 13.6$ μA; $I_C = 3.4$ mA; $V_C = 6.6$ V

9. $I_C = 3.67$ mA (saturation current); $V_{CE} = 0.1$ V

11. (a) decrease (to zero)   (b) remain the same
   (c) increase   (d) increase
   (e) increase

13. $I_C \cong I_E = 0.92$ mA; $V_{CE} = 8.34$ V

15. $I_{C(sat)} = 5.52$ mA; $V_{CE(cutoff)} = 15$ V

17. $P = (36.2$ mA$)(9.23$ V$) = 334$ mW

19. (a) $I_C = 36.2$ mA; $V_{CE} = 7.1$ V
   (b) $P_{R_C} = 432$ mW
   (c) $P_D = 256$ mW

21. $V_B = 2.64$ V; $V_E = 1.94$ V; $V_C = 10.0$ V

23. $A_{v(max)} = 123$; $A_{v(min)} = 2.9$

25. $R_{in(tot)} = 5.44$ kΩ; $A_i = 5.44$

27. $I_{c(sat)(ac)} = 32.1$ mA; $V_{ce(cutoff)(ac)} = 13.0$ V

29. Low input resistance

31. 45 Ω

33. $I_{C(sat)(Q1)} = 1.19$ mA; $I_{C(sat)(Q2)} \cong 10$ mA

35. See Figure ANS–8.

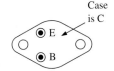

Case
is C

E B C

FIGURE ANS–8

37. With the positive probe on the emitter and negative probe on the base, the reading is an open (or extremely high resistance). With the leads reversed, the reading is much lower.

39. (a) 27.8   (b) 109

41. Q_2 shorted collector-to-emitter, Q_4 shorted collector-to-emitter, R_1 open, R_6 open, Q_1 open, or Q_3 open

Chapter 4

1. JFETs

3. (a) Depletion region widens (creating a narrower channel)
   (b) Increase
   (c) Less

5. +5.0 V

7. (a) 10 mA   (b) 4 GΩ   (c) R_{IN} drops

9. (a) Approximately +4 V
   (b) Approximately 2.5 mA
   (c) Approximately +15.8 V

11. (a) +2.1 V   (b) 2.1 mA   (c) +5.97 V

13. (a) $V_{DS} = +6.3$ V; $V_{GS} = -1.0$ V
   (b) $V_{DS} = +7.29$ V; $V_{GS} = -0.3$ V
   (c) $V_{DS} = -1.65$ V; $V_{GS} = +2.35$ V

15. $I_D = 0.514$ mA; $V_D = +6.86$ V

17. The gate is separated from the channel by a silicon-dioxide insulating layer.

19. +3 V

21. (a) Since $V_{GS} > V_{GS(th)}$, the device is on.
   (b) Since $V_{GS} < V_{GS(th)}$, the device is off.

23. See Figure ANS–9.

FIGURE ANS–9 I_D (mA)

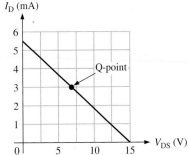

25. $A_v = 21.9$

27. $A_{v(min)} = 0.64$; $A_{v(max)} = 0.9$

29. Q_1 or Q_2 open, R_E open, no negative supply voltage, open path between transistors.

31. 0.953 mA

33. (a) $I_D = 4.85$ mA; $V_{DS} = 9.3$ V   (b) $A_v = 3.5$

35. It is a resistance that appears to be in series with the signal and indicates how much the FET departs from ideal.

37. Saturation and cutoff

Chapter 5

1. $A_{v(overall)} = 38.4$; $R_{in} \cong 1.0$ MΩ; $R_{out} = 2.7$ kΩ

3. $A_{v(overall)} = 812$; $A'_{v(overall)} = 58.2$ dB

5. (a) See Figure ANS–10.
   (b) $A_{v(overall)} = 6000$
   (c) $A_{v(overall)} = 3600$

FIGURE ANS–10

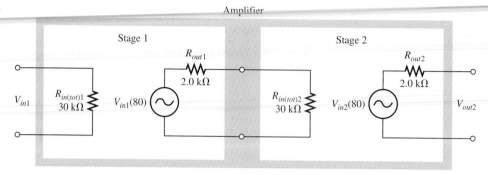

Amplifier

7. Since it increases the input resistance of stage 2, gain will be larger.

9. To prevent reflections from affecting the signal

11. An increase in the input signal will cause it to move over a larger portion of the load line. At the upper end, the transconductance is higher, thus the gain is higher. At the lower end, the opposite is true. The overall effect is increased distortion.

13. 10 kΩ

15. See Figure ANS–11.

FIGURE ANS–11

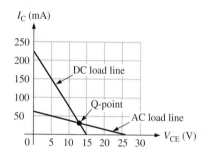

17. $Q = 79$; $A_{v(NL)} = 415$; $BW = 5.75$ kHz

19. (a) $I_{C(Q2)} = 5.3$ mA; $V_{B(Q3)} = +0.7$ V; $I_{C(Q3)} = 120$ mA; $V_{E(Q3)} = 0$ V
    (b) 0.25 W

21. See Figure ANS–12.

FIGURE ANS–12

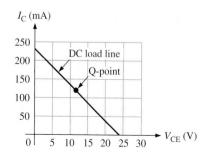

23. (a) None
    (b) Gain increases to 101
    (c) No noticeable effect

25. (a) $I_{CQ} = 68.4$ mA; $V_{CEQ} = 5.14$ V.
    (b) $A_v = 11.7$; $A_p = 263$

27. The changes are shown on Figure ANS–13. The advantage of this arrangement is that the load resistor is referenced to ground.

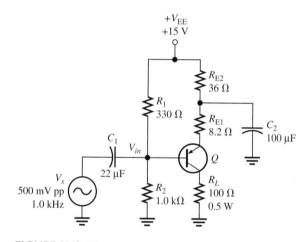

FIGURE ANS–13

29. (a) $V_{B(Q1)} = +0.7$ V; $V_{B(Q2)} = -0.7$ V; $V_E = 0$ V; $I_{CQ} = 8.3$ mA; $V_{CEQ(Q1)} = +9$ V; $V_{CEQ(Q2)} = -9$ V
    (b) 0.5 W

31. (a) $V_{B(Q1)} = +8.2$ V; $V_{B(Q2)} = +6.8$ V; $V_E = 7.5$ V; $I_{CQ} = 6.8$ mA; $V_{CEQ(Q1)} = +7.5$ V; $V_{CEQ(Q2)} = -7.5$ V
    (b) $P_L = 167$ mW

33. (a) C_2 open or Q_2 open
    (b) power supply off, open R_1, Q_1 base shorted to ground
    (c) Q_1 has collector-to-emitter short
    (d) one or both diodes shorted

35. 2 W

37. There would be no output because neither transistor can conduct.

Chapter 6

1. *Practical op-amp:* High open-loop gain, high input impedance, low output impedance, large bandwidth, high CMRR.
Ideal op-amp: Infinite open-loop gain, infinite input impedance, zero output impedance, infinite bandwidth, infinite CMRR.

3. **(a)** Single-ended input; differential output
   (b) Single-ended input; single-ended output
   (c) Differential input; single-ended output
   (d) Differential input; differential output

5. V1: differential output voltage
   V2: noninverting input voltage
   V3: single-ended output voltage
   V4: differential input voltage
   A1: bias current

7. 8.1 μA

9. 107.96 dB

11. 0.3

13. 40 μs

15. $V_f = 49.5$ mV, $B = 0.0099$

17. **(a)** 11   **(b)** 101   **(c)** 47.81   **(d)** 23

19. **(a)** 1.0   **(b)** −1.0   **(c)** 22.3   **(d)** −10

21. **(a)** 0.45 mA   **(b)** 0.45 mA
    (c) −10 V   **(d)** −10

23. **(a)** $Z_{in(VF)} = 1.32 \times 10^{12}$ Ω; $Z_{out(VF)} = 0.455$ mΩ
    (b) $Z_{in(VF)} = 5 \times 10^{11}$ Ω; $Z_{out(VF)} = 0.6$ mΩ
    (c) $Z_{in(VF)} = 40,000$ MΩ; $Z_{out(VF)} = 1.5$ mΩ

25. **(a)** R_1 open or op-amp faulty   **(b)** R_2 open

27. The closed-loop gain will become a fixed −100.

Chapter 7

1. 70 dB

3. 1.67 kΩ

5. **(a)** 79,603   **(b)** 56,569
   (c) 7960   **(d)** 80

7. **(a)** −0.67°   **(b)** −2.69°
   (c) −5.71°   **(d)** −45°
   (e) −71.22°   **(f)** −84.29°

9. **(a)** 0 dB/decade   **(b)** −20 dB/decade
   (c) −40 dB/decade   **(d)** −60 dB/decade

11. **(a)** 29.8 dB; closed-loop
    (b) 23.9 dB; closed-loop
    (c) 0 dB; closed-loop

13. 21.14 MHz

15. Circuit (b) has smaller *BW* (97.5 kHz).

17. **(a)** 150°   **(b)** 120°   **(c)** 60°
    (d) 0°   **(e)** −30°

19. **(a)** Unstable   **(b)** Stable
    (c) Marginally stable

21. 25 Hz

Chapter 8

1. 24 V, with distortion

3. $V_{UTP} = +2.77$ V; $V_{LTP} = −2.77$ V

5. See Figure ANS–14.

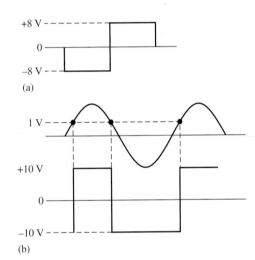

FIGURE ANS–14

7. +8.57 V and −0.968 V

9. **(a)** −2.5 V   **(b)** −3.52 V

11. 110 kΩ

13. $V_{OUT} = −3.57$ V; $I_f = 357$ μA

15. −4.46 mV/μs

17. 1 mA

19. See Figure ANS–15.

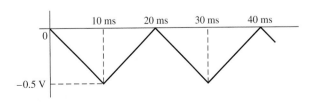

FIGURE ANS–15

21. See Figure ANS–16.

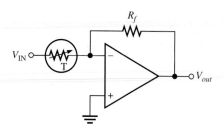

R_f

V_{IN} ⊙—◁▷⊿— T

V_{out}

FIGURE ANS–16

23. The output is not correct because the output should also be high when the input goes below +2 V. Possible faults: Op-amp A2 bad, diode D_2 open, noninverting (+) input of op-amp A2 not properly set at +2 V, or V_{in} is not reaching inverting input.

25. Output is not correct. R_2 is open.

Chapter 9

1. (a) Band pass   **(b)** High pass
   (c) Low pass   **(d)** Band stop

3. 48.2 kHz; No

5. 700 Hz, 5.05

7. (a) 1, not Butterworth
   (b) 1.44, approximate Butterworth
   (c) 1st stage: 1.67
       2nd stage: 1.67
       Not Butterworth

9. (a) Chebyshev **(b)** Butterworth
   (c) Bessel   **(d)** Butterworth

11. 190 Hz

13. Add another identical stage and change the ratio of the feedback resistors to 0.068 for first stage, 0.586 for second stage, and 1.482 for third stage.

15. Exchange positions of resistors and capacitors in the filter network.

17. (a) Decrease R_1 and R_2 or C_1 and C_2.
   (b) Increase R_3 or decrease R_4.

19. (a) f_0 = 4.95 kHz, BW = 3.84 kHz
   (b) f_0 = 449 Hz, BW = 96.5 Hz
   (c) f_0 = 15.9 kHz, BW = 838 Hz

21. Sum the low-pass and high-pass outputs with a two-input adder.

Chapter 10

1. An oscillator requires no input (other than dc power).

3. $^1/_{75}$

5. 733 mV

7. 50 kΩ

9. 2.34 kΩ

11. 136 kΩ; 691 Hz

13. Change R_1 to 3.54 kΩ

15. R_4 = 65.8 kΩ, R_5 = 47 kΩ

17. 3.33 V; 6.67 V

19. 0.0076 μF

21. 13.6 ms

23. 0.01 μF; 9.1 kΩ

Chapter 11

1. 0.033%

3. 1.01%

5. See Figure ANS–17.

7. 8.5 V

9. 9.57 V

11. 500 mA

13. 10 mA

15. $I_{L(max)}$ = 250 mA, P_{R1} = 6.25 W

17. 40%

19. Decreases

FIGURE ANS–17

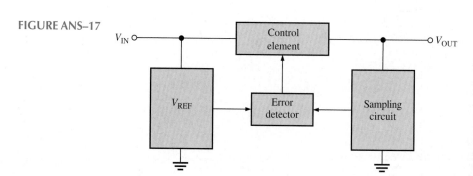

V_{IN} ⊙ — Control element — ⊙ V_{OUT}
V_{REF} Error detector   Sampling circuit

21. 14.25 V

23. 1.3 mA

25. 2.8 Ω

27. $R_{\lim} = 0.35 \, \Omega$

29. See Figure ANS–18.

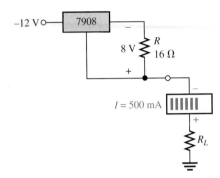

FIGURE ANS–18

Chapter 12

1. $A_{v(1)} = A_{v(2)} = 101$

3. 1.005 V

5. 51.5

7. Change R_G to 2.2 kΩ

9. 300

11. Change the 18 kΩ resistor to 68 kΩ.

13. Connect output (pin 15) directly to pin 14, and connect pin 6 directly to pin 10 to make $R_F = 0$.

15. 500 μA, 5 V

17. $A_v \cong 11.5$

19. See Figure ANS–19.

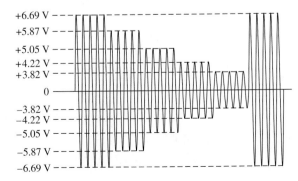

FIGURE ANS–19

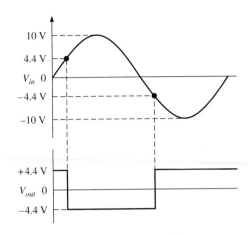

FIGURE ANS–20

21. See Figure ANS–20.

23. **(a)** −0.301   **(b)** 0.301
(c) 1.699   **(d)** 2.114

25. The output of a log amplifier is limited to 0.7 V because of the transistor's *pn* junction.

27. −157 mV

29. $V_{out(max)} = -147$ mV, $V_{out(min)} = -89.2$ mV; the 1 V input peak is reduced 85% whereas the 100 mV input peak is reduced only 10%.

31. Probe 1: ≈ -15 V
Probe 2: ≈ 0 V
Probe 3: 22 mV @ 1 kHz
Probe 4: +15 V
Probe 5: ≈ 0 V

Chapter 13

1. See Figure ANS–21.

3. 1135 kHz

5. RF: 91.2 MHz, IF: 10.7 MHz

7. 739 μA

9. −8.12 V

11. **(a)** 0.28 V   **(b)** 1.024 V
(c) 2.07 V   **(d)** 2.49 V

13. $f_{diff} = 8$ kHz, $f_{sum} = 10$ kHz

15. $f_{diff} = 1.7$ MHz, $f_{sum} = 1.9$ MHz, $f_1 = 1.8$ MHz

17. $f_c = 850$ kHz, $f_m = 3$ kHz

19. $V_{out} = 15$ mV cos[(1100 kHz)2πt] − 15 mV cos[(5500 kHz)2πt]

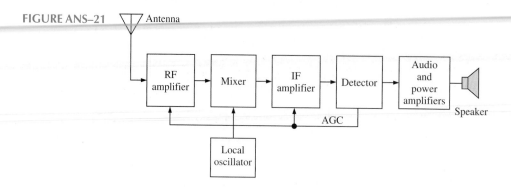

FIGURE ANS–21

21. See Figure ANS–22.

23. See Figure ANS–23.

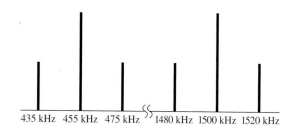

FIGURE ANS–22

FIGURE ANS–23

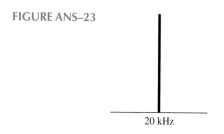

25. The IF amplifier has a 450 kHz–460 kHz passband. The audio/power amplifiers have a 10 Hz–5 kHz passband.

27. The modulating input signal is applied to the control terminal of the VCO. As the input signal amplitude varies, the output frequency of the VCO will vary proportionally.

29. Varactor

31. **(a)** 10 MHz   **(b)** 48.3 mV

33. 1005 Hz

35. $f_o = 233$ kHz, $f_{lock} = \pm 103.6$ kHz, $f_{cap} \cong \pm 4.56$ kHz

Chapter 14

1. See Figure ANS–24.

3. See Figure ANS–25.

5. **(a)** 1   **(b)** 3

7. See Figure ANS–26.

9. 5 kΩ, 2.5 kΩ, 1.25 kΩ

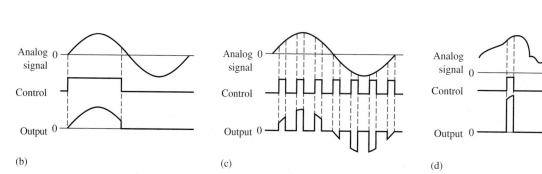

(b)   (c)   (d)

FIGURE ANS–24

13. (a) 16   **(b)** 32   **(c)** 256   **(d)** 65,536

15. 1 mV

17.

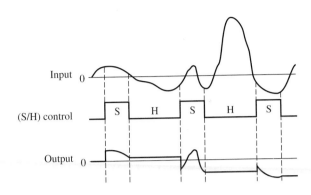

Input 0

(S/H) control

Output 0

FIGURE ANS–25

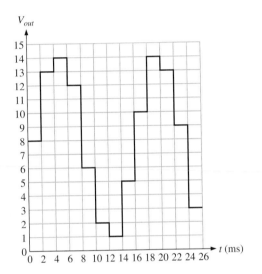

FIGURE ANS–26

Sampling Time (μs)	Binary Output
0	000
10	000
20	001
30	100
40	110
50	101
60	100
70	011
80	010
90	001
100	001
110	011
120	110
130	111
140	111
150	111
160	111
170	111
180	111
190	111
200	100

11.

D_3	D_2	D_1	D_0	V_{out}
0	0	0	0	0 V
0	0	1	1	-0.50 V $+ (-0.25$ V$) = -0.75$ V
1	0	0	0	-2.00 V
1	1	1	1	-2.00 V $+ (-1.00$ V$) + (-0.50$ V$) + (-0.25$ V$) = -3.75$ V
1	1	1	0	-2.00 V $+ (-1.00$ V$) + (-0.50$ V$) = -3.50$ V
0	1	0	0	-1.00 V
0	0	0	0	0 V
0	0	0	1	-0.25 V
1	0	1	1	-2.00 V $+ (-0.50$ V$) + (-0.25$ V$) = -2.75$ V
1	1	1	0	-2.00 V $+ (-1.00$ V$) + (-0.50$ V$) = -3.50$ V
1	1	0	1	-2.00 V $+ (-1.00$ V$) + (-0.25$ V$) = -3.25$ V
0	1	0	0	-1.00 V
1	0	1	1	-2.00 V $+ (-0.50$ V$) + (-0.25$ V$) = -2.75$ V
0	0	0	1	-0.25 V
0	0	1	1	$-0.50 + (-0.25$ V$) = -0.75$ V

19. f_{out} increases.

21. 691 pF (Use standard 680 pF).

23. $f_{out(min)} = 26.2$ kHz, $f_{out(max)} = 80.9$ kHz

25. The D_0 (LSB) is stuck high and the D_2 is stuck low.

Chapter 15

1. 5 V

3. 189.84°

5. 12 bits

7. Thermocouple C

9. -4.36 V

11. 540 Ω

13. The effects of the wire resistances are cancelled in the 3-wire bridge.

15. $\Delta R = 4.5$ mΩ

17. **(a)** absolute pressure transducer
(b) gage pressure transducer
(c) differential pressure transducer

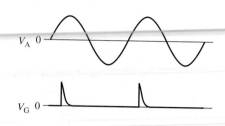

FIGURE ANS–27

19. See Figure ANS–27.

21. Reverse the comparator inputs.

GLOSSARY

ac beta (β_{ac})   The ratio of a change in collector current to a corresponding change in base current in a bipolar junction transistor.

Accuracy   In relation to DACs or ADCs, a comparison of the actual output with the expected output, expressed as a percentage.

Acquisition time   In an analog switch, the time required for the device to reach its final value when switched from hold to sample.

ac resistance   The ratio of a small change in voltage divided by a corresponding change in current for a given device; also called *dynamic, small-signal,* or *bulk resistance.*

Active filter   A frequency-selective circuit consisting of active devices such as transistors or op-amps coupled with reactive components.

A/D conversion   A process whereby information in analog form is converted into digital form.

Amplification   The process of producing a larger voltage, current, or power using a smaller input signal as a "pattern."

Amplifier   An electronic circuit having the capability of amplification and designed specifically for that purpose.

Amplitude modulation (AM)   A communication method in which a lower-frequency signal modulates (varies) the amplitude of a higher-frequency signal (carrier).

Analog signal   A signal that can take on a continuous range of values within certain limits.

Analog switch   A type of semiconductor switch that connects an analog signal from input to output with a control input.

Analog-to-digital converter (ADC)   A device used to convert an analog signal to a sequence of digital codes.

Anode (semiconductor diode definition)   The terminal of a semiconductor diode that is more positive with respect to the other terminal when it is biased in the forward direction.

Antilog   The number corresponding to a given logarithm.

Aperture jitter   In an analog switch, the uncertainty in the aperture time.

Aperture time   In an analog switch, the time to fully open after being switched from sample to hold.

Astable   Characterized by having no stable states; a type of oscillator.

Astable multivibrator   A type of circuit that can operate as an oscillator and produces a pulse waveform output.

Attenuation   The reduction in the level of power, current, or voltage.

Audio   Related to the range of frequencies that can be heard by the human ear and generally considered to be in the 20 Hz to 20 kHz range.

Automatic gain control (AGC)   A feedback system that reduces the gain for larger signals and increases the gain for smaller signals.

Balanced modulation   A form of amplitude modulation in which the carrier is suppressed; sometimes known as *suppressed-carrier modulation.*

Band-pass filter   A type of filter that passes a range of frequencies lying between a certain lower frequency and a certain higher frequency.

Band-stop filter   A type of filter that blocks or rejects a range of frequencies lying between a certain lower frequency and a certain higher frequency.

Bandwidth   The characteristic of certain types of electronic circuits that specifies the usable range of frequencies that pass from input to output. It is the upper critical frequency minus the lower critical frequency.

Barrier potential The inherent voltage across the depletion region of a *pn* junction.

Base One of the semiconductor regions in a BJT.

Base bias A form of bias in which a single resistor is connected between a BJT's base and V_{CC}.

Bessel A type of filter response having a linear phase characteristic and less than -20 dB/decade/pole roll-off.

Bias The application of dc voltage to a diode or other electronic device to produce a desired mode of operation.

Bipolar Characterized by two *pn* junctions.

Bipolar junction transistor (BJT) A transistor constructed with three doped semiconductor regions separated by two *pn* junctions.

Bounding The process of limiting the output range of an amplifier or other circuit.

Butterworth A type of filter response characterized by flatness in the passband and a -20 dB/decade/pole roll-off.

Bypass capacitor A capacitor connected in parallel with a resistor to provide the ac signal with a low impedance path.

Carrier The high frequency (RF) signal that carries modulated information in AM, FM, and other communications systems.

Cathode (semiconductor diode definition) The terminal of a doide that is more negative with respect to the other terminal when it is biased in the forward direction.

C_{iss} The common-source input capacitance of a FET as seen looking into the gate.

Center tap A connection at the midpoint of the secondary of a transformer.

Characteristic curve A plot which shows the relationship between two variable properties of a device. For most electronic devices, a characteristic curve refers to a plot of the current, I, plotted as a function of voltage, V.

Chebyshev A type of filter response characterized by ripples in the passband and a greater than -20 dB/decade/pole roll-off.

Clamper A circuit that adds a dc level to an ac signal; also called a *dc restorer.*

Class A An amplifier that operates in the active region at all times.

Class AB An amplifier that is biased into slight conduction. The Q-point is slightly above cutoff.

Class B An amplifier that has the Q-point located at cutoff, causing the output current to vary only during one-half of the input cycle.

Closed-loop An op-amp configuration in which the output is connected back to the input through a feedback circuit.

Closed-loop voltage gain The net voltage gain of an amplifier when negative feedback is included.

Coax A transmission line, principally used for high frequencies, in which a center conductor is surrounded by a tubular conducting shield.

Cold junction A reference thermocouple held at a fixed temperature and used for compensation in thermocouple circuits.

Collector One of the semiconductor regions in a BJT.

Collector characteristic curves A set of collector I-V curves that show how I_C varies with V_{CE} for a given base current.

Collector feedback bias A form of bias, used in CE and CB amplifiers, in which a single resistor is connected between a BJT's base and its collector.

Common-base (CB) A BJT amplifier configuration in which the base is the common terminal to an ac signal or ground.

Common-collector (CC) A BJT amplifier configuration in which the collector is the common terminal to an ac signal or ground.

Common-drain (CD) A FET amplifier configuration in which the drain is the grounded terminal to an ac signal or ground.

Common-emitter (CE) A BJT amplifier configuration in which the emitter is the common terminal to an ac signal or ground.

Common-gate (CG) A FET amplifier configuration in which the gate is the grounded terminal to an ac signal or ground.

Common mode A condition characterized by the presence of the same signal on both op-amp inputs.

Common-mode input impedance The ac resistance between each input and ground.

Common-mode input voltage range The range of input voltage, which when applied to both inputs, will not cause clipping or other output distortion.

Common-mode rejection ratio (CMRR) The ratio of open-loop gain to common-mode gain; a measure of an op-amp's ability to reject common-mode signals.

Common-source (CS) A FET amplifier configuration in which the source is the grounded terminal.

Comparator A circuit which compares two input voltages and produces an output in either of two states indicating the greater than or less than relationship of the inputs.

Compensation The process of modifying the roll-off rate of an amplifier to ensure stability.

Complementary symmetry transistors These are a matching pair of *npn/pnp* BJTs or a matching pair of *n*-channel/*p*-channel FETs.

Conduction electron An electron that has broken away from the valance band of the parent atom and is free to move from atom to atom within the atomic structure of a material; also called a *free electron.*

Constant-current region   The region on the drain characteristic of a FET in which the drain current is independent of the drain-to-source voltage.

Constant-current source   A circuit that delivers a load current that remains constant when the load resistance changes.

Coupling capacitor   A capacitor connected in series with the ac signal and used to block dc voltages.

Covalent bond   A type of chemical bond in which atoms share electron pairs.

Critical frequency   The frequency that defines the end of the passband of a filter; also called *cutoff frequency*.

Crossover distortion   Distortion in the output of a class B push-pull amplifier at the point where each transistor changes from the cutoff state to the on state.

Crystal   A solid in which the particles form a regular, repeating pattern.

Current mirror   A circuit that uses matching diode junctions to form a current source. The current in a diode junction is reflected as a matching current in the other junction (which is typically the base-emitter junction of a transistor). Current mirrors are commonly used to bias a push-pull amplifier.

Current-to-voltage converter   A circuit that converts a variable input current to a proportional output voltage.

Cutoff   The nonconducting state of a transistor.

Cycle   The complete sequence of values that a waveform exhibits before another identical pattern occurs.

D/A conversion   The process of converting a sequence of digital codes to an analog form.

Damping factor (DF)   A filter characteristic that determines the type of response.

dBm   Decibel power level when the reference is understood to be 1 mW (see Decibel).

dc beta (β_{DC})   The ratio of collector current to base current in a bipolar junction transistor.

Decibel   A dimensionless quantity that is 10 times the logarithm of a power ratio or 20 times the logarithm of a voltage ratio.

Decoupling network   A low-pass filter that provides a low-impedance path to ground for high-frequency signals.

Demodulation   The process in which the information signal is recovered from the IF carrier signal; the reverse of modulation.

Depletion mode   A class of FETs that is on with zero-gate voltage and is turned off by gate voltage. All JFETs and some MOSFETS are depletion-mode devices.

Depletion region   The area near a *pn* junction on both sides that has no majority carriers.

Differential amplifier (diff-amp)   An amplifier that produces an output voltage proportional to the difference of the two input voltages.

Differential input impedance   The total resistance between the inverting and the noninverting inputs.

Differential mode   The input condition of an op-amp in which opposite polarity signals are applied to the two inputs.

Differentiator   A circuit that produces an inverted output which approximates the rate of change of the input function.

Digital signal   A noncontinuous signal that has discrete numerical values assigned to the specific steps.

Digital-to-analog converter (DAC)   A device in which information in digital form is converted to an analog form.

Diode   An electronic device that permits current in only one direction.

Discrete device   An individual electrical or electronic component that must be used in combination with other components to form a complete functional circuit.

Discriminator   A type of FM demodulator.

Domain   The values assigned to the independent variable. For example, frequency or time are typically used as the independent variable for plotting signals.

Doping   The process of imparting impurities to an intrinsic semiconductive material in order to control its conduction characteristics.

Drain   One of the three terminals of a field-effect transistor; it is one end of the channel.

Droop   In an analog switch, the change in the sampled value during the hold interval.

Dynamic emitter resistance (r_e')   The ac resistance of the emitter; it is determined by the dc emitter current.

Efficiency (power)   The ratio of the signal power supplied to the load to the power from the dc supply.

Electroluminescence   The process of releasing light energy by the recombination of electrons in a semiconductor.

Electron   The basic particle of negative electrical charge in matter.

Electrostatic discharge (ESD)   The discharge of a high voltage through an insulating path that frequently destroys a device.

Emitter   One of the three semiconductor regions in a BJT.

Emitter bias   A very stable form of bias requiring two power supplies. The emitter is connected through a resistor to one supply; another resistor is connected between a BJT's base and ground.

Energy   The ability to do work.

Enhancement mode   A MOSFET in which the channel is formed (or enhanced) by the application of a gate voltage.

Feedback oscillator   A type of oscillator that returns a fraction of output signal to the input with no net phase shift resulting in a reinforcement of the output signal.

Feedforward   A method of frequency compensation in op-amp circuits.

Feedthrough   In an analog switch, the component of the output voltage which follows the input voltage after the switch opens.

Field-effect transistor (FET)   A voltage-controlled device in which the voltage at the gate terminal controls the amount of current through the device.

Filter   A type of electrical circuit that passes certain frequencies and rejects all others.

Flash   A method of A/D conversion.

Floating point   A point in the circuit that is not electrically connected to ground or a "solid" voltage.

Fold-back current limiting   A method of current limiting in voltage regulators.

Forward bias   The condition in which a *pn* junction conducts current.

Four-quadrant multiplier   A linear device that produces an output voltage proportional to the product of two input voltages.

Frequency   The number of repetitions per unit of time for a periodic waveform.

Frequency modulation (FM)   A communication method in which a lower-frequency intelligence-carrying signal modulates (varies) the frequency of a higher-frequency signal.

Full-wave rectifier   A circuit that converts an alternating sine wave into a pulsating dc voltage consisting of both halves of a sine wave for each input cycle.

Gage factor (*GF*)   The ratio of the fractional change in resistance to the fractional change in length along the axis of the gage.

Gain   The amount of amplification. Gain is a ratio of an output quantity to an input quantity (e.g., voltage gain is the ratio of the output voltage to the input voltage).

Gate   One of the three terminals of a field-effect transistor. A voltage applied to the gate controls drain current.

Germanium   A semiconductive material.

Half-wave rectifier   A circuit that converts an alternating sine wave into a pulsating dc voltage consisting of one-half of a sine wave for each input cycle.

Harmonics   Higher-frequency sinusoidal waves that are integer multiples of a fundamental frequency.

High-pass filter   A type of filter that passes frequencies above a certain frequency while rejecting lower frequencies.

Hole   A mobile vacancy in the electronic valence structure of a semiconductor. A hole acts like a positively charged particle.

Hysteresis   The property that permits a circuit to switch from one state to the other at one voltage level and switch back to the original state at another lower voltage level.

I_{DSS}   The drain current in a FET when the gate is shorted to the source. For JFETs, this is the maximum allowed current.

I_{GSS}   The gate-reverse current in a FET. The value is based on a specified gate-to-source voltage.

Input bias current   The average dc current required by the inputs of an op-amp to properly operate the device.

Input offset voltage (V_{OS})   The differential dc voltage required between the op-amp inputs to force the differential output to zero volts.

Input offset voltage drift   A parameter that specifies how much change occurs in the input offset voltage for each degree change in temperature.

Instrumentation amplifier   A differential voltage-gain device that amplifies the difference between the voltage existing at its two input terminals.

Intregrated circuit (IC)   A type of circuit in which all the components are constructed on a single chip of silicon.

Integrator   A circuit that produces an inverted output which approximates the area under the curve of the input function.

Intermediate frequency   A fixed frequency that is lower than the RF, produced by beating an RF signal with an oscillator frequency.

Intrinsic (pure)   An intrinsic semiconductor is one in which the charge concentration is essentially the same as a pure crystal with relatively few free electrons.

Inverting amplifier   An op-amp closed-loop configuration in which the input signal is applied to the inverting input.

Ion   An atom that has gained or lost a valence electron, resulting in a net positive or negative charge.

Isolation amplifier   An amplifier in which the input and output stages are not electrically connected.

Junction field-effect transistor (JFET)   A type of FET that operates with a reverse-biased *pn* junction to control current in a channel. It is a depletion-mode device.

Large-signal   A signal that operates an amplifier over a significant portion of its load line.

Light-emitting diode (LED)   A type of diode that emits light when there is forward current.

Limiter   A circuit that removes part of a waveform above or below a specified level; also called a *clipper*.

Linear component   A component in which an increase in current is proportional to the applied voltage.

Linear regulator   A voltage regulator in which the control element operates in the linear region.

Linearity   A straight-line relationship. A linear error is a deviation from the ideal straight-line output of a DAC.

Line regulation   The change in output voltage for a given change in line (input) voltage, normally expressed as a percentage.

Load line   A straight line plotted on a current versus voltage plot that represents all possible operating points for an external circuit.

Load regulation   The change in output voltage for a given change in load current, normally expressed as a percentage.

Logarithm   An exponent; the logarithm of a quantity is the exponent or power to which a given number called the base must be raised in order to equal the quantity.

Loop gain   An op-amp's open-loop voltage gain times the attenuation of the feedback network.

Low-pass filter   A type of filter that passes frequencies below a certain frequency while rejecting higher frequencies.

Mean   Average value.

Microcontroller   A specialized microprocessor designed for control functions.

Mixer   A nonlinear circuit that combines two signals and produces the sum and difference frequencies; a device for down-converting frequencies in a receiver system.

Modem   A device that converts signals produced by one type of device to a form compatible with another; *mo*dulator/*dem*odulator.

Modulation   The process in which a signal containing information is used to modify the amplitude, frequency, or phase of a much higher-frequency signal called the carrier.

Monostable   Characterized by having one stable state.

Monotonicity   In relation to DACs, the presence of all steps in the output when sequenced over the entire range of input bits.

MOSFET   Metal-oxide semiconductor field-effect transistor; one of two major types of FET. It uses a SiO_2 layer to insulate the gate lead from the channel. MOSFETs can be either depletion mode or enhancement mode.

Natural logarithm   The exponent to which the base e ($e =$ 2.71828) must be raised in order to equal a given quantity.

Negative feedback   The process of returning a portion of the output back to the input in a manner to cancel changes that may occur at the input.

Neutralization   A method of preventing unwanted oscillations by adding negative feedback to just cancel the positive feedback caused by internal capacitances of an amplifier.

Noise   An unwanted voltage or current fluctuation.

Noninverting amplifier   An op-amp closed-loop configuration in which the input signal is applied to the noninverting input.

Nonmonotonicity   In relation to DACs, a step reversal or missing step in the output when sequenced over the entire range of input bits.

Norton's theorem   An equivalent circuit that replaces a complicated two-terminal linear network with a single current source and a parallel resistance.

Nyquist rate   In sampling theory, the minimum rate at which an analog voltage can be sampled for A/D conversion. The sample rate must be more than twice the maximum frequency component of the input signal.

Ohmic region   The region on the drain characteristic of a FET with low values of V_{DS} in which the channel resistance can be changed by the gate voltage; in this region the FET can be operated as a voltage-controlled resistor.

One-shot   A monostable multivibrator that produces a single output pulse for each input trigger pulse.

Open-loop   A condition in which an op-amp has no feedback.

Open-loop voltage gain   The internal gain of an amplifier without external feedback.

Operational amplifier (op-amp)   A type of amplifier that has very high voltage gain, very high input impedance, very low output impedance, and good rejection of common-mode signals.

Operational transconductance amplifier   An amplifier in which the output current is the gain times the input voltage.

Order   The number of poles in a filter.

Oscillator   An electronic circuit that operates with positive feedback and produces a time-varying output signal without an external input signal.

Output impedance   The ac resistance viewed from the output terminal of an op-amp.

Passband   The region of frequencies that are allowed to pass through a filter with minimum attenuation.

Peak detector   A circuit used to detect the peak of the input voltage and store that peak value on a capacitor.

Period (*T*)   The time for one cycle of a repeating wave.

Periodic   A waveform that repeats at regular intervals.

Phase angle (in radians)   The fraction of a cycle that a waveform is shifted from a reference waveform of the same frequency.

Phase-locked loop (PLL)   A device for locking onto and tracking the frequency of an incoming signal.

Phase margin   The difference between the total phase shift through an amplifier and 180°, the additional amount of phase shift that can be allowed before instability occurs.

Phase shift   The relative angular displacement of a time-varying function relative to a reference.

Phase-shift oscillator   A type of sinusoidal feedback oscillator that uses three RC networks in the feedback loop.

Photodiode   A diode whose reverse resistance changes with incident light.

Pinch-off voltage   The value of the drain-to-source voltage of a FET at which the drain current becomes constant when the gate-to-source voltage is zero.

PN **junction**   The boundary between n-type and p-type materials.

Pole   A network containing one resistor and one capacitor that contributes -20 dB/decade to a filter's roll-off rate.

Positive feedback   A condition where an in-phase portion of the output voltage is fed back to the input.

Power gain   The ratio of the power delivered to the load to the input power of an amplifier.

Power supply   A device that converts ac or dc voltage into a voltage or current suitable for use in various applications to power electronic equipment. The most common form is to convert ac from the utility line to a constant dc voltage.

Push-pull   A type of class B amplifier with two transistors in which one transistor conducts for one half-cycle and the other conducts for the other half-cycle.

Quality factor (Q)   A dimensionless number that is the ratio of the maximum energy stored in a cycle to the energy lost in a cycle. The ratio of a band-pass filter's center frequency to its bandwidth.

Quantization   The determination of a value for an analog quantity.

Quantization error   The error resulting from the change in the analog voltage during the A/D conversion time.

Quantizing   The process of assigning numbers to sampled data.

Quiescent point   The point on a load line that represents the current and voltage conditions for a circuit with no signal (also called operating or Q-point). It is the intersection of a device characteristic curve with a load line.

$r_{DS(on)}$ The resistance of the channel of a FET measured between the drain and the source when the FET is fully on and only a small voltage is between the drain and the source.

Radio frequency interference (RFI)   High frequencies produced when high values of current and voltage are rapidly switched on and off.

Recombination   The process of a free electron in the conduction band falling into a hole in the valence band of an atom.

Rectifier   An electronic-circuit that converts ac into pulsating dc.

Regulator   An electronic circuit that is connected to the output of a rectifier and maintains an essentially constant output voltage despite changes in the input, the load current, or the temperature.

Relaxation oscillator   A type of oscillator that uses an RC timing circuit to generate a nonsinusoidal waveform.

Resistance temperature detector (RTD)   A type of temperature transducer in which resistance is directly proportional to temperature.

Resolution   In relation to DACs or ADCs, the number of bits involved in the conversion. Also, for DACs, the reciprocal of the maximum number of discrete steps in the output.

Resolver   A type of synchro.

Resolver-to-digital converter (RDC)   An electronic circuit that converts resolver voltages to a digital format which represents the angular position of the rotor shaft.

Reverse bias   The condition in which a *pn* junction blocks current.

Ripple voltage   The variation in the dc voltage on the output of a filtered rectifier caused by the slight charging and discharging action of the filter capacitor.

Roll-off   The rate of decrease in gain, below or above the critical frequencies of a filter.

Root mean square (RMS)   The value of an ac voltage that corresponds to a dc voltage that produces the same heating effect in a resistance.

Rotor   The part of a synchro that is attached to the shaft and rotates.

Sample-and-hold   The process of taking the instantaneous value of a quantity at a specific point in time and storing it on a capacitor.

Sampling   The process of breaking the analog waveform into time "slices" that approximate the original wave.

Saturation   The state of a BJT in which the collector current has reached a maximum and is independent of the base current.

Schmitt trigger   A comparator with hysteresis.

Semiconductor   A material that has a conductance value between that of a conductor and that of an insulator. Silicon and germanium are examples.

Settling time   The time it takes a DAC to settle within $\pm \frac{1}{2}$ LSB of its final value.

Shell   An energy level in which electrons orbit the nucleus of an atom.

Signal compression   The process of scaling down the amplitude of a signal voltage.

Silicon   A semiconductive material used in diodes and transistors.

Silicon-controlled rectifier   A type of three-terminal thyristor that conducts current when triggered on and remains on until the anode current falls below a specific value.

Single-ended mode The input condition of an op-amp in which one input is grounded and the signal voltage is applied only to the other input.

Skin effect The phenonenon at high frequencies which causes current to move to the outside surface of conductors.

Slew rate The rate of change of the output voltage of an op-amp in response to a step input.

Source One of the three terminals of a field-effect transistor; it is one end of the channel.

Spectrum A plot of amplitude versus frequency for a signal.

Stability A condition in which an amplifier circuit does not oscillate.

Stage Each transistor in a multistage amplifier that amplifies a signal.

Standing wave A stationary wave on a transmission line formed by the interaction of an incident and reflected wave.

Stator The part of a synchro that is fixed. The stator windings are located on the stator.

Strain The expansion or compression of a material caused by stress forces acting on it.

Strain gage A transducer formed by a resistive material in which a lengthening or shortening due to stress produces a proportional change in resistance.

Successive approximation A method of A/D conversion.

Summing amplifier A variation of a basic comparator circuit that is characterized by two or more inputs and an output voltage that is proportional to the magnitude of the algebraic sum of the input voltages.

Switch An electrical or electronic device for opening and closing a current path.

Switching regulator A voltage regulator in which the control element is a switching device.

Synchro An electromechanical transducer used for shaft angle measurement and control.

Synchro-to-digital converter (SDC) An electronic circuit that converts synchro voltages to a digital format which represents the angular position of the rotor shaft.

Terminal An external contact point on an electronic device.

Thermal overload A condition in a rectifier where the internal power dissipation of the circuit exceeds a certain maximum due to excessive current.

Thermistor A type of temperature transducer in which resistance is inversely proportional to temperature.

Thermocouple A type of temperature transducer formed by the junction of two dissimilar metals which produces a voltage proportional to temperature.

Thevinen's theorem An equivalent circuit that replaces a complicated two-terminal linear network with a single voltage source and a series resistance.

Thyristor A class of four-layer (*pnpn*) semiconductor devices, such as the silicon-controlled rectifier.

Transconductance The ratio of output current to input voltage; the gain of a FET; it is determined by a small change in drain current divided by a corresponding change in gate-to-source voltage. Is is measured in siemens or mhos.

Transducer A device that converts a physical quantity from one form to another; for example, a microphone converts sound into voltage.

Transfer curve A plot ot the output of a circuit or system for a given input.

Transistor A semiconductor device used for amplification and switching applications in electronic circuits.

Transducer A device that converts a physical parameter into an electrical quantity.

Triac A three-terminal thyristor that can conduct current in either direction when properly activated.

Trim To precisely adjust or fine tune a value.

Uncompensated op-amp An op-amp with more than one critical frequency.

$V_{\text{GS(off)}}$ The voltage applied between the gate and the source that is just sufficient to turn off a FET. The exact point is arbitrary; some manufacturers use a specific very small current to determine it.

Valence electron An electron in the outermost shell or orbit of an atom.

Varactor A diode that is used as a voltage-variable capacitor.

Vector Any quantity that has both magnitude and direction.

Voltage-controlled oscillator A type of relaxation oscillator whose frequency can be changed by a variable dc voltage; also known as a VCO.

Voltage-divider bias A very stable form of bias in which a voltage divider is connected between V_{CC} and ground; the output of the divider supplies bias current to the base of a BJT.

Voltage-follower A closed-loop, noninverting op-amp with a voltage gain of one.

Voltage regulation The process of maintaining an essentially constant output voltage over variations in input voltage or load.

Voltage-to-current converter A circuit that converts a variable input voltage to a proportional output current.

Wien-bridge oscillator A type of sinusoidal feedback oscillator that uses an *RC* lead-lag network in the feedback loop.

Zener diode A type of diode that operates in reverse breakdown (called zener breakdown) to provide voltage regulation.

Zero-voltage switching The process of switching power to a load at the zero crossings of an ac voltage to minimize RF noise generation.

INDEX